Immunity Against Mucosal Pathogens

Michael Vajdy
Editor

Immunity Against Mucosal Pathogens

Springer

Editor
Dr. Michael Vajdy
University of California, Davis
School of Medicine
Department of Medical Microbiology
and Immunology
Davis CA 95616
USA
michaelvajdy@comcast.net

ISBN: 978-1-4020-8411-9 e-ISBN: 978-1-4020-8412-6

Library of Congress Control Number: 2008926594

Cover Illustration: BIRD FLU REASSORTMENT GRAPHIC: The illustration shows reassortment of viral RNA segments in a cell (e.g. lung epithelial cell) infected by two strains of influenza virus (human and bird flu) leading to a new and potentially dangerous strain that could spread easily from human to human and so trigger a deadly worldwide epidemic. Such genetic mixing might occur in pigs, since a pig might be infected by both strains and then pass the new virus on to humans. Alternatively, a person might become infected with bird flu and human flu and start an epidemic of the novel virus.

Printed on acid-free paper

9 8 7 6 5 4 3 2 1

springer.com

Preface

We see ourselves enveloped by our skins. But what we don't see or usually think about is that most of our body is covered by a different type of epithelial cells than that of the skin. Surprisingly what separates us from the open environment all around us sometimes is a single layer of epithelial cells. It is at these seemingly fragile sites that most pathogens gain access to our inside milieu. These sites comprise most of the mucosal membranes of the gastrointestinal, respiratory, and the upper genitourinary tracts. It is also at some of these sites that we need to acquire food or air. The evolution has endowed us with the ability to distinguish between a myriad of food or air particles and pathogens. Thus, we are racing against time and evolution and to win this race to survive we need to better prepare our mucosal membranes to neutralize and eliminate mucosal pathogens preferably right at the portals of entry through vaccinations or treatments.

The mucosal membranes of the gastrointestinal and respiratory tracts together with the genitourinary tract comprise the overall mucosal membranes of our bodies. Given the uniqueness of these surfaces, it is not surprising that the immune system that evolved to protect these surfaces against pathogens is somewhat different than that of the systemic immune system. This is evident in the different types of the cells that make up the mucosal tissues, as well as the innate physiological barriers that exist there. While Mucosal Immunology has been generally thought of as a separate and smaller portion of the general, systemic, Immunology, it is in fact more comprehensive and general than systemic Immunology. This is because Mucosal Immunology includes not only the complex mucosal immune systems, but also because of its close interaction and link to the systemic immune system, the whole mammalian host's immune system.

While there are major similarities between the cells and the immune responses generated at the various mucosal tracts, there are also important differences. Knowledge of these differences and similarities is required in order to understand the interactions between us, as the host, and the pathogens that attack through each tract, and how our immune system reacts to each of them. Whether we want to devise rational prophylactic or therapeutic vaccines or treatments to either prevent or treat mucosal infections we must acquire such knowledge. This is the rationale behind putting this book together. This book will provide the readers in the areas of vaccinology, virology, bacteriology, epidemiology, immunology and mucosal immunology within academia (undergraduate, graduate, post doctoral fellows and professors), as well as preclinical and clinical scientists in vaccine and drug industries a thorough appreciation of the mucosal immune system and its importance in protecting humans against mucosal pathogens.

This book is divided into 6 parts. It starts with Part I, an introductory yet in depth study of structure and cells of mucosal tissues, giving the reader an understanding of the structure and cells of the gastrointestinal, respiratory and genitourinary tracts. Because most of what we know about the mucosal immune system stems from the studies of the gastrointestinal tract, this chapter is in more detail. This section thus provides a good background for better understanding of the subsequent chapters.

Part II, Immunology of Mucosal Tissues, is divided into 3 subsections. Mucosal B cells, the first subsection, is a detailed study of the role of mucosal B cells in protection of various mucosal tissues, their differences and similarities with systemic B cells and vaccine design for optimal antibody protection through various routes. Thus, this chapter provides a good example of how a mucosal lymphocyte population is different than its systemic sister population. The second subsection, "Bridging Mucosal Innate Immunity to the Adaptive Immune System", is an important and comprehensive study of current knowledge of the mucosal innate immune system and how it interacts and leads to adaptive immune responses. The mucosal innate immune system is the first line of defense against mucosal pathogens and thus an appreciation of its components and how it leads to adaptive and memory type responses is essential in any study of the mucosal immune system. A major issue on how the organized mucosal immune system develops is the interaction of its cells with the resident microbial flora. Thus, Chapter 5 explains the current knowledge and hypotheses on the importance of the microbial flora in the development of organized lymphoid tissues of the intestines and how they affect B cell development and antibody diversity.

While Parts I and II serve to give a thorough and necessary background about the structure and unique cells of the mucosal tissues, the next sections proceed to describe immunity against one or more representative bacterial or viral pathogens that infect the host through the gastrointestinal, respiratory and the genitourinary tracts. The choice of each of these pathogens was made due to either their importance in terms of the number of infections they cause and the severity of the disease or their intriguing nature and potential to cause pandemics, such as the SARS virus.

With regards to the recent concern for pathogens that can be used as biological weapons, the National Institutes of Allergy and infectious Diseases in USA, lists the subject of Chapter 13, Anthrax, as the top agent with bioterrorism potential in Category A. As a viral example for a high potential bioterrorism agent, Chapter 9, describes *Noroviruses*, in Category B. The final chapter lists mucosal immunopotentiaing adjuvants and delivery systems, without which there is little hope of developing licensed prophylactic or therapeutic vaccines or treatments for human consumption.

I am sincerely indebted to each and every of the authors for their contributions. Each of the authors was carefully selected because of his/her academic expertise/leadership in the area or relevant industry experience or in some cases both.

My hope is that at this critical juncture of biomedical history, with the threat of pandemic influenza and reality of HIV upon us, this book will increase the current awareness of the nature of the current and emerging pathogens that infect us, almost all of which use our mucosal membranes as their portals of entry. While relatively few efforts have been made to develop mucosal vaccines, awareness is gradually increasing. However, even for HIV, which is proven to be mostly transmitted through the genitourinary tracts, in the area of HIV vaccine development, thus far there have been no published clinical trials with mucosally administered vaccines. It remains to be seen whether systemically injectable vaccines that are currently used against almost all common pathogens against which we do have a level or pre-existing immunity, will be protective against emerging mucosal pathogens against which we do not have any pre-existing immunity.

Orinda, CA Michael Vajdy

Contents

Viral Genital Pathogens

Contributors

Marli S.P. Azevedo
Food Animal Health Research Program, Ohio Agricultural Research and Development Center, The Ohio State University, Wooster, OH, USA

Barbara Baudner
Novartis Vaccines and Diagnostics S.r.l., Siena, Italy

Zohar Biron
Novartis Vaccines and Diagnostics, Inc., Emeryville, CA, USA

Prosper N. Boyaka
Departments of Veterinary Biosciences, Internal Medicine and Center for Microbial Interface Biology, The Ohio State University, Columbus, OH, USA

Per Brandtzaeg
Laboratory for Immunohistochemistry and Immunopathology (LIIPAT), Institute and Division of Pathology, University of Oslo, Rikshospitalet University Hospital, Oslo, Norway

Lynn Bry
Department of Pathology, Brigham and Women's Hospital and Harvard Medical School, Boston, MA, USA

Jimna Cisto
Novartis Vaccines and Diagnostics, Inc., Emeryville, CA, USA

John D. Clemens
International Vaccine Institute (IVI), SNU Research Park, Kwanak-gu, Seoul, Korea

Jacinta E. Cooper
Division of Infectious Diseases and Department of Microbiology, Colorado Center for AIDS Research, University of Colorado at Denver and Health Sciences Center, Department of Veterans Affairs Eastern Colorado Health Care System, Denver, CO, USA

Estelle Cormet-Boyaka
Department of Internal Medicine, The Ohio State University, Columbus, OH, USA

Satya Dandekar
Department of Medical Microbiology and Immunology, University of California, Davis, CA, USA

Alexandra Duverger
Departments of Veterinary Biosciences, The Ohio State University, Columbus, OH, USA

Tom Evans
Novartis Institute of Biomedical Research, Cambridge, MA, USA

Tiffany Glavan
Department of Medical Microbiology and Immunology, University of California, Davis, CA, USA

Ana María González
Food Animal Health Research Program, Ohio Agricultural Research and Development Center, The Ohio State University, Wooster, OH, USA

Jan Holmgren
Gothenburg University Vaccine Research Institute (GUVAX), Gothenburg, Sweden

Edward N. Janoff
Department of Medicine and Microbiology, Mucosal and Vaccine Research Program Colorado (MAVRC), Division of Infectious Diseases, University of Colorado Denver, CO, USA

Finn-Eirik Johansen
Laboratory for Immunohistochemistry and Immunopathology (LIIPAT), Institute and Division of Pathology, University of Oslo, Rikshospitalet University Hospital, Oslo, Norway

Elaine Kan
Novartis Vaccines and Diagnostics, Inc., Emeryville, CA, USA

Katherine L. Knight
Department of Microbiology and Immunology, Stritch School of Medicine, Loyola University Chicago, Maywood, IL, USA

Dennis K. Lanning
Department of Microbiology and Immunology, Stritch School of Medicine, Loyola University Chicago, Maywood, IL, USA

Juan S. Leon
Hubert Department of Global Health, Rollins School of Public Health, Emory University, Atlanta, GA, USA

James W. Lillard, Jr.
Department of Microbiology and Immunology, James Graham Brown Cancer Center, University of Louisville, Louisville, Kentucky, USA

Nils Lycke
Department of Microbiology and Immunology, Mucosal Immunobiology and Vaccine Research Center (MIVAC), Institute of Biomedicine, The Sahlgrenska Academy at Gothenburg University, Gothenburg, Sweden

Padma Malyala
Novartis Vaccines and Diagnostics, Inc., Emeryville, CA, USA

Nicholas J. Mantis
Division of Infectious Disease, Wadsworth Center, New York State Department of Health, Albany, NY, USA

Ellen Marks
Department of Microbiology and Immunology, Mucosal Immunobiology and Vaccine Research Center (MIVAC), Institute of Biomedicine, The Sahlgrenska Academy at Gothenburg University, Gothenburg, Sweden

Christine L. Moe
Hubert Department of Global Health, Rollins School of Public Health, Emory University, Atlanta, GA, USA

Firdausi Qadri
International Centre for Diarrhoeal Disease Research [ICDDR, B], Dhaka, Bangladesh

Rino Rappuoli
Novartis Vaccines and Diagnostics S.r.l., Siena, Italy

Paolo Ruggiero
Novartis Vaccines and Diagnostics S.r.l., Siena, Italy

Linda J. Saif
Food Animal Health Research Program, Ohio Agricultural Research and Development Center (OARDC), Department of Veterinary Preventive Medicine, The Ohio State University, Wooster, OH, USA

Sumathi Sankaran
Department of Medical Microbiology and Immunology, University of California, Davis, CA, USA

Kari M. Severson
Department of Microbiology and Immunology, Stritch School of Medicine, Loyola University Chicago, Maywood, IL, USA

Manmohan Singh
Novartis Vaccines and Diagnostics, Inc., Emeryville, CA, USA

Rajesh Singh
Department of Microbiology and Immunology, James Graham Brown Cancer Center, University of Louisville, Louisville, KY, USA

Emily R. Smith
Hubert Department of Global Health, Rollins School of Public Health, Emory University, Atlanta, GA, USA

Menira Souza
Food Animal Health Research Program, Ohio Agricultural Research and Development Center (OARDC), Department of Veterinary Preventive Medicine, The Ohio State University, Wooster, OH, USA

Indresh K. Srivastava
Novartis Vaccines and Diagnostics, Inc., Emeryville, CA, USA

Isha N. Srivastava
Johns Hopkins University, Baltimore, MD, USA

Ann-Mari Svennerholm
Department of Microbiology and Immunology, Gothenburg University Vaccine Research Institute (GUVAX) and Department of Microbiology and Immunology, The Sahlgrenska Academy at University of Gothenburg, Gothenburg, Sweden

Jean-Nicolas Tournier
Department de Biologie des Agents Transmissibles, CRSSA, La Tronche, Franc and USAMRIID, Integrated Toxicology Division, Fort Detrick, MD, USA

Michael Vajdy
University of California, Davis, School of Medicine, Department of Medical Microbiology and Immunology, Davis CA 95616, USA

Qiuhong Wang
Department of Microbiology and Molecular Genetics, Medical College of Wisconsin, Milwaukee, WI, USA

Part I
Structure and Cells of Mucosal Tissues

Chapter 1
The Intestinal Epithelium: The Interface Between Host and Pathogen

Nicholas J. Mantis and Lynn Bry

Abstract The gastrointestinal (GI) epithelium consists of a single layer of epithelial cells whose primary function is digestion, management of fluid and electrolyte balance, and nutrient acquisition. Indeed, as will be discussed in this chapter, the cellular structure and function of the intestinal epithelium is optimally designed to undertake these tasks. However, the intestinal epithelium is also the first point of contact for toxins, viruses, bacteria, and parasites that are transmitted by the oral route. As such, the intestinal epithelium must constantly be on guard against enteric pathogens, and be equipped to respond in the event that infections do occur. Specialized cells within the epithelium function directly in innate immunity through the secretion of mucus and antimicrobial factors, while others function in antigen sampling and immune surveillance. The epithelium is also capable of secreting important cytokines and pro-inflammatory molecules that direct local immune responses and the subsequent development of adaptive immune responses in mucosal locations. As such the intestinal epithelium plays the conductor's role in orchestrating the immune response to mucosal infections.

1.1 Introduction

The gastrointestinal (GI) epithelium is the largest continuous mucosal surface in the human body. In addition to the GI tract's role in digestion and nutrient acquisition, the intestinal epithelium is also an exclusive barrier, which separates the microbial world of the intestinal lumen from the aseptic environment of the interstitium. Remarkably, these diametrically opposed tasks-sieve versus barrier – are achieved by a single layer of epithelial cells. Within the epithelium, however, there are a number of highly specialized cell types that are dedicated to epithelial regeneration, nutrient acquisition, innate defense, immunoglobulin transport, and immunological surveillance. This chapter provides an overview of the structure and function of the

N.J. Mantis
Division of Infectious Disease, Wadsworth Center, New York State Department of Health, Albany, NY, USA
email: nmantis@wadsworth.org

M. Vajdy (ed.), *Immunity Against Mucosal Pathogens*,

intestinal epithelium with particular regard to how it keeps pathogenic microbes, as well as the commensal microflora, at bay.

The stomach, small and large intestinal epithelium are characterized by a gastric pit or crypt-villus architecture (Fig. 1.1) [1, 2]. In general, the crypts are primarily involved in cell renewal, ion and water secretion, whereas the villus epithelium functions in nutrient acquisition. Multi-potent stem cells in the so-called crypts of Lieberkühn give rise to a series of committed progenitors, which subsequently undergo differentiation to defined cell types in each location [3, 4]. Differentiated cells migrate along each of the villus axis, ultimately undergoing programmed cell death and subsequent shedding into the gut lumen. Thus, the epithelium is in a constant state of renewal, undergoing complete turn-over of its surface every 5–10 days in most mammalian species. In fact, it is estimated that the intestinal epithelium sheds $\sim 10^{10}$ cells per day [5]. The capacity for rapid turnover and self-renewal is critical for maintaining epithelial integrity and barrier function, as it prevents microbial penetration of dead or damaged cells facing the luminal environment.

Indeed, the intestinal epithelium is under constant barrage from the microbial world. Within hours after parturition, the gut becomes colonized with a microbial

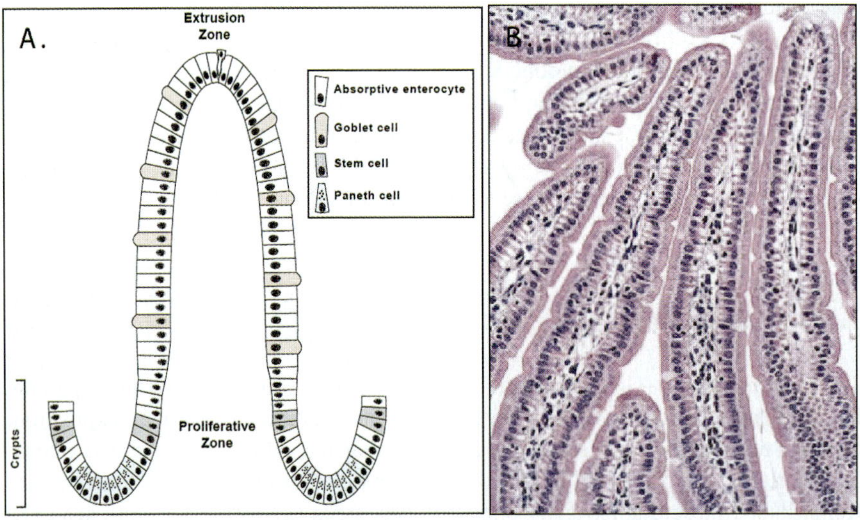

Fig. 1.1 The intestinal epithelium. (**A**) Multi-potent stem cells residing in the proliferative zone at the base of each villus, in the so-called crypts of Lieberkühn, give rise to four primary epithelial lineages: absorptive enterocytes, mucin-secreting goblet cells, enteroendocrine cells, and Paneth cells. The crypts function primarily in water and ion secretion, whereas the villi are responsible for fluid and nutrient absorption. The function of each cell type within the epithelium is described in the text. Cellular migration along the villus axis culminates in epithelial sloughing or "extrusion" through a process involving apoptosis. (**B**) A representative micrograph of mouse ileum stained with hematoxylin and eosin

flora that will eventually consist of more than 10 trillion organisms [6]. This microbial population plays an integral part in human metabolism and, is for the most part, considered beneficial (if not essential) for health [7]. However, a number of the bacterial species that cohabit the intestinal lumen are opportunistic pathogens, and are capable of causing disease if they gain access to the systemic compartment [8, 9, 10]. In addition to the normal microflora, the intestinal mucosa is also constantly exposed to pathogenic microorganisms and enterotoxins, as a consequence of the normal human diet. Many of these microbial pathogens are equipped with sophisticated molecular weaponry designed exclusively to aid in the colonization and invasion of intestinal epithelial cells [11]. It is for these reasons that the intestinal epithelium has a number of innate barriers in place to deter microbial attachment and colonization of the mucosal surfaces, and is highly coordinated with both the innate and adaptive immune systems (e.g., through the secretion of chemokines, cytokines and lipid mediators) in the event that infection or intoxication does occur [12, 13, 14].

1.2 Gastric Mucosa

The glandular mucosa of the stomach consists of three general regions, the pure parietal cell zone, muco-parietal cell zone and pure mucous zone, which is located adjacent to the pylorus. Each glandular unit of the stomach consists of pit, isthmus, neck and base regions that extend from the luminal surface to the region adjacent to the submucosa [15]. Multi-potent stem cells, located in the mid-region of the pits, give rise to all epithelial cell lineages [16]. Parietal cells, responsible for secretion of hydrochloric acid (HCl), undergo a bipolar migration within the gastric unit and are found from the surface pit region to the base of each gastric unit. The stomach's low pH provides an initial barrier for many pathogens to colonize subsequent locations in the gastrointestinal tract. Neck mucous cells in themed-region of the gastric pit secrete a variety of mucins [17], while pit and surface mucous cells that face the lumen of the stomach, secrete much of the protective mucous layer. Intrinsic-factor secreting zymogenic cells arise from the neck cell lineage and come to reside at the base of the gastric pits. Enteroendocrine cells, also residing in the base of gastric units, secrete various hormonal factors including gastrin that regulate epithelial secretion and muscular motility of the viscus.

However, certain bacterial species, *Helicobacter pylori* in particular, have evolved unique adaptations to survive in this environmental niche. *H. pylori* binds carbohydrate structures expressed on surface and neck mucous cells [18]. A potent urease is capable of sheathing the organism in a protective layer of ammonium ion to allow a short period of survival if exposed to acidic gastric juices. A type IV secretion system *H. pylori* is capable of delivering toxins to target epithelial cells [19], producing damage to the epithelium and chronic inflammation in the stomach [20].

1.3 Small Intestine

The small intestine is comprised of a tough outer serosa, three muscular layers, whose rhythmic contractions are responsible for peristalsis, the loose connective tissues of the submucosa, and the intestinal epithelium arranged in finger-like projections known as villi. Multi-potent stem cells residing at the base of each villus, in the so-called crypts of Lieberkühn, give rise to four primary epithelial lineages: absorptive enterocytes, mucin-secreting goblet cells, enteroendocrine cells, and Paneth cells (Fig. 1.1). Each of these cell types will be discussed in detail below. The epithelium is anchored on a continuous sheet of extracellular matrix, or basement membrane, consisting of a mixture of collagen, laminin, and fibronectin [21]. The actual composition of the basement membrane varies along the crypt-villus axis and this is thought to provide signals essential for enterocyte proliferation, survival and differentiation [21]. The cellular constituents of the crypts, notably stem cells, goblet cells, enteroendocrine cells, and Paneth cells, collectively are responsible for water and ion secretion, as well as exocrine, paracine and endocrine secretions, whereas the intestinal villi are primarily responsible for fluid and nutrient absorption [22, 23, 24]. A number of in vitro models have been developed that replicate (with varying degrees of success) villus enterocytes, goblet cells, and Paneth cells [25].

Migration of cells along the crypt-villus access takes 5–7 days, and ultimately results in the extrusion or "shedding" of apoptotic epithelial cells from the tips of villi. While this event has been well-defined histologically (for review see Mayhew and colleagues [26]), the exact molecular and cellular events associated with epithelial shedding are only partially understood, in part because there appears to be mechanistic differences between species [5]. Irrespective of these differences, extrusion of a single epithelial cell from the epithelial monolayer necessitates the loss of both cell-matrix, as well as, cell-cell interactions. Electron microscopy studies and biochemical analysis of murine intestinal tissues suggest that these events occur sequentially [27]. In human small intestine, cells shed from the tips of villi are caspase-3 positive indicating that apoptosis initiates just prior to (or simultaneously with) detachment from the extracellular matrix [5]. Although it has been proposed that sub-epithelial macrophages and lymphocytes may be involved in mediating epithelial shedding [26], in a recent report Bullen and colleagues found no evidence for this, at least in human small intestine [5]. Extrusion of apoptotic cells is achieved without compromising the epithelial barrier [28, 29]. Madara demonstrated by freeze-fracture, light and electron microscopy that as cells extrude from the epithelium, processes of adjacent cells extend under them. In particular, tight junction elements (see below) proliferate between extruding cells and their neighbors and appear to move down the lateral margin of the extruding cell as it extends into the lumen. Neighboring cells are generally positive for phosphorylated myosin light chain, suggesting that intestinal shedding shares similarities with oligocellular wound repair [5, 29]. The actual extrusion event involves actin-myosin rings formed around dying cells, which are pinched off like a purse string [30, 31]. Using high-resolution in vivo light microscopy Watson and colleagues propose that there are indeed transient gaps in the epithelium, but that they are filled by an impermeable substance that preserves integrity of the epithelium [29]. In vitro, apoptosis is initiated while cells are within the monolayer, and occurs concurrently with extrusion or shedding. Occasional neighboring cells (morphologically normal) were caspase-3 positive, indicating apoptosis and suggesting detachment and initiation of detachment occur simultaneously. In this regard, Fouquet and colleagues have detected loss of E-cadherin as being a key event [27].

1.3.1 Absorptive Enterocytes

Absorptive enterocytes are the predominant epithelial cell type in the small intestine, and are responsible for the bulk of nutrient absorption from the gut lumen [22, 23, 24]. Absorptive enterocytes are columnar in shape, being $\sim 25\,\mu m$ in height and $\sim 8\,\mu m$ in width. Enterocytes are joined side-by-side at their apical aspects by so-called tight junctions (TJ) or Zonula Occludens, a subcellular multiprotein complex that plays two important roles in epithelial biology. First, TJs serve as a "fence" which physically restricts the movement of proteins (e.g., membrane enzymes, channels and transporters) between the biochemically and morphologically distinct apical and basolateral plasma membranes [32, 33]. Second, TJs form a seal between cells, thereby limiting (and controlling) paracellular transport of solutes and macromolecules. Microbial pathogens and toxins are known to alter the integrity of the TJs, thereby enabling penetration of the epithelial barrier [34, 35, 36].

The apical surfaces of enterocytes are highly specialized sub-cellular structures responsible for terminal digestion and the uptake of proteins, lipids, sterols (e.g., cholesterol), carbohydrates, essential vitamins, and minerals from the intestinal lumen [32, 37, 38]). The apical surfaces consist of rigid, closely packed microvilli ($\sim 1\,\mu m$ in height, $\sim 0.1\,\mu m$ in width), which collectively form the so-called brush border membrane [37, 38, 39]. The brush border membrane contains stalked glycoproteins, including digestive enzymes, and is particularly rich in glycolipids. The brush border membranes are further proposed to be organized into at least two distinct types of lipid rafts that enable compartmentalization of enzymes and transporters, and thereby carry out specific subcellular functions [39].

In addition, the tips of enterocyte microvilli are coated with a 400–500 nm thick meshwork referred to as the filamentous brush border glycocalyx (FBBG) that is postulated to function primarily in host defense by limiting pathogen access to membrane bound receptors. The FBBG can be visualized by electron microscopy when samples are fixed simultaneously with osmium and gluteraldehyde [40, 41]. Horiuchi and colleagues recently analyzed the FBBG in mice by scanning electron microscopy and reported that the filamentous glycocalyx originating from each microvillus anastamoses with neighboring ones to form a finely porous blanket on the enterocyte surface [42]. This morphological assessment agrees with experimental work by Neutra and colleagues who demonstrated that the FBBG is a size-selective barrier that can prevent particles from gaining access to both glycolipid and glycoprotein receptors present on the tips of microvilli [41, 43, 44]. In rabbits, the FBBG consists of a mucin-like glycoprotein of apparent molecular mass of 400kDa that is postulated to be anchored in the brush border membrane by a peptide stalk [45]. It should be noted that polarized human intestinal epithelial cell lines grown in culture (e.g.,Caco-2) have poorly developed FBBG [41].

1.3.2 Goblet Cells

Goblet cells, also known as mucin-producing cells, are responsible for the production of the mucus gel that blankets the surface of the intestinal epithelium. Goblet cells, named for their characteristic goblet-like shape, are present in the small and large intestine, and are found along the entire crypt-villus axis. In villi of the small intestine, goblet cells are interspersed among absorptive enterocytes (Fig. 1.2) [1, 46]. Under normal conditions, individual goblet cells constitutively secrete mucins, high molecular weight glycoproteins that consist of core polypeptides heavily decorated with both N- and O-linked oligosaccharide side chains. This so-called baseline secretion of mucins is necessary for both maintenance and renewal of the mucus layer, which is important in both epithelial function and defense [47, 48, 49]. For example, the mucus layer aids in lubrication and protection of the intestinal mucosa, as well as serving as an important defense mechanism

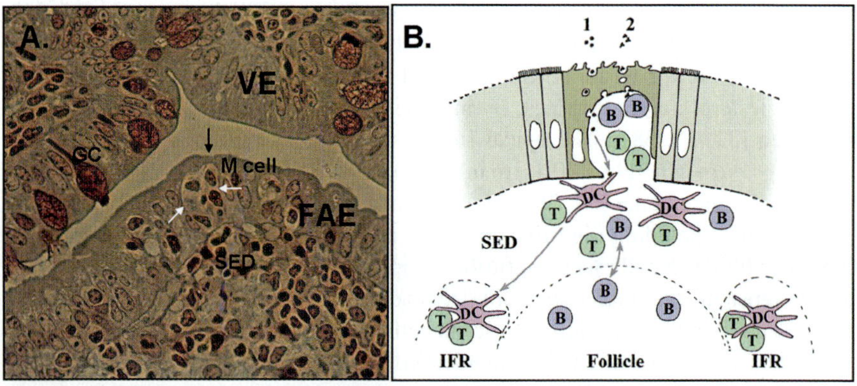

Fig. 1.2 The Follicle-Associated Epithelium and M cells. The FAE is found exclusively overlying organized lymphoid follicles situated throughout the small and large intestines. (**A**) A representative micrograph of a Peyer's patch showing both villus epithlieum (VE, upper half of image) and FAE (lower half of image). The VE consists of entrocytes and interspersed Goblet cells (GC); the latter being largely absent from the FAE. The FAE consists of enterocytes and M cells. A particularly good example of an M cell is shown in the center of the image. The unique M cell apical membrane is highlighted with a black arrow. The M cell pocket (white arrows) is occupied by at least four leukocytes. (**B**) Cartoon depicting antigen uptake and transport by M cells. The attachment of particulate antigens (1) or pathogens (2) to the apical surfaces of M cells results in their transepithelial transport and delivery into the intraepithelial pocket or SED. The intraepithelial pocket is populated by B cells, T cells and occasional DCs, and is proposed to be an extension of the underlying germinal center. Cartoon from [66]

against microbial pathogens and toxins. The viscous mucus gel limits diffusion of macromolecues and impedes motility of parasites and bacterial pathogens [50, 51]. Through heterogeneous N- and O-linked oligosaccharide side chains on mucins, the mucus layer also provides "decoy" ligands for lectin like adhesins expressed by microbial pathogens and toxins, thereby competitively inhibiting these agents from gaining access to their receptors on the apical surfaces of enterocytes. In situations where the intestinal epithelium is exposed to microbial pathogens, toxins or other intraluminal irritants, goblet cells can release additional mucins that are normally stored in apically-residing granules. Recently, Robbe and colleagues [52] used highly sensitive mass spectroscopy methods to begin to define the variations in structure of oligosaccharide side chains on mucins present throughout the gut.

1.3.3 Enteroendocrine Cells

Enteroendocrine cells are a specialized sub-population of "sensory" epithelial cells that serve as a link between the intestinal lumen and the enteric nervous system [53]. The primary function of this epithelial cell type is to secrete peptide hormones and transmitters. Enterochromaffin cells, for example, are responsible for the production and secretion of serotonin (5-hydroxytryptamine), a hormone which regulates (among other things) intestinal peristaltic and secretory reflexes. Recently, a sub-population of enteroendocrine cells have been proposed to serve as "taste receptors," based on the immunohistochemistry and real time PCR analysis demonstrating the presence of taste signaling molecules in the small intestine of mice [54].

1.3.4 Paneth Cells

Paneth cells migrate to the crypt base where these cells reside for an average of 21–28 days in most mammals [55, 56]. Paneth cells possess a well-developed apical secretory apparatus. Their location in the crypt base places them in an ideal position to deliver growth factors to dividing cells in higher compartments of the crypt, and to create gradients of anti-microbial factors which limit, or altogether prevent, microbial colonization of the small intestinal crypts. Key anti-microbial factors produced by Paneth cells include lysozyme, the cryptidins alpha-defensins and other digestive enzymes with known anti-microbial properties [56, 57, 58, 59, 60].

1.3.5 Other Cell Types

Undifferentiated crypt enterocytes are morphologically and biochemically distinct from villus enterocytes and constitute a minor population of cells within the crypts. A more compete review of their unique properties was recently published [61]. A cardinal feature of undifferentiated crypt enterocytes is their expression of the polymeric immunoglobulin receptor (pIgR) which is responsible for the unidirectional (basolateral to apical) transport of polymeric IgA and IgM from the interstitial fluids into the intestinal lumen [62, 63].

1.4 Colonic Epithelium

Unlike the small intestine with the crypt-villus structure, the colon consists of a flattened crypt-cuff unit. Migrating cells traverse the epithelial crypts to the cuff edge and surface epithelium facing the lumen, where they are subsequently lost through the same process of apoptosis. Epithelial crypts of Lieberkühn in the colon give rise to three colonic primary differentiated lineages. Absorptive colonocytes play specific roles, depending upon their location. Epithelium in the cecum and proximal colon expresses a basolaterial $H+/K+$ ATPase that drives significant absorption of water and electrolytes secreted during digestion. More distally, healthy epithelium is water and electrolyte impervious, assisting in the further compaction and passage of fecal material. Like in the small intestine, colonic goblet cells secrete a variety of mucins that create a viscous protective and lubricating layer over the epithelium, impenetrable to many microbial species, and also containing carbohydrate structures capable of saturating microbial adhesions. Paneth cells are present in the cecum and proximal colon of humans and other species, but absent in the mouse colon. Colonic epithelial cells are capable of secreting a number of anti-microbial peptides, including β-defensins and cathelicidins (see below), particularly during infection. It should be noted that the bacterial burden in post-natal mammals reaches greatest levels in the cecum and colon. Microbial degradation of residual foodstuffs allows the host to absorb additional calories that would otherwise remain unavailable. In many cases, the cecal epithelium is specialized to absorb microbially-produced products such as short- and medium-chain fatty acids [64].

1.5 Follicle-Associated Epithelium (FAE) and M Cells

The intestinal epithelium maintains a close collaboration with an underlying local immunological network, collectively referred to as the mucosal immune system [65, 66]. Nowhere is this collaboration more apparent than in organized lymphoid follicles, which are present throughout the small intestine and colon. In the small intestine, these macroscopic structures consist of aggregates of between 5 and10 lymphoid follicles known as Peyer's patches. As discussed in detail elsewhere in this book, these organized lymphoid follicles contain germinal centers which represent the primary sites of mucosal B cell differentiation and somatic cell hypermutation [67]. As mucosa-associated lymphoid tissues lack afferent lymphatics, germinal center activity is driven exclusively in response to antigens present in the intestinal lumen. Uptake and transepithelial transport of macromolecular antigens from the intestinal lumen to the organized lymphoid follicles is achieved the so-called follicle-associated epithelium (FAE).

1.5.1 FAE

The FAE is distinct from the villus epithelium in both structure and function (Fig. 1.2) [66]. Whereas the villus epithelium is specialized for digestion and absorption of nutrients and is dominated by absorptive enterocytes, mucin-secreting goblet cells and enteroendocrine cells, the FAE contains few or no goblet or enteroendocrine cells and has reduced levels of certain digestive enzymes [68]. There are also fewer defensin- and lysozyme-producing Paneth cells in follicle-associated crypts [69]. Enterocytes within the FAE, like their counterparts on villi, have well-developed microvilli and are coated with a thick FBBG, but they are not identical to villus cells [41, 45]. For example, FAE enterocytes express lower levels of the membrane associated hydrolases involved in digestive functions [68, 70]. It is also apparent that the glycosylation patterns of epithelial cells in the entire FAE differ from those on villi, indicating that glycosyltransferase expression in the FAE is distinct [43, 69, 71]. The FAE of Peyer's patches also express chemokines (e.g., CCL9, CCL20) involved in leukocyte homing, which are not expressed elsewhere in the small intestinal epithelium [72, 73]. Finally, the FAE is devoid of pIgR and therefore it is unable to transport IgA from the interstitium to the lumen [74, 75, 76].

1.5.2 M Cells

Probably the most distinguishing feature of the FAE is the presence of M cells, a unique epithelial cell type that is specialized in the uptake and transepithelial transport of particulate antigens, including particles and macromolecules [44, 77], viruses [78, 79, 80], bacteria [81, 82, 83], and parasites [84]. Indeed, M cells have been considered the "gateway" to the gut associated lymphoid tissue (GALT) [66]. The apical and basolateral surfaces of M cells have distinct features that enable them to rapidly and efficiently deliver mucosal antigens from the lumen to underlying lymphoid follicles (Fig. 1.2). For example, M cells lack the well-developed brush border and FBBG present on enterocytes. Consequently, M cell apical membranes are more accessible to particles, viruses and bacteria than adjacent enterocytes [41, 43]. In mice and humans, the apical surfaces of M cells have a pattern of glycosylation that is distinct from FAE enterocytes and villus enterocytes [69, 85, 86]. M cells also selectively express Toll-like receptors and pattern recognition receptors on their apical membrane that may facilitate antigen recognition and contribute to signaling in the local environment [87, 88, 89].

The M cell basolateral membrane is deeply invaginated to form a large intraepithelial "pocket" containing specific sub-populations of naive and memory B and T cells [90, 91, 92, 93, 94, 95], and occasional dendritic cells [96]. The pocket brings the M cell basolateral surface to within a few microns of the apical surface, shortening the distance that transcytotic vesicles must travel to cross the epithelium. Antigens transported by M cells are sampled by adjacent DCs [97, 98, 99]. The immunological function of the B and T cell populations within M cell pockets, however, is still unclear. The resident T cells are mostly CD4+ and in humans the majority display CD45RO, a surface marker typical of memory T cells [92], which distinguishes them from effector CD8+ intraepithelial lymphocytes. The B cell population is a mixture of naïve (SIgD+) and memory (SIgD−) cells that are proposed to originate from the underlying B cell follicles [90]. Based on co-stimulatory molecule expression and in vitro co-culture assays, Brandtzaeg and colleagues have proposed that M cells pockets are extensions of germinal centers [94, 95, 100] and that memory B cells in this niche may be actively engaged in sampling luminal antigens and presenting them to adjacent T cells [94, 95]. Activated T cells expressing CD40L could in turn induce CD40 positive memory B cell survival and proliferation.

1.6 Intraepithelial lymphocytes (IELs)

Intraepithelial lymphocytes (IELs) comprise a diverse population of T cells that closely associate with the gut epithelium. Most of these cells are thymic-independent γ/δ TCR+ CD8+ T cells that originate in gut-associated lymphoid tissues including sub-epithelial cryptopatches and isolated lymphoid follicles (ILFs) where they undergo selection and differentiation [101, 102, 103].

IELs commonly express mucosal specific integrins $\alpha_4\beta_7$ and $\alpha_E\beta_7$. The $\alpha_4\beta_7$ integrin mediates the attachment of T and B cells to MadCAM (Mucosal Addressin Cell Adhesion Molecule) on high endothelial venules (HEV) in Peyer's patches and mesenteric lymph nodes, facilitating entry of these cells into the mucosal compartment [104, 105]. The $\alpha_E\beta_7$ integrin mediates attachment of IELs to epithelial cells via its association with E-cadherin [106, 107]. $\alpha_E\beta_7$ is also expressed on populations of dendritic cells and mast cells [107]. Greater than 90% of IELs and 45–50% of lamina propria T cells found in mouse and human small intestine express the $\alpha_E\beta_7$ integrin [108]. In contrast, fewer than 5% of human peripheral blood mononuclear cells express the $\alpha_E\beta_7$ integrin, illustrating its preferential expression on mucosal T cells.

IELs intercalate within the epithelium and are believed to provide a means of immune surveillance. However, no primary role has been demonstrated for $\gamma\delta$ TCR+ T cells in host protection from Gram negative enteric pathogens including *Salmonella typhimurium, Citrobacter rodentium* or *Yersinia enterocolitica* [109, 110]. In contrast, defects in CD4+ T cell or $\alpha\beta$ TCR+ T cells are known to impact pathogen immunity to Gram negative organisms including *Salmonella* species, *Citrobacter rodentium*, and enteric *Yersinia* species [111, 112, 113, 114]. α_E-integrin deficient mice, the closest approximation to an IEL-deficient host, have to date demonstrated no phenotypic differences in survival, immune responses or infection-induced pathology in response to a variety of bacterial and viral pathogens [110, 115].

γδTCR+ T cells have been shown to improve maintenance of epithelial barrier function during in vitro *Toxoplasma gondii* and *Salmonella* infections [116]. In the absence of αβTCR+ CD4+ T cells, γδTCR+ T cells can provide some protection against *Salmonella* species and *Cryptosporidium parvum* [117, 118] and they have demonstrated protective functions in primary infections with intestinal parasites [119]. However, while γδTCR+ T cells have been shown to increase in number during a variety of infectious processes in vivo, from bacterial peritonitis to primary or secondary infections with viruses and bacterial pathogens, defining their role in vivo in host defense has remained elusive [120, 121].

1.7 Anti-Microbial Factors Produced by the Epithelium

Mucosal epithelia produce a diverse armamentarium of anti-microbial factors that play key roles in host defense against bacterial pathogens and commensal microflora. These factors include the alpha and beta defensins, cathelicidins, and larger LPS-neutralizing proteins such as bactericidal permeability increasing protein (BPI). In addition, many digestive enzymes, including lysozyme, proteases, and phospholipases with specificity for microbial phospholipids, have activity against commensal and pathogenic species. Gut epithelia are also capable of producing lactoferrin, a secreted iron-binding protein that is also found in milk and tears [122]. The protein's high iron-binding capacity limits availability of this nutrient for commensal or pathogenic species in the gut lumen.

1.7.1 Defensins

The defensins are a collection of small cationic peptides that bind and disrupt bacterial cell membranes [57, 59]. Defensins are divided into two main classes. Alpha-defensins include the cryptdins expressed primarily by small intestinal Paneth cells. Beta-defensins are produced by neutrophils and other mucosal epithelial cells including bronchial and colonic epithelium.

More than 17 cryptdin genes have been identified in the mouse [59]. These cysteine-rich peptides are believed to form pores in the outer-membranes of bacteria. Sequence differences among peptides likely provide specificity for activity against particular microbial populations. Expression occurs in skin, reproductive organs and in Paneth cells of the small intestinal epithelium. While some cryptdins are expressed throughout the length of the small bowel, others, notably cryptdin-4, demonstrate regional localization to the distal portion of the bowel in the jejunum and ileum [123]. Cryptdins also appear to modulate other epithelial responses including chloride secretion. Rapid secretion of Paneth cell cryptdins is known to occur in the presence of bacterial antigens, from commensal and pathogenic species

[124]. In vitro bactericidal activity in the micromolar range has been demonstrated against pathogens such as *Salmonella typhimurium, Listeria monocytogenes* and commensal strains of *E. coli* [59].

Beta-defensins have been detected in mucosal epithelia of the gut, respiratory and reproductive tracts [125, 126]. Human beta-1 defensin is expressed constitutively in gastric and colonic epithelium while expression of beta-2 defensin has been demonstrated in epithelia infected with *H. pylori* or in inflamed colon [126, 127]. Similarly, infection of human colonic xenografts into nude mice demonstrated up-regulation of HBD-2 during infection with *Salmonella*, a process dependent upon NF-kappaB activation [128].

1.7.2 Cathelicidins

Cathelicidins are also small cationic peptides with broad-spectrum anti-microbial activity that are found in mammals, birds and many fish species [129]. They are expressed in the secondary granules of neutrophils and in mucosal epithelia of the upper and lower gastrointestinal tract and in a variety of secretions including saliva [130, 131, 132]. Cathelicidins have a similar mechanism of action as the defensins, but demonstrate more activity in the presence of divalent cations.

Like the defensins, cathelicidins possess a net positive charge at neutral pH, a property that facilitates electrostatic interactions with negatively-charged bacterial surfaces. This interaction leads to subsequent insertion of the peptide in the bacterial membrane. Most cathelicidins demonstrate subsequent lytic activity, though some classes such as the bactenecins and proline-rich cathelicidins do not produce overt membrane lysis, but detectable effects on bacterial transcription [133, 134].

Unlike the defensins, cathelicidins cover a broader range of activity against different microbial species, including Gram positive and Gram negative species [135, 136, 137]. Some also have demonstrated effects in vitro against fungal and parasitic species [138].

1.7.3 Bactericidal Permeability Increasing Protein (BPI)

Bactericidal Permeability Increasing Protein (BPI) is expressed in neutrophil primary granules and in gut epithelial cells [139, 140]. At 55kDa, the protein is significantly larger than the peptide defensin and cathelicidin antimicrobials. It also has a different mechanism of action, containing structural motifs of a lipid binding protein that serves to bind and neutralize endotoxin from the outer membrane of Gram negative bacteria. Binding rapidly produces growth arrest in Gram negative species, followed by subsequent bactericidal activity [140, 141]. The protein is active against commensal and enteric pathogens at nanomolar concentrations.

In addition to providing a direct, innate defense against common bacterial pathogens including *Salmonella*, pathogenic *E. coli*, and other species, the protein's rapid neutralization of endotoxin is also thought to have immuno-modulatory properties by limiting local inflammatory responses triggered by the presence of un-neutralized LPS [142, 143]. Thus, BPI provides a potential example whereby an innate bactericidal molecule may also be capable of modifying subsequent host responses to infection. This activity is of particular use in the GI tract, providing one mechanism by which the host can avoid untoward stimulation of inflammatory responses against commensal organisms.

1.7.4 NOD/TLR

Epithelial cells also express a variety of pattern-recognition receptors including Toll-like receptors (TLRs) and Non-Obese Diabetic (NOD) family of proteins [144, 145]. Expression of these key receptors has importance both in host defense to infectious agents, as well as providing important functions in immune surveillance and tolerance to luminal antigens and commensal species.

The family of Toll-Like Receptors recognizes a variety of foreign pathogen-associated molecular patterns (PAMPs) [146]. Among these receptors TLR2 recognizes many bacterial lipopeptides [147]. TLR3 recognizes double-stranded RNA, a viral-specific PAMP [148, 149, 150]. TLR4, with the MD-2 accessory molecule, recognizes LPS [151] while TLR5 binds flagellins [152]. Bacterial peptidoglycan and lipotechoic acids are recognized by a combination of TLR2 and TLR6. TLR9 recognizes unmethylated CpG dinucleotides, found in bacterial DNA [153].

Activation of TLRs leads to a series of downstream signaling events, including recruitment of accessory molecules such as MyD88 [154], and in epithelial cells commonly results in activation of Mitogen Activated Protein kinases and nuclear translocation of the transcription factor NFκB [155, 156]. These cascades generally result in pro-inflammatory responses through the release of cytokines and chemokines such as interferon-gamma and IL-8, which serve to recruit inflammatory cells that comprise both innate and subsequent adaptive immune responses [157]. In epithelial cells, TLR signaling also up-regulates expression of many alpha- and beta-defensins [158, 159].

Many TLRs are constitutively expressed in intestinal epithelium while others are selectively up-regulated during infection or inflammatory states [144, 160]. Constitutive expression of TLR3 and TLR5 has been found in human colonic epithelium [161, 162] while TLR2 and TLR4 are expressed during infection or in inflammatory bowel diseases [161, 163]. Apical epithelial expression of TLR9 in mice has been associated with colonization of the flora, specifically in response to Gram negative commensal or pathogenic species, but not in response to the Gram positive commensal *Bifidobacterium breve* [164]. Interestingly, activation of apical TLR9 has demonstrated down-regulatory effects on NFkappaB signaling via the accumulation of ubiquitinated IkappaB, preventing activation of NFkappaB. In contrast, activation of basolateral TLR9 triggers IkappaB degradation and subsequent activation of NFkappaB [153]. This unique mechanism may provide a key mechanism by which the gut maintains "tolerance" to commensal bacteria and luminal antigens that might otherwise trigger TLR-induced pro-inflammatory responses.

Assessment of epithelial-expressed TLRs on host immunity is frequently confounded in vivo in mice fully deficient for one or more receptors or accessory molecules by the pleiotropic effects of TLR-expression in immune and other cell types. For instance, mouse strains lacking TLR4 are LPS non-responsive and frequently susceptible to a variety of pathogens [165, 166]. However, dissecting mucosal versus systemic effects of TLR activation, as well as activation by different cell types in mucosal compartments in knockout mice would be needed to define epithelial contributions that directly impact host survival, effects on local inflammatory responses and clearance of pathogens. To date studies in bone marrow chimeras have successfully delineated effects of TLR4 and TLR5 activation on immune cells during intraperitoneal *Salmonella* or *Pseudomonas aeruginosa* infection [167]. Similar types of studies adoptively transferring isogenic, TLR-sufficient immune cells into TLR-deficient recipients might assist in dissecting functions played by epithelial-expressed TLRs. However, to date no such studies have been published.

In contrast to the surface-expressed TLRs, intracellular NOD proteins provide a means for detecting intracellular PAMPs, a function important for host defense against invasive, intracellular pathogens. NOD1 and NOD2 recognize peptidoglycan motifs [148, 168]. As with TLRs, binding and activation of NOD proteins leads to activation of NFkappaB with downstream effects including up-regulation of defensins, and pro-inflammatory cytokines and chemokines [169].

NOD1 is constitutively expressed in human colon [145], while NOD2 has been found primarily in intestinal Paneth cells [170, 171]. Defects in NOD2 have also been shown to predispose to development of Crohn's disease [172]. In vitro stimulation of epithelial cells with NOD1 agonists leads to NFκB activation, an effect that is abolished in vivo in NOD1-deficient mice. Joint activation of NOD and TLR pathways has also been demonstrated to enhance TLR signaling [169, 173]. Thus activation of intracellular PAMPs, in the presence of extracellular TLR activation provides the host a means to better distinguish a likely bacterial pathogen from a commensal, and mount an effective inflammatory response.

1.8 Intestinal Epithelial Cells: Linking Innate and Adaptive Immunity

As discussed above, it is becoming increasingly apparent that epithelial cells are effective "sensors" of microbial pathogens that can orchestrate the innate and adaptive arms of the immune system in response to infection. A notable example of this is the role of the enterocytes and colonocytes in the localized recruitment and activation of leukocytes, including polymorphonuclear (PMN) cells, dendritic cells, macrophages, and lymphocytes to sites of epithelial damage through the release of pro-inflammatory chemokines and lipid mediators [12, 13, 174]. Equally important, however, is the role of the epithelium in regulating immunoglobulin transport. Activation of TLR3 and TLR4 by their respective ligands, for example, stimulates pIgR expression, suggesting intestinal epithelial cells can directly contribute to adaptive immunity in response to infection through elevated transepithelial transport of polymeric IgA [175]. A number of very recent studies have highlighted the potential of human intestinal epithelial cells to transport IgG into mucosal secretions in response to microbial infection, although the exact mechanism by which IgG neutralizes pathogens in the luminal environment remains ill-defined [110, 176]. Finally, it is now well recognized that intestinal M cells express an IgA receptor on their apical surfaces that may function in the uptake and (retro) transepithelial transport of SIgA-antigen complexes [74, 177]. The uptake and delivery of SIgA-antigen complexes into organized mucosa-associated lymphoid follicles is postulated to serve an immunoregulatory function (e.g., tolerance or immunity) [66, 178].

1.9 Conclusions

At first glance, the intestinal epithelium appears both delicate and vulnerable, and hardly capable of functioning in digestion, management of fluid and electrolyte balance, and nutrient acquisition while constantly being exposed to microbial pathogens and commensal microflora. However, closer inspection reveals that the intestinal epithelium is in fact an agglomeration of highly specialized cell types that are responsive to environmental stimuli and capable of responding in a coordinated manner to mucosal infection. Indeed, it is only recently that we have begun to appreciate the level at which the intestinal epithelium orchestrates both the innate and adaptive immune responses at mucosal surfaces.

References

1. Karam, S. M. 1999, Front Biosci. 4, D286–298
2. Roth, K. A., Hermiston, M. L. and Gordon, J. I. 1991, Proc. Natl. Acad. Sci. USA. 88, 9407–9411
3. Blanpain, C., Horsley, V. and Fuchs, E. 2007, Cell. 128, 445–458
4. Clevers, H. 2006, Cell. 127, 469–480
5. Bullen, T. F., Forrest, S., Campbell, F., Dodson, A. R., Hershman, M. J., Pritchard, D. M., Turner, J. R., Montrose, M. H. and Watson, A. J. 2006, Lab Invest. 86, 1052–1063
6. Gill, S. R., Pop, M., Deboy, R. T., Eckburg, P. B., Turnbaugh, P. J., Samuel, B. S., Gordon, J. I., Relman, D. A., Fraser-Liggett, C. M. and Nelson, K. E. 2006, Science. 312, 1355–1359
7. Turnbaugh, P. J., Ley, R. E., Mahowald, M. A., Magrini, V., Mardis, E. R. and Gordon, J. I. 2006, Nature. 444, 1027–1031
8. Johansen, F. E., Pekna, M., Norderhaug, I. N., Haneberg, B., Hietala, M. A., Krajci, P., Betsholtz, C. and Brandtzaeg, P. 1999, J. Exp. Med. 190, 915–922
9. Kelly, D., Conway, S. and Aminov, R. 2005 Trend. Immunol. 26, 326–333
10. Macpherson, A. J., Gatto, D., Sainsbury, E., Harriman, G. R., Hengartner, H. and Zinkernagel, R. M. 2000, Science. 288, 2222–2226
11. Cossart, P. and Sansonetti, P. J. 2004, Science. 304, 242–248
12. Colgan, S. P. 2002, Cell Mol. Life Sci. 59, 754–760
13. Dwinell, M. B., Johanesen, P. A. and Smith, J. M. 2003, Surgery. 133, 601–607
14. Kagnoff, M. F. 2006, Ann. N Y Acad. Sci. 1072, 313–320

15. Karam, S. M., Li, Q. and Gordon, J. I. 1997, Am. J. Physiol. 272, G1209–1220
16. Lorenz, R. G. and Gordon, J. I. 1993, J. Biol. Chem. 268, 26559–26570
17. Hanby, A. M., Poulsom, R., Playford, R. J. and Wright, N. A. 1999, J. Pathol. 187, 331–337
18. Boren, T., Falk, P., Roth, K. A., Larson, G. and Normark, S. 1993, Science. 262, 1892–1895
19. Odenbreit, S., Puls, J., Sedlmaier, B., Gerland, E., Fischer, W. and Haas, R. 2000, Science. 287, 1497–1500
20. Lax, A. J. and Thomas, W. 2002, Trend. Microbiol. 10, 293–299
21. Beaulieu, J.-F. 1999, Front. Bioscience. 4, 310–321
22. Daniel, H. 2004, Annu. Rev. Physiol. 66, 361–384
23. Marsh, M. N. 1971, Annals Royal Coll. Surg. England. 48, 356–368
24. Simon, T. C., Roth, K. A. and Gordon, J. I. 1993, J. Biol. Chem. 268, 18345–18358
25. Simon-Assmann, P., Turck, N., Sidhoum-Jenny, M., Gradwohl, G. and Kedinger, M. 2007, Cell Biol, Toxicol. 23, 241–256
26. Mayhew, T. M., Myklebust, R., Whybrow, A. and Jenkins, R. 1999, Histol. Histopath. 14, 257–267
27. Fouquet, S., Lugo-Martinez, V. H., Faussat, A. M., Renaud, F., Cardot, P., Chambaz, J., Pincon-Raymond, M. and Thenet, S. 2004, J. Biol. Chem. 279, 43061–43069
28. Madara, J. L. 1990, J. Membr. Biol. 116, 177–184
29. Watson, A. J., Chu, S., Sieck, L., Gerasimenko, O., Bullen, T., Campbell, F., McKenna, M., Rose, T. and Montrose, M. H. 2005 Gastro. 129, 902–912
30. Florian, P., Schoneberg, T., Schulzke, J. D., Fromm, M. and Gitter, A. H. 2002, J Physiol. 545, 485–499
31. Rosenblatt, J., Raff, M. C. and Cramer, L. P. 2001, Curr. Biol. 11, 1847–1857
32. Turner, J. R. 2006, Am. J. Pathol. 169, 1901–1909
33. Van Itallie, C. M. and Anderson, J. M. 2006, Annu. Rev. Physiol. 68, 403–429
34. Di Pierro, M., Lu, R., Uzzau, S., Wang, W., Margaretten, K., Pazzani, C., Maimone, F. and Fasano, A. 2001. J. Biol. Chem. 276, 19160–19165
35. Nusrat, A., von Eichel-Streiber, C., Turner, J. R., Verkade, P., Madara, J. L. and Parkos, C. A. 2001, Infect. Immun. 69, 1329–1336
36. Shifflett, D. E., Clayburgh, D. R., Koutsouris, A., Turner, J. R. and Hecht, G. A. 2005, Lab. Invest. 85, 1308–1324
37. Mooseker, M. S. 1985, Annu. Rev. Cell Biol. 1, 209–241
38. Semenza, G. 1986, Annu. Rev. Cell Biol. 2, 255–313
39. Danielsen, E. M. and Hansen, G. H. 2003, Biochim. Biophys Acta. 1617, 1–9
40. Bye, W., Allan, C. and Trier, J. 1984, Gastroenterology. 86, 789–801
41. Frey, A., Giannasca, K. T., Weltsin, R., Giannasca, P. J., Reggio, H., Lencer, W. I. and Neutra, M. R. 1996, J. Exp. Med. 184, 1045–1059
42. Horiuchi, K., Naito, I., Nakano, K., Nakatani, S., Nishida, K., Taguchi, T. and Ohtsuka, A. 2005, Arch. Histol. Cytol. 68, 51–56
43. Mantis, N. J., Frey, A. and Neutra, M. R. 2000, Am. J Phys. 278, G915–924
44. Neutra, M., Phillips, T., Mayer, E. and Fishkind, D. 1987, Cell Tissue Res. 247, 537–546
45. Maury, J., Bernadac, A., Rigal, A. and Maroux, S. 1995, J. Cell Sci. 108, 2705–2713
46. Iwakiri, D. and Podolsky, D. K. 2001, Gastroenterology. 120, 1372–1380
47. Deplancke, B. and Gaskins, H. R. 2001, Am. J. Clin. Nutri. 73, 1131S–1141S
48. Forstner, J. F., Oliver, M. G. and Sylvester, F. A. 1995, In Infections of the gastrointestinal tract (Blaser, M., Smith, P. D., Ravdin, J. I., Greenberg, H. B. and Guerrant, R. L., eds.). pp. 71–88, Raven Press, New York
49. Lievin-Le Moal, V. and Servin, A. L. 2006, Clin. Microbiol. Rev. 19, 315–337
50. Lai, S. K., O'Hanlon, D. E., Harrold, S., Man, S. T., Wang, Y. Y., Cone, R. and Hanes, J. 2007, Proc. Natl. Acad. Sci. USA. 104, 1482–1487
51. Nielsen, A. T., Dolganov, N. A., Otto, G., Miller, M. C., Wu, C. Y. and Schoolnik, G. K. 2006, PLoS Pathog. 2, e109
52. Robbe, C., Capon, C., Coddeville, B. and Michalski, J. C. 2004, Biochem J. 384, 307–316
53. Flemstrom, G. and Sjoblom, M. 2005, Am. J. Physiol. Gastrointest. Liver Physiol. 289, G377–380

54. Bezencon, C., le Coutre, J. and Damak, S. 2007, Chemical Senses. 32, 41–49
55. Andreu, P., Colnot, S., Godard, C., Gad, S., Chafey, P., Niwa-Kawakita, M., Laurent-Puig, P., Kahn, A., Robine, S., Perret, C. and Romagnolo, B. 2005, Development. 132, 1443–1451
56. Bry, L., Falk, P., Huttner, K., Ouellette, A., Midtvedt, T. and Gordon, J. I. 1994, Proc. Natl. Acad. Sci. USA. 91, 10335–10339
57. Bevins, C. L. 2005, Am. J. Physiol. Gastrointest. Liver Physiol. 289, G173–176
58. Keshav, S. 2006, J. Leukoc. Biol. 80, 500–508
59. Ouellette, A. J. 2005, Springer Semin Immunopathol. 27, 133–146
60. Wehkamp, J., Chu, H., Shen, B., Feathers, R. W., Kays, R. J., Lee, S. K. and Bevins, C. L. 2006, FEBS Lett. 580, 5344–5350
61. Turner, J. R. 2003, In Microbial pathogenesis of the intetsinal epithelial cell (Hecht, G., ed.). pp. 1–22, ASM Press, Washington D.C.
62. Brandtzaeg, P. 1978, Scand. J. Immunol. 8, 39–52
63. Phalipon, A. and Corthesy, B. 2003, Trend. Immunol. 24, 55–58
64. Scheppach, W. 1994, Gut. 35, S35–38
65. Holmgren, J. and Czerkinsky, C. 2005, Nat. Med. 11, S45–53
66. Neutra, M., Mantis, N. and Kraehenbuhl, J. P. 2001, Nat. Immunol. 2, 1004–1009
67. Brandtzaeg, P. and Johansen, F. E. 2005, Immunol. Rev. 206, 32–63
68. Owen, R. and Bhalla, D. 1983, Am. J. Anat. 168, 199–212
69. Giannasca, P. J., Giannasca, K. T., Falk, P., Gordon, J. I. and Neutra, M. R. 1994, Am. J. Physiol. 267, G1108–G1121
70. Savidge, T. C. and Smith, M. W. 1995, Adv. Exp. Med. Biol. 371A, 239–241
71. Gebert, A. and Hach, G. 1993, Gastroenterology. 105, 1350–1361
72. Iwasaki, A. and Kelsall, B. L. 2000, J. Exp. Med. 191, 1381–1394
73. Zhao, X., Sato, A., Dela Cruz, C. S., Linehan, M., Luegering, A., Kucharzik, T., Shirakawa, A. K., Marquez, G., Farber, J. M., Williams, I. and Iwasaki, A. 2003, J. Immunol. 171, 2797–2803
74. Mantis, N. J., Cheung, M. C., Chintalacharuvu, K. R., Rey, J., Corthesy, B. and Neutra, M. R. 2002, J. Immunol. 169, 1844–1851
75. Pappo, J. and Owen, R. L. 1988, Gastroenterology. 95, 1173–1174
76. Weltzin, R., Lucia-Jandris, P., Michetti, P., Fields, B. N., Kraehenbuhl, J. P. and Neutra, M. R. 1989, J. Cell Biol. 108, 1673–1685
77. Pappo, J. and Ermak, T. H. 1989, Clin. Exp. Immunol. 76, 144–148
78. Amerongen, H. M., Weltzin, R., Farnet, C. M., Michetti, P., Haseltine, W. A. and Neutra, M. R. 1991, J. Acq. Immun. Def. Synd. 4, 760–765
79. Sicinski, P., Rowinski, J., Warchol, J. B., Jarzabek, Z., Gut, W., Szczygiel, B., Bielecki, K. and Koch, G. 1990, Gastroenterology. 98, 56–58
80. Wolf, J., Rubin, D., Finberg, R., Kauffman, R., Sharpe, A., Trier, J. and Fields, B. 1981, Science. 212, 471–472
81. Jones, B., Ghori, N. and Falkow, S. 1994, J. Exp. Med. 180, 15–23
82. Sansonetti, P. J., Arondel, J., Cantey, J. R., Prevost, M. C. and Huerre, M. 1996, Infect. Immun. 64, 2752–2764
83. Owen, R., Pierce, N., Apple, R. and Cray, W., Jr. 1986, J. Infect. Dis. 153, 1108–1118
84. Marcial, M. A. and Madara, J. L. 1986, Gastroenterology. 90, 583–594
85. Clark, M. A., Jepson, M. A., Simmons, N. L. and Hirst, B. H. 1993, J. Histochem. Cytochem. 41, 1679–1687
86. Giannasca, P. J., Giannasca, K. T., Leichtner, A. M. and Neutra, M. R. 1999, Infect Immun. 67, 946–953
87. Chabot, S., Wagner, J. S., Farrant, S. and Neutra, M. R. 2006, J. Immunol. 176, 4275–4283
88. Lo, D., Tynan, W., Dickerson, J., Mendy, J., Chang, H. W., Scharf, M., Byrne, D., Brayden, D., Higgins, L., Evans, C. and O'Mahony, D. J. 2003, Cell Immunol. 224, 8–16
89. Tohno, M., Shimosato, T., Kitazawa, H., Katoh, S., Iliev, I. D., Kimura, T., Kawai, Y., Watanabe, K., Aso, H., Yamaguchi, T. and Saito, T. 2005, Biochem Biophys Res Commun. 330, 547–554

90. Brandtzaeg, P., Baekkevold, E. S., Farstad, I. N., Jahnsen, F. L., Johansen, F. E., Nilsen, E. M. and Yamanaka, T. 1999, Immunol. Today. 20, 141–151
91. Ermak, T. and Owen, R. L. 1986, Anat. Rec. 215, 144–152
92. Farstad, I. N., Halstensen, T. S., Fausa, O. and Brandtzaeg, P. 1994, Immunology. 83, 457–464
93. Mantis, N. J. and Wagner, J. S. 2004, J. Drug Targeting. 12, 79–87
94. Yamanaka, T., Straumfors, A., Morton, H., Fausa, O., Brandtzaeg, P. and Farstad, I. 2001, Eur. J. Immunol. 31, 107–117
95. Yamanaka, T., Straumfors, A., Morton, H. C., Rugtveit, J., Fausa, O., Brandtzaeg, P. and Farstad, I. N. 1999, Immunol. Lett. 69, 42
96. Iwasaki, A. and Kelsall, B. 2001, J. Immunol. 166, 4884–4890
97. Hopkins, S. A. and Kraehenbuhl, J. P. 1997, Adv. Exp. Med. Biol. 417, 105–109
98. Niedergang, F., Sirard, J. C., Blanc, C. T. and Kraehenbuhl, J. P. 2000, Proc. Natl. Acad. Sci. USA. 97, 14650–14655
99. Pron, B., Boumaila, C., Jaubert, F., Berche, P., Milon, G., Geissmann, F. and Gaillard, J. L. 2001, Cell Microbiol. 3, 331–340
100. Yamanaka, T., Helgeland, L., Farstad, I. N., Fukushima, H., Midtvedt, T. and Brandtzaeg, P. 2003, J Immunol. 170, 816–822
101. Nonaka, S., Naito, T., Chen, H., Yamamoto, M., Moro, K., Kiyono, H., Hamada, H. and Ishikawa, H. 2005, J Immunol. 174, 1906–1912
102. Pabst, O., Herbrand, H., Worbs, T., Friedrichsen, M., Yan, S., Hoffmann, M. W., Korner, H., Bernhardt, G., Pabst, R. and Forster, R. 2005, Eur. J. Immunol. 35, 98–107
103. Suzuki, K., Oida, T., Hamada, H., Hitotsumatsu, O., Watanabe, M., Hibi, T., Yamamoto, H., Kubota, E., Kaminogawa, S. and Ishikawa, H. 2000, Immunity. 13, 691–702
104. Berlin, C., Berg, E. L., Briskin, M. J., Andrew, D. P., Kilshaw, P. J., Holzmann, B., Weissman, I. L., Hamann, A. and Butcher, E. C. 1993, Cell. 74, 185–195
105. Picarella, D., Hurlbut, P., Rottman, J., Shi, X., Butcher, E. and Ringler, D. J. 1997, J. Immunol. 158, 2099–2106
106. Higgins, J. M., Mandlebrot, D. A., Shaw, S. K., Russell, G. J., Murphy, E. A., Chen, Y. T., Nelson, W. J., Parker, C. M. and Brenner, M. B. 1998, J. Cell Biol. 140, 197–210
107. Taraszka, K. S., Higgins, J. M., Tan, K., Mandelbrot, D. A., Wang, J. H. and Brenner, M. B. 2000, J. Exp. Med. 191, 1555–1567
108. Schon, M. P., Arya, A., Murphy, E. A., Adams, C. A., Strauch, U. G., Agace, W. W., Marsal, J., Donohue, J. P., Her, H., Beier, D. R., Olson, S., Lefrancois, L., Brenner, M. B., Grusby, M. J. and Parker, C. M. 1999, J. Immunol. 162, 6641–6649
109. Autenrieth, I. B., Tingle, A., Reske-Kunz, A. and Heesemann, J. 1992, Infect. Immun. 60, 1140–1149
110. Bry, L. and Brenner, M. B. 2004, J. Immunol. 172, 433–441
111. Bry, L., Brigl, M. and Brenner, M. B. 2006, Infect. Immun. 74, 673–681
112. Hormaeche, C. E., Mastroeni, P., Arena, A., Uddin, J. and Joysey, H. S. 1999, Immunology. 70, 247–250
113. Kempf, V. A., Bohn, E., Noll, A., Bielfeldt, C. and Autenrieth, I. B. 1998, Clin. Exp. Immunol. 113, 429–437
114. McSorley, S. J., Asch, S., Costalonga, M., Reinhardt, R. L. and Jenkins, M. K. 2002, Immunity. 16, 365–377
115. Lefrancois, L., Parker, C. M., Olson, S., Muller, W., Wagner, N., Schon, M. P. and Puddington, L. 1999, J. Exp. Med. 189, 1631–1638
116. Dalton, J. E., Cruickshank, S. M., Egan, C. E., Mears, R., Newton, D. J., Andrew, E. M., Lawrence, B., Howell, G., Else, K. J., Gubbels, M. J., Striepen, B., Smith, J. E., White, S. J. and Carding, S. R. 2006, Gastroenterol. 131, 818–829
117. Eichelberger, M. C., Suresh, P. and Rehg, J. E. 2000, Comp Med. 50, 270–276
118. Mixter, P. F., Camerini, V., Stone, B. J., Miller, V. L. and Kronenberg, M. 1994, Infect. Immun. 62, 4618–4621

119. Bozic, F., Forcic, D., Mazuran, R., Marinculic, A., Kozaric, Z. and Stojcevic, D. 1998, Comp. Immunol. Microbiol. Infect. Dis. 21, 201–214
120. Ramsburg, E., Tigelaar, R., Craft, J. and Hayday, A. 2003, J. Exp. Med. 198, 1403–1414
121. Rosat, J. P., MacDonald, H. R. and Louis, J. A. 1993, J. Immunol. 150, 550–555
122. Ward, P. P., Paz, E. and Conneely, O. M. 2005, Cell Mol. Life Sci. 62, 2540–2548
123. Tanabe, H., Qu, X., Weeks, C. S., Cummings, J. E., Kolusheva, S., Walsh, K. B., Jelinek, R., Vanderlick, T. K., Selsted, M. E. and Ouellette, A. J. 2004, J. Biol. Chem. 279, 11976–11983
124. Ayabe, T., Satchell, D. P., Wilson, C. L., Parks, W. C., Selsted, M. E. and Ouellette, A. J. 2000, Nat. Immunol. 1, 113–118
125. Meyerholz, D. K., Gallup, J. M., Grubor, B. M., Evans, R. B., Tack, B. F., McCray, P. B., Jr. and Ackermann, M. R. 2004, Dev. Comp. Immunol. 28, 171–178
126. Wehkamp, J., Fellermann, K., Herrlinger, K. R., Baxmann, S., Schmidt, K., Schwind, B., Duchrow, M., Wohlschlager, C., Feller, A. C. and Stange, E. F. 2002, Eur. J. Gastroenterol. Hepatol. 14, 745–752
127. Boughan, P. K., Argent, R. H., Body-Malapel, M., Park, J. H., Ewings, K. E., Bowie, A. G., Ong, S. J., Cook, S. J., Sorensen, O. E., Manzo, B. A., Inohara, N., Klein, N. J., Nunez, G., Atherton, J. C. and Bajaj-Elliott, M. 2006, J. Biol. Chem. 281, 11637–11648
128. O'Neil, D. A., Porter, E. M., Elewaut, D., Anderson, G. M., Eckmann, L., Ganz, T. and Kagnoff, M. F. 1999, J. Immunol. 163, 6718–6724
129. Lehrer, R. I. and Ganz, T. 2002, Curr. Opin. Hematol. 9, 18–22
130. Song, J. J., Hwang, K. S., Woo, J. S., Chae, S. W., Cho, J. G., Kang, H. J., Hwang, S. J. and Lee, H. M. 2006, Inter. J. Ped. Otorhinolaryngol. 70, 487–492
131. Woo, J. S., Jeong, J. Y., Hwang, Y. J., Chae, S. W., Hwang, S. J. and Lee, H. M. 2003, Arch. Otolaryngol. Head Neck Surg. 129, 211–214
132. Wehkamp, J., Schauber, J. and Stange, E. F. 2007, Curr. Opin. Gastroenterol. 23, 32–38
133. Anderson, R. C., Hancock, R. E. and Yu, P. L. 2004, Antimicrob. Agents. Chemother. 48, 673–676
134. Shinnar, A. E., Butler, K. L. and Park, H. J. 2003, Bioorganic chem. 31, 425–436
135. Bals, R., Weiner, D. J., Moscioni, A. D., Meegalla, R. L. and Wilson, J. M. 1999, Infect. Immun. 67, 6084–6089
136. Lee, P. H., Ohtake, T., Zaiou, M., Murakami, M., Rudisill, J. A., Lin, K. H. and Gallo, R. L. 2005, Proc. Natl. Acad. Sci. USA. 102, 3750–3755
137. Wu, H., Zhang, G., Minton, J. E., Ross, C. R. and Blecha, F. 2000, Infect. Immun. 68, 5552–5558
138. Skerlavaj, B., Scocchi, M., Gennaro, R., Risso, A. and Zanetti, M. 2001, Antimicrob. Agents Chemother. 45, 715–722
139. Canny, G., Levy, O., Furuta, G. T., Narravula-Alipati, S., Sisson, R. B., Serhan, C. N. and Colgan, S. P. 2002, Proc. Natl. Acad. Sci. USA. 99, 3902–3907
140. Levy, O., Canny, G., Serhan, C. N. and Colgan, S. P. 2003, Biochem. Soc. Trans. 31, 795–800
141. Canny, G., Cario, E., Lennartsson, A., Gullberg, U., Brennan, C., Levy, O. and Colgan, S. P. 2006, Am. J. Physiol. Gastrointest. Liver Physiol. 290, G557–567
142. Azuma, M., Matsuo, A., Fujimoto, Y., Fukase, K., Hazeki, K., Hazeki, O., Matsumoto, M. and Seya, T. 2007, Biochem. Biophys. Res. Commun. 354, 574–578
143. Fang, W. H., Yao, Y. M., Shi, Z. G., Yu, Y., Wu, Y., Lu, L. R. and Sheng, Z. Y. 2001, Crit. Care Med. 29, 1452–1459
144. Abreu, M. T., Fukata, M. and Arditi, M. 2005, J. Immunol. 174, 4453–4460
145. Kim, J. G., Lee, S. J. and Kagnoff, M. F. 2004, Infect. Immun. 72, 1487–1495
146. Takeda, K., Kaisho, T. and Akira, S. 2003, Annu. Rev. Immunol. 21, 335–376
147. Kirschning, C. J. and Schumann, R. R. 2002, Curr. Top. Microbiol Immunol. 270, 121–144
148. Matsumoto, M., Funami, K., Oshiumi, H. and Seya, T. 2004, Microbiol. Immunol. 48, 147–154
149. Salio, M. and Cerundolo, V. 2005, Curr. Biol. 15, R336–339
150. Sen, G. C. and Sarkar, S. N. 2005, Cytokine Growth Factor Rev. 16, 1–14

151. Neal, M. D., Leaphart, C., Levy, R., Prince, J., Billiar, T. R., Watkins, S., Li, J., Cetin, S., Ford, H., Schreiber, A. and Hackam, D. J. 2006, J. Immunol. 176, 3070–3079
152. Rumbo, M., Nempont, C., Kraehenbuhl, J. P. and Sirard, J. C. 2006, FEBS Lett. 580, 2976–2984
153. Lee, J., Mo, J. H., Shen, C., Rucker, A. N. and Raz, E. 2007, Curr Opin Gastroenterol. 23, 27–31
154. Dauphinee, S. M. and Karsan, A. 2006, Lab. Invest. 86, 9–22
155. Akira, S. and Takeda, K. 2004, Comptes Rendus Biologies. 327, 581–589
156. Vogel, S. N., Fitzgerald, K. A. and Fenton, M. J. 2003, Mol. Interventions. 3, 466–477
157. Zhao, L., Kwon, M. J., Huang, S., Lee, J. Y., Fukase, K., Inohara, N. and Hwang, D. H. 2007, J. Biol. Chem. 282, 11618–11628
158. Hertz, C. J., Wu, Q., Porter, E. M., Zhang, Y. J., Weismuller, K. H., Godowski, P. J., Ganz, T., Randell, S. H. and Modlin, R. L. 2003, J. Immunol. 171, 6820–6826
159. Vora, P., Youdim, A., Thomas, L. S., Fukata, M., Tesfay, S. Y., Lukasek, K., Michelsen, K. S., Wada, A., Hirayama, T., Arditi, M. and Abreu, M. T. 2004, J. Immunol. 173, 5398–5405
160. Gewirtz, A. T. 2003, Curr. Pharmaceutical Design. 9, 1–5
161. Furrie, E., Macfarlane, S., Thomson, G. and Macfarlane, G. T. 2005, Immunology. 115, 565–574
162. Ortega-Cava, C. F., Ishihara, S., Rumi, M. A., Aziz, M. M., Kazumori, H., Yuki, T., Mishima, Y., Moriyama, I., Kadota, C., Oshima, N., Amano, Y., Kadowaki, Y., Ishimura, N. and Kinoshita, Y. 2006, Clin. Vaccine Immunol. 13, 132–138
163. Swerdlow, M. P., Kennedy, D. R., Kennedy, J. S., Washabau, R. J., Henthorn, P. S., Moore, P. F., Carding, S. R. and Felsburg, P. J. 2006, Vet. Immunol. Immunopath. 114, 313–319
164. Ewaschuk, J. B., Backer, J. L., Churchill, T. A., Obermeier, F., Krause, D. O. and Madsen, K. L. 2007, Infect. Immun. 75, 2572–2579
165. Miller, S. I., Ernst, R. K. and Bader, M. W. 2005, Nat. Rev. Microbiol. 3, 36–46
166. Roy, M. F., Lariviere, L., Wilkinson, R., Tam, M., Stevenson, M. M. and Malo, D. 2006, Gene Immunity. 7, 372–383
167. Feuillet, V., Medjane, S., Mondor, I., Demaria, O., Pagni, P. P., Galan, J. E., Flavell, R. A. and Alexopoulou, L. 2006, Proc. Natl. Acad. Sci. USA. 103, 12487–12492
168. Uehara, A., Sugawara, Y., Kurata, S., Fujimoto, Y., Fukase, K., Kusumoto, S., Satta, Y., Sasano, T., Sugawara, S. and Takada, H. 2005, Cell Microbiol. 7, 675–686
169. Kobayashi, K. S., Chamaillard, M., Ogura, Y., Henegariu, O., Inohara, N., Nunez, G. and Flavell, R. A. 2005, Science. 307, 731–734
170. Ogura, Y., Lala, S., Xin, W., Smith, E., Dowds, T. A., Chen, F. F., Zimmermann, E., Tretiakova, M., Cho, J. H., Hart, J., Greenson, J. K., Keshav, S. and Nunez, G. 2003, Gut. 52, 1591–1597
171. Wehkamp, J., Harder, J., Weichenthal, M., Schwab, M., Schaffeler, E., Schlee, M., Herrlinger, K. R., Stallmach, A., Noack, F., Fritz, P., Schroder, J. M., Bevins, C. L., Fellermann, K. and Stange, E. F. 2004, Gut. 53, 1658–1664
172. Rosenstiel, P., Fantini, M., Brautigam, K., Kuhbacher, T., Waetzig, G. H., Seegert, D. and Schreiber, S. 2003, Gastroenterol. 124, 1001–1009
173. Chamaillard, M., Hashimoto, M., Horie, Y., Masumoto, J., Qiu, S., Saab, L., Ogura, Y., Kawasaki, A., Fukase, K., Kusumoto, S., Valvano, M. A., Foster, S. J., Mak, T. W., Nunez, G. and Inohara, N. 2003, Nat. Immunol. 4, 702–707
174. Rimoldi, M., Chieppa, M., Salucci, V., Avogadri, F., Sonzogni, A., Sampietro, G. M., Nespoli, A., Viale, G., Allavena, P. and Rescigno, M. 2005, Nat. Immunol. 6, 507–514
175. Schneeman, T. A., Bruno, M. E., Schjerven, H., Johansen, F. E., Chady, L. and Kaetzel, C. S. 2005, J. Immunol. 175, 376–384
176. Yoshida, M., Masuda, A., Kuo, T. T., Kobayashi, K., Claypool, S. M., Takagawa, T., Kutsumi, H., Azuma, T., Lencer, W. I. and Blumberg, R. S. 2006, Springer Semin Immunopathol. 28, 397–403
177. Rey, J., Garin, N., Spertini, F. and Corthesy, B. 2004, J. Immunol. 172, 3026–3033
178. Corthesy, B. 2007, J. Immunol. 178, 27–32

Chapter 2
Structure of the Respiratory and Female Genitourinary Tracts

Michael Vajdy

Abstract The respiratory and genitourinary tracts comprise two major mucosal tissues. There are some similarities between the structure and cells of these sites, and there are also important differences. The respiratory tract is arguably more exposed to the outside environment than the genitourinary tract, as it is essential for continuously acquiring oxygen. Therefore, it is also more accessible, and socially acceptable, as a vaccination route. The genitourinary tract, together with the rectal mucosa, is a major site of a variety of sexually transmitted diseases, and as such they comprise the most important mucosal routes for transmission of HIV and many other pathogens. In this chapter, a description of the structure and cells of the upper and lower respiratory tract as well as the genitourinary and rectal mucosa is given.

2.1 Introduction

By far, the majority of pathogens enter the mammalian hosts through mucosal surfaces. The mucosal immune system in mammalian species covers a surprisingly large surface area. The mucosal immune system mainly comprises the gastrointestinal, respiratory and genitourinary tracts. While there are major similarities between these tissues, there are also considerable differences between them. The similarities generally pertain to the antigenic load (gastrointestinal and respiratory tracts) and the one layer of epithelial cells separating the lumen from the interior lamina propria (the intestine, the lower respiratory tract and the upper genitourinary tracts). The differences are mainly in the expression of cellular homing receptors, antigen sampling, and unique cell types. These differences and similarities play a major role in the design of effective vaccines as well as therapeutics. Thus, while it has generally been difficult to devise vaginally administered vaccines, it has been easier to design intranasal (IN) vaccines. Moreover, while oral vaccines require

M. Vajdy
University of California, Davis, School of Medicine, Department of Medical Microbiology and Immunology, Davis CA 95616, USA
e-mail: michaelvajdy@comcast.net

M. Vajdy (ed.), *Immunity Against Mucosal Pathogens*,

high antigenic concentrations, IN vaccines require several folds less. Therefore, knowledge of the cellular structure, the immune microenvironment and homing properties of each mucosal tissue is required to make a rational design of vaccines through each route. Given the notion that rapid local immune responses are required to induce optimal protection against new mucosal pathogens, knowledge about the cellular structure and the related mucosal immune system at each mucosal tract is essential for development of prophylactic and therapeutic protective strategies.

2.2 Structure of the Upper and Lower Respiratory Tract

The upper respiratory tract includes the nose and the larynx, which separates the upper and lower respiratory tracts. Except the anterior nares, which are lined with squamous epithelium, most of the upper respiratory tract epithelia are ciliated, pseudostratified columnar cells [1].

In primates, the palatine, lingual and nasopharyngeal tonsils form the Waldeyer's ring, which when compared to similar structures in rodents is collectively referred to as nasal associated lymphoid tissue (NALT). NALT is strategically placed for sampling at the portals of both gastrointestinal and respiratory tracts [2]. The pharynx is divided into three parts, the nasopharynx (posterior to the nose and superior to the soft palate), the oropharynx (posterior to the mouth) and the laryngopharynx (posterior to the larynx) [3]. The lymphoid tissue in the pharynx forms an incomplete circular lymphoid structure called the Waldeyer's ring. This lymphoid tissue is aggregated to form masses of LN called tonsils. The pharyngeal tonsil, known as adenoids, is in the mucous membrane of the roof and posterior wall of the nasopharynx [3]. The palatine tonsils are LN at each side of the oropharynx between the palatine arches [3]. Unlike peripheral LN, which are not directly associated with the mucosal lumen, the surface epithelium of the tonsils, similar to the mucosal associated lymphoid tissue (MALT) of the gastrointestinal tract (eg. Peyer's patches), is in direct contact with the lumen. The palatine tonsils and adenoids are covered with lymphoepithelium consisting of ciliary and non-ciliary epithelial cells, goblet cells and M cells, the latter showing many invaginating lymphoid cells. [4]. Direct uptake of antigens through the epithelial cells of the tonsils has been demonstrated and suggests that the tonsils play a major role as local inductive sites for mucosal immunity.

In addition to the lymphoid aggregates in the epithelium, local draining lymph nodes also represent important inductive sites for local and systemic immunity following application of antigens to the oral cavity. In the oral cavity, the lymphatic vessels of the parotid and submandibular glands drain to superficial and deep cervical LN. [3] Lymph from the end of the tongue drains to the superior deep cervical lymph node (LN), whereas lymph from the tip of the tongue drains to the submental LN. Lymph from the sides and the middle of the tongue drain to the inferior deep cervical LN and to the submandibular LN respectively [3].

2.3 Cells Involved in Immune Responses in Respiratory Tract

Dendritic cells (DC) are numerous within and underneath the epithelial layer of the tonsils and are in close contact with the neighboring B and T cells. [5] Alveolar macrophages are derived form bone marrow as well as by local division. However, alveolar macrophages have been shown to be involved in down regulation of antigen presenting function of pulmonary DC [6]. Approximately 10% of the bronchoalveolar cells consist of lymphocytes. Mast cells and basophils also exist within the lower respiratory tract, and can be directly involved in inflammatory responses such as asthma under the influence of T cells as well as other factors. Several studies suggest that the adenoids contain IgG, IgA and IgM secreting cells in lower numbers than palatine tonsils, with the predominant isotype being IgG (62% in adenoids and 73% in palatine tonsils). However, the relative numbers of IgA and IgM secreting cells appear to be similar [7, 8]. The predominant IgG subtype in both palatine tonsils and the adenoids are IgG1 followed by IgG3, with low numbers of IgG2 and few if any IgG4 [8]. The number of IgD+ and IgE+ Ig-containing cells are very low [9]. While the majority of IgA+ cells are in extra-follicular, non-GC areas, most IgM+ cells are in the follicles [9]. The germinal center (GC) in tonsils gives rise to mainly IgG (55–72%) and IgA (13–18%) producing B cell blasts/plasma cells [10], however IgM is mainly confined to the GC regions [11]. The tonsilar GC contains follicular dendritic cells (FDC), which are known to bind antibody-antigen immune complexes and play a role in the generation of secondary or memory B cell responses. The tonsilar FDC have been shown to be in association with IgM, IgG and IgA and the majority of the cells within their dendrites are B cells [12]. Importantly, it has been shown that memory-type B cells from human tonsil lie within the mucosal epithelium overlying the tonsils and directly present antigen to T cells [13].

The adenoids have patches of epithelium expressing polymeric Ig receptor (pIgR), whereas the palatine tonsils are covered with squamous epithelium and lack a secretory IgA system [14]. Polymeric IgA, produced in the adenoids is destined to populate mucosal effector sites of the upper airway mucosa as well as the salivary and lacrimal glands, where they produce secretory IgA (SIgA) [15]. Two subtypes of IgA exist in humans, IgA1 and IgA2. The latter is most frequent in the upper aerodigestive tracts, while the former predominates in the large intestine [16]. In this regard, tonsilar IgA+ cells are predominantly of the IgA1 subtype [8, 17], providing further evidence that they function as local inductive sites that seed the mucosal effector sites of the upper aerodigestive tracts. The palatine tonsils and the adenoids comprise 30–35% CD3+ cells, 20–28% CD4+ cells, and 5–6% CD8+ cells and the majority of the T cells appear to express TCR$\gamma\delta$. While the activation marker IL2R (CD25) is upregulated on only 3–8% of the cells, CD28 is expressed on 23–36% of the cells and mostly in the adenoids [8]. T cells comprise about half of tonsilar intraepithelial mononuclear cells, with equal ratios of CD4+ and CD8+ cells. In the deeper inter-follicular regions the ratio of $\alpha\beta$TCR+ to $\gamma\delta$TCR+ cells is 10:1, whereas in the superficial areas a reduction in the number of $\alpha\beta$TCR+ cells reduces this ratio to 2:1 [18].

In the adenoids, the predominant cytokine released in response to in vitro restimulation with tetanus and diphtheria toxoids, influenza virus or *Candida albicans* was IFNγ, followed by IL-5 [8]. However, polyclonal activation of tonsilar and adenoidal cells with phytohaemagglutinin induced proliferative responses and many cytokines, including IFNγ, IL-4, IL-5, IL-10, TNFα and TGFβ [8]. It is noteworthy that the adenoids of children aged 1–12 years contained a relatively large number of CD5+, possibly of the B-1 lineage, with a greater capacity of Ig production than conventional (B-2) CD5− cells [19]. Taken together, the presence of a lymphoepithelial structure, professional antigen presenting cells e.g. DC, FDC and the germinal center machinery, as well as functional CD4+ and CD8+ T cells within the epithelial layers as well as deep in the LN suggest that the oro/nasopharyngeal lymphoid tissues act as both immune inductive and effector sites for the upper aerodigestive tract.

The lower respiratory tract includes the trachea, the bronchi, bronchioles and the alveoli. The trachea and the major bronchi are covered by pseudostratified, columnar epithelia, whereas the bronchioles and alveoli are covered with simple columnar epithelia. Mucus-secreting cells are dispersed throughout the tracheobronchial tree, but are only found sparsely in small bronchioli. Simple, pseudostratified epithelial cells that are mostly ciliated, as well as mucus-secreting goblet cells are found in the lower respiratory tract [1]. An organized lymphoid tissue, bronchus associated lymphoid tissue (BALT), has been demonstrated in rabbits, rats and guinea pigs, but rarely in humans. Air from the trachea enters the two lung lobes via the two main bronchi, which are each subdivided into smaller bronchioles, forming the bronchial tree. Each terminal bronchiole ends in several alveolar sacs lined by alveoli where CO_2 and O_2 gas exchanges occurs.

Many of the proteins associated with innate immunity in the upper respiratory tract are to be found localized into mucus gels and the mucin-rich surface layers of the epithelium and the cilia. Mucus, produced by mucus-secreting cells in the respiratory tract, is rich in mucin and covers the epithelia. The gel-forming mucins, MUC5AC and MUC5B, are in charge of defending the epithelia from pathogens. Mucus moves upwards from the lung by mucociliary and through coughing mechanisms [20].

2.4 Structure and Cells of the Female Genitourinary Tract and Rectum

The vaginal mucosa is covered with multi-layered squamous epithelia, while the uterus, cervix and fallopian tubes are covered with pseudo-squamous and simple columnar epithelia. Underneath the epithelial layers of the vagina, uterus and fallopian tubes is the lamina propria compartment comprising a large array of B cells, CD4+ and CD8+ T cells, and antigen-presenting cells (APC) [21]. The presence of lymphoid aggregates in the female genital tract has also been reported, although whether these aggregates have follicle associated epithelium, as is the case with nasal-associated lymphoid tissue (NALT) and Peyer's patches, remains to be elucidated [21, 22]. DC and CD8+ T with cytotoxic activity cells are found interspersed within the squamous epithelium of the vagina [23, 24, 25]. Thus, the vaginal mucosa contains DC as well as CTL and can mount anti-viral cytotoxic T cell responses that can be protective. The vagina is considered to be a component of the common mucosal immune system and oral immunization in mice with microparticles has been shown to induce a vaginal antibody response [26]. In addition, IN immunization with microparticles also induced antibodies in the lower genital tract of mice [27]. Although there is no evidence to indicate the presence of lymphoid follicles or M cells in the vaginal mucosa [28], intra-vaginal (IVAG) immunization in humans induced local antibody responses [29]. However, IVAG immunization protocols in small animal models have not normally met with great success, despite the use of novel delivery systems and adjuvants [30, 31, 32], although a more recent report showed that vaginal or rectal but not IN or intra-muscular immunizations with alpha virus based replicon particles encoding HIV-1 gag protected against IVAG challenge with vaccinia virus encoding HIV-1 gag [33]. Moreover, the local immune response in the vagina is subject to significant hormonal regulation, with major changes in local antibodies at different stages of the menstrual cycle [34]. A study in mice showed that the IN route of immunization was more effective than the IVAG route for the induction of immune responses in the vagina [35]. In female humans the IN route of immunization may be exploited for the induction of genital tract antibody response [36]. Thus, although the vaginal mucosa contains the necessary immunological machinery to mount a local immune response, the IN immunization appears to be a more suitable route. However, it remains to be seen whether in the resting memory phase a more rapid local response is induced in the vaginal mucosa following IVAG immunization compared to IN immunization.

The rectal mucosa of several mammalian species, including humans, contain macroscopically invisible solitary lymphoid nodules that resemble Peyer's patches of the small intestine in their cellular structure and phenotype. These structures are overlaid with microfold (M) cells that are specialized in antigen uptake [37]. Of note, both the rectal and vaginal mucosa are drained by the iliac lymph nodes and there is indirect evidence that SIgA-secreting cells in the vaginal mucosa originate from the solitary lymphoid nodules of rectum [38, 39, 40]. In non-human primates [41] as well as in humans, the rectal and small intestinal lamina propria (LP) contain high numbers of CD69+ macrophages that are concentrated under the single layer of epithelial cells (enterocytes) whereas cells with denderites, that are far fewer in number (most likely DC), form a reticular frame work throughout the LP. The rectal mucosa may serve as a vaccines delivery route and because the vaccine does not have to go through the entire digestive tract and the intestine, lower amounts of antigen are required for intra-rectal compared to oral immunizations. However, it may not be an attractive route of immunization for socio-ethical reasons. Except for what is described above, the rectal mucosa's structure and cells generally resemble that of the small intestine, described in Chapter 1.

References

1. Jeffrey, P.K., and Corrin, B. 1984, Immunology of the Lung and Upper Respiratory Tract, J. Bienenstock (Ed.), McGraw-Hill, 1.
2. Kuper, C.F., Koornstra, P.J., Hameleers, D.M., Biewenga, J., Spit, B.J., Duijvestijn, A.M., Breda Vriesman, P.J., and Sminia, T. 1992, Immunol. Today, 13, 219.
3. Moore, K.L., and Dalley, A.F. 1999, Clinically Oriented Anatomy, P.J. Kelly (Ed.), Lippincott Williams & Wilkins, Chapter 8, 1062.
4. Fujimura, Y. 2000, Virchows Arch., 436, 560.
5. Pope, M. 1999, J. Infect. Dis., 179, S427.
6. Holt, P.G., Oliver, J., Bilyk, N., McMenamin, C., McMenamin, P.G., Kraal, G., and Thepen, T. 1993, J. Exp.Med., 177, 397.
7. Nadal, D., Soh, N., Schlapfer, E., Bernstein, J.M., and Ogra, P.L. 1992, Int. J. Pediatr. Otorhinolaryngol., 24, 121.
8. Boyaka, P.N., Wright, P.F., Marinaro, M., Kiyono, H., Johnson, J.E., Gonzales, R.A., Ikizler, M.R., Werkhaven, J.A., Jackson, R.J., Fujihashi, K., Di Fabio, S., Staats, H.F., and McGhee, J.R. 2000, Am. J. Pathol., 157, 2023.
9. Matthews, J.B., and Basu, M.K. 1982, Int. Arch. Allergy Appl. Immunol., 69, 21.
10. Brandtzaeg, P. 1996, Acta Otolaryngol. Suppl., 523, 55.
11. Franek, J., Kubin, V., Novak, V., and Bedka, J. 1981, Folia Microbiol. (Praha). 26, 253.
12. Heinen, E., Lilet-Leclercq, C., Mason, D.Y., Stein, H., Boniver, J., Radoux, D., Kinet-Denoel, C., and Simar, L.J. 1984, Eur. J. Immunol., 14, 267.
13. Liu, Y.J., Barthelemy, C., de Bouteiller, O., Arpin, C., Durand, I., and Banchereau, J. 1995, Immunity, 2, 239.
14. Brandtzaeg, P. 1999, Immunol. Today, 20, 383.
15. Brandtzaeg, P. 2003, Int. J. Pediatr. Otorhinolaryngol., 67, S69.
16. Brandtzaeg, P. 1992, J. Infect. Dis., 165, S167.
17. Brandtzaeg, P., Baekkevold, E.S., Farstad, I.N., Jahnsen, F.L., Johansen, F.E., Nilsen, E.M., and Yamanaka, T. 1999, Immunol. Today, 20, 141.
18. Graeme-Cook, F., Bhan, A.K., and Harris, N.L. 1993, Am. J. Pathol., 143, 1416.
19. Arita, M., Kodama, S., Suzuki, M., and Mogi, G. 2003, Laryngoscope, 113, 484.
20. Sheehan, J.K., Kesimer, M., and Pickles, R. 2006, Novartis Found Symp., 279, 155.
21. Yeaman, G.R., White, H.D., Howell, A., Prabhala, R., and Wira, C.R. 1998, AIDS Res. Hum. Retroviruses, 14 Suppl 1, S57.
22. Johansson, E.L., Rudin, A., Wassen, L., and Holmgren, J. 1999, Immunology, 96, 272.
23. Hu, J., Gardner, M.B., and Miller, C.J. 2000, J. Virol., 74, 6087.
24. Lohman, B.L., Miller, C.J., and McChesney, M.B. 1995, J. Immunol., 155, 5855.
25. McChesney, M.B., Collins, J.R., and Miller, C.J. 1998, AIDS Res. Hum. Retroviruses, 14 Suppl 1, S63.
26. Challacombe, S.J., Rahman, D., and O'Hagan, D.T. 1997, Vaccine, 15, 169.
27. Ugozzoli, M., O'Hagan, D.T., and Ott, G.S. 1998, Immunology, 93, 563.
28. Parr, M.B., and Parr, E.L. 1985, J. Reprod. Fertil., 74, 361.
29. Johansson, E.L., Rask, C., Fredriksson, M., Eriksson, K., Czerkinsky, C., and Holmgren, J. 1998, Infect. Immun., 66, 514.
30. Thaparr, M.A., Parr, E.L., Bozzola, J.J., and Parr, M.B. 1991, Vaccine, 9, 129.
31. O'Hagan, D.T., Rafferty, D., McKeating, J.A., and Illum, L. 1992, J. Gen. Virol., 73, 2141.
32. O'Hagan, D.T., Rafferty, D., Wharton, S., and Illum, L. 1993, Vaccine, 11, 660.
33. Vajdy, M., Gardner, J., Neidleman, J., Cuadra, L., Greer, C., Perri, S., O'Hagan, D., and Polo, J.M. 2001, J. Infect. Dis., 184, 1613.
34. Wira, C.R., Richardson, J., and Prabhala, R. 1994, Handbook of mucosal immunology, P.L. Ogra, M.E. Lamm, J.R. Mcghee, J. Mestecky, W. Strober and J. Bienenstock (Eds.), Academic Press, San Diego, 705.

35. Di Tommaso, A., Saletti, G., Pizza, M., Rappuoli, R., Dougan, G., Abrignani, S., Douce, G., and De Magistris, M.T. 1996, Infect. Immun., 64, 974.
36. Bergquist, C., Johansson, E.L., Lagergard, T., Holmgren, J., and Rudin, A. 1997, Infect. Immun., 65, 2676.
37. Neutra, M.R., Frey, A., and Kraehenbuhl, J.P. 1996, Cell, 86, 345.
38. Kutteh, W.H., Moldoveanu, Z., and Mestecky, J. 1998, AIDS Res. Hum. Retroviruses, 14 Suppl 1, S51.
39. Crowley-Nowick, P.A., Bell, M.C., Brockwell, R., Edwards, R.P., Chen, S., Partridge, E.E., and Mestecky, J. 1997, J. Clin. Immunol., 17, 370.
40. Mestecky, J., and Fultz, P.N. 1999, J. Infect. Dis., 179, S470.
41. Vajdy, M., Veazey, R.S., Knight, H.K., Lackner, A.A., and Neutra, M.R. 2000, Am. J. Pathol., 157, 485.

Part II
Immunology of Mucosal Tissues

Chapter 3
The Mucosal B-Cell System

Per Brandtzaeg and Finn-Eirik Johansen

Abstract The mucosal B-cell system forms the adaptive basis for humoral immune defense of the extensive mucosae. Such antibody protection depends on a complex cooperation between local B cells and secretory epithelia. Mucosa-associated lymphoid tissue (MALT) gives rise to activated B cells with striking J-chain expression that are seeded to local and distant secretory effector sites. Such homing is the biological prerequisite for local plasma cell (PC) production of polymeric immunoglobulin A (pIgA, mainly dimers) and pentameric IgM with high affinity to the epithelial pIg receptor that readily can export these antibodies to the mucosal surfaces. The J chain is also produced by IgG- and IgD-producing PCs occurring at secretory tissue sites; these PC isotypes may be considered as 'spin-offs' from early effector clones that through class switch are on their way to pIgA production.

Abundant evidence supports the notion that intestinal PCs are largely derived from B cells initially activated in gut-associated lymphoid tissue (GALT). Nevertheless, insufficient knowledge exists concerning the relative importance of M cells, major histocompatibility complex class II-expressing epithelial cells, and professional antigen-presenting cells for the uptake, processing, and presentation of luminal antigens in GALT to accomplish the extensive and sustained priming and expansion of mucosal B cells. Also, it is unclear how the germinal center reaction in GALT so strikingly can promote class switch to IgA and expression of J chain, but the commensal microbiota appears to contribute to both the diversification and memory of MALT responses.

Although B-cell migration from GALT to the intestinal lamina propria is guided by rather well-defined adhesion molecules and chemokines/chemokine receptors, the cues directing preferential homing to different segments of the gut require better definition. This is even more so for the molecules involved in homing of mucosal B cells to secretory effector sites beyond the gut. In this respect, the role of Waldeyer's

P. Brandtzaeg
Laboratory for Immunohistochemistry and Immunopathology (LIIPAT), Institute and Division of Pathology, University of Oslo, Rikshospitalet University Hospital, N-0027 Oslo, Norway
e-mail: per.brandtzaeg@medisin.uio.no

M. Vajdy (ed.), *Immunity Against Mucosal Pathogens*,

ring (including the palatine tonsils and adenoids) as a regional MALT in humans needs further characterization, although the balance of evidence suggests that it functions as nasopharynx-associated lymphoid tissue (NALT) identified in rodents. Altogether, data suggest a remarkable compartmentalization of the mucosal immune system that must be taken into account in the development of effective local vaccines to protect specifically the airways, eyes, oral cavity, small and large intestines, and female genital tract.

3.1 Introduction

Mucosal epithelia comprise an extensive and vulnerable physical barrier, which is reinforced by numerous innate defense mechanisms cooperating intimately with adaptive immunity, particularly the generation of secretory immunoglobulin A (SIgA) antibodies. Local formation and export of SIgA constitute the largest humoral immune system of the body, being involved in both the control of commensal bacteria and resistance against pathogens.

Prevention of infectious disease by exploiting the SIgA system is a compelling goal in an effort to improve public health in industrialized and developing societies. The rapid expansion of genome-based biotechnology provides a variety of new avenues for mucosal vaccine development, but it is nevertheless essential to learn more about unique features of the mucosal immune system, and characteristics shared with the systemic immune system. Thus, both mucosal and parenteral vaccination relies on immunological memory, yet our understanding of this fundamental characteristic of adaptive immunity remains incomplete.

Most infections involve the mucosae with regard to initial microbial colonization and entry into the body. In fact, diarrheal disease is ranked by WHO as the second most common lethal infection in children under 5 years of age – accounting for at least 20% of the 10.6 million annual deaths in this age group [1]. Repeated episodes of diarrhea, especially when long-lasting and associated with growth failure, also contribute significantly to malnutrition in developing countries. Rotavirus, diarrheagenic *Escherichia coli*, including enterotoxigenic (ETEC) and enteroaggregative strains, *Shigella* spp. and *Cryptosporidium parvum* are among the worst killers. Cholera is an important cause of diarrhea in the Bengal delta and occasionally causes epidemics with devastating effects even outside of Asia. Despite much intensive research, the development of vaccines against many of these important diarrheal pathogens has yet to be successful, and the same is true for common airway infections [2].

Vaccines applied directly to mucosal surfaces would make immunization procedures easier and better suited for mass administration; in poor countries, the avoidance of horizontal spread of infections with contaminated needles would also be a significant advantage [3]. Mucosal vaccination should, moreover, most efficiently induce immune exclusion [4] – a term coined for non-inflammatory antibody shielding at internal body surfaces – mediated principally by SIgA in co-operation with innate non-specific defenses, thus referring to 'first line' protection against micro-organisms [5]. This immune mechanism has a formidable task because the total mucosal surface area of an adult human body amounts to some 400 m^2. With the exception of certain sites such as the oral cavity and the vagina, the mucosae are generally covered only by a monolayered epithelium and therefore quite vulnerable.

This review discusses basic cellular and molecular mechanisms underlying strategies to enhance secretory immunity. The mucosal immune system is complex and involves dynamic interactions between structurally different tissue compartments. Moreover, important structural and functional species differences exist in this system [6]. Here, we will focus on human immunity but have to rely on information obtained in mice to provide a better functional picture.

3.2 Secretory Immunity

3.2.1 Mechanism of Epithelial Antibody Export

The cellular basis for the remarkable generation of SIgA is found locally in secretory mucosae and associated exocrine glands which harbor most (perhaps 90%) of the body's activated B cells – terminally differentiated to IgA-producing plasmablasts and PCs. SIgA antibodies are remarkably stable hybrid molecules – they consist of PC-derived IgA polymers (pIgA, dimers and some trimers) with one (or more) 'joining' (J) chain(s) and an epithelial portion called bound secretory component (SC) [7], which is disulfide-linked to one of the IgA subunit heavy chains (Fig. 3.1a). Most mucosal PCs (70–90%) normally produce pIgA [8, 9] (Fig. 3.1b) which, together with J chain-containing pentameric IgM, are exported by secretory epithelial cells to provide SIgA and secretory IgM (SIgM) antibodies [10, 11]. The intestinal IgA system is the best understood contributor to mucosal immunity. In fact, the gut mucosa contains at least 80% of the body's PCs, and some 90% of these lamina propria cells normally produce pIgA [12].

During the late 1960's and 70's at least eight different models were proposed to explain how IgA selectively could reach external secretions. One of us suggested in 1973–1974 that the epithelial glycoprotein identified in 1965 by Tomasi et al. [13] as part of SIgA, and later on named SC by WHO [6], could act as a membrane receptor for pIgA and pentameric IgM on secretory epithelia [9, 14]. This formed the basis for a common epithelial transport mechanism to generate SIgA and SIgM in which SC and the J chain of pIgs constitute 'key and lock' molecules [15]. The model is now generally accepted (Fig. 3.2) and membrane SC is known as the pIg receptor (pIgR).

SC thus occurs in three forms: membrane-associated (same as the 100-kD pIgR), bound (SIg-associated) and free (unassociated) [6]. The two latter (both 80 kD) represent the cleaved ectodomain of pIgR (Fig. 3.1a), and free SC is generated from unoccupied receptor. There is always a substantial excess of free SC in the secretions – on average approximately 50% of that produced [16]. Active epithelial pIg

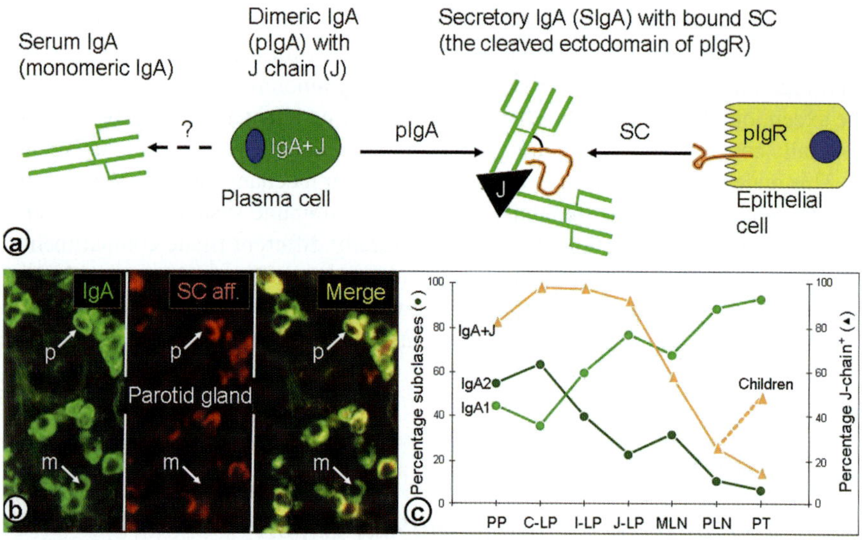

Fig. 3.1 Secretory IgA (SIgA) generation. (**a**) Depiction of how SIgA is formed as a hybrid antibody molecule stabilized by a disulfide bridge between the two cell products. The amount of dimeric IgA (pIgA) produced by a plasma cell depends on its level of J-chain expression, which generally is high in mucosal and glandular tissue, particularly in the gut. (**b**) Direct demonstration of abundant cytoplasmic expression of pIgA (p) in most parotid plasma cells, with only few producing monomers (m), obtained by in vitro affinity test with free SC on tissue section as described [8]. (**c**) Organ distribution of plasma cells expressing IgA associated with J chain (IgA+J), and the two IgA subclasses, based on studies in the authors' laboratory [65]. PP, Peyer's patch; C-LP, colonic lamina propria; I-LP, ileal lamina propria; J-LP, jejunal lamina propria; MLN, mesenteric lymph node; PLN, peripheral lymph node; PT, palatine tonsil from adults (or children with healthy tonsils: broken line)

export requires abundant expression of pIgR because it is continuously sacrificed at the luminal face to release bound or free SC by cleavage from its transmembrane form. SC has several natural antimicrobial properties [17], suggesting that pIgR is derived from the innate immune system.

The 15-kD J chain was identified independently in 1970 by Halpern and Koshland [18] and Mestecky et al. [19] as an Ig-associated peptide common to IgM and SIgA. J chain-mediated Ig polymerization provides antibodies with high avidity – a property particularly useful for agglutination of microbes in external secretions. As alluded to above, its incorporation into pIgA and pentameric IgM is mandatory for their binding to SC/pIgR [7, 15, 20, 21]. The receptor carries its two ligands to mucosal surfaces by a vesicular transport process termed transcytosis. The crucial importance of pIgR and J chain for such epithelial transport of pIgA is highlighted by the absence of SIgA in knockout (KO) mice deficient for either of these genes [22, 23, 24, 25].

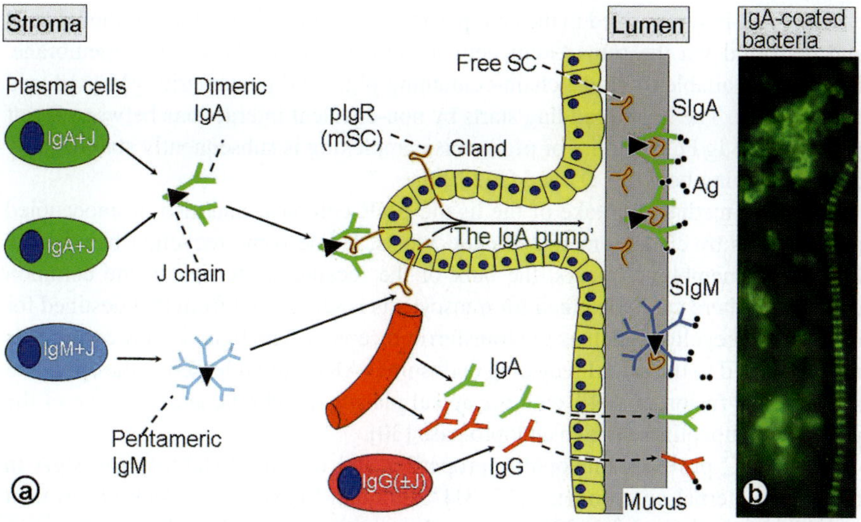

Fig. 3.2 Model for receptor-mediated export of J chain-containing dimeric IgA and pentameric IgM and their function in immune exclusion. (**a**) Polymeric Ig receptor (pIgR) is expressed basolaterally as membrane secretory component (mSC) on secretory epithelial cells and mediates transcytosis of dimeric IgA and pentameric IgM, which are produced with incorporated J chain (IgA + J and IgM + J) by mucosal plasma cells. The resulting secretory Ig molecules (SIgA and SIgM) act in a first line of defense by performing immune exclusion of antigens in mucus layer on epithelial surface. In addition, pIgR-mediated export of immune complexes from the lamina propria and intraepithelial compartment may contribute to noninflammatory mucosal defense (not shown). Although J chain is often (70–90%) produced by mucosal IgG plasma cells, it does not combine with this isotype and is therefore degraded intracellularly as denoted (±J). Locally produced (and serum-derived) IgG is not subjected to active external transport, but can be transmitted paracellularly to the lumen as indicated. Free SC (depicted in mucus) is generated when pIgR in its unoccupied state (top basolateral symbol) is cleaved at the apical face of the epithelium like bound SC in SIgA and SIgM. While bound SC is covalently linked to one subunit in SIgA, providing protection against degradation (Fig. 3.1a), SIgM contains only noncovalently bound SC in dynamic equilibrium with free SC in the secretion (not shown). (**b**) Commensal salivary bacteria coated in vivo with IgA. Adapted from Brandtzaeg et al. [199]

3.2.2 Structure and Regulation of pIgR

Intracellular trafficking of pIgR has been studied in great detail and represents the paradigm for transcytosis across a polarized epithelium [26]. The pIgR has an N-terminal extracellular ligand-binding domain, a single membrane-spanning α helix, and a 103 amino-acid cytoplasmic C-terminal tail that contains all the information for proper intracellular routing [27]. The extracellular region is comprised of five Ig-like domains (D1–D5); solution structure analysis has suggested a bend between D3 and D4, forming a hook-shaped molecule separated from the transmembrane domain by a flexible region, which also contains the receptor cleavage site [28].

The pIgR is synthesized in the endoplasmic reticulum of secretory epithelial cells and delivered via the trans-Golgi network directly to the basolateral membrane. Here, it is available to bind J chain-containing pIgA and pentameric IgM produced by local PCs. The ligand binding starts by non-covalent interactions between D1 of pIgR and the Ig Fc portion. For pIgA, this complexing is subsequently stabilized by covalent linking between D5 and Cα2 [29].

Receptor-mediated uptake of the ligand-pIgR complex, and also of unoccupied pIgR, occurs by clathrin-mediated endocytosis. While some recycling back to the basolateral membrane occurs, the bulk of the receptor is ferried to the common endosome where cargo destined for transcytosis is segregated from that destined for basolateral recycling (such as the transferrin receptor). The ligand-pIgR complex is next delivered to the apical recycling endosome (ARE) and released at the epithelial surface after fusion of ARE with the apical plasma membrane and cleavage of the pIgR by a leupeptin-sensitive endoprotease [30].

Bound SC provides mucophilic properties and increased stability for SIgA in the harsh external environment [17, 31]. Although SIgM, due to lack of covalent stabilization by bound SC [32], is not such a resistant antibody in the secretions as SIgA, it nevertheless can provide compensatory mucosal defense in newborns and subjects with selective IgA deficiency [33]. In the airways, SIgM compensation is less consistent than in the gut, and this variable apparently contributes to susceptibility of IgA-deficient individuals to infections in the respiratory tract [34].

Although epithelial membrane SC/pIgR expression is constitutively regulated, it can be enhanced at the transcriptional level by the immunoregulatory cytokines interferon-γ and interleukin (IL)-4, as well as by the proinflammatory cytokines tumor necrosis factor and IL-1 [29, 35, 36, 37, 38]. Both constitutive and cytokine-enhanced pIgR expression appears to depend on adequate presence of vitamin A (retinoic acid) and the nutritional state of the subject [39, 40].

3.2.3 Development and Functional Significance of Secretory Antibodies

In adult humans, more IgA is transported to the gut lumen every day (∼40 mg/kg body weight) than the total IgG production (∼ 30 mg/kg) – equalling an SIgA export of approximately 3 g/day [41]. Therefore, the intestinal mucosa is quantitatively the most important effector organ of antibody-mediated immunity (Table 3.1). Moreover, SIgA is the most stable antibody of the immune system [5, 10, 17, 29, 42], and its specific quantification should in theory represent a convenient read-out for mucosal vaccine responses. However, there are many pitfalls inherent in SIgA antibody measurements [43].

During the first postnatal period, only traces of SIgA and SIgM occur in human external secretions, whereas some IgG is often present, mainly as a result of 'leakage' from the mucosal lamina propria (Fig. 3.2) which – because of placental transfer – contains readily detectable maternal IgG as early as at 34 weeks of

Table 3.1 Key features of the intestinal SIgA system

• More IgA is produced every day than all other antibody classes collectively (IgG, IgM, IgD, and IgE)
• Daily adult intestinal SIgA export, > 40 mg/Kg (~ 3 g/day)
• Determination of specific SIgA is the best 'read-out' for mucosal vaccines but not easy to perform (saliva? feces? whole gut lavage?)

gestation [33]. Therefore, an adequate mucosal barrier function of neonates depends on a supply of SIgA antibodies from breast milk. In developing countries, infants are particularly dependent on milk antibodies to protect their mucosae, and this is essential in the face of rotavirus and *Vibrio cholerae* infections. Epidemiological data suggest that the risk of dying from diarrhea is reduced 14–24 times in breast-fed babies [44]. Indeed, breastfeeding currently represents the most efficient feasible intervention measure, in theory being able to prevent 13% of all deaths in children below 5 years of age [45, 46]. Moreover, experiments in neonatal rabbits clearly show that SIgA is the crucial antimicrobial component of breast milk [47].

Although the antimicrobial value of breastfeeding in countries with a modern hygiene is best seen in preterm infants [48], exclusively breast-fed infants are better protected against a variety of infections, also quite common ones [49, 50, 51, 52]. It should be noted, however, that intestinal uptake of maternal SIgA antibodies is of no importance for systemic immunity in breast-fed babies [53, 54], except perhaps to some extent in the preterm infant [55]. 'Gut closure' normally occurs in humans mainly before birth, but an adequate mucosal barrier may not be established until after 2 years of age; the different variables involved in this process are poorly defined [56] although secretory immunity appears to be a major component [33]. Interestingly, in mice the postnatal colonization of commensal bacteria is important both for establishing [57] and regulating [58] an appropriate epithelial barrier function – the effect of which could be both directly on the epithelium via microbial pattern recognition receptors (PRRs) and indirectly by driving the development of the mucosal immune system [59, 60, 61, 62].

3.3 Mucosal B-Cell Induction

3.3.1 Immune-Inductive Lymphoid Tissue

The mucosal B-cell system can principally be divided into two functionally distinct compartments (Fig. 3.3). First, the inductive sites consisting of organized mucosa-associated lymphoid tissue (MALT) together with mucosa-draining lymph nodes; here, antigens sampled from mucosal surfaces activate naïve T and B cells by complex stimulatory mechanisms [6, 11, 63]. Second, the effector sites consisting of secretory epithelium and its underlying connective tissue stroma such as the gut lamina propria; here, B cells become terminally differentiated to PCs that produce pIgs for epithelial export in a pIgR-dependent fashion as described above.

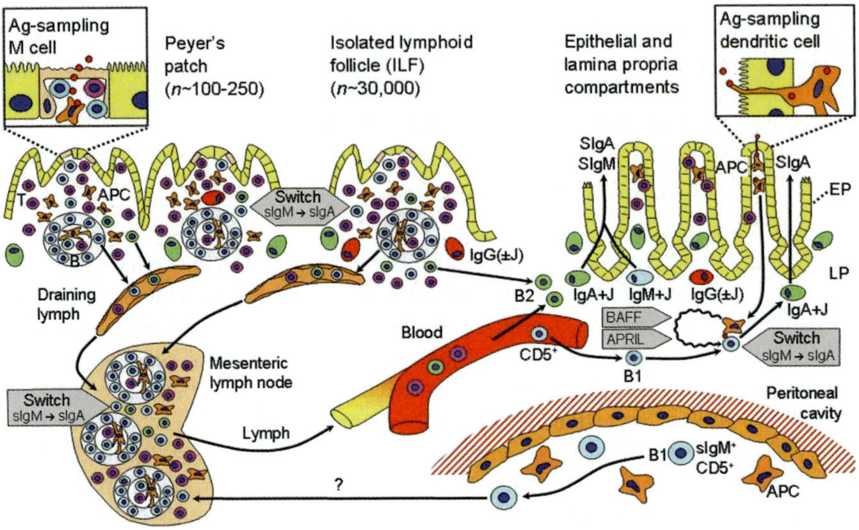

Fig. 3.3 Antigen-sampling and potential B-cell Ig-class switch sites for induction of intestinal antibody responses. The classical mucosal immune-inductive sites are constituted by gut-associated lymphoid tissue (GALT), which is equipped with antigen-sampling M cells, T-cell areas (T), B-cell follicles (B), and antigen-presenting cells (APC). Red dots denote luminal antigens. Switch of conventional B2 cells from surface (s)IgM to SIgA expression occurs in GALT and mesenteric lymph nodes; from here the activated B and T cells home to lamina propria (LP) via lymph and blood. T cells mainly end up in the epithelium (EP), whereas SIgA$^+$ cells differentiate to LP plasma cells that produce dimeric IgA with J chain (IgA + J). Dimeric IgA is next exported as secretory IgA (SIgA) to the lumen (see Fig. 3.2). Activated B cells may also migrate from PPs and isolated lymphoid follicles directly into the LP as indicated, but those differentiating to plasma cells just outside follicles often show reduced J-chain expression and a propensity for IgG production (IgG±J). B2 cells also give rise to plasma cells that produce pentameric IgM with J chain (IgM + J), which becomes secretory IgM (SIgM). B1 cells (CD5$^+$) from the murine peritoneal cavity reach the LP by an unknown route (?), perhaps via mesenteric lymph nodes. According to some data, these sIgM$^+$ cells may switch to SIgA within the LP under the influence of APCs that have sampled microbial antigens as dendritic cells within the epithelium and become activated to secrete stimulatory factors (waved arrow) such as BAFF and APRIL. In mice it appears that SIgA$^+$ B1 cells differentiate to plasma cells that provide SIgA mainly directed against the commensal gut microbiota. Reproduced from Brandtzaeg and Johansen [11] with permission from Blackwell Publishing

The various secretory effector sites receive their activated B cells from MALT by compartmentalized homing mechanisms (see below). Therefore, the old term 'a common mucosal immune system' is no longer valid, although considerable integration of the effector sites exists [64]. The homing to these sites appears to be antigen-independent, but locally available antigen contributes to the retention, proliferation and differentiation of the extravasated memory/effector B cells [6, 10, 11].

Gut-associated lymphoid tissue (GALT) – including Peyer's patches in the small intestine (mainly in the distal ileum), the appendix, and numerous isolated lymphoid follicles (Figs. 3.3 and 3.4) – constitutes the major part of human MALT. However,

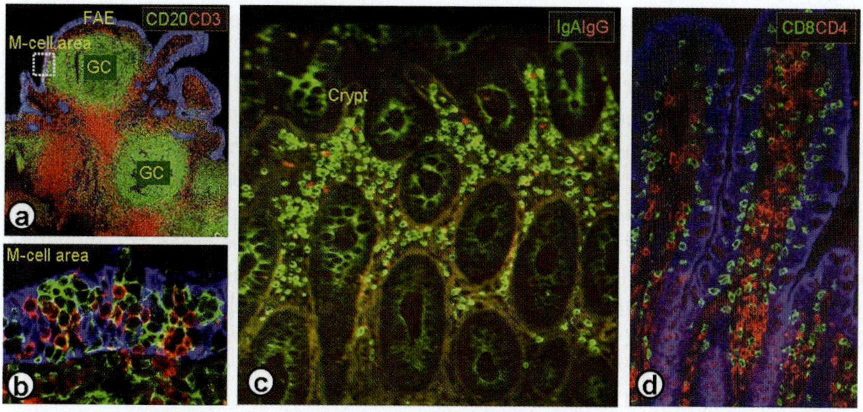

Fig. 3.4 Immunomorphological features of human mucosal inductive and effector sites. (**a, b**) Three-color immunofluorescence staining of B cells (CD20, green), T cells (CD3, red) and epithelium (cytokeratin, blue) in cryosection of human Peyer's patch. (*b*) Details from the M-cell areas framed on the left in the follicle-associated epithelium (**a**) covering a B-cell follicle. (**c**) Two-color immunofluorescence staining for IgA (green) and IgG (red) in section from normal human large bowel mucosa. Crypt epithelium shows selective transport of IgA and only a few scattered IgG-producing cells are seen in the lamina propria together with numerous IgA plasma cells. (**d**) Three-color immunofluorescence staining for $CD4^+$ (red) and $CD8^+$ (green) T cells in normal human duodenal mucosa. The epithelium of the villi is blue (cytokeratin). Note that most of the elements with weak CD4 expression seen in the background are either macrophages or dendritic cells. The pictures are original immunofluorescence images from the authors' laboratory. Reproduced from Brandtzaeg and Johansen [11] with permission from Blackwell Publishing

induction of mucosal immune responses can take place also in the paired palatine tonsils and other lymphoepithelial structures of Waldeyer's pharyngeal ring, including nasopharynx-associated lymphoid tissue (NALT) such as the unpaired adenoids [5, 6, 10, 65, 66, 67]. Bronchus-associated lymphoid tissue (BALT) may also contribute, but this type of MALT is not present in normal lungs of adults and only in approximately 40% of adolescents [68].

Peyer's patches are the best studied MALT structures. Their domes are covered by a characteristic follicle-associated epithelium (FAE), which contains variable numbers of antigen-sampling 'microfold' or 'membrane' (M) cells – depending both on the species and the degree of stimulatory activity [6]. These very thin and bell-shaped specialized epithelial cells transport effectively live and non-replicating antigens (especially particles) from the gut lumen into the lymphoid tissue [63].

Many enteropathogenic infectious agents use the M cells as portals of entry, so they represent extremely vulnerable parts of the surface epithelium. Such 'gaps' in the epithelial barrier are nevertheless needed to facilitate efficient induction of mucosal immunity. The antigen uptake mechanisms are poorly characterized, but are apparently both independent and dependent on various types of receptors [63]. Murine M cells express PRRs, including the Toll-like receptors (TLRs) TLR2 and TLR4 [69, 70, 71]. The former is reportedly of considerable interest to enhance the uptake of particulate microbial immunogens such as proteosomes [71].

An IgA-specific receptor has also been characterized – potentially enhancing the uptake of SIgA-antigen complexes from the gut [72].

All MALT structures resemble lymph nodes with B-cell follicles, intervening T-cell zones, and a variety of professional antigen-presenting cells (APCs), including both dendritic cells (DCs) and macrophages – but, notably, there are no afferent lymphatics (Fig. 3.3). Therefore, immunogenic stimuli are fully dependent on antigen and mitogen uptake via FAE – mainly by the M cells but also aided by DCs, which may penetrate the epithelium with their processes [73, 74]. Induction of intestinal immunity hence takes place primarily in the GALT structures and to a variable extent also in the gut-draining mesenteric lymph nodes (MLNs), as further discussed below [6, 10, 11].

3.3.2 Priming of Mucosal B and T Cells

B and T cells arrive in MALT via high endothelial venules as in other secondary lymphoid tissue [11]. Here, antigens are presented to the naïve T cells as immunogenic peptides by APCs. In addition, luminal peptides may be taken up and presented by B cells and epithelial cells to subsets of intra- and subepithelial T lymphocytes [65, 75]. Both professional mucosal APCs, B cells, and the small-intestinal villous epithelium, as well as FAE except for the M cells [76], express antigen-presenting MHC class II molecules – in humans particularly HLA-DR – in addition to classical and nonclassical MHC class I molecules [77, 78].

Interestingly, MHC class II-expressing naïve and memory B cells aggregate together with T cells in the M-cell pockets, which thus may represent the first contact site between immune cells and luminal antigens [10, 11, 65, 75]. The development of such M cell-associated B- and T-cell clusters depends on microbial colonization and are absent in germ-free (GF) animals [60]. The activated B cells probably perform important antigen-presenting functions in this compartment, perhaps promoting antibody diversification and immunological memory [65]. Other types of professional APCs – namely DCs and macrophages – are mainly located below the FAE and between the follicles [11, 60, 75].

$CD4^+$ helper T (Th) cells activated in GALT are known to release cytokines such as transforming growth factor-β and IL-10, which drive the class switch and differentiation of mucosal B cells to predominantly IgA-committed plasmablasts with J-chain expression [9, 10, 11, 62, 79, 80], although their detailed regulation is complex and still remains unclear (see below). Both naïve and primed B cells migrate rapidly from GALT via draining lymphatics to MLNs where they may be further stimulated; they next reach thoracic duct lymph and peripheral blood to become seeded by selective homing mechanisms into distant mucosal effector sites (Fig. 3.3), particularly the intestinal lamina propria where they finally differentiate to PCs [11, 66, 79, 80, 81]. This terminal differentiation is modulated by 'second signals' from local antigen-sampling DCs, lamina propria $CD4^+$ T cells, and available cytokines [6, 10, 11].

3.3.3 Divergent Microbial Activation of Mucosal B Cells

Stimulation of B cells may occur by various mechanisms. Some microbial substances are superantigens which interact with the B-cell receptor (BCR) outside the antigen binding site – the prototype being *Staphylococcus aureus* protein A. When B cells are exposed to cross-linked protein A, providing multiple binding sites like a micro-organism, they show a strong polyclonal response. B cells of MALT could be expanded in such a manner both by exogenous and microbe-induced endogenous superantigens [82].

Other microbial substances, called Type 1 T cell-independent (TI-1) antigens, are directly mitogenic for B cells – including sugars, lipid structures, and certain nucleic acids. Type 2 T cell-independent (TI-2) antigens, on the other hand, are not by themselves mitogenic but cause extensive cross-linking of BCR by repeating epitopes; their B-cell activation is exerted through synergy with soluble factors (e.g. cytokines) and interaction with various types of accessory cells. There are notable species differences in the mechanisms involved; for instance, while bacterial endotoxin or lipopolysaccharide (LPS) acts as a TI-1 antigen on mouse B cells, it acts as a TI-2 antigen on human B cells.

Microbial polysaccharides may also function as T cell-dependent antigens and exert profound effects on the immune system by stimulating $CD4^+$ Th cells after presentation by DCs on MHC class II molecules. Thus, a single bacterial polysaccharide was shown to have a striking impact both on lymphoid organogenesis and immune modulation in GF mice monocolonized with the ubiquitous gut commensal *Bacteroides fragilis* [83].

Although Peyer's patches containing primary follicles with naïve B lymphocytes are present at birth both in humans and mice, it takes some days before memory-generating secondary follicles with germinal centers (GCs) are induced by microbial stimulation [33, 59]. As alluded to earlier, studies in GF and conventional (CVN) specific pathogen-free rats demonstrated that bacterial colonization drives intraepithelial T- and B-cell accumulation with formation of M-cell pockets after an initial GC reaction – apparently induced by DC-mediated antigen transport from the gut lumen [60].

While TI antibody responses are restricted with regard to affinity and memory, the GC reaction is chiefly driven by BCR competition for a limited amount of T cell-dependent antigens – resulting in memory/effector B cells that undergo BCR affinity maturation and class switch [84]. As reportedly shown for GALT of experimental animals, however, IgA differentiation stimulated by commensal gut bacteria may, under certain conditions, bypass the usual BCR requirement [62, 85, 86]. Nevertheless, there always appears to be a dependency on some follicle-like aggregates of B cells, which interestingly, may lack antigen-retaining follicular DCs and GCs.

It follows from the above that if there is a sufficient innate drive of the immune system, B cells may survive with a restricted repertoire and rather low affinity. Thus, in the GF-appendix model in rabbits, it was shown that certain commensal bacteria are quite efficient in promoting GALT development, and that such ability depends on stress responses in the same bacteria – suggesting a non-specific (superantigen?)

impact on GALT [87]. The independency of BCR engagement for GALT development was confirmed in mice where the Epstein-Barr virus protein LMP2A was transgenically introduced as a constitutive BCR surrogate providing a weak signaling pathway [88]. It was concluded that commensal bacteria – by interacting with innate immune receptors such as PRRs – can promote the GC reaction in GALT.

Altogether, it is well established that the intestinal microbiota is required for activation of GALT and a normal mucosal PC development [33, 59, 60], and that commensal bacteria can shape the BCR repertoire of the host [82]. Perhaps the GC reaction originated in GALT during evolution to generate a protective antibody repertoire that is cross-reactive rather than antigen-specific. The indigenous microbiota might contribute to polyclonal B-cell activation through several mechanisms, including PRR signalling [89]. This notion is supported by an accumulating body of evidence showing that TLRs can be involved in polyclonal B-cell stimulation and maintenance of memory [90], not only indirectly via follicle-associated DC activation but also directly within the GCs [91, 92]. In humans, B cells consistently express various TLRs including TLR2, TLR4 and TLR9, which are up-regulated during the differentiation to memory cells [92, 93, 94]. The precise role of PRRs in mucosal IgA responses nevertheless remains elusive.

3.4 B-Cell Induction to Generate Dimeric IgA

3.4.1 IgA Class Switch Recombination and Plasma-Cell Development

In most mammals there is a single germline gene encoding the α constant heavy chain (CHα) domains at the 3′ end of the Ig heavy chain locus. In humans and higher primates, however, there has been a duplication of the $\gamma \rightarrow \alpha$ CH region, resulting in two functional CHα genes encoding the IgA1 and IgA2 subclasses whose PC contributions are disparate throughout the mucosal and systemic immune systems (Fig. 3.1c).

Generally, mammalian CHα loci have 3 exons corresponding to the heavy chain domains, with the region encoding the 19 amino acid extension of the secreted form at the 3′ end of the 3rd exon, and another short exon encoding the transmembrane segment and cytoplasmic tail [62]. Class switch recombination (CSR) occurs following transcription from the intronic germline Iα promoter 5′ of the CHα locus [95], and requires the activation-induced cytidine deaminase (AID) and other CSR enzymes [94, 96, 97]. The circular spliced segment is usually from the region between the switch segments upstream of the CHμ and CHα loci (Fig. 3.5), although trans-splicing between different chromosomes is also possible [98].

In general, CSR has been found to require two signals. The first is through cytokines which target transcription of the intronic promoters upstream of the CH exons and the switch region preceeding them. The second is delivered by ligation of CD40 on B cells with its ligand CD40L on T cells. However, in mice lacking

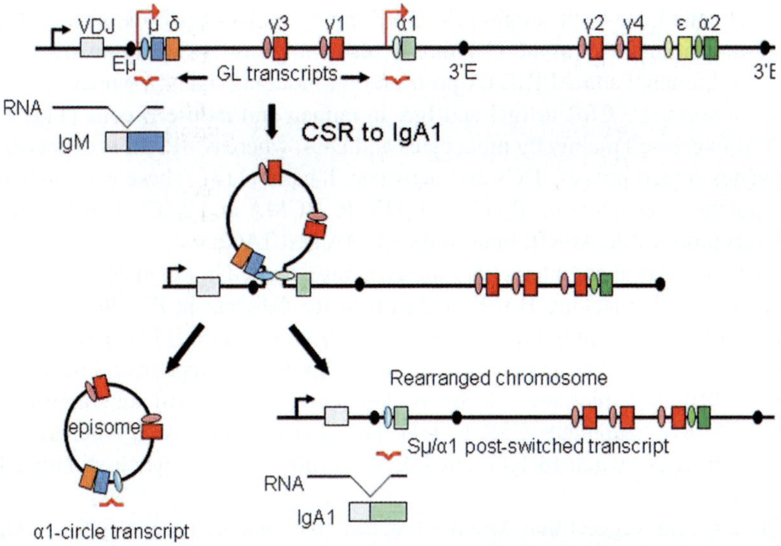

Fig. 3.5 Class-switch recombination (CSR) from IgM to IgA1 at the Ig heavy-chain locus. The variable region of the Ig heavy chain is assembled from the variable (V), diversity (D), and joining (J) gene segments by VDJ recombination during B-cell development. Transcription across the locus is driven by a promoter upstream of the rearranged VDJ segment (bent black arrow), which facilitates the synthesis of a μ heavy chain (CH). The latter associates with a light chain, thereby forming a complete IgM molecule, which is displayed on the surface of the B cell as part of the B-cell receptor. Secondary isotypes are produced by CSR, and is initiated by germ line (GL) transcription of the $CH\mu$ and the $CH\alpha 1$ genes (for switch to IgA1; bent red arrows). Products of recombination at the *CH* locus are the rearranged chromosome and an episomal circle, from which circle transcripts are derived. Cytokines stimulate transcription (bent red arrows) through the *CH* gene and determine the Ig isotype that the B cell will switch to. The constant region of downstream isotypes are denoted by their corresponding Greek letters and the 3′ enhancers (3′E), which influence GL transcription and thereby the CSR, are indicated below the schematic chromosome

CD40L or CD40, IgA production is not abrogated in the gut [99, 100], and this may also be the case in CD40L-deficient humans with the hyper-IgM syndrome [101].

The major cytokine signal for CSR to IgA is TGF-β with contributions from IL-2, IL-4, IL-5, IL-6 and IL-10; these stimulatory conditions were initially demonstrated by studying IgA production in non-specifically stimulated B-cell cultures [11, 102]. Evidence for the role of TGF-β in vivo came from an almost complete lack of IgA in mice deficient for the TGF-β receptor TβRII [103]. Mice deficient for the TGF-β negative regulator SMAD7 were found to have increased CSR to IgA but reduced proliferation to LPS [104], whereas mice deficient for the positive signal SMAD2 lack IgA [105]. IL-4-deficient mice show normal total IgA levels, but an impaired specific response to mucosal immunization with cholera toxin [106], while IL-6-deficient mice have defective IgA responses in some systems [107, 108, 109].

Because intestinal IgA production in mice is partially independent of T cells [110] and CD40–CD40L engagement [95], other costimulatory signals for CSR to

IgA in B cells have been sought. The TNF family includes two members – BAFF [B cell-activating factor of the TNF family, also known as BlyS (B lymphocyte stimulator) in humans] and APRIL (A proliferation-inducing ligand), which have been shown to stimulate CSR to IgG and IgA in human and mouse B cells [111, 112]. BAFF is expressed mainly by monocytes and DCs, whereas APRIL is expressed by monocytes, macrophages, DCs and activated T cells [113]. These cytokines have three potential receptors on B cells – BAFF-R, BCMA and TACI; BAFF binds all three receptors while APRIL binds only BCMA and TACI.

The function of these interactions has been investigated in KO mice with selective deficiencies. Mice lacking BAFF or BAFF-R are deficient in B cells, so they are not useful for determining the requirements of IgA induction [114]. TACI-deficient mice have low IgA and poor immunization responses to repetitive TI-2 antigens [115]. APRIL KO mice are selectively deficient for IgA [116]. In sorted murine B cells, APRIL stimulates CSR to IgG, IgA and IgE via TACI, whereas TACI-deficient B cells switch to IgG and IgE – but not to IgA – when stimulated by BAFF [112].

These results suggest that APRIL–TACI interactions are necessary for IgA production in mice, and that the CD40–CD40L interaction may be redundant for CSR to IgA. But why is the IgA class switch so prominent in the intestine (and other mucosal sites) when APRIL and BAFF can stimulate CSR also to other isotypes? Intestinal DCs from Peyer's patches and MLNs constitutively secrete retinoic acid that synergises with IL-6 or IL-5 to induce IgA production in purified B cells, both from mice and humans [117]. Retinoic acid secretion alone can also induce small intestinal homing molecules on purified B cells [117]. Thus, the environment of cytokines and DC phenotype in the regional lymphoid tissue may largely explain the dominating IgA response generated in the gut.

In addition to the complexity of CSR, an elaborate transcriptional program controls the lineage development from B cells to PCs [94]. BSAP (B-cell lineage-specific activator protein) is encoded by the *pax-5* gene, and is also called PAX5 (paired box protein 5). This transcription factor is required to establish and maintain B-cell identity, and it activates AID in the GC reaction. However, the terminal B-cell differentiation to PCs depends on inhibition of PAX5 by an increasing level of the transcription factor PRDI-BF (positive regulator domain I-binding factor) or BLIMP1 (B-lymphocyte-induced maturation protein 1), the latter designation referring to the mouse homologe.

3.4.2 Regulation of J-Chain Expression

For the mouse, J chain is a commonly used marker of activated B cells in general, and its expression is induced by treatments that crosslink BCR, in addition to various cytokines such as IL-2, IL-5, and IL-6 [118, 119]. In humans, the J chain is mainly a characteristic of mucosal PCs (Fig. 3.1c), regardless of concurrent production of the various Ig isotypes, but there is limited information regarding its regulatory elements [120]. This polypeptide is degraded in the cytoplasm of IgG- and IgD-producing

Table 3.2 J-chain positivity (%) of mucosal plasma cells belonging to major Ig classes*

Exocrine tissue site	Ig class expression			
	IgA	IgM	IgG	IgD
Mammary glands	94	100	56	100
Salivary and lacrimal glands	92	100	44 (72)**	95
Normal nasal mucosa	98	100	69	100
Normal small intestinal mucosa	99	100	87	N.D.

*Summary of LIIPAT's immunohistochemical studies of cytoplasmic J-chain expression in mucosal/glandular plasma cells (for references, see text).

**Data from IgA-deficient individuals.

PCs because it can only be secreted from the cell when it is associated with IgA or IgM [20, 38]. We consider J chain-expressing PCs that concurrently produce IgG or IgD at secretory effector sites (Table 3.2) as 'spin-offs' from early effector B-cell clones in the CSR process to generate pIgA [10, 11].

Several promoter elements and transcriptional factors important for antigen-induced J-chain expression have been identified in the mouse, and IL-2-induced chromatin remodeling of the gene locus is necessary [121, 122]. The best characterized promoter elements involved in the regulation of the murine *J-chain* gene have been designated JA, JB, JC and JE [123, 124, 125]. The transcription factors B-MEF2 (myocyte enhancer factor 2-related nuclear factor), PU.1 and USF-1 (upstream stimulatory factor) act positively on the transcription by binding to the JA, JB and JE elements in this promoter, respectively [122, 126, 127].

PAX5 binds the JC element and thereby represses transcription from the murine J-chain promoter, apparently by preventing the binding of the activators USF-1 and B-MEF2 [122]. PU.1 is also involved in BSAP-mediated repression of J-chain transcription by recruiting the co-repressor Groucho-4 [128] (Fig. 3.6). As mentioned above, terminal maturation of murine B cells requires repression of PAX5

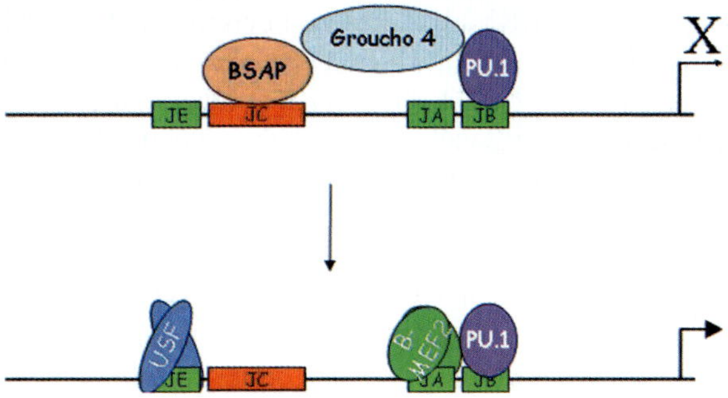

Fig. 3.6 Regulation of the murine J-chain gene promoter. B-cell lineage-specific activator protein (BSAP) acts as a repressor of J-chain transcription by recruitment of Groucho-4 and preventing DNA binding of upstream stimulatory factor (USF) and B-cell myocyte enhancer factor 2-related nuclear factor (B-MEF2). In terminally differentiated B cells, BSAP expression is reduced, and the *J-chain* gene is thus poised for activation (bottom)

by BLIMP1 that binds a target site in the *Pax-5* promoter. At the same time, the inhibitory effect of PAX5 on J-chain expression is relieved [129].

In mice, J-chain expression is only initiated during the last stages of antigen-driven B-cell differentiation, while in humans the transcription of J chain is active already during early stages of the B-cell development [130]. However, regulation of J-chain expression in the human mucosal immune system has apparently not been studied in any detail. Binding sites for BSAP, USF (E-Box), PU.1 and B-MEF2 are conserved in the human promoter [120], but their functionality has not been reported.

3.5 Homing of Mucosal B Cells

3.5.1 Distribution of B Cells for Priming in MALT

Several homeostatic chemokines have been identified as major cues for lymphocyte trafficking and positioning in organized lymphoid tissue [131, 132]. The B cell-attracting chemokine (BCA)-1, also designated chemokine ligand CXCL13, attracts naïve human B cells in vitro and has been shown to be produced in human lymph node follicles [133]. This chemokine is called B-lymphocyte chemoattractant (BLC) in mice [134]. Evidence suggests that CXCL13/BLC and its receptor CXCR5 are directly involved in the formation of organized murine lymphoid tissue [135, 136]. Thus, this chemokine upregulates the organogenic lymphotoxin $\alpha 1\beta 2$ on B cells, and a positive feedback loop may thereby be established [137].

The follicular expression of murine CXCL13/BLC is reportedly more consistent in murine Peyer's patches than in peripheral lymph nodes [134]. Alternative chemokines acting on B cells may also operate in human lymphoid tissue. For instance, stromal cell-derived factor 1 (SDF-1, CXCL12), which appears to be produced by cells lining tonsillar GCs, has been shown to attract naïve and memory B cells expressing CXCR4 in vitro [138].

CXCL13/BCA-1 and CXCR5 are expressed in normal human GALT structures, both in Peyer's patches and colonic ILFs [139]. Expression of CXCR5 in follicular mantle zones (Fig. 3.7) at a relatively low level agrees with the fact that CXCL13/BCA-1 attracts naïve B cells in vitro with moderate effect [132]. Scattered T cells with strong CXCR5 expression occur within the same follicles [139]. This $CD4^+CXCR5^+$ phenotype is functionally defined as 'follicular B-helper T cells', or T_{FH} cells (Fig. 3.7), and shows all the characteristics required for efficient B-cell help [84, 140, 141, 142], while being distinct from Th1 and Th2 cells [143]. A small T_{FH}-cell subset, identified as $CXCR5^+CD57^+$ and termed GC T-helper (GC-Th) cells, was initially described as quite essential for B-cell differentiation and antibody production [144] but an even more restricted phenotype has subsequently been described [145].

The partial overlap produced by immunostaining for CXCL13/BCA-1 and several traditional follicular DC markers in human lymphoid tissue [139], suggests that this chemokine is deposited on peripheral extensions of these antigen-retaining

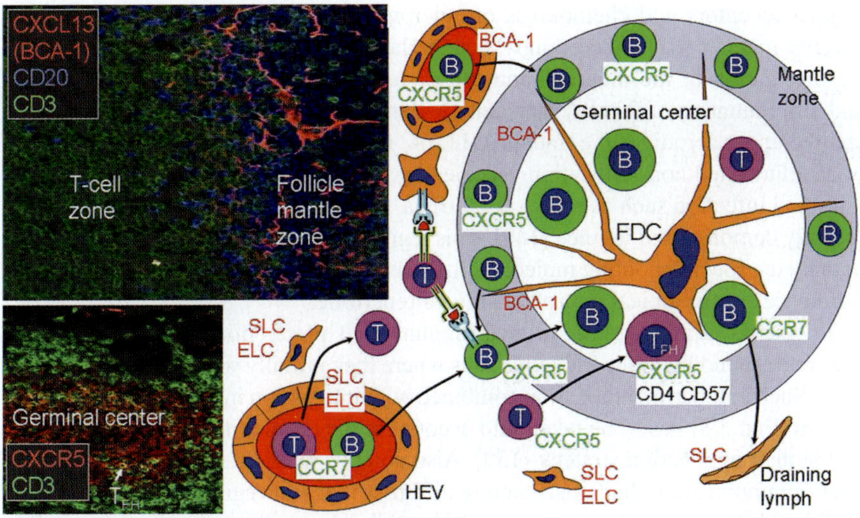

Fig. 3.7 Main adhesion molecule- and chemokine-regulated steps of T- and B-cell migration to, and positioning within, gut-associated lymphoid tissue (GALT). Naïve T and B cells enter both GALT and mesenteric lymph node via high endothelial venules (HEVs). The chemokines involved (right panel) are SLC (CCL21) and ELC (CCL19), produced by stromal cells and redistributed to the luminal face of HEVs as indicated to attract preferentially $CCR7^+$ naïve T cells and perhaps less actively B cells. SLC may also be involved in the exit of lymphoid cells from GALT via draining lymphatics as depicted. Naïve B cells are $CXCR5^+$ and extravasate in mice mainly via modified HEVs presenting CXCL13 (also called BCA-1 in humans) juxtaposed to, or inside of, lymphoid follicles; they are next attracted to the mantle zone where CXCL13 (BCA-1) is deposited on dendritic elements such as the follicular-dendritic cell (FDC) tips. The distribution of this chemokine, together with B and T cells, is shown by three-color immunofluorescence staining in the upper left panel, and the receptor distribution is similarly shown in the lower left panel. Also follicular B-helper T (T_{FH}) cells ($CXCR5^+CD4^+CD57^+$) are attracted to the follicle by similar interactions. B cells are primed just outside the lymphoid follicle by interaction with cognate T cells and antigen-presenting cells as indicated; they then re-enter the follicle and end up as $CCR7^+$ germinal-center cells after interactions with FDCs and T_{FH} cells. The B cells may thereafter leave the follicle as memory/effector cells. Immunofluorescence pictures adapted from Carlsen et al. [139]

DCs after secretion by another cell type (Fig. 3.7). The main source of CXCL13/BCA-1 does in fact appear to be the GCDCs previously reported to stimulate T cells in GCs [146]. Notably, we have observed that both GCDCs and large CXCL13/BCA-1-expressing cells in inflammation-associated lymphoid neogenesis with B-cell aggregates exhibit a phenotype compatible with macrophage derivation [147].

3.5.2 Compartmentalized Dispersion of Activated Mucosal B Cells

The homing of memory/effector lymphocytes depends on their surface expression of adhesion molecules and chemokine receptors that bind complementary counter-receptors and chemokines on microvascular endothelial cells in a tissue-specific manner [81]. Accordingly, accumulating evidence suggests a remarkable regionalization in the mucosal immune system with regard to homing properties and differentiation of B cells, especially with regard to a dichotomy between the gut and the upper aerodigestive tract [10, 11, 64, 66, 148]. Thus, intranasal application of an adjuvanted non-replication vaccine (e.g., virus-like particles) against a small-intestinal infection such as rotavirus diarrhea, cannot be expected to be effective, as recently demonstrated in mice [149]. This regional disparity is apparently explained both by the specific homing molecules involved in local leukocyte extravasation and differences in the topical microbial antigen repertoire.

In general, it appears that primed immune cells preferentially home to effector sites corresponding to the inductive sites where they initially were triggered by antigens. Such compartmentalization combined with integration in the mucosal immune system (Fig. 3.8) has to be taken into account in the development of local vaccines and in their application strategy [150]. Also notably, in addition to the variable cellular communication that exists among different mucosal regions, the mucosal and systemic lymphoid cell systems are not completely segregated – particularly not

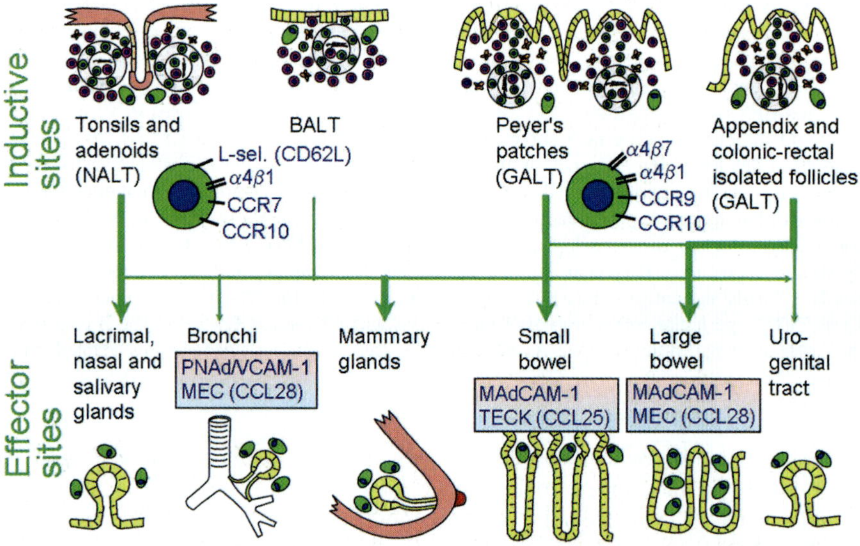

Fig. 3.8 Putative scheme for compartmentalized mucosal B-cell homing from inductive (top) to effector (bottom) sites in humans. Depicted are more or less preferred pathways (graded arrows) presumably followed by mucosal B cells of any isotype activated in nasopharynx-associated lymphoid tissue (NALT) represented by human Waldeyer's lymphoid ring (including tonsils and adenoids), and bronchus-associated lymphoid tissue (BALT), *versus* gut-associated lymphoid tissue (GALT) represented by Peyer's patches, appendix, and colonic-rectal isolated lymphoid follicles. The principal homing receptor profiles of the respective B-cell populations, and compartmentalized adhesion/chemokine cues directing extravasation at different effector sites, are indicated (pink and blue panels)

in the upper airways where several homing molecules are shared between the two systems as discussed below. In both systems memory/effector B cells express mainly L-selectin and also CCR7 but little of the gut homing receptor α4β7 [64, 151], and they may be guided to the bone marrow by CXC3 and CXCR4 [152].

3.5.3 Homing Molecules Operating at Mucosal Effector Sites

Homing of GALT-induced B cells to the gut lamina propria relies mainly on a high surface level of the integrin α4β7 and expression of the chemokine receptors CCR9 or CCR10 (see below), while a combined expression of integrin α4β1, L-selectin (CD62L) and CCR7 appears to direct mucosal B cells to the human upper airways and perhaps to the uterine cervix mucosa [64]. The counter receptor for the α4β7 integrin is the mucosal addressin cell adhesion molecule-1 (MAdCAM-1), expressed apically on endothelial cells of the intestinal lamina propria microvasculature [65, 81, 153], while α4β1 integrin binds vascular cell adhesion molecule (VCAM)-1, expressed on the endothelium in human bronchial and nasal mucosa [154, 155].

The microenvironment of the mucosal inductive sites and draining lymph nodes regulate integrin expression pattern on lymphocytes, thus imprinting the homing capacity of the memory/effector immune cells [156]. This imprinting is at least partially provided by regional DCs. Such cells isolated from Peyer's patches and MLNs are unique in their capacity to increase the expression of the mucosal homing receptor α4β7 integrin on lymphocytes [117, 157]. Acquisition of the gut-homing properties apparently depends on retinoic acid derived by oxidative conversion from vitamin A – an enzymatic process preferentially mediated by the regional DCs and acting on both B and T cells [117, 157, 158].

To attract the appropriately activated lymphocytes, expression of integrin counter-receptors on the microvascular endothelium is regulated by cell-intrinsic and micro-environmental factors at the mucosal effector sites. Additionally, compartmentalized production of specific chemokines that activate integrins on the lymphocytes and promote chemotaxis directs selective extravasation (Fig. 3.8).

In the gut, expression of different chemokines in the small and large intestine explains the selective recruitment of lymphocytes activated by the oral and rectal route [11, 148]. The chemokine CCL25 (TECK) is selectively produced by the crypt epithelium in small intestine [11, 159] and attracts B-cell blasts expressing CCR9 [160]. During an immune response to fed antigens, DCs in Peyer's patches and MLNs imprint lymphocytes with high expression of CCR9 together with α4β7 integrin, while L-selectin is down-regulated [117, 157, 161] – the combined effect being extravasation in the lamina propria of the small intestine. In the large intestine, CCL28 (MEC) expression appears to be a decisive cue for attracting IgA^+ plasmablasts that express high levels of CCR10 as well as α4β7 [11, 162, 163]. However, CCL28 is upregulated in the inflamed small intestine, such as during rotavirus infection in mice [164]. Because this chemokine is a cue also for CCR10-expressing

NALT-derived B-cell blasts [64], one may not exclude some protective effect from a nasal vaccine during a small intestinal infection, but experimental evidence speaks against this possibility [149].

The interaction between CCR10 and CCL28 may furthermore attract circulating mucosal B cells to the bone marrow where stromal cells secrete CCL28 [165]. Thus, we found that human NALT-derived plasmablasts homed to the bone marrow [64]. Nevertheless, most long-lived PCs in the bone marrow produce IgG and are of systemic origin, the precursors being attracted by more generalized chemokine-chemokine receptor pair interactions such as CXCR4-CXCL12 [152, 165].

3.5.4 Local Microbiota's Impact on B-Cell Regulation

A regionalized impact on mucosal immune regulation is also reflected by the highly disparate subclass distribution of IgA-producing PCs (Fig. 3.1c). Relatively more IgA2 is generally produced in the mucosal immune system than in peripheral lymphoid tissue – reaching a clear dominance over the IgA1 subclass in the large bowel mucosa [10, 11, 65]. An increase of the IgA2 isotype in secretions compared with serum can be important for the stability of secretory antibodies because SIgA2, in contrast to SIgA1, is resistant to several proteases synthesized by a variety of potentially pathogenic bacterial species [166].

However, production of IgA1 is dominating both in the nasal (93%) and bronchial (75%) mucosa [10, 11], and bacteria such as *Haemophilus influenzae, Streptococcus pneumoniae* and *Neisseria meningitides* show frequent synthesis of IgA1-specific proteases; this could contribute to the fact that those three bacterial species are prone to produce invasive disease of the upper respiratory tract, apparently by being able to exploit SIgA1-derived Fab fragments for epithelial adhesion [167]. The same bacteria may express an IgD-binding factor, which facilitates their interaction with the surface of naïve B cells; this could favor excessive B-cell proliferation-driving non-classical CSR to IgD-producing PCs as well as downstream CSR to IgA1 expression [10, 11, 33].

3.5.5 Mucosa-Draining Lymph Nodes as Amplifiers of Mucosal Immunity

MLNs and cervical lymph nodes apparently share immune-inductive properties with the related MALT structures – that is, GALT and NALT, respectively. The local/regional lymph nodes receive antigens in both MALT-derived and mucosa-derived lymph as well as via antigen-transporting DCs that continuously migrate through lymph from the same surface sites [168]. Thus, it has been shown in mice that T and B cells obtain gut- or skin-homing properties during antigen priming in MLNs or peripheral lymph nodes, respectively [11, 169]. Although DC-derived retinoic acid contributes to the homing instruction imprinted in GALT and MLNs [117, 157, 158, 170], all responsible inductive factors are not fully known. Nevertheless, similar mediators can probably induce homing capacity for the upper aerodigestive tract when T and B cells are stimulated by antigens in Waldeyer's ring and/or cervical lymph nodes [64].

Antigens reaching MALT-associated lymph nodes may hence elicit or amplify mucosal immunity in the same region. Importantly, the human nasal mucosa is extremely rich in various DC types, both within and beneath the epithelium [171]. In a rat model such airway DCs have been shown to extend their processes between epithelial cells and sample bacteria from the mucosal surface [168]. To be highly successful, an intranasally applied vaccine (see later) should therefore aim to target both mucosal DCs in the nasal cavity and FAE (including M cells) in the crypts of the MALT structures in Waldeyer's ring.

3.6 Induction of Cross-Reactive Antibodies and Immunological Memory

3.6.1 Role of the Commensal Microbiota

The peritoneal cavity is a substantial source of intestinal B cells in normal mice, reportedly giving rise to 40–50% of the lamina propria IgA-producing PCs [62, 172], which produce 'natural' cross-reactive SIgA antibodies directed mainly against microbial TI antigens with no clear dependency on a GC reaction [111]. The peritoneal precursors are self-renewing IgM^+B1 ($CD5^+$) cells (Fig. 3.3), but it remains controversial where they switch to the IgA^+ phenotype and how they reach the gut mucosa [6, 79, 99]. Also notably, rather than being encoded by germline sequences, murine B1 cells may show hypermutation of Ig heavy-chain V-region (V_H) genes as a sign of selection [173].

Several studies have analyzed the intestinal immune system of GF mice after monoassociation with a variety of non-invasive, commensal bacteria [174, 175]. In general, these microbes were found to induce a GC reaction in GALT with generation of plasmablasts that accumulated in the lamina propria as IgA-producing PCs and secreted 'natural' as well as specific antibodies. Individual bacterial species were shown to differ both with regard to the induced maximal amount of total 'natural' IgA and the fraction found to be specific for the colonizer [176]. The bacteria elicited a waxing followed by a long-term waning IgA response, accompanied by a GC reaction that both developed and declined much more rapidly. Such kinetics could be attributed to the 'shielding' of GALT from inductive microbial antigens by the lasting specific SIgA response generated because of relatively long-term persistence of both specific and 'natural' IgA-producing PCs in the lamina propria.

Mucosal immune modulation has been particularly well documented with the segmented filamentous bacterium (SFB, related to *Clostridia*), which is a major gut colonizer of the distal ileum in mice after weaning. Colonization of formerly GF weanlings resulted in a transient GC reaction of GALT and accumulation of IgA-producing PCs in the lamina propria, thereby providing an SIgA level comprising 50–70% of that seen normally in CVN mice; only about 1% of this IgA was found to be specific for SFB [177, 178].

Interestingly, super-colonization with *Morganella morganii* 100 days following monoassociation with SFB, induced little change in the production of total intestinal IgA although the specific response to *M. morganii* increased 20-fold compared to that against SFB [177]. This implied that the chronic GC reaction observed in GALT of CVN mice must be caused by the continuous exposure to novel microbial antigens. The sustained gut colonization of commensal bacteria therefore appears to provide the necessary chronic stimulation of previously induced 'natural' as well as specific IgA production, thereby promoting both cross-reactivity and memory. Thus, in contrast to the systemic immune system, MALT structures have a unique innate strategy to diversify and maintain the B-cell repertoire.

3.6.2 B1 and B2 Cells and Intestinal Antibody Production

The relative contribution of B1 and conventional bone marrow-derived B2 cells in murine IgA responses to commensal bacteria remains elusive [62, 86]; when one of these subsets is deleted in genetically manipulated mice, the other subset apparently occupies the whole intestinal B-cell compartment [99]. Lamina propria IgA-producing PCs of both B1 and the B2 origin showed quite restricted (oligodisperse) usage of V_H genes and multiple clonally related sequences when the repertoire was analyzed by complementarity-determining region (CDR)3 spectrotyping, cloning and sequencing [179]. Out of 15–20 sequences examined from various types of mice, there were two or more likely clonal relatives based on identical V/D/J junctional sequences. Such restricted V_H-gene usage was seen whether monoassociated immunodeficient recipients of B cells or CVN immunocompetent mice of several common strains were analyzed.

One could expect that the polyclonal microbial stimuli in the gut should induce polydisperse B-cell responses. It appears, instead, that both B1 and B2 cells generate an oligodisperse population of IgA-producing PCs that produce 'natural' cross-reactive antibodies, probably after BCR stimulation by TI-1 and/or TI-2 bacterial antigens. In this process, B cells with randomly recombined V/D/J segments – but without appreciable N-additions or point mutations in the V_H regions – might be selected by relatively few TI antigens in the gut.

Such observations in mice constitute useful explanation for the enormous IgA drive provided by the gut microbiota in the absence of a high-affinity BCR development. It appears that the production of large amounts of IgA with a restricted or oligodisperse repertoire provides antibody capacity to bind with relatively low affinity to the numerous redundant epitopes of commensal bacteria; to maintain host–parasite mutualism, the homeostatic balance thus relies mainly on the large quantities of cross-reactive SIgA produced in the gut [62, 86].

Altogether, the induction of intestinal IgA by the indigenous microbiota appears to be a rather primitive system that limits microbial colonization and penetration through the epithelial barrier without eliminating the bacteria from the gut lumen. Superimposed on this 'innate-like' antibody defense mechanism, the B2 system has the property to undergo GC-driven high-affinity BCR selection for particular virulence factors to provide more powerful adaptive protection against exogenous infections and expel overt pathogens from the host. These observations suggest that innate signals imitating the stimulation provided by commensal bacteria might be exploited to promote diversity and memory in response to vaccination against mucosal pathogens [92].

3.6.3 Intestinal B-Cell Responses Differ Between Humans and Mice

It remains elusive whether such a two-layered SIgA-mediated antibody strategy operates in humans where peritoneal B1 cells apparently do not contribute significantly to the intestinal PC population [79, 180]. Human lamina propria PCs of both the IgA and IgM classes, have highly mutated V_H-region genes even from childhood – consistent with precursor selection in GCs [181, 182, 183]. The level of mutations is significantly higher in the intestinal B cells than that seen in splenic PCs [184]. Moreover, spectrotyping of CDR3 variability shows a rather restricted repertoire for circulating human IgA^+ cells compared with V_H transcripts from the colon, which are quite diverse, particularly for the V_H1–V_H5 regions [185].

The IgM V_H-region genes in the human peritoneal cavity likewise exhibit fewer mutations than the corresponding genes from intestinal B cells [180]. Also, the V_H4-34 genes used by IgG and IgA in human peritoneal B cells show significantly lower numbers of mutations than their mucosal counterparts. Of further note, V_H-gene sequences from human Peyer's patch B cells are clonally related to ileal lamina propria PCs, in accordance with a predominant derivation from GCs of GALT [186]. Finally, clonally related V_H transcripts are widely distributed along the colon [184], likewise suggesting B-cell seeding from GALT mainly via peripheral blood (Fig. 3.3).

Altogether, there is no convincing evidence to suggest that mucosal B cells normally switch outside of GCs [6]; two independent studies in mice [99, 187] and one in humans [188] have quite convincingly rejected this possibility. However, it was recently observed that IgA^+ B cells in the colon of OcaB KO mice, which lack GCs, expressed low levels of AID [97], perhaps reflecting lamina propria CSR involving the B1 subset (Fig. 3.3).

Conversely, in humans there appears to be no reason to question that intestinal IgA-producing PCs are chiefly generated by precursor induction in GCs of GALT; B-cell migration from the peritoneal cavity [6, 10, 11, 180] or the greater omentum [189] to the lamina propria, and switching to IgA there [188, 190], are negligible or absent. Nevertheless, in the large bowel, bacteria could induce switch factors in epithelial cells and lamina propria DCs that may drive TI B-cell development towards the IgA2 subclass [191]. The first evidence for such a role of the gut microbiota was obtained by showing that jejunal segments with bacterial overgrowth had a $\sim 50\%$ reduction of the IgA1:IgA2 mucosal PC ratio [192].

Also notably, considerable levels of cross-reactive 'natural' SIgA antibodies directed against self as well as microbial antigens do occur in human external secretions [193]. The reason for this could be stimulation via BCR by microbial TI antigens and/or innate PRR-driven microbial activation of GALT – independent of BCR-mediated antigen recognition as discussed earlier. The principle of translating innate and polyclonal commensal signals into immunological diversity and memory [92] should therefore be a promising research avenue to be explored also for mucosal vaccination in humans.

3.7 Antibody-Mediated Mucosal Defense

3.7.1 Functions of Secretory IgA Antibodies and SC

In cooperation with a variety of innate mucosal defenses [194], secretory antibodies perform immune exclusion of exogenous antigens (Table 3.3 and Fig. 3.9) [5]. SIgA is a remarkably stable hybrid molecule (Fig. 3.1a) [41] and can therefore retain its antibody activity well in hostile environments such as the gut lumen [195] and oral cavity [196]. The function of SIgA is probably enhanced significantly by the high level of cross-reactivity (see earlier) as also detected in human secretions [197]. These 'natural' antibodies are apparently designed for urgent protection before an adaptive specific immune response is elicited; they are therefore reminiscent of innate immunity [193, 198]. In the lumen, SIgA will coat commensal bacteria, apparently without stopping their growth; this has been demonstrated both in saliva [199] and in feces [200] – most likely reflecting cross-reactive antibodies, although microbial binding via the Fc portion of IgA cannot be excluded [201]. Regardless of mechanism, such coating reduces bacterial access to the epithelial surface and protects against bacterial overgrowth and invasion, containing the indigenous microbiota within the lumen to maintain host–parasite mutualism [86].

Table 3.3 Antimicrobial effects of SIgA antibodies

- Provide efficient microbial agglutination and virus neutralization
- Perform anti-inflammatory extracellular and intracellular immune exclusion by inhibiting epithelial adherence and invasion
- Exhibit extensive cross-reactive ('innate-like') activity which provides cross-immunity and herd protection
- Especially the SIgA2 isotype is quite resistant against proteolysis (bound SC stabilizes both isotypes)
- Exhibit mucophilic and lectin-binding properties (via bound SC in both isotypes, and mannose in SIgA2)

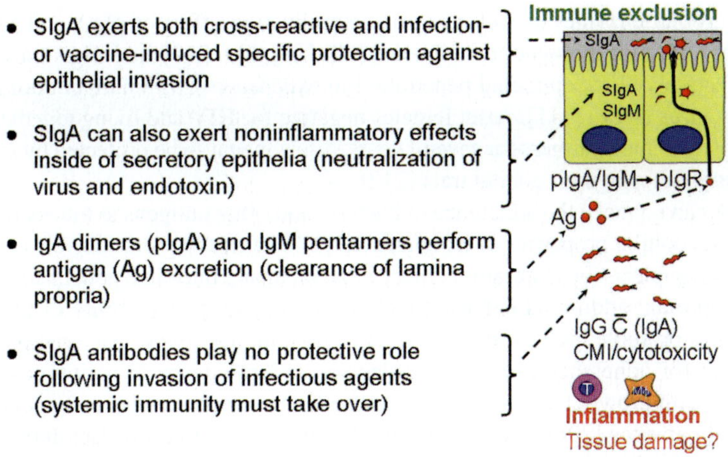
• SIgA exerts both cross-reactive and infection- or vaccine-induced specific protection against epithelial invasion
• SIgA can also exert noninflammatory effects inside of secretory epithelia (neutralization of virus and endotoxin)
• IgA dimers (pIgA) and IgM pentamers perform antigen (Ag) excretion (clearance of lamina propria)
• SIgA antibodies play no protective role following invasion of infectious agents (systemic immunity must take over)
Immune exclusion
SIgA
SIgA
SIgM
pIgA/IgM→pIgR
Ag
IgG C̄ (IgA)
CMI/cytotoxicity
T
Inflammation
Tissue damage?

Fig. 3.9 Different principles of SIgA-mediated contribution to mucosal homeostasis. In addition to immune exclusion at the epithelial surface, the pIgR-mediated external transport of dimeric IgA and pentameric IgM (pIgA/IgM) may be exploited for intraepithelial virus and toxin neutralization, as well as antigen excretion from the lamina propria. However, when infection with pathogen invasion occurs, systemic immunity must take over to save life. This involves potent proinflammatory mechanisms such as complement activation (C) by IgG antibodies, cell-mediated immunity (CMI), and cytotoxicity, all of which may cause tissue damage

After the induction of a secretory immune response in the newborn's gut, GALT will temporarily be 'shielded' by SIgA – apparently reflecting an antibody-mediated negative feed-back mechanism for homeostatic regulation of the mucosal B-cell system as discussed above [174]. Uptake of SIgA-coated bacteria by M cells may nevertheless be mediated by a receptor specific for IgA identified on these cells [72, 202, 203]. SIgA may exploit this M-cell receptor for delivery of antigens to DCs in the domes of Peyer's patches [72, 204]. This putative positive feedback mechanism could particularly target relevant environmental antigens to GALT of breast-fed infants by means of cognate maternal SIgA antibodies. In this manner, SIgA might act as an instructive immunological enhancer – accompanied by a balanced cytokine pattern that promotes homeostasis [72, 205]. Many reports show that breast-fed babies over time develop enhanced secretory immunity [206].

Whether SIgA-coated or not, commensal bacteria sampled by the M cells mainly become destroyed by subepithelial macrophages in murine Peyer's patches; only tiny amounts ($\sim$0.0001%) are targeted to DCs, but this is sufficient to induce an immune response in GALT and MLNs [86, 207]. The systemic immune system remains untriggered in CVN clean mice – demonstrating a compartmentalization of the response, which contributes to peripheral tolerance, or rather ignorance, of the indigenous microbiota under normal conditions – although this barrier does not appear so complete under natural conditions in humans [86].

It has been traditionally believed that SIgA can also efficiently inhibit epithelial colonization and invasion of overt pathogens (Table 3.3). The agglutinating and microbial enzyme- as well as virus-neutralization effect of pIgA and SIgA is superior to that of monomeric antibodies [42, 208, 209, 210], and SIgA antibodies can effectively block epithelial penetration in experiments with human immunodeficiency virus (HIV) [211]. Also, females negative for HIV and living together with HIV-positive male partners for several years, often appear to be protected by specific SIgA antibodies in their genital tract [212].

SIgA can enhance the adherence of bacteria and other antigens to mucus because of the mucophilic properties of its bound SC [17] – thereby promoting clearance of immune complexes by respiratory ciliary movement and intestinal peristalsis. Potentially important additional defense functions are the suggested ability of SIgA antibodies to promote biofilm formation [213, 214], induce loss of bacterial plasmids that code for adherence-associated molecules and resistance to antibiotics [215], interfere with growth factors (e.g. iron) and enzymes necessary for pathogenic bacteria and parasites [208], and enhance the bacteriostatic effects of lactoferrin [216] and the peroxidase system [217].

Animal and cell culture experiments have suggested that secretory antibodies can perform immune exclusion not only at the epithelial luminal surface (Fig. 3.9). During the pIgR-mediated transport of pIgA and pentameric IgM, such antibodies may even inactivate viruses (e.g. rotavirus, influenza virus and HIV) inside of secretory epithelial cells and carry the pathogens and their products back to the lumen, thus avoiding cytolytic damage to the epithelium [218, 219, 220, 221, 222]. It has also been suggested that pIgA antibodies can remove antigens from the lamina propria [223, 224] and neutralize bacterial LPS within intestinal epithelial cells [225] – implying putative novel anti-inflammatory and non-cytotoxic roles for this antibody class during its export to the lumen (Fig. 3.9).

In order to pump pIgs actively out to epithelial surfaces for efficient first-line defense, SC/pIgR might have originated from the innate defense system like several other proteins exploited for adaptive immunity. This evolutionary possibility is supported by the fact that free SC – generated by epithelial cleavage and apical release of unoccupied pIgR (Fig. 3.2) – possesses several natural protective properties [17, 226] such as binding to ETEC and reducing the effect of *Clostridium difficile* toxin [227, 228]. Moreover, a pneumococcal surface protein (SpsA/PspC) has been shown to interact directly with both free and bound SC [229]. Fc receptors for IgA (FcαRI, CD89) are expressed on human monocytes and granulocytes, particularly the neutrophils [230]. Eosinophils may even interact with, and become activated by, SIgA via its inherently bound SC [10, 231]. Interestingly, however, it has been shown that free SC can bind and inhibit the proinflammatory cytokine/chemokine IL-8, which is a potent chemotactic factor for neutrophilic granulocytes [232].

3.7.2 Functions of Secretory IgM and Local IgG Antibodies

Although SIgA is the chief effector of immune exclusion at mucosal surfaces, SIgM also contributes – particularly in newborns and subjects with selective IgA deficiency [33, 34]. In addition, there may be a significant contribution to immune exclusion by serum-derived or locally produced IgG antibodies (Fig. 3.2) transferred passively to the lumen by paracellular leakage [233] – or perhaps to some extent exported actively by the neonatal Fc receptor (FcRn) expressed on the gut epithelium [234]. IgG and monomeric IgA antibodies, when cross-linked via antigen with pIgA of the same specificity, might contribute to pIgR-mediated epithelial excretion of foreign material from the intestinal lamina propria [223]. Notably, however, because IgG is complement-activating, its contribution to surface defense is potentially pro-inflammatory, which could jeopardize the epithelial barrier function [235]. Nevertheless, such deterioration of local homeostasis is most likely counteracted by competition with anti-inflammatory IgA and by a variety of complement regulatory factors produced by mucosal epithelia [236].

It should also be noted that when overt infection with microbial invasion occurs, SIgA antibodies will no longer determine the fate of the host as experimentally documented in pIgR KO mice compared with wild-type mice [237, 238]. Moreover, studies in mucosally vaccinated wild-type mice challenged with live influenza virus intranasally, have suggested that while SIgA antibodies are essential to control virus replication locally, serum IgG antibodies protect against clinical illness [239]. Nevertheless, although systemic immunity may be considered a life-saving layer of defense, it operates at the risk of causing inflammation and tissue damage (Fig. 3.9). Thus, it has been clinically documented that SIgA antibodies prevent virally induced pathology in the upper airways, whereas IgG antibodies neutralize newly replicated influenza virus after the initiation of infection [240]. The lung parenchyma, which lacks a SIgA system, fully depends on serum-derived IgG and monomeric IgA for antibody protection [239, 241].

Therefore, in the face of most infections, it is important that vaccination induces both secretory and serum antibodies. The nasal route of immunization appears to be particularly advantageous to this end because of the shared homing molecules (L-selectin and CCR7) expressed on NALT-derived B-cell blasts [11, 64].

3.8 Clinical Value and Vaccine Stimulation of Mucosal Antibodies

3.8.1 Determining Protective Effects of Secretory Immunity

It is difficult to evaluate the defense functions of SIgA (and SIgM) antibodies during mucosal infection due to concurrent stimulation of systemic immunity; this may also be the case after local immunization with an attenuated live vaccine and when non-replicating virus-like particles or subunit vaccines are applied together with a mucosal adjuvant [242, 243]. As alluded to above, a protective effect of serum antibodies (mainly IgG) can contribute to immune exclusion (Fig 3.2), particularly in the respiratory tract – in addition to the general importance of systemic immunity to inhibit spread of the infectious agent throughout the body and performing antigen elimination within the mucosa [10] (Fig. 3.9).

An essential protective role of SIgA has been questioned by observations in IgA KO mice [244]. This strain remains healthy under ordinary laboratory conditions; and when challenged with influenza virus, it shows similar pulmonary virus levels and mortality as wild-type mice. According to the principle of 'pathotopic potentiation of local immunity' established in a rabbit model with influenza virus [245], leakage of serum IgG antibodies through an irritated mucosal surface epithelium [233] can play a significant protective role in the airways, as discussed earlier. Thus, antibody defense of the lung parenchyma fully depends on serum IgG [239, 241]. In addition, the IgA KO mice usually show a compensatory SIgM response and – compared with wild-type mice – they are less prone to inflammation-induced pathology, apparently because of reduced antigen-presenting ability [246].

Wild-type mice on exclusive parenteral nutrition show markedly reduced IgA anti-influenza virus titers in the upper respiratory tract with no significant compensatory SIgM response, which together probably explain their impaired mucosal immunity against experimental influenza [247]. The fact that humans with selective IgA deficiency do not suffer significantly more than others from intestinal virus infections, may largely be ascribed to their consistently enhanced SIgM and IgG1 responses in the gut, as well as increased numbers of intraepithelial lymphocytes [34].

The notion that secretory antibodies are essential for an optimal mucosal barrier function is supported by the systemic IgG antibody response against *E. coli* [23] and other gut bacteria [248] observed in our pIgR KO mice – reflecting excessive uptake of microbial antigens from the gut lumen. Moreover, the lamina propria population of IgA-producing PCs shows a three-fold increase, suggesting overstimulation of the mucosal B-cell system [249].

A similar pIgR KO strain has been used to test the role of SIgA antibodies in the airway defense against influenza virus [250]. After intranasal immunization, wild-type mice exhibited complete protection or partial cross-protection when challenged with live A/PR8 virus; lack of secretory antibodies in the airways caused a decrease of both protection and cross-protection, despite leakage of IgG and some IgA from serum into nasal secretions. These results emphasize the importance of secretory immunity in virus defense. Earlier reports likewise demonstrated that secretory antibodies induced by intranasal vaccination in humans, show a wider spectrum of activity against influenza virus than comparable serum (mainly IgG) antibodies [251, 252]. Furthermore, a correlation between influenza-specific SIgA in nasal lavage and protection has been suggested in vaccinated subjects [253].

In harmony with these results, SIgA responses have been shown to interfere significantly with mucosal uptake of macromolecules in experimental animals without inducing immunopathology [254], in contrast to the adverse effects on the epithelial barrier caused by serum IgG antibodies [235]. SIgA antibodies are also essential in protection against cholera toxin (CT), as demonstrated in mice deficient for J chain [24] or for pIgR [238] – two central components of the secretory immune system [15]. The same is probably true for heat-labile *E. coli* enterotoxin, which binds with high affinity to the same ganglioside receptor as CT. Notably, in a clinical setting there is significant cross-protection against ETEC diarrhea after vaccination against cholera with a killed whole-cell/CT B-subunit vaccine [255].

The initial epithelial colonization of bacterial pathogens is apparently also largely controlled by SIgA antibodies, even in naïve animals [238]. Most importantly, the natural horizontal fecal–oral spread of gut pathogens is significantly diminished by secretory antibodies [256], apparently because the SIgA coating of bacteria (see above) reduces both their mucosal colonization and fecal shedding. This explanation has likewise been given for the significant herd protection observed in unvaccinated subjects (up to 80% indirect protection) associated with the concurrent coverage level of the same population with inactivated oral cholera vaccine in a residential cluster of Bangladesh [257].

3.8.2 Mucosal Vaccines for Induction of Secretory Immunity

Mucosal infections represent a challenge for the development of vaccines that can either prevent the pathogen from colonizing the surface epithelium (noninvasive bacteria), inhibit its epithelial penetration and replication within the body (invasive bacteria and viruses), and/or block the binding of microbial toxins and neutralize them. In most cases, it would seem desirable to induce specific SIgA antibodies associated with immunological memory, in addition to systemic immunity. Such infections include in the gastrointestinal tract: *Helicobacter pylori, V. cholerae*, ETEC, *Salmonella, Shigella* spp., *C. parvum, Campylobacter jejuni, C. difficile, Giardia*, rotaviruses and calici viruses; in the respiratory tract: *Mycoplasma pneumoniae*, influenza virus and respiratory syncytial virus; in the genital tract: HIV, *Chlamydia, Neisseria gonorrhoeae* and herpes simplex virus; and in the urinary tract: selected strains of *E. coli* [150, 255, 258].

Some mucosal vaccines have been approved for human use, mostly being attenuated live variants (Table 3.4). With the exception of the oral killed whole-cell/CT B-subunit cholera vaccine (Dukoral), it has proven difficult to stimulate SIgA responses by peroral administration of non-replicating immunogens in humans

Table 3.4 Mucosal vaccines approved for human use

- Oral live attenuated (Sabin) polio vaccine
- Oral killed whole-cell/B subunit cholera vaccine
- Oral live attenuated cholera vaccine
- Oral live attenuated typhoid vaccine
- Oral live attenuated adenovirus vaccine (restricted to military personell)
- Oral live attenuated rotavirus vaccine (RotaShield); withdrawn 1999 because of reported intussusception in 15 out of 800,000 vaccinated American children
- New oral live attenuated rotavirus vaccines: Rotarix and RotaTeq
- Nasal enterotoxin-adjuvanted inactivated influenza vaccine (Nasalflu); withdrawn in 2001 because of reported facial paresis
- Nasal live attenuated influenza vaccine (FluMist)

[2, 255], and some controversy exists concerning the side effects and long-term efficacy of this and similar vaccines [259]. Many of the somatic microbial antigens may best be delivered by means of recombinant live attenuated carrier strains. Adenoviral vectors have been particularly studied as vaccine platforms because of their tropism for mucosal surfaces [260]. A platform can incorporate several microbial antigens in a single bacterial host to overcome the possibility of overgrowth of one live vaccine strain over others in a mixed vaccine.

Live vaccines are difficult to produce for small production facilities in developing countries, thus counteracting adequate pricing in a global perspective. An additional advantage of killed or subunit vaccines is that they are not jeopardized with the possibility of single-strain overgrowth or the potential problem of pathogenic reversion by recombination or other genetic modification in the vaccines or contacting microorganisms.

Two mucosal vaccines, an oral attenuated Rhesus monkey–human recombinant rotavirus vaccine (RotaShield) and a nasal enterotoxin-adjuvanted inactivated influenza vaccine (Nasalflu), were withdrawn from the market after a short time due to serious adverse reactions (Table 3.4) – illustrating the complexity of mucosal vaccine development [2, 150, 261]. The withdrawal of RotaShield has been estimated to put some 2.5–4 million potential vaccinees in developing countries at unnecessary risk to die from rotavirus infection over the following 5–8 years [262]. Hopefully, the two more recently marketed attenuated oral vaccines for rotavirus diarrhea – Rotarix (human) and RotaTeq (bovine-human) – will prove sufficiently effective in field trials without complications in industrialized as well as in a variety of developing countries, and also become affordable in that part of the world [263, 264]. As yet, there is no convincing evidence that these two vaccines increase the risk of life-threatening intussusception (mainly ileocecal 'telescoping'), but this conclusion may be premature [261].

Oral delivery of non-replicating vaccines is difficult due to poor stability of immunogenic proteins, peptides and DNA in the acidic and enzyme-rich gastrointestinal environment, although many strategies – including administration of particulates – have been attempted to avoid such problems [2], in addition to the use of a variety of adjuvants. CT and *E. coli* heat-labile toxin are considered as the most potent mucosal adjuvants – being well-known for their costimulatory properties from numerous animal studies. However, a high degree of toxicity limits their use in humans without extensive molecular modifications.

After nasal application in mice, CT was shown to reach the olfactory bulb and could thus adversely affect the central nervous system [265]. Although the wild-type *E. coli* heat-labile holotoxin has been used in humans [266], problems with cases of facial paresis (Bell's palsy) was noted and led to withdrawal of the Swiss inactivated influenza vaccine (Nasalflu) after 1 year as mentioned above [267]. Therefore, it is important to develop genetically detoxified mutants of mucosal adjuvants that are tolerated by humans. Animal experiments have shown that such mutants can retain their capacity to enhance humoral as well as cell-mediated immune responses to a mucosally applied inactivated influenza vaccine [268]. Alternatively, there are completely non-toxic oil-based mucosal adjuvants available (Eurocine) that have been shown to be efficient for nasal application both with DNA and peptide vaccines [269].

For peroral vaccination, some of the problems mentioned above may in the future be overcome by introducing edible plant vaccines [270, 271, 272]; a number of transgenic plants have been tested, but there are many unsolved problems related to such biofarming, and a shift to plant bioreactors may make the vaccines excessively expensive. Also, a general problem with non-replicating oral vaccines is poor induction of serum antibodies, which are needed – in addition to SIgA antibodies – to combat most mucosal infections (see earlier). This is one reason why there is an increasing focus on the development of nasal vaccines, even for tuberculosis [273] – and particularly against influenza virus as illustrated in the next section. However, the lack of consistent homing of NALT-derived B cells to the small intestine probably excludes this route of vaccine administration for many gut diseases in humans [11, 64, 150]. Such compartmentalization in vaccine effects may be masked in rodents because of the massive supply of serum-derived pIgA as SIgA via the bile to the lumen of the upper small intestine – amounting to 90% in the rat [274].

3.8.3 Nasal Influenza Vaccination

At present, parenteral vaccination is generally recommended for protection against influenza in vulnerable subjects. However, as discussed above, this approach induces little or no cross-protection, as reflected in the inter-pandemic manufacturing of the actual vaccines; when a genomic drift occurs in a virus, the vaccine strain has to be replaced, and it usually takes at least 6 months before a new vaccine is available. Conversely, many studies in experimental animals and humans have demonstrated that nasal vaccination gives rise to cross-protection against drifted strains. With the available mucosal influenza vaccine (Table 3.4), good protection was achieved despite the fact that the epidemic strain was not part of the vaccine [275]. Cross-clade immunity against HIV has also been reported after experimental nasal DNA prime in mice, followed by nasal peptide boost; the vaccine contained epitopes of clade B, but high and long-lasting serum antibody titers against the neutralizing gp41 ELDKWAS epitopes from both clades A, B, C and D were observed [276].

In nasal mucosa of unvaccinated adult subjects, antibody-producing PCs with specificity for influenza virus are present [277], but it remains unknown whether they are induced by previous (subclinical?) infection or reflect cross-reactivity of the mucosal IgA system. Notably, parenteral immunization with an inactivated trivalent virus vaccine did not result in an increase of influenza-specific PCs in nasal mucosa [278], although an IgA response was elicited in tonsils and saliva [279, 280].

These results suggest that to stimulate a nasal immune response, the vaccine should be targeted both against mucosal DCs migrating to cervical lymph nodes and against the crypts with M cells characteristic of the local MALT structures – probably the adenoids in particular (Fig. 3.8). Such regional mucosal immune induction apparently imprints the necessary homing properties of primed B cells to extravasate efficiently in the airway mucosa and associated glands [64, 151].

The amount of antigen reaching the lymphoid tissue of Waldeyer's ring and cervical lymph nodes after parenteral immunization is clearly insufficient to induce a nasal immune response, although some SIgA antibodies may occur in nasopharyngeal secretions. This most likely reflects local production in the adenoids where there is epithelial expression of pIgR, in contrast to the palatine tonsils [67, 281]. Not unexpectedly, some communication exists in terms of memory/effector B-cell distribution between the mucosal and the systemic immune systems [282], particularly so in cervical lymph nodes and Waldeyer's lymphoid ring because of shared homing molecules as described earlier [64].

Recent reports have documented the efficacy and effectiveness of the trivalent, live attenuated nasal spray influenza vaccine (CAIV-T), both in healthy children and adults [283]. Although nasal vaccination also efficiently induces systemic immunity, a combination of intranasal and parenteral immunization may be preferable for optimal protection when an inactivated influenza vaccine is used [284]. Alternatively, the effect of subunit vaccines applied topically can be enhanced by incorporation into liposomes or with the addition of a nontoxic mucosal adjuvant (Eurocine). The initial optimism regarding *E. coli* heat-labile enterotoxin as a nasal adjuvant in humans [285] vanished with the occurrence of Bell's palsy [267]. Other adjuvants such as the hydrophobic outer-membrane protein preparations (proteosomes) from *N. meningitidis* may be an efficient alternative [286, 287, 288]. In mice, the uptake of proteosomes can be enhanced by incorporation of TLR2 ligand (PorB) [71], but it remains to be shown whether TLRs are expressed on human M cells.

The only licensed commercial mucosal influenza vaccine (FluMist) is live attenuated and quite effective as mentioned earlier [275], but it remains too expensive for poor societies. Production of a non-proliferating vaccine would be cheaper, although such preparations are still generally believed to require repeated applications of quite high doses combined with adjuvants. Avoidance of immunogen treatment with formalin or β-propiolactone appears to be preferable; and particulates are more stimulatory for the mucosal immune system than soluble proteins. Thus, virus-derived particles may function without adjuvants as demonstrated for a trivalent inactivated whole-cell influenza vaccine [289] – probably because of better targeting to M cells and DCs. Protection against avian influenza viruses are facing similar scientific challenges [290].

An inactivated whole-virus monovalent influenza vaccine was recently tested with different devices for intranasal or intraoral spray application, and exhibited promising results for induction of antibodies both in serum and nasal secretions (Table 3.5). Mild side effects were deemed to be acceptable [291, 292]. Most importantly, the same intranasal vaccine also induced cellular immunity in addition to the desirable two-tiered antibody response – mucosal and systemic. A serum hemagglutination inhibition titer of 40 or higher – which is considered a protective level – was obtained in most volunteers after two vaccine doses given 1 week

Table 3.5 Vaccinated subjects (% of n = 15–19) with hemagglutination inhibition serum antibody titer ≥ 40 before and after 2 doses (3 wks) or 4 doses (6 wks) of inactivated whole-virus influenza vaccine, and showing a highly significant nasal IgA response after 4 doses

Vaccine delivery	Serum: pre-immunization (%)	Serum: after 2 doses (%)	Serum: after 4 doses (%)	Nasal fluidIgA response
Nasal spray (OptiMist™)	32	89	100	p = 0.0003
Nasal spray (conventional)	32	79	94	p = 0.0006
Nasal drops	28	94	100	P = 0.0006
Oral spray	29	76	87	No response*

Based on data from [291, 292].
*No increase of IgA antibody titer in nasal fluid, but a slight increase in whole saliva.

apart; and an additional memory effect was revealed after three or four doses in that 100% of the individuals had acquired protective antibody titers. This result was best achieved by vaccine application in nasal drops or with a special breath-activated device (OptiMist™, Optinose), whereas oral spray only induced serum antibodies.

3.9 Summary and Conclusions

3.9.1 Characteristics of Secretory Immunity

Secretory antibodies apparently function both by performing antigen exclusion at mucosal surfaces and by virus and endotoxin neutralization within epithelial cells. SIgA is thus persistently containing commensal bacteria outside the epithelial barrier but can also target invasion of pathogens and penetration of other harmful antigens. Importantly, in contrast to IgG antibodies, IgA antibodies do not activate complement and are therefore generally considered as anti-inflammatory.

Several studies have shown that natural infections and mucosal immunization are more effective in giving rise to SIgA antibodies than parenteral vaccination. The latter appears to induce an SIgA response mainly when the mucosal immune system has been primed locally in advance [293]. Mucosal stimulation is generally much more efficient with live than with inactivated vaccines. However, various types of mucosal adjuvants and particulate immunogens can enhance the stimulatory effect of non-replicating vaccines.

Like natural infections, live mucosal vaccines or adequate combinations of non-replicating vaccines and mucosal adjuvants, give rise not only to SIgA antibodies but also to longstanding serum IgG and IgA responses, especially when applied by the nasal route. Vaccine targeting of NALT may therefore be advantageous to protect against many infections, but only if successful stimulation is achieved without the use of toxic adjuvants that might reach the central nervous system. There is considerable disparity with regard to the migration of memory/effector B cells from mucosal inductive sites to secretory effector sites, and NALT-derived cells do not home consistently to the gut [64, 150]. Also, although immunological memory is generated after mucosal priming in the gut, this may be masked by a self-limiting response temporarily shielding the GALT.

The degree of protection obtained after mucosal vaccination ranges from reduction of symptoms to complete inhibition of re-infection. In this scenario, it is often difficult to determine the relative importance of SIgA *versus* serum antibodies. Nevertheless, infection models in KO mice strongly support the notion that SIgA exerts a decisive role in protection and cross-protection against a variety of infectious agents. Resistance to toxin-producing bacteria such as *V. cholera* and ETEC appears to depend largely on SIgA, and so does herd protection against horizontal fecal–oral spread of enteric pathogens under naïve or immunized conditions – with a substantial innate impact both on cross-reactivity and memory. However, relatively few mucosal vaccines have been approved for human use, and more basic work is needed in vaccine and adjuvant design, including particulate or live-vectored combinations.

3.9.2 Species- and Site-Specific Features of Mucosal Immune-Inductive Sites

Considerable levels of cross-reactive SIgA antibodies directed against self as well as against microbial antigens occur in human external secretions [193]. As in mice, the reason for this may be polyclonal microbial activation of MALT, independently of BCR-mediated antigen recognition. This is an important distinction between systemic and mucosal immunity, which can be exploited by vaccine targeting of both GALT and NALT – the latter approach at the same time providing substantial stimulation of the systemic immune system because of shared B-cell homing molecules [64, 150].

The enormous innate drive of the mucosal immune system does not only enhance diversity but also memory. However, with regard to NALT, the situation in humans and mice appears to be significantly different. First, NALT of rodents is an organized bell-shaped tissue structure situated in the floor of the nasal cavity, and its organogenesis is different from that of GALT and human palatine and nasopharyngeal tonsils [6, 294]. The tonsils have deep and branched antigen-retaining crypts while rodent NALT has a smooth surface – like GALT in all mammalian species [6]. This disparity probably explains that GCs develop shortly after birth in tonsils as in the heavily microbe-exposed GALT structures, whereas rodent NALT requires an infection or another danger signal such as CT to drive GC formation [187, 295].

The relative lack of innate stimulation is therefore probably the reason that a non-replicating rotavirus vaccine adjuvanted with a mutant *E. coli* toxin did not induce a substantial memory response in murine NALT [149]. By contrast, human NALT is equipped with antigen- and microbe-retaining crypts and is hence liable to polyclonal stimulation for enhanced development of B-cell diversity and memory.

In addition, the crypt epithelium is activated via PRRs to secrete BAFF and the thymic stromal lymphopoietin (TSLP) – a cytokine that further promotes CSR and broad reactivity of local B cells by activating BAFF-producing DCs [296]. Together, these features of human NALT constitute an intriguing basis for the current interest in exploiting the nasal route for vaccine administration to combat a variety of diseases [297].

Acknowledgments Studies in the authors' laboratory are supported by the Research Council of Norway, the University of Oslo and Rikshospitalet University Hospital. Hege Eliassen and Erik K. Hagen provided excellent assistance with the manuscript and figures, respectively.

References

1. World Health Report 2005, Annex Tables 3.3 and 3.4 (World Health Organization).
2. Mitragotri, S. 2005, Nat. Rev. Immunol., 5, 905.
3. Levine, M.M. 2003, Nat. Med., 9, 99.
4. Neutra, M.R. and Kozlowski, P.A. 2006, Nat. Rev. Immunol., 6, 148.
5. Brandtzaeg, P. 2003, Int. J. Med. Microbiol., 293, 3.
6. Brandtzaeg, P. and Pabst, R. 2004, Trends Immunol., 25, 570.
7. Brandtzaeg, P. 1985, Scand. J. Immunol., 22, 111.
8. Brandtzaeg, P. 1973, Nature New Biol., 243, 142.
9. Brandtzaeg, P. 1974, Nature, 252, 418.
10. Brandtzaeg, P., Farstad, I.N., Johansen, F.-E., Morton, H.C., Norderhaug, I.N., and Yamanaka, T. 1999, Immunol. Rev., 171, 45.
11. Brandtzaeg, P., and Johansen, F.-E. 2005, Immunol. Rev., 206, 32.
12. Brandtzaeg, P., Halstensen, T.S., Kett, K., Krajci, P., Kvale, D., Rognum, T.O., Scott, H., and Sollid, L.M. 1989, Gastroenterology, 97, 1562.
13. Tomasi, T.B., Tan, E.M., Solomon, A., and Prendergast, R.A. 1965, J. Exp. Med., 121, 101.
14. Brandtzaeg, P. 1974, J. Immunol., 112, 1553.
15. Brandtzaeg, P., and Prydz, H. 1984, Nature, 311, 71.
16. Brandtzaeg, P. 1973, Ann. Inst. Pasteur Immunol., 124C, 417.
17. Phalipon, A., and Corthésy, B. 2003, Trends Immunol., 24, 55.
18. Halpern, M.S., and Koshland, M.E. 1970, Nature, 228, 1276.
19. Mestecky, J., Zikan, J., and Butler, W.T. 1971, Science, 171, 1163.
20. Johansen, F.-E., Braathen, R., and Brandtzaeg, P. 2000, Scand. J. Immunol., 52, 240.
21. Johansen, F.-E., Braathen, R., and Brandtzaeg, P. 2001, J. Immunol., 167, 5185.
22. Hendrickson, B.A., Conner, D.A., Ladd, D.J., Kendall, D., Casanova, J.E., Corthesy, B., Max, E.E,. Neutra, M.R., Seidman, C.E., and Seidman, J.G. 1995, J. Exp. Med., 182, 1905.
23. Johansen, F.E., Pekna, M., Norderhaug, I.N., Haneberg, B., Hietala, M.A., Krajci, P., Betsholtz, C., and Brandtzaeg, P. 1999, J. Exp. Med. 190, 915.
24. Lycke, N., Erlandsson, L., Ekman, L., Schèon, K., and Leanderson, T. 1999, J. Immunol., 163, 913.
25. Shimada, S., Kawaguchi-Miyashita, M., Kushiro, A., Sato, T., Nanno, M., Sako, T., Matsuoka, Y., Sudo, K., Tagawa, Y., Iwakura, Y., and Ohwaki, M. 1999, J. Immunol., 163, 5367.
26. Verges, M., Sebastian, I., and Mostov, K.E. 2007, Exp. Cell. Res., 313, 707.
27. Mostov, K.E., Friedlander, M., and Blobel, G. 1984, Nature, 308, 37.
28. Bonner, A., Perrier, C., Corthesy, B., and Perkins, S.J. 2007, J. Biol. Chem., 282, 16969.
29. Norderhaug, I.N., Johansen, F.-E., Schjerven, H., and Brandtzaeg, P. 1999, Crit. Rev. Immunol., 19, 481.

30. Rojas, R., and Apodaca, G. 2002, Nat. Rev. Mol. Cell. Biol., 3, 944.
31. Crottet, P., and Corthesy, B. 1998, J. Immunol., 161, 5445.
32. Brandtzaeg, P. 1975, Immunology, 29, 559.
33. Brandtzaeg, P., Nilssen, D.E., Rognum, T.O., and Thrane, P.S. 1991, Gastroenterol. Clin. North Am., 20, 397.
34. Brandtzaeg, P., and Nilssen, D.E. 1995, Curr. Opin. Gastroenterol., 11, 532.
35. Brandtzaeg, P., Halstensen, T.S., Huitfeldt, H.S., Krajci, K., Kvale, D., Scott, H., and Thrane, P.S. 1992, Ann. NY Acad. Sci., 664, 157.
36. Schjerven, H., Brandtzaeg, P., and Johansen, F.-E. 2000, J. Immunol., 165, 3898.
37. Schjerven, H., Brandtzaeg, P., and Johansen, F.-E. 2001, J. Immunol., 167, 6412.
38. Johansen, F.-E., and Brandtzaeg, P. 2004, Trends Immunol., 25, 150.
39. Ha, C.L., and Woodwar,d B. 1998, Lab. Invest., 78, 1255.
40. Sarkar, J., Gangopadhyay, N.N., Moldoveanu, Z., Mestecky, J., and Stephensen, C.B. 1998, J. Nutr., 128, 1063.
41. Conley, M.E., and Delacroix, D.L. 1987, Ann. Intern. Med., 106, 892.
42. Berdoz, J., Blanc, C.T., Reinhardt, M., Kraehenbuhl, J.P., and Corthesy, B. 1999, Proc. Natl. Acad. Sci. USA, 96, 3029.
43. Brandtzaeg, P. 2007, Ann. NY Acad. Sci., 1098, 288.
44. Anonymous. 1994, Lancet, 344, 1239.
45. Jones, G., Steketee, R.W., Black, R.E., Bhutta, Z.A., Morris, S.S.; Bellagio Child Survival Study Group. 2003, Lancet, 362, 65.
46. Jong-wook, L. 2003, Lancet, 362, 2083.
47. Dickinson, E.C., Gorga, J.C., Garrett, M., Tuncer, R., Boyle, P., Watkins, S.C., Alber, S.M., Parizhskaya, M., Trucco, M., Rowe, M.I., and Ford, H.R. 1998, Surgery, 124, 284.
48. Hylander, M.A., Strobino, D.M., and Dhanireddy, R. 1998, Pediatrics, 102, E38.
49. Pisacane, A., Graziano, L., Zona, G., Granata, G., Dolezalova, H., Cafiero, M., Coppola, A,, Scarpellino, B., Ummarino, M., and Mazzarella, G. 1994, Acta Paediatr., 83, 714.
50. Wold, A.E., and Hanson, L. Å. 1994, Curr. Opin. Gastroenterol., 10, 652.
51. Newman, J. 1995, Sci. Am., 273, 58.
52. Wright, A.L., Bauer, M., Naylor, A., Sutcliffe, E., and Clark, L. 1998, Pediatrics, 101, 837.
53. Ogra, S.S., Weintraub, D., and Ogra, P.L. 1977, J. Immunol., 119, 245.
54. Klemola, T., Savilahti, E., and Leinikki, P. 1986, Acta Paediatr. Scand., 75, 230.
55. Weaver, L.T., Wadd, N., Taylor, C.E., Greenwell, J., and Toms, G.L. 1991, Pediatr. Allergy Immunol., 2, 2.
56. van Elburg, R.M., Uil, J.J., de Monchy, J.G., and Heymans, H.S. 1992, Scand. J. Gastroenterol., 194(suppl), 19.
57. Hooper, L.V., Wong, M.H., Thelin, A., Hansson, L., Falk, P.G., and Gordon, J.I. 2001, Science, 291, 881.
58. Neish, A.S., Gewirtz, A.T., Zeng, H., Young, A.N., Hobert, M.E., and Karmali, V., Rao, A.S., and Madara, J.L. 2000, Science, 289, 1560.
59. Crabbé, P.A., Nash, D.R., Bazin, H., Eyssen, H., and Heremans, J.F. 1970, Lab. Invest., 22, 448.
60. Yamanaka, T., Helgeland, L., Farstad, I.N., Midtvedt, T., Fukushima, H., and Brandtzaeg, P. 2003, J. Immunol., 170, 816.
61. Macpherson, A.J., and Harris, N.L. 2004, Nat. Rev. Immunol., 4, 478.
62. Macpherson, A.J., McCoy, K.D., Johansen, F.-E., and Brandtzaeg, P. 2008, Mucosal Immunol., 1, 11.
63. Neutra, M.R., Mantis, N.J., and Kraehenbuhl, J.P. 2001, Nat. Immunol., 2, 1004.
64. Johansen, F.-E., Baekkevold, E.S., Carlsen, H.S., Farstad, I.N., Soler, D., and Brandtzaeg, P. 2005, Blood, 106, 593.
65. Brandtzaeg, P., Baekkevold, E.S., Farstad, I.N., Jahnsen, F.L., Johansen, F.-E., Nilsen, E.M., Yamanaka, T. 1999, Immunol. Today, 20, 141.
66. Brandtzaeg, P., Farstad, I.N., and Haraldsen, G. 1999, Immunol. Today, 20, 267.

67. Brandtzaeg, P. 2003, International Congress Series (ICS), 1254, 89 (Elsevier)/Int. J. Pediatr. Otorhinolaryngol., 67(Suppl. 1), S69.
68. Hiller, A.S., Tschernig, T., Kleemann, W.J., and Pabst, R. 1998, Scand. J. Immunol., 47, 159.
69. Chabot, S., Wagner, J.S., Farrant, S., and Neutra, M.R. 2006, J. Immunol., 176, 4275.
70. Tyrer, P., Foxwell, A.R., Cripps, A.W., Apicella, M.A., and Kyd, J.M. 2006, Infect. Immun., 74, 625.
71. Chabot, S.M., Chernin, T.S., Shawi, M., Wagner, J., Farrant, S., Burt, D.S., Cyr, S., and Neutra, M.R. 2007, Vaccine, 25, 5348.
72. Corthesy, B. 2007, J. Immunol., 178, 27.
73. Rescigno, M., Urbano, M., Valzasina, B., Francolini, M., Rotta, G., Bonasio, R., Granucci, F., Kraehenbuhl, J.P., and Ricciardi-Castagnoli, P. 2001, Nat. Immunol., 2, 361.
74. Milling, S.W., Cousins, L., and MacPherson, G.G. 2005, Trends Immunol., 26, 349.
75. Yamanaka, T., Straumfors, A., Morton, H., Fausa, O., Brandtzaeg, P., and Farstad, I. 2001, Eur. J. Immunol., 31, 107.
76. Bjerke, K., and Brandtzaeg, P. 1988, Clin. Exp. Immunol., 71, 502.
77. Christ, A.D., and Blumberg, R.S. 1997, Springer Sem. Immunopathol., 18, 449.
78. Perera, L., Shao, L., Patel, A., Evans, K., Meresse, B., Blumberg, R., Geraghty, D., Groh, V., Spies, T., Jabri, B., and Mayer, L. 2007, Inflamm. Bowel Dis., 13, 298.
79. Brandtzaeg, P., Baekkevold, E.S., and Morton, H.C. 2001, Nat. Immunol., 2, 1093.
80. Fagarasan, S., and Honjo, T. 2003, Nat. Rev. Immunol., 3, 63.
81. Kunkel, E.J., and Butcher, E.C. 2002, Immunity, 16, 1.
82. Lanning, D.K., Rhee, K.J., and Knight, K.L. 2005, Trends Immunol., 26, 419.
83. Mazmanian, S.K., Liu, C.H., Tzianabos, A.O., and Kasper, D.L. 2005, Cell, 122, 107.
84. Manser, T. 2004, J. Immunol., 172, 3369.
85. Macpherson, A.J., Lamarre, A., McCoy, K., Harriman, G.R., Odermatt, B., Dougan, G., Hengartner, H., and Zinkernagel, R.M. 2001, Nat. Immunol., 2, 625.
86. Macpherson, A.J., Geuking, M.B., and McCoy, K.D. 2005, Immunology, 115, 153.
87. Rhee, K.J., Sethupathi, P., Driks, A., Lanning, D.K., Knight, K.L. 2004, J. Immunol., 172, 1118.
88. Casola, S., Otipoby, K.L., Alimzhanov, M., Humme, S., Uyttersprot, N., Kutok, J.L., Carroll, M.C., and Rajewsky, K. 2004, Nat. Immunol., 5, 317.
89. Beisner, D.R., Ch'en, I.L., Kolla, R.V., Hoffmann, A., and Hedrick, S.M. 2005, J. Immunol., 175, 3469.
90. Bernasconi, N.L., Traggiai, E., and Lanzavecchia, A. 2002, Science, 298, 2199.
91. Pasare, C., and Medzhitov, R. 2005, Nature, 438, 364.
92. Pulendran, B., and Ahmed, R. 2006, Cell, 124, 849.
93. Ganley-Leal, L.M., Liu, X., and Wetzler, L.M. 2006, Clin. Immunol., 120, 272.
94. Shapiro-Shelef, M., and Calame, K. 2005, Nat. Rev. Immunol., 5, 230.
95. Chaudhuri, J., and Alt, F.W. 2004, Nat. Rev. Immunol., 4, 541.
96. Muramatsu, M., Kinoshita, K., Fagarasan, S., Yamada, S., Shinkai, Y., and Honjo, T. 2000, Cell, 102, 553.
97. Crouch, E.E., Li, Z., Takizawa, M., Fichtner-Feigl, S., Gourzi, P., Montano, C., Feigenbaum, L., Wilson, P., Janz, S., Papavasiliou, F.N., and Casellas, R. 2007, J. Exp. Med., 204, 1145.
98. Kingzette, M., Spieker-Polet, H., Yam, P.C., Zhai, S.K., and Knight, K.L. 1998, Proc. Natl. Acad. Sci. USA, 95, 11840.
99. Bergqvist, P., Gardby, E., Stensson, A., Bemark, M., and Lycke, N.Y. 2006, J. Immunol., 177, 7772.
100. Renshaw, B.R., Fanslow, W.C., Armitage, R.J., Campbell, K.A., Liggitt, D., Wright, B., Davison, B.L., and Maliszewski, C.R. 1994, J. Exp. Med., 180, 1889.
101. Stensvold, K., Brandtzaeg, P., Kvaløy, S., Seip, M., and Lie, S.O. 1984, Br. J. Haematol., 56, 417.

102. Sonoda, E., Hitoshi, Y., Yamaguchi, N., Ishii, T., Tominaga, A., Araki, S., and Takatsu, K. 1992, Cell. Immunol., 140, 158.
103. Cazac, B.B., and Roes, J. 2000, Immunity, 13, 443.
104. Li, R., Rosendahl, A., Brodin, G., Cheng, A.M., Ahgren, A., Sundquist, C., Kulkarni, S., Pawson, T., Heldin, C.H., and Heuchel, R.L. 2006, J. Immunol., 176, 6777.
105. Klein, J., Ju, W., Heyer, J., Wittek, B., Haneke, T., Knaus, P., Kucherlapati, R., Bottinger, E.P., Nitschke, L., and Kneitz, B. 2006, J. Immunol., 176, 2389.
106. Vajdy, M., Kosco-Vilbois, M.H., Kopf, M., Kohler, G., and Lycke, N. 1995, J. Exp. Med., 181, 41.
107. Kopf, M., Le Gros, G., Coyle, A.J., Kosco-Vilbois, M., and Brombacher, F. 1995, Immunol. Rev., 148, 45.
108. Bromander, A.K., Ekman, L., Kopf, M., Nedrud, J.G., and Lycke, N.Y. 1996, J. Immunol., 156, 4290.
109. Ramsay, A.J., Husband, A.J., Ramshaw, I.A., Bao, S., Matthaei, K.I., Koehler, G., and Kopf, M. 1994, Science, 264, 561.
110. Macpherson, A.J., Gatto, D., Sainsbury, E., Harriman, G.R., Hengartner, H., and Zinkernagel, R.M. 2000, Science, 288, 2222.
111. Litinskiy, M.B., Nardelli, B., Hilbert, D.M., He, B., Schaffer, A., Casali, P., and Cerutti, A. 2002, Nat. Immunol., 3, 822.
112. Castigli, E., Wilson, S.A., Scott, S., Dedeoglu, F., Xu, S., Lam, K.P., Bram, R.J., Jabara, H., and Geha, R.S. 2005, J. Exp. Med., 201, 35.
113. Mackay, F., Schneider, P., Rennert, P., and Browning, J. 2003, Annu. Rev. Immunol., 21, 231.
114. Rolink, A.G., and Melchers, F. 2002, Curr. Opin. Immunol., 14, 266.
115. von Bulow, G.U., van Deursen, J.M., and Bram, R.J. 2001, Immunity, 14, 573.
116. Castigli, E., Scott, S., Dedeoglu, F., Bryce, P., Jabara, H., Bhan, A.K., Mizoguchi, E., and Geha, R.S. 2004, Proc. Natl. Acad. Sci. USA, 101, 3903.
117. Mora, J.R., Iwata, M., Eksteen, B., Song, S.Y., Junt, T., Senman, B., Otipoby, K.L., Yokota, A., Takeuchi, H., Ricciardi-Castagnoli, P., Rajewsky, K., Adams, D.H., and von Andrian, U.H. 2006, Science, 314, 1157.
118. Randall, T.D., Heath, A.W., Santos-Argumedo, L., Howard, M.C., Weissman, I.L., and Lund, F.E. 1998, Immunity, 8, 733.
119. Tigges, M.A., Casey, L.S., and Koshland, M.E. 1989, Science, 243, 781.
120. Max, E.E., and Korsmeyer, S.J. 1985, J. Exp. Med., 161, 832.
121. Kang, C.J., Sheridan, C, Koshland, M.E. 1998, Immunity, 8, 285.
122. Wallin, J.J., Rinkenberger, J.L., Rao, S., Gackstetter, E.R., Koshland, M.E., and Zwollo, P. 1999, J. Biol. Chem., 274, 15959.
123. Lansford, R.D., McFadden, H.J., Siu, S.T., Cox, J.S., Cann, G.M., and Koshland, M.E. 1992, Proc. Natl. Acad. Sci. USA, 89, 5966.
124. Rinkenberger, J.L., Wallin, J.J., Johnson, K.W., and Koshland, M.E. 1996, Immunity, 5, 377.
125. Wallin, J.J., Gackstetter, E.R., and Koshland, M.E. 1998, Science, 279, 1961.
126. Rao, S., Karray, S., Gackstetter, E.R., and Koshland, M.E. 1998, J. Biol. Chem., 273, 26123.
127. Shin, M.K., and Koshland, M.E. 1993, Genes Dev., 7, 2006.
128. Linderson, Y., Eberhard, D., Malin, S., Johansson, A., Busslinger, M., and Pettersson, S. 2004, EMBO Rep., 5, 291.
129. Lin, K.I., Angelin-Duclos, C., Kuo, T.C., and Calame, K. 2002, Mol. Cell. Biol., 22, 4771.
130. Bertrand, F.E., Billips, L.G., Gartland, G.L., Kubagawa, H., and Schroeder, H.W. 1996, J. Immunol., 156, 4240.
131. Cyster, J.G. 1999, Science, 286, 2098.
132. Moser, B., and Loetscher, P. 2001, Nat. Immunol., 2, 123.
133. Legler, D.F., Loetscher, M., Roos, R.S., Clark-Lewis, I., Baggiolini, M., and Moser, B. 1998, J. Exp. Med., 187, 655.
134. Gunn, M.D., Ngo, V.N., Ansel, K.M., Ekland, E.H., Cyster, J.G., and Williams, L.T. 1998, Nature, 391, 799.

135. Förster, R., Mattis, A.E., Kremmer, E., Wolf, E., Brem, G., and Lipp, M. 1996, Cell, 87, 1037.
136. Luther, S.A., Lopez, T., Bai, W., Hanahan, D., and Cyster, J.G. 2000, Immunity, 12, 471.
137. Ansel, K.M., Ngo, V.N., Hyman, P.L., Luther, S.A., Forster, R., Sedgwick, J.D., Browning, J.L., Lipp, M., and Cyster, J.G. 2000, Nature, 406, 309.
138. Bleul, C.C., Schultze, J.L., and Springer, T.A. 1998, J. Exp. Med., 187, 753.
139. Carlsen, H.S., Baekkevold, E.S., Johansen, F.-E., Haraldsen, G., and Brandtzaeg, P. 2002, Gut, 51, 364.
140. Schaerli, P., Willimann, K., Lang, A.B., Lipp, M., Loetscher, P., and Moser, B. 2000, J. Exp. Med., 192, 1553.
141. Breitfeld, D., Ohl, L., Kremmer, E., Ellwart, J., Sallusto, F., Lipp, M., and Forster, R. 2000, J. Exp. Med., 192, 1545.
142. Moser, B., Schaerli, P., and Loetscher, P. 2002, Trends Immunol., 23, 250.
143. Chtanova, T., Tangye, S.G., Newton, R., Frank, N., Hodge, M.R., Rolph, M.S., and Mackay, C.R. 2004, J. Immunol., 173, 68.
144. Kim, C.H., Rott, L.S., Clark-Lewis, I., Campbell, D.J., Wu, L., and Butcher, E.C. 2001, J. Exp. Med., 193, 1373.
145. Rasheed, A.U., Rahn, H.P., Sallusto, F., Lipp, M., and Muller, G. 2006, Eur. J. Immunol., 36, 1892.
146. Grouard, G., Durand, I., Filgueira, L., Banchereau, J., Liu, Y.-J. 1996, Nature, 384, 364.
147. Carlsen, H.S., Baekkevold, E.S., Morton, H.C., Haraldsen, G., and Brandtzaeg, P. 2004, Blood, 104, 3021.
148. Kiyono, H., and Fukuyama, S. 2004, Nat. Rev. Immunol., 4, 699.
149. Ogier, A., Franco, M.A., Charpilienne, A., Cohen, J., Pothier, P., and Kohli, E. 2005, Eur. J. Immunol., 35, 2122.
150. Holmgren, J., and Czerkinsky, C. 2005, Nat. Med., 11(4 Suppl), S45.
151. Quiding-Järbrink, M., Nordström, I., Granström, G., Kilander, A., Jertborn, M., Butcher, E.C., Lazarovits, A.I., Holmgren, J., and Czerkinsky, C. 1997, J. Clin. Invest., 99, 1281.
152. Odendahl, M., Mei, H., Hoyer, B.F., Jacobi, A.M., Hansen, A., Muehlinghaus, G., Berek, C., Hiepe, F., Manz, R., Radbruch, A., and Dorner, T. 2005, Blood, 105, 1614.
153. Butcher, E.C., and Picker, L.J. 1996, Science, 272, 60.
154. Bentley, A.M., Durham, S.R., Robinson, D.S., Menz, G., Storz, C., Cromwell, O., Kay, A.B., and Wardlaw, A.J. 1993, J. Allergy Clin. Immunol., 92, 857.
155. Jahnsen, F.L., Haraldsen, G., Aanesen, J.P., Haye, R., and Brandtzaeg, P. 1995, Am. J. Respir. Cell Mol. Biol., 12, 624.
156. Campbell, D.J., and Butcher, E.C. 2002, J. Clin. Invest., 110, 1079.
157. Mora, J.R., Bono, M.R., Manjunath, N., Weninger, W., Cavanagh, L.L., Rosemblatt, M., and Von Andrian, U.H. 2003, Nature, 424, 88.
158. Iwata, M., Eshima, Y., and Kagechika, H. 2003, Int. Immunol., 15, 1017.
159. Kunkel, E.J., Campbell, J.J., Haraldsen, G., Pan, J., Boisvert, J., Roberts, A.I., Ebert, E.C., Vierra, M.A., Goodman, S.B., Genovese, M.C., Wardlaw, A.J., Greenberg, H.B., Parker, C.M., Butcher, E.C., Andrew, D.P., and Agace, W.W. 2000, J. Exp. Med., 192, 761.
160. Pabst, O., Ohl, L., Wendland, M., Wurbel, M.A., Kremmer, E., Malissen, B., and Forster, R. 2004, J. Exp. Med., 199, 411.
161. Johansson-Lindbom, B., Svensson, M., Wurbel, M.A., Malissen, B., Marquez, G., and Agace, W. 2003, J. Exp. Med., 198, 963.
162. Hieshima, K., Kawasaki, Y., Hanamoto, H., Nakayama, T., Nagakubo, D., Kanamaru, A., and Yoshie, O. 2004, J. Immunol., 173, 3668.
163. Kunkel, E.J., Campbell, D.J., and Butcher, E.C. 2003, Microcirculation, 10, 313.
164. Feng, N., Jaimes, M.C., Lazarus, N.H., Monak, D., Zhang, C., Butcher, E.C., and Greenberg, H.B. 2006, J. Immunol., 176, 5749.
165. Nakayama, T., Hieshima, K., Izawa, D., Tatsumi, Y., Kanamaru, A., and Yoshie, O. 2003, J. Immunol., 170, 1136.

166. Kilian, M., Reinholdt, J., Lomholt, H., Poulsen, K., and Frandsen, E.V. 1996, APMIS, 104, 321.
167. Weiser, J.N., Bae, D., Fasching, C., Scamurra, R.W., Ratner, A.J., and Janoff, E.N. 2003, Proc. Natl. Acad. Sci. USA, 100, 4215.
168. Jahnsen, F.L., Strickland, D.H., Thomas, J.A., Tobagus, I.T., Napoli, S., Zosky, G.R., Turner, D.J., Sly, P.D., Stumbles, P.A., and Holt, P.G. 2006, J. Immunol., 177, 5861.
169. Campbell, D.J., and Butcher, E.C. 2002, J. Exp. Med., 195, 135.
170. Iwata, M., Hirakiyama, A., Eshima, Y., Kagechika, H., Kato, C., and Song, S.Y. 2004, Immunity, 21, 527.
171. Jahnsen, F.L., Gran, E., Haye, R., and Brandtzaeg, P. 2004, Am. J. Respir. Cell Mol. Biol., 30, 31.
172. Kroese, F.G., Butcher, E.C., Stall, A.M., Lalor, P.A., Adams, S., and Herzenberg, L.A. 1989, Int. Immunol., 1, 75.
173. Bos, N.A., Bun, J.C., Popma, S.H., Cebra, E.R., Deenen, G.J., van der Cammen, M.J., Kroese, F.G., and Cebra, J.J. 1996, Infect. Immun., 64, 616.
174. Shroff, K.E., Meslin, K., and Cebra, J.J. 1995, Infect. Immun., 63, 3904.
175. Umesaki, Y., Okada, Y., Matsumoto, S., Imaoka, A., and Setoyama, H. 1995, Microbiol. Immunol., 39, 555.
176. Bos, N.A., Jiang, H.Q., and Cebra, J.J. 2001, Gut, 48, 762.
177. Talham, G.L., Jiang, H.Q., Bos, N.A., and Cebra, J.J. 1999, Infect. Immun., 67, 1992.
178. Jiang, H.-Q., Bos, N.A., and Cebra, J.J. 2001, Infect. Immun., 69, 3611.
179. Stoel, M., Jiang, H.Q., van Diemen, C.C., Bun, J.C., Dammers, P.M., Thurnheer, M.C., Kroese, F.G., Cebra, J.J., and Bos, N.A. 2005, J. Immunol., 174, 1046.
180. Boursier, L., Farstad, I.N., Mellembakken, J.R., Brandtzaeg, P., and Spencer, J. 2002, Eur. J. Immunol., 32, 2427.
181. Dunn-Walters, D.K., Boursier, L., and Spencer, J. 1997, Eur. J. Immunol., 27, 2959.
182. Fischer, M., and Kuppers, R. 1998, Eur. J. Immunol., 28, 2971.
183. Boursier, L., Dunn-Walters, D.K., and Spencer, J. 1999, Immunology, 97, 558.
184. Dunn-Walters, D.K., Hackett, M., Boursier, L., Ciclitira, P.J., Morgan, P., Challacombe, S.J., and Spencer, J. 2000, J. Immunol., 164, 1595.
185. Holtmeier, W., Hennemann, A., and Caspary, W.F. 2000, Gastroenterology, 119, 1253.
186. Dunn-Walters, D.K., Isaacson, P.G., and Spencer, J. 1997, Eur. J. Immunol., 27, 463.
187. Shikina, T., Hiroi, T., Iwatani, K., Jang, M.H., Fukuyama, S., Tamura, M., Kubo, T., Ishikawa, H., and Kiyono, H. 2004, J. Immunol., 172, 6259.
188. Boursier, L., Gordon, J.N., Thiagamoorthy, S., Edgeworth, J.D., and Spencer, J. 2005, Gastroenterology, 128, 1879.
189. Boursier, L., Montalto, S.A., Raju, S., Culora, G., and Spencer, J. 2006, Immunology, 119, 90.
190. Farstad, I.N., Carlsen, H., Morton, H.C., and Brandtzaeg, P. 2000, Immunology, 101, 354.
191. He, B., Xu, W., Santini, P.A., Polydorides, A.D., Chiu, A., Estrella, J., Shan, M., Chadburn, A., Villanacci, V., Plebani, A., Knowles, D.M., Rescigno, M., and Cerutti, A. 2007, Immunity, 26, 812.
192. Kett, K., Baklien, K., Bakken, A., Kral, J.G., Fausa, O., and Brandtzaeg, P. 1995, Gastroenterology, 109, 819.
193. Bouvet, J.P., and Fischetti, V.A. 1999, Infect. Immun., 67, 2687.
194. Muller, C.A., Autenrieth, I.B., and Peschel, A. 2005, Cell. Mol. Life Sci., 62, 1297.
195. Haneberg, B. 1974, Scand. J. Immunol., 3, 191.
196. Ma, J.K., Hikmat, B.Y., Wycoff, K., Vine, N.D., Chargelegue, D., Yu, L., Hein, M.B., and Lehner, T. 1998, Nat. Med., 4, 601.
197. Quan, C.P., Berneman, A., Pires, R., Avrameas, S., and Bouvet, J.P. 1997, Infect. Immun., 65, 3997.
198. Bouvet, J.P., and Dighiero, G. 1998, Infect. Immun., 66, 1.
199. Brandtzaeg, P., Fjellanger, I., and Gjeruldsen, S.T. 1968, J. Bacteriol., 96, 242.

200. van der Waaij, L.A., Limburg, P.C., Mesander, G., and van der Waaij, D. 1996, Gut, 38, 348.
201. Kronvall, G., Simmons, A., Myhre, E.B., and Jonsson, S. 1979, Infect. Immun., 25, 1.
202. Roy, M.J., and Varvayanis, M. 1987, Cell. Tissue Res., 248, 645.
203. Mantis, N.J., Cheung, M.C., Chintalacharuvu, K.R., Rey, J., Corthesy, B., and Neutra, M.R. 2002, J. Immunol., 169, 1844.
204. Rey, J., Garin, N., Spertini, F., and Corthesy, B. 2004, J. Immunol., 172, 3026.
205. Favre, L., Spertini, F., and Corthesy, B. 2005, J. Immunol., 175, 2793.
206. Brandtzaeg, P. 2002, Integrating Population Outcomes, Biological Mechanisms and Research Methods in the Study of Human Milk and Lactation, M.K. Davis, C.E. Isaacs, L. Å. Hanson, and A.L. Wright (Ed.), Kluwer Academic/Plenum Publishers, New York,. Advanc. Exp. Med., 503, 1.
207. Macpherson, A.J., and Uhr, T. 2004, Science, 303, 1662.
208. Goldblum, R.M., Hanson, L. Å., and Brandtzaeg, P. 1996, Immunologic Disorders in Infants, Children, 4th Ed., E.R. Stiehm (Ed.), W.B. Saunders Co., Philadelphia, 159.
209. Renegar, K.B., Jackson, G.D., and Mestecky, J. 1998, J. Immunol., 160, 1219.
210. Berdoz, J., and Corthesy, B. 2004, Mol. Immunol., 41, 1013.
211. Hocini, H., Bélec, L., Iscaki, S., Garin, B., Pillot, J., Becquart, P., and Bomsel, M. 1997, AIDS Res. Hum. Retrovirus, 13, 1179.
212. Mazzoli, S., Trabattoni, D., Lo Caputo, S., Piconi, S., Blé, C., Meacci, F., Ruzzante, S., Salvi, A., Semplici, F., Longhi, R., Fusi, M.L., Tofani, N., Biasin, M., Villa, M.L., Mazzotta, F., and Clerici, M. 1997, Nat. Med., 3, 1250.
213. Bollinger, R.R., Everett, M.L., Palestrant, D., Love, S.D., Lin, S.S., and Parker, W. 2003, Immunology, 109, 580.
214. Bollinger, R.R., Everett, M.L., Wahl, S.D., Lee, Y.H., Orndorff, P.E., and Parker, W. 2006, Mol. Immunol., 43, 378.
215. Porter, P., and Linggood, M.A. 1983, Ann. NY Acad. Sci., 409, 564.
216. Rogers, H.J., and Synge, C. 1978, Immunology, 34, 19.
217. Tenovuo, J., Moldoveanu, Z., Mestecky, J., Pruitt, K.M., and Rahemtulla, B.M. 1982, J. Immunol., 128, 726.
218. Burns, J.W., Siadat-Pajouh, M., Krishnaney, A.A., and Greenberg, H.B. 1996, Science, 272, 104.
219. Bomsel, M., Heyman, M., Hocini, H., Lagaye, S., Belec, L., Dupont, C., Desgranges, C. 1998, Immunity, 9, 277.
220. Fujioka, H., Emancipator, S.N., Aikawa, M., Huang, D.S., Blatnik, F., Karban, T., DeFife, K., and Mazanec, M.B. 1998, J. Exp. Med., 188, 1223.
221. Mazanec, M.B., Coudret, C.L., and Fletcher, D.R. 1995, J. Virol., 69, 1339.
222. Huang, Y.T., Wright, A., Gao, X., Kulick, L., Yan, H., and Lamm, M.E. 2005, J. Immunol., 174, 4828.
223. Mazanec, M.B., Nedrud, J.G., Kaetzel, C.S., and Lamm, M.E. 1993, Immunol. Today, 14, 430.
224. Robinson, J.K., Blanchard, T.G., Levine, A.D., Emancipator, S.N., and Lamm, M.E. 2001, J. Immunol., 166, 3688.
225. Fernandez, M.I., Pedron, T., Tournebize, R., Olivo-Marin, J.C., Sansonetti, P.J., and Phalipon, A. 2003, Immunity, 18, 739.
226. Kaetzel, C.S. 2005, Immunol. Rev., 206, 83.
227. Giugliano, L.G., Ribeiro, S.T.G., Vainstein, M.H., and Ulhoa, C.J. 1995, J. Med. Microbiol., 42, 3.
228. Dallas, S.D., and Rolfe, R.D. 1998, J. Med. Microbiol., 47, 879.
229. Hammerschmidt, S., Talay, S.R., Brandtzaeg, P., and Chhatwal, G.S. 1997, Mol. Microbiol., 25, 1113.
230. Hamre, R., Farstad, I.N., Brandtzaeg, P., and Morton, H.C. 2003, Scand. J. Immunol., 57, 506.
231. Lamkhioued, B., Gounni, A.S., Gruart, V., Pierce, A., Capron, A., and Capron, M. 1995, Eur. J. Immunol., 25, 117.
232. Marshall, L.J., Perks, B., Ferkol, T., and Shute, J.K. 2001, J. Immunol., 167, 2816.

233. Persson, C.G., Erjefält, J.S., Greiff, L., Erjefält, I., Korsgren, M., Linden, M., Sundler, F., Andersson, M., and Svensson, C. 1998, Scand. J. Immunol., 47, 302.
234. Yoshida, M., Claypool, S.M., Wagner, J.S., Mizoguchi, E., Mizoguchi, A., Roopenian, D.C., Lencer, W.I., and Blumberg, R.S. 2004, Immunity, 20, 769.
235. Brandtzaeg, P., and Tolo, K. 1977, Nature, 266, 262.
236. Berstad, A.E., and Brandtzaeg, P. 1998, Gut, 42, 522.
237. Sun, K., Johansen, F.-E., Eckmann, L., and Metzger, D.W. 2004, J. Immunol., 173, 4576.
238. Uren, T.K., Wijburg, O.L.C., Simmons, C., Johansen, F.-E., Brandtzaeg, P., and Strugnell, R.A. 2005, Eur. J. Immunol., 35, 180.
239. Renegar, K.B., Small, P.A., Boykins, L.G., and Wright, P.F. 2004, J. Immunol., 173, 1978.
240. Bižanov, G., Janakova, L., Knapstad, S.E., Karlstad, T., Bakke, H., Haugen, I.L., Haugan, A., Samdal, H.H., and Haneberg, B. 2005, Scand. J. Immunol., 61, 503.
241. Daniele, R.P. 1990, Annu. Rev. Physiol., 52, 177.
242. Velazquez, F.R., Matson, D.O., Calva, J.J., Guerrero, L., Morrow, A.L., Carter-Campbell, S., Glass, R.I., Estes, M.K., Pickering, L.K., and Ruiz-Palacios, G.M. 1996, N. Engl. J. Med., 335, 1022.
243. O'Neal, C.M., Clements, J.D., Estes, M.K., and Conner, M.E. 1998, J. Virol., 72, 3390.
244. Mbawuike, I.N., Pacheco, S., Acuna, C.L., Switzer, K.C., Zhang, Y., and Harriman, G.R. 1999, J. Immunol.,162, 2530.
245. Fazekas de St. Groth, S. 1951, Aust. J. Exp. Biol. Med. Sci., 29, 339.
246. Arnaboldi, P.M., Behr, M.J., and Metzger, D.W. 2005, J. Immunol., 175, 1276.
247. Renegar, K.B., Johnson, C.D., Dewitt, R.C., King, B.K., Li, J., Fukatsu, K., and Kudsk, K.A. 2001, J. Immunol., 166, 819.
248. Sait, L.C., Galic, M., Price, J.D., Simpfendorfer, K.R., Diavatopoulos, D.A., Uren, T.K., Janssen, P.H., Wijburg, O.L., and Strugnell, R.A. 2007, Int. Immunol., 19, 257.
249. Uren, T.K., Johansen, F.E., Wijburg, O.L., Koentgen, F., Brandtzaeg, P., and Strugnell, R.A. 2003, J. Immunol., 170, 2531.
250. Asahi, Y., Yoshikawa, T., Watanabe, I., Iwasaki, T., Hasegawa, H., Sato, Y., Shimada, S., Nanno, M., Matsuoka, Y., Ohwaki, M., Iwakura, Y., Suzuki, Y., Aizawa, C., Sata, T., Kurata, T., and Tamura, S. 2002, J. Immunol., 168, 2930.
251. Waldman, R.H., Wigley, F.M., and Small, P.A. 1970, J. Immunol., 105, 1477.
252. Shvartsman, Y.S., Agranovskaya, E.N., and Zykov, M.P. 1977, J. Infect. Dis., 135, 697.
253. Clements, M.L., and Murphy, B.R. 1986, J. Clin. Microbiol., 23, 66.
254. Lim, P.L., and Rowley, D. 1982, Int. Arch. Allergy Appl. Immunol., 68, 41.
255. Sanchez, J., and Holmgren, J. 2005, Curr. Opin. Immunol., 17, 388.
256. Wijburg, O.L.C., Uren, T.K., Simpfendorfer, K., Johansen, F.E., Brandtzaeg, P., and Strugnell, R.A. 2006, J. Exp. Med., 203, 21.
257. Ali, M., Emch, M., von Seidlein, L., Yunus, M., Sack, D.A., Rao, M., Holmgren, J., and Clemens, J.D. 2005, Lancet, 366, 44.
258. Eriksson, K., and Holmgren, J. 2002, Curr. Opin. Immunol., 14, 666.
259. Kabir, S. 2005, Rev. Med. Microbiol., 16, 101.
260. Santosuosso, M., McCormick, S., and Xing, Z. 2005, Viral Immunol., 18, 283.
261. Angel, J., Franco, M.A., and Greenberg, H.B. 2007, Nat. Rev. Microbiol., 5, 529.
262. Glass, R.I., Bresee, J.S., Parashar, U.D., Jiang, B., and Gentsch, J. 2004, Lancet, 363, 1547.
263. Ruiz-Palacios, G.M., Perez-Schael, I., Velazquez, F.R., Abate, H., Breuer, T., Clemens, S.C., Cheuvart, B., Espinoza, F., Gillard, P., Innis, B.L., Cervantes, Y., Linhares, A.C., López, P., Macías-Parra, M., Ortega-Barría, E., Richardson, V., Rivera-Medina, D.M., Rivera, L., Salinas, B., Pavía-Ruz, N., Salmerón, J., Rüttimann, R., Tinoco, J.C., Rubio, P., Nuñez, E., Guerrero, M.L., Yarzábal, J.P., Damaso, S., Tornieporth, N., Sáez-Llorens, X., Vergara, R.F., Vesikari, T., Bouckenooghe, A., Clemens, R., De Vos, B., O'Ryan, M.; Human Rotavirus Vaccine Study Group. 2006, N. Engl. J. Med., 354, 11.
264. Vesikari, T., Matson, D.O., Dennehy, P., Van Damme, P., Santosham, M., Rodriguez, Z., Dallas, M.J., Heyse, J.F., Goveia, M.G., Black, S.B., Shinefield, H.R., Christie, C.D.,

Ylitalo, S., Itzler, R.F., Coia, M.L., Onorato, M.T., Adeyi, B.A., Marshall, G.S., Gothefors, L., Campens, D., Karvonen, A., Watt, J.P., O'Brien, K.L., DiNubile, M.J., Clark, H.F., Boslego, J.W., Offit, P.A., Heaton, P.M.; Rotavirus Efficacy and Safety Trial (REST) Study Team. 2006, N. Engl. J. Med., 354, 23.
265. van Ginkel, F.W., Jackson, R.J., Yuki, Y.,. and McGhee,, J.R. 2000, J. Immunol., 165, 4778.
266. Glück, R., Mischler, R., Durrer, P., Furer, E., Lang, A.B., Herzog, C., and Cryz, S.J. Jr. 2000, J. Infect. Dis., 181, 1129.
267. Mutsch, M., Zhou, W., Rhodes, P., Bopp, M., Chen, R.T., Linder, T., Spyr, C., and Steffen, R. 2004, N. Engl. J. Med., 350, 896.
268. Lu, X., Clements, J.D., and Katz, J.M. 2002, Vaccine, 20, 1019.
269. Hinkula, J., Devito, C., Zuber, B., Benthin, R., Ferreira, D., Wahren, B., and Schröder, U. 2006, Vaccine, 24, 4494.
270. Mason, H.S., Warzecha, H., Mor, T., and Arntzen, C.J. 2002, Trends Mol. Med., 8, 324.
271. Yusibov, V., Rabindran, S., Commandeur, U., Twyman, R.M., and Fischer, R. 2006, Drugs R. D., 7, 203.
272. Schillberg, S., Twyman, R.M., and Fischer, R. 2005, Vaccine, 23, 1764.
273. Källenius, G., Pawlowski, A., Brandtzaeg, P., and Svenson, S. 2007, Tuberculosis, 87, 257.
274. Lemaitre-Coelho, I., Jackson, G.D., and Vaerman, J.P. 1978, Scand. J. Immunol., 8, 459.
275. Belshe, R.B., Gruber, W.C., Mendelman, P.M., Cho, I., Reisinger, K., Block, S.L., Wittes, J., Iacuzio, D., Piedra, P., Treanor, J., King, J., Kotloff, K., Bernstein, D.I., Hayden, F.G., Zangwill, K., Yan, L., Wolff, M. 2000, J. Pediatr., 136, 168.
276. Devito, C., Zuber, B., Schröder, U., Benthin, R., Okuda, K., Broliden, K., Wahren, B., and Hinkula, J. 2004, J. Immunol., 173, 7078.
277. Brokstad, K.A., Cox, R.J., Eriksson, J.C., Olofsson, J., Jonsson, R., and Davidsson, A. 2001, Scand. J. Immunol., 54, 243.
278. Brokstad, K.A., Eriksson, J.C., Cox, R.J., Tynning, T., Olofsson, J., Jonsson, R., and Davidsson, A. 2002, J. Infect. Dis., 185, 878.
279. Brokstad, K.A., Cox, R.J., Olofsson, J., Jonsson, R., and Haaheim, L.R. 1995, J. Infect. Dis., 171, 198.
280. El-Madhun, A.S., Cox, R.J., Søreide, A., Olofsson, J., and Haaheim, L.R. 1998, J. Infect. Dis., 178, 933.
281. Brandtzaeg, P. 1998, The Nose, P. van Cauwenberge, D.-Y. Wang, K. Ingels, and C. Bachert (Eds.), Kugler Publications, The Hague, 233.
282. McCaughan, G.W., Adams, E., and Basten, A. 1984, J. Immunol., 132, 1190.
283. Glezen, W.P. 2002, Curr. Opin. Infect. Dis., 15, 283.
284. Keitel, W.A., Cate, T.R., Nino, D., Huggins, L.L., Six, H.R., Quarles, J.M., and Couch, R.B. 2001, J. Infect. Dis., 183, 329.
285. de Bernardi di Valserra, M., Zanasi, A., Ragusa, S., Glück, R., and Herzog, C. 2002, Clin. Ther., 24, 100.
286. Berstad, A.K., Andersen, S.R., Dalseg, R., Dromtorp, S., Holst, J., Namork, E., Wedege, E., and Haneberg, B. 2000, Vaccine, 18, 1910.
287. Plante, M., Jones, T., Allard, F., Torossian, K., Gauthier, J., St-Felix, N., White, G.L., Lowell, G.H., and Burt, D.S. 2001, Vaccine, 20, 218.
288. Treanor, J., Nolan, C., O'Brien, D., Burt, D., Lowell, G., Linden, J., and Fries, L. 2006, Vaccine, 24, 254.
289. Greenbaum, E., Furst, A., Kiderman, A., Stewart, B., Levy, R., Schlesinger, M., Morag, A., and Zakay-Rones, Z. 2002, Vaccine, 20, 1232.
290. Subbarao, K., and Joseph, T. 2007, Nat. Rev. Immunol., 7, 267.
291. Bakke, H., Samdal, H.H., Holst, J., Oftung, F., Haugen, I.L., Kristoffersen, A.C., Haugan, A., Janakova, L., Korsvold, G.E., Krogh, G., Andersen, E.A., Djupesland, P., Holand, T., Rappuoli, R., and Haneberg, B. 2006, Scand. J. Immunol., 63, 223.

292. Bakke, H., and Haneberg, B. 2006, Tidsskr. Nor. Laegeforen., 126, 2818.
293. Herremans, T.M.P.T., Reimerink, J.H.J., Buisman, A.M., Kimman, T.G., and Koopmans, M.P. 1999, J. Immunol., 162, 5011.
294. Kunisawa, J., Fukuyama, S., and Kiyono, H. 2005, Curr. Mol. Med., 5, 557.
295. Henriksson, G., Helgeland, L., Midtvedt, T., Stierna, P., and Brandtzaeg, P. 2004, Am. J. Respir. Cell Mol. Biol., 31, 657.
296. Xu, W., He, B., Chiu, A., Chadburn, A., Shan, M., Buldys, M., Ding, A., Knowles, D.M., Santini, P.A., and Cerutti, A. 2007, Nat. Immunol., 8, 294.
297. Vajdy, M., and Singh, M. 2006, Expert Opin. Drug Deliv., 3, 247.

Chapter 4
Bridging Mucosal Innate Immunity to the Adaptive Immune System

Rajesh Singh and James W. Lillard

Abstract The human adult mucosa covers immense surface area 200 times greater than that of the skin, i.e., 400 m^2. The mucosal immune system mediates the symbiotic relationship between hosts and endogenous microorganisms (commensal bacteria) and serves as a physical and immunological defense against invading pathogens and/or toxins. To accomplish this task, the mucosal immune system, consisting of an integrated network of tissues, lymphoid and non-lymphoid cells, and effector molecules such as antibodies, chemokines, and cytokines for host protection, tightly controls the balance between immune responsiveness (i.e., immunity and inflammation) and non-responsiveness (i.e., tolerance). The mucosal immune system, antigen-presenting cells, lymphocytes, and the factors they produce are essential for orchestrating specific mucosal immune responses. Cells of the mucosa initiate and support the development of both innate and adaptive immunity. Here, we discuss how the innate mucosal immune responses lead to host adaptive immunity.

4.1 Introduction

Starting at birth, we live in a sea of microorganisms that colonize our skin, nose, throat, etc. It is, however, quite rare that these microbes are able to breach these natural barriers and invade host bloodstream and tissues. Therefore, we are endowed with multiple defense mechanisms that promptly detect and kill these intruders. The microorganisms that manage to cross physical barriers, such as the skin, will initially face the agents of innate immunity. Phagocytosis seems to be the most basic defense mechanism, present in many multi-cellular organisms. The cells that perform this function are called phagocytes and are not only the major

J.W. Lillard
Senior Scientist, James Graham Brown Cancer Center, Associate Professor of Microbiology & Immunology, Smith & Lucille Gibson Chair in Medicine, University of Louisville, Louisville, Kentucky 40202
e-mail: James.Lillard@louisville.edu

M. Vajdy (ed.), *Immunity Against Mucosal Pathogens*,

effector cells of innate immunity, but connect innate and acquired (adaptive) responses. All vertebrates, starting with jawed fish, are endowed with an adaptive immune system. The defining feature of the adaptive immune system is specificity, which is an inducible response to specific antigens. Accordingly, it has been demonstrated that the body fluids of an immunized animal can contain cells and/or soluble molecules that react to a single immunogen, but not to other antigens. This response is also considered inducible, because the response to foreign antigen increases with subsequent challenge(s). Thus, the major distinction between innate and acquired immunity is that of scope. Phagocytic cells are general-purpose effector cells that can kill a wide variety of microbes, whereas lymphocytes, the agents of acquired immunity, are specific to a single microbe, and perhaps homologous antigens. The capacity of lymphocytes to discriminate and distinguish between different, yet similar, microbes from the components of the host is unique indeed. In fact, microbes have evolved mechanisms to mutate and evade the adaptive immune response. Nonetheless, the innate and adaptive mucosal immune response is critical to allow coexistence with commensal organisms and robust protective responses to pathogens.

4.1.1 Innate versus Adaptive Immune Systems

The innate immune system uses a limited number of genetically encoded receptors (e.g., pattern recognition receptors and toll-like receptors) to detect foreign organisms. In contrast, the adaptive immune system employs subsets of an elite set of leukocytes called **lymphocytes**, which have the capacity to generate a large number of antigen-specific cell surface receptors by random gene rearrangement. Bone marrow-derived lymphocytes (**B lymphocytes** or **B cells**) generate **B cell receptors (BCRs)**, and thymus-derived lymphocytes **(T lymphocytes or T cells)** generate T cell receptors (**TCRs**) that are varied specificities to antigen due to **somatic recombination** of the antigen-binding regions. Diversity is added by variability **(junctional diversity)** in the cutting and joining of these genes. The affinity of BCRs can be enhanced with the accumulation of mutations (**somatic hypermutation or affinity maturation**) as B cells undergo periods of intense proliferation [1]. Although less appreciated, evidence suggests that TCRs might also undergo somatic hypermutation events, which affects their affinity [2, 3]. The number of possible receptors becomes immense, with estimates exceeding 10^{15} possible specificities.

Another key difference between the two systems is the way they respond to a foreign antigen or pathogen. Unlike cells of the innate immune system, those of the adaptive immune system (Fig. 4.1), once stimulated to respond, have the ability to recall previous exposures to the same stimulus and to modify their response accordingly, a process called **memory**. Additional features that distinguish innate and adaptive immune responses are summarized in Table 4.1.

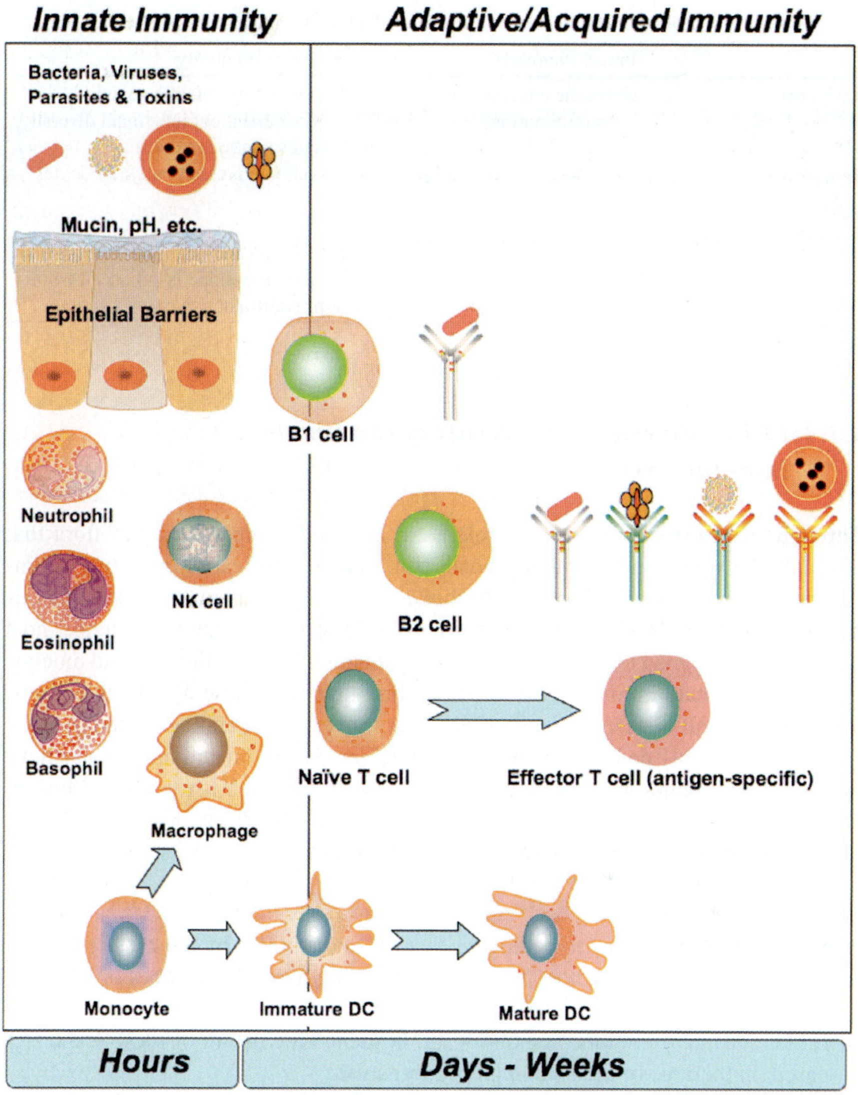

Fig. 4.1 Temporal and physical transition of the innate and adaptive immunity. The innate immune system consists of formidable physical and chemical mucin, pH, etc.) barriers to pathogens and toxins. Phagocytic cells, e.g., neutrophils, eosinophils, basophils, B1 cells, dendritic cells and macrophages, are present to engulf or destroy pathogens circumventing the epithelium. Within hours, B1, B2, and T cells become activated by these foreign intruders and transit from effector sites to inductive sites, where the adaptive immune response develops in days and matures over weeks to efficiently clear foreign antigen

Table 4.1 Hallmarks of innate and adaptive immune systems

	Innate immunity	Adaptive immunity
Receptors	Germline encoded (no recombination)	Generated by somatic genetic recombination, junctional diversity, and somatic mutation
Receptor distribution	Expressed by many cells	Clonally expressed on lymphocytes
Receptor repertoire	Limited	Vast
Memory	None	Magnitude and quality of response to specific epitope is enhanced from previous encounter with antigen

4.1.2 Functions and Components of the Innate Immune System

Throughout life the body surfaces tolerate a rich and complex array of flora that is usually harmless and frequently beneficial, but can also cause opportunistic infections. In contrast, domains of the body just a few microns beneath mucous membranes (e.g., lamina propria) are usually free of microorganisms. Hence, host defenses are designed to intensify as microorganisms encounter the skin and mucous membranes. Innate immunity is the body's first line of defense against infections. It is present in all metazoans, and in plants and invertebrate animals – and in these species it is the only defense against infections. Vertebrates, including humans, have not only innate immunity but also adaptive immunity, which is the second line of defense.

The following are the functions of innate immunity:

- Recognize and respond to pathogens that are encountered by the host.
- Prevent infection of the host by either eliminating harmful microbes or allowing normal flora to exist on mucosal surfaces.
- Initiate adaptive immune responses and influence the nature of these responses based on the type of the invading microorganism.
- Serve, in some cases, as an effector mechanism for adaptive immune responses.

The components of the innate immune system are either cellular, consisting of various types of cells, or humoral, consisting of various secreted soluble peptides, proteins, and other mediators (Table 4.2). The interactions of these components in the host with the invading microbe often, but not always, lead to a constellation of responses called inflammation. The purpose of inflammation is to amplify the body's defenses by increasing the number of leukocytes and blood supply to the site of infection, bringing antimicrobial proteins and white cells to defend the host and eliminate the pathogen or foreign antigen encountered.

Table 4.2 Components of the innate immune system

Component	Function(s)
Cellular	
Skin and Mucous membrane (epithelial cells)	Mechanical and chemical defenses
Phagocytic cells (neutrophils, macrophage)	Ingest and kill pathogens
Proinflammatory cells (macrophages, mast cells, eosinophils)	Innate defense and inflammation
Natural killer (NK) cells	Kill infected cells and tumor cells
Antigen-presenting cells (dendritic cells, macrophages, and B cells)	Recognize, process and present antigen to lymphocytes and initiate adaptive cellular immune response
Humoral	
Antibodies	bind microbes and activates complement
Complement	Enhances phagocytosis (opsonization), induces inflammation, and destroys some pathogens.
Cytokines	Secreted by many cells and activate innate and adaptive defenses
Chemokines	Secreted by many cells and attract as well as activate leukocytes
Kinins	Induce inflammation
Acute phase proteins	Enhance cellular and humoral defense
Enzymes	Kill and digest microorganisms
Inflammation	
Increase Blood supply (erythema)	Brings more antimicrobial cells and proteins to the site of infection
Induce vascular permeability (edema)	Brings more antimicrobial protein to the site of infection

4.2 The Mucosal Barrier

4.2.1 Physiologic and Molecular Barriers

The mucosal barrier is a complex structure composed of both cellular and non-cellular components [4]. Probably the most significant barrier to antigen entry into the mucosa-associated lymphoid tissue (**MALT**) is the presence of enzymes starting in the mouth (and eyes) and extending down to the stomach, small bowel, and colon. **Proteolytic enzymes** in the stomach (pepsin, papain, etc.) and small bowel (trypsin, chymotrypsin, and pancreatic proteases) digest polypeptides and proteins down to smaller peptide fragments. The breakdown of large polypeptides into small dipeptides and tripeptides accomplishes two tasks. First, it allows for the process of digestion and absorption of nutrients. Second, it renders potentially immunogenic proteins non-immunogenic (peptides 8–10 amino acids in length or smaller are poor immunogens). Coupled with the effects of these proteases are the emulsifying effect(s) of bile salts and enzymes that break down carbohydrates. Hence, these innate molecules are a potent system to alter antigen (Ag) exposure. Considering the **pH extremes** of the stomach and proximal small bowel as well as commensal products in the colon, it is amazing that immune responses to oral antigens are necessary or can occur at all. However, they do, and many of these responses provide protection against potential pathogens. The decision to respond or suppress a response may relate to the pathway used by the antigen to gain access to the host. Invasive pathogens, with the ability to break down the mucosal barrier, elicit aggressive host responses, whereas luminal colonizers require a more tolerant response.

A key component of the mucosal barrier is the production of **mucin** or mucus. These glycoproteins line the surface epithelium from the nasal cavity/oropharynx to the rectum [5, 6, 7, 8, 9]. Mucus-producing goblet cells continuously produce a thick barrier covering adjacent epithelium. Particles, bacteria, and viruses become trapped in the mucus layer and are expelled by the peristaltic processes of the gut, nasal tract and lungs. Hence, mucus prevents potential pathogens and antigens from gaining access to the underlying epithelium, a process called non-immune exclusion. Mucins or mucus also serve as a reservoir for secretory immunoglobulin A (**SIgA**). This antibody traverses the epithelium and is secreted into the lumen.

The SIgA, present in the mucus layer, binds pathogens and prevents their attachment to the epithelium. An associated family of factors, called **trefoil factors**, helps to strengthen this obstacle to pathogens and toxins as well as promote restoration of the mucosal barrier if any defect occurs. In the absence of mucin gene products or trefoil factors, the host becomes more susceptible to inflammation and less capable of repairing breeches in the barrier [10, 11]. Whether such defects exist in food allergy, inflammatory bowel disease or other mucosal inflammatory diseases are not known, but would be worth studying.

Several investigators have demonstrated that the rodent (rat and mouse) neonate has increased intestinal permeability, allowing for the passage of dietary and possibly bacterial antigens into the underlying lymphoid rich lamina propria region of the gastrointestinal tract [12, 13, 14, 15, 16]. This would effectively bypass mechanisms involved in tolerance induction and could promote an active immune response. Along these lines, it has been well documented that oral tolerance cannot be induced in the neonate. The mechanism(s) underlying this observation has not been elucidated, although the onset of tolerance seems to correlate with "gut closure". However, immaturity of the MALT is also a factor and studies to characterize these processes have not been performed.

4.2.2 Skin and Mucous Membrane Defenses

Before microorganisms can enter the normally aseptic regions of the body, they must pass through the barriers of the skin, the conjunctivae of the eyes, or the mucous membrane of the respiratory, alimentary, or urogenital tracts. Each barrier has its own protective mechanisms, which can be broadly classified as mechanical, chemical, molecular, and microbial (Table 4.3). Mechanical barriers are highly

Table 4.3 Defenses of the skin and mucous membranes

Component	Main function
Mechanical	
Keratinized epithelium	Offers protection against microorganisms
Desquamation of stratified epithelium	Removes pathogens attached to its surface
Epithelial cells joined by tight junctions	Prevents entry of microbes
Mucus-coated hair in the nose	Trap particles
Mucus-coated ciliated epithelium	Removes particles from the respiratory tract
Coughing and sneezing	Expel particles
Flow of Urine	Cleanses the urethra
Chemical and molecular	
Low pH in the stomach	Inactivates many bacteria, fungi, parasites and viruses
Antimicrobial peptides (constitutive and induced)	Kill many microorganisms
Defensins (Skin, Intestine, mucous membranes)	Lyses Bacteria and fungi
Phospholipases (in intestinal paneth cells and tears)	Kill Bacteria
Dermacidin (in sweat glands)	Kill Bacteria and fungi
Ribonucleases (on skin)	Kill Bacteria and fungi
Fatty acids (from sebaceous glands of the skin	Inhibits bacterial growth
Enzymes	
Lysozyme (in tears, sweat, saliva and serum)	Hydrolyses bacterial cell wall peptidoglycan
Amidase (in the skin and serum)	
Microbial antagonism	
Normal flora bacteria and fungi	Compete for nutrients with potential pathogen Produce antimicrobial factors (bactriocins)

effective, especially the skin, which is covered with a thick layer of **keratinized epithelium**. The mucous membranes of the mouth, pharynx, esophagus, and lower urinary tract are composed of several layers of epithelial cells, whereas those of the lower respiratory, gastrointestinal, and upper urinary tracts are delicate single layers of epithelial cells, often endowed with specialized functions. Membranes of the alveoli and the intestine are thin because they serve as exchangers of gases, fluids, and solutes. Although epithelia have protective tight junctions that do not allow passage of microorganisms, these cells are easily perturbed, especially when subjected to high pressures or abrasions. In fact, trauma occurs daily in the colon during defecation and in the mouth (e.g., during eating).

As mentioned previously, many mucous membranes are covered by a protective layer of mucus that provides a formidable mechanical and chemical barrier. Mucus is a giant cross-linked gel-like structure made of glycoprotein subunits. It entraps particles and prevents them from reaching the mucosa. Mucus is hydrophilic and allows diffusion of many substances produced by the body, including antimicrobial peptides and enzymes such as **lysozyme** and **peroxidase**. Its viscous properties enable it to bear substantial encounters and yet be readily moved by the motion of the cilia of the underlying cells.

To afford additional protection, skin and mucous membranes are rich in chemical and molecular antimicrobial factors. Some of these factors are constitutively expressed, e.g., low pH in the stomach, fatty acids on the skin, and lysozyme (a bacteriolytic enzyme in tears, sweat, and saliva). The most potent of these factors are antimicrobial peptides (Table 4.4). One of the most active peptides is **phospholipase A_2**, which is an enzyme that destroys bacterial cytoplasmic membrane. **Defensins** are cysteine-rich small peptides (3 to 5 kDa), which are ubiquitous in all vertebrates and invertebrates, and in humans are subdivided into two families, α- and β-defensins based on their structure. Six α-defensins and four β-defensins have been well characterized in humans, but recent analysis of the human genome revealed 34 defensin genes, which highlights the significance of this family of antimicrobial peptides. Defensins, like most other antimicrobial peptides, are highly cationic, enabling them to bind to the negatively charged cell walls of bacteria

Table 4.4 Antimicrobial peptides

Peptide	Location	Active against
Phospholipases	• PMN primary granules • Intestine Paneth cells • Tears • Serum	• Gram positive bacteria • Gram-negative bacteria together with BPI
α-Defensins	• PMN primary granules • Intestinal Paneth cells	• Bacteria • Fungi • Parasites • Viruses
β-Defensins	• Crypt and epithelial cells • Skin, respiratory tract, genitourinary tract	• Bacteria • Fungi
Ribonucleases	• PMNs, monocytes • Epithelial cells (skin, genitourinary tract, intestinal tract)	• Bacteria • Fungi
Dermcidin	• Sweat glands	• Bacteria • Fungi
BPI	• PMN primary granule	• Gram negative bacteria
Cathelicidins (CAP18 in humans)	• PMN secondary granule	• Bacteria • Fungi
Serprocidins (Cationic serine proteases: elastase, proteinase 3, cathepsin G)	• PMN primary granule	• Bacteria • Fungi
Lysozyme	• Phagocyte granules • Tears, saliva • Serum • Intestinal Paneth cells	• Gram positive bacteria peptidoglycan)

and fungi and to kill them by pore formation and permeabilization of their cell membranes [17].

Many antimicrobial peptides are constitutively present on the skin and mucous membranes, such as phospholipase A_2, α-defensins, and ribonucleases (**RNases**). Other antimicrobial peptides, such as β-defensins and RNases, are present in low levels on epithelial surfaces of the skin as well as mucous membranes and their increased production can be induced by microbial signatures (e.g., **lipopolysaccharides**) or by proinflammatory cytokines released by cells of the mucosa following exposure to bacteria. Lysozyme is present in large amounts in tears, sweat, saliva, and serum. It hydrolyses peptidoglycan, the main structural component of bacterial cell walls. Lysozyme acts mainly on Gram-positive bacteria, although many bacterial species have evolved resistant modifications of their cell wall chemistry. Gram-negative bacteria, however, are resistant because their peptidoglycan substrate is shielded by an outer membrane. In this case, lysozyme can work synergistically with other antimicrobial peptides, e.g., **complement**, that can damage the outer membrane of bacteria and allow lysozyme to access its substrate.

The innate immune system prevents colonization of the skin and mucous membranes by potential pathogens, but allows colonization by a limited number of the nonpathogenic bacteria and fungi of the normal flora. The **flora** is unique for each area of the body and is beneficial to the host because it protects the skin and mucous membranes from colonization by pathogenic microorganisms. This protective mechanism is likely accomplished by competition for nutrients with potential pathogens and by production of antimicrobial factors (e.g., bacteriocins) by the normal flora. However, in immunocompromised patients, the normal flora can cause opportunistic infections that are quite severe or even life threatening.

The obstructive function of skin and mucous membranes is seldom breached except by injuries, such as burns, cuts, or wounds. Once microorganisms have traversed the skin, they encounter powerful defenses in the underlying soft tissues. However, these defenses do not work at full capacity under all conditions. For instance, abrasions or lacerations impair the local vascular and lymphatic circulation and interfere with soluble and cellular defense mechanisms, thus rendering the underlying connective tissue vulnerable. When this occurs, substantially fewer microorganisms are required to cause infection. For example, many chronically debilitated patients suffer from decubitus ulcers (bed sores) that become contaminated and are constantly infected with normally harmless organisms on the skin. When a microorganism crosses the protective epidermis of the skin or the epithelia of mucous membranes, it encounters the next line of defense mechanisms: phagocytic cells, proinflammatory cells, natural killer (NK) cells, and antigen-presenting cells as well as humoral defenses.

4.3 Innate Immune Responses at the Epithelium

Innate mechanisms include the physical barrier provided by epithelial cells and cilia movement, mucus production, and secreted molecules with antibacterial activity.

Epithelial cells in the GI tract form a monolayer of cells tightly connected by apical tight junctions, which allow for intercellular transport of molecules, but not the trans-epithelial passage of macromolecules. Unlike the GI tract, uterine epithelium and lower respiratory tracts, the epithelium of other mucosal surfaces such as the nasal tract and vagina are made of stratified, squamous epithelium that are devoid of tight junctions. However, the multi-layer structure of these epithelia and the layer of mucus that covers them contribute to the mucosal barrier. A unique population of lymphoid cells is closely associated with epithelial cells of the GI tract and other mucosal tissues. This population of intraepithelial lymphocytes contains an unusually high proportion of $\gamma\delta TCR^+$ cells. Several questions remain unanswered about these cells and their role in host defense. However, their high frequency in exposed mucosal tissues, ability to recognize antigen presented by non-classical MHC pathways, and the presence of a large subpopulation of cytotoxic cells in this population argue for an essential role of $\gamma\delta TCR^+$ cells for innate and adaptive immunity at mucosal surfaces.

Numerous endogenous molecules with anti-microbial activity are present at the epithelium. The best-described molecules include: defensins, lactoferrin, lysozyme, lactoperoxidase, secretory phospholipase A2 and cathelin-associated peptides. Some of these molecules are secreted by intestinal Paneth cells, but all are produced by neutrophils present in the lamina propria during inflammation and/or infection, suggesting that both cell types play important roles in the innate mucosal immune defenses of mucosal surfaces. Lactoferrin is a member of the transferrin family, which is found in mucosal exocrine secretions, e.g., milk, and exhibits antimicrobial properties. Epithelial cells do not produce lactoferrin; however, they express the lactoferrin receptor. Gastric pepsin hydrolysis of either human or bovine lactoferrin generates a 25-residue peptide called lactoferricin with potent anti-microbial activity [18]. Further, an 11-residue peptide containing five of the eight basic residues in bovine lactoferricin displays anti-microbial properties similar to those of the 25-residue peptide [19].

Defensins generally contain between 30 and 40 amino acid residues and can be divided into α-defensins comprised of two contiguous cysteine residues and β-defensins whose cysteine residues are separated by six amino acids. Defensins are produced by epithelial cells including crypt cells and Paneth cells, which contain granules rich in α-defensins that are subsequently secreted into the lumen of the small intestine [20, 21]. Two of these human intestinal α-defensins (HD), HD-5 and HD-6, have been identified in Paneth cells and human reproductive tract epithelial cells [22]. α-defensins are also secreted by tracheal epithelial cells and homologous to peptides that function as mediators of nonoxidative microbial cell killing in human neutrophils (termed human neutrophil peptides; HNPs). β-defensins and in particular human β-defensin-1 (HBD-1), are expressed in the epithelial cells of the oral mucosa, trachea, bronchi, as well as mammary and salivary glands [23, 24, 25].

HNP-1, -2, -3 share more than 60% homology and 40% identity with murine crypt cell defensins (cryptins) [26, 27, 28]. In addition to being antimicrobial, defensins have been shown to be chemotactic for T cells and monocytes via

CCR6 [29, 30, 31] and may therefore be capable of activating T cell dependent immune responses. Lillard et al. fully characterized the adaptive immune response promoted by these innate peptides and found that HNPs significantly increased antigen-specific IgG and IgM antibody levels in the serum [32]. These serum Ab responses were associated with the induction of Ag-specific $CD4^+$ T cell proliferative responses and IFN-γ, IL-5, IL-6 and IL-10 secretion. Nasal immunization of mice with the model vaccine antigen (chicken egg ovalbumin, OVA) in the presence of HNPs stimulated OVA-specific serum IgG antibody responses. However, neither IgA nor IgE Abs was induced when HNPs were used as adjuvant. The pattern of IgG subclass antibody responses consisted of IgG1, but also IgG2a and suggested the involvement of a mixed Th1/Th2 response. This finding was confirmed by the high level of IFN-γ, IL-5, IL-6 and IL-10 secreted by Ag-specific $CD4^+$ T cells. It is important to note that HNPs failed to induce SIgA antibody responses in mucosal tissue secretions. To this end, α-defensins enhanced IL2R and CD28 expression by splenic T cells more dramatically than by Peyer's patch-derived T lymphocytes, which suggests a tissue-specific effect by these classical defensins. HNPs enhanced the expression of CD40 by spleen- and Peyer's patch-derived B cells. While the expression of CD80 was unaffected, HNPs down-regulated the expression of CD86 by LPS-stimulated B cells. These effects were more pronounced on Peyer's patch- than on splenic-B cells.

It is interesting that these innate peptides initiated significant systemic- but not mucosal- humoral immunity after nasal delivery. Several reasons could have accounted for the low or minimal IgA Ab levels in mucosal secretions. It is possible that the cytokines induced by HNPs do not support the switch to or synthesis of IgA Abs. It is unlikely that HNPs disrupted the cell integrity of the mucosa, which could have allowed the nasally administered antigen to be delivered directly to the periphery. Potentially, these characteristics would allow crypt and Paneth cell defensins to contain pathogens and commensals in mucosal compartments without invoking brisk mucosal immune responses whenever a foreign substance is encountered in the mucosa. Correspondingly, these mechanisms would permit defensins to markedly enhance adaptive immune responses to microbes that enter the sterile peripheral environment to thereby bridge innate with adaptive immunity mechanisms.

4.4 Mucosal Phagocytes and Epithelial Cells

4.4.1 Neutrophils

Of all the constitutive antimicrobial defenses of the body, the most potent is the cellular response. It consists of the influx of neutrophils, eosinophils, and monocytes into infected tissues. Neutrophils are actively motile phagocytic cells produced in the bone marrow. They differentiate from stem cells over 2 weeks. Together with eosinophils and basophils, they are also known as **polymorphonuclear leukocytes** because of their segmented nucleus, or as **granulocytes** because of their granules.

Neutrophils migrate to sites of infection and, once activated, initiate a cascade of defense mechanisms. Neutrophils recognize and engulf microorganisms by a process known as phagocytosis and release of reactive oxygen species (e.g., superoxide, hydrogen peroxide, hypochlorous acid, etc.). They also produce four kinds of granules-primary, secondary, tertiary, and secretory vesicles of which azurophil (primary) and specific (secondary) granules are the largest and most easily visible by microscopy. These granules have many antimicrobial and chemotactic factors (e.g., defensins). Together, these factors can kill and degrade microorganisms as well as damage host tissue.

Lung inflammation of the respiratory mucosal surface in diseases contributes significantly to the pathology. Dysentery caused by *Shigella* and other bacterial species is characterized by infiltration of PMNs into the colonic mucosa. A large population of these cells also infiltrate the vaginal mucosa following intravaginal HSV-2 change to control disease [33]. PMNs have also been shown to enhance Th1 immune responses and down-regulate IL-10 as well as protection against Candida-induced tissue damage [34]; thus bridging innate and adaptive immune responses. Perhaps due to their short lifespan, the phenotype of systemic PMNs, other than activation, does not change after entry into mucosal sites. Clear evidence was found of neutrophil infiltration into the intestinal lamina propria and epithelium through epithelial breaks as well as intact epithelium [35].

Neutrophils and monocytes can be enticed into the foci of infection by a number of chemoattractant gradients, such as the C5a, formylated proteins, and chemokines. But what insures that these phagocytes will arrive precisely where they are needed? The short answer to this question is neutrophils and monocytes, as well as endothelial cells to which they must adhere, become sticky as the result of expression of sugars on glycosylated surface proteins. These glycoproteins on the endothelial cells are called **selectins** (which mediate weak binding) and **integrin ligands** or **addressins** (which mediate strong binding) and bind to **integrins** on leukocytes. However, blood cells that originate in the marrow must enter the bloodstream and be able to circulate there without adhering too firmly. For example, neutrophils enter the capillary bed where they adhere to the endothelial venules in response to distress signals, e.g., complement components, IL-l, TNF-α, etc., from tissues, whereby, the endothelium in the involved region expresses a set of adhesion molecules that bind more firmly to neutrophils. When summoned by chemotaxins, the neutrophils, like other leukocytes, cross the endothelium by **diapedesis** through the cell junctions, traverse the basement membrane, and enter the extravascular tissue spaces. At the proper time, both leukocytes and endothelial cells must be able to stick to each other. The transition from loosely adherent to tightly adherent is critical for the leukocytes to leave the circulation and move through the tissues. Clearly, leukocytes and endothelial cells are subject to regulatory mechanisms that induce adherence of these cells at the proper time. Complement components, such as C3a, chemokines and cytokines direct and regulate leukocyte chemotaxis and adherence to endothelial cells.

The importance of glycoprotein receptors for endothelial cells that are present on neutrophils is illustrated in people with congenital defects in these proteins. Neutrophils from these patients are unable to pass through the vascular endothelium and fail to orient, bind, and ingest particles. Patients with this condition are said to have congenital **leukocyte adhesion deficiency (LAD)** and suffer recurrent, often fatal infections. The reason for failure to bind and to ingest bacteria is a defect in one of the cellular receptors for C3b (called Mac-1or CD11b, which is one of the integrins). In a healthy person, adherence of leukocytes to endothelial surfaces is very intense in the submucosa of the alimentary tract. The large intestine has an enormous microbial population just a cell layer away from the host's aseptic tissues. This abundant flora generates large amounts of chemotaxins that recruit the bulk of neutrophils. Thus, the submucosa of the gut is in a constant state of low-level inflammation, which keeps the microbial flora at the lumen in check.

4.4.2 Eosinophils

Granulocytes, especially eosinophils, differentiation occurs within the bone marrow in response to eosinopoietic cytokines, particularly IL-5. The mechanisms underlying mucosal tissue basophil/mast cell or eosinophil accumulation and/or differentiation are unclear. Eosinophil precursors (CD34/IL-5Rα^+ cells) and IL-5 mRNA$^+$ cells have been identified in the lungs of asthmatics, indicating that a population of eosinophils or basophil/mast cells may differentiate in situ [36]. Nasal epithelial cells from atopic, as opposed to nonatopic, patients produce basophil/mast cell and eosinophil colony-stimulating activities with an enhanced effect of the former on atopic of similar peripheral blood cell growth. These studies support the hypothesis that basophil/mast cells and eosinophils accumulate in allergic inflammation as a result of proliferation and differentiation [37]. To this end, epithelial cells from respiratory mucosa increase eosinophil survival and activation [38]. Taken together, a subset of basophil, mast cells, and/or eosinophils may differentiate in the mucosa, in a IL-5-dependent fashion.

The eosinophils parallel the neutrophils in many of their functions. However, they are not efficient in phagocytosis, but can readily release their contents of their granules to the outside. Their targets are usually parasites rather than bacteria. Indeed, the increase of these cells in the circulation, **eosinophilia**, is the hallmark of multicellular parasitic diseases such as schistosomiasis or trichinosis. It has been shown that the cytoplasmic granules of the eosinophils carry large amounts of an enzyme known as **eosinophil peroxidase**, as well as specific cationic proteins. These compounds have the power to kill certain parasites. Thus, eosinophils have an anti-infectious armamentarium similar to that of neutrophils but specifically target multi-cellular pathogens, which are too large to be phagocytized by neutrophils.

4.4.3 Monocytes and Macrophages

Terminal maturation of macrophages is an important step for cell diversity among mucosal subpopulations and their functional competence in vivo. These activated macrophages to hold two functions, either to facilitate tissue remodeling or clear infection. Optimal phagocytosis of mucosal pathogens is achieved by mature macrophages but not blood-derived monocytes. Studies of factors controlling macrophage differentiation in tissues may help to explain the great macrophage, DC and neutrophil heterogeneity in the intestinal during disease. The phenotype and function of monocytes recruited to revealed $CD68^{Hi}$ $Gr\text{-}1^{+/-}$ monocytes, along with $CD68^{+/-}$ $Gr\text{-}1^{Hi}$ neutrophils, rapidly accumulate in Peyer's patches and mesenteric lymph nodes after oral *Salmonella* infection. These monocytes have increased MHC-II and costimulatory molecule expression; in contrast, neutrophils and DC produce inducible nitric oxide synthase [39]. In the normal intestinal lamina propria, only few macrophages express calprotectin and CD86; however, the number of $calprotectin^{+}$ $CD80^{Hi}$ $CD86^{Hi}$ macrophages from inflamed mucosa increase with the degree of inflammation [40, 41]. Similar mechanisms of monocyte-macrophage recruitment, activation, and maturation exist in other mucosal tissue. Human upper airway epithelial cells from normal and inflamed, allergic rhinitis and nasal polyp tissues induce monocytic differentiation of hemopoietic progenitors into mucosal macrophages that largely required GM-CSF, G-CSF and IL-6 [42].

Monocytes reach the sites of microbial invasion after arrival of neutrophils. These circulating members of the mononuclear family eventually settle in tissues and become known as resident **tissue macrophages**. Although monocytes and macrophages share a common progenitor with neutrophils, their kinetics of maturation and function are substantially different. Unlike neutrophils, monocytes continue to differentiate after they leave the bone marrow. Most importantly, monocytes and macrophages function in both innate and adaptive immune responses.

In general, monocytes and macrophages come into play slowly, often hours after neutrophils have been actively combating invading microorganisms. Neutrophils play a role in recruiting mononuclear cells because they release a specific, potent chemoattractant for monocytes, **azurocidin** or CAP37. Tissue or resident macrophages exist through out the body and have different names and functions, depending on the tissue. Thus, they are called **Kupffer cells** in the liver, **alveolar macrophages** in the lungs, **osteoclasts** in the bone, and **microglia** in the brain. Tissue macrophages contribute greatly to the inflammatory response by releasing IL-1, which enhances the adherence of neutrophils to endothelia, and TNF-α, which activates newly arrived neutrophils and monocytes. In addition, tissue macrophages release IL-6 and many chemokines that attract other leukocytes to bridge the innate with the adaptive immune system.

4.4.4 Natural Killer Cells

NK cells are large granular lymphocytes that have cytoxic activity. In this aspect they resemble cytotoxic T cells. However, NK cells are neither T cells nor B cells because they do not have the markers characteristic of T or B-lymphocytes. They also do not have T or B cell antigen receptors and the associated signal transduction molecules that make B and T cells antigen recognition specific. Therefore, they do not recognize antigens through the types of receptors that T and B cells use, and they do not need to be selected and generated by adaptive immune responses. The name natural killer thus stems from their readiness to kill other cells without the need for time-consuming antigen-driven selection that is required for adaptive immunity.

NK cells have several types of **natural killer receptors (NKRs)**, which are lectins that bind to various glycoproteins present on many host cells. However, this would potentially induce the killing of many healthy cells. Therefore, NK cells also have **killer inhibitory receptors (KIRs)**, also called killer Ig-like receptors (e.g., CD158), because they inhibit the killing of normal healthy cells by NK cells. However, virus-infected cells or some tumor cells often have missing MHC class I molecules and therefore do not engage KIRs but still have glycoproteins that engage NKRs. In the absence of inhibition by KIRs, NKRs trigger the killing of these target cells.

NK cells are stimulated by macrophage-derived IL-15 and IL-12. **IL-15** is a growth factor for NK cells, and IL-12 induces IFN-γ production by NK cells and enhances their cytolytic activity. IFN-γ in turn activates macrophages and inhibits growth of viruses in host cells. NK cells also have **Fc-γRIIIa receptors** (CD16), which bind the Fc portion of IgG antibodies when complexed with antigen. This allows NK cells to function by **antibody-dependent cell-mediated cytotoxicity (ADCC)**, which is another important mechanism that bridges humoral adaptive responses with this innate immunity cell.

4.4.5 Dendritic Cell Activation, Maturation and Role in Bridging Innate with Adaptive Mucosal Immunity

Specialized antigen presenting cells, such as Dendritic cells (DCs), are sensors of "microbes" [43]. Their activation leads to an increase in immunogenicity with delivery of these signals: peptide loading on MHC class I and II, increased levels of costimulatory molecules, and polarizing cytokines. Direct activation of DCs via TLRs has been well described. TLRs expressed by DCs, likely cooperating to widen the repertoire of recognition specificity and triggering DC maturation in a nuclear factor (NF)-κB–dependent manner [43, 44]. Heat shock proteins released by necrotic cells may also mature DCs and be considered as danger signals. In contrast to LPS-signaling, triggering receptor expressed on myeloid cells 2 (TREM-2)/DNAX-activating protein of 12 kDa (DAP12)-induced DC maturation [45], like that initiated by FcRs [46], is dependent on protein tyrosine kinase (PTKs) and extracellular signal–regulated kinase (ERK) signaling. TREM2/DAP12 engagement by DCs leads to upregulation of chemokine receptor 7 (CCR7), CD40, CD86, and DC survival but not IL-12 or TNF-α production. In the absence of microbial stimulus, activated NK cells, secrete IL-12 and TNF-α to trigger activation and maturation (i.e., upregulation of CD80, CD86, CD83, HLA-DR, CCR7) of immature DC [47, 48].

At the onset of infection and before antigen-specific cognate T cells are expanded, NK cells become activated and amplify the maturation of DCs induced by microbial products or by virus-induced IFNs. Activated NK cells, by lysing target cells or surrounding immature DCs that have phagocytized and processed foreign antigen, provide antigenic cellular debris internalized by maturing DCs that can be presented to T cells in lymph nodes. Thus, NK cells likely participate in DC-mediated cognate T cell responses. At later stages of immune responses, activated NK cells overwhelm surrounding DCs and the cross talk between activated NK cells and DCs leads to NK cell–mediated DC death shutting off the antigen presentation. Hence, the interaction between immature DCs and activated NK cells results in either DC maturation or cell death. The mechanisms that determine the outcome between death and maturation depend on the dynamics between DC and NK cell density and the stage of DC maturation. In addition, the cytokine secreted by DC activation after interaction with NK cells in response to various microbial stimuli could influence polarization of T cell responses.

DCs are generated from bone marrow precursors and populate non-lymphoid tissues via blood, where they remain in a state defined as immature. DCs produced in long term culture are partially mature, and quite distinct from DCs in peripheral tissue sites, which clearly express MHC-II and CD40 as well as migrate to lymph nodes to present antigen to T cells. Antigen presentation by DCs activates T helper cells to express CD40 ligand, which in turn activates $CD40^+$ DCs [49]. DCs undergo maturation, with subsequent trafficking from non-lymphoid to lymph nodes where they may encounter and stimulate naïve T cells [50]. Maturation of DCs occurs at peripheral inflammatory sites and likely to be induced by inflammatory cytokines such as IL-1β or TNF-α the release that is stimulated by bacterial products, e.g., LPS. This maturation process coincides with profound functional changes [51]. The activation of DCs induces the production of pro-inflammatory cytokines and chemokines that induce the development of immature to mature DCs [52].

4.4.6 Epithelial Barrier Function

The formation of a regulated selectively permeable epithelial cell barrier is essential to prevent the uncontrolled passage into the host of partially digested components of ingested food as well as bacterial products and to regulate fluid and electrolyte absorption and secretion (Fig. 4.2). Tight adheren junctions are located near the apical part of columnar epithelial cells and separate the paracellular space between those cells from the intestinal lumen [53]. Cell-cell adhesion at these junctions is maintained by a tightly regulated complex of proteins, such as occludin, zonula occludens-(ZO) 1 and –2, and members of the claudin family. However, IFN-γ can down-regulate the expression of ZO-1 [54]. This time-dependent decrease corresponds with a significant decrease in trans epithelial resistance as well as an increase in mannitol flux. In vivo, expression of the tight junction protein occluding appears to be diminished in IBD, suggesting that down-regulation of epithelial occludin may

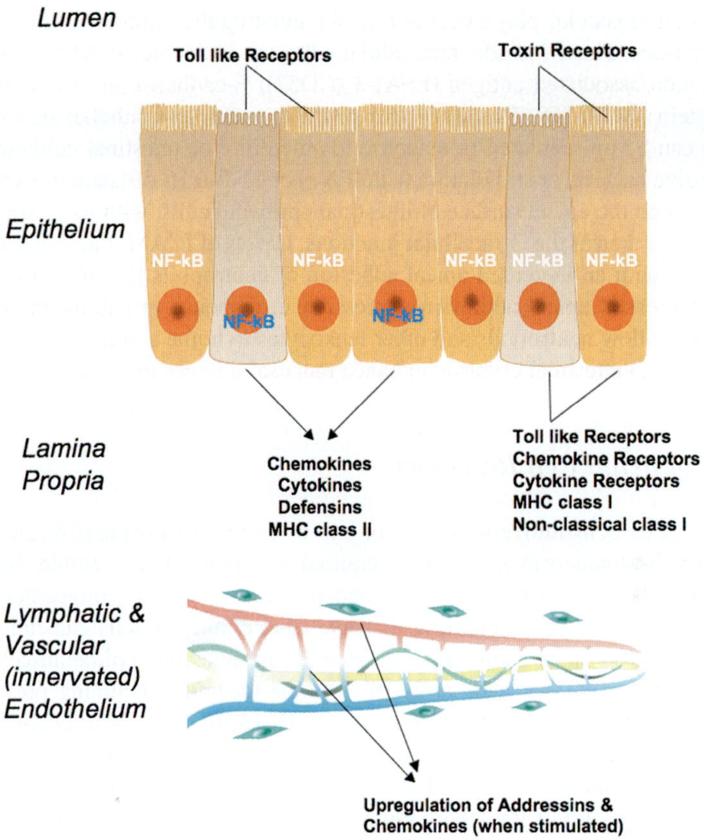

Fig. 4.2 Epithelial barrier function and signaling to promote innate as well as adaptive immunity. Epithelial cells of the mucosa (e.g., gastrointestinal, nasopharynx, lung, and reproductive tracts) provide a selectively permeable barrier to protect the host from the outside environment. The epithelium has innate receptors that can detect toll receptor ligands or toxins to lead to the activation of transcription factors, e.g., nuclear factor kappa B. Subsequent translocation of these transcriptional factors to the nucleus can promote the expression and secretion of chemokines and cytokines to recruit and activate leukocytes in the lamina propria that traffic by lymphatic and vascular endothelial cells from this effector site to inductive sites (e.g., lymph nodes), where adaptive immune responses can develop further

enhance paracellular permeability and neutrophil transmigration that is observed in active disease [55]. When the epithelial cell layer is disrupted, several epithelial cell produced factors are known to influence epithelial cell proliferation, migration and wound healing. These include growth factors such as epidermal growth factor (EGF) that exerts its effect partially through an increase in production of transforming growth factor (TGF)-β [56], fibroblast growth factor (FGF), hepatocyte growth factor (HGF), and intestinal trefoil factor (ITF). ITF is secreted towards the apical side of the epithelium and has been shown to increase the migration of cells towards sites of injury [57].

Adhesion molecules play a central role in regulating the trafficking of leukocytes through tissues. These include intracellular adhesion molecule (ICAM)-1, lymphocyte function associated antigen (LFA)-3 (CD58), E-cadherin, gp180, and biliary glycoprotein (BGP) [58, 59, 60, 61, 62, 63]. For example, epithelial ICAM-1 expression can be up-regulated in response to co-culture of intestinal epithelial cells with invasive bacteria or stimulation with IFN-γ or TNF-α [6, 59], and its expression is polarized on the apical surface of intestinal epithelial cells, with its density being greatest in the area of the intracellular junctions. Levels of ICAM-1 have been shown to correlate with an increased apical adhesion of neutrophils [59, 60]. It is plausible that increased apical addressin expression in response to pathogenic bacteria functions to allow neutrophils and other leukocytes to home and migrate across the epithelium into intestinal crypts to enhance mucosal defense mechanisms.

4.4.7 Epithelial Cell Receptors

Epithelial cells constitutively express or can be induced to express receptors important for host mucosal innate and acquired immunity. For example, intestinal epithelial cells express MHC class II and non-classical MHC molecules (e. g., CD1d, MICA) [64, 65], suggesting that intestinal epithelial cells can function as antigen-presenting cells and/or costimulatory cells. Intestinal epithelial cells have been shown to express a wide array of cytokine receptors including putative receptors for IL-1, IL-4, IL-6, IL-7, IL-9, IL-10, IL-15, IL-17, IFN-γ, GM-CSF, and TNFα [57, 66, 67, 68, 69, 70, 71, 72, 73, 74] as well as receptors for several chemokines including CXCR4, CCR5, CCR6, and CX3CR1 [75, 76, 77]. Several of these receptors are expressed on the basolateral surface of the epithelium, while others are expressed apically or both apically and basolaterally. This indicates that epithelial cell signaling and function can be influenced not only by luminal antigens, but also by cytokines and chemokines released from local immune cell populations in the mucosa thereby allowing epithelial cells to detect and subsequently respond to immunological changes within the subepithelial compartment.

Some pathogenic bacteria (e.g., *Salmonella, Shigella*, enteroinvasive *E. coli, Yersinia*, and *Listeria*) have been shown to invade the epithelium and alter intracellular signaling pathways, either directly or indirectly through the secretion of bacterial products into the cell [78, 79]. To mitigate these molecular pathogenic mechanisms, intestinal epithelial cells express certain TLRs that represent an evolutionarily conserved family of receptors that function in innate immunity via recognition of conserved patterns in microbial molecules [80]. Epithelial cells express TLR2, TLR3, TLR4 and TLR5 that recognize LPS and other bacterial products like flagellin, etc. [81, 82, 83].

4.5 Mucosal Cytokines and Other Innate Factors

A number of reasons make mucosal delivery of cytokines and innate factors an attractive strategy for manipulation of mucosal immunity. Unlike other molecules that regulate mucosal immunity through unknown mechanisms, cytokines primarily interact with their corresponding receptors on target cells. Significant levels of IL-12 and subsequent IFN-γ responses were achieved after oral delivery of this regulatory molecule [84]. Even though significant serum IL-12 levels were achieved following nasal delivery, these levels were equivalent to approximately one-tenth of those achieved by parenteral injection of the same dose of IL-12 [84]. Interestingly, both orally and nasally administered IL-12 could exert biological activity and regulate mucosal immune responses. Thus, mucosal delivery of cytokines could represent a safer alternative to systemic cytokine treatment, since the large and repeated doses generally needed for systemic cytokine treatment are often associated with toxicity.

A number of chemokines have been tested as adjuvants for nasal vaccines since chemokines are produced by a variety of mucosa-associated cells, including epithelial cells and immune cells. For example, $\gamma\delta TCR^+$ intraepithelial lymphocytes (IELs) produce large amounts of XCL1, a C chemokine with chemotactic activity for T and NK cells [85, 86]. XCL1 enhanced both mucosal and systemic immunity to a nasally co-administered protein antigen [87]. Further, analysis of antigen-specific $CD4^+$ T cell responses showed that the adjuvant effect of this chemokine involved both Th1 and Th2 responses. The fact that XCL1 induced mucosal SIgA antibody responses is consistent with the reported impairment of SIgA antibody responses in mice lacking $\gamma\delta$ T cells [88].

Epithelial cells also produce significant amounts of CCL5, CCL4, and CCL3 when stimulated by TLR ligands. CCL5 displays mucosal adjuvant activity for a nasally co-administered protein [89]. Analysis of antigen-specific mucosal and systemic $CD4^+$ T cell and cytokine responses showed that nasal CCL5 as adjuvant promotes strong IFN-γ and predominantly Th1 responses. Systemic immunization with a plasmid DNA expressing CCL5 was also reported to enhance antigen-specific immunity confirming the adjuvant activity of this chemokine by nonmucosal routes [90]. CCL3 and CCL4 are distinct, but highly homologous CC chemokines, that share an affinity for CCR5. Lillard et al., 2003 showed that these chemokines differentially affect adaptive mucosal and systemic immunity to nasally co-administered antigens [91]. CCL3 stimulated strong serum IgG with IgG1, IgG2b followed by IgG2a and IgG3 subclass antibody responses. On the other hand, CCL4 promoted lower IgG antibodies of IgG1 and IgG2b subclasses with high IgA and IgE antibody responses. Correspondingly, CCL3 produced higher titers of antigen-specific mucosal SIgA antibodies when compared with CCL3. CCL3 as adjuvant promoted Th1-biased $CD4^+$ T cell responses while CCL4 triggered Th2-type responses with higher levels of IL-4.

4.5.1 Secreted Epithelial Cell Mediators

In the course of infection or insult of the epithelial layer, there is an increased influx of immune and inflammatory cells into the subepithelial compartment towards the site of injury. For example, in response to bacterial infection, human intestinal epithelial cells produce mediators essential for the onset of acute mucosal inflammation. In this regard, a number of enteroinvasive bacteria, and some non-invasive bacterial pathogens, interact with the epithelial cell membrane and induce the release of potent chemoattractants for neutrophils, e.g., CXCL1, CXCL2, CXCL5, CXCL8 [92, 93, 94], monocytes/macrophages, e.g., CCL2 [94, 95], and immature CCR6-expressing DCs, e.g., CCL20 [96]. Furthermore, under inflammatory conditions, human intestinal epithelial cells can produce IFN-γ-inducible chemokines (e.g., CXCL9, CXCL10, CXCL11) [97, 98] that attract $CXCR3^+$ T cells that have a Th1 phenotype.

The variance in the kinetics of chemokine expression, function and production by intestinal epithelial cells may be due to the different spatial and temporal chemokine gradients and expression, respectively, which would modulate the chemoattraction of target cells within the intestinal mucosa [94]. Recent studies have begun to elucidate the role the epithelium plays in providing important chemotactic signals for the development of mucosal adaptive immunity. In this regard, small intestinal epithelial cells constitutively produce CCL25, whose cognate receptor CCR9, is expressed by $\alpha_4\beta_7$-expressing T cells that preferentially home to the small intestinal mucosa [99, 100]. In contrast, human colon epithelial cells do not produce CCL25 but do produce CCL28, whose cognate ligand is CCR10, which allows $CCR10^+$ cells to preferentially migrate to sites of CCL28 production in the small intestine [101, 102]. Hence, CCL25 and CCL28 appear to be important for selective migration of lymphocyte subsets to the small intestine and colon, respectively.

In addition to chemokine expression, intestinal epithelial cells are capable of producing a number of cytokines in response to infection, thereby possessing the ability to affect local immune responses. Some of the cytokines expressed by intestinal epithelial cells include: EBI3, TNF-α, IL-12, IL-15, GM-CSF, and MIF [66, 93, 103, 104]. Besides affecting gene expression patterns of immune cells in the mucosa, factors released by epithelial cells upon microbial challenge, or under the influence of proinflammatory cytokines, might directly induce endothelial expression of adhesion molecules such as ICAM-1 and VCAM-1, which could lead to an increased mucosal influx of leukocytes [105]. The release of proinflammatory cytokines and chemokines by epithelial cells can also be induced by IL-1α or TNF-α (or infection), return to baseline within hours after induction, suggesting that epithelial cells initiate mucosal inflammatory responses [92, 94, 106]. Similarly, other factors, e.g., CXCL5 and CCL5, that are expressed relatively late in response to microbial challenge indicates that epithelial cells might also participate in the later phases of the mucosal inflammatory or immunity process [94].

4.6 Recognition Strategies of the Innate Immune System

Signaling within the intestinal epithelial cells that is initiated by a broad array of cytokines, chemokines, and microbial antigens frequently converge to activate the transcription factor NF-κB and its target genes [107]. This has led to the concept that NF-κB is a central regulator of epithelial cell signaling pathways essential for initiating host innate immune responses to microbial infection. Activation of the

Iκ kinase (IKK) complex, and notably the IKKβ subunit is an essential step in the activation of NF-κB by a number of enteric pathogens and by proinflammatory cytokines (e.g., TNF-α, IL-1, etc.). Indeed, some pathogens (e.g., *Yersinia*), through their type III secretory proteins, have developed strategies to prevent the activation of NF-κB and other signal transduction pathways and consequently down-modulate the host inflammatory response [108, 109].

One of the central questions in immunology is how the immune system is able to discriminate between pathogens or toxins and harmless (or beneficial) commensal bacteria and food antigens. During its long evolution and coexistence with microorganisms, the innate immune system has developed three recognition strategies to detect microbes and distinguish dangerous from harmless agents in the absence of MHC through pattern recognition receptors. The first and most frequently used strategy is recognition through receptors specific for microbial molecules that are not present in the host. These **pattern recognition receptors**, detect unique structures present in microbes, but not in mammals (Table 4.5). These structures are found in microbial products that are essential for their survival and thus cannot be easily changed, dispensed with or substituted. Several pattern recognition receptors are located on the cell membrane of phagocytic cells, lymphocytes, endothelial cells, and epithelial cells, although not all of these cells express all pattern recognition receptors.

The largest family of pattern recognition are classified as toll-like receptors (**TLRs**), because of their homology to a fruit fly *Drosophila* receptor TLRs recognize many essential microbial molecules, including peptidoglycan, lipoteichoic acids, lipopeptides, lipopolysaccharide, flagella, bacterial cytidine phosphate guanosine (CpG) nucleotide motifs, and viral double-stranded or single stranded RNA. Some of the same molecules are also recognized by another pattern recognition receptor, **CD14**. TLRs and CD14 allow host cells, especially macrophages and dendritic cells, to recognize most microbes (bacteria, fungi, viruses), alert the host to their presence, initiate the innate or early host response, and bridge with the adaptive immune response. Other pattern recognition receptors, such as the **scavenger receptor**, mannose receptor and β-**glucan receptor**, are present primarily on phagocytic cells and function to recognize various bacterial and fungal surface molecules to enhance phagocytosis.

Some pattern recognition receptors are intracellular. One such receptor, **protein kinase R**, recognizes intracellular viruses and induces the synthesis of antiviral interferons. Another family of intracellular pattern recognition receptors includes proteins with a **nucleotide oligomerization domain (NOD)**. NOD proteins recognize fragments of peptidoglycan from bacterial cell walls and allow the infected or antigen-containing cells to produce proinflammatory cytokines, which enhance antimicrobial responses. The need for intracellular recognition of bacteria stems from the ability of some bacteria (e.g., *Salmonella, Shigella* and *Listeria*) to enter host cells and escape into the cytoplasm. This strategy allows these bacteria to avoid recognition and destruction by phagocytic cells and other extracellular antimicrobial mechanisms.

Table 4.5 Pattern recognition receptors

Receptor	Recognition OF	Main effect
Cell membrane		
TLRs	Bacterial, viral and fungal components	Activation of macrophages and dendritic cells
TLR2 (plus TLR1 or TLR6)	Peptidoglycan (all bacteria)	
	Lipoteichoic acid (Gram-positive bacteria)	
	Lipopeptides (many bacteria)	
	Lipoarabinomannan (Mycobacterium)	
	Zymosan (fungal cell wall)	
TLR3	Double-strand RNA (viruses)	
TLR4	Lipopolysacchride (Gram negative bacteria)	
TLR5	Flagella (many bacteria)	
TLR7 and TLR8	Single strand RNA (viruses) and rRNA (bacteria)	
TLR9	Bacterial DNA (CpG)	
CD14	Lipopolysachhride (Gram-negative bacteria)	
		Activation of macrophages
	Peptidoglycan (all bacteria)	
	Lipoteichoic acid (Gram-positive bacteria)	
Scavenger receptors	Negative charge polysaccharides (bacteria and fungi)	Phagocytosis by macrophages
Mannose receptor		Phagocytosis by macrophages
β-glucan receptor (dectin-1)		Phagocytosis by macrophages
N-formylmethionyl peptide receptor		Chemoattraction of PMNs and macrophages
Intracellular		
Protein kinase R	Double strand RNA (viruses)	Induces expression of a viral-inhibitory protein, interferon
NOD1 and NOD2	Peptidoglycan (bacteria)	Activation of macrophages and epithelial cells to produce proinflammatory mediators
Secreted		
Soluble CD14	Lipopolysachhride (Gram-negative bacteria)	Activation of $CD14^{-}$ cells (e.g., Endothelial cells)
Lipopolysaccharide-binding protein	Lipopolysachhride (Gram-negative bacteria)	Activation of macrophages
Mannose-binding lectin (collectin)	Mannose-containing polymers (bacteria and fungi)	Activation of complement
Surfactant proteins (collectin)	Bacteria, fungi, viruses	Opsonization
Serum amyloid P (pentraxin)	Bacteria	Activation of complement
C-reactive protein (pentraxin)	Bacteria polysaccharides	Activation of complement

CD, cluster of differentiation; NOD, nucleotide oligomerization domain; PMN, polymorphonuclear leukocyte; TLR, toll-like receptor

The clinical significance of these pattern recognition receptors has only recently been realized through the identification of patients who suffer from severe recurrent pyogenic bacterial infections resulting from genetic defect in signal transduction pathways activated by TLRs. Recently, polymorphisms in the gene encoding the intracellular recognition protein, **NOD2**, has been correlated with patients with **Crohn's disease**, one of the chronic inflammatory bowel diseases (IBD) associated with an inability to properly control the intestinal inflammatory responses [110, 111, 112, 113, 114, 115]. In general, IBD arises from a dysregulated mucosal immune response to luminal bacteria. It has been suggested that TLR4 recognizes LPS and transduces a proinflammatory signal through the adapter molecule myeloid differentiation marker 88 (**MyD88**). Mesenteric lymph nodes and intestinal lamina propria from $TLR4^{-/-}$ and $MyD88^{-/-}$ mice frequently develop colitis due to high bacterial levels [116]. Hence, a defective innate immune response may result in diminished bacterial clearance and dysregulated responses to normal flora.

The innate immune system detects danger signals produced by tissue damage, which would usually be found in tissue-damaging infections. For example, uric acid crystals released from damaged cells can induce proinflammatory responses in DCs. However, it is still unclear what receptors recognize these danger signals. Recent molecular biologic advances in this field have given us a clear picture of the mechanistic basis of gouty inflammation due to deposits of monosodium urate (MSU) crystals [117]. MSU crystals promote inflammation by stimulating cells via TLR signaling and cleaving of C5 that leads to the formation of the membrane attack complex to culminate in secretion of cytokines, chemokines, and other inflammatory mediators. To this end, $TLR2^{-/-}$, $TLR4^{-/-}$, and $MyD88^{-/-}$ bone marrow derived-DCs demonstrate impaired MSU crystals uptake and associated inflammatory cytokine release, when compared to similar cells from wild type mice [118]. Dying mammalian cells have also been found to release uric acid due to injury. Uric acid enhances the adaptive immune response to antigens associated with injured cells [119]. Taken together, these data suggest the innate immune system is not only capable of detecting pathogens, but also cell death and metabolic stress.

4.6.1 TLR-Regulated Cell Recruitment to Infection Sites

One of the distinctive features of the immune system is that it relies on cell migration for surveillance, attack, containment and clearance of invading pathogens. Two types of cell migrations are use by the cells for both innate and adaptive immune systems: inducible and homeostatic. Inducible cell migration is generally triggered as a result of the sensing of pathogens through PRRs, as in neutrophil recruitment to the local site of infection. Steady state, or homeostatic, cell migration allows naïve lymphocytes to circulate between the secondary lymphoid tissues throughout the body to enhance their chances of encountering the cognate antigen. This pathway is constitutive and does not require induction through TLRs. Innate recognition of pathogen-associated molecular patterns (PAMPs) through TLRs initiates an inflammatory response characterized by the recruitment of cells to sites of infection to augment the killing of invading pathogens and to halt their spread. Cell migration from peripheral blood into the inflamed tissue involves a tightly controlled sequence of events [120].These events are mediated by two types of signals: diffusible chemotactic factors and cell surface adhesion molecules.

Activation of TLRs induces the expression of selectin, chemokine and chemokine receptor genes that regulate cell migration to the sites of inflammation [121]. Sensing inflammatory signals, leukocytes first roll on the vascular endothelial cells, a process mediated by selectins expressed by the endothelium and the carbohydrate ligands on leukocytes. The selectins are rapidly displayed on the endothelial cell surface either after direct recognition of pathogens through TLRs or by tumor necrosis factor secreted from activated tissue macrophages. Next, PAMP recognition through the TLR induces chemokine secretion. Key inflammatory chemokines produced during acute microbial infection include CXCL1, CCL2, CXCL8, CCL8, CCL3, CCL4, CCL5, CCL7, and CCL13 [122]. These chemokines bind to the luminal surface of the vascular endothelium and trigger activation of leukocytes and induce conformational changes in the integrins [120]. The activated integrins on the leukocytes allow firm adhesion to the vascular endothelium. Integrin ligands on the endothelium, such as intercellular cell adhesion molecule (ICAM-1), are also induced either directly by TLRs expressed on the endothelium, or indirectly, through TNF and IL-1, which are induced by TLR-activated resident macrophages. Finally, the adherent leukocytes migrate between the endothelial cells and extravasate into tissue interstitium. To provide maximum surveillance for infectious agents, in addition to the inflammation-induced cell recruitment, most tissues of the body, particularly at the mucosal surfaces that represent portals of pathogen entry, are interlaced with resident innate leukocytes such as DCs, macrophages and mast cells. Pathogen recognition through TLRs orchestrates the recruitment of leukocytes to the sites of infection by activating the tissue stromal cells, tissue-resident innate cells and the circulating leukocytes.

4.6.2 TLR Activation of Leukocytes

Acute inflammatory cellular infiltrate consists of innate immune cells such as monocytes, neutrophils, basophils, eosinophils and NK cells. Neutrophils and NK cells are critical effector cells that protect the host by killing pathogenic microbes and infected cells, respectively. In addition, expression of some TLRs is regulated, both positively and negatively, in several cell types. Human neutrophils express mRNA for TLR1 through TLR10, except for TLR3, and secrete inflammatory chemokines after stimulation with LPS, zymosan and R848 [123]. However, neutrophils are not stimulated by CpG [124]. Similarly, NK cells express TLR9 mRNA, but are not responsive to CpG stimulation [125]. The potent ability of CpG DNA to stimulate NK cells was most likely attributed by the contamination of plasmacytoid DCs (pDCs), which after being stimulated with CpG secrete large amounts of type I interferons and induce NK cell activation [125].

Eosinophils, despite their constitutive expression of TLR1, TLR4, TLR7, TLR9 and TLR10 mRNA, only respond to stimulation with R848 through TLR7, which results in their activation and prolonged survival [126]. Mast cells express TLR1, TLR2, TLR4 and TLR6 and respond to TLR2 and TLR4 agonists [127, 128, 129]. Monocytes express a variety of TLRs that induce their activation (discussed below). The monocytes that infiltrate the inflamed tissue can differentiate into either tissue macrophages or DCs that contribute to the generation of adaptive immunity. TLR signals not only regulate chemokine and chemokine receptor gene expression but also control cell surface chemokine receptor expression by inhibiting chemokine receptor desensitization. For example, neutrophils stimulated through TLR4 transcriptionally down regulate the expression of G protein-coupled receptor kinases [130].

4.6.3 TLR Activation

The epithelial layer provides the first line of defense against invading pathogens. Consequently, many TLRs are expressed by epithelial cells of the intestinal, respiratory and urogenital tracts [131, 132]. Pathogen recognition by TLRs expressed by these cells leads to the production of cytokines, chemokines and antimicrobial peptides [133]. Epithelial chemokines can be displayed on the luminal surface of both the local tissue and draining lymph node vascular endothelium.

Cell migration is also regulated by direct recognition of PAMPs by the vascular endothelium. In a mouse model of septic shock, systemic LPS injection induces rapid neutrophil sequestration into the lungs. In one study of bone marrow- chimeric mice that were TLR4-deficient in either their leukocytes or in the stromal cell compartment, TLR4 responsiveness by the vascular endothelial cells, but not neutrophils, was required for migration of these cells into the lung tissue [134]. In another study, TLR4 expression in both the hematopoietic and the stromal compartments was required for the recruitment of leukocytes after challenge with the uropathogenic *Escherichia coli* [135]. Thus, the recognition of PAMPs by epithelial, endothelial and hematopoietic cells through TLRs are integral in the innate immune defense against microbial pathogens at sites of infection.

4.6.4 TLRs and DCs

DCs are central to T lymphocyte activation and differentiation into effector T helper cells and CTLs [136]. DCs take up antigens, become activated and migrate to local lymphoid tissues to present the antigen to lymphocytes. This process involves phagocytosis, the upregulation of costimulatory and MHC molecules, a switch in chemokine receptor expression, secretion of cytokines and chemokines and the presentation of antigens by DCs [136]. All of these events are regulated through the recognition of pathogens via the pathogen-recognition receptors (**PRRs**) expressed by DCs and are influenced by the microenvironment in which they reside. Many types of PRRs are expressed by DCs, including C-type lectins, mannose receptors and TLRs. However, TLR family members are critical PRRs, whose signals lead to the generation of effector responses including Th1 [135] and CTL [137] responses. Distinct DC subsets occupy special niches defined by their anatomical location and their ability to respond to certain types of pathogens through the expression of distinct sets of TLRs.

Studies of DC subsets isolated from humans and mice have revealed that TLRs have distinct expression patterns in these important antigen-presenting cells. Human blood monocytes express TLR1, TLR2, TLR4 and TLR5, but progressively lose these receptors as they differentiate into immature DCs in the presence of granulocyte-macrophage colony-stimulating factor and IL-4. These cells instead acquire the expression of TLR3 [138]. Notably, human myeloid DCs and in vitro-differentiated immature DCs express TLR3 in their intracellular compartments, unlike human fibroblasts, which express TLR3 on the cell surface. Human Langerhans cells do not express TLR3 [139]. Freshly isolated human plasmacytoid ($CD8\alpha^+$) DCs express TLR7 and TLR9, whereas myeloid ($CD8\alpha^-$) DCs express TLR1, TLR2, TLR3, TLR5, TLR6 and TLR8 [125, 140, 141]. In some studies, TLR7 expression was detected on both plasmacytoid and myeloid DCs [142, 143], whereas others found TLR7 was exclusively expressed by plasmacytoid DCs [140, 141]. In mice, plasmacytoid DCs express TLR1, TLR2, TLR4, TLR6, TLR8 and TLR9 [144, 145]. There are also distinct differences in the expression and responsiveness of certain TLRs in freshly isolated DCs versus in vitro-derived DCs. Although TLR4 is expressed at low amounts on splenic DCs, freshly isolated mouse DC subsets do not consistently respond to LPS stimulation [146]. In contrast, DCs derived from bone marrow precursors in the presence of granulocyte-macrophage colony-stimulating factor (GM-CSF) have high expression of TLR4 and respond robustly to LPS, a characteristic shared with murine macrophages. Inhibitors of endosomal pathways such as chloroquine and bafilomycin have been used to delineate the functional location of certain TLRs within the DC subsets [147]. These studies showed that TLR3, TLR7, TLR8 and TLR9 all require an acidic environment for activation of the endosomes, whereas TLR1, TLR2, TLR4, TLR5, and TLR6 are expressed by DCs and do not require endosomal maturation.

TLR expression profiles of distinct DC subsets raise some interesting questions about the function of these DCs. For example, TLR3 is lacking in both human and mouse plasmacytoid DC populations. Yet these cells are efficient detectors of viral infections, recognizing the viral genomes through TLR9 (double-stranded DNA) [142] and TLR7 (single-stranded RNA). Another point that arises is the lack of responsiveness of all human and mouse DC subsets to LPS. Only monocytes and in vitro-differentiated DCs express TLR4 and respond notably to LPS stimulation. These data suggest that some, if not all, of the immunostimulatory effects induced by injected LPS on DCs are perhaps mediated indirectly by non-DC or DC precursors in vivo. Whichever is the case, the accumulating evidence points to a differential function for DC subsets in pathogen recognition and influencing adaptive immune responses.

4.6.5 TLR-Dependent DC-Mediated Control of T Cell Activation

The generation of adaptive immunity can begin with DCs capturing microbial antigens in the peripheral tissues. Subsequently, DCs migrate to the draining lymph nodes to present the processed peptides to naive T lymphocytes in the context of MHC molecules. This migration is mediated by TLR-induced down-regulation of inflammatory chemokine receptors and up-regulation of the receptors for lymphoid chemokines. The stimulation of immature DCs with TLR ligands results in CCR6 down-regulation and CCR7 up-regulation [148, 149], which enhances the ability of DCs to migrate from the peripheral tissues to the draining lymph node [150, 151]. In transit, DCs also undergo a maturation program that endows the cells with the ability to stimulate naïve T lymphocytes. It is only after encountering microbial pathogens that DCs begin the process of maturation.

Once inside the lymph nodes, some DCs migrate to the T cell zones, seek out antigen-specific T cells and induce their activation and differentiation into effector cells. DCs provide the naïve T cell with two signals required for their activation. The first signal is the antigen-specific signal received as a result of binding of the TCR to peptide presented by MHC molecules. The second signal is provided by costimulatory molecules such as B7-1 (CD80) and B7-2 (CD86), which are expressed by the DCs and trigger CD28 expressed on naïve T cells. Depending on the density of the peptides presented, the costimulatory molecules expressed and cytokines secreted by the DCs motivate naïve $CD4^+$ T cells to differentiate into either Th1 or Th2 cells [152]. Primary activation of $CD8^+$ T cells requires similar signals from the DCs, whereas memory $CD8^+$ T cell responses require $CD4^+$ T cell help [153, 154, 155, 156].

A given DC population will respond only to the pathogens for which they have appropriate TLRs. Furthermore, stimulation through a given TLR, depending on which DCs express it, can result in a distinct outcome, even when triggered by the same ligand. Stimulation of human plasmacytoid and myeloid DCs with synthetic TLR7 agonists induces the secretion of IFN-γ and IL-12, respectively [143]. Similarly, stimulation with certain synthetic oligonucleotides containing CpG motifs induces plasmacytoid DCs to secrete IFN-γ and myeloid DCs to secrete IL-12 [157].

Despite these differences in cytokine secretion by DC subsets, signals through TLRs generally result in the activation and maturation of all DCs, as measured by enhanced expression of the costimulatory molecules CD80, CD86 and CD40. Induction of the costimulatory molecules CD80 and CD86 on the DC surface is a particularly important step in the initiation of adaptive immunity. Indeed it is the coupling of microbial recognition with the induction of the costimulators that allows activation of pathogen-specific T cells, as was originally proposed by

Janeway [158]. CD80 and CD86 costimulators “flag” the DCs as a mature antigen-presenting cell for naïve T cells, which can only receive costimulatory signals in an antigen-specific, cognate interaction with these cells. Hence, TLR-induced expression of costimulators translates the nonclonal pattern recognition signal into clonal antigen-specific (adaptive) immune responses.

Once $CD4^+$ T cells are activated, their differentiation into effector cells is controlled by a variety of factors, including cytokines produced by DCs. TLR agonists induce DCs to secrete cytokines, such as IL-12, generally specialized in the induction of Th1 responses [159]. Accordingly, the immunization of mice with adjuvants containing various TLR ligands, including LPS, CpG DNA and complete Freund’s adjuvant, results in Th1 responses and the induction of Th1-dependent antibody isotypes. In contrast, MyD88 deficient mice are severely compromised in Th1 differentiation and instead generate Th2 responses [135, 160]. These data are consistent with a view that TLRs control the induction of Th1, but not Th2, responses. However, several reports have demonstrated that triggering TLR2 can induce Th2 differentiation [161, 162, 163]. The physiological importance of this is unclear because pathogens that trigger TLR2, such as Gram-positive bacteria, do not generally induce Th2 responses in vivo. However, stimulation of lung DCs with low doses of inhaled LPS induces Th2 responses and allergic inflammation, whereas high dose of LPS in the same conditions induce Th1 responses. Physiologically, the low dose of LPS presumably corresponds to the amounts available after inhalation of LPS associated with environmental antigens, whereas high dose LPS corresponds to the amount present during infections with Gram-negative bacteria. It is therefore likely that the TLR2-dependent induction of Th2 responses after inhalation of environmental LPS represents a mechanism for the pathogenesis of asthma, whereas the TLR4-dependent induction of a Th1 response after Gram-negative infection represents a mechanism for protection from infection.

4.6.6 TLRs in B Cell Activation and Antibody Production

B lymphocytes have a somewhat unusual status in the immune system because they express both clonally rearranged antigen receptors (i.e., BCRs) and non clonal PPRs, most notably TLRs. Stimulation of TLRs expressed on B cells can lead to polyclonal activation and production of low-affinity IgM antibodies. B1 cells of the peritoneal cavity and marginal zone B cells are particularly sensitive to TLR ligands and participate in T cell-independent immediate antibody responses that are typically directed against common bacterial components, such as phosphatidylcholine [164].

Only B Cells and plasma cells synthesize Igs. Developing B cells undergo dramatic changes in their responses to chemokines [165]. Pre-B cells are capable of making pro-B colonies in the presence of IL-7 and Flt3 ligand by migration towards CCL25. This response is lost in later stages of development in exchange for the ability to migrate towards CXCL13 via CXCR5 expression. Arising from pluripotent hematopoietic stem cells in the fetal liver and bone marrow, B cells are divided into two distinct lineages, B1 and B2 cells. Because they are the first to embryologically develop, B1 cells are a self-renewing population that predominate the plural and peritoneal cavities. Although the function of B1 cells is not fully understood, as a population they constitutively secrete IgM antibodies of limited diversity that react strongly with carbohydrates. A large portion of serum IgM in normal individuals is of B1 origin and is often referred to as "natural antibodies." As a cellular class, B1 cells may bridge innate and adaptive immunity by producing antibodies that react with carbohydrates expressed by infectious organisms. In this context, the splenic marginal zone B (innate-like) cells, by virtue of their pre-activated state collaborate with B1 cells to generate significant levels of IgM antibodies as plasmablasts during the innate response to bacterial antigens [166]. Together B1 and innate-like marginal zone B cells have been proposed to possess "natural memory" to bridge the innate immune response with the adaptive immune response.

Antigen, LPS, and cytokines have been shown to activate B cells and induce their differentiation into plasma cells via stimulation of TLR1, TLR2, TLR4, TLR6, TLR7, TLR9, and TLR10 expression [125, 167]. Signals directed through TLR1, TLR4, and TLR9 modulate integrins and CD9 on B1 cells, which is required for detachment from local matrix in preparation to respond to chemotactic signals [168, 169]. NF-kB is critical in the response of B1 cells. The peritoneal cavity of mice with a mutation in the NF-kB-inducing kinase gene, contain more B1 cells than normal mice [170]. These mice fail to migrate lymphoid tissues via CCL21 and CXCL13. Indeed, CXCL13 is highly chemotactic for B1 cells, which express significantly higher surface CXCR5 (the ligand for CXCL13) than B2 cells [171].

B2 or conventional follicular B cells express many of the same innate receptors as B1 or innate-like marginal zone B cells. However, conventional or B2 cells arise during and after the neonatal period, are continuously replaced from the bone marrow, and are widely distributed throughout the lymphoid organs and tissues. Somatic gene rearrangements are more extensive in B2 cells than in B1 cells, giving them a broader repertoire of BCR specificities and affinities. Unstimulated B2 cells, unlike B1 cells, produce low levels of Igs that increase only when they are activated by helper T cells that cause them to differentiate into antibody-secreting cells. B2 pathway-committed cell precursors graduate though several levels that involve changes in gene activity. Hence, the cytokines and chemokines that are generated early on in response to infection contribute to immunoglobulin (Ig) isotype switching, clonal expansion and differentiation into memory or Ig-secreting plasma cells.

Because the polyclonal activation of B2 cells may potentially result in the production of autoreactive antibodies, several mechanisms have evolved to regulate B cell activation, depending on the specificity of their BCR. Thus, B cells that express autoreactive membrane Ig induce a tolerogenic signaling pathway that inhibits TLR-induced B cell activation [172]. In contrast, the physiological importance of polyclonal activation by TLR ligands has been demonstrated for memory B cells. Human memory B cells produced antibodies in response to CpG stimulation independently of antigen-specific cognate T cell help [173]. It seems that naïve and memory B cells may respond to TLR stimulation differently to ensure both self-tolerance and a rapid response to reinfection: Because memory B cells have already been selected for their specificity, their polyclonal activation is not expected to result in autoimmunity.

4.6.7 TLR and Chemokines: Key Regulators in Bridging Innate and Adaptive Immunity

TLRs differentially regulate chemokine and chemokine receptor expression by a variety of host cells for induction of an innate response. TLR4 agonists promote IL-12 p70 and CXCL10, which are associated with Th1 responses [174]. In contrast, TLR2 stimulation does not induce IL-12 p70 or CXCL10 expression, but does support secretion of the IL-12 inhibitory p40 homodimer promoting conditions that favor Th2 adaptive immune responses. Correspondingly, activation of either TLR2 (by Pam (3) CysSerLys (4)) or TLR4 (by purified LPS) results in a decrease in both CCR1 and CCR2 expression by human monocytes [175]. Interestingly, TLR2 and TLR5, but not TLR4, are required for *Helicobacter pylori*-induced NF-κB activation and chemokine expression by gastric epithelial cells [176]. These findings demonstrate that the gastric mucosa recognizes and responds to *H. pylori* infection at least in part by sensing TLR2 and TLR5 ligands, which highlights the unique characteristics of *H. pylori* LPS that is unable to act as a TLR4 agonist. Nonetheless, TLR2 and TLR4 are crucial for mucosal innate responses. TLR2 and TLR4 have been shown to be express by lymphatic capillaries in the lamina propria of mucosal tissues that results in CCL21 expression [177]. Hence, the lymphatic endothelium contributes to DC migration into secondary lymphoid tissues (i.e., effector sites) through the expression of TLRs, the ligand engagements of which result in the induction of chemokines.

IFN-γ and TLR3 ligation induces the expression of CXCL9 and CXCL11 by leukocytes and fibroblasts [178]. In addition, stimulation of uterine epithelial cells with TLR3 agonist, poly(I:C), induces the expression of TNF-α, IL-6, GM-CSF, G-CSF, CXCL8, CCL2, and CCL4 as well as defensins to further fuel the adaptive and innate host responses, respectively [179]. Following vaginal inoculation of monkeys with TLR7 and TLR9 agonists, a massive influx of mucosal monocytes, $CD4^+$ T cells, DC infiltrates has been documented due to beta-chemokine-expression [180].

Innate immune stimuli, through activation of TLRs, set in motion a genetic program that induces the expression of chemokines by epithelium, endothelium, resident tissue macrophages and DCs. Changes in chemokine/chemokine-receptor expression orchestrate the movement of antigen-loaded DCs from the tissue into lymphoid tissue to activate naïve T and B cells to initiate the adaptive immune response. Chemokines expressed after TLR activation also help to guide the newly activated T cells back into the tissue where the innate immune system first sensed the foreign challenge. During secondary immune responses, chemokines induced by antigen-specific lymphocyte responses recruit innate immune cells into sites of inflammation, serving to amplify the adaptive response with innate immune effector cells. Thus, chemokines and their receptors serve a critical function to coordinate and link the interdependent innate and adaptive immune responses.

4.6.8 Detection of Toxins

The relationship in the intestinal mucosa between the GM1 ganglioside concentration and the number of binding sites for cholera toxin (CT) supported the concept that the GM1 is the intestinal binding receptor for CT [181]. While this binding provides a pathogenic mechanism to support the spread of *Vibrio cholera*, this activity initiates a robust innate response. Earlier studies have shown that CT induces IL-1 and IL-6 production [182, 183] by cultured epithelial cell lines. In addition, macrophages incubated with CT produce IL-1 [184]. CCL20, CCL25 and CCL28, which mediate mucosal immunity in part by recruiting DCs and IgA plasma cells (ASCs) into mucosal lamina propria, are expressed in response to CT stimulation of both small and large intestinal cells [185, 186]. This reveals a mechanism by which host cells can invoke innate responses that initiate the adaptive mucosal immune response to toxins. As a result, CT mixed with or conjugated to unrelated protein antigens, are known to enhance mucosal IgA and T cell adaptive responses to CT and unrelated antigens. While, CTB is responsible for binding the mucosal adjuvant activity is largely mediated by CT's, a subunit responsible for the ADP-ribosylating activity of CT and related toxins, since derivatives that bind GM1 but lack cAMP activity still induce mucosal humoral and cellular immunity [187].

It is proposed that Shiga toxins influence dysentery severity by inducing colonic vascular damage, which accounts for bloody stools, intestinal ischemia, and inflation of a polymorphonuclear intestinal compartment during the infectious process [188]. Many of these affects are due to Shiga toxin's ability to induce the secretion of significant amounts of chemokines [189, 190]. Indeed, children with dysentery caused by *Shigella dysenteriae* express elevated serum and mucosal IL-1β, IL-6, TNF-α, and IFN-γ [191]. The secretion of these cytokines appears to be regulated by host cell expression of glycolipid globotriaosylceramide (Gb-3), which serves as the Shiga toxin receptor [192, 193]. Like CT, Shiga toxin derivatives, e.g., its B subunit or nontoxic mutants harboring amino acid substitutions in the A subunit, possess adjuvant activity via activation of DCs [194].

Similarly, anthrax edema toxin is another AB-type toxin that binds to anthrax toxin receptors on target cells via its B subunit. Mucosal delivery of this toxin enhanced both systemic and mucosal immunity to co-administered antigens and resulted in high antigen-specific antibody and T helper cell cytokine responses [195]. An attenuated non-toxinogenic non-encapsulated *Bacillus anthracis* spore vaccine expressing mutant anthrax edema toxin, when mucosally administered is effective at providing protection against a lethal anthracis challenge (bacterial or toxin challenge) [196]. Taken together, these data provide important examples of how innate responses to mucosal toxins lead to adaptive immune response, namely antigen-specific secretory (s) IgA, which effectively neutralizes the attachment of toxins (e.g., cholera toxin, shiga toxin, etc.) to the surface epithelial cells.

4.7 Consequences of Recognition: Effector Mechanisms

The consequences of recognition by the innate immune system depend on the type of tissue and the recognition systems that are present. Constitutive innate immunity is an immediate response based on pre-synthesized antimicrobial factors and effector cells ready to be set into motion. Keratinocytes, epithelial cells, and several specialized cells (e.g., Paneth cells present in the crypts of the intestine, or cells in sweat and sebaceous glands) produce many antimicrobial peptides [197]. When a microbe comes in contact with the skin or mucous membranes, it is confronted by mechanical barriers and antimicrobial factors. If the microbe escapes these mechanisms and penetrates the skin or mucous membranes, it is phagocytozed and killed by phagocytic cells (neutrophils and macrophages) present in blood and tissues. The presence of microbes on the skin or mucous membranes or in tissues also triggers innate responses, which amplify and supplement the initial constitutive response. This response consists of several changes. There is an increased production of **antimicrobial peptides** by keratinocytes, epithelial cells, and specialized cells in the tissues; this action enhances the killing of microorganisms. There is secretion of cytokines, chemokines, and vasoactive products by macrophages, mast cells, eosinophils, basophils, and platelets. Complement is activated, which enhances phagocytosis (opsonization), chemotaxis, and direct killing of some microorganisms. Activation of clotting cascade and generation of **bradykinin** increases vascular permeability, vasodilation, hypotension, pain, and smooth muscle contraction.

In summary, the extent of the innate response is usually proportional to the need, i.e., the location and the extent of the infection. External tissues (skin and mucous membranes) constantly come in contact with potential immune stimulants (e.g., bacteria). The response is usually controlled by constitutively produced antimicrobial peptides. Therefore, only high numbers of microorganisms may induce mucosal immune responses, because no response is necessary when low numbers of microorganisms are present. By contrast, the immune cells in tissues that are normally sterile, especially macrophages and leukocyte, are highly sensitive to very low amounts of microbial products. The reason for this high responsiveness is the need for detection of low numbers of microorganisms in normally sterile blood and tissues, because any appearance of microorganisms in these tissues indicates that the first line of defense (skin and mucous membranes) is breached.

4.7.1 Type I IFNs and Adaptive Immunity

The type I interferon family consists of IFN-α, IFN-β, IFN-ω and IFN-λ [198]. Whereas the IFN-α gene family consists of many members, there is a single gene encoding IFN-β. Stimulation of many cell types with viruses, bacteria or TLR ligands, such as poly (I) · poly(C), LPS, CpG and imiquimod, results in the production of IFN-β and IFN-α [199]. Factors involved in the transcription of IFN-β have been well characterized [199] and involve a complex containing the interferon regulatory factor (IRF) family member IRF-3. Once secreted, IFN-β can bind to the type I interferon receptor, a heterodimeric molecule consisting of IFNAR-1 and IFNAR-2, to activate the Janus kinase (Jak) protein tyrosine kinases Jak1 and Tyk2. This step leads to the phosphorylation of signal transducer and activator of transcription 1 (STAT1) and STAT2. STAT1–STAT2-IRF-9 forms a complex known as interferon-stimulated gene factor 3 (ISGF3), which binds to interferon-stimulated responsive element (ISRE) and turns on genes such as those encoding 2′–5′-oligoadenylate. Thus, the initial IFN-β and IFN-α secretion induced by viral infection results in expression of more IFN-α gene members. In contrast, LPS stimulation induces mainly IFN-β production without inducing IFN-α.

Although the function of type I interferons is most closely associated with their antiviral activities, these cytokines also have diverse effector functions in the development of adaptive immunity that are not restricted to antiviral defense. Type I interferons promote the proliferation of memory T cells and prevent T cell apoptosis [200]. Type I interferons can activate STAT4 directly and induce IFN-γ secretion from $CD8^+$ T cells in mice [201] and $CD4^+$ T cells in humans [202, 203]. Further, cross-presentation of viral antigens occurs via a mechanism dependent on type I interferons. One study found that the activation of cross-priming by type I interferons was independent of $CD4^+$ T cell help but involved the direct stimulation of DCs [204]. IFN-α/β also enable B cells to undergo isotype switching and differentiation into plasma cells through the activation of DCs [205]. They are also critical activators of NK cells [206]. Finally, IFN-α/β induce DC maturation following stimulation via CpG, poly (I: C), or LPS treatment or viral infection [207, 208, 209].

Given the multitude of genes that are responsive to type I interferons, it is not unexpected that they are essential for diverse aspects of adaptive immunity. In this context, TLRs differ in their ability to induce IFN-α/β expression. TLR3, TLR4, TLR7 and TLR9 can all induce IFN-α/β production, albeit through different signaling pathways. TLRs 1, 2, 5 and 6 do not induce IFN-α/β. Because IFN-α/β functions are important in the control of adaptive immunity, it is curious that these particular TLRs do not induce their expression. It is tempting to speculate, therefore, that for any given pathogen, a combination of TLRs that becomes activated always includes at least one TLR that can induce the IFN response.

4.7.2 Complement

The complement system has many components (Tables 4.6 and 4.7), mediates a large number of biological effects, and interacts with other complex systems, such as blood clotting and adaptive immune responses. Complement derives its name from the original belief that it complements, or completes, the action of antibodies. Only later did researchers realize that it plays a crucial role in body defenses even in the absence of antibodies. The complement system is normally barely "on" and must be activated to become a significant part of the defense mechanisms [210, 211].

Table 4.6 Components of the complement system

Component	Role in the complement cascade
Classical pathway	
C1q	Bind to Fc region of Ig in antigen-antibody complexes; this binding leads to the activation of C1r
C1r	C1r is cleaved on activation to generate C1r, a serine protease that cleaves C1s
C1s	C1s is cleaved to produce, a serine protease; which, in turn, cleaves C4 and C2
C4	Cleaved by C1s into C4a and C4b; C4b binds to the surface membrane and becomes part of the C3 convertase
C2	Binds to C4b and cleaved by C1s into C2a, which is a serine protease component of the C3/C5 convertase.
C3 (also part of the lectin and alternative pathways)	Cleaved by C4bC2a into the anaphylatoxin C3a and C3b which is an opsonin and part of the C3/C5 convertase
Lectin pathway	
MBL	Binds to mannose-containing carbohydrates on bacterial and fungal cells, analogous to C1q in the classical pathway, associated with MASP-1 and MASP-2
MBL-associated proteases (MASP-1 and MASP-2)	Serine protease that activates C4 and C2 by cleaving them similar to C1s
Alternative pathway	
Factor B	Analogous to C2 in the classical pathway
Factor D	A serine protease that activates factor B by cleaving it
Properdin	Binds to the C3-C5 convertase of the alternative pathway and stabilizes it
Membrane attack complex	
C5	Cleaved by the convertase complex; C5a is an anaphylatoxin, C5b is the anchoring protein for C6
C6	Binds to C5b and this complex becomes the anchor for C7
C7	Binds to the C5bC6 complex and then C5bC6C7 inserts into the membrane and becomes an anchor for C8
C8	Attaches to C5bC6C7 and produces a stable membrane-associated complex that can bind C9
C9	Polymerizes at the site of the C5/C8 complex; this completes formation of the membrane attack complex.

Table 4.6 (continued)

Component	Role in the complement cascade
Complement receptors	
Complement receptor type 1 (CR1, CD35)	Accelerates dissociation of the C3 convertases, enhances phagocytosis of C3b-or C4b-coated microorganisms
Complement receptor type 2 (CR2, CD21)	Activate B cells
Complement receptor type 3 (CR3, CD11b/CD18)	Adhesion protein, important for phagocytosis of C3b-coated microbes.
Complement receptor type 4 (CR4, CD11c/CD18)	Important in phagocytosis of C3b-coated microorganisms
C3a receptor	Signals chemotaxis and release of vasoactive products from mast cells
C5a receptor	Signals chemotaxis and release of vasoactive products from mast cells

Once activated, it functions to enhance the antimicrobial defenses in several ways. Complement makes invading microorganisms susceptible to phagocytosis through opsonization. Activated complement lyses some microorganisms directly and/or activates antimicrobial systems of phagocytes. Certain complements cause the chemotaxis of white blood cells to promote the inflammatory response.

Table 4.7 Proteins that regulate complement activity

Component	Function
Soluble serum proteins	
C1 inhibitor	Inhibit enzymatic activity of C1r and C1s of the classical pathway and MASPs of the lectin pathway by suicide substrate mechanism.
C4 binding protein (C4bp)	Binds to C4b and enhance the decay of C4bC2a classical and lectin pathway C3 convertase
Factor H	Binds to C3b and enhance the decay of the alternative pathway C3 convertase
Factor I	Proteolytically inactivates C4b and C3b
Anaphylatoxin inactivator	Proteolytically inactivates the anaphylatoxins, C3a and C5a
S-Protein (vitronectin)	Inhibits insertion of the MAC into the lipid bilayer by binding to the C5b/C7 complex
Membrane proteins	
Complement receptor type-1 (CR1, CD35)	Accelerates dissociation of C3 convertases
Membrane cofactor protein	Cofactor for Factor I-mediated cleavage of C3b
Decay accelerating factor (DAF, CD55)	Accelerates dissociation of C3 convertases
Membrane inhibitor of reactive lysis (MIRL, CD59)	Bind to C9 and prevents MAC formation
Homologous restriction factor (HRF)	Blocks MAC insertion into the surface of cells

The complement system can be activated in one of three ways. The three activation pathways are known as **classical**, **lectin**, and **alternative**. Each pathway starts out separately, but the three eventually converge to make the same products. The classical pathway is usually set in motion by the presence of antigen-antibody complexes. The lectin and alternative pathways are elicited independent of antibodies and activated by microbial components like mannose-containing polysaccharides (lectin pathway) or bacterial lipopolysaccharides (alternative pathway). In either case, activation results from proteolytic cleavage of inert larger proteins.

Activation of the complement system is involved in several important aspects of host defenses. Two activities of complement are specifically directed toward enhancing phagocytosis, which is probably the most effective of the constitutive defenses against microorganisms. These activities are the recruitment of white cells by **chemotactic proteins**, such as C5a, and the facilitation of phagocytosis by proteins called **opsonins**. Complement activation also induces acute inflammation by generating **anaphylatoxins** (C3a and C5a), which act by causing release of vasoactive mediators from mast cells. This activity can be considered beneficial, insofar as the inflammatory response helps fight invading microorganisms.

Other components of complement make the so-called **membrane attack complex (MAC)**, which is responsible for the lysis of Gram-negative bacteria, enveloped viruses, and foreign cells. They may even lyse infected tissue cells that appear alien because they display viral or other foreign proteins in their cell membranes. This activity is particularly important in killing bacteria that resist phagocytosis, such as meningococci and gonococci. Indeed, genetic deficiencies of the proteins involved in the formation of the MAC result in increased susceptibility to infections by these particular organisms.

The three pathways of complement activation converge at a key biochemical step, the cleavage of component C3. From that point the remaining steps of all three pathways are the same. The enzymes responsible for this activity, **C3 convertases**, yield fragments C3a and C3b. Both components are pharmacologically active. C3a, which diffuses into the surrounding fluid, is an anaphylatoxin. C3b, which has an active thioester group that covalently binds to hydroxyl groups of carbohydrates or proteins on the antigen (e.g., bacterium), is an **opsonin**. C3b binds to **complement receptor type 1 (CRI)** on phagocytic cells and thus greatly enhances binding of the opsonized bacteria to phagocytic cells. C3b also binds to CR2 on B cells and enhances B cell activation and production of antibodies in the adaptive immune response. In addition, C3b becomes part of the C3 convertase of the alternative pathway, and a part of C5 convertase. The formation of this complex can have a direct impact on mucosal homeostasis. Inhibition of the C3 convertase has been shown to prevent ischaemia-reperfusion stress in the rat gastric mucosa [212].

4.7.3 C3 Convertase

Activation of complement by the classical pathway is usually initiated by the presence of antigen-antibody complexes. A protein complex called C1, composed of three proteins, C1q, Clr, and C1s, recognizes the immune complexes. Clq is made up of six subunits, each shaped like a tulip. One IgM molecule is sufficient for Clq binding, but two closely adjacent IgG molecules are needed. For this reason IgM is a very potent C activator, whereas only high concentrations of IgG can activate C. Clq binds only to antibodies bound to antigens, which prevents C activation by free antibodies always present in the serum, and ensures that only antigen-antibody complexes activate the classical pathway. Following binding to the antigen-antibody complex, C1q converts C1r into protease that carries out the next step in the pathway, the cleavage of C1s. The activated enzyme C1s in turn cleaves C2 and C4 in to fragment C4b2a, which bind covalently to the antigen and become the C3 convertase of the classical pathway and C4a and C2b fragments that are released.

Eventhough IgA has been shown to be a relatively weak activator of complement, respiratory SIgA has been shown to activate the alternative complement pathway [213]. Regardless of SIgA's role in complement activation, the role of the classical complement pathway in initiating mucosal adaptive response is well documented. IgA nephropathy or Bürger's disease is a form of renal disease characterized by IgA with C3 deposition in the mesangium of kidney [214]. Dermatitis herpetiformis [215] is a skin condition associated with celiac disease and is also associated with the deposition of IgA and C3.

Relatively low levels of diffusible complement activity and the C3 component of complement are present in the mucosa of the cervix, uterus, and fallopian tubes, whereas C1q, C1r, and C4 components are either absent or present only in trace quantities [216]. The fact that IgG and IgA are released in measurable quantities from these sites suggests that complement activation may be an important mediator of mucosal immunity of the reproductive tract. There is also evidence that complement components may form locally in inflamed mucosal tissues containing macrophages. Jejunal secretion of C4, but not C3 or factor B, parallel with the clinical activity of IBD [217]. Indeed, the increased secretion of complement by clinically unaffected jejunal tissue in patients with Crohn's disease suggests the synthesis of complement by activated intestinal macrophages. Human IgG1 and activated complement are co-localized along the colonic epithelium in active ulcerative colitis [218]. Granular deposition of C3b and terminal complement complex were observed at the luminal face of the surface epithelium in 10 of 18 patients with colitis suggesting complement activation takes place at the luminal face of the epithelium. Factors regulating proximal small-intestinal luminal concentrations of human IgG3 are not well established, but the intestinal mucosa is immunoreactive for C4 as C4 significantly correlates with IgG3 concentrations in proximal small-intestinal luminal and caecal secretions.

The lectin pathway is activated by binding of **mannose-binding-lectin (MBL)**, which is a mannose-containing polysaccharide, and a C1q analog, to mannose- containing polysaccharides, frequently present in the cell walls of many bacteria and fungi. The result is activation and complexing of MBL with the MBL associated proteases **MASP-1** and **MASP-2**, which then cleave C4 and C2 to generate C4b2a, which is the same **C3 convertase** as in the classical pathway. Interestingly, MBL deficiency is associated with diarrhea. To this end, individuals homozygous for MBL mutations have an increased risk of cryptosporidiosis and lectin-mediated binding of MBL to activate C4 that was detected on sporozoites, but not oocytes [219, 220]. MBL and C3, but not C1q, present in the vaginal cavity help to recognize infectious agents (e.g., bacteria and fungi) that colonize the vaginal mucosa [221] suggesting lectin- and classical-mediated pathways of complement activation may play a role in host defense in the genital tract.

4.7.4 Complement: Chemotaxins and Opsonins

Several of the complement components are important chemoattractants that can mobilize cells of both innate and adaptive immune systems. For example, C5a is an attractant for neutrophils and monocytes. While phagocytic cells alone can engulf bacteria, without enhancing cofactors this process is inefficient. In the presence of complement **opsonins**, neutrophils and macrophages will phagocytose pathogens more efficiently. Several substances normally serve as opsonins, including the C3b component of complement. C3b binds to the surface of bacteria and forms a ligand that is recognized by receptors on neutrophils, monocytes, and macrophages. Microorganisms coated with C3b become anchored to the surface of phagocytes, which facilitates their uptake. Three receptors for C3b and its various cleavage products are located on white blood cells: CR1, CR3 (CD11b/CD18), and CR4 (CD11c/CD18). While CR1 is localized on the ciliary surface of epithelium in both normal subjects and patients with chronic sinusitis, CR1-binding to C3b is associated with chronic sinusitis [222]. Similarly, elevated CR1 expression by granulocytes in the bronchial mucosa is involved in prolonged inflammatory responses in asthma and bronchial hyperresponsiveness [223]. Interestingly, intestinal macrophages, which are believed to orchestrate mucosal inflammatory responses, do not express CR3 or CR4 [224]. These findings suggest peripheral monocytes recruited to the intestinal mucosa eventually become anergic to inflammatory stimuli to presumably limit responses to potentially immunostimulatory commensal bacteria.

4.8 Linking Innate and Adaptive Immunity

The innate immune response not only provides the first line of defense against microorganisms, but also provides the "danger signal" – that instructs the adaptive immune system to mount a response [225]. On the other hand, adaptive immune response calls on the innate immune system to provide the professional phagocytes (e.g. macrophages and neutrophils) and specialized granulocytes (e.g. eosinophils and basophils) necessary to engulf small pathogens and contain larger parasites. The primary adaptive immune response takes place in the draining lymph nodes and not in the tissue itself. Antigen is taken up by dendritic cells in the mucosa and carried into regional lymph nodes (i.e., inductive sites), where the DCs activate naive T and

B cells. Activated, T and B cells then leave the lymph node and find their way back to the site of insult or inflammation in the tissue to carry out their effector function.

Although the trafficking patterns of DCs and lymphocytes have been appreciated for some time, the molecules that control their movement in vivo have only recently been identified as members of the chemokine family. These chemotactic cytokines orchestrate the trafficking of DCs as well as T and B cells needed to generate an adaptive immune response. By activating TLRs, pathogens stimulate the local release of chemokines. In general, immature DCs constitutively express CCR1, CCR5 and CCR6 that mediate their attraction to sites of lymphocyte activation. However, the immature DCs are inefficient at picking-up antigen and must mature and differentiate into cells that can activate naïve T cells. During this process, the DC down regulates expression of CCR1, CCR5 and CCR6 and up-regulates expression of CCR7, causing its migration into the afferent lymphatic system to eventually migrate to mucosal inductive sites or organs. The CCR7 ligands, CCL19 and CCL21, are expressed by lymphatic endothelium and direct the migration of the antigen-loaded mature DCs.

Chemokines are also involved in recruiting naïve T and B cells from the periphery via high endothelial venules (HEVs) into the inductive sites and into contact with activated DCs. The molecular details remain to be elucidated, but it is likely that chemokines such as CCL18 and CCL19 are important in juxta-positioning these cells in lymph nodes. CCL19 and CCL21 are produced by stromal cells in the T-cell zone and CXCL13 is secreted by stromal cells in B cell follicles, helping to guide cells to both T and B cell areas, respectively. Indeed, some activated T cells down regulate expression of CCR7 and up regulate CXCR5 to be better directed to follicles to deliver help to B cells, whereas other activated effector T cells up regulate CXCR3 and are attracted into mucosal tissues (i.e., effector sites.) T cells that are activated in regional lymph nodes, after encountering antigen-loaded DCs, subsequently return to sites of inflammation by sensing chemokine gradients. CXCR3 ligands, CXCL9, CXCL10 and CXCL11 are highly expressed by activated Th1 cells and are believed to be important in adaptive immunity and mucosal inflammation.

4.8.1 Chemokines that Lead to Th1-Driven Responses

The chemokine system has emerged, as an innate response that is crucial to regulate DC and lymphocyte trafficking that is needed to initiate the mucosal adaptive immune response as well as inflammation. As mentioned above, chemokines also link activation of the Th1 adaptive immunity and the innate immune system. Chemokines are released from CTLs after antigen-specific activation. Indeed, CTLs specific for HIV-1, hepatitis B virus (HBV) or HCV release large amounts of CCL3, CCL4, and CCL5 from their cytotoxic granules [226]. These chemokines adhere to infected cells and recruit other inflammatory cells, including professional phagocytes, to the site of viral replication, amplifying the response to include innate immune effector cells.

Th1 and Th2 cells were found to express different sets of chemokine receptors, enabling the differential regulation of their trafficking and localization during different types of immune responses [227, 228]. Th1 cells preferentially express CCR5, CXCR3, and CCR1. This is similar to the expression pattern found on monocytes and macrophages, which allows them to be localized in inflammatory sites and co-localize with Th1 cells. Th1, but not Th2, cells also express CCR7 and home to the periarteriolar lymphoid sheathes. Retroviral transduction of Th2 cells with CCR7 forces them to localize in a Th1-like pattern and modulates their function in vivo [229].

The first hint of possible chemokine involvement in T cell polarization and the progression of Th1 type diseases was that the expression of some chemokines, such as CXCL10 and CXCL9, are induced by a Th1 cytokine, i.e., IFN-γ [230, 231]. CXCL11 is also expressed in response to IFN-γ [232]. These three chemokines share CXCR3 as their major receptor [233], which is expressed by virtually all Th1 cells. CCL4 is a chemokine that binds CCR5, another Th1 cell-expressed chemokine receptor. Expression of CCL4 is not induced by IFN-γ, but is inhibited by a Th2 cytokine IL-4 in LPS-activated monocytes [234]. IFN-γ also greatly increases the expression of CCL2 that can bind CCR5, which is preferentially express by Th1 cells [235]. Interestingly, CCL3, CCL4, and CCL5 are produced by Th1, but not Th2, cells [236]. CCL5 production is induced by TNF-α and IFN-γ in airway smooth muscle cells and is partially inhibited by Th2-derived cytokines, IL-4, IL-10, IL-13, and by dexamethasone [237]. In addition to CXCR3 and CCR5, CXCR6, the receptor for CXCL16, is also preferentially expressed by a subset of Th1 cells [238]. Taken together, these findings suggest that chemokines are important for directing and amplifying Th1 polarization and expansion, respectively.

Lillard et al., 2001 demonstrated that nasal immunization of mice with chicken egg ovalbumin (OVA) and CCL5 enhanced OVA-specific antibody as well as $CD4^+$ T cell proliferative and cytokine responses in both systemic and mucosal compartments. Our results show that both Th1- and Th2-type pathways can be induced by mucosally administered CCL5 even though this molecule primarily induced Th1-type responses. Interestingly, CCL5 enhances the expression of CD28, CD40L, and CD80 to mediate T cell responses [89]. Palaniappan et al., 2006 also demonstrated that CCL5 blockade resulted in a decrease of $CD4^+$ and $CD8^+$ T cells as well as $CD11b^+$ cells in mucosal and systemic compartments during the recognition phase of the pneumococcal adaptive immune response [239]. CCL5 blockade significantly reduced the antigen-specific IgG2a and IgG1 Abs in serum and IgA antibody levels in nasal washes, corresponding to reductions in antigen-specific T cell (mucosal and systemic) responses. CCL5 inhibition decreased the quantity of IL-4- and IFN-γ-secreting CD4 T cells and increased the number of antigen-specific IL-10-producing CD4 T cells. When combined, this also corresponded with the transition from pneumococcal carriage to lethal pneumonia. These data suggest that CCL5 is an essential factor for the induction and maintenance of protective pneumococcal immunity. Lillard et al., 2003 showed that immunization of mice with OVA plus CCL3 or CCL4, which bind CCR5, enhances antigen-specific serum and mucosal IgG2a and IgA, respectively, as well as antibody and $CD4^+$ T cell proliferation and

Th1 cytokine responses in systemic and mucosal compartments [91]. The cytokine responses promoted by these chemokines, particularly CCL3, can induce significant antigen-specific cytotoxic T lymphocyte (CTL) responses in both systemic and mucosal tissues and can up-regulate CD28, CD40L, and CD80 expression.

Several IBD models indicate $CD4^+$ T cells are major components in the induction of inflammation and much of the intestinal damage that this disease causes [240]. Singh, et al., 2003 have shown that CXCL9, CXCL10, and CXCL11 are up-regulated at sites of colitis [241], and the serum levels of CXCL9, CXCL10, and CXCL11 are significantly increased in Crohn's disease patients when compared to normal healthy donors, regardless of ethnicity (i.e., African-, European/Caucasian-, Ashkenazi Jewish-, and Asian-American). CXCR3-expressing T cells have been shown to produce predominantly Th1 cytokines, supporting the hypothesis that CXCR3 ligands selectively mobilize Th1 cells and leukocytes to support mucosal inflammation. Indeed, studies from our laboratory demonstrated that CXCL10 blockade abrogated spontaneous colitis in IL-$10^{-/-}$ mice mediated predominantly by Th1 $TCR\alpha\beta^+$ $CXCR3^+$ cells [242].

4.8.2 Chemokines that Lead to Th2-Driven Responses

Th2 effector cells secrete IL-4 and IL-13, and amplify Th2 cell recruitment by inducing the release of STAT6-inducible chemokines. Interestingly, many of these same chemokines are also active on eosinophils, basophils and mast cells. In this way, Th2 cells control the trafficking of eosinophils, basophils and perhaps mast cells into sites of allergic inflammation. Human Th2 cells preferentially express the CCR3, CCR4, and CCR8 and migrate to their ligands: CCL1, CCL11, CCL17, CCL22, CCL24 [243, 244]. Endobronchial biopsies from six asthmatics, taken 24 hours after allergen challenge, demonstrates that virtually all T cells express IL-4 after stimulation and are responsive to these Th2-associated chemokines. Similarly, the cytokine milieu in the nasal mucosa of patients with active Wegener's granulomatosis are associated with $CCR3^+$ IL-4^+ Th2 cells [245] that respond to CCL17 [246]. CCR3 is also expressed by eosinophils and mast cells to allow recruitment to the skin, lung and nasal tract during in the development of mucosal hyper-responsiveness [247, 248]. This chemokine axis with support of IL-13 and IL-4 secretion by a number of host cells following Th2-inducing agents (e.g., *Mycobacterium bovis, Schistosoma mansoni* egg antigens, etc.) initiates responses required to remove large extracellular pathogens [249, 250]. The ability of certain chemokines to induce Th2 adaptive responses has perhaps influenced the evolution of viral chemokines that target Th2-associated chemokine receptors. In this regard, viral macrophage-inflammatory protein-II (MIP-II) is a viral chemokine homologue of CCL2 that is encoded by human herpes virus 8. vMIP-II is selectively chemotactic for Th2 lymphocytes. Singh et al., 2004 have shown that CCL1, CCL2, and vMIP-II significantly enhanced antigen-specific serum and mucosal mouse IgG1 antibodies by increasing Th2 cytokine (e.g., IL-10) secretion by $CD4^+$ T cells [251].

Although CCL1 increased the number of CD28-, CD40L-, and CD30-positive, antigen-stimulated naïve T helper cells, vMIP-II and CCL1 decreased the number of CD28-, CD40L-, and CD30-positive, resting naive CD4+ T cells. Taken together, these studies suggest that CCR8 ligands (I-309) and its mouse counterparts T cell activation gene 3 (TCA-3) generated by the innate response direct Th2 adaptive mucosal responses and vMIP-II limits costimulatory molecule expression to mitigate host immunity.

It has also been shown that chemokines, like defensins, can mediate anti-microbial clearance of microbes. The Th2-associated chemokine, CCL20 binds to and activates CCR6. While CCL20 does not share the binding site of CCR6 with any other chemokine, β-defensin-1 and -2 (hBD-1 and hBD-2), small cationic antimicrobial peptides, have also been found to bind to and activate CCR6. Conversely, it was found that CCL20 possesses antibacterial activity of greater potency than the β-defensins against *Escherichia coli* and *Staphylococcus aureus* [252]. CCL20 and HBD-2 possess antibacterial activity against *Streptococcus mutans* and *Lactobacillus casei*, which are involved in caries [253]. While CXCL9, CXCL10 and CXCL11, attract $CXCR3^+$ T cells, B cells, and NK cells, all three chemokines possess antimicrobial activities [254]. Hence, chemokines are versatile factors that can mediate leukocyte migration and activation as well as killing of bacteria.

4.9 Conclusion

It is apparent that innate immunity is not a static host defense system without specificity per se. The adaptive and innate immune system(s) are intimately integrated and connected to better handle microbial invaders. Innate immunity is a complex and dynamic web of microbial and other antigenic recognition systems that are specific for various microorganisms and can swiftly generate various antimicrobial responses precisely tailored to the microbial invader that challenges the host. Moreover, innate immunity is intricately connected with adaptive immunity because it serves as the initial recognition system for antigens and is required for efficient initiation of adaptive immune responses. It also often directs which type of adaptive immune response is induced and in some cases serves as the effector mechanism of adaptive immunity.

The adaptive immune system builds and amplifies upon the defensive foundation established by the innate immune system. It adds several defensive features including: the random generation of antigen-specific receptors with far greater variety than is available in the innate system and the ability to intensely focus a variety of destructive mechanisms (including those of the innate system) to narrowly defined targets. The property of memory allows the adaptive system to modify its response to stimuli that are encountered repeatedly. In concert with the innate immune system, the adaptive immunity provides a versatile and coordinated set of defensive responses capable of dealing with the invasive styles of a broad array of mucosal pathogens and toxins.

References

1. Brenner S, Milstein C. 1966. Origin of antibody variation. *Nature* 211: 242–3
2. Zheng B, Xue W, Kelsoe G. 1994. Locus-specific somatic hypermutation in germinal centre T cells. *Nature* 372: 556–9
3. Tjoa B, Kranz DM. 1992. Diversity of T cell receptor-alpha chain transcripts from hyperimmune alloreactive T cells. *Journal of Immunology* 149: 253–9
4. Nagler-Anderson C. 2001. Man the barrier! Strategic defences in the intestinal mucosa. *Nature Reviews. Immunology* 1: 59–67
5. Renes IB, Verburg M, Van Nispen DJ, Buller HA, Dekker J, Einerhand AW. 2002. Distinct epithelial responses in experimental colitis: implications for ion uptake and mucosal protection. *American Journal of Physiology – Gastrointestinal & Liver Physiology* 283: G169–79
6. Kelly CP, O'Keane JC, Orellana J, Schroy PC, 3rd, Yang S, LaMont JT, Brady HR. 1992. Human colon cancer cells express ICAM-1 in vivo and support LFA-1-dependent lymphocyte adhesion in vitro. *American Journal of Physiology* 263: G864–70
7. McNabb PC, Tomasi TB. 1981. Host defense mechanisms at mucosal surfaces. *Annual Review of Microbiology* 35: 477–96
8. Belley A, Keller K, Gottke M, Chadee K. 1999. Intestinal mucins in colonization and host defense against pathogens.[erratum appears in Am J Trop Med Hyg 1999 Jun;60(6):1062 Note: Goettke M [corrected to Gottke M]]. *American Journal of Tropical Medicine & Hygiene* 60: 10–5
9. Deplancke B, Gaskins HR. 2001. Microbial modulation of innate defense: goblet cells and the intestinal mucus layer. *American Journal of Clinical Nutrition* 73: 1131S–41S
10. Sands BE, Podolsky DK. 1996. The trefoil peptide family. *Annual Review of Physiology* 58: 253–73
11. Podolsky DK. 1997. Lessons from genetic models of inflammatory bowel disease. *Acta Gastroenterologica Belgica* 60: 163–5
12. Walker WA. 1975. Antigen absorption from the small intestine and gastrointestinal disease. *Pediatric Clinics of North America* 22: 731–46
13. Walker WA, Isselbacher KJ. 1974. Uptake and transport of macromolecules by the intestine. Possible role in clinical disorders. *Gastroenterology* 67: 531–50
14. Kuvaeva IB. 1979. Permeability of the gastronintestinal tract for macromolecules in health and disease. *Human Physiology* 4: 272–83
15. Weaver LT, Laker MF, Nelson R. 1984. Intestinal permeability in the newborn. *Archives of Disease in Childhood* 59: 236–41
16. Vukavic T. 1984. Timing of the gut closure. *Journal of Pediatric Gastroenterology and Nutrition* 3: 700–3
17. Ganz T. 2003. Defensins: antimicrobial peptides of innate immunity. *Nature reviews. Immunology* 3: 710–20
18. Bellamy W, Takase M, Wakabayashi H, Kawase K, Tomita M. 1992. Antibacterial spectrum of lactoferricin B, a potent bactericidal peptide derived from the N-terminal region of bovine lactoferrin. *Journal of Applied Bacteriology* 73: 472–9
19. Kang JH, Lee MK, Kim KL, Hahm KS. 1996. Structure-biological activity relationships of 11-residue highly basic peptide segment of bovine lactoferrin. *International Journal of Peptide & Protein Research* 48: 357–63
20. Porter EM, Liu L, Oren A, Anton PA, Ganz T. 1997. Localization of human intestinal defensin 5 in Paneth cell granules. *Infection and Immunity* 65: 2389–95
21. Porter EM, van Dam E, Valore EV, Ganz T. 1997. Broad-spectrum antimicrobial activity of human intestinal defensin 5. *Infection & Immunity* 65: 2396–401
22. Quayle AJ., Porter EM, Nussbaum AA, Wang YM, Brabec C, Yip KP, Mok SC. 1998. Gene expression, immunolocalization, and secretion of human defensin-5 in human female reproductive tract. *American Journal of Pathology* 152: 1247–58

23. Zhao X, Deak E, Soderberg K, Linehan M, Spezzano D, Zhu J, Knipe DM, Iwasaki A. 2003. Vaginal submucosal dendritic cells, but not Langerhans cells, induce protective Th1 responses to herpes simplex virus-2. *Journal of Experimental Medicine* 197: 153–62
24. Singh PK, Jia HP, Wiles K, Hesselberth J, Liu L, Conway BA, Greenberg EP, Valore EV, Welsh MJ, Ganz T, Tack BF, McCray PB. Jr. 1998. Production of beta-defensins by human airway epithelia. *Proceedings of the National Academy of Sciences of the United States of America*. 95: 14961–6
25. Mathews M, Jia HP, Guthmiller JM, Losh G, Graham S, Johnson GK, Tack BF, McCray PB, Jr. 1999. Production of beta-defensin antimicrobial peptides by the oral mucosa and salivary glands. *Infection & Immunity* 67: 2740–5
26. Jones DE, Bevins CL. 1992. Paneth cells of the human small intestine express an antimicrobial peptide gene. *Journal of Biological Chemistry* 267: 23216–25
27. Eisenhauer PB, Harwig SS, Lehrer RI. 1992. Cryptdins: antimicrobial defensins of the murine small intestine. *Infection and Immunity* 60: 3556–65
28. Chertov O, Michiel DF, Xu L, Wang JM, Tani K, Murphy WJ, Longo DL, Taub DD, Oppenheim JJ. 1996. Identification of defensin-1, defensin-2, and CAP37/azurocidin as T-cell chemoattractant proteins released from interleukin-8-stimulated neutrophils. *Journal of Biological Chemistry* 271: 2935–40
29. Yang D, Chertov O, Bykovskaia SN, Chen Q, Buffo MJ, Shogan J, Anderson M, Schroder JM, Wang JM, Howard OM, Oppenheim JJ. 1999. Beta-defensins: linking innate and adaptive immunity through dendritic and T cell CCR6. *Science* 286: 525–8
30. Ganz T. 1999. Defensins and host defense.[comment]. *Science* 286: 420–1
31. Territo MC, Ganz T, Selsted ME, Lehrer R. 1989. Monocyte-chemotactic activity of defensins from human neutrophils. *Journal of Clinical Investigation* 84: 2017–20
32. Lillard JW, Jr., Boyaka PN, Chertov O, Oppenheim JJ, McGhee JR. 1999. Mechanisms for induction of acquired host immunity by neutrophil peptide defensins. *Proceedings of the National Academy of Sciences of the United States of America* 96: 651–6
33. Milligan GN, Bourne N, Dudley KL. 2001. Role of polymorphonuclear leukocytes in resolution of HSV-2 infection of the mouse vagina. *Journal of Reproductive Immunology* 49: 49–65
34. Schaller M, Boeld U, Oberbauer S, Hamm G, Hube B, Korting HC. 2004. Polymorphonuclear leukocytes (PMNs) induce protective Th1-type cytokine epithelial responses in an in vitro model of oral candidosis. *Microbiology* 150: 2807–13
35. Lewis DC, Walker-Smith JA, Phillips AD. 1987. Polymorphonuclear neutrophil leucocytes in childhood Crohn's disease: a morphological study. *Journal of Pediatric Gastroenterology & Nutrition* 6: 430–8
36. Cameron L, Christodoulopoulos P, Lavigne F, Nakamura Y, Eidelman D, McEuen A, Walls A, Tavernier J, Minshall E, Moqbel R, Hamid Q. 2000. Evidence for local eosinophil differentiation within allergic nasal mucosa: inhibition with soluble IL-5 receptor. *Journal of Immunology* 164: 1538–45
37. Denburg JA, Otsuka H, Ohnisi M, Ruhno J, Bienenstock J, Dolovich J. 1987. Contribution of basophil/mast cell and eosinophil growth and differentiation to the allergic tissue inflammatory response. *International Archives of Allergy and Applied Immunology* 82: 321–6
38. Mullol J, Xaubet A, Lopez E, Roca-Ferrer J, Carrion T, Rosello-Catafau J, Picado C. 1997. [Eosinophil activation by epithelial cells of the respiratory mucosa. Comparative study of normal mucosa and inflammatory mucosa]. *Medicina clínica (Barc)* 109: 6–11
39. Rydstrom A, Wick MJ. 2007. Monocyte recruitment, activation, and function in the gut-associated lymphoid tissue during oral Salmonella infection. *Journal of Immunology* 178: 5789–801
40. Rugtveit J, Bakka A, Brandtzaeg P. 1997. Differential distribution of B7.1 (CD80) and B7.2 (CD86) costimulatory molecules on mucosal macrophage subsets in human inflammatory bowel disease (IBD). *Clinical and Experimental Immunology* 110: 104–13

41. Rugtveit J, Brandtzaeg P, Halstensen TS, Fausa O, Scott H. 1994. Increased macrophage subset in inflammatory bowel disease: apparent recruitment from peripheral blood monocytes. *Gut* 35: 669–74
42. Ohtoshi T, Vancheri C, Cox G, Gauldie J, Dolovich J, Denburg JA, Jordana M. 1991. Monocyte-macrophage differentiation induced by human upper airway epithelial cells. *American Journal of Respiratory Cell and Molecular Biology* 4: 255–63
43. Reis e Sousa C. 2001. Dendritic cells as sensors of infection. *Immunity* 14: 495–8
44. Medzhitov R, Preston-Hurlburt P, Janeway CA, Jr. 1997. A human homologue of the Drosophila Toll protein signals activation of adaptive immunity.[see comment]. *Nature* 388: 394–7
45. Bouchon A, Hernandez-Munain C, Cella M, Colonna M. 2001. A DAP12-mediated pathway regulates expression of CC chemokine receptor 7 and maturation of human dendritic cells. *Journal of Experimental Medicine* 194: 1111–22
46. Regnault A, Lankar D, Lacabanne V, Rodriguez A, Théry C, Rescigno M, Saito T, Verbeek S, Bonnerot C, Ricciardi-Castagnoli P, Amigorena S. 1999. Fc R-mediated induction of dendritic cell maturation and MHC classI–restricted antigen presentation after immune complex internalization. *Journal of Experimental Medicine* 189: 371–80
47. Piccioli D, Sbrana S, Melandri E, Valiante NM. 2002. Contact-dependent stimulation and inhibition of dendritic cells by natural killer cells. *Journal of Experimental Medicine* 195: 335–41
48. Gerosa F, Baldani-Guerra B, Nisii C, Marchesini V, Carra G, Trinchieri G. 2002. Reciprocal activating interaction between natural killer cells and dendritic cells. *Journal of Experimental Medicine* 195: 327–33
49. Cella M, Scheidegger D, Palmer-Lehmann K, Lane P, Lanzavecchia A, Alber G. 1996. Ligation of CD40 on dendritic cells triggers production of high levels of interleukin-12 and enhances T cell stimulatory capacity: T-T help via APC activation. *Journal of Experimental Medicine* 184: 747–52
50. Cella M, Sallusto F, Lanzavecchia A. 1997. Origin, maturation and antigen presenting function of dendritic cells. *Current Opinion in Immunology* 9: 10–6
51. Cumberbatch M, Kimber I. 1995. Tumour necrosis factor-alpha is required for accumulation of dendritic cells in draining lymph nodes and for optimal contact sensitization. *Immunology* 84: 31–5
52. Melchers F, Rolink AG, Schaniel C. 1999. The role of chemokines in regulating cell migration during humoral immune responses. *Cell* 99: 351–4
53. Anderson JM, Van Itallie CM. 1999. Tight junctions: closing in on the seal. *Current Biology* 9: R922–4
54. Youakim A, Ahdieh M. 1999. Interferon-gamma decreases barrier function in T84 cells by reducing ZO-1 levels and disrupting apical actin. *American Journal of Physiology* 276: G1279–88
55. Kucharzik T, Walsh SV, Chen J, Parkos CA, Nusrat A. 2001. Neutrophil transmigration in inflammatory bowel disease is associated with differential expression of epithelial intercellular junction proteins. *American Journal of Pathology* 159: 2001–9
56. Dignass AU, Podolsky DK. 1993. Cytokine modulation of intestinal epithelial cell restitution: central role of transforming growth factor beta. *Gastroenterology* 105: 1323–32
57. Dignass A, Lynch-Devaney K, Kindon H, Thim L, Podolsky DK. 1994. Trefoil peptides promote epithelial migration through a transforming growth factor beta-independent pathway. *Journal of Clinical Investigation* 94: 376–83
58. Dogan A, Wang ZD, Spencer J. 1995. E-cadherin expression in intestinal epithelium. *Journal of Clinical Pathology* 48: 143–6
59. Huang GT, Eckmann L, Savidge TC, Kagnoff MF. 1996. Infection of human intestinal epithelial cells with invasive bacteria upregulates apical intercellular adhesion molecule-1 (ICAM)-1 expression and neutrophil adhesion. *Journal of Clinical Investigation* 98: 572–83

60. Kvale D, Krajci P, Brandtzaeg P. 1992. Expression and regulation of adhesion molecules ICAM-1 (CD54) and LFA-3 (CD58) in human intestinal epithelial cell lines. *Scandinavian Journal of Immunology* 35: 669–76
61. Morales VM, Christ A, Watt SM, Kim HS, Johnson KW, Utku N, Texieira AM, Mizoguchi A, Mizoguchi E, Russell GJ, Russell SE, Bhan AK, Freeman GJ, Blumberg RS. 1999. Regulation of human intestinal intraepithelial lymphocyte cytolytic function by biliary glycoprotein (CD66a). *Journal of Immunology* 163: 1363–70
62. Parkos CA, Colgan SP, Diamond MS, Nusrat A, Liang TW, Springer TA, Madara JL. 1996. Expression and polarization of intercellular adhesion molecule-1 on human intestinal epithelia: consequences for CD11b/CD18-mediated interactions with neutrophils. *Molecular Medicine* 2: 489–505
63. Yio XY, Mayer L. 1997. Characterization of a 180-kDa intestinal epithelial cell membrane glycoprotein, gp180. A candidate molecule mediating t cell-epithelial cell interactions. *Journal of Biological Chemistry* 272: 12786–92
64. Colgan SP, Hershberg RM, Furuta GT, Blumberg RS. 1999. Ligation of intestinal epithelial CD1d induces bioactive IL-10: critical role of the cytoplasmic tail in autocrine signaling. *Proceedings of the National Academy of Sciences of the United States of America* 96: 13938–43
65. Steinle A, Groh V, Spies T. 1998. Diversification, expression, and gamma delta T cell recognition of evolutionarily distant members of the MIC family of major histocompatibility complex class I-related molecules. *Proceedings of the National Academy of Sciences of the United States of America* 95: 12510–5
66. Reinecker HC, MacDermott RP, Mirau S, Dignass A, Podolsky DK. 1996. Intestinal epithelial cells both express and respond to interleukin 15. *Gastroenterology* 111: 1706–13
67. Varilek GW, Neil GA, Bishop WP. 1994. Caco-2 cells express type I interleukin-1 receptors: ligand binding enhances proliferation. *American Journal of Physiology* 267: G1101–7
68. Colgan SP, Resnick MB, Parkos CA, Delp-Archer C, McGuirk D, Bacarra AE, Weller PF, Madara JL. 1994. IL-4 directly modulates function of a model human intestinal epithelium. *Journal of Immunology* 153: 2122–9
69. Reinecker HC, Podolsky DK. 1995. Human intestinal epithelial cells express functional cytokine receptors sharing the common gamma c chain of the interleukin 2 receptor. *Proceedings of the National Academy of Sciences of the United States of America* 92: 8353–7
70. Yamada K, Shimaoka M, Nagayama K, Hiroi T, Kiyono H, Honda T. 1997. Bacterial invasion induces interleukin-7 receptor expression in colonic epithelial cell line, T84. *European Journal of Immunology* 27: 3456–60
71. Panja A, Goldberg S, Eckmann L, Krishen P, Mayer L. 1998. The regulation and functional consequence of proinflammatory cytokine binding on human intestinal epithelial cells. *Journal of Immunology* 161: 3675–84
72. Fish SM, Proujansky R, Reenstra WW. 1999. Synergistic effects of interferon gamma and tumour necrosis factor alpha on T84 cell function. *Gut* 45: 191–8
73. Awane M, Andres PG, Li DJ, Reinecker HC. 1999. NF-kappa B-inducing kinase is a common mediator of IL-17-, TNF-alpha-, and IL-1 beta-induced chemokine promoter activation in intestinal epithelial cells. *Journal of Immunology* 162: 5337–44
74. Denning TL, Campbell NA, Song F, Garofalo RP, Klimpel GR, Reyes VE, Ernst PB. 2000. Expression of IL-10 receptors on epithelial cells from the murine small and large intestine. *International Immunology* 12: 133–9
75. Delezay O, Koch N, Yahi N, Hammache D, Tourres C, Tamalet C, Fantini J. 1997. Coexpression of CXCR4/fusin and galactosylceramide in the human intestinal epithelial cell line HT-29. *AIDS* 11: 1311–8
76. Dwinell MB, Eckmann L, Leopard JD, Varki NM, Kagnoff MF. 1999. Chemokine receptor expression by human intestinal epithelial cells. *Gastroenterology* 117: 359–67
77. Jordan NJ, Kolios G, Abbot SE, Sinai MA, Thompson DA, Petraki K, Westwick J. 1999. Expression of functional CXCR4 chemokine receptors on human colonic epithelial cells. *Journal of Clinical Investigation* 104: 1061–9

78. Brumell JH, Steele-Mortimer O, Finlay BB. 1999. Bacterial invasion: Force feeding by Salmonella. *Current Biology* 9: R277–80
79. Galan JE. 2001. Salmonella interactions with host cells: type III secretion at work. *Annual Review of Cell & Developmental Biology* 17: 53–86
80. Kaisho T, Akira S. 2000. Critical roles of Toll-like receptors in host defense. *Critical Reviews in Immunology* 20: 393–405
81. Cario E, Rosenberg IM, Brandwein SL, Beck PL, Reinecker HC, Podolsky DK. 2000. Lipopolysaccharide activates distinct signaling pathways in intestinal epithelial cell lines expressing Toll-like receptors. *Journal of Immunology* 164: 966–72
82. Gewirtz AT, Simon PO, Jr., Schmitt CK, Taylor LJ, Hagedorn CH, O'Brien AD, Neish AS, Madara JL. 2001. Salmonella typhimurium translocates flagellin across intestinal epithelia, inducing a proinflammatory response.[see comment]. *Journal of Clinical Investigation* 107: 99–109
83. Gewirtz AT, Navas TA, Lyons S, Godowski PJ, Madara JL. 2001. Cutting edge: bacterial flagellin activates basolaterally expressed TLR5 to induce epithelial proinflammatory gene expression. *Journal of Immunology* 167: 1882–5
84. Marinaro M, Boyaka PN, Jackson RJ, Finkelman FD, Kiyono H, Jirillo E, McGhee JR. 1999. Use of intranasal IL-12 to target predominantly Th1 responses to nasal and Th2 responses to oral vaccines given with cholera toxin. *Journal of Immunology* 162: 114–21
85. Kelner GS, Kennedy J, Bacon KB, Kleyensteuber S, Largaespada DA, Jenkins NA, Copeland NG, Bazan JF, Moore KW, Schall TJ, et al. 1994. Lymphotactin: a cytokine that represents a new class of chemokine. *Science* 266: 1395–9
86. Hedrick JA, Zlotnik A. 1997. Lymphotactin: a new class of chemokine. *Methods in Enzymology* 287: 206–15
87. Lillard JW, Jr., Boyaka PN, Hedrick JA, Zlotnik A, McGhee JR. 1999. Lymphotactin acts as an innate mucosal adjuvant. *Journal of Immunology* 162: 1959–65
88. Fujihashi K, McGhee JR, Kweon MN, Cooper MD, Tonegawa S, Takahashi I, Hiroi T, Mestecky J, Kiyono H. 1996. gamma/delta T cell-deficient mice have impaired mucosal immunoglobulin A responses. *Journal of Experimental Medicine* 183: 1929–35
89. Lillard JW, Jr., Boyaka PN, Taub DD, McGhee JR. 2001. RANTES potentiates antigen-specific mucosal immune responses. *Journal of Immunology* 166: 162–9
90. Xin KQ, Lu Y, Hamajima K, Fukushima J, Yang J, Inamura K, Okuda K. 1999. Immunization of RANTES expression plasmid with a DNA vaccine enhances HIV-1-specific immunity. *Clinical Immunology* 92: 90–6
91. Lillard JW, Jr., Singh UP, Boyaka PN, Singh S, Taub DD, McGhee JR. 2003. MIP-1alpha and MIP-1beta differentially mediate mucosal and systemic adaptive immunity. *Blood* 101: 807–14
92. Eckmann L, Kagnoff MF, Fierer J. 1993. Epithelial cells secrete the chemokine interleukin-8 in response to bacterial entry. *Infection and Immunity* 61: 4569–74
93. Jung HC, Eckmann L, Yang SK, Panja A, Fierer J, Morzycka-Wroblewska E, Kagnoff MF. 1995. A distinct array of proinflammatory cytokines is expressed in human colon epithelial cells in response to bacterial invasion.[see comment]. *Journal of Clinical Investigation* 95: 55–65
94. Yang SK, Eckmann L, Panja A, Kagnoff MF. 1997. Differential and regulated expression of C-X-C, C-C, and C-chemokines by human colon epithelial cells. *Gastroenterology* 113: 1214–23
95. Reinecker HC, Loh EY, Ringler DJ, Mehta A, Rombeau JL, MacDermott RP. 1995. Monocyte-chemoattractant protein 1 gene expression in intestinal epithelial cells and inflammatory bowel disease mucosa. *Gastroenterology* 108: 40–50
96. Izadpanah A, Dwinell MB, Eckmann L, Varki NM, Kagnoff MF. 2001. Regulated MIP-3alpha/CCL20 production by human intestinal epithelium: mechanism for modulating mucosal immunity. *American Journal of Physiology – Gastrointestinal & Liver Physiology* 280: G710–9

97. Dwinell MB, Lugering N, Eckmann L, Kagnoff MF. 2001. Regulated production of interferon-inducible T-cell chemoattractants by human intestinal epithelial cells.[see comment]. *Gastroenterology* 120: 49–59
98. Shibahara T, Wilcox JN, Couse T, Madara JL. 2001. Characterization of epithelial chemoattractants for human intestinal intraepithelial lymphocytes.[see comment]. *Gastroenterology* 120: 60–70
99. Kunkel EJ, Campbell JJ, Haraldsen G, Pan J, Boisvert J, Roberts AI, Ebert EC, Vierra MA, Goodman SB, Genovese MC, Wardlaw AJ, Greenberg HB, Parker CM, Butcher EC, Andrew DP, Agace WW. 2000. Lymphocyte CC chemokine receptor 9 and epithelial thymus-expressed chemokine (TECK) expression distinguish the small intestinal immune compartment: Epithelial expression of tissue-specific chemokines as an organizing principle in regional immunity. *Journal of Experimental Medicine* 192: 761–8
100. Wurbel MA, Philippe JM, Nguyen C, Victorero G, Freeman T, Wooding P, Miazek A, Mattei MG, Malissen M, Jordan BR, Malissen B, Carrier A, Naquet P. 2000. The chemokine TECK is expressed by thymic and intestinal epithelial cells and attracts double- and single-positive thymocytes expressing the TECK receptor CCR9. *European Journal of Immunology* 30: 262–71
101. Pan J, Kunkel EJ, Gosslar U, Lazarus N, Langdon P, Broadwell K, Vierra MA, Genovese MC, Butcher EC, Soler D. 2000. A novel chemokine ligand for CCR10 and CCR3 expressed by epithelial cells in mucosal tissues. *Journal of Immunology* 165: 2943–9
102. Wang W, Soto H, Oldham ER, Buchanan ME, Homey B, Catron D, Jenkins N, Copeland NG, Gilbert DJ, Nguyen N, Abrams J, Kershenovich D, Smith K, McClanahan T, Vicari AP, Zlotnik A. 2000. Identification of a novel chemokine (CCL28), which binds CCR10 (GPR2). *Journal of Biological Chemistry* 275: 22313–23
103. Eckmann L, Reed SL, Smith JR, Kagnoff MF. 1995. Entamoeba histolytica trophozoites induce an inflammatory cytokine response by cultured human cells through the paracrine action of cytolytically released interleukin-1 alpha. *Journal of Clinical Investigation* 96: 1269–79
104. Maaser C, Eckmann L, Paesold G, Kim HS, Kagnoff MF. 2002. Ubiquitous production of macrophage migration inhibitory factor by human gastric and intestinal epithelium. *Gastroenterology* 122: 667–80
105. Maaser C, Schoeppner S, Kucharzik T, Kraft M, Schoenherr E, Domschke W, Luegering N. 2001. Colonic epithelial cells induce endothelial cell expression of ICAM-1 and VCAM-1 by a NF-kappaB-dependent mechanism. *Clinical and Experimental Immunology* 124: 208–13
106. Eckmann L, Jung HC, Schurer-Maly C, Panja A, Morzycka-Wroblewska E, Kagnoff MF. 1993. Differential cytokine expression by human intestinal epithelial cell lines: regulated expression of interleukin 8.[see comment][comment]. *Gastroenterology* 105: 1689–97
107. Elewaut D, DiDonato JA, Kim JM, Truong F, Eckmann L, Kagnoff MF. 1999. NF-kappa B is a central regulator of the intestinal epithelial cell innate immune response induced by infection with enteroinvasive bacteria. *Journal of Immunology* 163: 1457–66
108. Meijer LK, Schesser K, Wolf-Watz H, Sassone-Corsi P, Pettersson S. 2000. The bacterial protein YopJ abrogates multiple signal transduction pathways that converge on the transcription factor CREB. *Cellular Microbiology* 2: 231–8
109. Schesser K, Spiik AK, Dukuzumuremyi JM, Neurath MF, Pettersson S, Wolf-Watz H. 1998. The yopJ locus is required for Yersinia-mediated inhibition of NF-kappaB activation and cytokine expression: YopJ contains a eukaryotic SH2-like domain that is essential for its repressive activity. *Molecular Microbiology* 28: 1067–79
110. Hugot JP, Chamaillard M, Zouali H, Lesage S, Cezard JP, Belaiche J, Almer S, Tysk C, O'Morain CA, Gassull M, Binder V, Finkel Y, Cortot A, Modigliani R, Laurent-Puig P, Gower-Rousseau C, Macry J, Colombel JF, Sahbatou M, Thomas G. 2001. Association of NOD2 leucine-rich repeat variants with susceptibility to Crohn's disease.[see comment]. *Nature* 411: 599–603

111. Ogura Y, Bonen DK, Inohara N, Nicolae DL, Chen FF, Ramos R, Britton H, Moran T, Karaliuskas R, Duerr RH, Achkar JP, Brant SR, Bayless TM, Kirschner BS, Hanauer SB, Nunez G, Cho JH. 2001. A frameshift mutation in NOD2 associated with susceptibility to Crohn's disease.[see comment]. *Nature* 411: 603–6
112. Hampe J, Cuthbert A, Croucher PJ, Mirza MM, Mascheretti S, Fisher S, Frenzel H, King K, Hasselmeyer A, MacPherson AJ, Bridger S, van Deventer S, Forbes A, Nikolaus S, Lennard-Jones JE, Foelsch UR, Krawczak M, Lewis C, Schreiber S, Mathew CG. 2001. Association between insertion mutation in NOD2 gene and Crohn's disease in German and British populations.[see comment][erratum appears in Lancet 2002 Sep 7;360(9335):806]. *Lancet* 357: 1925–8
113. Ahmad T, Armuzzi A, Bunce M, Mulcahy-Hawes K, Marshall SE, Orchard TR, Crawshaw J, Large O, de Silva A, Cook JT, Barnardo M, Cullen S, Welsh KI, Jewell DP. 2002. The molecular classification of the clinical manifestations of Crohn's disease.[see comment][erratum appears in Gastroenterology. 2003 Jul;125(1):281]. *Gastroenterology* 122: 854–66
114. Cuthbert AP, Fisher SA, Mirza MM, King K, Hampe J, Croucher PJ, Mascheretti S, Sanderson J, Forbes A, Mansfield J, Schreiber S, Lewis CM, Mathew CG. 2002. The contribution of NOD2 gene mutations to the risk and site of disease in inflammatory bowel disease.[see comment]. *Gastroenterology* 122: 867–74
115. Lesage S, Zouali H, Cezard JP, Colombel JF, Belaiche J, Almer S, Tysk C, O'Morain C, Gassull M, Binder V, Finkel Y, Modigliani R, Gower-Rousseau C, Macry J, Merlin F, Chamaillard M, Jannot AS, Thomas G, Hugot JP, Group E-I, Group E, Group G. 2002. CARD15/NOD2 mutational analysis and genotype-phenotype correlation in 612 patients with inflammatory bowel disease. *American Journal of Human Genetics* 70: 845–57
116. Fukata M, Michelsen KS, Eri R, Thomas LS, Hu B, Lukasek K, Nast CC, Lechago J, Xu R, Naiki Y, Soliman A, Arditi M, Abreu MT. 2005. Toll-like receptor-4 is required for intestinal response to epithelial injury and limiting bacterial translocation in a murine model of acute colitis. *American Journal of Physiology – Gastrointestinal & Liver Physiology* 288: G1055–65
117. Cronstein BN, Terkeltaub R. 2006. The inflammatory process of gout and its treatment. *Arthritis Research & Therapy* 8 Suppl 1: S3
118. Liu-Bryan R, Scott P, Sydlaske A, Rose DM, Terkeltaub R. 2005. Innate immunity conferred by Toll-like receptors 2 and 4 and myeloid differentiation factor 88 expression is pivotal to monosodium urate monohydrate crystal-induced inflammation. *Arthritis & Rheumatism* 52: 2936–46
119. Shi Y, Evans JE, Rock KL. 2003. Molecular identification of a danger signal that alerts the immune system to dying cells.[see comment]. *Nature* 425: 516–21
120. Laudanna C, Kim JY, Constantin G, Butcher E. 2002. Rapid leukocyte integrin activation by chemokines. *Immunological Reviews* 186: 37–46
121. Huang Q, Liu D, Majewski P, Schulte LC, Korn JM, Young RA, Lander ES, Hacohen N. 2001. The plasticity of dendritic cell responses to pathogens and their components.[see comment]. *Science* 294: 870–5
122. Mantovani A. 1999. The chemokine system: redundancy for robust outputs.[see comment]. *Immunology Today* 20: 254–7
123. Hayashi F, Means TK, Luster AD. 2003. Toll-like receptors stimulate human neutrophil function. *Blood* 102: 2660–9
124. Neufert C, Pai RK, Noss EH, Berger M, Boom WH, Harding CV. 2001. Mycobacterium tuberculosis 19-kDa lipoprotein promotes neutrophil activation. *Journal of Immunology* 167: 1542–9
125. Hornung V, Rothenfusser S, Britsch S, Krug A, Jahrsdorfer B, Giese T, Endres S, Hartmann G. 2002. Quantitative expression of toll-like receptor 1–10 mRNA in cellular subsets of human peripheral blood mononuclear cells and sensitivity to CpG oligodeoxynucleotides. *Journal of Immunology* 168: 4531–7

126. Nagase H, Okugawa S, Ota Y, Yamaguchi M, Tomizawa H, Matsushima K, Ohta K, Yamamoto K, Hirai K. 2003. Expression and function of Toll-like receptors in eosinophils: activation by Toll-like receptor 7 ligand. *Journal of Immunology* 171: 3977–82
127. McCurdy JD, Olynych TJ, Maher LH, Marshall JS. 2003. Cutting edge: distinct Toll-like receptor 2 activators selectively induce different classes of mediator production from human mast cells. *Journal of Immunology* 170: 1625–9
128. Supajatura V, Ushio H, Nakao A, Akira S, Okumura K, Ra C, Ogawa H. 2002. Differential responses of mast cell Toll-like receptors 2 and 4 in allergy and innate immunity. *Journal of Clinical Investigation* 109: 1351–9
129. Supajatura V, Ushio H, Nakao A, Okumura K, Ra C, Ogawa H. 2001. Protective roles of mast cells against enterobacterial infection are mediated by Toll-like receptor 4. *Journal of Immunology* 167: 2250–6
130. Fan J, Malik AB. 2003. Toll-like receptor-4 (TLR4) signaling augments chemokine-induced neutrophil migration by modulating cell surface expression of chemokine receptors.[erratum appears in Nat. Med. 2003 Apr; 9(4):477]. *Nature Medicine* 9: 315–21
131. Ashkar AA, Bauer S, Mitchell WJ, Vieira J, Rosenthal KL. 2003. Local delivery of CpG oligodeoxynucleotides induces rapid changes in the genital mucosa and inhibits replication, but not entry, of herpes simplex virus type 2. *Journal of Virology* 77: 8948–56
132. Zhang D, Zhang G, Hayden MS, Greenblatt MB, Bussey C, Flavell RA, Ghosh S. 2004. A toll-like receptor that prevents infection by uropathogenic bacteria.[see comment]. *Science* 303: 1522–6
133. Kagnoff MF, Eckmann L. 1997. Epithelial cells as sensors for microbial infection. *Journal of Clinical Investigation* 100: 6–10
134. Andonegui G, Bonder CS, Green F, Mullaly SC, Zbytnuik L, Raharjo E, Kubes P. 2003. Endothelium-derived Toll-like receptor-4 is the key molecule in LPS-induced neutrophil sequestration into lungs.[erratum appears in J Clin Invest. 2003 Oct;112(8):1264]. *Journal of Clinical Investigation* 111: 1011–20
135. Schnare M, Barton GM, Holt AC, Takeda K, Akira S, Medzhitov R. 2001. Toll-like receptors control activation of adaptive immune responses. *Nature Immunology* 2: 947–50
136. Banchereau J, Steinman RM. 1998. Dendritic cells and the control of immunity. *Nature* 392: 245–52
137. Palliser D, Ploegh H, Boes M. 2004. Myeloid differentiation factor 88 is required for cross-priming in vivo. *Journal of Immunology* 172: 3415–21
138. Visintin A, Mazzoni A, Spitzer JH, Wyllie DH, Dower SK, Segal DM. 2001. Regulation of Toll-like receptors in human monocytes and dendritic cells. *Journal of Immunology* 166: 249–55
139. Muzio M, Bosisio D, Polentarutti N, D'Amico G, Stoppacciaro A, Mancinelli R, van't Veer C, Penton-Rol G, Ruco LP, Allavena P, Mantovani A. 2000. Differential expression and regulation of toll-like receptors (TLR) in human leukocytes: selective expression of TLR3 in dendritic cells. *Journal of Immunology* 164: 5998–6004
140. Jarrossay D, Napolitani G, Colonna M, Sallusto F, Lanzavecchia A. 2001. Specialization and complementarity in microbial molecule recognition by human myeloid and plasmacytoid dendritic cells. *European Journal of Immunology* 31: 3388–93
141. Kadowaki N, Ho S, Antonenko S, Malefyt RW, Kastelein RA, Bazan F, Liu YJ. 2001. Subsets of human dendritic cell precursors express different toll-like receptors and respond to different microbial antigens. *Journal of Experimental Medicine* 194: 863–9
142. Krug A, Towarowski A, Britsch S, Rothenfusser S, Hornung V, Bals R, Giese T, Engelmann H, Endres S, Krieg AM, Hartmann G. 2001. Toll-like receptor expression reveals CpG DNA as a unique microbial stimulus for plasmacytoid dendritic cells which synergizes with CD40 ligand to induce high amounts of IL-12. *European Journal of Immunology* 31: 3026–37
143. Ito T, Amakawa R, Kaisho T, Hemmi H, Tajima K, Uehira K, Ozaki Y, Tomizawa H, Akira S, Fukuhara S. 2002. Interferon-alpha and interleukin-12 are induced differentially by Toll-like receptor 7 ligands in human blood dendritic cell subsets. *Journal of Experimental Medicine* 195: 1507–12

144. Edwards AD, Diebold SS, Slack EM, Tomizawa H, Hemmi H, Kaisho T, Akira S, Reis e Sousa C. 2003. Toll-like receptor expression in murine DC subsets: lack of TLR7 expression by CD8 alpha+ DC correlates with unresponsiveness to imidazoquinolines. *European Journal of Immunology* 33: 827–33
145. Doxsee CL, Riter TR, Reiter MJ, Gibson SJ, Vasilakos JP, Kedl RM. 2003. The immune response modifier and Toll-like receptor 7 agonist S-27609 selectively induces IL-12 and TNF-alpha production in CD11c + CD11b + CD8− dendritic cells. *Journal of Immunology* 171: 1156–63
146. Boonstra A, Asselin-Paturel C, Gilliet M, Crain C, Trinchieri G, Liu YJ, O'Garra A. 2003. Flexibility of mouse classical and plasmacytoid-derived dendritic cells in directing T helper type 1 and 2 cell development: dependency on antigen dose and differential toll-like receptor ligation. *Journal of Experimental Medicine* 197: 101–9
147. Ahmad-Nejad P, Hacker H, Rutz M, Bauer S, Vabulas RM, Wagner H. 2002. Bacterial CpG-DNA and lipopolysaccharides activate Toll-like receptors at distinct cellular compartments. *European Journal of Immunology* 32: 1958–68
148. Sallusto F, Schaerli P, Loetscher P, Schaniel C, Lenig D, Mackay CR, Qin S, Lanzavecchia A. 1998. Rapid and coordinated switch in chemokine receptor expression during dendritic cell maturation. *European Journal of Immunology* 28: 2760–9
149. Dieu MC, Vanbervliet B, Vicari A, Bridon JM, Oldham E, Ait-Yahia S, Briere F, Zlotnik A, Lebecque S, Caux C. 1998. Selective recruitment of immature and mature dendritic cells by distinct chemokines expressed in different anatomic sites. *Journal of Experimental Medicine* 188: 373–86
150. Forster R, Schubel A, Breitfeld D, Kremmer E, Renner-Muller I, Wolf E, Lipp M. 1999. CCR7 coordinates the primary immune response by establishing functional microenvironments in secondary lymphoid organs. *Cell* 99: 23–33
151. Gunn MD, Kyuwa S, Tam C, Kakiuchi T, Matsuzawa A, Williams LT, Nakano H. 1999. Mice lacking expression of secondary lymphoid organ chemokine have defects in lymphocyte homing and dendritic cell localization.[see comment]. *Journal of Experimental Medicine* 189: 451–60
152. Constant SL, Bottomly K. 1997. Induction of Th1 and Th2 CD4+ T cell responses: the alternative approaches. *Annual Review of Immunology* 15: 297–322
153. Janssen EM, Lemmens EE, Wolfe T, Christen U, von Herrath MG, Schoenberger SP. 2003. CD4+ T cells are required for secondary expansion and memory in CD8+ T lymphocytes. *Nature* 421: 852–6
154. Shedlock DJ, Shen H. 2003. Requirement for CD4 T cell help in generating functional CD8 T cell memory.[see comment]. *Science* 300: 337–9
155. Sun JC, Bevan MJ. 2003. Defective CD8 T cell memory following acute infection without CD4 T cell help.[see comment]. *Science* 300: 339–42
156. Bourgeois C, Veiga-Fernandes H, Joret AM, Rocha B, Tanchot C. 2002. CD8 lethargy in the absence of CD4 help. *European Journal of Immunology* 32: 2199–207
157. Hemmi H, Kaisho T, Takeda K, Akira S. 2003. The roles of Toll-like receptor 9, MyD88, and DNA-dependent protein kinase catalytic subunit in the effects of two distinct CpG DNAs on dendritic cell subsets. *Journal of Immunology* 170: 3059–64
158. Janeway CA, Jr., Bottomly K. 1994. Signals and signs for lymphocyte responses. *Cell* 76: 275–85
159. Trinchieri G. 2003. Interleukin-12 and the regulation of innate resistance and adaptive immunity. *Nature Reviews Immunology* 3: 133–46
160. Kaisho T, Hoshino K, Iwabe T, Takeuchi O, Yasui T, Akira S. 2002. Endotoxin can induce MyD88-deficient dendritic cells to support T(h)2 cell differentiation. *International Immunology* 14: 695–700
161. Redecke V, Hacker H, Datta SK, Fermin A, Pitha PM, Broide DH, Raz E. 2004. Cutting edge: activation of Toll-like receptor 2 induces a Th2 immune response and promotes experimental asthma. *Journal of Immunology* 172: 2739–43

162. Pulendran B, Kumar P, Cutler CW, Mohamadzadeh M, Van Dyke T, Banchereau J. 2001. Lipopolysaccharides from distinct pathogens induce different classes of immune responses in vivo. *Journal of Immunology* 167: 5067–76
163. Agrawal S, Agrawal A, Doughty B, Gerwitz A, Blenis J, Van Dyke T, Pulendran B. 2003. Cutting edge: different Toll-like receptor agonists instruct dendritic cells to induce distinct Th responses via differential modulation of extracellular signal-regulated kinase-mitogen-activated protein kinase and c-Fos. *Journal of Immunology* 171: 4984–9
164. Martin P, Del Hoyo GM, Anjuere F, Arias CF, Vargas HH, Fernandez LA, Parrillas V, Ardavin C. 2002. Characterization of a new subpopulation of mouse CD8alpha+ B220+ dendritic cells endowed with type 1 interferon production capacity and tolerogenic potential. *Blood* 100: 383–90
165. Bowman EP, Campbell JJ, Soler D, Dong Z, Manlongat N, Picarella D, Hardy RR, Butcher EC. 2000. Developmental switches in chemokine response profiles during B cell differentiation and maturation. *Journal of Experimental Medicine* 191: 1303–18
166. Martin F, Oliver AM, Kearney JF. 2001. Marginal zone and B1 B cells unite in the early response against T-independent blood-borne particulate antigens. *Immunity* 14: 617–29
167. Saikh KU, Kissner TL, Sultana A, Ruthel G, Ulrich RG. 2004. Human monocytes infected with Yersinia pestis express cell surface TLR9 and differentiate into dendritic cells. *Journal of Immunology* 173: 7426–34
168. Ha S-a, Tsuji M, Suzuki K, Meek B, Yasuda N, Kaisho T, Fagarasan S. 2006. Regulation of B1 cell migration by signals through Toll-like receptors. *Journal of Experimental Medicine* 203: 2541–50
169. Dasari P, Nicholson IC, Hodge G, Dandie GW, Zola H. 2005. Expression of toll-like receptors on B lymphocytes. *Cellular Immunology* 236: 140–5
170. Fagarasan S, Shinkura R, Kamata T, Nogaki F, Ikuta K, Tashiro K, Honjo T. 2000. Alymphoplasia (aly)-type nuclear factor kappaB-inducing kinase (NIK) causes defects in secondary lymphoid tissue chemokine receptor signaling and homing of peritoneal cells to the gut-associated lymphatic tissue system. *Journal of Experimental Medicine* 191: 1477–86
171. Ishikawa S, Sato T, Abe M, Nagai S, Onai N, Yoneyama H, Zhang Y, Suzuki T, Hashimoto S, Shirai T, Lipp M, Matsushima K. 2001. Aberrant high expression of B lymphocyte chemokine (BLC/CXCL13) by C11b+CD11c+ dendritic cells in murine lupus and preferential chemotaxis of B1 cells towards BLC.[see comment]. *Journal of Experimental Medicine* 193: 1393–402
172. Rui L, Vinuesa CG, Blasioli J, Goodnow CC. 2003. Resistance to CpG DNA-induced autoimmunity through tolerogenic B cell antigen receptor ERK signaling. *Nature Immunology* 4: 594–600
173. Bernasconi NL, Traggiai E, Lanzavecchia A. 2002. Maintenance of serological memory by polyclonal activation of human memory B cells. *Science* 298: 2199–202
174. Re F, Strominger JL. 2001. Toll-like receptor 2 (TLR2) and TLR4 differentially activate human dendritic cells. *Journal of Biological Chemistry* 276: 37692–9
175. Parker LC, Whyte MKB, Vogel SN, Dower SK, Sabroe I. 2004. Toll-like receptor (TLR)2 and TLR4 agonists regulate CCR expression in human monocytic cells. *Journal of Immunology* 172: 4977–86
176. Smith MF, Jr., Mitchell A, Li G, Ding S, Fitzmaurice AM, Ryan K, Crowe S, Goldberg JB. 2003. Toll-like receptor (TLR) 2 and TLR5, but not TLR4, are required for Helicobacter pylori-induced NF-kappa B activation and chemokine expression by epithelial cells. *Journal of Biological Chemistry* 278: 32552–60
177. Kuroshima S, Sawa Y, Kawamoto T, Yamaoka Y, Notani K, Yoshida S, Inoue N. 2004. Expression of Toll-like receptors 2 and 4 on human intestinal lymphatic vessels. *Microvascular Research* 67: 90–5
178. Proost P, Verpoest S, Van de Borne K, Schutyser E, Struyf S, Put W, Ronsse I, Grillet B, Opdenakker G, Van Damme J. 2004. Synergistic induction of CXCL9 and CXCL11 by Toll-like receptor ligands and interferon-gamma in fibroblasts correlates with elevated levels of CXCR3 ligands in septic arthritis synovial fluids. *Journal of Leukocyte Biology* 75: 777–84

179. Schaefer TM, Fahey JV, Wright JA, Wira CR. 2005. Innate immunity in the human female reproductive tract: antiviral response of uterine epithelial cells to the TLR3 agonist poly(I:C). *Journal of Immunology* 174: 992–1002
180. Wang Y, Abel K, Lantz K, Krieg AM, McChesney MB, Miller CJ. 2005. The Toll-like receptor 7 (TLR7) agonist, imiquimod, and the TLR9 agonist, CpG ODN, induce antiviral cytokines and chemokines but do not prevent vaginal transmission of simian immunodeficiency virus when applied intravaginally to rhesus macaques.[erratum appears in J Virol. 2006 Sep;80(17):8846]. *Journal of Virology* 79: 14355–70
181. Holmgren J, Lonnroth I, Mansson J, Svennerholm L. 1975. Interaction of cholera toxin and membrane GM1 ganglioside of small intestine. *Proceedings of the National Academy of Sciences of the United States of America* 72: 2520–4
182. Bromander AK, Kjerrulf M, Holmgren J, Lycke N. 1993. Cholera toxin enhances alloantigen presentation by cultured intestinal epithelial cells. *Scandinavian Journal of Immunology* 37: 452–8
183. McGee DW, Elson CO, McGhee JR. 1993. Enhancing effect of cholera toxin on interleukin-6 secretion by IEC-6 intestinal epithelial cells: mode of action and augmenting effect of inflammatory cytokines. *Infection and Immunity* 61: 4637–44
184. Bromander A, Holmgren J, Lycke N. 1991. Cholera toxin stimulates IL-1 production and enhances antigen presentation by macrophages in vitro. *Journal of Immunology* 146: 2908–14
185. Hieshima K, Kawasaki Y, Hanamoto H, Nakayama T, Nagakubo D, Kanamaru A, Yoshie O. 2004. CC chemokine ligands 25 and 28 play essential roles in intestinal extravasation of IgA antibody-secreting cells. *Journal of Immunology* 173: 3668–75
186. Anjuere F, Luci C, Lebens M, Rousseau D, Hervouet C, Milon G, Holmgren J, Ardavin C, Czerkinsky C. 2004. In vivo adjuvant-induced mobilization and maturation of gut dendritic cells after oral administration of cholera toxin. *Journal of Immunology* 173: 5103–11
187. Yamamoto S, Takeda Y, Yamamoto M, Kurazono H, Imaoka K, Yamamoto M, Fujihashi K, Noda M, Kiyono H, McGhee JR. 1997. Mutants in the ADP-ribosyltransferase cleft of cholera toxin lack diarrheagenicity but retain adjuvanticity. *Journal of Experimental Medicine* 185: 1203–10
188. Fontaine A, Arondel J, Sansonetti PJ. 1988. Role of Shiga toxin in the pathogenesis of bacillary dysentery, studied by using a Tox- mutant of Shigella dysenteriae 1. *Infection and Immunity* 56: 3099–109
189. Miyamoto Y, Iimura M, Kaper JB, Torres AG, Kagnoff MF. 2006. Role of Shiga toxin versus H7 flagellin in enterohaemorrhagic Escherichia coli signalling of human colon epithelium in vivo. *Cellular Microbiology* 8: 869–79
190. Thorpe CM, Hurley BP, Lincicome LL, Jacewicz MS, Keusch GT, Acheson DW. 1999. Shiga toxins stimulate secretion of interleukin-8 from intestinal epithelial cells. *Infection and Immunity* 67: 5985–93
191. de Silva DG, Mendis LN, Sheron N, Alexander GJ, Candy DC, Chart H, Rowe B. 1993. Concentrations of interleukin 6 and tumour necrosis factor in serum and stools of children with Shigella dysenteriae 1 infection.[see comment]. *Gut* 34: 194–8
192. Harrison LM, van den Hoogen C, van Haaften WCE, Tesh VL. 2005. Chemokine expression in the monocytic cell line THP-1 in response to purified shiga toxin 1 and/or lipopolysaccharides. *Infection and Immunity* 73: 403–12
193. Sasaki S, Omoe K, Tagawa Yi, Iwakura Y, Sekikawa K, Shinagawa K, Nakane A. 2002. Roles of gamma interferon and tumor necrosis factor-alpha in shiga toxin lethality. *Microbial Pathogenesis* 33: 43–7
194. Ohmura M, Yamamoto M, Tomiyama-Miyaji C, Yuki Y, Takeda Y, Kiyono H. 2005. Nontoxic Shiga toxin derivatives from Escherichia coli possess adjuvant activity for the augmentation of antigen-specific immune responses via dendritic cell activation. *Infection and Immunity* 73: 4088–97
195. Duverger A, Jackson RJ, van Ginkel FW, Fischer R, Tafaro A, Leppla SH, Fujihashi K, Kiyono H, McGhee JR, Boyaka PN. 2006. Bacillus anthracis edema toxin acts as an

adjuvant for mucosal immune responses to nasally administered vaccine antigens. *Journal of Immunology* 176: 1776–83
196. Aloni-Grinstein R, Gat O, Altboum Z, Velan B, Cohen S, Shafferman A. 2005. Oral spore vaccine based on live attenuated nontoxinogenic Bacillus anthracis expressing recombinant mutant protective antigen. *Infection and Immunity* 73: 4043–53
197. Ghosh D, Porter E, Shen B, Lee SK, Wilk D, Drazba J, Yadav SP, Crabb JW, Ganz T, Bevins CL. 2002. Paneth cell trypsin is the processing enzyme for human defensin-5.[see comment]. *Nature Immunology* 3: 583–90
198. Katze MG, He Y, Gale M, Jr. 2002. Viruses and interferon: a fight for supremacy. *Nature Reviews Immunology* 2: 675–87
199. Taniguchi T, Takaoka A. 2002. The interferon-alpha/beta system in antiviral responses: a multimodal machinery of gene regulation by the IRF family of transcription factors. *Current Opinion in Immunology* 14: 111–6
200. Tough DF, Sun S, Zhang X, Sprent J. 1999. Stimulation of naive and memory T cells by cytokines. *Immunological Reviews* 170: 39–47
201. Nguyen KB, Watford WT, Salomon R, Hofmann SR, Pien GC, Morinobu A, Gadina M, O'Shea JJ, Biron CA. 2002. Critical role for STAT4 activation by type 1 interferons in the interferon-gamma response to viral infection. *Science* 297: 2063–6
202. Sareneva T, Matikainen S, Kurimoto M, Julkunen I. 1998. Influenza A virus-induced IFN-alpha/beta and IL-18 synergistically enhance IFN-gamma gene expression in human T cells. *Journal of Immunology* 160: 6032–8
203. Rogge L, D'Ambrosio D, Biffi M, Penna G, Minetti LJ, Presky DH, Adorini L, Sinigaglia F. 1998. The role of Stat4 in species-specific regulation of Th cell development by type I IFNs. *Journal of Immunology* 161: 6567–74
204. Le Bon A, Etchart N, Rossmann C, Ashton M, Hou S, Gewert D, Borrow P, Tough DF. 2003. Cross-priming of CD8+ T cells stimulated by virus-induced type I interferon.[see comment]. *Nature Immunology* 4: 1009–15
205. Le Bon A, Schiavoni G, D'Agostino G, Gresser I, Belardelli F, Tough DF. 2001. Type i interferons potently enhance humoral immunity and can promote isotype switching by stimulating dendritic cells in vivo. *Immunity* 14: 461–70
206. Biron CA, Nguyen KB, Pien GC, Cousens LP, Salazar-Mather TP. 1999. Natural killer cells in antiviral defense: function and regulation by innate cytokines. *Annual Review of Immunology* 17: 189–220
207. Hoshino K, Kaisho T, Iwabe T, Takeuchi O, Akira S. 2002. Differential involvement of IFN-beta in Toll-like receptor-stimulated dendritic cell activation. *International Immunology* 14: 1225–31
208. Honda K, Sakaguchi S, Nakajima C, Watanabe A, Yanai H, Matsumoto M, Ohteki T, Kaisho T, Takaoka A, Akira S, Seya T, Taniguchi T. 2003. Selective contribution of IFN-alpha/beta signaling to the maturation of dendritic cells induced by double-stranded RNA or viral infection. *Proceedings of the National Academy of Sciences of the United States of America* 100: 10872–7
209. Hoebe K, Janssen EM, Kim SO, Alexopoulou L, Flavell RA, Han J, Beutler B. 2003. Upregulation of costimulatory molecules induced by lipopolysaccharide and double-stranded RNA occurs by Trif-dependent and Trif-independent pathways.[see comment]. *Nature Immunology* 4: 1223–9
210. MacKay I, Rosen F. 2000. Innate immunity. *The New England Journal of Medicine* 343: 338–44
211. Parkin J, Cohen B. 2001. An overview of the immune system. *Lancet* 357: 1777–89
212. Iwata F, Joh T, Tada T, Okada N, Morgan BP, Yokoyama Y, Itoh M. 1999. Role of complement regulatory membrane proteins in ischaemia-reperfusion injury of rat gastric mucosa. *Journal of Gastroenterology & Hepatology* 14: 967–72
213. Robertson J, Caldwell JR, Castle JR, Waldman RH. 1976. Evidence for the presence of components of the alternative (properdin) pathway of complement activation in respiratory secretions. *Journal of Immunology* 117: 900–3

214. Emancipator SN, Gallo GR, Lamm ME. 1985. IgA nephropathy: perspectives on pathogenesis and classification. *Clinical Nephrology* 24: 161–79
215. Katz SI, Hall RP, 3rd, Lawley TJ, Strober W. 1980. Dermatitis herpetiformis: the skin and the gut. *Annals of Internal Medicine* 93: 857–74
216. Tauber PF, Wettich W, Nohlen M, Zaneveld LJ. 1985. Diffusable proteins of the mucosa of the human cervix, uterus, and fallopian tubes: distribution and variations during the menstrual cycle. *American Journal of Obstetrics & Gynecology* 151: 1115–25
217. Laufer J, Oren R, Goldberg I, Horwitz A, Kopolovic J, Chowers Y, Passwell JH. 2000. Cellular localization of complement C3 and C4 transcripts in intestinal specimens from patients with Crohn's disease. *Clinical and Experimental Immunology* 120: 30–7
218. Andoh A, Fujiyama Y, Bamba T, Hosoda S. 1993. Differential cytokine regulation of complement C3, C4, and factor B synthesis in human intestinal epithelial cell line, Caco-2. *Journal of Immunology* 151: 4239–47
219. da Rosa Utiyama SR, da Silva Kotze LM, de Messias Reason IT. 2005. Complement factor B allotypes in the susceptibility and severity of coeliac disease in patients and relatives. *International Journal of Immunogenetics* 32: 307–14
220. Pellis V, De Seta F, Crovella S, Bossi F, Bulla R, Guaschino S, Radillo O, Garred P, Tedesco F. 2005. Mannose binding lectin and C3 act as recognition molecules for infectious agents in the vagina. *Clinical and Experimental Immunology* 139: 120–6
221. Kelly P, Jack DL, Naeem A, Mandanda B, Pollok RC, Klein NJ, Turner MW, Farthing MJ. 2000. Mannose-binding lectin is a component of innate mucosal defense against Cryptosporidium parvum in AIDS. *Gastroenterology* 119: 1236–42
222. Smythies LE, Sellers M, Clements RH, Mosteller-Barnum M, Meng G, Benjamin WH, Orenstein JM, Smith PD. 2005. Human intestinal macrophages display profound inflammatory anergy despite avid phagocytic and bacteriocidal activity. *Journal of Clinical Investigation* 115: 66–75
223. Lundahl J, Skedinger M, Hed J, Johansson SG, Zetterstrom O. 1992. Lability in complement receptor mobilization of granulocytes in patients with bronchial hyperreactivity. *Clinical & Experimental Allergy* 22: 834–8
224. Miyaguchi M, Uda H, Sakai S, Kubo T, Matsunaga T. 1988. Immunohistochemical studies of complement receptor (CR1) in cases with normal sinus mucosa and chronic sinusitis. *Archives of Oto-Rhino-Laryngology* 244: 350–4
225. Gallucci S, Matzinger P. 2001. Danger signals: SOS to the immune system. 13: 114–9
226. Wagner L, Yang OO, Garcia-Zepeda EA, Ge Y, Kalams SA, Walker BD, Pasternack MS, Luster AD. 1998. Beta-chemokines are released from HIV-1-specific cytolytic T-cell granules complexed to proteoglycans. *Nature* 391: 908–11
227. Rossi D, Zlotnik A. 2000. The biology of chemokines and their receptors. *Annual Review of Immunology* 18: 217–42
228. Sallusto F, Mackay CR, Lanzavecchia A. 2000. The role of chemokine receptors in primary, effector, and memory immune responses. *Annual Review of Immunology* 18: 593–620
229. Randolph DA, Huang G, Carruthers CJ, Bromley LE, Chaplin DD. 1999. The role of CCR7 in TH1 and TH2 cell localization and delivery of B cell help in vivo. *Science* 286: 2159–62
230. Luster AD, Unkeless JC, Ravetch JV. 1985. Gamma-interferon transcriptionally regulates an early-response gene containing homology to platelet proteins. *Nature* 315: 672–6
231. Farber JM. 1990. A macrophage mRNA selectively induced by gamma-interferon encodes a member of the platelet factor 4 family of cytokines. *Proceedings of the National Academy of Sciences of the United States of America* 87: 5238–42
232. Cole KE, Strick CA, Paradis TJ, Ogborne KT, Loetscher M, Gladue RP, Lin W, Boyd JG, Moser B, Wood DE, Sahagan BG, Neote K. 1998. Interferon-inducible T cell alpha chemoattractant (I-TAC): a novel non-ELR CXC chemokine with potent activity on activated T cells through selective high affinity binding to CXCR3. *The Journal of Experimental Medicine* 187: 2009–21

233. Loetscher M, Gerber B, Loetscher P, Jones SA, Piali L, Clark-Lewis I, Baggiolini M, Moser B. 1996. Chemokine receptor specific for IP10 and mig: structure, function, and expression in activated T-lymphocytes. *The Journal of Experimental Medicine* 184: 963–9
234. Ziegler SF, Tough TW, Franklin TL, Armitage RJ, Alderson MR. 1991. Induction of macrophage inflammatory protein-1 beta gene expression in human monocytes by lipopolysaccharide and IL-7. *Journal of Immunology* 147: 2234–9
235. van der Velden VH, Verheggen MM, Bernasconi S, Sozzani S, Naber BA, van der Linden-van Beurden CA, Hoogsteden HC, Mantovani A, Versnel M. 1998. Interleukin-1beta and interferon-gamma differentially regulate release of monocyte chemotactic protein-1 and interleukin-8 by human bronchial epithelial cells. *European Cytokine Network* 9: 269–77
236. Schrum S, Probst P, Fleischer B, Zipfel PF. 1996. Synthesis of the CC-chemokines MIP-1alpha, MIP-1beta, and RANTES is associated with a type 1 immune response. *Journal of Immunology* 157: 3598–604
237. John M, Hirst SJ, Jose PJ, Robichaud A, Berkman N, Witt C, Twort CH, Barnes PJ, Chung KF. 1997. Human airway smooth muscle cells express and release RANTES in response to T helper 1 cytokines: regulation by T helper 2 cytokines and corticosteroids. *Journal of Immunology* 158: 1841–7
238. Kim CH, Kunkel EJ, Boisvert J, Johnston B, Campbell JJ, Genovese MC, Greenberg HB, Butcher EC. 2001. Bonzo/CXCR6 expression defines type 1-polarized T-cell subsets with extralymphoid tissue homing potential. *Journal of Clinical Investigation* 107: 595–601
239. Palaniappan R, Singh S, Singh UP, Singh R, Ades EW, Briles DE, Hollingshead SK, Royal W, 3rd, Sampson JS, Stiles JK, Taub DD, Lillard JW, Jr. 2006. CCL5 modulates pneumococcal immunity and carriage. *Journal of Immunology* 176: 2346–56
240. Elson CO, Beagley KW, Sharmanov AT, Fujihashi K, Kiyono H, Tennyson GS, Cong Y, Black CA, Ridwan BW, McGhee JR. 1996. Hapten-induced model of murine inflammatory bowel disease: mucosa immune responses and protection by tolerance. *Journal of Immunology* 157: 2174–85
241. Singh UP, Singh S, Taub DD, Lillard JW, Jr. 2003. Inhibition of IFN-gamma-inducible protein-10 abrogates colitis in IL-10-/- mice. *Journal of Immunology* 171: 1401–6
242. Singh UP, Singh S, Iqbal N, Weaver CT, McGhee JR, Lillard JW, Jr. 2003. IFN-gamma-inducible chemokines enhance adaptive immunity and colitis. *Journal of Interferon & Cytokine Research* 23: 591–600
243. Lezcano-Meza D, Davila-Davila B, Vega-Miranda A, Negrete-Garcia MC, Teran LM. 2003. Interleukin (IL)-4 and to a lesser extent either IL-13 or interferon-gamma regulate the production of eotaxin-2/CCL24 in nasal polyps. *Allergy* 58: 1011–7
244. Panina-Bordignon P, Papi A, Mariani M, Di Lucia P, Casoni G, Bellettato C, Buonsanti C, Miotto D, Mapp C, Villa A, Arrigoni G, Fabbri LM, Sinigaglia F. 2001. The C-C chemokine receptors CCR4 and CCR8 identify airway T cells of allergen-challenged atopic asthmatics. *Journal of Clinical Investigation* 107: 1357–64
245. Balding CE, Howie AJ, Drake-Lee AB, Savage CO. 2001. Th2 dominance in nasal mucosa in patients with Wegener's granulomatosis. *Clinical and Experimental Immunology* 125: 332–9
246. Terada N, Nomura T, Kim WJ, Otsuka Y, Takahashi R, Kishi H, Yamashita T, Sugawara N, Fukuda S, Ikeda-Ito T, Konno A. 2001. Expression of C-C chemokine TARC in human nasal mucosa and its regulation by cytokines.[see comment]. *Clinical & Experimental Allergy* 31: 1923–31
247. Miyazaki E, Nureki S-i, Fukami T, Shigenaga T, Ando M, Ito K, Ando H, Sugisaki K, Kumamoto T, Tsuda T. 2002. Elevated levels of thymus- and activation-regulated chemokine in bronchoalveolar lavage fluid from patients with eosinophilic pneumonia. *American Journal of Respiratory & Critical Care Medicine* 165: 1125–31
248. Ma W, Bryce PJ, Humbles AA, Laouini D, Yalcindag A, Alenius H, Friend DS, Oettgen HC, Gerard C, Geha RS. 2002. CCR3 is essential for skin eosinophilia and airway hyperresponsiveness in a murine model of allergic skin inflammation. *Journal of Clinical Investigation* 109: 621–8

249. Chiu B-C, Freeman CM, Stolberg VR, Komuniecki E, Lincoln PM, Kunkel SL, Chensue SW. 2003. Cytokine-chemokine networks in experimental mycobacterial and schistosomal pulmonary granuloma formation. *American Journal of Respiratory Cell and Molecular Biology* 29: 106–16
250. Zimmermann N, Hershey GK, Foster PS, Rothenberg ME. 2003. Chemokines in asthma: cooperative interaction between chemokines and IL-13. *The Journal of Allergy and Clinical Immunology* 111: 227–42; quiz 43
251. Singh UP, Singh S, Ravichandran P, Taub DD, Lillard JW, Jr. 2004. Viral macrophage-inflammatory protein-II: a viral chemokine that differentially affects adaptive mucosal immunity compared with its mammalian counterparts. *Journal of Immunology* 173: 5509–16
252. Hoover DM, Boulegue C, Yang D, Oppenheim JJ, Tucker K, Lu W, Lubkowski J. 2002. The structure of human macrophage inflammatory protein-3alpha/CCL20. Linking antimicrobial and CC chemokine receptor-6-binding activities with human beta-defensins. *Journal of Biological Chemistry* 277: 37647–54
253. Shiba H, Mouri Y, Komatsuzawa H, Ouhara K, Takeda K, Sugai M, Kinane DF, Kurihara H. 2003. Macrophage inflammatory protein-3alpha and beta-defensin-2 stimulate dentin sialophosphoprotein gene expression in human pulp cells. *Biochemical & Biophysical Research Communications* 306: 867–71
254. Petkovic V, Moghini C, Paoletti S, Uguccioni M, Gerber B. 2004. I-TAC/CXCL11 is a natural antagonist for CCR5. *Journal of leukocyte biology* 76: 701–8

Chapter 5
Intestinal Bacteria: Mucosal Tissue Development and Gut Homeostasis

Dennis K. Lanning, Kari M. Severson and Katherine L. Knight

Abstract Intestinal microbiota affect multiple aspects of the mucosal immune system, including promotion of mucosal organogenesis, T_H cytokines balance, somatic diversification of Ig genes, and production of IgA. The mechanism by which microbiota interacts with vertebrate hosts and mediates these processes remains largely unexplored. Here we discuss the organogenesis of mucosal-associated lymphoid tissues, mechanisms by which microbiota promote MALT development, and how IgA can maintain a peaceful coexistence between commensals and their host.

5.1 Introduction

The mucosal immune system protects the body's mucosal surfaces from invasion by microbial pathogens. Paradoxically, however, the body requires interaction with resident microorganisms for the development of its mucosal immune system. The developmental importance of host interaction with commensal microbes is most pronounced in mucosa-associated lymphoid tissues in the gut, where the body harbors an enormous and complex resident microbiota. The intestinal microbiota influences many aspects of mucosal and systemic immunity, including secondary lymphoid tissue organogenesis [1, 2], mucosal IgA production [3], intraepithelial lymphocyte (IEL) development [4], B cell expansion and Ig gene diversification [5], and the systemic T_H1/T_H2 cytokine balance [6].

Interruption of the host-microbial interactions required for normal development of the mucosal immune system can result in significant pathology. The marked increase in incidence of allergies and asthma in the industrialized countries over the past 50 years, for example, has been linked to overly hygienic practices and excessive antibiotic use, particularly during infancy [7]. The hygiene hypothesis, proposed to explain this linkage, suggests that excessive hygiene and antibiotic use during early childhood reduce exposure to microbes that help the mucosal

K.L. Knight
Department of Microbiology and Immunology, Stritch School of Medicine, Loyola University Chicago, Maywood, IL, USA
e-mail: kknight@lumc.edu

immune system develop tolerance to harmless environmental antigens [8]. Although controversial, this idea has recently found experimental support. Asthma and allergies are characterized by overproduction of T_H2 cytokines and IgE [9]. Mazmanian et al. (2005) demonstrated that colonization of germfree mice, which exhibit a T_H2-skewed cytokine profile, with *Bacteroides fragilis* restored the T_H1/T_H2 cytokine balance [6]. This change was due to a single molecule, the capsular polysaccharide PSA, which activated $CD4^+$ T cells in an MHC II-dependent manner and promoted differentiation of T_H1 cells. PSA thus appears to be an immunomodulatory molecule that can shift the systemic T_H1/T_H2 balance away from the T_H2-skewed cytokine profile associated with allergy and asthma. Another group of diseases caused by disruption of host-microbial interactions, inflammatory bowel diseases (IBDs), result from impaired homeostasis with the intestinal microbiota. IBDs, which are also on the rise in industrialized countries, are caused by defects in the complex innate, effector and/or regulatory mechanisms that mediate beneficial host-microbial interactions in the intestine [10]. In this chapter, we explore the host-microbial interactions known to be required for normal development and function of the mucosal immune system and, also, how its major effector molecule, IgA, regulates gut homeostasis.

5.2 MALT Organogenesis

Mucosa-associated lymphoid tissue, or MALT, is a general term comprising a collection of lymphoid tissues located at multiple sites throughout the body. Three sites of MALT have been closely studied: bronchus-associated lymphoid tissue (BALT), nasopharynx-associated lymphoid tissue (NALT) and gut-associated lymphoid tissue (GALT). These tissues share, to varying degrees, a common characteristic structure. The simplest example of MALT structure is the isolated lymphoid follicle (ILF). ILFs contain a single B cell follicle covered by a specialized follicle-associated epithelium (FAE) containing M cells [11]. M cells are specialized epithelial cells that can selectively transcytose particles and molecules across their apical membrane into a pocket formed by invagination of their basolateral membrane [12, 13]. Immune cells, including B and T lymphocytes and dendritic cells (DCs), transiently populate the M cell pockets, where they interact with transcytosed material and with each other. More elaborate MALT structures, such as Peyer's patches (PPs), contain multiple B cell follicles and specialized interfollicular T cell areas. Cryptopatches, another MALT structure, contain large T cell aggregates [14]. MALTs appear to form a common mucosal immune system in which lymphocytes primed at one mucosal site selectively home to other mucosal locations [15].

5.2.1 *Bronchus-Associated Lymphoid Tissue (BALT)*

Structures possessing characteristic MALT features are found in the epithelium of the airways in the upper two-thirds of the lung. These include ILFs in the bronchial epithelium and larger aggregates of B cell follicles, resembling PPs, found primarily at bifurcations of the bronchial tree [16]. BALT development can clearly occur in the absence of microbial stimulation, as BALT has been observed in germfree rats and mice [16], and also in human fetuses during the 2nd trimester of gestation [17]. However, microbial exposure strongly enhances, and can induce, BALT development. Moyron-Quiroz et al. (2004) demonstrated that BALT was induced in mice by inhalation of influenza virus [18], and further, that strong anti-viral B and T cell primary responses were initiated in this induced BALT (iBALT). Interestingly, iBALT was induced in mice lacking both tumor necrosis factor (TNF) and lymphotoxin α (LTα), molecules required for development of lymph nodes, PPs and organized spleen [18]. The capacity to develop in the absence of TNF and LTα establishes iBALT as a unique mucosal lymphoid tissue.

5.2.2 Nasal-Associated Lymphoid Tissue (NALT)

Although the nasal mucosa is one of the first sites to contact inhaled antigens, comparatively little research has addressed the local sites at which immune responses are induced in this tissue. In humans and some other species, oropharyngeal lymphoid tissues, including the adenoids, bilateral tubule, and palatine and lingual tonsils (Waldeyer's ring), probably contribute to respiratory and gastrointestinal immunity. However, lymphoid structures in the nasal mucosa have been clearly identified only in rodents [19]. In these species, NALT consists of paired lymphoid structures located on either side of the nasal septum at the entrance to the bifurcated pharygeal duct. Rodent NALT contains multiple B cell follicles and interfollicular T cell areas, lying beneath an FAE containing M cells [19]. In contrast to most other secondary lymphoid tissues, NALT development does not begin until after birth, reaching maximal size at about 8 weeks of age [20]. Although structurally less well-organized than in wild type mice, NALT develops upon exposure to antigen in LTα-deficient mice, as well as in a number of other genetically-deficient mouse strains that lack PPs and/or lymph nodes [20, 21]. NALT and iBALT thus appear to develop through a different pathway than that used by PPs and most other secondary lymphoid tissues, perhaps initiated by interaction with commensal or pathogenic microorganisms. Research addressing this possibility will likely reveal novel mechanisms of lymphoid tissue development, driven by host-microbial interactions.

5.2.3 Gut-Associated Lymphoid Tissue (GALT)

The GALT of primates and rodents includes PPs serially distributed along the small intestine, ILFs located throughout the antimesenteric wall of the small intestine and cryptopatches found in the crypt lamina propria of the alimentary tract [14, 22]. In a number of other species, specialized GALT serves as a site for proliferative expansion of the B cell population and diversification of the primary antibody repertoire. These include the bursa of Fabricius in chickens, the ileal Peyer's patch in sheep, cattle and pigs, and the sacculus rotundus and appendix in rabbits. Although some GALT is fully developed, or nearly so, by birth (e.g. the bursa in chickens and the ileal PP in sheep and cattle), most GALT is immature at birth and undergoes significant postnatal development. Much of GALT development thus occurs following microbial colonization of the gastrointestinal tract and, in many cases, is dependent on interaction with intestinal commensals. Unlike BALT and NALT, GALT resides in an environment dominated by a resident microbial community of enormous complexity and density. The mammalian intestinal microbiota comprises 500–1000 microbial species colonizing at a density of about 10^{12} viable bacteria per gram of colonic content [23]. Through millions of years of co-evolution with their intestinal microbiota, mammals have come to not only tolerate, but also require developmental cues derived from their intestinal commensals.

Studies of germfree and gnotobiotic animals have clearly demonstrated the importance of intestinal commensals for GALT development. Germfree mice have reduced numbers of PPs, with fewer lymphoid follicles, reduced cellularity and fewer M cells than those of conventional mice [24]. Normal numbers of PPs, structurally identical to those of conventional mice, can be restored by transfer to conventional specific pathogen-free conditions, or by mono-association with a single commensal isolate [24, 25]. In rats, PP B cells do not express Bcl-2, an anti-apoptotic molecule associated with cellular differentiation and longevity, in the absence of intestinal colonization [26]. Intestinal commensals probably also regulate FAE function through stimulation of FAE-expressed TLRs, which recognize specific microbial components and subsequently induce upregulation of inflammatory and immunomodulatory genes [27]. For example, Chabot et al. (2006) demonstrated that the mouse PP FAE expresses several TLRs and that TLR2 stimulation induced rapid TLR2 redistribution, enhanced M cell transcytosis and subepithelial DC migration into M cells [28]. Yamanaka et al. (2003) found that the PP FAE of germfree rats, unlike that of conventional rats, was populated primarily by immature DCs instead of B and T cells [26]. These DCs rapidly disappeared from the FAE after conventionalization, likely due to migration to the interfollicular T cell areas, where they induce proinflammatory differentiation of naïve T cells [26, 29, 30]. The intestinal microbiota thus influences a wide range of developmental events in GALT.

A number of studies have shown that members of the intestinal microbiota differ in their ability to stimulate GALT development. Rhee et al. (2004) demonstrated that some, but not all, intestinal isolates induced robust follicle development after being introduced into sterile rabbit appendix [31]. Co-introduction of *Bacteroides fragilis* and *Bacillus subtilis*, for example, induced robust follicle development, while the introduction of other commensals (e.g. *Clostridium subterminale*) induced little or none. CCL21 expression and T cell area formation in the rabbit appendix are also dependent on the intestinal microbiota, and as in the induction of follicle formation, intestinal commensals differ strikingly in their ability to induce these developmental events [32]. Similarly, Maeda et al. (2001) reported that some, but not all, intestinal commensals upregulated CCL21 expression in murine PPs, which resulted in T cell accumulation and the induction of oral tolerance [33]. Others have also noted differential impacts of commensal species on GALT development. For example, Shroff et al. (1995) and Snel et al. (1998) reported that, unlike most other intestinal commensals, *Morganella morganii* and segmented filamentous bacteria (SFB), respectively, potently induced PP germinal center formation and specific IgA responses when introduced into germfree mice [34, 35]. Collectively, these studies demonstrate that the intestinal microbiota stimulates numerous morphological, functional and immunological aspects of GALT development and is thus critical for the development and optimal functioning of the mucosal immune system.

5.3 How does the Microbiota Drive Development of GALT?

Rabbits provide a useful model for investigating GALT development because their GALT is extraordinarily large. Together, the appendix, the Peyer's patches, and the sacculus rotundus, an organ lying at the ileal-cecal junction, contain more than 10^{10} lymphocytes, most of which are B lymphocytes. GALT development results in the formation of B cell follicles separated by T cell areas. This process occurs in two stages, the first beginning shortly after birth with an influx of B cells into the tissue mediated, at least in part, by the interaction of CD62L expressed on the B cells and peripheral lymph node addressin (PNAd) expressed by high endothelial venules (HEVs) [36]. This influx of B cells declines rapidly after the first week of life due to a sharp decline in total area of $PNAd^+$ HEVs, and by 6 weeks of age, 100-fold fewer B cells are recruited into the appendix [37]. Despite the decrease in B cells entering GALT, B cell follicles continue to grow at a rapid pace due to extensive B cell proliferation within the follicles. During this second stage of development, the Ig genes are somatically diversified, and B cell selection occurs [5, 38, 39]. Although, in the first stage of development, the influx of B cells can occur in the absence of the intestinal microbiota, B cell proliferation, follicle formation, and Ig gene diversification in the second stage cannot [39].

The mechanisms by which intestinal commensals induce stage two processes are not known. One possibility is that B cell proliferation is induced in response to antigen in a T cell-dependent manner. Such B cell stimulation in GALT could be initiated by transepithelial or subepithelial DCs that capture and internalize microbes [40] and present antigen to T cells, which in turn activate B cells (Fig. 5.1A). Although polyclonal development of B cell follicles and diversification of Ig genes in GALT could occur in this manner in response to the enormous number of epitopes present in the intestinal microbiota, we do not think this is the case. Instead, we think that GALT in rabbit plays a role similar to that of the bone marrow in mouse and human, where the diversified preimmune B cell repertoire develops in an antigen-independent manner [41, 42]. Consistent with this idea, somatic mutations in VDJ genes in GALT are targeted to different locations than those in splenic VDJ genes induced in response to immunization with antigen [43]. Further, transgenic mice in which all B cells were specific for the chicken protein HEL developed normal-sized PPs, even though the spleen and MLN were significantly decreased in size [44]. This finding indicates that B cells in PPs proliferated in the absence of antigen. Similarly,

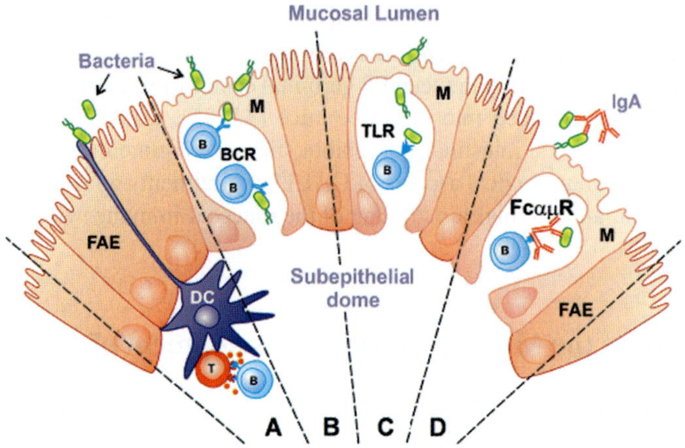

Fig. 5.1 Mechanisms by which the commensal microbiota might mediate development of rabbit GALT, Ig diversification, and selective expansion of B cells. Bacteria may enter GALT through M cells or through transepithelial DCs. B cells may be stimulated to proliferate in a T cell- and antigen-dependent manner (**A**) in which DCs present antigen to T cells that, in turn, provide help to B cells; or in a T cell- and antigen-independent manner (**B–D**). In the T cell-independent mechanism, B cells within M cell pockets could interact with bacteria and be stimulated through the BCR (**B**) or TLR (**C**); additionally, IgA bound to Fcα/μR could provide a co-stimulatory signal (**D**). DC= dendritic cell; FAE = follicle-associated epithelium; M = M cell

we think B cells in rabbit GALT are stimulated in an antigen-independent manner, giving rise to a diversified preimmune repertoire. How then are B cells stimulated in GALT, and how does the microbiota contribute to this process? Below, we discuss the polyclonal antigen-independent mechanisms by which commensals might stimulate B cells in stage two of GALT development (Fig. 5.1).

5.3.1 Polyclonal Antigen-Independent Stimulation of B Cells

We consider several mechanisms by which intestinal commensals could contribute to polyclonal B cell stimulation and somatic diversification of Ig genes in GALT (Fig. 5.1B–D). Our models for antigen-independent stimulation of B cells are based in part on the possibility that commensals interact directly with B cells either within M cell pockets or in the SED region. It is known that commensal bacteria can be transcytosed across the apical membrane of M cells into the M cell pockets [45]. There, the bacteria can interact with, and potentially stimulate, B cells. Evidence that B cells are found near the M cell pockets during the initial stages of GALT development was reported recently [46]. In immunodeficient IgH transgenic rabbits, in which the appearance of peripheral B cells and development of GALT are greatly delayed, a few B cells can be found in the SED region of the appendix, close to the location of M cells, before B cell follicles form. This observation suggests to us that the B cells entering the appendix migrate into the SED region, and likely

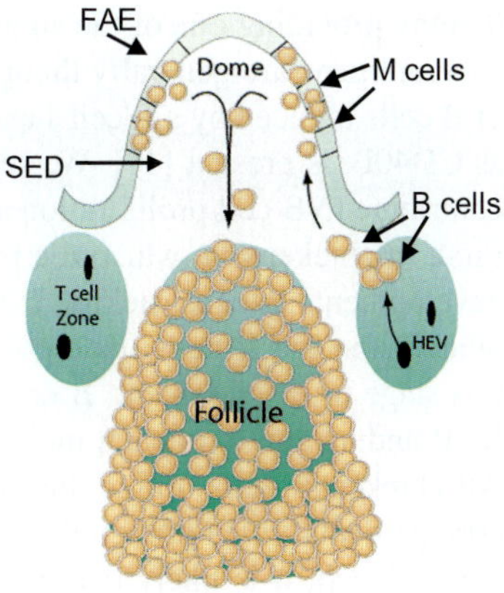

Fig. 5.2 Model of B cell development in GALT. B cells enter GALT from blood through the HEVs and migrate into the subepithelial dome (SED) region. As shown in Fig. 5.1, the B cells may be stimulated through interaction with bacteria that have been transcytosed across the follicle-associated epithelium (FAE) through M cells. (Adapted from [46] with permission from Elsevier Limited publishers)

enter M cell pockets, where they may encounter bacteria and become stimulated (Fig. 5.2). Below, we consider the possibility that the B cells are activated through Toll-like receptors (TLRs) or through the BCR. Then, we discuss the possibility that IgA-coated microbes stimulate B cells through the Fc$\alpha\mu$R.

5.3.1.1 TLRs

TLRs recognize microbial molecular patterns and this interaction leads to recruitment of adaptor molecules and initiation of an intracellular signaling cascade that results in activation of NF-κB [27]. B cells can be stimulated through TLRs in a polyclonal, antigen-independent manner [47]. After bacteria are taken up by M cells, they could potentially interact with TLRs on B cells located within the M cell pockets or in the SED region (Fig. 5.1C) and stimulate B cells to proliferate and somatically diversify their Ig genes. Alternatively, the bacteria may not interact directly with B cell TLRs but instead may interact with TLRs on DCs found in the M cell pockets and SED regions of the appendix. Activated DCs could then release stimulatory molecules such as B cell activating factor (BAFF), which in turn would activate B cells. Because TLRs are present on multiple cell types in the mucosa, and because the microbiota is required for GALT development, we assume that microbial stimulation of cells through TLRs contributes, in some manner, to B cell proliferation and GALT development. A combination of in vitro and in vivo experiments with TLR agonists may help elucidate how TLRs contribute to GALT development.

5.3.1.2 BCR

The microbiota may also polyclonally stimulate B cells to proliferate and somatically diversify their Ig genes by the interaction of a microbial superantigen with the BCR (Fig. 5.1B). We envision that bacterial molecules of some, but not all, bacteria serve as B cell superantigens and bind V_H framework regions (FR) of most B cells, thereby providing one of the signals required for B cell activation. Although B cell superantigens are generally thought to delete B cells to which they bind, apoptosis of B cells induced by a B cell superantigen can be overcome if a second signal, such as CD40L, is present [48]. We suggest that microbial-derived B cell superantigens contribute to B cell proliferation and GALT development for the following reasons. First, in chickens, in which the bursa of Fabricius plays an essential role in B cell development, B cells lacking a functional BCR develop normally until hatch, at which time the B cells undergo apoptotic death [49]. These data suggest that, at this stage of development, B cell survival is dependent on interaction between the BCR and an environmental molecule, perhaps expressed by intestinal commensals. All chicken B cells utilize the same V_H and V_L gene segment in V(D)J gene rearrangements. Even though the BCR is somatically diversified by gene conversion before hatch, it is likely that FRs are sufficiently conserved among the donor V_H or V_L gene segments that they remain available for binding to a microbial B cell superantigen. We propose that post-hatch bursal B cell development is driven by interaction between the BCR and a microbial B cell superantigen, or an endogenous bursal superantigen induced by the intestinal microbiota.

Our studies in rabbit also lead us to propose that GALT development is driven by a B cell superantigen [5]. Most B cells in rabbit utilize the V_H gene segment, $V_H 1$, in their VDJ gene rearrangements. These B cells are designated V_Ha B cells because the $V_H 1$ genes of different H-chain haplotype rabbits encode the V_Ha allotype. A small percentage of B cells utilize other V_H gene segments, designated V_Hn (V_Ha-negative), which encode V_H regions that lack the V_Ha allotype. However, the number of V_Hn B cells exiting the bone marrow is low, and these B cells proliferate less than V_Ha B cells in GALT [5, 50]. Rhee et al. (2005) examined mutant Alicia rabbits, in which GALT is initially populated primarily by V_Hn B cells, and later, by V_Ha B cells [5]. By ligating the appendix at birth, thereby preventing the interaction of intestinal commensals with B cells in the appendix, the authors showed that V_Ha B cells did not accumulate in the appendix and, instead, the appendix continued to be populated predominantly by V_Hn B cells, albeit at low numbers. These data show that the proliferation of V_Ha B cells in GALT is dependent on the intestinal microbiota. This finding taken together with the finding that V_Hn B cells do not expand significantly in number with or without the microbiota, leads to the question of how V_Ha and V_Hn B cells intrinsically differ from each other such that V_Ha, but not V_Hn, B cells are stimulated to proliferate in the presence of intestinal microbiota.

V_Ha and V_Hn B cells appear to differ only in the V_H regions of their VDJ genes. Comparison of the amino acid sequences of V_Ha and V_Hn regions led to the identification of seven FR1 and FR3 amino acids that distinguish V_Ha molecules from V_Hn molecules (Fig. 5.3). These seven amino acids are located on the β-strands on the exterior face of the V_H domain and could serve as a ligand-binding site for a superantigen-like molecule. We hypothesize that V_Ha B cells bind a microbial superantigen or an endogenous superantigen upregulated by the microbiota, and further, that this binding leads to proliferation and expansion of these B cells. We suggest that V_Hn B cells, on the other hand, have low affinity for potential superantigens and are therefore not expanded in GALT. Identification of a B cell superantigen,

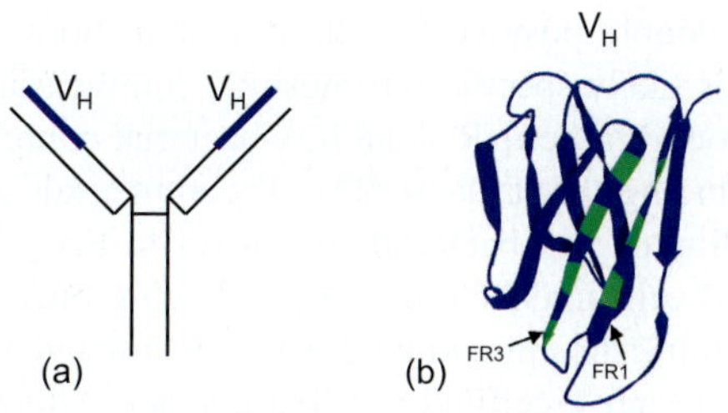

Fig. 5.3 (**a**) Drawing of an antibody molecule, highlighting the V_H region in blue. (**b**) Ribbon drawing of V_H region showing seven amino acids (*green*) in FR1 and FR3 that distinguish V_Ha from V_Hn molecules and may form a binding site for a microbial superantigen. (Adapted from [79])

whether of microbial or microbial-induced endogenous origin, would enhance our understanding of the multiple dimensions by which the commensal microbiota contributes to the well being of the host.

5.3.1.3 Fc$\alpha\mu$R

B cells in the follicles of rabbit appendix express both IgM and IgA on the cell surface [51]. In fact, by 4 weeks of age, more than half (58%) of B cells in the rabbit appendix are IgM^+IgA^+ double positive (Fig. 5.4a); in contrast, only 1% of B cells in spleen express IgA. By single cell RT-PCR, we found μ-chain, but not α-chain, mRNA in appendix B cells, demonstrating that the IgM^+IgA^+ double positive cells do not synthesize IgA. Accordingly, the IgA must be passively adsorbed, presumably through one of the IgA receptors: pIgR, FcαR (CD89), or Fc$\alpha\mu$R. As determined by RT-PCR, the IgA^+IgM^+ B cells do not express pIgR (Severson and Knight, unpublished data), and FcαR (CD89) is reportedly not present in rabbits [52]. We found by RT-PCR that rabbit appendix cells express Fc$\alpha\mu$R, a receptor shown to bind IgA and also IgM [53, 54]. We hypothesize that the IgA associated

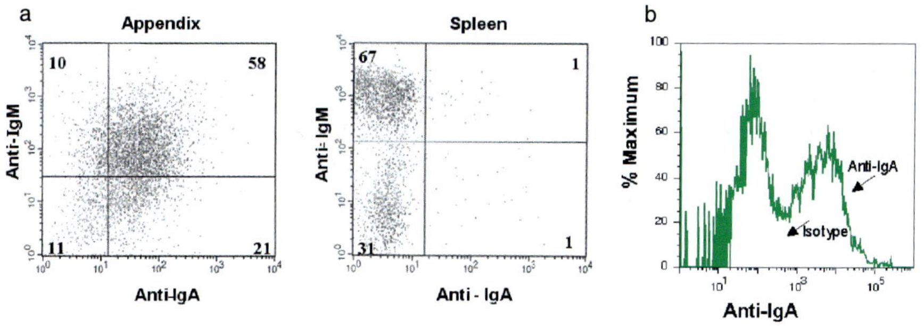

Fig. 5.4 IgA bound to IgM^+ B cells in GALT, and to intestinal commensals. (**a**) Cells from appendix and spleen of a 4-week-old rabbit were stained with FITC-anti-α-chain mAb and PE-anti-μ-chain mAb; (**b**) Lumenal bacteria from 6-week-old rabbit were stained with anti-rabbit IgA mAb or isotype control

with the IgM^+IgA^+ double positive B cells is bound through Fc$\alpha\mu$R. The IgA bound to the B cells is likely specific for intestinal commensals [55], and we suggest that the IgA is bound to Fc$\alpha\mu$R as an IgA-bacterial complex. Consistent with this proposal is the finding that at least 50% of commensals are coated with IgA (Fig. 5.4b). The binding of IgA-bacterial complexes to Fc$\alpha\mu$R may deliver a co-stimulatory signal to B cells in GALT, leading to B cell proliferation.

In conclusion, we think the microbiota drives polyclonal development of B cells in GALT in an antigen- and T cell-independent manner. Although we think B cell stimulation in GALT is antigen-independent, we assume that in other tissues, such as the LP, B cell stimulation is antigen-dependent. We have focused on three mechanisms by which the microbiota could contribute to stimulation of B cells in GALT, and if two signals are required for activation, one or both of these could occur through TLR, BCR and/or Fc$\alpha\mu$R. Other co-stimulatory signals such as those provided by CD40/CD40L interactions, or cytokines such as BAFF, may also contribute to B cell stimulation in GALT. While many of the studies described above were performed with rabbits and chickens, species in which GALT is required for development of the antibody repertoire, we suggest the mechanisms by which the microbiota induces B cell proliferation are not unique to these species, but instead, likely contribute to development of GALT in many species, including humans and mice.

5.4 IgA Production and Regulation of Gut Homeostasis

5.4.1 Production of IgA

The importance of the intestinal microbiota in promoting mucosal tissue organogenesis has been demonstrated in various species. Once mucosal lymphoid tissues are established, the mucosal immune system functions to protect mucosal surfaces against invading microbes. This protection is mediated largely through IgA. As demonstrated in humans, each day more IgA is produced than all other isotypes combined [56]. Very few IgA antibody secreting cells and low levels of IgA are found in germfree mice and neonates, indicating that IgA is induced in response to the intestinal microbiota [3, 57, 58]. The mechanisms by which commensal bacteria induce IgA are not well defined (reviewed by Cebra et al. (2005)) [59]. As we described above for mucosal organogenesis, bacteria differ in their capacity to promote class switch to IgA and subsequent antibody secretion [4], and in this section, we discuss mechanisms of IgA induction in the intestinal mucosa. We also speculate as to how the intestinal microbiota may contribute to these mechanisms.

5.4.1.1 Peyer's Patches

B cells in PPs undergo class switch recombination (CSR) to IgA during typical germinal center (GC) reactions [60, 61, 62]. Unlike transient GCs in the spleen and lymph nodes formed in response to antigen, PP GCs are constitutive, which is likely due to the continuous supply of antigens from the intestinal microbiota [61]. Below, we discuss possible mechanisms by which intestinal commensals may stimulate continual IgA CSR in the PPs.

Casola et al. (2004) demonstrated that the intestinal microbiota induced PP GCs in a T cell-dependent, but BCR-independent, manner [44]. The authors predicted that PP B cells enter GCs following microbial stimulation through innate immune receptors instead of through the BCR. Alternatively, Macpherson and Uhr (2004) showed that, upon commensal bacterial translocation through the epithelium, PP DCs took up the bacteria, became activated, and induced the production of IgA^{+}B cells in a T cell-independent manner [63]. Following the uptake of intestinal bacteria, DCs may stimulate IgA class switch via the production of BAFF and APRIL, molecules secreted by DCs that induce IgA class switch [64]. Another mechanism of bacterial-induced IgA^{+} B cell production in the PPs may involve M cells, which can transcytose intestinal microbes and thus, perhaps, facilitate direct stimulation of B cells to undergo IgA class switch in PP GCs.

5.4.1.2 Lamina Propria (LP)

Although PPs are considered inductive sites and the majority of IgA-expressing B cells are generated during GC reactions in the PPs, there is evidence that B cells can also class switch to IgA independently of GCs and PPs, in gut LP–a mucosal effector site. Naïve IgM^{+} B cells can be recruited directly to the gut LP and subsequently class switch to IgA in situ [65, 66]. The microenvironment established by LP stromal cells, specifically, LTβR signaling in the LP microenvironment, can promote IgM^{+} B cell-class switch to IgA [66, 67].

Both bone marrow-derived and peritoneal B1 cells can be recruited to the gut LP where they are induced to undergo IgA class switch [68]. Macpherson et al. (2000) demonstrated that the majority of IgA produced in response to commensal bacteria was produced in the gut LP by B1 cells, and that the IgA produced was directed against cell surface proteins of intestinal commensals [55]. Further, these authors showed that the IgA was produced in a T cell- and GC-independent manner. Ha et al. (2006) suggested that the intestinal microbiota contributes to IgA production in the LP by stimulating peritoneal B1 cells through TLRs and downregulating the expression of genes to promote detachment and emigration from the peritoneum [69]. These authors suggest that, following contact with gut microbial antigens, peritoneal B1 cells subsequently migrate to the LP where they can then be induced to undergo IgA class switch.

Naïve B cells within the LP may also receive signals directly from the intestinal microbiota in situ to class switch to IgA in a GC- and PP-independent manner. Bacteria may enter the LP through two distinct mechanisms involving DCs or villus M cells. LP DCs can penetrate epithelial cell tight junctions to sample lumenal bacteria [40]. After taking up intestinal bacteria, the DCs may become activated to secrete molecules, such as BAFF and APRIL, which can promote IgA production in the B cells within the LP. Alternatively, M cells present in the intestinal villus epithelium have been shown to take up bacteria and promote antigen-specific immune responses in the absence of PPs, ILFs, and MLN [70]. We speculate that this mechanism may facilitate bacterial-induced IgA production. Bacteria may also induce IgA production in the LP without gaining entry into the tissue. In the gut, the intestinal microbiota is in intimate contact with epithelial cells, and as shown by He et al. (2007), intestinal bacteria can promote IgA class switch via TLRs on intestinal epithelial cells [71]. In response to TLR engagement by select commensal bacterial species, intestinal epithelial cells produced APRIL, which in turn induced IgA class switch. A similar mechanism has recently been reported in oropharyngeal lymphoid tissues, in which tonsillar epithelial cells promoted IgA class switch by producing BAFF and IL-10 in response to engagement of TLR3 [72].

Taken together, these data indicate that intestinal commensals induce IgA class switch and likely, the production of IgA, through a variety of T cell-dependent and -independent mechanisms. The bacterial-induced IgA subsequently binds to the commensals to prevent bacterial translocation and invasion of host tissues.

5.4.2 IgA Regulation of Gut Homeostasis

While commensals are required for many physiologic processes, including mucosal lymphoid tissue organogenesis and IgA production, the microbiota outnumbers the host cells by 100-fold, presenting a serious risk of the commensals invading and infecting the host. It seems paradoxical that the host is dependent on its intestinal microbiota for the development of mucosal lymphoid tissues in the gut, because the major function of these tissues is to protect mucosal surfaces against invading microbes. This situation might have arisen evolutionarily from the need for the host and its intestinal microbiota to strike a peaceful coexistence, or homeostasis. It has become clear that the principal effector molecule of mucosal immunity, IgA, is also a major arbiter of host-microbial homeostasis in the gut.

By binding to commensals, secretory IgA (SIgA) prevents their translocation into host tissues and subsequent initiation of infection and host immune responses [73]. During early postnatal development, this protection is provided by maternal SIgA present in the nursing mother's breast milk [74]. With age, the neonate increasingly produces its own SIgA, induced in GALT by intestinal commensals. While some of the SIgA produced in mice colonized with a single microorganism is specific for that particular organism, the specificity of much of the synthesized IgA is not known and is considered "natural" IgA [75]. This "natural" IgA may cross-react with other microbial species due to common molecules on the cell surface, such as cell-surface polysaccharides [76], or similar to bacterial superantigens, intestinal bacterial molecules may bind to common regions of SIgA, such as the FRs or constant regions [76]. The induction of both high-affinity, organism-specific SIgA and "natural" polyreactive IgA provides broad protection against bacterial translocation and colonization of host tissues, and thus helps maintain a peaceful coexistence between commensals and their host [75].

Striking evidence for the regulation of gut homeostasis by IgA was reported by Suzuki et al. (2004) and Fagarasan et al. (2002), who found that the number of anaerobic bacteria, particularly segmented filamentous bacteria (SFB), expanded 100-fold in the small intestine of activation-induced cytidine deaminase null ($AID^{-/-}$) mice, in which B cells can neither class switch nor somatically hypermutate their Ig genes [77, 78]. These animals also displayed extensive B cell activation in peripheral lymphoid tissues and increased numbers of ILFs along the small intestine, likely due to unregulated activation of B cells by SFB. Introduction of IgA into these mice reestablished normal SFB numbers and decreased the abnormally high level of B cell activation in peripheral lymphoid organs, as well as the high numbers of ILFs. These findings show that IgM, somatically undiversified in the $AID^{-/-}$ mice, cannot compensate for the loss of IgA. It is likely that diversified Igs are required for regulating the growth of anaerobes in the small intestine, ensuring that B cell activation and GC formation are not rampant in secondary lymphoid organs.

Bacterial induction of host IgA synthesis is a good example of how mutualism is achieved and maintained between commensal bacteria and their host. The induced IgA maintains homeostasis between the host and the intestinal microbiota by preventing commensal translocation into host tissues and the initiation of host immune responses against the bacteria. This IgA helps maintain a peaceful coexistence between the host and the intestinal microbiota so that both benefit from their interaction. The intestinal commensals gain a stable, nutrient-rich environment, and among the many benefits gained by the host, commensal-induced IgA provides a potent defense against opportunistic pathogens.

5.5 Concluding Statement

In summary, as demonstrated by early studies in germ-free and gnotobiotic mice, the intestinal microbiota is required for mucosal immune system development, from the initiation of mucosal tissue development and establishment of mucosal immune response effector functions, such as IgA production, to the generation of a systemic T_H1/T_H2 balance. While the importance of the interaction between intestinal bacteria and its host has been known for years, the mechanisms underlying these interactions are only now being elucidated. Continued research in the area of microbial-mucosal immune system interactions will help us gain insight into these mechanisms and may also provide a general and better understanding of pathogen-host interactions at mucosal surfaces.

References

1. Tlaskalova-Hogenova, H., and Stepánková, R. 1980, Folia Biologica, 26, 81.
2. Kim, Y.B. 1979, Microbiology, D. Schlessinger (Ed.), American Society for Microbiology, 343.

3. Crabbe, P.A., Bazin, H., Eyssen, H., and Heremans, J.F. 1968, Int. Arch. Allergy Appl. Immunol., 34, 362.
4. Umesaki, Y., Setoyama, H., Matsumoto, S., Imaoka, A., and Itoh, K. 1999, Infect. Immun., 67, 3504.
5. Rhee, K.J., Jasper, P.J., Sethupathi, P., Shanmugam, M., Lanning, D., and Knight, K.L. 2005, J. Exp. Med., 201, 55.
6. Mazmanian, S.K., Liu, C.H., Tzianabos, A.O., and Kasper, D.L. 2005, Cell, 122, 107.
7. Gent, A.E., Berry, M.L., Kim, A., Nelson, A.J., Welch, M.P., and Aruffo, A. 1994, Lancet, 343, 766.
8. Strachan, D.P. 1989, BMJ, 299, 1259.
9. Umetsu, D.T., McIntire, J.J., Akbari, O., Macaubas, C., and DeKruyff, R.H. 2002, Nat. Immunol., 3, 715.
10. Elson, C.O., Cong, Y., McCracken, V.J., Dimmitt, R.A., Lorenz, R.G., and Weaver, C.T. 2005, Immunol. Rev., 206, 260.
11. Sminia, T., van der Brugge-Gamelkoorn, G.J., and Jeurissen, S.H. 1989, Crit. Rev. Immunol., 9, 119.
12. Bockman, D.E., and Cooper, M.D. 1973, Am. J. Anat., 136, 455.
13. Owen, R.L. 1977, Gastroenterology, 72, 440.
14. Saito, H., Kanamori, Y., Takemori, T., Nariuchi, H., Kubota, E., Takahashi-Iwanaga, H., Iwanaga, T., and Ishikawa, H. 1998, Science, 280, 275.
15. McDermott, M.R., and Bienenstock, J. 1979, J. Immunol., 122, 1892.
16. Beinenstock, J., and Clancy, R. 2005, Bronchus-associated lymphoid tissue in mucosal immunology, Elsevier, New York.
17. Gould, S.J., and Isaacson, P.G. 1993, J. Pathol., 169, 229.
18. Moyron-Quiroz, J.E., Rangel-Morneo, J., Kusser, K., Hartson, L., Sprague, F., Goodrich, S., Woodland, D.L., Lund, F.E., and Randall, T.D. 2004, Nat. Med., 10, 927.
19. Spit, B.J., Hendriksen, E.G., Bruijntjes, J.P., and Kuper, C.F. 1989, Cell Tissue Res., 255, 193.
20. Kiyono, H., and Fukuyama, S. 2004, Nat. Rev. Immunol., 4, 699.
21. Harmsen, A., Kusser, K., Hartson, L., Tighe, M., Sunshine, M.J., Sedgwick, J.D., Choi, Y., Littman, D.R., and Randall, T.D. 2002, J. Immunol., 168, 986.
22. Hamada, H., Hiroi, T., Nishiyama, Y., Takahashi, H., Masunaga, Y., Hachimura, S., Kaminogawa, S., Takahashi-Iwanaga, H., Iwanaga, T., Kiyono, H., Yamamoto, H., and Ishikawa, H. 2002, J. Immunol., 168, 57.
23. Hooper, L.V., and Gordon, J.I. 2001, Science, 292, 1115.
24. Smith, M.W., James, P.S., and Tivey, D.R. 1987, Am. J. Pathol., 128, 385.
25. Neutra, M.R., Pringault, E., and Kraehenbuhl, J.P. 1996, Annu. Rev. Immunol., 14, 275.
26. Yamanaka, T., Helgeland, L., Farstad, I.N., Fukushima, H., Midtvedt, T., and Brandtzaeg, P. 2003, J. Immunol., 170, 816.
27. Takeda, K., Kaisho, T., and Akira, S. 2003, Annu. Rev. Immunol., 21, 335.
28. Chabot, S., Wagner, J.S., Farrant, S., and Neutra, M.R. 2006, J. Immunol., 176, 4275.
29. Iwasaki, A., and Kelsall, B.L. 2000, J. Exp. Med., 191, 1381.
30. Dieu, M.C., Vanbervliet, B., Vicari, A., Bridon, J.M., Oldham, E., Ait-Yahia, S., Briere, F., Zlotnik, A., Lebecque, S., and Caux, C. 1998, J. Exp. Med., 188, 373.
31. Rhee, K.-J., Sethupathi, P., Driks, A., Lanning, D.K., and Knight, K.L. 2004, J. Immunol., 172, 1118.
32. Hanson, N.B., and Lanning, D.K. 2008, Dev. Comp. Immunol., 32, 980.
33. Maeda, Y., Noda, S., Tanaka, K., Sawamura, S., Aiba, Y., Ishikawa, H., Hasegawa, H., Kawabe, N., Miyasaka, M., and Koga, Y. 2001, Immunobiology, 204, 442.
34. Shroff, K.E., Meslin, K., and Cebra, J.J. 1995, Infect. Immun., 63, 3904.
35. Snel, J., Hermsen, C.C., Smits, H.J., Bos, N.A., Eling, W.M., Cebra, J.J., and Heidt, P.J. 1998, Can. J. Microbiol., 44, 1177.
36. Sinha, R.K., and Mage, R.G. 2004, Dev. Comp. Immunol., 28, 829.
37. Sinha, R.K., Alexander, C., and Mage, R.G. 2006, Vet. Immunol. Immunopathol., 110, 97.

38. Weinstein, P.D., Anderson, A.O., and Mage, R.G. 1994, Immunity, 1, 647.
39. Vajdy, M., Sethupathi, P., and Knight, K.L. 1998, J. Immunol., 160, 2725.
40. Rescigno, M., Urbano, M., Valzasina, B., Francolini, M., Rotta, G., Bonasio, R., Granucci, F., Kraehenbuhl, J.P., and Ricciardi-Castagnoli, P. 2001, Nat. Immunol., 2, 361.
41. Reynolds, J.D., and Morris, B. 1984, Eur. J. Immunol., 14, 1.
42. Weill, J.C., and Reynaud, C.A. 2005, J. Exp. Med., 201, 7.
43. Sehgal, D., Obiakor, H., and Mage, R.G. 2002, J. Immunol., 168, 5424.
44. Casola, S., Otipoby, K.L., Alimzhanov, M., Humme, S., Uyttersprot, N., Kutok, J.L., Carroll, M.C., and Rajewsky, K. 2004, Nat. Immunol., 5, 317.
45. Neutra, M.R., Phillips, T.L., Mayer, E.L., and Fishkind, D.J. 1987, Cell Tissue Res., 247, 537.
46. Jasper, P.J., Rhee, K.J., Kalis, S.L., Sethupathi, P., Yam, P.-C., Zhai, S.-K., and Knight, K.L. 2007, Eur. J. Immunol., 37, 2290.
47. Martin, F., Oliver, A.M., and Kearney, J.F. 2001, Immunity, 14, 617.
48. Goodyear, C.S., and Silverman, G.J. 2003, J. Exp. Med., 197, 1125.
49. Sayegh, C.E., and Ratcliffe, M.J. 2000, J. Immunol., 164, 5041.
50. Pospisil, R., Young-Cooper, G.O., and Mage, R.G. 1995, Proc. Natl. Acad. Sci. USA, 92, 6961.
51. Dasso, J.F., Obiakor, H., Bach, H., Anderson, A.O., and Mage, R.G. 2000, Dev. Comp. Immunol., 24, 797.
52. Abi-Rached, L., Dorighi, K., Norman, P.J., Yawata, M., and Parham, P. 2007, J. Immunol., 178, 7943.
53. Shibuya, A., Sakamoto, N., Shimizu, Y., Shibuya, K., Osawa, M., Hiroyama, T., Eyre, H.J., Sutherland, G.R., Endo, Y., Fujita, T., Miyabayashi, T., Sakano, S., Tsuji, T., Nakayama, E., Phillips, J.H., Lanier, L.L., and Nakauchi, H. 2000, Nat. Immunol., 1, 441.
54. Sakamoto, N., Shibuya, K., Shimizu, Y., Yotsumoto, K., Miyabayashi, T., Sakano, S., Tsuji, T., Nakayama, E., Nakauchi, H., and Shibuya, A. 2001, Eur. J. Immunol., 31, 1310.
55. Macpherson, A.J., Gatto, D., Sainsbury, E., Harriman, G.R., Hengartner, H., and Zinkernagel, R.M. 2000, Science, 288, 2222.
56. van der Heijden, P.J., Stok, W., and Bianchi, A.T. 1987, Immunology, 62, 551.
57. Crabbe, P.A., Nash, D.R., Bazin, H., Eyssen, H., and Heremans, J.F. 1970, Lab. Invest., 22, 448.
58. Moreau, M.C., Ducluzeau, R., Guy-Grand, D., and Muller, M.C. 1978, Infect. Immun., 21, 532.
59. Cebra, J.J., Jiang, H.-Q., Boiko, N., and Tlaskalova-Hogenova, H. 2005, Handbook of Mucosal Immunology, J. Mestecky, M.W. Lamm, W. Strober, J. Bienenstock, J.R. McGhee, and L. Mayer (Eds.), Elsevier Academic Press, San Diego, 487.
60. Craig, S.W., and Cebra, J.J. 1971, J. Exp. Med., 134, 188.
61. Weinstein, J., and Cebra, J. 1991, J. Immunol., 147, 4126.
62. Butcher, E.C., Rouse, R.V., Coffman, R.L., Nottenburg, C.N., Hardy, R.R., and Weissman, I.L. 1982, J. Immunol., 129, 2698.
63. Macpherson, A.J., and Uhr, T. 2004, Science, 303, 1662.
64. Litinskiy, M.B., Nardelli, B., Hilbert, D.M., He, B., Schaffer, A., Casali, P., and Cerutti, A. 2002, Nat. Immunol., 3, 822.
65. Suzuki, K., Meek, B., Doi, Y., Honjo, T., and Fagarasan, S. 2005, Proc. Natl. Acad. Sci. USA, 102, 2482.
66. Fagarasan, S., Kinoshita, K., Muramatsu, M., Ikuta, K., and Honjo, T. 2001, Nature, 413, 639.
67. Kang, H.S., Chin, R.K., Wang, Y., Yu, P., Wang, J., Newell, K.A., and Fu, Y.X. 2002, Nat. Immunol., 3, 576.
68. Suzuki, K., Ha, S.A., Tsuji, M., and Fagarasan, S. 2007, Semin. Immunol., 19, 127.
69. Ha, S.A., Tsuji, M., Suzuki, K., Meek, B., Yasuda, N., Kaisho, T., and Fagarasan, S. 2006, J. Exp. Med., 203, 2541.
70. Jang, M.H., Kweon, M.N., Iwatani, K., Yamamoto, M., Terahara, K., Sasakawa, C., Suzuki, T., Nochi, T., Yokota, Y., Rennert, P.D., Hiroi, T., Tamagawa, H., Iijima, H., Kunisawa, J., Yuki, Y., and Kiyono, H. 2004, Proc. Natl. Acad. Sci. USA, 101, 6110.

71. He, B., Xu, W., Santini, P.A., Polydorides, A.D., Chiu, A., Estrella, J., Shan, M., Chadburn, A., Villanacci, V., Plebani, A., Knowles, D.M., Rescigno, M., and Cerutti, A. 2007, Immunity, 26, 812.
72. Xu, W., He, B., Chiu, A., Chadburn, A., Shan, M., Buldys, M., Ding, A., Knowles, D.M., Santini, P.A., and Cerutti, A. 2007, Nat. Immunol., 8, 294.
73. Kramer, D.R., and Cebra, J.J. 1995, J. Immunol., 154, 2051.
74. Harris, N.L., Spoerri, I., Schopfer, J.F., Nembrini, C., Merky, P., Massacand, J., Urban, J.F., Jr., Lamarre, A., Burki, K., Odermatt, B., Zinkernagel, R.M., and Macpherson, A.J. 2006, J. Immunol., 177, 6256.
75. Talham, G.L., Jiang, H.Q., Bos, N.A., and Cebra, J.J. 1999, Infect. Immun., 67, 1992.
76. Bos, N.A., Bun, J.C., Popma, S.H., Cebra, E.R., Deenen, G.J., van der Cammen, M.J., Kroese, F.G., and Cebra, J.J. 1996, Infect. Immun., 64, 616.
77. Suzuki, K., Meek, B., Doi, Y., Muramatsu, M., Chiba, T., Honjo, T., and Fagarasan, S. 2004, Proc. Natl. Acad. Sci. USA, 101, 1981.
78. Fagarasan, S., Muramatsu, M., Suzuki, K., Nagaoka, H., Hiai, H., and Honjo, T. 2002, Science, 298, 1424.
79. Lanning, D.K., Rhee, K.J., and Knight, K.L. 2005, Trends Immunol., 26, 419.

Part III
Gastrointestinal Pathogens

Chapter 6
Mucosal Immune Responses Against Enterotoxigenic *Escherichia coli* [ETEC] in Humans

Ann-Mari Svennerholm and Firdausi Qadri

Abstract Enterotoxigenic Escherichia coli [ETEC] is among the most important causes of diarrhoea morbidity and mortality, particularly in children below 5 years of age in developing countries. The bacteria are also a common cause of diarrhoea outbreaks as well as of diarrhoeal disease in travellers to Africa, Asia and Latin America. ETEC is a classical non-invasive mucosal pathogen that colonizes the epithelium in the small intestine by means of different colonization factors [CFs] that usually are fimbriae. The bacteria produce either or both of a heat-labile [LT] and a heat-stable toxin [ST] that bind to the epithelial cells and cause secretion of water and electrolytes resulting in watery diarrhoea. ETEC infection results in intestinal secretory IgA and also systemic IgA and IgG antibody responses against the main virulence factors, i.e. the CFs and LT and also against the O antigen of the infecting strain. Protective immunity is most likely provided by such locally produced antibodies that prevent binding of bacteria and toxin action on the epithelial cells. Different live as well as inactivated candidate vaccines against ETEC have been developed based on toxin and/or CF antigens and administered to induce mucosal immune responses, predominantly by the oral, but recently also by the transcutaneous route. Several of these candidate vaccines have been tested for safety and immunogenicity in ETEC endemic as well as non-endemic countries and in a few instances also for protective efficacy in travellers and in infants in the developing world. In this chapter we describe mucosal immune responses against both clinical ETEC disease and current ETEC candidate vaccines.

Keywords ETEC · Enterotoxins · Colonization Factors · Mucosal Immune Responses · SIgA · ETEC Vaccines · Protective Efficacy

---

A.-M. Svennerholm
Department of Microbiology and Immunology, Gothenburg University Vaccine Research Institute [GUVAX], the Sahlgrenska Academy at University of Gothenburg, Box 435 S-40530, Gothenburg, Sweden
e-mail: ann-mari.svennerholm@microbio.gu.se

6.1 Introduction

Enteric infections resulting in diarrhoeal disease remain a leading global health problem. It has been estimated that approximately 2–4 billion episodes of diarrhoea, resulting in more than 2 million deaths occur annually in children below 5 years in developing countries [1, 2]. Diarrhoeal diseases also constitute a major health problem in travellers to developing countries in Africa, Asia and Latin America [2, 3, 4]. Almost half of all diarrhoeas are due to bacteria that cause watery diarrhoeal disease by producing one or more enterotoxins. Among these bacteria enterotoxigenic *Escherichia coli* [ETEC] cause the largest number of cases, and based on recent estimates ETEC infections are responsible for ca 20% of all diarrhoeas in children in endemic countries and for 30–50% of all cases in travellers [3, 4]. Based on the great health impact of infections with ETEC, there is a great interest in developing an effective ETEC vaccine.

In this chapter we describe current knowledge of immune responses against ETEC as an important background for developing an effective vaccine against this common pathogen.

6.2 ETEC

Importance. ETEC is the most common cause of bacterial induced diarrhoea in the developing world. In these areas, on an annual basis, ETEC results in 280–400 million diarrhoeal episodes in children below 5 years and an additional 100 million episodes in children aged 5–14 years [1, 2, 3, 4]. Repeated episodes of ETEC diarrhoea have also been found to be an important cause of malnutrition in poor children in developing countries [2, 5, 6, 7] leading to growth retardation and stunting in later life [6]. Moreover, ETEC causes substantial disease in adults in developing countries with an estimated number of 400 million cases per year in persons above 15 years [3, 4]. ETEC may also be an important cause of diarrhoea during natural disasters such as flooding, tsunamis, earthquakes as well as man-made calamities and upheavals. We have recently shown that ETEC was as common as *V. cholerae* during a recent flooding in Bangladesh and responsible for ca 25% of all severely dehydrated diarrhoea cases [8].

ETEC cause disease by colonizing the small intestine and elaborating one or two different enterotoxins [3]. The resulting illness usually lasts from 3–5 days and ranges from mild diarrhoea without dehydration to severe cholera-like disease. Though the illness is typically mild, especially in children, it can result in severe dehydration, which non-treated may result in fatal outcomes [3]. The clinical picture of ETEC diarrhoea is very similar to that of cholera and in many patients it may be difficult to discriminate diarrhoea caused by ETEC and *V. cholerae* without immunological or molecular biological diagnostic procedures [3]. The high incidence of ETEC diarrhoea in inhabitants in developing countries, in travellers to these areas and during outbreaks makes ETEC an important target for immunoprophylaxis.

In regions of the world where ETEC is highly endemic there is a decline in ETEC diarrhoea incidence with age with peaks observed in the age groups 6–18 months [9, 10], whereas there is no evident association with age in short time visitors to endemic areas [2, 3, 4]. However, the incidence of ETEC rapidly decreases also in persons from industrialized countries during prolonged stay in ETEC endemic areas [3]. The rate of isolation of ETEC from asymptomatic children has varied between 0% and 20% in numerous studies carried out in children in different settings but has in most instances been significantly lower than the rates in children with diarrhoea [11, 12, 13, 14, 15, 16]; on average ETEC is seen at least 2–3 times more frequently in symptomatic as compared to in asymptomatic children [7, 11]. These observations strongly support the belief that effective immunity may develop after repeated infections and consequently protection by way of an effective ETEC vaccine is achievable. The design of such a vaccine should be based on the knowledge of mechanisms of disease and immunity in ETEC infections.

6.3 Pathogenesis

The major virulence mechanisms in ETEC include production of a heat-labile [LT] and/or a heat-stable [ST] enterotoxin that both have been fully characterized, cloned, sequenced, and their genetic control in transmissible plasmids identified [17, 18, 19, 20, 21]. Other important virulence factors in ETEC include production of one or more colonization factors [CFs] that also are under plasmid control [3, 22].

Enterotoxins. LT has been shown to be, structurally, functionally and antigenically very similar to cholera toxin [CT]. Thus, LT consists of an active [A] subunit and five identical binding [B] subunits which both share about 80% homology with corresponding CT subunits [20]. Immunity against LT is predominantly directed against the B subunit portion of the toxin molecule [20]. ST, which differs extensively from LT or CT, exists in two variants, STp and STh, which are very small molecular weight peptides, consisting of 18 and 19 amino acids, respectively; these peptides are not antigenic unless coupled to a carrier protein [21]. Hence, immune responses to ST are not induced after infection with ST producing ETEC. The relative proportion of strains producing LT alone, ST alone or LT/ST seems to vary from one geographic area to another and in patients with ETEC diarrhoea and in asymptomatic carriers [3, 4, 9, 10].

Colonization factors. There are more than 22 ETEC CFs that have been recognized among human ETEC and additional ones are likely to be identified [3, 22]. The CFs are mainly fimbrial or fibrillar proteins, although some CFs are not fimbrial in structure, notable among these is CS6 which has been increasingly isolated in recent studies [3, 23, 24]. The CFs promote colonization of the ETEC organisms in the small bowel, thus allowing expression of either or both of LT and ST in close proximity to the intestinal epithelium. Studies in humans as well as in experimental animals have shown that CF positive bacteria, but not their isogenic CF negative mutants, colonize and induce diarrhoea [3, 22]. Of the wide range of CFs, the most commonly present on diarrheaogenic strains include CFA/I, CS1, CS2, CS 3, CS4, CS5, CS6, and in some studies also CS7, CS14, CS17 and CS21 [3, 9, 10, 22]. Several of these CFs may be co-expressed on the same bacteria, e.g. CS1 + CS3, CS2 + CS3, CS4 + CS6, CS5 + CS6 are usually co-expressed although some strains may express CS3 alone and an increasing number CS6 alone [22]. The different CFs have been found on ETEC strains worldwide in varying frequencies in different geographic areas and settings as well as in various categories of patients/subjects [10, 23, 25, 26, 27, 28]. However, the CFs identified so far have not been detected on all ETEC, irrespective of whether genotypic or phenotypic methods have been used, and on roughly 20–50% of strains worldwide no known CFs could be detected [3, 22]. This could be due either to the absence of CFs, to loss of CF property on subculture of strains or that new, as yet not identified CFs are expressed. Interestingly, in a diarrhoeal epidemic where ETEC was detected in high frequency, over 70% of the strains were CF positive compared to <50% seen in strains isolated during seasonal epidemics and during the rest of the year. In addition, some CFs such as CFA/I, are more common during the spring diarrhoeal peak than during other times of the year, suggesting a relationship between expression of CFs and high diarrhoeal load in the community [7].

Some of the better characterized CFs can be subdivided into different families [29], i.e. the colonization factor I group [including CFA/I, CS1, CS2, CS4, CS14, CS17, CS22 and PCFO71] and the coli surface 5 group [CS5, CS7, CS18 and CS20] and those that are unique [CS3, CS6, CS10–12]. Most of the CFs are composed of up to 100 identical structural subunits [22] and several of the CFs also express distinct tip proteins [22, 29, 30]. For example, in the CFA/I family there are cross-reactive epitopes, both among the structural subunits (particularly in the N-terminal region; [31]) and the tip proteins [29], which have been considered as candidates for vaccine development. Both the tip proteins and the structural subunits have been shown to bind specifically to intestinal epithelial cells or glycoconjugates derived from such cells [29, 32]. Thus, we could recently demonstrate that CFA/I binds to certain glycosphingolipids which are present in human small intestine, e.g. the Lewis a blood group antigen, through the major subunit protein CfaB [32]. Similar binding patterns were observed for immunologically related CFs, e.g. CS1 and CS4. Binding to epithelium may also be mediated by the single tip subunit CfaE of CFA/I, but specific binding structures for the tip have not yet been identified ([29], *Savarino S*, personal communication). Protective immune responses against CFs may either be directed against the major subunit protein or the tip protein by preventing adhesion. Indeed, recent studies [*Savarino S* et al., Vaccines for enteric diseases, VED, conference, Lisbon, April 2007] have shown that antibody preparations against whole CFA/I fimbriae as well as against the tip protein in milk formulas have afforded significant protection in human volunteers challenged with virulent CFA/I ETEC.

ETEC serogroups. More than 100 different O-groups of *E. coli* have been identified among large strain collections of ETEC [23, 33]. In addition, rough strains which are non-typeable with regard to O antigen, are not uncommon [3]. Although there are some ETEC serogroups that are more prevalent than others, e.g. O6, O8,

O78, O128 and O153, and there may be regional clustering of O-groups [25], there are large geographic differences and there are no distinct serogroups that can be used for ETEC diagnostics or used as basis for a world-wide ETEC vaccine. Thus, immunity against O-antigens is serogroup-specific. We have previously shown that anti-O antibodies can afford passive protection against ETEC induced fluid secretion in experimental animals, but only against ETEC expressing homologous O antigens [34].

6.4 Immune Responses Against Natural ETEC Disease

Immune responses against ETEC infections in humans have predominantly been studied in hospitalized patients or in the community, both in children and adults, in different countries in Asia, Africa and South America. Due to the mucosal nature of ETEC disease there has often been a focus on determining mucosal immune responses, in particular immune responses locally in the intestine [Table 6.1]. Such intestinal immune responses have often been determined as specific immunoglobulin A [IgA] per total IgA responses in intestinal lavage fluid or as specific IgA antibody secreting cell [ASC] per total mononuclear cells [MNC] in blood or in intestinal biopsies [35, 36, 37, 38]; ASC responses of other isotypes have often also been determined. Alternative approaches have been to determine specific antibodies in saponin-extracted intestinal biopsies [37] or by determining circulating ASC or antibodies produced by such cells in blood using the antibody in lymphocyte supernatant [ALS] method [39].

In a series of studies in Bangladesh we have shown that patients hospitalized with acute, watery diarrhoea due to ETEC develop significant IgA antibody responses against the key antigens of the infecting strain in intestine. Thus, in an early study during the 1980's [38] we demonstrated that 80% of the patients infected with LT-positive ETEC responded to LT, 63% of those infected with CF positive ETEC responded to the corresponding CFs and 78% to homologous lipopolysaccharide O-antigen in intestinal lavage fluid. Furthermore, the intestinal immune responses were associated with immune responses in breast-milk, saliva and serum [38]. All these responses were optimal already within a week of hospitalization.

Table 6.1 Methods to assess mucosal immune responses against ETEC

Responses against CFs
–Specific IgA/total IgA in intestinal lavage is "golden standard"
–Circulating IgA ASC responses reflect lavage responses well
–IgA responses in stool reflect lavage responses well but are nonsensitive
–Intestinal IgA ASC or IgA in saponin-extracted biopsies reflect well, but are invasive
–IgA responses in serum comparatively infrequent and weak
–IgG responses in serum infrequent and very weak
Responses against LT/CTB
–Serum IgG and IgA reflect lavage responses very well
–Circulating IgA ASC reflect intestinal lavage responses very well

Immune responses locally in the intestine. In a subsequent study in Bangladeshi patients we confirmed the mucosal nature of ETEC immune responses after infection by showing that more than 90% of adult patients responded with high specific B cell responses in intestine, determined as specific ASC per total MNC, against LT and the CFs of the infecting strain in duodenal biopsies [37]. Optimal immune responses were recorded either at the acute stage or within a week of hospitalisation. The infection induced comparable levels of IgM ASC response against CFs in the duodenum at these time points, but increased levels of IgG ASC only in specimens collected in early convalescence. Similarly, the patients infected with LT producing ETEC responded with significant IgA ASC responses against the LT enterotoxin in intestine in most [83%] of the cases. Marked increases of IgA responses both against CFs and enterotoxin were also observed in fecal specimens collected at early convalescence [37]. Natural ETEC infection was also shown to induce high IgA ASC responses against the homologous CFs and more modest antitoxic responses in peripheral blood; these responses peaked already around 3 days after admission to hospital, i.e. ca 5–6 days after onset of infection [37]. Interestingly, there was a significant correlation [$p = 0.03$] between numbers of CF-specific B cells in blood sampled from patients early after infection and numbers of CF-specific B cells in duodenal biopsies collected 1 week later. These findings support the notion that determination of circulating ASC responses may be used to estimate B cell responses locally in the intestine, but only early after onset of disease. Thus, specific circulating ASC are only found ca 5–9 days after antigen stimulation locally in the gut and decline to baseline levels thereafter, whereas elevated intestinal ASC responses may remain for several weeks [36, 37].

Immune responses against CS6. In a recent study in Bangladesh, we were particularly interested in analyzing immune responses against the non-fimbrial colonization factor CS6, since a substantial portion of ETEC strains isolated from patients in endemic areas and in travellers to ETEC endemic countries express CS6, either alone or in combination with CS5 or CS4 [3, 23, *Torres O* et al., to be published]. Furthermore, CS6 is expressed in considerably lower quantities on the bacterial surface than the fimbrial CFs [*Svennerholm A-M*, unpublished]. Both adults [mean age 37 years] and children [mean age 1.9 years] hospitalized with diarrhoea caused by CS6 expressing ETEC [CS6 alone or together with CS5] and who had a disease spectrum ranging from mild to moderate to severe dehydration were enrolled for these studies [40]. More than 90% of the patients had significant mucosal immune responses to CS6, assessed either as IgA ASC or ALS in peripheral blood, and these responses peaked around a week after hospitalisation. Interestingly, the magnitude of IgA ASC response to CS6 were comparable to those we have previously found against fimbriated CFs, i.e. CFA/I, CS1, CS2, CS3 or CS5, in peripheral blood of Bangladeshi ETEC patients [37]. Both children and adults [altogether 57%] responded with rises in CS6-specific IgA in their stool specimens, which showed a similar kinetics of response as that seen in blood. These faecal responses declined in around 21 days. More than 80% of the patients responded with significant IgA and 70% with IgG responses in serum. There were no significant differences in magnitude of responses in patients infected with ETEC expressing CS6 alone as compared to with ETEC expressing CS6 together with CS5, nor between responses in children and adults; however, antibody levels recorded early after onset of diarrhoea and during early convalescence were higher in the adults than in the children [40].

Immune responses against cross-reactive CF antigens. In all the studies described so far only immune responses to antigens expressed by the infecting strain were assessed, To evaluate whether the CFs may induce immune responses not only against homologous but also against heterologous CFs sharing cross-reactive epitopes, e.g. in the CFA/I and CS5 groups, we have studied immune responses in Bangladeshi children 2–5 years of age with diarrhoea due to ETEC expressing CFs of the CFA/I and the CS5 groups, respectively [41]. These studies revealed that patients infected with ETEC belonging to the CFA/I group also responded with IgA ASC as well as plasma IgA responses to the cross-reacting CFs, although at lower levels than against the homologous CF; however no significant responses were observed against heterologous CFs. Similarly, children infected with ETEC of the CS5 group also responded both with IgA ASC and plasma antibody responses to the cross-reactive CF, but not significantly to CFs from the heterologous CF group. Also, these cross-reactive immune responses were observed early, with peak response already 1–2 weeks after onset of infection [41]. These findings give hope that a CF based ETEC vaccine composed of maybe the 5–7 most common CFs may provide cross-protection also against a larger mixture of heterologous ETEC strains expressing cross-reactive CF epitopes.

Immune responses against ETEC during infancy and early childhood. In a recent study we evaluated the natural history of ETEC infections and development of immune responses during early life in a birth cohort in an urban slum area in Dhaka, Bangladesh [7]. Over 300 children were followed up for studying incidence of ETEC infections, both symptomatic and asymptomatic, during their first two years of life with emphasis on the rate of initial and repeated infections with the same or heterologous phenotypes of the bacteria. In this setting, ETEC was the most common diarrhoeal pathogen with an incidence of 0.5 episodes/child/year and was significantly associated with diarrhoea when compared to the infection rate in healthy children [19.5% vs. 8% $P < 0.001$]. Among children with diarrhoea, ETEC expressing ST alone was predominant [49%] followed by the LT + ST phenotype [30%], and LT only expressing ETEC [21%]; ETEC from healthy children showed a different pattern with LT only strains being more common [35%] than in diarrhoea cases [7]. ETEC isolated from diarrhoeal cases were positive for one or more colonization factors in significantly higher frequency [66%] than ETEC from healthy children [33%] [$P < 0.001$], supporting the notion that the CFs are indeed significant virulence factors.

Interestingly, in this birth cohort study we found that none of the children who had experienced diarrhoea due to CFA/I, CS1 + CS3, CS2 + CS3 or CS5 + CS6 expressing ETEC had a repeat episode of diarrhoea from a homologous CF type of ETEC strain, but frequently episodes caused by heterologous ETEC strains during the two year study period, serving as supporting data that development of immunity against the CFs does occur [7]. Similarly, asymptomatic infections with ETEC expressing these CFs were not followed by symptomatic ETEC expressing homologous CFs. Repeated infections with ETEC expressing only CS6 were, however, not uncommon both after symptomatic and asymptomatic initial infections with CS6- positive ETEC. Likewise, symptomatic infections with LT only ETEC did not seem to protect against LT-producing *E. coli*, since they were often followed by symptomatic or asymptomatic infections of the same toxin type [7]. These data suggest that LT only ETEC strains are less protective, and while this is supported by recent findings from Egypt [27], it is not supported by findings from Guinea Bissau [42] [see below]. The poor protection induced by LT in some different studies may be explained by the fact that LT is usually produced in comparatively low amounts during clinical infection and immune responses are often rather modest, e.g. as compared to after administering CTB or LTB orally in different candidate vaccines [4, 43]. Recent studies carried out in Brazil have also shown that production and release of LT by wild-type human-derived ETEC strains are heterogeneous traits under both *in vitro* and *in vivo* growth conditions, which may affect the outcome of the disease and also the immune responses that are stimulated by LT [*Ferreira L*, personal communication].

The capacity of the major CFs on ETEC such as CFA/I and CS1–3 and CS5+CS6 strains to prevent re-infection with ETEC expressing corresponding CFs is encouraging and strengthens the present strategy of including these CFs in ETEC vaccine development programs [43]. However, more studies are needed to evaluate the duration of protection against reinfection, at least during childhood up to five years of age.

Immune responses against ETEC in infancy in Egypt and Guinea-Bissau. Support for a protective effect of anti-CF immunity has also been obtained from a case-control study in Egyptian infants in which 397 children <36 months of age were tested for a possible association between antitoxin and anti-colonization titres and risk of developing ETEC diarrhoea. In that study, serum antibody titres against CFA/I were inversely related to the risk of developing CFA/I ETEC diarrhoea in children <18 months [44]. However, no association was found with anti-LT titres and the risk of developing LT ETEC disease similar to findings from Bangladesh with a lack of protection induced by LT disease [7]. In a study from Guinea Bissau, on the other hand, in which 200 neonates were followed up to 2 years of age, infection with LT positive ETEC conferred 45% protection against symptomatic infections with LT-ETEC [42]. In the latter study, the authors concluded that the CFs did not seem to contribute to protection against reinfection with ETEC expressing homologous CFs; however, this conclusion was only based on estimates of protection against asymptomatic infections, since protection against symptomatic infections was not possible to determine due to the low number of cases [42]. Thus, there seems to be evidence that immunity against CFs, at least when induced by clinical disease, is related to protection against diarrhoea caused by ETEC expressing homologous CFs, whereas variable results have been achieved with regard to the protective effect induced by LT after natural infection.

Effect of breast-feeding. Since secretory IgA [SIgA] antibodies against CFs and LT are present in breast-milk from mothers in developing countries, it would be expected that breast-fed children are protected from ETEC diarrhoea [38, 45].

However, a possible relationship between breast-feeding and the prevalence of ETEC diarrhoea has varied in different studies; in studies in Egypt an association was found between severe diarrhoea and the breast-feeding pattern [44], whereas no such relation was observed in different studies in Bangladesh [3, 45]. Difficulties in proving an effect of breast-feeding could partly be explained by the fact that in ETEC endemic areas most women are breast-feeding during the first 2 years of life but this is very seldom an 'exclusive breast-feeding' pattern. Thus, children are given liquids including sugar syrups and water very early on in life to supplement breast milk, which may be a cause of infection. However, lower rates of ETEC diarrhoea have been observed during the first 6 months of life of breastfed children as compared to after weaning in most studies. The lower infection rates observed in the youngest infants may be attributed to SIgA antibodies and/or receptor analogues etc. in breast-milk [46], although it may also partly be explained by less exposure to contaminated foods and drinks.

6.5 Immune Responses Against ETEC in Challenged Human Volunteers

Immune responses against ETEC infection have been determined in human volunteers after challenge with fully virulent ETEC strains. Such studies were initiated already during the late 70's at the Center for Vaccine Development [CVD] in Baltimore [47] as a background for the development of ETEC vaccines. Different doses of various ETEC strains, e.g. ETEC expressing CFA/I and CFA/II [CS1 + CS3] were tested and the dose giving significant clinical symptoms in a majority of the volunteers, which usually corresponded to 10^8–10^9 colony forming units [cfu], was determined. In these early studies, mucosal immune responses in the challenged volunteers were assessed as specific ETEC antibodies in jejunal aspirates [47].

In subsequent studies at Johns Hopkins University, Baltimore and Walter Reed Army Institute of Research, Silver Spring, Maryland, several ETEC challenge studies have been undertaken to assess protective efficacy of different ETEC candidate vaccines. In these studies, that have included determination of immune responses that may be associated with protection against ETEC challenge [48, 49, *Savarino S* et al., unpublished], correlations were found between the magnitude of serological as well as mucosa-derived immune responses against CFs and protection against ETEC expressing corresponding CFs. It was shown that human volunteers having a reciprocal ELISA antibody titre of $\geq$155 against CS3 before challenge with CS1 + CS3 positive ETEC bacteria were protected against symptoms induced by the challenge organisms [48, *Bourgeois* et al., unpublished]. Serological antibody levels against LT, which seemed to be associated with protection, have also been identified. Immunological correlates of protection have also been searched for in field studies in Egyptian children; in those studies ELISA IgG titres against CFA/I of $\geq$76 were reported to reflect protection against CFA/I positive ETEC [44]. Hence, determination of immune responses in challenged volunteers as well as in field studies in endemic areas may help in identifying the nature and levels of protective immune responses, but more studies are clearly needed to identify reliable correlates of protection.

6.6 Protective Immunity Against ETEC

Studies both in experimental animals and in human volunteers have shown that an initial ETEC infection may provide highly significant protection against reinfection with the homologous strain or strains expressing the same or related CFs. Thus, infection in a rabbit non-ligated intestine [RITARD] model with a CS positive ETEC strain provided highly significant protection against subsequent challenge with heterologous strains expressing the same CS factors [50]. Furthermore, *Levine* et al. [47] showed that initial infection with a CS1+CS3 positive, toxin-negative mutant ETEC strain provided complete protection against subsequent challenge with a toxin positive CS1 + CS3 strain. In a study in Mexico, *Cravioto* et al. [13] found that during the first year of life, the risk of developing diarrhoea due to reinfection with ETEC strains expressing the same CF was significantly lower [$p < 0.01$] than to be re-infected and develop symptoms due to ETEC expressing heterologous CFs. Similarly, results from the birth cohort studies in Bangladesh and Egypt [7, 44] further support the notion that immunity against CFs is important for protection against ETEC disease.

Previous studies in animals and humans have shown that also antitoxic immunity may be effective against ETEC. Thus, anti-LT antibodies cooperated synergistically with anti-CF antibodies in protecting animals against ETEC producing LT and expressing homologous CFs [43] and immunization with an oral inactivated

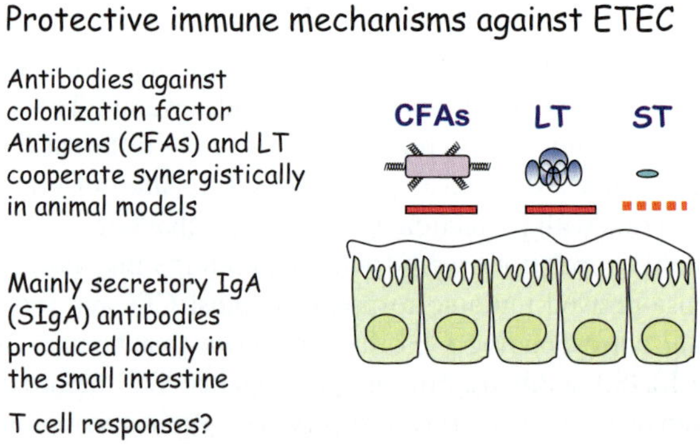

Fig. 6.1 Postulated protective immune mechanisms against ETEC. SIgA antibodies can prevent binding of ETEC CFs to the epithelial surface [different glycoconjugates] and antibodies against LTB prevent binding to surface receptors [GM1 and glycoproteins]. Antibodies against ST, if induced by immunization with an ST-protein conjugate, can also block receptor-binding and toxin action

cholera vaccine containing CTB, which cross-reacts immunologically with LTB, has afforded statistically significant, 50–70%, protection against LT ETEC both in travellers and children in an ETEC endemic area [51, 52]. These findings, together with the low rates of ETEC infections in children above 2 years in high endemic areas are all in favour of development of protective immune responses following ETEC infection. Such protective immunity appears to be mediated by SIgA antibodies directed against the CFs and LT [43]; ST, which is a small peptide, does not elicit neutralizing antibodies following natural infection [43; Fig. 6.1].

6.7 ETEC Vaccines

Based on the proven efficacy of CF antigens and the LT antigen, an ETEC vaccine should most likely contain fimbrial antigens present on the most prevalent ETEC pathogens, in particular on strains producing LT + ST or ST alone, to provide broad spectrum protection. Thus, a multivalent ETEC vaccine containing CFA/I, CS1–6 and an LT toxoid may provide protection against ca 80% of ETEC strains world-wide [43, 53, 54]. The vaccine should also provide strong mucosal immune responses against the key protective antigens locally in the small intestine. Different strategies have been undertaken to deliver ETEC fimbriae and toxin antigens to the human immune system to elicit protective immune responses in the form of inactivated or live candidate vaccines [53, 54, 55]. A number of different live and attenuated candidate vaccines have been develop based on such strategies [Table 6.2] and are presented below.

6.7.1 Inactivated ETEC Vaccines

6.7.1.1 Purified CFs and Enterotoxoids

Purified CFs have drawbacks as oral immunogens since they are expensive to prepare and sensitive to proteolytic degradation [54, 55]. To protect the fimbriae from degradation in the stomach, immunization with purified CFs incorporated into biodegradeable microspheres has been attempted. However, no significant protection was induced against subsequent challenge with ETEC expressing the homologous CFs, either when immunizing with high doses of a combination of CS1 and CS3 or recombinantly produced CS6 [53, 55]. Alternative approaches include to present the CFs transcutaneously. Thus, the possibility of immunizing transcutaneously with *E. coli* CS6, alone or together with non-mutated LT, has been evaluated in human volunteers [56]. The rational for such administration is that the skin is a highly active immunologic organ in which induction of immune responses can be initiated if antigens can penetrate the keratinized layer to reach the sub-epidermal region where Langerhans cells [a dendritic cell] are in abundance. When administering comparatively high doses of CS6 incorporated into patches which are applied on

Table 6.2 Current ETEC vaccines and vaccines under development [producers]

Inactivated ETEC vaccines
Toxoids
–CTB in killed Dukoral® cholera vaccine [SBLVaccin/Crucell]
–LT in transcutaneous patches [IOMAI Corporation]
Purified CFs
–CS6 [CS1 + CS3] in microcapsules [Walter Reed Army Research Institute, WRAIR]
–CfaE [CFA/I tip protein] and other tip proteins recombinantly produced [WRAIR]
–Transcutaneous CS6 together with LT adjuvant [IOMAI]
Inactivated whole bacteria
–ETEC expressing CFs together with rCTB [rCTB-CF ETEC; SBL/Crucell]
–Recombinant E. coli overexpressing CFs together with rCTB or CTB/LTB hybrid [University of Gothenburg and SBL/Crucell]
Live ETEC vaccine candidates
Toxin-deleted ETEC strains expressing CFs
–Attenuated wild type ETEC expressing CS1 and CS3 [Camebridge biostability]
–A mixture of attenuated ETEC strains expressing CFA/I, CS1-6 and LTB [Camebridge biostability]
Attenuated Shigellae expressing ETEC antigens
–*S. flexneri 2a* T32 strains expressing CFA/I, CS2, CS3, CS4 or mutated LT [Center for Vaccine Development, CVD]
–A mixture of 5 different *Shigellae* strains expressing CFA/I, CS1-CS6 and LTB [CVD]
Attenuated V. cholerae or Salmonellae expressing ETEC antigens [different companies/ universities]

the surface of the skin, no response to CS6 was observed in the absence of LT [which is also a strong adjuvant]; however, combined administration of CS6 and LT induced ASC as well as serum antibody responses against CS6 in about half of the volunteers and anti-LT responses in serum in all of the vaccinees [56].

Since LTB as well as the immunologically cross-reactive CTB, are strongly immunogenic, but lack toxicity, are stable in the gastrointestinal milieu, are capable of binding to the intestinal epithelium, they are both suitable candidate antigens to provide anti-LT immunity. Although CTB has afforded significant protection against *E. coli* LT disease both in endemic areas and in travellers [51, 52] an LT toxoid may be slightly more effective than CTB in inducing protective anti-LT immunity [54]. Alternatively, hybrid CTB/LTB molecules [57] may be used to provide protection both against cholera and ETEC since recent studies have shown that these infections are both prevalent in young children less than 2 years of age, e.g. in West Bengal in India [58].

The approach to administer *E. coli* LT as a candidate immunogen transcutanously has recently also been tested for protective efficacy in human volunteers [49]. Adult Americans were given three transcutaneous immunizations with active LT in a patch at four week intervals. Following challenge, the attack rate for diarrhoea was not significantly diminished in the vaccinated volunteers versus controls, but the vaccinees had a significantly longer time to onset of diarrhoea, significantly fewer and smaller stools and significantly fewer of the vaccinated volunteers needed intravenous rehydration [49]. Collectively, these results suggest a biological effect of transcutaneous immunization with LT, which has been shown to induce systemic as well as mucosal antitoxic immune responses.

6.7.1.2 Inactivated Whole Bacteria

A simple approach has been to prepare killed ETEC bacteria that express the most important CFs in immunogenic form on the bacterial surface. Such inactivated organisms may be combined with an appropriate LT toxoid, i.e. CTB or LTB. Attempts to prepare a suitable ST toxoid have not been successful, partly due to the small size and high content of cysteine residues of the ST molecule [21]. Inactivation of the bacteria may be achieved by treatment with formalin or colicin E1, which has resulted in killing of the bacteria without significant loss in antigenicity of the different CFs and O-antigens [43, 54]. CFs on ETEC bacteria, inactivated by mild formalin-treatment, have been shown to retain their antigenicity, the fimbrial structure as well as the capacity to bind to eukaryotic cells; they are also more stable than purified fimbriae in the gastrointestinal milieu.

The ETEC vaccine that has been most extensively studied in clinical trials consists of a combination of recombinantly produced CTB and formalin-inactivated ETEC bacteria expressing CFA/I and CS1-CS5 as well as some of the most prevalent O-antigens of ETEC [rCTB-CF ETEC vaccine] [43, 54]. This vaccine has the potential to provide broad protective coverage against ETEC disease. The rCTB-CF ETEC vaccine has been shown to be safe and give rise to significant SIgA immune responses in intestinal lavage fluid in a majority [70–90%] of Swedish vaccinees [35]. Phase I and II trials of the CTB-CF ETEC vaccine in Swedish, Bangladeshi and Egyptian adult volunteers have all shown that the vaccine is well tolerated and gives rise to mucosal immune responses, i.e. peripheral blood ASCs, against the different vaccine CFs in 70–100% of the adult volunteers [35, 36, 37, 59]. Studies in Bangladeshi adults have also shown that the vaccine was capable of inducing comparable immune responses against the CFs and LT locally in the intestine, i.e. ASCs in duodenal biopsies, as clinical ETEC disease [37].

Analogous safety/immunogenicity clinical trials have been carried out with the rCTB-WC ETEC vaccine in Egyptian and Bangladeshi children down to 6 months of age. In these children, the vaccine was equally immunogenic as in the adults and also well tolerated [60, 61, 62] except in the youngest infants aged 6–17 months [63]. Since increased frequency of vomiting was observed in vaccinated Bangladeshi infants, a dose-finding study was initiated showing that a quarter of a full dose of rCTB-CF ETEC vaccine did not give rise to any vomiting and was equally safe and almost as immunogenic in the youngest infants as a full dose in older children and adults [63].

A number of different studies to evaluate the protective efficacy of the CTB-CF ETEC vaccine have been initiated in adult travellers going from industrialized areas to different countries in Asia, Africa and Latin America. In an initial pilot study, the vaccine was tested for protective efficacy in European travellers going to 20 different countries in Africa, Asia and Latin America [64]. This study revealed very promising results in that the ETEC vaccine conferred 82% protective efficacy [$p < 0.05$]

against ETEC disease. However, the number of cases fulfilling the inclusion criteria was low.

This study was followed by a larger placebo-controlled Phase III vaccine trial in nearly 700 American travellers going to Mexico and Guatemala to assess the protective efficacy of the rCTB-CF ETEC vaccine [65]. In this study, the vaccine was shown to provide significant protection [protective efficacy 77%, $p = 0.039$] against non-mild ETEC diarrhoeal illness, defined as symptoms that interfered with daily activities or more than five loose stools in a day; however, no significant protection was observed against ETEC diarrhoea of any severity, including very mild cases. Another trial subsequently undertaken in the same setting in American travellers has similarly shown that the vaccine protected against more severe symptoms in those vaccinees in whom vaccine intake could be documented, but no significant protection against mild diarrhoeal symptoms, i.e. three loose stools in a day [*Bourgeois A* et al., VED, Lisbon, April 2007].

The only paediatric study to assess efficacy of the rCTB-CF ETEC vaccine has been undertaken in rural Egypt [43, 53, 54]. In that double-blind, placebo controlled trial conducted in about 350 6–18 months old children, disease detection was based on active surveillance through semi-weekly household visits and cultures of faecal specimens from children with diarrhoea. Unfortunately, no significant protection was induced by the vaccine in this trial. This could partly be due to the fact that active surveillance for diarrhoea was undertaken and hence most cases were relatively mild, which usually results in lower protective effects as compared to in passive surveillance studies. Furthermore, it has been well documented that several vaccines are considerably less effective in the developing world, particularly in infants and young children, who may already have developed pre-existing immunity that may prevent vaccine "take rates" e.g. in the intestine. However, these are areas where the vaccine is needed the most.

Based on the disappointing results from testing of the vaccine in children in developing countries, studies to improve vaccine efficacy have been initiated. These studies include attempts to increase the amounts of the protective antigens, in particular the CFs, on the bacterial surface by recombinant technology [54, 66]. They also include evaluation of putative adjuvants or modified delivery forms, the effects of breast-milk withdrawal at the time for vaccination [Ahmed et al., Submitted for publication] as well as of pre-treating the children with micronutrients, e.g. zinc [67], before and during vaccination. The possible benefits of such interventions, to increase the immunogenicity of inactivated whole cell vaccines, are being explored in different ongoing studies in Sweden and Bangladesh.

6.7.2 Live Oral ETEC Vaccines

The potential of live ETEC vaccines as a tool for protection against diarrhoea has been demonstrated by findings in human volunteers that a live vaccine strain expressing CS1 and CS3 fimbriae, but lacking the genes encoding LT and ST, induced 75% protection against challenge with wild type ETEC expressing corresponding CS factors as well as LT and ST [47]. Different strategies have thereafter been attempted in which non-pathogenic *E. coli*, attenuated *Shigellae, V. cholerae or Salmonellae* expressing different CF components alone or in combination with an LT toxoid have been constructed as putative vaccine candidates [53, 68].

6.7.2.1 Genetically Attenuated ETEC Strains as Live Oral Vectors

One promising approach that has been attempted is to utilize attenuated ETEC strains as vectors of key protective antigens. Such strains were developed by researchers and vaccine manufacturers [69], e.g. the strains PTL002 and PTL003 that are derivatives of strain E1392-75-2A, the O6:H16 prototype CS1 + CS3 vaccine. The latter prototype strain had spontaneously lost the genes encoding LT and ST but retained genes for expression of CS1 and CS3 and had been shown to stimulate specific anti-CF immunity and to confer significant protection against experimental challenge with an ETEC strain expressing homologous CFs [O139:H28, LT/ST, CS1,CS3]. Both PTL002 and 003 strains harbour a mutation in *aroC*; PTL002 has an additional mutation in *ompR*, while PTL003 has mutations in *ompC* and *ompF* [68]. Evaluation of these two strains in human volunteers has shown that they are safe and immunogenic when given in a single dose of 5×10^9 cfu. In a Phase 1 clinical trial, both strains were well tolerated but significantly more recipients of PTL003 exhibited faecal shedding and mounted more robust serum antibody and IgA anti-CF ASC responses; PTL003, was even shown to induce immune responses against both CS1 and CS3 of comparable magnitudes as the wild type strain [69].

An initial experimental challenge study with virulent ETEC was carried out in North American volunteers immunized with two 2×10^9 cfu doses of PTL003 vaccine or placebo to assess preliminarily the ability of the strain to protect against ETEC diarrhea [48]. Subjects were challenged with a virulent LT/ST strain expressing CS1 and CS3. Neither the attack rate for diarrhea nor total stool volume was significantly diminished in vaccinees versus placebo recipients. It was hypothesized that the reason for the lack of protection may have been the fact that PTL003 was given at too low a dose and that too high a challenge dose was used [49]. Further development of the approach of expressing CFs in nonpathogenic *E. coli* include construction of ACAM2010, which is a similar vaccine candidate strain derived from wild type WS-1858B [O71 : H^-, LT-/ST+, CFA/I, *astA*] by deleting STh [encoded by *estA*], EAST1 [enteroaggregative *E. coli* enterotoxin 1 encoded by *astA*], *aroC, ompC* and *ompF* [68]. In a small Phase 1 clinical trial, ACAM2010 was well tolerated and most recipients of a 10^9 cfu dose excreted the vaccine strain and had serum IgG and peripheral blood IgA ASC responses to CFA/I. The constructs clinically tested so far have not contained genes encoding the expression of LTB subunit or a mutant LT to stimulate LT antitoxin immune responses, although such further modifications are planned [68].

6.7.2.2 Multivalent Shigella/ETEC Vaccines

Considerable success has been achieved recently in engineering strains of *Shigella* that are attenuated compared with their wild-type parent and in using these strains as live vectors to express ETEC antigens [70, 71]. Building upon this strategy, different live *Shigella* based multivalent *Shigella*/ETEC hybrid vaccines have been constructed wherein the important fimbrial CFs are expressed along with mutated LT in attenuated *Shigella* [70]. An ambitious project has been underway for several years in which five attenuated *Shigella* serotypes [*S. dysenteriae* 1, *S. flexneri* 2a, *S. flexneri* 3a, *S. flexneri* 6 and *S. sonnei*] have each been engineered to carry stable expression plasmids that encode various ETEC fimbriae and LTh B subunit [68]. The five *Shigella* live vector strains in the final vaccine will collectively express CFA/I, CS1-6 and LTB subunit. Combinations of several attenuated *Shigella* strains expressing various ETEC fimbriae and LTB subunit or a mutant, toxin-negative LT molecule, LTK63, which is a putative mucosal adjuvant, have been shown in pre-clinical models [guinea pigs and Rhesus macaque non-human primates] to be well tolerated and to stimulate systemic IgG and mucosal IgA immune responses to both the ETEC and *Shigella* antigens. In a pre-clinical, non-intestinal model, the responses to the different fimbrial antigens are as robust when the strains are administered in combination as when they are administered as individual vaccines. One of the five live vector strains, attenuated *S. flexneri* 2a strain CVD 1208S expressing CFA/I and LTh B subunit, is currently under evaluation as a prototype in dose-escalating Phase 1 clinical trials to assess the live vector's clinical acceptability and mucosal immunogenicity in humans [68].

V. cholerae and Salmonellae expressing ETEC antigens. A number of other live oral vaccine prototypes have been developed and evaluated for safety and immunogenicity in Phase 1 clinical trials in recent years. These include attenuated *Salmonella* serovar Typhi and *V. cholerae* expressing LTB subunit and different ETEC CFs [53, 68]. It is unclear if any of these is progressing to industrial product development.

6.8 Status of ETEC Vaccine Development

Only one vaccine, rCTB-CF ETEC, has been evaluated for protective efficacy in a field trial in children in endemic areas. Unfortunately, this vaccine did not induce significant protection in this important target group. Against this background considerable work has been initiated to try to improve the immunogenicity of this vaccine, especially for use in children in developing countries. Intense efforts are also in progress in many laboratories to try to develop alternative candidate vaccines, both for travellers and for use in endemic populations. Initially candidate vaccines should be assessed for safety and capacity to induce mucosal immune responses that are likely to reflect protection, before being tested for protective efficacy against primarily moderate to severe ETEC disease; such studies should preferably be conducted in passive surveillance studies in ETEC endemic areas.

References

1. Kosek, M., Bern, C., and Guerrant, R.L. 2003, Bull. World Health Organ., 81, 197.
2. Black, R.E., Brown, K.H., and Becker, S. 1984, Pediatrics, 73, 799.
3. Qadri, F., Svennerholm, A.M., Faruque, A.S.G., and Sack, R.B. 2005, Clin. Microbiol. Rev.18, 465.
4. Wennerås, C., and Erling, V. 2004, J. Health Popul. Nutr., 22, 370.
5. Mata, L. 1992, Am. J. Trop. Med. Hyg., 47, 16.
6. Brown, K.H. 2003, J. Nutr., 133, 328S.
7. Qadri, F., Saha, A., Ahmed, T., Tarique, A., Begum, Y., and Svennerholm, A. M. 2007, Infect. Immun. 75, 3961.
8. Qadri, F., Khan, A.I., Abu Syed, G., Faruque, A.S.G., Begum, Y.A., Chowdhury, F., Nair, G.B., Salam, M.A., Sack, D.A., and Svennerholm, A.M. 2005, Emerg. Infect. Dis., 11, 1104.
9. Abu-Elyazeed, R., Wierzba,T.F., Mourad, A.S., Peruski L.F., Kay B.A., Rao, M., Churilla, A.M., Bourgeois, A.L., AMortagy, .K.S., Kamal, M., Savarino, S.J.J., Campbell, R.J., Murphy, R., Naficy, A., and Clemens, J.D. 1999, J. Infect. Dis.,179, 382.
10. Qadri, F., Das, S.K., Faruque, A.S., Fuchs, G.J., Albert, M.J., Sack, R.B., and Svennerholm, A.M. 2000, J. Clin. Microbiol., 38, 27.
11. Albert, M.J., Faruque, A.S., Faruque, S.M., Sack, R.B., and Mahalanabis. D. 1999, J. Clin. Microbiol., 37, 3458.
12. Black, R.E. 1993, Pediatr. Infect. Dis. J.,12, 751.
13. Cravioto, A., Reyes, R.E., Truilla, F., Furiba, M.M., Navarro, A., De La Roco, Hernandez Rand, J.M., and Perez, G. 1990, Am. J. Epidemiol., 131, 886.
14. Guerrant, R.L., Kirchhoff, L.V., Shields, D.S., Nations, M.K., Leslie, J., de Sousa, M.A., Araujo, J.G., Correia, L.L. Sauer, K.T., and McClelland, K.E. 1983, J. Infect. Dis., 148, 986.
15. Huilan, S., Zhen, L.G., Mathan, M.M., Mathew, M.M., Olarte, J., Espejo, R., Khin Maung, U., Ghafoor, M.A., Khan, M.A., and Sami, Z. 1991, Bull. World Health Organ., 69, 549.
16. Reis, M.H., Guth, B.E., Gomes, T.A., Murahovschi, J., and Trabulsi. L.R. 1982, J. Clin. Microbiol., 15, 1062.
17. So, M., Dallas, W., and Falkow, S. 1978, Infect. Immun., 21, 405.
18. Gill, D.M., and Richardson, S.H. 1980, J. Infect. Dis., 141, 64.
19. Rao, M.C., 1985, Ciba Found. Symp., 12, 74.
20. Holmgren, J. 1981, Nature, 292, 413.
21. Nair, G.B., and Takeda Y. 1998, Microb. Pathog. 24, 123.
22. Gaastra, W., and Svennerholm, A.M. 1996, Trends Microbiol., 4, 444.
23. Shaheen, H., Khalil, S.B., Rao, M.R., Abu Elyazeed, R., Wierzba, T.F., Peruski, L.F., Putnam, S., Navarro, A., Badria, Z.M., Cravioto, A., Clemens, J.D., Svennerholm, A.M., and Savarino, S.J. 2004, J. Clin. Microbiol., 42, 5588.
24. Steinsland, H., Valentiner-Branth, P., Perch, M., Dias, F., Fischer, T.K., Aaby, P., Molbak, K., and Sommerfelt, H. 2002, J. Infect. Dis., 186, 1740.
25. Begum, Y.A., Talukder, K.A., Nair, G.B., Khan, S.I., Svennerholm, A.M., Sack, R.B., and Qadri, F. 2007. Can. J. Microbiol. 53, 19.
26. Sommerfelt, H., Steinsland, H., Grewal, H.M., Viboud, G.I., Bhandari, N.,, Gaastra, W., Svennerholm, A.M., and Bhan, M.K. 1996, J. Infect. Dis.,174, 768.
27. Rao, M.R., Abu-Elyazeed, R., Savarino, S.J., Naficy, A.B., Wierzba, T.F., Abdel-Messih, I., Shaheen, H., Frenck, R.W., Svennerholm, A.M., and Clemens, J.D. 2003, J. Clin. Microbiol., 41, 4862.
28. Valvatne, H., Steinsland, H., and Sommerfelt, H. 2002, APMIS, 110, 665.
29. Anantha, R.P., McVeigh, A.L., Lee, L.H., Agnew, M.K., Cassels, F.J., Scott, D.A., Whittam, T.S., and Savarino, S.J. 2004, Infect. Immun., 72, 7190.
30. Scott, J.R., Wakefield, J.C., Russell, P.W., Orndorff, P.E., and Froehlich, B.J. 1992, Mol. Microbiol., 6, 293.
31. Rudin, A., McConnell, M.M., and Svennerholm, A.M. 1994, Infect. Immun., 62, 4339.

32. Jansson, L., Tobias, J., Lebens, M., Svennerholm, A.M., and Teneberg, S. 2006, Infect. Immun., 76, 3488.
33. Wolf, M.K. 1997, Clin. Microbiol. Rev., 10, 569.
34. Svennerholm, A.M., and Åhren, C. 1982, Acta Pathol. Microbiol. Immunol. Scand. [C], 90, 1.
35. Åhren, C., Jertborn, M., and Svennerholm, A.M. 1998, Infect. Immun., 66, 3311.
36. Jertborn, M., Åhren, C., and Svennerholm, A.M. 2001, Clin. Diagn. Lab. Immunol., 8, 424.
37. Wennerås, C., Qadri, F., Bardhan, P.K., Sack, R.B., and Svennerholm, A.M. 1999, Infect. Immun., 67, 6234.
38. Stoll, B.J., Svennerholm, A.M., Gothefors, L., Barua, D., Huda, S., and Holmgren, J. 1986, J. Infect. Dis., 153, 527.
39. Carpenter, C.M., Hall, E.R., Randall, R., McKenzie, R., Cassels, F., Diaz, N., Thomas, N., Bedford, P., Darsley, M., Gewert, C., Howard, C., Sack, R.B., Sack, D.A., Chang, H.S., Gomes, G., and Bourgeois, A.L. 2006, Vaccine, 24, 3709.
40. Qadri, F., Ahmed, T., Ahmed, F., Bhuiyan S., Mostafa, M.G., Cassels, F.J., Helander, A., and Svennerholm, A.M. 2007, Infect. Immun., 75, 2269.
41. Qadri, F., Ahmed, F., Ahmed, T., and Svennerholm, A.M. 2006, Infect. Immun., 74, 4512.
42. Steinsland, H., Valentiner-Branth, P., Gjessing, H.K., Aaby, P., Mølbak, K., and Sommerfelt, H. 2003, Lancet, 362, 286.
43. Svennerholm, A.M., and Steele, D. 2004, Best Pract. Res. Clin. Gastroenterol., 18, 42.
44. Rao, M.R., Wierzba, T.F., Savarino, S.J., Abu-Elyazeed, R., El-Ghoreb, N., Hall, E.R., Naficy, A., Abdel-Messih, I., Frenck, R.W., Svennerholm, A.M., and Clemens J.D. 2005, J. Infect. Dis., 91, 562.
45. Clemens, J.D., Rao, M.R. Chakraborty J., Yunus, M., Ali, M., Kay, B., van Loon, F.P.L., Naficy, A., and Sack, D.A. 1997, Pediatrics, 100, E2.
46. Holmgren, J., Svennerholm, A.M., and Lindblad, M. 1983, Infect. Immun., 39, 147.
47. Levine, M.M., Nalin, D.R., Hoover, D.L., Bergquist, E.J., Hornick, R.B., and Young, C.R., 1979, Infect. Immun., 23, 729.
48. McKenzie, R., Bourgeois, A.L., Engstrom, F., Hall, E., Chang, H.S., Gomes, J.G., Kyle, J.L., Cassels, F., Turner, A.K., Randall, R., Darsley, M., Lee, C., Bedford, P., Shimko, J., and Sack, D.A. 2006, Infect. Immun., 74, 994.
49. McKenzie, R., Bourgeois, A.L., Frech, S.A., Flyer, D.C., Bloom, A., Kazempour, K., et al. 2007, Vaccine, 25, 3684.
50. Svennerholm, A.M., Wennerås, C., Holmgren, J., McConnell, M.M., and Rowe, B. 1990, Infect. Immun., 58, 341.
51. Clemens, J.D., Sack, D.A. Harris, J.R., Chakraborty, J., Neogy, P.K., Stanton, B., Huda, N., Khan, M.U., Kay, B.A., Khan, M.R., et al. 1988, J. Infect. Dis., 158, 372.
52. Peltola, H., Siitonen, A., Kyronseppa, H., Simula, I., Mattila, L. Oksanen, P.. Kataja, M. J, and Cadoz, M. 1991, Lancet, 338, 1285.
53. Walker, R.I., Steele, D., and Aguado, T. 2007, Vaccine, 25, 2545.
54. Svennerholm, A.M., and Savarino, S. 2004, New Generation Vaccines 3rd ed, Levine M.M. et al [eds], p 737, Marcel Decker, Inc. New York.
55. Katz, D.E., DeLorimier, A.J., Wolf, M.K., Hall, E.R., Cassels, F.J., van Hamont, J.E., Newcomer, R.L., Davachi, M.A., Taylor, D.N., and McQueen, C.E. 2003, Vaccine, 21, 341.
56. Guerena-Burgueno, F., Hall, E.R., Taylor, D.N., Cassels, F.J., Scott, D.A., Wolf, M.K., Roberts, Z.J., Nesterova, G.V., Alving, C.R., and Glenn, G.M. 2002, Infect. Immun., 70, 1874.
57. Lebens, M., Shahabi, V., Backstrom, M., Houze, T., Lindblad, M., and Holmgren, J. 1996, Infect. Immun., 64, 2144.
58. Sur, D., Deen, J.L., Manna, B., Niyogi, S.K., Deb, A.K., Kanungo, S., Sarkar, B.L., Kim, D.R., Danovaro-Holliday, M.C., Holliday, K., Gupta, V.K., Ali, M., von Seidlein, L., Clemens, and J.D., Bhattacharya, S.K. 2005, Arch Dis. Child. 90, 1175.
59. Savarino, S.J., Brown, F.M., Hall, E., Bassily, S., Youssef, F., Wierzba, T., Peruski, L., El-Masry, N.A., Safwat, M., Rao, M., Jertborn, M., Svennerholm, A.M., Lee, Y.J., and Clemens, J.D. 1998, J. Infect. Dis., 177, 796.

60. Savarino, S.J., Hall, E.R., Bassily, S., Brown, F.M., Youssef, F., Wierzba, T.F. Peruski, L., El-Masry, N.A., Safwat, M., Rao, M., El Mohamady, H., Abu-Elyazeed, R., Naficy, A., Svennerholm, A.M., Jertborn, M., Lee, Y.J., Clemens, J.D. and PRIDE Study Group. 1999, J. Infect. Dis., 179, 107.
61. Savarino, S.J., Hall, E.R., Bassily, S., Wierzba, T.F., Youssef, F.G., Peruski, L.F., Abu-Elyazeed, R., Rao, M., Francis, W.M., El Mohamady, H., Safwat, M., Naficy, A.B., Svennerholm, A.M., Jertborn, M., Lee, Y.J., and Clemens, J.D. 2002, Pediatr. Infect. Dis. J., 21, 322.
62. Qadri, F., Ahmed, T., Ahmed, F., Sack, R.B., Sack, D. A, and Svennerholm, A.M. 2003, Vaccine, 21, 2394.
63. Qadri, F., Ahmed, T., Ahmed, F., Begum, Y.A., Sack, D.A., and Svennerholm, A.M. 2006, Vaccine, 24, 1726.
64. Wiedermann, G., Kollaritsch, H., Kundi M., Svennerholm, A.M., and Bjare, U., J. Travel Med., 7, 27.
65. Sack, D.A., Shimko, J., Torres, O., Bourgeois, A., Gustafsson, B., Kärnell, A., Nyquist, I., and Svennerholm, A.M. 2007, Vaccine, 25, 4392.
66. Tobias, J., Lebens, M., Bölin I., Wiklund, G., and Svennerholm A.M. 2008, Vaccine, 26, 743.
67. Albert, M.J., Qadri, F., Wahed, M.A., Ahmed, T., Rahman, A.S., Ahmed, F., Bhuiyan, N.A., Zaman, K., Baqui, A.H., Clemens, J.D., and Black, R.E. 2003, J. Infect. Dis.,187, 909.
68. Levine, M.M., and Svennerholm, A.M. 2007, Travelers' Diarrhea 2nd edition, C. Ericsson et al [Eds.], B.C. Decker, Inc. Hamilton, Canada, Chapter 24.
69. Turner, A.K., Beavis, J.C., Stephens, J.C., Greenwood, J., Gewert, C., Thomas, N., Deary, A., Casula, G., Daley,A., Kelly, P., Randall, R., and Darsley, M.J. 2006, Infect. Immun., 74, 1062.
70. Barry, E.M., Altboum, Z., Losonsky, G., and Levine, M.M. 2003, Vaccine, 21, 333.
71. Barry, E.M., Wang, J., Wu, T., Davis, T., and Levine, M.M. 2006, Vaccine, 24, 3727.

Chapter 7
Cholera Immunity and Cholera Vaccination

Jan Holmgren and John D. Clemens

Abstract Cholera remains an important global health problem. Studies in the 1970ies, identifying the AB_5 molecular structure of cholera toxin, the GM1 ganglioside receptor for the toxin, and the toxin's ADP-ribosylating effect on the intestinal epithelium resulting in life-threatening dehydrating diarrhoea, made the pathogenesis of cholera the best understood of all infectious diseases. In the 1980's, the toxin coregulated pilus (TCP) was identified as being a critical factor for the colonization of *Vibrio cholerae* in the intestine, and through immunological studies cholera also emerged as a prototype infection in which protective immunity is mediated by a local mucosal immune response. Locally produced secretory IgA antibodies to cell-wall lipopolysaccharide and to the cholera toxin B subunit (CTB) moiety were found to be important for protection against cholera by inhibiting colonization and toxin binding, respectively, and when present together in the intestine these two antibodies were found to exert a synergistic cooperative protective effect. Studies of the intestinal immune response after immunization by different routes showed that oral but not parenteral immunization with a vaccine containing inactivated whole-cell cholera bacteria together with CTB could effectively stimulate local intestinal production of protective IgA antibacterial and antitoxin antibodies and also a long-lasting intestinal immunologic memory. Based on this, safe and effective oral cholera vaccines have been developed, and cholera is one of still very few infections for which there exists a licensed effective mucosal vaccine for human use.

7.1 Introduction

Cholera remains an important global health challenge. It is the most severe of all diarrheal diseases and the archetype for the large group of infectious diarrheas caused

J. Holmgren
Gothenburg University Vaccine Research Institute (GUVAX), P.O. Box 435, S-405 30 Göteborg, Sweden
e-mail: Jan.Holmgren@microbio.gu.se

This paper is dedicated to our close friend, late Professor Dang Duc Trach, who "fathered" the VietNam oral cholera vaccine and was a great inspiration for the whole cholera vaccine community.

M. Vajdy (ed.), *Immunity Against Mucosal Pathogens*,

by bacteria that produce one or more enterotoxins. Collectively these "enterotoxic enteropathies" [1] may account for one-half of all infection-related diarrheas in the world. Cholera may account for at least 3–5 million cases and 120 000–200 000 deaths annually, hitting both children and adults. These figures are probably significant underestimates of the true burden of cholera disease due to notorious underreporting [2]. Enterotoxigenic *Escherichia coli* (ETEC) is the most prevalent pathogen in the target groups afflicted by diarrhea, estimated to cause almost half a billion cases and half a million deaths each year, mainly in infants and young children [3].

Pioneering studies in the 1970ies, defining the molecular structure of cholera toxin and its mode of action on the intestinal epithelium that leads to diarrhoea and dehydration, made the pathogenesis of cholera the best understood of all infectious diseases [4, 5]. In the 1980ies cholera also emerged as the best studied example of an infection in which protective immunity is mainly or exclusively mediated by a local mucosal immune response [4, 6, 7]. Cholera and cholera toxin have thereafter played pivotal roles in elucidating fundamental aspects of mucosal immunity and mucosal responses to immunization [8]. Based on this, effective oral cholera vaccines have been developed, and cholera is one of very few infections for which there exists a licensed effective mucosal vaccine for human use [9, 10, 11, 12].

7.2 *Vibrio Cholerae* and Cholera Pathogenesis and Disease

7.2.1 The Pathogen and its Epidemiology

Vibrio cholerae of serogroup O1 is the prototype for the enterotoxin-producing bacteria. The organism was first isolated by Robert Koch in 1884. *V. cholerae* O1 can appear as either of two main different serotypes, Inaba and Ogawa. Until the beginning of the 20th century all examined *V. cholerae* O1 isolates were of the same "classical" biotype. In 1906, however, vibrios of a new biotype, El Tor, were isolated, and for many years vibrios of either the classical or El Tor biotype were isolated from cholera cases [2].

A much feared characteristic of cholera, distinguishing it from most other enteric pathogens, is the propensity for causing big epidemic outbreaks and even large pandemics. This became known in the early part of the 19th century, when cholera started to spread from its ancient home in Bengal. Since 1817, seven major pandemics have been described which have affected large parts of the world [13]. The latest and still ongoing pandemic took its departure from Celebes in 1961 and has since then spread to and even made cholera endemic in many countries in Asia and Africa. From 1991, for the first time in more than 100 years, cholera caused by the same El Tor pandemic strain as that isolated from Asia and Africa, has also reappeared as a significant health problem in much of South and Central America. When an epidemic strikes an area where hygiene is poor and health care is not adequate the results can be disastrous. This was illustrated during the refugee crisis in Goma, Zaire in 1994, when an estimated 58,000–80,000 cases and 23,800 deaths occurred within 1 month [14]. In endemic areas the highest incidence of cholera is seen in children below 5 years of age; in the age group 1–5 years annual cholera incidence rates above 10 cases per 1000 children have been reported in high-endemicity locations. However, still more than half of all *V. cholerae* O1 cases occur in older children and adults in these areas. When, on the other hand, cholera has spread to new countries, all age groups have been affected to the same degree. This pattern is probably due to the fact that natural immunity normally develops by age in endemic countries, which is lacking in the newly attacked/infected [2, 15, 16].

Since 1993, as yet limited to a few countries in South-East Asia, a new *V. cholerae* serogroup, O139 Bengal, has been found to cause a variable percentage, usually below 5%, of all cholera cases [2]. If increasing, or even worse, if also spreading to other areas, this new organism may cause an 8th cholera pandemic in parallel with the still ongoing 7th *V. cholerae* O1 El Tor pandemic.

As mentioned, the total number of cholera cases annually in the world is uncertain since several heavily affected countries do not monitor or report the disease accurately, in part due to surveillance difficulties, but also for fear of economic and social consequences [2]. The recent outbreaks of *V. cholerae* O1 in Latin America as well as of *V. cholerae* O139 in Asia have probably resulted in a substantial increase in the number of cholera cases in the last 10–15 years. Therefore, the often cited figures of 3–5 million cases and 120,000–200,000 deaths from cholera annually are likely underestimates of the present situation.

7.2.2 Cholera Disease

The disease caused by *V. cholerae* is characterized by watery diarrhea without blood and mucus [2]. In the majority of cases, cholera is characterized by acute, profuse watery diarrhoea lasting for one or a few days. In a substantial proportion, however, cholera causes dehydrating disease and acidosis. In its extreme manifestations, cholera is one of the most rapidly fatal infectious illnesses known. Within 3–4 h after the onset of symptoms, a previously healthy person may become hypotensive and may die within 6–8 h. More commonly, fatal cases progress to shock within 6–12 h with death following in the next one to several days of dehydrating disease unless proper rehydration treatment is provided. The most severe cholera cases can purge as much as 15–25 l of water and electrolytes per day, and the mortality rate in severe, non-treated cholera is 30–50%. Persons of blood group O are known to have an increased risk of developing severe cholera (cholera gravis) when infected [17, 18, 19]. Interestingly, this correlation between increased severity and blood group O appears restricted to cholera caused by the *V. cholerae* O1 El Tor biotype (and the novel O139 serogroup which is a mutated form of O1 El Tor), and was not seen for cholera caused by the O1 classical biotype when this form coincided with large numbers of infections in the 1980's in Bangladesh [18].

The recommended primary treatment for cholera today is simple, oral rehydration solutions containing salts and glucose, which will save lives when properly administered [2]. If the patient is severely dehydrated on arrival to the treatment center, is unable to drink, or if the rate of fluid loss by diarrhea exceeds what can be compensated by oral rehydration, which is often the case with cholera cases, aggressive intravenous fluid rehydration is necessary. With effective rehydration treatment, the case fatality rate may go down to below 1% even in severe cases of cholera. Antibiotics can be used to shorten the duration of cholera disease and decrease the risk for further spread of the infection. WHO recommends that antibiotics be used only in the treatment of cholera cases with signs of severe dehydration. Whenever possible, the sensitivity of the cholera isolate to antibiotics should be assessed. Antibiotics are not indicated in the treatment of milder cases of cholera or for mass prophylaxis.

7.2.3 Virulence Factors and Pathogenesis

Cholera is to the largest extent a disease originating from the upper part of the small intestine. Intestinal perfusion studies have revealed that as much as 90% of the intestinal secretion occurs in the uppermost one meter of the intestine in adult cholera patients. *V. cholerae* O1 (and O139) bacteria have developed special, highly efficient means to colonize and multiply to excessive numbers in the small intestine, and in this process they also efficiently produce and release the cholera toxin (CT) molecule [2, 5]. Through its high-affinity binding to the gut epithelium and its subsequent cellular action (to be described further below), CT is directly responsible for most, if not all, of the pathogenic effects on intestinal ion and water secretion processes that may lead to life-threatening diarrhea and dehydration. The most important attributes of *V. cholerae* allowing it to efficiently colonize the small intestine include: (i) the toxin-coregulated pilus (TCP), which has been found to be a critical attachment fimbriae in at least the early stage of colonization [20, 21, 22]; (ii) mucinase (soluble hemagglutin), a metalloprotease which both facilitates bacterial penetration through the intestinal mucus layer and helps to activate CT by "nicking" the peptide bond between CTA1 and CTA2 [23]; (iii) and the single flagellum which operates in concert with chemotactic receptors and intracellular sensor molecules to allow directed motility towards the intestinal cell wall [24].

The *V. cholerae* genome is made up of two chromosomes, one of approximately 3 Mb in size (chromosome I) and another of around 1 Mb in size (chromosome II), both of which have been completely sequenced [25]. Chromosome I harbours all of the known virulence factors, including CT and TCP.

The identification of the subunit structure and function of CT was of pivotal importance for clarifying the pathogenesis of cholera [4].

CT is composed of two types of subunit, a single toxin-active A subunit (CTA), which is embedded in a circular B subunit homopentamer (CTB) responsible for toxin binding to cells [4, 5]. CTA comprises 240 amino acids and has a molecular weight of 28 kDa, whereas the 11.6 kDa B subunit monomers each comprise 103 amino acids. Although CTA is synthesized as a single polypeptide chain, it is post-translationally modified through the action of a *V. cholerae* protease (soluble hemagglutinin/mucinase, see above). This generates two fragments, CTA1 and

CTA2, which, although functionally distinct, remain linked by a disulfide bond. The enzymatic ADP-ribosylating "toxic" activity of CTA resides in CTA1, whereas CTA2 inserts CTA into the CTB pentamer. The interactions between CTA2 and the CTB pentamer are noncovalent, and the last four amino acids (lysine-aspartate-glutamate-leucine; KDEL) at the carboxy- terminal tail of CTA2 are not engaged in the association to CTB but rather protrude from the associated toxin [26]. The CTB pentamer attaches CT to the intestinal epithelial cell through its high-affinity binding to cell-surface receptors. These were already identified more than 30 years ago as the monosialoganglioside GM1, which was then the first biologic receptor that was fully defined chemically [20, 27]. Both the specific sugar residues in GM1 and the amino acid residues in CTB interacting with each other have been defined [26]. Although there is one GM1-binding site in each B subunit monomer, a single amino acid from the neighboring CTB monomer also has a role in the binding. This explains the much higher binding strength of the CTB pentamer compared with that (practically negligible) of individual B subunit monomers. GM1 is present on most mammalian cell types, and CT has been demonstrated to bind to (and intoxicate) many different types of cells experimentally.

After binding to GM1, which appears to be localized mainly in lipid rafts on the cell surface, CT is endocytosed by the cell [28]. After endocytosis, CT travels to the endoplasmic reticulum (ER) via a retrograde transport pathway, possibly via the Golgi-system. It is thought that the actin cytoskeleton has a role in CT trafficking from the plasma membrane to the Golgi–ER. After CT has reached the ER, CTA dissociates from CTB to enter the cytosol through a process which is still not fully resolved [5]. The arrival of CTA1 in the cytosol is the crucial step for intoxication because this polypeptide catalyzes the ADP ribosylation of the trimeric Gsα component of adenylate cyclase (AC). This makes AC to remain in its GTP-bound state, resulting in enhanced AC activity and increased intracellular cyclic adenosine mono-phosphate (cAMP) production. The resulting higher levels of cAMP then cause an imbalance in electrolyte transport across the epithelial cell, with a decrease in sodium uptake and an increase in chloride and bicarbonate export [29]. Decreased sodium uptake reduces water absorption by the enterocyte (mainly villus cells), and, at the same time, increased chloride and bicarbonate extrusion causes water secretion. The combined effect is an abundant net fluid loss from the intestine that will soon lead to dehydration. In addition to the direct effect of CT on AC activity and cAMP production in the enterocytes, it has been proposed that the diarrheal response to CT might have a significant (perhaps up to 50%) neurological component [30]. The experimental evidence for the involvement of the enteric nervous system in the pathophysiology of cholera has been obtained mainly in vivo and on extrinsically denervated intestinal segments of cats and rats. One way in which CT could then be acting is through stimulation of enterochromaffin cells to release serotonin, the latter substance in turn stimulating and inducing the release of vasointestinal peptide from local neural networks.

TCP, a type 4 pilus, which is closely related to other enterobacterial type 4 pili, has been shown to be essential for colonization and virulence in humans and in an infant mouse model [21, 22]. Passive immunization studies of infant mice using polyclonal and monoclonal antibodies directed against TCP suggests that the adhesive moiety lies within the major pilin subunit itself (TcpA). The detailed mechanisms by which TCP promotes colonization has remained elusive. It has been proposed [31] that the initial attachment of bacteria to the epithelium is mediated by an outer membrane protein, rather than by TCP. Instead, it is proposed, TCP together with a soluble protein (TcpF), secreted via the TCP biogenesis apparatus, mediates bacterial microcolony and biofilm formation on the epithelial surface as critical events in the colonization process.

Although CT and TCP are undoubtedly the main virulence factors of *V.cholerae*, several accessory factors have also been described (reviewed in [5]). These range from additional pilus structures, such as the fucose-binding and mannose-binding hemagglutinins, which may or may not contribute to colonization of classical and El Tor biotypes, respectively, to various soluble factors. In addition to mucinase and neuraminidase, the latter group includes a variety of "minor toxins" that might contribute to cholera diarrhea. Among these are the *ctx*-encoded Zonula occludens toxin (zot), which might be both a morphogenetic phage protein and an enterotoxin; the actin-crosslinking RTX toxin; the S-CEP (Chinese hamster cell elongating protein) cytotonic protein and the pore-forming and vacuolating hemolysin A. Even though the role of all of these factors in the virulence of *V. cholerae* is probably minor relative to CT and TCP, they might explain, singly or jointly, why live attenuated vaccines based on the removal of the ctxAB operon have usually been found to cause mild disease, including low-volume diarrhea, when orally administered to volunteers.

An intriguing novel finding is that the O1 El Tor [32] as well as the O139 (J.Sanchez and J.Holmgren, unpublished) *V. cholerae* strains isolated after 2001 produce CT of the classical biotype rather than of the El Tor type as was the case in 100% of strains isolated before 1995. There is evidence that this shift in CT type is associated with increased virulence resulting in a higher hospitalization rate and more severe dehydration (A.K. Siddique et al. unpublished).

7.3 Innate Immune Mechanisms in Cholera Infection

Susceptibility to infection with *V. cholerae* is dependent on both adaptive immune responses induced by previous infection or vaccination and on innate host factors. In contrast to the adaptive immune mechanisms, which have been studied extensively in cholera, little is known about the innate immune components and mechanisms.

Among the intrinsic host factors that influence susceptibility to cholera, stomach acidity and ABO blood groups are the most studied. In Bangladesh, patients with cholera (or ETEC diarrhea) had significantly lower gastric acid levels than other groups studied, and low gastric acid level was also associated with more severe cholera disease [33]. It is also well known that neutralization of stomach acid dramatically reduces the minimal infectious dose of *V. cholerae* [2]. In healthy North-American volunteers given different doses of *V. cholerae* O1 bacteria, the average minimal effective dose was 10^6–10^8 when given without bicarbonate but a 1000-fold less when the bacteria were given together with 2 g of sodium bicarbonate. In endemic situations it has been calculated that the average infectious dose of cholera vibrios when ingested together with rice or other food resulting in transient neutralization of stomach acidity, is in the order of 10^3 organisms. Likewise, it is well established that people who had undergone surgical removal of the acid-producing part of the stomach had a far increased risk of attracting cholera infection and disease.

When it comes to ABO blood groups, several case-control studies found that individuals with blood group O are at increased risk of hospitalization due to both *V. cholerae* O1 and *V. cholerae* O139 (reviewed in [19]). In a prospective study of household contacts of (O1 El Tor) cholera patients in Bangladesh, Glass et al. [17] found that contacts with blood group O had 5–10 times increased risk of getting moderate or severe cholera compared to contacts with blood group A or B, and a more than 20-fold increased risk compared to blood group AB contacts. It was suggested [17] that *V. cholerae* infection may be an evolutionary force behind the exceptionally low prevalence of persons with blood group O in the Ganges delta, where historic reports indicate that cholera has been a killing disease for thousands of years. In a later study from the same area performed between 1985–1987, at which time cholera was caused at about the same frequency by *V. cholerae* O1 of El Tor and the classical biotypes, Clemens et al. [18] made the important observation that the link between blood group O and cholera severity appeared to be restricted to El Tor cholera but was not seen for cholera caused by the classical biotype. Although the classical and El Tor biotypes may well have undergone significant changes over time influencing their virulence, as illustrated by the recently observed shift in cholera toxin type in O1 El Tor and O139 strains in recent years (see above [32], these data may suggest that cholera due to the El Tor biotype rather than the classical biotype historically may have been an evolutionary force away from blood group O in the Ganges delta.

The mechanism by which blood group influences susceptibility to cholera remains to be defined. The specific relationship to the El Tor biotype would suggest a mechanisms affecting colonization rather than CT action, since CT of classical and El Tor type are functionally indistinguishable despite small amino acid and epitope differences. Further, analyses of intestinal mucosal specimens from (Swedish) people of O and non-O blood groups did not reveal any differences in either GM1 ganglioside contents or capacity to bind CT, and CT produced by classical and El Tor *V. cholerae* O1 were found to be indistinguishable in binding specificity and affinity for GM1 ganglioside in human intestine (J. Holmgren and L. Svennerholm, unpublished).

Many factors in addition to gastric acidity and ABO blood group related factors may contribute to the innate defence against cholera. Although there is no pronounced inflammation during cholera, several studies have noted increased infiltration of neutrophils, degranulation of mast cells and eosinophils, and production of some innate defence molecules during acute cholera (reviewed in [34]. CT can also induce the migration of $CD8^+$ intraepithelial lymphocytes (IEL) from the epithelium to the lamina propria region in the small intestine with possible immune defence consequences. To further the search for factors that are activated in the early stages of cholera infection, a whole-genome microarray screening was recently used to study gene expression in duodenal mucosa during acute cholera [34]. Biopsies were taken from the duodenal mucosa of cholera patients 2 and 30 days after the onset of diarrhea, and the gene expression pattern in the acute and convalescent phase samples was compared pair-wise. Of about 21,000 genes expressed in the intestinal epithelium, 29 were defined as being up-regulated and 33 as down-regulated during acute cholera. The majority of the up-regulated genes were noted to have a previously described role in the innate defence against infections These genes include a number of proteins with antibacterial activity e.g. growth differentiation factor 15, matrix metalloproteinases 1 and 3, and a-antichymotrypsin, plus various other genes that appeared to have been indirectly activated through the (CT mediated) activation of IL-1 e.g. the von Ebner minor salivary gland protein and lactoferrin. Confirmative PCR demonstrated good correlation with the micro-array data. The data indicate that during acute cholera infection, innate defence mechanisms are switched on to an extent previously not known. Both direct effects of CT on the epithelial cells and changes in the lamina propria cells appear to contribute to the up-regulation of the innate immune system.

7.4 Adaptive Immune Mechanisms

The best-studied correlate of adaptive immunity to *V. cholerae* is serum vibriocidal antibody titer. Seroepidemiologic studies in cholera endemic areas have shown that vibriocidal antibodies increase with age and that the risk of disease is inversely proportional to the vibriocidal antibody titer [15, 35]. Natural serum antibodies to CT, which may largely be induced by exposure to cross-reactive *E. coli* LT, on the other hand do not appear to correlate with protection against cholera [35]. However, parenteral vaccination, which induces extremely high vibriocidal antibody titers, confers only limited and short-lived protection [36]. Thus, vibriocidal antibodies in unvaccinated individuals are only a surrogate marker for the intestinal mucosal immune status. It is notable, however, that breast-milk IgA antibody titers to either *V. cholerae* LPS or CT in breast-feeding women in Bangladesh were inversely related to the risk of the breast-fed child to develop cholera [37], consistent with a closer link between the gut mucosal immune status with breast milk IgA antibodies than with systemic antibodies [38].

In parallel with the rapid clarification of the mechanisms of disease in cholera, especially as relating to the structure and function of CT as described above, there was also a rapid increase in understanding the mechanisms of protective immunity following clinical infection or vaccination (Fig. 7.1, and reviewed in [6]). The main findings, which directly guided the development of more effective cholera vaccines may be summarized as follows:

(1) In accordance with the findings that the pathogenesis of cholera depends critically on bacterial colonization and the production and action of CT in the

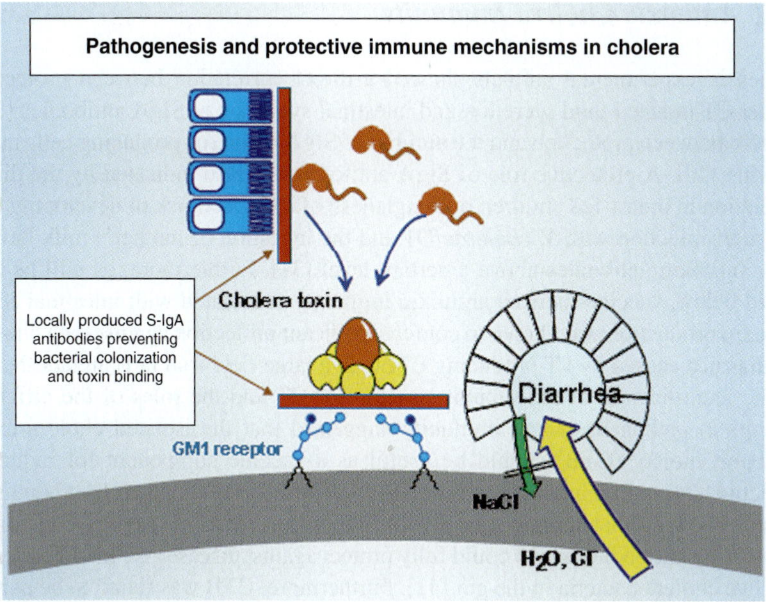

Fig. 7.1 Pathogenesis and protective immune mechanisms in cholera. After ingestion of an infective dose of *V. cholerae* O1 or O139 bacteria, the bacteria can penetrate the mucus layer, attach to the epithelium and colonize and grow to large numbers in the small intestine. The toxin coregulated pilus (TCP) has been shown to be important for the colonization process, and various additional bacterial components including cell-associated factors such as the flagellum, the cell wall lipopolysaccharide (LPS) and fimbrial or outer membrane adhesins as well as mucinase/soluble hemagglutinin and other secreted factors may also partake in these processes. In the intestine, the bacteria produce the main pathogeneic molecule, cholera toxin, which bind to intestinal GM1 ganglioside epithelial receptors through its five B subunits and then through its toxic-active A subunit inhibits NaCl and water uptake by the villus cells and induces active chloride, bicarbonate and water secretion by the crypt cells leading to severe diarrheal fluid loss. Protective immunity depends on locally produced secretory IgA antibodies that can prevent bacterial colonization and/or cholera toxin binding and action in the intestine. The two most important antibodies for protection are antibodies to LPS and antibodies to the cholera toxin B subunits, which can independantly protect against cholera infection and disease and which, when present together in the intestine, also cooperative synergistically in their protective action

small intestine, it was found in animal models that antibacterial and antitoxic antibodies capable of preventing these different events could protect against experimentally induced *V. cholerae* infection and disease. It was further noted that in the intestine antibacterial and antitoxic antibodies produce a synergistic cooperative effect in protection against disease.

(2) Both antibacterial and antitoxic immunity were found to depend mainly if not exclusively on locally produced mucosal antibodies of the secretory IgA (S-IgA) type.
(3) To stimulate a protective gut mucosal immune response immunization by the oral route was found to be superior to parenteral immunization.

7.4.1 Antitoxic Cholera Immunity

Studies in experimental animals showed a direct correlation between protection against CT-induced fluid secretion and intestinal synthesis of SIgA antibodies [39], and also between protection and the number of SIgA antitoxin-producing cells in the intestine [40]. A protective role of SIgA antitoxin was also indicated by the direct correlation in breast-fed children in Bangladesh of a reduced risk of developing disease after infection with *V. cholerae* O1 and the ingestion of mother's milk having SIgA antitoxin antibodies above a certain level [37]. Furthermore, as will be discussed below, vaccine-induced antitoxic immunity associated with intestinal SIgA antitoxin production was shown to confer significant protection against both cholera and diarrhea caused by LT producing ETEC in a large field trial in Bangladesh.

The identification of the subunit structure of CT and the roles of the different subunits in pathogenesis and immunity suggested that the isolated cholera toxin B-subunit moiety (CTB) should be useful as a vaccine component for inducing protective antitoxic immunity. This notion was further strengthened by findings in animals showing that immunization with highly purified CTB gave rise to toxin-neutralizing antibodies which could fully protect against disease also after challenge with live cholera bacteria in the gut [41]. Furthermore, CTB was found to be particularly well suited as an oral immunogen, because it is stable in the intestinal milieu and capable of binding to the intestinal epithelium, including the M-cells of the Peyer's patches, properties which are critically important for stimulating mucosal immunity and local immunological memory.

7.4.2 Colonization Factors and Antibacterial Immunity

It is well established that *V. cholerae* O1 lipopolysaccharide (LPS) is the predominant antigen inducing protective antibacterial immunity against experimental cholera caused by O1 bacteria [6]. Indeed, when protective hyperimmune sera raised against either live or killed *V. cholerae* O1 were absorbed on an immunosorbent column, which completely and specifically removed antibodies to LPS, and then tested for residual passive protective activity against cholera infection all such activity had disappeared. Further, although it was shown that during infection in the gut *V. cholerae* can in vivo develop some surface antigens that are not expressed in vitro, also immune sera against in vivo grown organisms completely lost their passive protective activity when absorbed on LPS immunosorbent columns [42]. Also for protection against *V. cholerae* O139, immunity appears to a large extent to be mediated by specific antibodies to (0139) LPS [43].

Immunologically, the O1 LPS contains group-specific epitopes shared between the Inaba and Ogawa serotypes and additional serotype-specific epitopes. Studies have shown that both antibodies against the main shared epitope(s) and serotype specific antibodies can protect against experimental *V. cholerae* O1 infection in accordance [6]. Thus, it is advantageous but not absolutely critical for a cholera vaccine to contain both Inaba and Ogawa LPS to protect against both serotypes of *V. cholerae*.

In *V. cholerae* O1 bacteria of the classical biotype, a toxin-coregulated pilus (TCP) has been shown to be of importance for colonization of the small intestine [20, 22]. There is evidence that also for *V. cholerae* O1 El Tor and O139, an antigenically distinct form of TCP is in an analogous way important for colonization and disease [21]. In addition, *V. cholerae* bacteria have been found to express a number of other fimbrial structures. One of these is the mannose-sensitive hemagglutinin (MSHA), which is found on O1 El Tor and O139 *V. cholerae* but not on O1 bacteria of classical biotype, and which can mediate bacterial attachment to epithelial cells [44]. Specific antibody to MSHA also protects against cholera infection and disease in an animal model [45]. However, it was recently shown that the expression of MSHA pilus is specifically repressed in vivo and that this may be an important mechanism for bacteria to avoid to be blocked by secretory IgA in a mannose-sensitive, non-antibody specific manner [46]. The role of MSHA and other non-TCP attachment factors for colonization and infection in humans, however, remains to be defined [47].

The identification of TCP as an important colonization factor on *V. cholerae* suggests that it should be possible to raise protective antibacterial immunity also against these fimbrial antigens. Indeed, in experimental systems it has been found that both monoclonal antibodies and polyclonal antisera against TCP can protect against infection and disease [20, 21]. However, following infection in human volunteers, little if any anti-TCP immunity was seen [22], while it was reported that the majority of patients with natural cholera infection developed both a mucosal and a systemic IgA response to the TcpA subunit [48]. Therefore, as an overall conclusion, it remains to be defined whether mucosal immune responses against TCP and other surface antigens on *V. cholerae* could add significantly to the protective action mediated by antibodies to O1 (or O139) LPS antigen.

An important observation guiding the design of new cholera vaccines is the synergistic cooperation between antitoxic and antibacterial immune mechanisms in mediating protection. Either of the two main protective antibodies against cholera which, as mentioned above, are directed against the cell wall LPS and CTB, can confer strong protection against disease by inhibiting bacterial colonization and toxin binding, respectively [6, 41, 49]. When present together in the gut, these antibodies have been shown to have a strongly synergistic protective effect [41, 49].

7.4.3 Immunologic Memory

The acute local antibody response to the immunologically best studied of the new generation of cholera vaccines, the so-called whole cell-B subunit (WC-CTB) vaccine, which is administered orally in two doses ca 2 weeks apart (an interval between doses of 1–6 weeks have been found to give rise to similar immune responses, normally lasts for up to 6–9 months in unprimed individuals [7, 50]. However, in accordance with studies in animals showing almost life-long gut mucosal immunologic memory after oral vaccination with CT [51, 52] it was shown that a long-lasting immunological memory is also induced in humans by the oral vaccine [53, 54, 55].

Thus, healthy Swedish volunteers who received a booster vaccination 5 years after an initial two-dose cholera vaccination responded with significantly higher antitoxin and antibacterial serum IgG and IgA antibody levels than previously unvaccinated controls [55]. In another study, the frequencies of CTB-specific antibody secreting cells (ASCs) extracted from duodenal mucosal biopsies of vaccinated Swedish volunteers were found to be considerably higher after the second than after the first vaccine dose, and the response to a booster dose at 5 months (when the response after the two first doses had largely vanished) fully matched the response after the second dose at 2 weeks [50]. Almost identical results were obtained in Swedish volunteers given a booster dose 10 months after either two or three initial vaccinations [54].

Taken together these studies suggest that the gut mucosal immune system has capacity to mount a very long-lasting immunological memory. A stronger and faster immune response than after an initial vaccination can thus be obtained by a renewed oral antigen exposure (or a booster vaccination) at any time from 2 weeks up to at least 5 years after the primary immunization. Based on the very rapid development of the IgA gut mucosal immune response, noted within 12–16 h after renewed oral antigen exposure even at very long time after primary immunization in experimental animals [51], the immunologic memory appears to usually allow an individual to mount a protective mucosal immune response to a renewed exposure to *V. cholerae* infection before any symptoms have developed.

7.5 Oral Cholera Vaccines

Parenteral vaccines based on inactivated *V. cholerae* O1 bacteria have not been recommended by WHO since the late 1970's. Before that they were available and used for more than 70 years. Their protective efficacy was modest and of short duration (less than 50% for less than 6 months), local side-effects made the vaccines not well accepted, and the vaccines did not prevent transmission of the infective agent [9, 36].

As mentioned, over the last 20 years new oral cholera vaccines have been developed, and two such vaccines are registered internationally [9, 10, 11]. One of these vaccines (WC-rCTB; Dukoral®) consists of killed whole *V. cholerae* O1 cells in combination with recombinantly produced CTB. This vaccine is well tolerated and confers a high level (85%) of protection for 6 months after two or three immunizations in both adults and children. The level of protection is still about 50% 3 years after immunization in vaccinees who were >5 years old at the time of vaccination [9, 11]. Modelled on this vaccine, an inexpensive, locally produced WC-only vaccine (*i.e.* lacking the B subunit) has been developed in Viet Nam and also found to give good protection in a field trial [56].

The other internationally licensed oral cholera vaccine is a live, attenuated vaccine based on the genetically manipulated *V. cholerae* O1 strain CVD103-HgR [10]. A single dose of this vaccine conferred good protection in adult volunteers in USA challenged 3 months after vaccination. However, this vaccine did not provide detectable protection when tested in an endemic setting (Indonesia) [57] and the vaccine has therefore not been produced in the last several years. Additional live-attenuated vaccines, such as the Peru-15 strain, are currently in phase 2 trials [58].

These different types of oral cholera vaccine are described in some greater detail below, especially the killed WC-rCTB vaccine, which is the only widely registered cholera vaccine internationally available at the present time. Recently, a few anti-O139 vaccine candidates have also been developed, but their efficacy is not yet documented.

7.5.1 Killed WC-rCTB Vaccine

This vaccine, consisting of killed whole-cell *V. cholerae* O1 bacteria (classical and El Tor biotypes, Inaba and Ogawa serotypes) in combination with recombinantly manufactured [59] B-subunit of cholera toxin has been marketed (Dukoral®) since the early 1990's. The vaccine is now licensed in more than 50 countries world-wide and recommended and pre-qualified by WHO for UN agency purchasing [9, 11]. After having been given more than 10 million doses together with a bicarbonate buffer to people in countries with good systems for registration of adverse reactions, the conclusion is that this vaccine is exceptionally safe and well tolerated. The only adverse effect reported is occasional mild gastrointestinal disturbances in persons sensitive to the bicarbonate buffer. The WC-rCTB vaccine has also been found to be well tolerated by HIV-positive individuals, by pregnant or breast-feeding women and by children and infants down to at least 6 months of age.

In a large randomized field trial in Bangladesh in the late 1980's, two or three doses of WC-CTB vaccine were tested in comparison with either WC vaccine alone or placebo (killed *E. coli* K12 bacteria) [60, 61]. At this time CTB was not made recombinantly but was prepared and extensively purified after first isolating CT on a GM1-affinity column and then separating the A and B subunits by repeated chromatographic steps [62]. In comparison with the placebo group, the WC-CTB vaccine gave 85 and 60% cholera-specific protection when assessed after 4–6 months and 3 years, respectively, in all age groups; two doses were similarly effective as three doses. Protection was similar against cholera caused by both classical and El Tor biotypes and by Inaba and Ogawa serotypes, the only differences noted being a slightly lesser efficacy against El Tor compared to classical cholera and in blood group O versus non-O individuals. In children older than 5 years and in adults, protective efficacy remained high for the first two years of follow-up and was evident also during the third year (Table 7.1). In children 2–5 years of age, on the other hand, the 100% efficacy noted during the first 4–6 months of surveillance [60] then declined substantially to be ca 40% during the first two years of follow-up, and then being absent during the third year after vaccination [61], Table 7.1).

In a placebo-controlled phase 3 trial in Peru in 1993 in military recruits [63], two doses of the WC-rCTB vaccine given 1–2 weeks apart induced 86% protection in the vaccinees against an outbreak of cholera occurring ca 3–5 months after

Table 7.1 Protection against cholera by the oral B subunit-killed whole cell (CTB-WC) and WCV-only vaccines in the Bangladesh field trial (60, 61)

Follow-up period	Protective efficacy % (95% confidence interval)					
	CTB-WC vaccine			WC-only vaccine		
	All ages	2–5 y	>5 y			
4–6 months	85% (56–95%)	100%	76%	58% (14–79%)	35%	71%
1st year	64% (50–74%)	38%	78%	56% (39–76%)	31%	67%
2nd year	52% (30–76%)	47%	63%	55% (33–69%)	24%	73%
3rd year	19% (Nil–46%)	Nil	41%	41% (7–62%)	2%	61%

vaccination, *i.e.* very similar to the 85% protection after 4–6 months noted in the previous Bangladesh trial. Importantly, this high level of vaccine-induced protection was achieved against cholera due to *V. cholerae* O1 El Tor in a previously unexposed population of individuals who were almost exclusively of the high-risk blood group O. In a later study in Peru [64], two doses of the vaccine did not appear to give significant protection during the first year of surveillance, but the study was criticized for lack of rigor during this observation period [65]. In the second year of follow-up, when a booster dose had also been given the efficacy was 80% [64].

Recently, the O1 WC/rCTB vaccine was also tested with very good results in a large effectiveness trial in Mozambique [66]. The vaccine was administered through the normal public health system to a high-endemicity population with a high sero-prevalence also of HIV infection (estimated to be circa 30% in the adult population). Despite the potential immunocomprising impact of HIV infection the protection was found to be very high, 80% against all hospital admittance for cholera and 90% against severe cholera, thus confirming the previous phase 3 efficacy trial findings in both Asia and Latin America.

7.5.2 Killed WC-Only Vaccines

When tested side-by-side in Bangladesh with the WC-CTB vaccine, also the same WC vaccine without any CTB provided significant short-term as well as long-term protection against cholera (Table 7.1, [60, 61]). The short-term protection, 58% for the initial 4–6 month period, was however significantly lower than the 85% efficacy of the combined WC-CTB vaccine during the same period, which indicates an independent protective immunogenic effect of the CTB component in the latter vaccine. Indeed, if for the calculation of the protective efficacy of the CTB component separate from that of the WC component, one compares the rate of cholera in the WC-CTB group with that in the WC group, the protective efficacy of the CTB over the WC component was 73% during this period [60]. The WC-CTB vaccine continued to be significantly more protective than the WC-alone vaccine for the first 9 months after vaccination [61]. Thereafter, the overall efficacy was similar, approximately 55% during the second year of follow-up and 20–40% during the third year.

As mentioned, a killed *V. cholerae* O1 whole-cell vaccine, modelled on the Swedish WC-rCTB vaccine but lacking the B-subunit, has been locally produced, tested and licensed in Viet Nam. Immunization by two oral doses of this vaccine resulted in 66% protective efficacy during a local cholera outbreak that occurred 8–10 months after vaccination [56]. The protective efficacy was similar among children aged 1–5 years as in older children and adults. A second-generation bivalent vaccine (O1/O139-WC), containing killed vibrios of the serogroup O139 in addition to the O1 WC vaccine, has also been developed in Viet Nam. With support from the International Vaccine Institute (IVI), this vaccine was recently tested for safety and immunogenicity in a large phase 2 trial side-by-side with the international O1 WC-rCTB vaccine [67]. Results showed that the Viet Nam-produced O1/O139-WC vaccine was safe and immunogenic, that it could be administered without buffer, and that it could elicit robust antibacterial immune responses in both adults and children. The vibriocidal antibody responses to O1 *V. cholerae* were similar to those obtained with the international O1 WC-rCTB vaccine. In accordance with the different compositions of the two vaccines, only the Viet Nam vaccine induced detectable vibriocidal antibodies to O139 *V. cholerae* (although lower than against O1) and only the Swedish O1 WC-rCTB reference vaccine induced antitoxic antibody responses. Notably, when there was a cholera outbreak 3 or 5 years after the locally produced vaccine had been given in a two-dose schedule under public health conditions, there was a ca 50% protection in comparison with unvaccinated individuals [68].

Based on this, with the continued support from IVI, South–South technology transfer and collaboration have been initiated to facilitate local production, evaluation and introduction of the O1/O139 WC oral cholera vaccine also in other South-East Asian countries. In India this has now led to national registration of the vaccine (based on safety and immunogenicity data in phase 2 studies) and currently a large phase 3 placebo-controlled trial is going on in Kol Kata to assess the protective efficacy.

7.5.3 Live, Attenuated CVD 103-HgR Vaccine and Related Vaccines

A live, attenuated oral cholera vaccine containing the genetically manipulated classical *V. cholerae* strain CVD 103-HgR (10) was marketed (Orochol®) in several countries for use in travellers. The vaccine was well-tolerated and immunogenic in infants as young as 3 months of age as well as in HIV-infected individuals. Except for transient mild diarrhoea in 2–5% of the vaccinees, and some cases of nausea and abdominal cramps, adverse reactions to the vaccine were not reported.

Experimental challenge studies in volunteers demonstrated protection as early as 1 week after a single-dose vaccination. A high level of protection was conferred against moderate and severe cholera caused by challenge with *V. cholerae* O1 of either El Tor (60–70% protection) or classical biotype (80–100% protection). Efficacy in volunteers lasted for at least 6 months; data beyond that time are not available [9, 10]. However, when tested in a randomized placebo-controlled field trial in Indonesia, a single dose of CVD 103-HgR did not confer any significant protection during any of 4 years of follow-up [57]. The low number of cholera cases identified during the first year of the trial made the estimates of protection imprecise, which might explain why it was difficult to demonstrate significant protection in this trial even at a time when protection would have been expected based on the studies in challenged volunteers. The CVD103-HgR vaccine has not been produced since 2004.

Several other live, attenuated newer-generation cholera vaccines have also been developed and a few of them have even been tested for protective efficacy in the challenged volunteer model. The best studied among these newer vaccine strains is Peru-15, which was derived from a non-motile *V. cholerae* O1 Inaba strain by a series of gene deletions and modifications including deletion of the entire *ctx* genetic element [69]. When given to North-American volunteers in a single dose, the Peru-15 vaccine strain was well tolerated and gave better than 90% protection against challenge with (a low dose of) a virulent *V. cholerae* O1 (Inaba, El Tor) strain [58]. In recent phase 2 studies in Bangladesh [70], the Peru-15 vaccine was also reported to be safe and immunogenic, giving rise to vibriocidal antibodies in serum in the majority of vaccinated adults as well as children. A slightly different, live attenuated O1 El Tor cholera vaccine strain named 638 developed in Cuba has also shown promise in volunteer studies [71].

7.5.4 O139 Cholera Vaccine Candidates

Both killed bivalent O1/O139 WC-rCTB and O1/O139 WC-only vaccines modelled on the licensed O1 WC-rCTB vaccine were developed in Sweden and found to be safe and immunogenic in volunteers [72]. However, in expanded phase 2 studies vibriocidal responses against the O139 component were less frequent and at lower titres than against *V. cholerae* O1 (P.Askelöf, personal communication). When the O1/O139 WC-rCTB vaccine was tested in North-American volunteers, the protective efficacy against challenge with a virulent O139 strain was also lesser than that against *V. cholerae* O1 El Tor Inaba challenge (DA Sack et al., unpublished).

Also a few genetically engineered, live, attenuated O139 vaccine candidate strains have been generated, and two such strains were also reported to protect against homologous O139 challenge in human volunteers [73, 74]. The possibility of combining live attenuated O1 and O139 strains into a bivalent vaccine has not been assessed.

7.6 Immunological Cross-Reactivity Between Cholera and *E. coli* Enterotoxins

There is convincing evidence from preclinical studies that antibodies to CTB can effectively neutralize *E. coli* LT, and that antitoxic immunity induced by either CT or CTB can protect against LT ETEC diarrhea [75, 76].

The first evidence that oral immunization with CTB provides significant cross-protection against disease caused by LT-producing ETEC also in humans was demonstrated in the Bangladesh field trial, where two or three doses of WC-CTB cholera vaccine were found to give significant protection against LT-ETEC disease [77]. The protective efficacy against ETEC diarrhea during the first three months after vaccination when there was much ETEC disease was 67%, including both strains that produced LT alone and strains producing LT together with heat-stable ST (Please define!) toxin. Protection was stronger against severe dehydrating disease (86%) than against milder illness (54%), and there were no age-dependent differences in protection. There was no protection against ETEC strains that only produced the ST toxin, and the concurrently tested WC vaccine lacking CTB did not protect against any form of ETEC. These findings strongly indicate that the protection achieved against *E. coli*-LT disease by the WC-CTB vaccine was specifically mediated by the CTB vaccine component. It was not possible to determine the protective efficacy for the following 6 months as there were no ETEC cases in this period. When the next ETEC season occurred 9–12 months after the last vaccination, no protection was evident. Thus, in endemic areas, a significant short-term cross-protection against LT-ETEC disease by the oral WC-CTB cholera vaccine, specifically mediated by the CTB component, was shown.

As ETEC is the single most common cause of travellers diarrhea, it was of interest to study protection in travellers. This has been done in two randomized, placebo-controlled studies. The first of these studies was undertaken in Finnish travellers going to Morocco [78]. In this study a protective efficacy of 60% was seen against LT-producing ETEC. A second traveller study was conducted among US students going to Mexico [79]. The design differed from that of the Finnish study in that the US students were vaccinated immediately after arrival in Mexico, while the Finnish tourists were vaccinated before departure. The majority of diarrhea cases among the US students occurred during the first two weeks after arrival, when the vaccine could not be expected to protect. When considering only cases occurring more than 7 days after the second vaccination, there was significant 50% protection against all ETEC diarrhea. The conclusion of these studies is that to protect travellers, the vaccine needs to be given so that the second dose is taken at least one week before arrival. Under these conditions, the WC-rCTB vaccine can be expected to give significant, ca 50–60%, protection against diarrhea caused by LT-producing ETEC.

7.7 Effects on Diarrhea Morbidity and Mortality

Consistent with the demonstrated efficacy of oral WC-CTB and WC-only cholera vaccines and the importance of cholera and ETEC as causes of severe watery diarrhoea in Bangladesh, both vaccines were found to substantially reduce the overall diarrhea morbidity [80]. Admissions for severe watery diarrhea were significantly reduced, by 51 and 32%, in the WC-CTB and WC vaccinated groups, as compared with the placebo group during the first year after vaccination. In the first year of follow-up after vaccination there was also a dramatic effect of cholera vaccines on total mortality [80]. Thus, as compared with placebo, overall mortality rates were 26% lower in the WC-CTB group and 23% lower in the WC group during the first year. Among women vaccinated at ages >15 years, those who received CTB-WC vaccine had 45% fewer deaths ($p < 0.01$) and those who received WC vaccine 33% fewer deaths ($p < 0.05$) than placebo recipients. Several findings suggested that this reduction in overall mortality was a specific effect rather than a statistical coincidence: (i) the effect was restricted to the cholera season, (ii) it was correlated with deaths associated with or preceded by watery diarrheal disease according to "verbal autopsy" reports by household members, and (iii) as mentioned, it was limited to the underprivileged group of women and was not seen in the children also participating in the study (80). The effect on total mortality was restricted to the first year, while reduction in diarrhea morbidity was seen also during the second year of follow-up. The results suggest that even in a "well-treated" area such as the field site in Matlab, Bangladesh there may be a significant number of hidden cholera and severe ETEC diarrhea deaths that might be averted by effective cholera and/or ETEC vaccination programs.

One special aspect studied in the Bangladesh field trial of oral cholera vaccines was the association between breast feeding and the risk of severe cholera among children under 36 months of age [81]. Maternal vaccination with oral cholera vaccines was associated with a 50% reduced risk of their non-vaccinated children to develop severe cholera as compared to the risk in children of mothers given placebo vaccination. These results suggest that vaccination of mothers may provide specific protection to their young children in endemic settings. This could be due to either or both of interrupting maternal-child transmission of cholera and inducing specific antibodies increasing the specific immune protective activity of the mothers' breast milk [37].

7.8 Vaccine-Induced Herd Immunity in Cholera

The public-health importance of a vaccine depends on both the direct protection of vaccinated individuals and on possible additional indirect protection, so-called herd immunity, which results from a reduction in transmission of the pathogen within the vaccinated population. Phase III efficacy trials have typically used individually randomized designs to ensure that measurements of vaccine protective efficacy reflect only the direct effects of the vaccine. As a result, decisions about introduction of newly licensed vaccines into publichealth programs often do not consider the substantially greater protection that can occur when a vaccine is deployed in practice than when it is assessed in a trial setting [82].

These aspects were recently addressed [83] in a reanalysis of indirect and direct effects of vaccination with killed oral cholera WC-CTB and WC vaccines in the large field trial in Bangladesh described above. It was examined whether there was an inverse relationship between the level of vaccine coverage in a residential cluster (a "bari") and the incidence of cholera in individual placebo recipients residing in the bari after controlling for potential confounding variables. If found, such a relationship would support a herd immunity effect. Vaccine (including placebo) coverage of the targeted population ranged from 4 to 65%. Incidence rates of cholera among placebo recipients were inversely related to levels of vaccine coverage (7·01 cases per 1000 in the lowest quintile of coverage *vs* 1·47 cases per 1000 in the highest quintile corresponding to a 79% indirect protection for the placebo recipients in the baris with the highest coverage; $p < 0{\cdot}0001$ for trend). After adjustment for the level of vaccine coverage of the cluster, specific vaccine protective efficacy remained significant (55%, $p < 0{\cdot}0001$). These results indicate that in addition to providing direct protection to vaccine recipients, killed oral cholera vaccines confer significant herd protection to neighboring non-vaccinated individuals. Use of these vaccines could have a major effect on the burden of cholera in endemic settings. Indeed, the latter conclusion was further emphasized by mathematical modelling of the calculated effect of public health use of killed cholera vaccines in a setting such as that in Bangladesh [84]. The results indicate that through the combination of direct (vaccine-specific) and indirect (herd immunity) protection, routine immunization with oral cholera vaccine could even at a moderate coverage where only 50% of the target population was immunized practically eliminate cholera in high-endemic settings and thus be an important tool in the public health control of cholera.

Acknowledgments We acknowledge with gratitude the invaluable contributions by our many co-workers in the cited studies of our own published work. We also gratefully acknowledge the financial support of GUVAX and IVI from many sources. We also thank Ms. Helene Eliasson for skilled secretarial assistance in the preparation of the chapter.

References

1. Craig J.P. 1980, A survey of the enterotoxic enteropathies, in Ouchterlony Ö and Holmgren J (eds). Cholera and related diarrheas. 43rd Nobel Symposium, Stockholm 1978, Karger, Basel, pp 15–25.
2. Sack D.A., Sack R.B., Nair G.B., and Siddique A.K. 2004, Lancet, 363, 223.
3. Svennerholm A.M. and Steele D. 2004, Best. Pract. Res. Clin. Gastroenterol., 18, 421.
4. Holmgren J. 1981, Nature, 292, 413.
5. Sanchez J. and Holmgren J. 2005, Curr. Opin. Immunol., 17, 388.
6. Holmgren J. and Svennerholm A.M. 1983, Prog. Allergy, 33, 106.
7. Svennerholm A.M., Jertborn M., Gothefors L., Karim A.M., Sack D.A., and Holmgren J. 1984, J. Infect. Dis., 149, 884.
8. Holmgren J., Harandi A.M., and Czerkinsky C. 2005, Mucosal Immunol. Update, 13, 3.
9. WHO. 2001, WHO position paper. Weekly Epidemiol. Record, 76, 117.
10. Levine M.M. and Kaper J.B. 1995, Bull. Inst. Pasteur 93, 243.
11. Holmgren J. and Bergquist C. 2004, In New Generation Vaccines, edn 3. Levine M.M., Kaper J.B., Rappuoli R., Liu M.A., and Good M.F. (eds), Marcel Decker, New York, p 499.
12. Holmgren J. and Czerkinsky C. 2005, Nature Med., 11, S45.
13. Blake P.A. 1994, Historical perspectives on pandemic cholera; in IK W, PA B and Olsvik O (eds). Washington, ASM Press, pp 293.
14. Goma Epidemiology Group 1995. Lancet, 345, 339.

15. Mosley W.H., Ahmad S., Benenson A.S., and Ahmed A.,1968, Bull. World Health Organ, 38, 777.
16. Tauxe R.V., Seminario L., Tapia R., and Libel M. 1994, The Latin American Epidemic; in Wachsmuth IK, PA B and Olsvik O (eds). *Vibrio cholerae* and cholera: Molecular to global perspectives. Washington, ASM Press, p 321.
17. Glass R.I., Holmgren J., Haley C.D., Khan M.R., Svennerholm A.-M., Stoll B.J. Hossain K.M.B., Black R.E., Yunus M., and Barva D. 1985, Am. J. Epid., 121, 791.
18. Clemens J.D., Sack D.A., Harris J.R., Chakraboryt J., Khan M.R., Huda S., Ahmed F., Gomes J., Rao M.R., Svennerholm A.-M., and Holmgren J. 1989, J. Inf. Dis., 159, 770.
19. Harris J.B., Khan A.I., LaRocque R.C., Dorer D.J., Chowdhury F., Faruque A.S., Sack D.A., Ryan E.T., Qadri F., and Calderwood S.B. 2005,. Infect. Immun. 73, 7422.
20. Taylor R.K., Miller V.L., Furlong D.B., and Mekalanos J.J. 1987, Proc. Natl. Acad. Sci. USA, 84, 2833.
21. Voss E., Manning P.A., and Attridge S.R. 1996, Microb. Pathog. 20, 141.
22. Herrington D.A., Hall R.H., Losonsky G., Mekalanos J.J., Taylor R.K., and Levine M.M. 1988, J. Exp. Med. 168, 1487.
23. Silva A.J., Pham K., and Benitez J.A. 2003, Microbiology, 149, 1883.
24. Richardson K. 1991, Infect. Immun., 59, 2727.
25. Heidelberg J.F., Eisen J.A., Nelson W.C., Clayton R.A., Gwinn M.L., Dodson R.J., Haft D.H., Hickey E.K., Peterson J.D., Umayam L. et al. 2000, Nature, 406, 477.
26. Merritt E.A., Sarfaty S., van den Akker F., L'Hoir C., Martial J.A., and Hol W.G. 1994, Protein Sci., 3, 166.
27. Holmgren J., Lönnroth I., and Svennerholm L. 1973, Infect. Immun., 8, 208.
28. Hansen G.H., Dalskov S.-M., Rasmussen C.R., Immerdal L., Niels-Christiansen L.-L., and Danielsen E.M. 2005, Biochemistry, 44, 873.
29. Field M., Fromm D., al-Awqati Q., and Greenough W.B. 1972, J. Clin. Invest., 51, 796.
30. Lundgren O. 2002, Pharmacol. Toxicol., 90, 109.
31. Kirn T.J. and Taylor R.K. 2005, Infect. Immunity, 4461.
32. Nair G.B., Qadri F., Holmgren J., Svennerholm A.M., Safa A., Bhuiyan N.A., Ahmad Q.S., Faruque A.S., Takeda Y., and Sack D.A. 2006, J. Clin. Microbiol., 44, 4211.
33. Evans C.A., Gilman R.H., Rabbani G.H., Salazar, and Ali A.1997, Trans. R. Soc. Trop. Med. Hyg., 91, 681.
34. Flach C.F., Qadri F., Bhuiyan T., Alam N., Jennische E., Lönnroth I., and Holmgren J. 2007, A broad up-regulation of innate defense factors during acute cholera. Infect. Immun., 75, 2343.
35. Clemens J.D., van Loon F., Sack D.A., Chakraborty J., Rao M.R., Ahmed F., Harris J.R., Khan M.R., Yunus M., Huda S., Kay B.A., Svennerholm A.-M., and Holmgren J. 1991, J. Infect. Dis., 163, 1235.
36. Feeley J., and Gangarosa E. 1978, Field trials of cholera vaccines; in Ouchterlony O and Holmgren J (eds). Cholera and related diarrheas. 43rd Nobel Symposium. Stockholm. Basel, Karger, 204.
37. Glass R.I., Svennerholm A.M., Stoll B.J., Khan M.R., Hossain K.M., Huq M.I., and Holmgren J. 1983, N. Engl. J. Med., 308, 1389.
38. Jertborn M., Svennerholm A.M., and Holmgren J. 1986, J. Clin. Microbiol., 24, 203.
39. Svennerholm A.-M., Lange S., and Holmgren J. 1978, Infect. Immun., 21, 1.
40. Pierce N.F., Cray W.C., Jr., Sacci J.B., Jr., Craig J.P., Germanier R., and Furer E. 1983, Infect. Immun., 40, 1112.
41. Holmgren J., Svennerholm A.M., Lonnroth I., Fall-Persson M., Markman B., and Lundbeck H. 1977, Nature, 269, 602.
42. Jonson G., Svennerholm A.-M., and Holmgren J. 1989, Infect. Immun., 57, 1809.
43. Jonson G., Osek J., Svennerholm A.-M., and Holmgren J. 1998, Infect. Immun., 64, 3778.
44. Jonson G., Holmgren J., and Svennerholm A.M. 1991, Microb. Pathog., 11, 433.
45. Osek J., Svennerholm A.M., and Holmgren J. 1992, Infect. Immun., 60, 496.
46. Hsiao A., Liu Z., Joelsson A., and Zhu J. 2006, Proc. Natl. Acad. Sci., 103, 14542.

47. Manning P.A. 1994, Curr. Top. Microbiol. Immunol., 192, 265.
48. Asaduzzaman M., Ryan E.T., John M., Hang L., Khan A.I., Faruque A.S., Taylor R.K., Calderwood S.B., and Qadri F. 2004, Infect. Immun., 72, 4448.
49. Svennerholm A.M. and Holmgren J. 1976, Infect. Immun., 13, 735.
50. Quiding M., Nordstrom I., Kilander A., Andersson G., Hanson L.A., Holmgren J. and Czerkinsky C. 1991, J. Clin. Invest., 88, 143.
51. Lycke N. and Holmgren J. 1987, Scand. J. Immunol., 25, 407.
52. Vajdy M. and Lycke N. 1993, Immunology, 80, 197.
53. Svennerholm A.M., Gothefors L., Sack D.A., Bardhan P.K., and Holmgren J. 1984, Bull. World Health Organ., 62, 909.
54. Jertborn M., Svennerholm A.M., and Holmgren J. 1994, Vaccine, 12, 1078.
55. Jertborn M., Svennerholm A.M., and Holmgren J. 1988, J. Infect. Dis., 157, 374.
56. Trach D.D., Clemens J.D., Ke N.T., Thuy H.T., Son N.D., Canh D.G., Hang P.V., and Rao M.R. 1997, Lancet, 349, 231.
57. Richie E.E., Punjabi N.H., Sidharta Y.Y., Peetosutan K.K., Sukandar M.M., Wasserman S.S., Lesmana M.M., Wangsasaputra F.F., Pandam S.S., Levine M.M. et al. 2000, Vaccine, 18, 2399.
58. Cohen M.B., Giannella R.A., Bean J., Taylor D.N., Parker S., Hoeper A., Wowk S., Hawkins J., Kochi S.K., Schiff G., and Killeen K.P. 2002, Infect. Immun., 70, 1965.
59. Sancehz J. and Holmgren J. 1989, Proc. Natl. Acad. Sci. USA, 86, 481.
60. Clemens J.D., Sack D.A., Harris J.R., Chakraborty J., Khan M.R., Stanton B.F., Kay B.A., Khan M.U, Yunus M.D., Atkinson W., Svennerholm A.-M., and Holmgren J. 1986, Lancet, 2, 124.
61. Clemens J.D., Sack D.A., Harris J.R., Van Loon F., Chakraborty J., Ahmed F., Rao M.R., Khan M.R., Yunus M., Huda N., Stanton B.F., Kay B.A., Walter S., Eekels R., Svennerholm A.-M., and Holmgren J. 1990, Lancet, 335, 270.
62. Tayot J.L., Holmgren J., Svennerholm L., Lindblad M., and Tardy M.1981, Eur. J. Biochem., 113, 249.
63. Sanchez J.L., Vasquez B., Begue R.E., Meza R., Castellares G., Cabezas C., Watts D.M., Svennerholm A.-M., Sadoff J.C., and Taylor D.N. 1994, Lancet, 344, 1273.
64. Taylor D.N., Cardenas V., Sanchez .JL., Begue R.E., Gilman R., Bautista C., Perez J., Puga R., Gaillour A., Meza R., Echeverria P., and Sadoff J. 2000, J. Infect. Dis., 181, 1667.
65. Clemens J.D., Sack D.A., and Ivanoff B. 2001, J. Infect. Dis., 183, 1306.
66. Lucas M.E., Deen J.L., von Seidlein L., Wang X.Y., Ampuero J., Puri M., Ali M., Ansaruzzaman M., Amos J., Macuamule A., Cavailler P., Guerin P.J., Mahoudeau C., Kahozi-Sangwa P., Chaignat C.L., Barreto A., Songane F.F., and Clemens J.D. 2005, N. Engl. J. Med., 352, 757.
67. Anh D., Lopez A., Thiem V., Canh G., Long P., Son N., Deen J., von Seidlegin L., Carbis R., Han S., Shin S., Attridge S., Holmgren J., and Clemens J. 2007, Vaccine, 25,1149.
68. Thiem V., Deen J., von Seidlein L., Canh do G., Anh D., Park J., Ali M., Danovaro-Holliday M., Son N., Hoa N., Holmgren J., and Clemens J. 2006, Vaccine, 24, 4297.
69. Kenner J.R., Coster T.S., Taylor D.N., Trofa A.F., Barrera-Oro M., Hyman T., Adams J.M., Beattie D.T., Killeen K.P., Spriggs D.R., et al. 1995, J. Infect. Dis., 172, 1126.
70. Qadri F., Chowdhury M.I., Farugue S.M., Salam M.A., Ahmed T., Begum Y.A., Saha A., Al Tarique A., Seidlein L.V., Park E., Killeen K.P., Mekalanos J.J., Clemens J.D., Sack D.A.2007, Vaccine, 25, 231.
71. Garcia L., Jidy M.D., Garcia H., Rodriquez B.L., Fernandez R., Ario G., Cedre B., Valmaseda T., Suzarte E., Ramirez M., Pino Y., Campos J., Menendez J., Valera R., Gonzales D., Gonzalez I., Perez O., Serrano T., Lastre M., Miralles F., Del Campo J., Maestre J.L., Perez J.L., Talavera A., Perez A., Marrero K., Ledon T., and Fando R. 2005, Infect. Immun., 73, 3018.
72. Jertborn M., Svennerholm A.-M. and Holmgren J. 1996, Vaccine, 14, 1459.
73. Coster T.S., Killeen K.P., Waldor M.K., Beattie D.T., Spriggs D.R., Kenner J.R., Trofa A., Sadoff J.C., Mekalanos J.J., and Taylor D.N. 1995, Lancet, 345, 949.

74. Tacket C.O., Losonsky G., Nataro J.P., Comstock L., Michalski J., Edelman R., Kaper J.B., and Levine M.M. 1995, J. Infect. Dis., 172, 883.
75. Holmgren J., Söderlind O., and Wadström T. 1973, Acta Path. Microbial. Scand., 81, 757.
76. Pierce N.F. 1977, Infect. Immun., 18, 338.
77. Clemens J.D., Sack D.A., Harris J.R., Chakraborty J., Neogy P.K., Stanton B., Huda N., Khan M.U., Kay B.A., Khan M.R., Ansaruzzaman M., Yunus M., Rao M.R., Svennerholm A.-M., and Holmgren J. 1988, J. Infect. Dis., 158, 372.
78. Peltola H., Siitonen A., Kyronseppa H., Simula I., Mattila L., Oksanen P., Kataja M.J., and Cadoz M. 1991, Lancet, 338, 1285.
79. Scerpella E.G., Sanchez J.L., Mathewson J.J., Torres-Cordero J.V., Sadoff J.C., Svennerholm A.-M., DuPont H.L., Taylor D.N., and Ericsson C.D. 1995, J. Travel Med., 2, 22.
80. Clemens J.D., Sack D.A., Harris R., Chakraborty J., Khan M.R., Stanton B.F., Ali M., Ahmed F., Uynus M., Kay B.A., Khan M.V., Rao M.R., Svennerholm A.-M., and Holmgren J. 1988, Lancet, i, 1375.
81. Clemens J.D., Sack D.A., Harris J.R., Khan M.R., Chakraborty J., Chowdhury S., Rao M.R., van Loon F..P, Stanton B.F., Yunus M., Ali M.D., Ansaruzzaman M., Svennerholm A-M., and Holmgren J. 1990, Am. J. Epidemiol., 131, 400.
82. Clemens J. and Jodar L. 2005, Nat. Med., 11, S12.
83. Ali M., Emch M., von Seidlein L., Yunus M., Sack D.A., Rao M., Holmgren J., and Clemens J.D. 2005, Lancet, 366, 44.
84. Longhini I.M., Nizam A., Ali M., Yonus M., Shenvi N., and Clemens J.D. 2007, PLoS Medicine, 4, e336.

Chapter 8
Helicobacter pylori Pathogenesis and Vaccines

Paolo Ruggiero

Abstract *Helicobacter pylori* is a spiral-shaped, Gram-negative bacillus that chronically infects the gastric mucosa of more than 50% of the human population worldwide. *H. pylori* infection causes chronic gastritis, peptic ulcer and MALT lymphoma, and constitutes a risk factor for developing gastric adenocarcinoma. The pathological outcome of *H. pylori* infection is determined by complex host-pathogen interactions, which have been elucidated only in part. Antibiotic-based therapies against *H. pylori* are generally effective, but can fail due to antibiotic resistance or poor patient compliance. Vaccination would represent a valid alternative approach to fight this pathogen; however, presently, there are not licensed anti-*H. pylori* vaccines. Observations made in both animals and humans have provided evidence that the natural response to *H. pylori* infection is of Th1 type; nevertheless, this response, although strong, is rarely able to clear the infection. On the other hand, the inflammatory response to the infection can contribute to the development of the gastric pathologies. Studies have been performed, mostly in animals, to elucidate the nature of protective immune response elicited by vaccination; in spite protection has been obtained in various animal models by both prophylactic and therapeutic vaccination, mechanisms of protection still need to be clarified, and correlates of protection have to be identified.

8.1 Introduction

Bacteria in mammalian stomach had been already observed since the late 19th century, but only in 1982 a bacterium was isolated from human gastric biopsies and cultured, and then its relationship with gastritis in humans was proposed [1, 2]. In 1989 this microorganism was finally classified as *Helicobacter pylori* [3]. For the revolutionary discovery that gastritis and peptic ulcer are caused by an infectious agent, in

P. Ruggiero
Novartis Vaccines and Diagnostics S.r.l, Via Fiorentina 1, 53100 Siena, Italy
e-mail: paolo.ruggiero@novartis.com

2005 Barry Marshall and Robin Warren received the Nobel Prize in Physiology or Medicine.

H. pylori is a spiral-shaped, flagellated Gram-negative bacillus which colonizes human gastric mucosa. It has been estimated that humans have been hosting *H. pylori* for at least 100,000 years [4]. Presently, more than 50% of human population is infected. Prevalence of *H. pylori* infection is much higher in developing than in developed countries [5, 6], most probably as a consequence of different hygiene, sanitation, and living conditions; however, population-based long-term follow-up studies in large groups should be performed to clarify the current development of *H. pylori* infection incidence in children [5].

Commonly, *H. pylori* infections are acquired during the first years of life [7], mostly within the family [8]. The exact way of transmission is still to be understood: it is supposed to be oro-fecal and/or oro-oral [9, 10, 11], and it could also imply contaminated food or water [12, 13]. A role of pets [14, 15, 16] or houseflies in *H. pylori* transmission seems to be unlikely [17]. *H. pylori* can assume a non-cultivable coccoid form, that has been proposed to be either a degenerative or a dormant form; in the latter case, it could represent an important way of dissemination of the microorganism [18, 19, 20]. *H. pylori* can be also found in the oral cavity, but the relationship between its presence in this site and the gastric infection is doubtful [21, 22].

Although mostly asymptomatic, the *H. pylori* infection causes chronic gastritis, peptic ulcer [23], and gastric mucosa-associated lymphoid tissue (MALT) lymphoma [24]. Moreover, *H. pylori* infection plays a role in gastric carcinogenesis, and WHO has included this pathogen among the category 1 carcinogens [25].

8.2 *H. pylori*-Related Diseases

Both direct bacterial action and host response concur to originate chronic inflammation of the stomach in the presence of *H. pylori* infection. Bacterial products act on the gastric tissues, weakening the mucus layer protection, and damaging the gastric epithelial cells, also inducing apoptosis [26, 27]. *H. pylori* adheres to the epithelial tight junctions, increasing the paracellular barrier permeability and changing the distribution of the tight-junction related molecules, conceivably to have access to nutrients and ions [28]. The strong immune response elicited by the *H. pylori* infection, even though unable to confer protection against the bacterium, may contribute to the local inflammation [29]. Moreover, the mimicry of the *H. pylori* LPS with human Lewis blood antigens [30] can elicit in the infected host production of autoantibodies reacting with the same Lewis antigens expressed by the gastric epithelial cells, thus exacerbating the inflammation [31]. Recently, it has been proposed the relevance of molecular mimicry between *H. pylori* antigens and H^+, K^+-ATPase in development of human gastric autoimmunity [32]; however, the *H. pylori*-induced chronic inflammation by itself might be sufficient to partially breakdown gastric mucosal tolerance and initiate gastric autoimmunity [33].

H. pylori infection causes gastritis, peptic ulcer [2, 23], and gastric mucosa-associated lymphoid tissue (MALT) lymphoma [24, 34]. It has been observed that

H. pylori-positive human population is exposed to a significantly higher risk of gastric cancer than non-infected population [35, 36]. The causative effect of *H. pylori* infection on gastric adenocarcinoma development has been proven in animal models [37, 38, 39]. Possible association of *H. pylori* infection with T cell lymphoma has also been reported [40].

H. pylori isolates have been divided into two main categories: type I and type II strains. Type I *H. pylori* strains contain in their genome the cytotoxin-associated gene pathogenicity island (*cag* PAI) and express CagA protein [41, 42], which will be described more in detail below. Type I strains are often associated with the most severe outcome of the infection, in particular with gastric cancer, as demonstrated both in experimentally infected animals [37] and in naturally infected individuals [43, 44, 45].

Host's genetic background can influence the severity of the *H. pylori*-associated pathologies. For example, polymorphisms in the human IL-1β and IL-1 receptor antagonist genes can determine the outcome of infection [46, 47, 48, 49], possibly related to the acid suppressive effect of IL-1β. In fact, upon *H. pylori* infection, people having a high gastric acid secretion are likely to develop antral gastritis and duodenal ulceration, while people having lower acid output can develop corpus gastritis, atrophy of the fundic mucosa and gastric ulcer, with possible malignant outcome [46, 50] (Fig. 8.1). This is consistent with the previous observation that individuals with duodenal ulcer disease are less likely to develop gastric adenocarcinoma than the average population [35, 51]. Polymorphisms of other human genes, such as TNF-α [52] and IL-10 [53], could influence the host response to *H. pylori* infection; enhanced production of TNF-α, as well as insufficient IL-10 expression, could exacerbate the inflammatory response. Polymorphism 251A of IL-8 has been reported to be associated with progression of gastric atrophy in patients with *H. pylori* infection, thus possibly increasing the risk of gastric ulcer and gastric cancer [54]. Further studies on association of host genetic polymorphisms

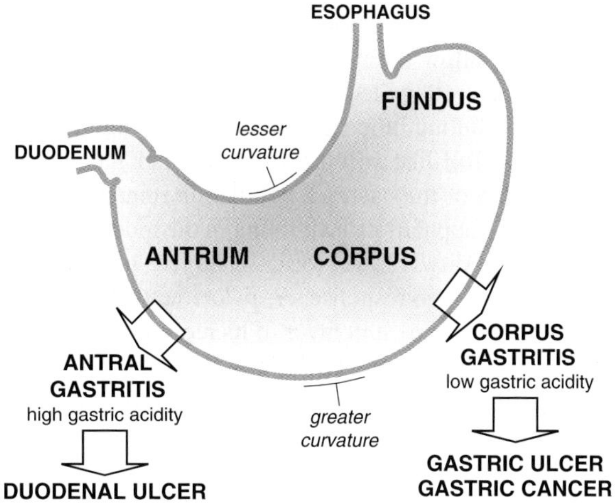

Fig. 8.1 Schematic representation of the stomach anatomy and of the most severe pathological outcomes of *H. pylori* infection

with the type and severity of *H. pylori*-associated disease could be useful not only to better understand the pathogenesis, but also to develop predictive analyses of the pathological outcome of *H. pylori* infection.

Risk of development of gastric cancer is also linked with family history of gastric cancer, low consumption of fruits and vegetables, consumption of salted, smoked or poorly preserved foods and cigarette smoking [55].

The association of *H. pylori* infection with some extra-gastric pathologies, and in particular, but not only, with cardiovascular diseases, has been suggested [56, 57, 58, 59, 60, 61]. Clarifying the mechanisms by which the *H. pylori* infection causes or contributes to the development of pathologies at sites apparently unrelated to the site of infection could help to understand the action of *H. pylori* on the host immune system, both at local and systemic level.

There is evidence that eradication of *H. pylori* infection can cause regression of the related diseases [38, 62, 63, 64, 65, 66]. By preventing mortality from peptic ulcer disease and gastric cancer, *H. pylori* eradication can produce important cost-effective health benefits [67, 68, 69]. Eradicating *H. pylori* can be also advantageous in non-ulcer dyspepsia [70].

It has been hypothesized that *H. pylori* infection could protect against gastroesophageal reflux disease (GERD), Barrett's esophagus, and esophageal adenocarcinoma [71, 72, 73, 74]. Nevertheless, in the case of GERD, this hypothesis has not been confirmed so far in controlled clinical trials or population-based studies [75, 76, 77, 78]. *H. pylori* eradication, although cannot cure GERD, has been reported to prolong the disease-free interval when compared with the anti-secretory therapy alone [76]. Also, no evidence has been found that *H. pylori* infection reduces the risk of esophageal adenocarcinoma [79].

8.3 *H. pylori* Factors Relevant to Virulence and Pathogenesis

The presence of flagella at one end of the spiral-shaped bacillus confers to *H. pylori* a screw-like movement that allows penetration of the gastric mucus layer, required for the gastric colonization [80]. After crossing the mucus layer, a redundant series of adhesins allow the microorganism to adhere to the gastric epithelial cells [81]. Several host cell molecules have been proposed to act as receptor for bacterial binding, including blood group antigens, class II Major histocompatibility (MHC) molecules, Toll-like receptors (TLR), and gastric trefoil factor TFF1 (reviewed in [82]). Part of the bacterial population remains non-adherent, in equilibrium with the adherent population, swimming in the mucus layer close to the epithelial surface, where the acidity is lower [83]. To survive into the hostile gastric environment and to escape immune response, *H. pylori* activates a series of mechanisms to modify the microenvironment in favor of its requirements, and to impair the phagocytic and antigen presenting activities [28]. In the following paragraphs of this section some of the major *H. pylori* colonization/virulence factors will be described more in detail, with particular attention to those that have already been involved in clinical vaccine trials.

8.3.1 Urease

The *H. pylori* urease catalyzes the conversion of urea to CO_2 and ammonia, neutralizing the acidic environment and allowing *H. pylori* to survive in the gastric niche in the presence of acidic juices. It is required for bacterial colonization [84]. Urease consists of two moieties, UreA (27 kDa), and UreB (62 kDa). It is released by bacteria through autolysis, and constitutes up to 10% of the total bacterial protein content. Urease seems to be involved also in the activation and adhesion of inflammatory cells at the site of gastric lesions [85], and to stimulate macrophage inducible nitric oxide synthase [86], thus contributing to mucosal damage and, possibly, to the development of gastric pathology.

8.3.2 Neutrophil-Activating Protein (HP-NAP)

HP-NAP is a 17 kDa protein which assembles to form dodecamers [87, 88]. It induces chemotaxis and direct activation of neutrophils and monocytes, and stimulates production of reactive oxygen intermediates [89]. In rat mesenteric venules, HP-NAP crosses the endothelium and promotes polymorphonuclear cell (PMN) adhesion [90]. The contribution of HP-NAP to the recruitment of neutrophils and monocytes is also indirect, through the stimulation of mast cells [91] and consequent IL-6 production. These effects can induce local gastric inflammation, and are potentiated by IFN-γ and TNF-α [89]. HP-NAP has iron-binding capacity, suggesting its role in iron uptake; however, its expression is not regulated by iron, thus its main function could be to maintain an inflammation level that causes the release of nutrients from the tissue, thus promoting *H. pylori* growth [88].

8.3.3 Cytotoxin-Associated Gene (cagA) and cag Pathogenicity Island

As mentioned above, *H. pylori* isolates belonging to type I, which are associated with the most severe pathological outcome of infection, differently from those of type II, contain in their genome a 40 kb region called *cag* pathogenicity island (*cag* PAI). Similarly to that observed in other pathogens [92], *H. pylori* has acquired this pathogenicity island during its evolution. This may have represented for *H. pylori* one of the most important steps in changing from a commensal or moderately dangerous guest to an insidious pathogen. The *cag* PAI encompasses several genes including the cytotoxin-associated gene *cagA*, which encodes CagA protein [42]. The majority of clinical isolates are CagA-positive. CagA is highly immunogenic at both antibody and cellular level; this immune response has been considered as a marker of severe infection. Other genes in the *cag* PAI encode proteins that constitute a type IV secretion system, able to translocate CagA into the eukaryotic cell, where it is phosphorylated by the host's enzymes [93]. The translocation and phosphorylation of CagA are followed by induction of IL-8 production by epithelial cells, activation of NF-kB, remodeling of the cytoskeleton and formation of cellular pedestals [41]; it has been shown in vitro that cellular shape and motility dramatically change following co-culture with *H. pylori* strains, strictly depending on the CagA expression and its phosphorylation [94]. It has been proposed that the alteration of cell-cell and cell matrix interactions, induced in vitro by CagA, acts as an early event in *H. pylori*-induced carcinogenesis [95]. CagA activity may explain the association of infection by CagA-positive strains with the development of severe gastric pathology [96]. In agreement with epidemiological observations made in infected humans, in Mongolian gerbils *H. pylori* strains containing an intact *cag* PAI have been shown to cause more severe gastric pathology than strains in which *cag* PAI was absent or incomplete [37, 97].

8.3.4 Vacuolating Cytotoxin (VacA)

The vacuolating cytotoxin VacA is one of the best characterized *H. pylori* factors. The mature VacA protein contains a moiety of 37 kDa essential for the toxic activity, and a 58 kDa portion that binds to target cells. VacA monomers oligomerize to form esa- and eptameric structures [98]. Its sequence is well conserved among different isolates, except for the mid-region of the 58 kDa moiety which expresses allelic variation. Clinical and experimental data clearly show that both alleles are equally toxic [99]; they could well represent the evolutionary expression at the bacterial level of a genetic polymorphism at the level of the specific host cell receptor for this toxin. In vitro, VacA induces cell vacuolation and rearrangements of late endosomes and lysosomes, interfering with their functions [97]. VacA forms oligomeric, anion-selective, urea-permeable channels in artificial and plasma membranes [88, 100], and promotes urea diffusion across the epithelium [101], thus providing the *H. pylori* urease with its substrate. It has been shown in vivo that VacA damages murine gastric epithelium [102]. VacA could support the *H. pylori* survival by inhibiting antigen processing and presentation [103]. Moreover, the observation that VacA favors the intracellular survival of *H. pylori* within AGS cells [104] suggests its possible role in vivo in maintaining an intracellular reservoir of live bacteria, protected from immune response and antibiotic action. VacA blocks proliferation of T lymphocytes and inhibits their activation [105, 106], has chemotactic activity on mast cells, and induces production of pro-inflammatory cytokines [107, 108]. VacA seems to confer an advantage in the first steps of colonization, as a VacA isogenic mutant strain can be outcompeted by wild type strain; however, VacA isogenic mutant is able to infect mice, and, once the infection is established, the bacterial load and degree of inflammation induced are similar to those associated with the wild-type strain [109].

8.3.5 Outer Membrane Proteins

Some *cag*-negative *H. pylori* strains are still able to induce IL-8 production: this ability has been correlated with Outer Inflammatory Protein (OipA or HopH) expression, leading to hypothesize that it is an important virulence factor in relation to the risk of clinically significant outcomes of *H. pylori* infection [110]. More recently, it has been reported that OipA is involved in bacterial adherence, but its role in directly influencing proinflammatory signaling was put in doubt, as *oip*A isogenic mutants were still able to induce IL-8 expression in vitro [111]. Other outer membrane proteins that could be relevant for virulence and pathogenesis are adhesins, such as the Blood group Antigen Binding adhesin (BabA), whose alleles are distributed reflecting adaptation to the types of receptors available in local populations [112] and the *H. pylori* adhesin A (HpaA), which has been reported to be essential for colonization in mice [113].

8.4 The Fight Against *H. pylori*: Current Status and Perspectives

Current therapies against *H. pylori* are based on the use of one proton pump inhibitor (PPI) plus two or more antibiotics [114] for one–two weeks. These therapies are generally effective, eradicating the bacterium from the infected individual in 80–90% of the cases; however, the real efficacy at the level of general practitioners is not well known, and could be significantly lower that that reported in controlled studies. Moreover, since only symptomatic patients are treated, a large number of infected – but asymptomatic – subjects would still remain at risk of developing severe complications of *H. pylori* infection.

The main reason of therapy failure is the lack of compliance due to the side effects of the treatment, such as nausea and general discomforts that can lead the patient to discontinue the treatment. Another relevant reason of failure of the first-line therapy is the antibiotic resistance, which is a general problem that progressively limits the efficacy of the antibiotics. Prevalence of *H. pylori* antibiotic resistance can reach values up to 95% for nitroimidazoles, 50% for macrolides, and 30% for penicillins, varying according geographic regions and being higher in developing than in developed countries [115].

Recurrence of *H. pylori* infection after a successful therapy is generally uncommon in developed countries, while can reach high values in developing countries and in particular geographic areas. The recurrence rate is highly variable among different studies, ranging from 0 to more than 50% [116].

There is a growing field of investigation on anti-*H. pylori* activity of non-antibiotic substances, encompassing phytomedicines, probiotics, and antioxidants [117]. The additional use of lactoferrin and probiotics has been recently proposed to increase the efficacy of current therapies [118]. However, while the inclusion of non-antibiotic substances in the therapies against *H. pylori* appears interesting and it is likely to be further developed in the near future, at present there are not validated therapies as an alternative to the current ones.

Vaccines represent the most cost-effective and successful approach to prevent infectious diseases. In the case of *H. pylori*, the drawbacks of antibacterial therapy would be overcome by the use of an efficacious vaccine, which would be cost-effective [119], preventing the insurgence of peptic ulcer, gastric cancer, and other *H. pylori*-related pathologies. Based on a mathematical model that compartmentalizes the population according to age, infection status and clinical state, and also takes into account the progressive, natural decrease of the incidence of infection, it has been calculated that even a 50%-effective vaccination campaign of infants with a prophylactic vaccine lasting only 10 years would be cost-effective in developed countries such as United States or Japan [120]. The same model indicates that in developing countries, due to the higher incidence of infection, the vaccination effort should be more prolonged to eliminate the pathogen and its associated disease. However, the cost-effectiveness of a vaccination against *H. pylori* in developing countries would increase considerably when taking into account other clinical consequences of the infection with *H. pylori*, that are much more evident in developing countries. For example, studies from Africa, Asia, and South America have reported that early acquisition of *H. pylori* may cause growth retardation [121, 122] and that this infection, by causing hypochlorhydria, can favor an increase of diarrheal diseases [123] and enteric infections such as cholera [124] and typhoid fever [125].

8.5 Animal Models of *H. pylori* Infection

Most of the knowledge of the *H. pylori* infection in humans comes from the first investigations in patients with chronic infection, in whom, however, it was impossible to get information on the initial stages of stomach colonization by the microorganism. Therefore, the early phases of infection in humans remain largely unknown, with the exception of some data obtained through few cases of self-infection [126, 127] and of accidental infections for professional reasons [128], and by examining volunteers experimentally infected with a CagA-negative *H. pylori* strain [129]. The limited possibility of following all the aspects of the pathogenesis of *H. pylori* infection in humans makes necessary the use of animal models to understand the bacterium-host interactions, and to study the mechanisms of immune response during the infection or after vaccination. Although so far we do not have an ideal animal model reproducing all numerous aspects of infection and disease observed in humans, the available models represent a precious tool for the development of vaccination strategies, including the selection of bacterial molecules as potential candidates for vaccine development.

Mouse models [130, 131, 132] have been extensively used for studies with *H. pylori*. Other models have been developed in non-mouse rodents [133, 134, 135], gnotobiotic piglets [84], monkeys [136], and dogs [137, 138].

8.6 Vaccination Against *H. pylori* in Animal Models

That protection against *H. pylori* infection can be achieved in animals, both prophylactically and therapeutically, has been proven by various groups (reviewed in [139]).

Since the microorganism is localized extracellularly at the mucosal level, particular emphasis has been given to oral immunization, although other mucosal routes of immunization [140], as well as the parenteral route [139] or prime-boost regimens [141, 142] have been considered, with interesting protective results obtained by mucosal priming followed by systemic boost [143]. Mucosal immunization requires the concomitant use of strong mucosal adjuvants, as proteins are poor immunogens when given mucosally. Some of the strongest mucosal adjuvants presently known are bacterial toxins, such as cholera toxin (CT) and the *E. coli* heat-labile enterotoxin (LT), which induce severe diarrhea in humans, seriously limiting their use. Thus, non-toxic mutants of these molecules are being generated and tested as mucosal adjuvants [144].

The feasibility of mucosal immunization was first demonstrated in mice immunized orally with bacterial lysates or chemically inactivated whole-cell bacteria together with CT, LT, or their non-toxic mutants (reviewed in [139]), obtaining a high rate of protection against *H. pylori*. Presently, the use of whole-cell based vaccines is unlikely to be accepted for humans, facing with problems related with the quality and reproducibility of preparations, and in particular to the presence of unwanted contaminants. For *H. pylori*, the presence of LPS with Lewis antigens, which are expressed on the surface of the majority of human cells, including gastric epithelial cells [30], could seriously hamper the use of a whole-cell vaccine in humans. Thus, the current research on *H. pylori* vaccine is focused on recombinant protein antigens.

A crucial step in selecting vaccine candidates is to test their protective efficacy. The animal models, and the murine one in particular, which represents a valid model to screen antigens for human vaccination [145], have greatly helped in antigen selection. The efficacy of prophylactic and therapeutic immunization against *H. pylori* has now been demonstrated for a variety of native or recombinant antigens, such as urease, heat shock proteins, VacA, CagA, HP-NAP, catalase (reviewed in [139], and HpaA [146]. Also, DNA vaccines and different delivery systems such as live *Samonella* or Polio Virus vectors have been used in animals (reviewed in [147]). Although in most cases the immunization resulted in a strong reduction of bacterial colonization rather than a sterilizing immunity, the data obtained in animal models indicate the feasibility of vaccination against *H. pylori*.

The information derived by studies on *H. pylori* genomics and proteomics could lead to identify further vaccine candidates [148, 149, 150].

8.7 *H. pylori* and the Immune System: What Is the Protective Immune Response?

H. pylori infection elicits a strong immune response, at both B and T cell level, as indicated by both antibody and cytokine and chemokine production. Nevertheless, the infection is seldom cleared.

Presently, although important observations on immune response to *H. pylori* have been made in both animal models and humans, it is not well understood how mucosal or systemic immunization can induce protection against *H. pylori*. Some data appear to be controversial being obtained by using different animal species or strains and/or bacterial strains; moreover, observations made in animals, although extremely valuable, cannot be automatically extended to humans.

Specific IgA [151] or IgG [152] antibodies produced locally in the gastric compartment have been originally hypothesized to mediate protection by favoring killing by neutrophils or monocytes through bacterial opsonization [153], or by neutralizing the activity of bacterial toxin VacA [154]. The need of specific antibodies in the effector mechanisms against *H. pylori* became doubtful when it was observed that humans with congenital IgA deficiency do not suffer from a *H. pylori* gastric pathology more severe than that observed in normal individuals [155], and it was proven that protection against *H. pylori* was attainable by vaccinating immunoglobulin-deficient mice [156, 157]. Protection against *H. pylori* was also achieved in MHC class I knockout mice, indicating that MHC class I-restricted $CD8^+$ T lymphocytes do not play a major role in such protection [158]. On the contrary, unsuccessful vaccination of MHC class II knockout mice demonstrated a crucial role of MHC class II-restricted $CD4^+$ T cells in achieving protection against *H. pylori* [156, 157].

Predominant Th1 response to *H. pylori* infection has been observed in animal models [159, 160, 161] and in humans [162, 163]. Consistent with this observation, some *H. pylori* factors including HP-NAP [164], LPS [165] and Outer Membrane Protein (OMP) 18 [166] have been shown to elicit a Th1 response in animals. After *H. pylori* infection, a long-lasting Th1-type response can be established, contributing to worsening of the disease. In a subset of infected people, the infection can affect the development or the expression in gastric T cells of regulatory cytotoxic mechanisms on B-cell proliferation, making easier the possibility of neoplastic transformation [167]; as already mentioned, host's genetic factors could be relevant in determining the outcome of *H. pylori*-associated pathologies [46, 47].

In the past, it was proposed that a protective response following vaccination should have been of Th2-type [168]. This hypothesis was supported, for example, by the finding that oral immunization failed to confer protection to IL-4 (Th2) knockout mice [169]. In addition, after therapeutic immunization with urease B subunit plus CT as adjuvant, the levels of IL-4 were found to increase, while those of IFN-γ (Th1) decreased [170]. Furthermore, systemic vaccination with *H. pylori* lysate and aluminum hydroxide was able to confer protection against *H. pylori* challenge [171]. Finally, helminth infection, which is known to elicit strong Th2-type immune responses, was shown to reduce gastric atrophy in *H. pylori*-infected mice [172]. On the other hand, more recent observations indicate that the regulatory roles of Th1 and Th2 cells in protective immunity against *H. pylori* are still to be clarified, as the crucial role of IL-12 and Th1 responses in achieving protection has been demonstrated [173, 174, 175]. Moreover, IL-12 has been recently proposed as a possible correlate of protection [142]. Also, IL-18 seems to be involved in protective Th1-type immune response, as vaccination against *H. pylori* does not confer protection in IL-18 knockout mice [176]. In mice, both IFN-γ and TNF-α have been proposed to contribute to protection, IFN-γ being also involved in gastric inflammation development [177]; on the other hand, significant correlation between gastritis severity, *H. pylori* colonization, apoptosis of epithelial cells and local secretion of IFN-γ and TNF-α has been found in humans [178].

In patients with peptic ulcer, it has been found that the large majority of *H. pylori*-specific clones, upon stimulation with the specific *H. pylori* antigen, showed a polarized Th1 profile, with high production of IFN-γ; conversely, in uncomplicated gastritis, the majority of Th clones were Th0 [179]. Inhibition of Th1 or activation of Th2 responses reduce dyspeptic symptoms, balancing and quenching some detrimental effect of Th1 responses [179]. Based on these and other similar observations, it can be concluded that a prolonged Th1-driven inflammation would result in immunopathology, while a polarized Th2 response alone would not guarantee protection. Thus, specific Th1 response appropriately tuned by Th2 cells would lead to a properly balanced, protective response [167, 179, 180, 181].

The difficulty to understand the protective immune response to *H. pylori* and why the natural response seems to be ineffective also resides in the incomplete knowledge of *H. pylori* immunopathogenesis (for a review, see [82]), and in particular of the mechanisms by which *H. pylori* disturbs and/or escapes the host immune response.

For example, it has been reported that *H. pylori* can either impair phagocytosis or survive in the phagosome after phagocytosis [182, 183]. However, efficient killing by phagocytes have been shown in vitro in the presence of specific antibodies and in absence of complement [184]. A powerful feature by which *H. pylori* could become protected from the immune response is its ability, observed in vitro, to penetrate epithelial cells and survive within them [185]; if this mechanism is efficient in vivo, it can create a reservoir of live bacteria in the gastric epithelium, and this could explain at least in part the difficulties in eradicating the *H. pylori* infection.

Regulatory $CD4^+$ $CD25^{high}$ T cells have been found to increase in *H. pylori*-infected individuals, and their involvement in suppressing mucosal immune responses has been proposed [186].

It has been shown in vitro that *H. pylori* factors decrease mouse dendritic cell IL-12 secretion [187, 166], while inducing dendritic cell maturation and function [166]. In contrast to data obtained in mice, human dendritic cells have been reported to preferentially produce IL-12 in the presence of *H. pylori* [188], and to be impaired by chronic exposure to *H. pylori* [189]. Remarkably, *H. pylori* Lewis antigens are able to bind dendritic cells, and decreasing IL-6 production, thus suggesting a possible mechanism of immune response suppression [190].

NK cells produce high levels IFN-γ upon co-stimulation with *H. pylori* and IL-12, thus an NK role in the local immune response to *H. pylori* infection has been hypothesized [191]. Interestingly, a *H. pylori* proinflammatory, cecropin-like peptide has been identified that enhances monocyte-induced NK cell suppression, and could be involved in the strategy *H. pylori* employs to protect itself from their action [192].

Mast cells are critical mediators of vaccine-induced *H. pylori* clearance [193]. It has been hypothesized that crosstalk between mast cells, $CD4^+$ T cells and possibly other unidentified mediators would make the environmental conditions of the stomach unfavorable to *H. pylori* survival and favor the bacterial clearance after vaccination [193]. One of the mechanisms by which *H. pylori* could counteract this activity of the immune system could include VacA toxin, which, as aforementioned, inhibits proliferation and activation of T lymphocytes [105, 106]; on the other hand, VacA has chemotactic activity on mast cells, and induces production of pro-inflammatory cytokines [107, 108].

The study of signaling pathways activated during *H. pylori* infection of human cells could contribute to a better understanding of the mechanisms by which *H. pylori* infection induces an inflammatory but functionally ineffective host response. It has been shown in vitro that the expression and activation of c-*fos* and c-*jun* via mitogen-activated protein kinase pathways [194] and NF-κB [195] in response to infection by *cag*-positive *H. pylori* strains lead to IL-8 expression. Further in vitro studies have shown that Toll-Like Receptor (TLR) 2 and TLR5 are involved in *H. pylori*-induced NF-κB activation and IL-8 expression [196], but also that an alternative TLR4 and TLR5-independent pathway exists [197]. Interestingly, in *H. pylori*-infected duodenal ulcer patients, the local IL-8 response is down-regulated by host factors, including the possible action of regulatory T cells [198].

A deeper knowledge of the immunological parameters of protection against *H. pylori* following immunization will be of critical importance to interpret and evaluate the results obtained with vaccines presently under clinical trials and for developing future vaccines.

Based on the numerous studies of vaccination both in animal models and human volunteers, it appears that the route of administration, as well as the adjuvant used, can play a crucial role in the achievement of protective immune response. Conceivably, as in the case of other bacterial vaccines, a combination of two or more antigens could constitute an efficacious vaccine. Finally, it has been demonstrated that vaccination is able to induce strong immune responses in the *H. pylori*-infected mucosa; thus, a therapeutic vaccine may lead to success, as previous *H. pylori*-induced inflammation could have efficiently primed the immune responses against bacterial antigens [199].

There are some indications that protective immunity can produce a transient enhancement of gastric inflammation, as observed in animals vaccinated and then challenged with *H. pylori*, and recently confirmed by microarray analysis on gastric tissue that revealed upregulation of IFN-γ and T cell markers in mice immunized prior to challenge as compared to non-immunized, challenged mice [200]. Thus it can be hypothesized that post-immunization gastritis may be part of the protective response, being involved in bacterial clearance [180, 181, 201, 202]. Recently, a further piece of this puzzle has been identified, by the demonstration that Peyer's patches in the small intestine are critical for priming $CD4^+$ T cells and essential in the development of *H. pylori*-induced gastritis in mice [203]. Moreover, it has been shown that coccoid forms of *H. pylori* are phagocytosed by dendritic cells in Peyer's patches [203].

8.8 Vaccination Against *H. pylori* in Humans

Despite the large body of information that has been acquired on *H. pylori* vaccination in animals, few clinical studies have been carried out so far in humans, and no vaccines are currently licensed. Several of the trials conducted have focused on the use of the most abundant *H. pylori* antigen, i.e. urease. In addition, with one exception, all trials have been carried out using the mucosal route of immunization. The need for strong and safe mucosal adjuvants and the issues inherent to the formulation of vaccines to be given mucosally may have limited the number of studies with these vaccines. A model of experimental *H. pylori* infection in human volunteers has been recently reported, which could be useful for studying *H. pylori* pathogenesis in humans in addition to studies on vaccine efficacy [129].

8.8.1 Inactivated Whole-Cell Vaccines

The efficacy of inactivated whole-cell vaccines against *H. pylori* has been shown in a large number of studies carried out in animals immunized either mucosally or parenterally (reviewed in [139]). Whole-cell vaccines offer the advantage of eliciting immune responses against a wide variety of antigens; however they may include potentially dangerous components of the bacterium, such as those sharing homologies with the self antigens, and able to induce autoimmune responses, *e.g. H. pylori* LPS [31].

A formalin-inactivated whole-cell *H. pylori* vaccine was evaluated in a phase I trial in both *H. pylori*-negative and *H. pylori*-positive subjects [204]. The vaccine, containing various amounts of bacterial cells, was given orally three times, on days 0, 14, and 28, together with 25 μg of the LTR192G mutant of LT toxin. The first part of the trial was an open-label, dose-response study in which *H. pylori*-infected or -uninfected individuals received 2.5×10^6 to 2.5×10^{10} bacterial cells together with the LT mutant. Vaccination elicited *H. pylori*-specific antibody responses only in subjects receiving the highest dose of the vaccine. A marginal increase of IgA and IgG titers was observed only in *H. pylori*-infected patients. The number of specific antibody-secreting cells induced by the vaccine remained negligible; however, some detectable responses were observed at duodenal level in *H. pylori*-negative subjects [205]. While some antibody response was induced only in *H. pylori*-infected patients, proliferative responses of peripheral blood mononuclear cells and production of IFN-γ were observed only in uninfected volunteers (5 and 7 volunteers out of 10, respectively) who had received the highest dose of the vaccine, following in vitro re-stimulation with a bacterial sonicate (and not with a purified antigen). The second part of the trial was a randomized, double-blind study in which *H. pylori*-infected individuals received either 2.5×10^{10} bacteria plus 25 μg of LTR192G, or placebo plus 25 μg of LTR192G. Vaccinated subjects had significantly higher IgA antibody titers in the stools than subjects receiving the placebo. The co-administration of the vaccine with the adjuvant influenced only marginally the serum antigen-specific IgA antibody response. Vaccination of *H. pylori*-infected patients did not achieve bacterial eradication since in both parts of the trial [^{13}C]urea breath test remained positive up to 7.5 months after vaccination [204]. It is not known, however, whether vaccination affected the degree of *H. pylori* gastric colonization, since no microbiological data were reported in this study. Finally, diarrhea occurred in 5 out of 18 subjects vaccinated with the highest dose of bacteria plus the LTR192G mutant, and in one out of three who received the LTR192G mutant. Conceivably, diarrhea was due to the LT mutant, that retains most of its toxic activity in vitro and in vivo [144].

8.8.2 Vaccination with Purified Recombinant Urease

A detailed study was conducted in *H. pylori*-infected volunteers, which were orally immunized with high doses of recombinant urease (20, 60, or 180 mg once weekly for 4 weeks) together with 5 μg of LT. Although the urease was well tolerated, 16 out 24 of the vaccinees had diarrhea, regardless of the dose of urease, thus conceivably due to LT toxicity. Volunteers receiving the highest doses of urease (60 or 180 mg) showed high levels of anti-urease IgA serum antibodies. Neither eradication nor decrease of the gastric inflammation was achieved, although a reduction of bacterial burden in the stomach was observed [206]. The reasons of the partial results of this trial in terms of both immunogenicity and efficacy may be linked to the vaccination regimen, to the vaccine formulation, to the urease antigen used, to the low dose of the mucosal adjuvant employed (higher doses, such as 10 μg, induced too severe diarrhea), or to difficulties inherent to the therapeutic vaccination itself that are not understood yet. Indeed, despite all efforts, so far there are no therapeutic vaccines available on the market.

Following these results, the same group attempted to ameliorate the safety profile of this vaccine by reducing the amount of LT [207]. In this study, 42 healthy *H. pylori*-negative subjects were immunized orally with 60 mg of recombinant urease either in soluble or acid-resistant, encapsulated form together with 2.5, 0.5, or 0.1 μg of LT. The vaccine was given on days 1, 8, 28, and 57. The subjects who received the highest dose of LT showed a slightly better urease-specific antibody response and an increase of $CD4^+$, $CD45RO^+$, $CD69^+$ cells; however, diarrhea was evident in 50% of the subjects of this group. These data confirm, as already known from in vitro assays and animal studies, that LT toxicity, immunogenicity, and adjuvanticity are dose-dependent, and that a fine tuning is required in order to induce protective immune responses against the vaccine without causing unacceptable side effects, such as diarrhea.

Another attempt, made to circumvent the safety issues inherent to the use of the oral administration of wild type LT, has been done by replacing the oral with the rectal route of vaccination, which had been previously successfully employed in mice [208]. Also in this study recombinant urease was used at the dose of 60 mg, administered as a rectal enema to 18 healthy, *H. pylori*-negative adults, together with either 5 or 25 μg of LT, given three times on days 0, 14, and 28 [209]. A strong systemic antibody response to LT was detectable in the majority of the vaccinees, mainly in the group of subjects receiving the lowest dose of LT (5 μg). Only a small minority of subjects developed systemic anti-urease IgG or IgA antibody responses. However, no anti-urease or anti-LT IgA were detectable in stool or in salivary samples. Finally, the urease-driven proliferative response and IFN-γ production were negligible.

All these studies clearly show that oral administration of recombinant urease in particular, and likely of other antigens, requires the use of strong and safe mucosal adjuvants to be given at doses sufficiently high to exert their adjuvant effects, but still unable to induce unwanted effects, such as diarrhea. Both the oral and the rectal routes require the development of appropriate formulations able to induce adequate protective immune responses to *H. pylori*.

8.8.3 Salmonella-*Vectored Urease*

The attempts to develop *H. pylori* vaccines based on the administration of live recombinant *Salmonella* strains expressing urease have resulted in poor immunogenicity and efficacy so far, in marked contrast with the results previously obtained in mice.

In one study, one or two oral administrations of 10^{10} CFU of a *Salmonella enterica* serovar Typhi strain, attenuated by deletion of the *phoP/phoQ* virulence regulon and expressing *H. pylori* urease, induced serum antibody responses to salmonella antigens, while no detectable immune response to urease was observed, even after a booster dose of recombinant urease plus wild type LT [210].

Slightly better results were reported in a subsequent study, in which six volunteers were immunized orally (5 to 8 $\times$ 10^7 CFU) with *S. enterica* serovar Typhimurium harboring the same *phoP/phoQ* deletion [211]. Only one of the six volunteers had detectable urease-specific IgA antibody secreting cells; two others had slight amounts of urease-specific antibodies produced in vitro by cultured peripheral blood mononuclear cells; two subjects had some specific serum IgA antibodies detectable by ELISA, but not by western blotting.

Another study was carried out in volunteers immunized orally with *S. enterica* serovar Typhi Ty21a expressing *H. pylori* urease. None of the 9 vaccinated volunteers showed any detectable antibody response against urease, but 5 subjects developed cellular response [212]. Similar results were obtained in a further study, in which 5 out of 9 vaccinees that were pre-immunized with the carrier strain developed cellular, but not antibody, response against urease; however, in this study, vaccinees that were not pre-immunized with the carrier strain did not develop any kind of specific immune response [213]. More recently, it has been reported that, with a similar vaccination, three out of nine vaccinees cleared *H. pylori* after experimental infection [214]; however, this study has not been reported in detail yet.

It appears that *Salmonella*-vectored *H. pylori* urease vaccines still require optimization, in particular in terms of increase of urease expression [213]; also, coexpression of an adjuvant or immunostimulatory, as shown in mice [215], could increase the efficacy of this kind of vaccine.

8.8.4 Parenteral, Multi-Component Vaccines

Parenteral vaccination has been shown to confer protective immunity against *H. pylori* in various animal models (reviewed in [139]). In most cases, however, the experimental vaccine consisted of an uncharacterized bacterial lysate, instead of well-defined recombinant antigens. Based on previous experience with the acellular pertussis vaccine [216], a multi-component vaccine against *H. pylori* was developed, consisting of VacA, CagA, and HP-NAP. Previous work in mice had shown that immunization with VacA, CagA, or HP-NAP protected animals against challenge with *H. pylori* [139, 217]. The effect of immunization with the combination of these three antigens was then tested in an experimental beagle dog model of infection with *H. pylori* [138]. Both prophylactic (unpublished data) and therapeutic [218] protection against *H. pylori* was observed after parenteral immunization with the three antigens formulated with aluminium hydroxide, the adjuvant most widely utilized for human use. Then, the safety and the immunogenicity of a similar aluminium hydroxide-adjuvanted, multicomponent vaccine was evaluated in human volunteers. *H. pylori*-uninfected individuals were immunized intramuscularly three times, following three different immunization regimes, with a vaccine consisting of either 10 or 25μg each of CagA, VacA, and HP-NAP, plus aluminium hydroxide [219]. This vaccine was extremely safe and highly immunogenic, inducing antibody responses to the three antigens in almost all vaccinees. Months after the last immunization, most of the subjects had still detectable antibody responses to each of the three antigens. Interestingly, parenteral vaccination with VacA, CagA, and HP-NAP induced very strong and sustained antigen-driven cellular proliferative responses and IFN-γ production, as well as strong and long lasting immunological memory.

8.9 Conclusions

As the relationshipbetween *H. pylori* infection and peptic ulcer and increased risk of gastric cancer is well established, the need of blocking colonization or eradicating this pathogen appears to be the logical consequence. The introduction of a vaccine against *H. pylori* could be cost-effective [119, 120] and overcome the limits of the current antibiotic-based therapies. Several potential vaccine candidates have been identified through the studies on *H. pylori* and the mechanisms of its virulence and pathogenesis, and on the bacterial genome. The work on animal models of *H. pylori* infection has allowed not only a further selection of the more promising vaccines, but also the investigation on mechanisms of pathogenesis and on the immune response to both infection and vaccination. Most of the experimental vaccinations against *H. pylori* in animal models did not achieve a complete eradication of the bacterium; however, a strong reduction in bacterial burden was observed, which in some cases was accompanied by an amelioration of gastric pathology. This indicates that further investigation is required on antigens, routes of immunization and adjuvants to achieve sterilizing protection. Moreover, it is clear that further studies are required to better clarify the mechanisms of pathogenesis and to identify immunological correlates of protection.

Some vaccines underwent early phases of clinical trials, giving encouraging results of safety and immunogenicity, but the available results in terms of protection against *H. pylori* infection are poor or have not been produced yet. In the absence of immunological correlates of protection, only large phase III clinical trials can provide evaluation of the efficacy of these vaccines.

Bibliography

1. Marshall, B.J., and Warren, J.R. 1984, Lancet, 1, 1311.
2. Goodwin, C.S., Armstrong, J.A., and Marshall, B.J. 1986, J. Clin. Pathol., 39, 353.
3. Goodwin, C.S., Armstrong, J.A., Chilvers, T., Peters, M., Collins, M.D., Sly, L., McConnel, W., and Harper, W.E.S. 1989, Int. J. Syst. Bacteriol. 39, 397.
4. Covacci, A., Telford, J.L., Del Giudice, G., Parsonnet, J., and Rappuoli, R. 1999, Science, 284, 1328.
5. Rothenbacher, D., and Brenner, H. 2003, Microbes Infect., 5, 693.
6. Frenck, R.W., and Clemens, J. 2003, Microbes Infect., 5, 705.
7. Malaty, H.M., El-Kasabany, A., Graham, D.Y., Miller, C.C., Reddy, S.G., Srinivasan, S.R., Yamaoka, Y., and Berenson, G.S. 2002, Lancet 359, 931.
8. Rothebacher, D., Winkler, M., Gonser, T., Adler, G., and Brenner, H. 2002, Pediatr. Infect. Dis. J. 21, 674.
9. Parsonnet, J., Shmuely, H., and Haggerty, T. 1999, JAMA, 282, 2240.
10. Mitchell, H., and Megraud, F. 2002, Helicobacter, 7(suppl. 1), 8.
11. Dowsett, S.A., Archila, L., Segreto, V.A., Gonzalez, C.R., Silva, A., Vastola, K.A., Bartizek, R.D., and Kowolik, M.J. 1999, J. Clin. Microbiol., 37, 2456.
12. Lu, Y., Redlinger, T.E., Avitia, R., Galindo, A., and Goodman, K. 2002, Appl. Environ. Microbiol., 68, 1436.
13. Klein, P.D., Graham, D.Y., Gaillour, A., Opekun, A.R., and Smith, E.O. 1991, Lancet, 337, 1503.
14. El-Zaatari, F.A.K., Woo, J.S., Badr, A., Osato, M.S., Serna, H., Lichtenberger, L.M., Genta, R.M., and Graham, D.Y. 1997, J. Med. Microbiol., 46, 372.
15. Bode, G., Rothenbacher, D., Brenner, H., and Adler, G. 1998, Pediatr. Infect. Dis. J., 17, 909.
16. Brown, L.M., Thomas, T.L., Ma, J.L., Chang, Y.S., You, W.C., Liu, W.D., Zhang, L., and Gail, M.H. 2001, Scand. J. Infect. Dis., 33, 686.
17. Osato, M.S., Ayub, K., Le, H.H., Reddy, R., and Graham, D.Y. 1998, J. Clin. Microbiol., 36, 2786.
18. Andersen, L.P., Dorland, A., Karacan, H., Colding, H., Nilsson, H.O., Waldström, T., and Blom, J. 2000, Scand. J. Gastroenterol., 9, 898.
19. Cellini, L., Allocati, N., Angelucci, D., Iezzi, T., Di Campli, E., Marzio, L., and Dainelli, B. 1994, Microbiol. Immunol., 38, 843.
20. She, F.F., Lin, J.Y., Liu, J.Y., Huang, C., and Su, D.H. 2003, World J, Gastroenterol., 9, 516.
21. Karczewska, E., Konturek, J.E., Konturek, P.C., Czesnikiewicz, M., Sito, E., Bielanski, W., Kwiecien N., Obtulowicz, W., Ziemniak, W., Majka, J., Hahn, E.G., and Konturek, S.J. 2002, Di. Dis. Sci., 47, 978.
22. Czesnikiewicz-Guzik, M., Loster, B., Bielanski, W., Guzik, T.J., Konturek, P.C., Zapala, J., and Konturek, S.J. 2007, J. Clin. Gastroenterol., 41, 145.
23. Sontag, S.J. 1997, Am. J. Gastroenterol., 92, 1255.
24. Du, M.Q., and Isaccson, P.G. 2002, Lancet Oncol., 3, 97.
25. International Agency for Research on Cancer, World Health Organization. 1994, Schistosomes, liver flukes and Helicobacter pylori. IARC Monographs on Evaluation of Carcinogenic Risks to Humans, IARC Press, Lyon, 60, 177.
26. Wagner, S., Beil, W., Westermann, J., Logan, R.P., Bock, C.T., Trautwein, C., Bleck, J.S., Manns, M.P. 1997, Gastroenterology, 113, 1836.
27. Souza, H.S., Neves, M.S., Elia, C.C., Tortori, C.J., Dines, I., Martinusso, C.A., Madi, K., Andrade, L., Castelo-Branco, M.T. 2006, World J. Gastroenterol., 12, 6133.
28. Montecucco, C., and Rappuoli, R. 2001, Nat. Rev. Mol. Cell. Biol., 2, 457.
29. Ernst, P.B., and Gold, B.D. 2000, Annu. Rev. Microbiol., 54, 615.
30. Appelmelk, B.J., Negrini, R., Moran, A.P., and Kuipers, E.J. 1997, Trends Microbiol., 5, 70.
31. Moran, A.P., and Prendergast, M.M. J. 2001, Autoimmun., 16, 241.

32. Amedei, A., Bergman, M.P., Appelmelk, B.J., Azzurri, A., Benagiano, M., Tamburini, C., Van der Zee, R., Telford, J.L., Vandenbroucke-Grauls, C.M., D'Elios, M.M., and Del Prete, G. 2003, J. Exp. Med., 198, 1147.
33. Bergman, M., Del Prete, G., Van Kooyk, Y., and Appelmelk, B. 2006, Nat. Rev. Microbiol., 4, 151.
34. Wotherspoon, A.C., Ortiz-Hidalgo, C., Falzon, M.R., and Isaacson, P.G. 1991, Lancet, 338, 1175.
35. Uemura, N., Okamoto, S., Yamamoto, S., Matsumura, N., Yamaguchi, S., Yamakido, M., Taniyama, K., Sasaki, N., and Schlemper, R.J. 2001, N. Eng. J. Med., 345, 784.
36. Forman, D., Webb, P., and Parsonnet, J. 1994, Lancet, 343, 243.
37. Ogura, K., Maeda, S., Nakao, M., Watanabe, T., Tada, M., Kyutoku, T., Yoshida, H., Shiratori, Y., and Omata, M. 2000, J. Exp. Med., 192, 1601.
38. Watanabe, T., Tada, M., Naga, i H., Sasaki, S., and Nakao, M. 1998, Gastroenterology, 115, 642.
39. Han, S.U., Kim, Y.B., Joo, H.J., Hahm, K.B., Lee, W.H., Cho, Y.K., Kim, D.Y., and Kim, M.W. 2002, J. Gastroenterol. Hepatol., 17, 253.
40. Bariol, C., Field, A., Vickers, C.R., and Ward, R. 2001, Gut, 48, 269.
41. Covacci, A., and Rappuoli, R. 2000, J. Exp. Med., 191, 587.
42. Censini, S., Lange, C., Xiang, Z., Crabtree, J.E., Ghiara, P., Borodovsky, M., Rappuoli, R., and Covacci, A. 1996, PNAS, 93, 14648.
43. Parsonnet, J., Friedman, G.D., Orentreich, N., and Vogelman, H. 1997, Gut, 40, 297.
44. Nomura, A.M., Lee, J., Stemmermann, G.N., Nomura, R.Y., Perez-Perez, G.I., and Blaser, M.J. 2002, J. Infect. Dis., 186, 1138.
45. Palli, D., Masala, G., Del Giudice, G., Plebani, M., Basso, D., Berti, D., Numans, M.E., Ceroti, M., Peeters, P.H.M, De Mesquita, H.B., Buchner, F.L., Clavel-Chapelon, F., Boutron-Ruault, M.C., Krogh, V., Saieva, C., Vineis, P., Panico, S., Tumino, R., Nyren, O., Siman, H., Berglund, G., Hallmans, G., Sanchez, M.J., Larranaga, N., Barricarte, A., Navarro, C., Quiros, J.R., Key, T., Allen, N., Bingham, S., Khaw, K.T., Boeing, H., Weikert, C., Linseisen, J., Nagel, G., Overvad, K., Thomsen, R.W., Tjonneland, A., Olsen, A., Trichoupoulou, A., Trichopoulos, D., Arvaniti, A., Pera, G., Kaaks, R., Jenab, M., Ferrari, P., Nesi, G., Carneiro, F., Riboli, E., and Gonzalez, C.A. 2007, Int. J. Cancer, 120, 859.
46. El-Omar, E.M., Carrington, M., Chow, W.H., McColl, K.E., Bream, J.H., Young, H.A., Herrera, J., Lissowska, J., Yuan, C.C., Rothman, N., Lanyon, G., Martin, M., Fraumeni, J.F. Jr., and Rabkin, C.S. 2000, Nature, 404, 398.
47. Hwang, I.R., Kodama, T., Kikuchi, S., Sakai, K., Peterson, L.E., Graham, D.Y., and Yamaoka, Y. 2002, Gastroenterology, 123, 1793.
48. Figueiredo, C., Machado, J.C., Pharoah, P., Seruca, R., Sousa, S., Carvalho, R., Capelinha, A.F., Quint, W., Caldas, C., Van Doorn, L.J., Carneiro, F., and Sobrinho-Simoes, M. 2002, J. Natl. Cancer Inst., 94, 1680.
49. Leung, W.K., Chan, M.C., To, K.F., Man, E.P., Ng, E.K., Chu, E.S., Lau, J.Y., Lin, S.R., and Sung, J.J. 2006, Am J Gastroenterol., 101, 714.
50. Konturek, S.J., Konturek, P.C., Konturek, J.W., Plonka, M., Czesnikiewicz-Guzik, M., Brzozowski, T., Bielanski, W. 2006 J. Physiol. Pharmacol., 57(Suppl. 3), 29.
51. Hansson, L.E., Nyren, O., Hsing, A.W., Bergstrom, R., Josefsson, S., Chow, W.H., Fraumeni, J.F. Jr., and Adami, H.O. 1996, N. Engl. J. Med., 335, 242.
52. Yea, S.S., Yang, Y.I., Jang, W.H., Lee, Y.J., Bae, H.S., and Paik, K.H. 2001, J. Clin. Pathol., 54, 703.
53. Wu, M.S., Wu, C.Y., Chen, C.J., Lin, M.T., Shun, C.T., and Lin, J.T. 2003, Int. J. Cancer., 104, 617.
54. Ohyauchi, M., Imatani, A., Yonechi, M., Asano, N., Miura, A., Iijima, K., Koike, T., Sekine, H., Ohara, S., and Shimosegawa, T. 2005, Gut, 54, 330.
55. Fuchs, C.S., and Mayer, R.J. 1995, N. Engl. J. Med. 333, 32.
56. Konturek, S.J., Konturek, P.C., Pieniazek, P., and Bielanski, W. 1999, J. Physiol. Pharmacol., 50, 683.

57. Gasbarrini, A., Franceschi, F., Armuzzi, A., Ometti, V., Candelli, M., Torre, E.S., De Lorenzo, A., Anti, M., Pretolani, S., and Gasbarrini, G. 1999, Gut, 4(Suppl. 1), I9.
58. Davydov, L., and Cheng, J.W. 2000, Investig. Drugs., 9, 2505.
59. Wedi, B., and Kapp, A., 2002, Am. J. Clin. Dermatol., 3, 273.
60. Mayr, M., Kiechl, S., Mendall, M.A., Willeit, J., Wick, G., and Xu, Q. 2006, Stroke, 2003, 34, 610.
61. Ponzetto, A., Cardaropoli, S., Piccoli, E., Rolfo, A., Gennero, L., Kanduc, D., and Todros, T. 2006, J. Hypertens., 24, 2445.
62. Sung, J.J., Lin, S.R., Ching, J.Y., Zhou, L.Y., To, K.F., and Wang, R.T. 2000, Gastroenterology, 119, 7.
63. Papa, A., Cammarota, G., Tursi, A., Gasbarrini, A., and Gasbarrini, G. J. 2000, Clin. Gastroenterol., 31, 169.
64. Sugiyama, T., Sakaki, N., Kozawa, H., Sato, R., Fujioka, T., Satoh, K., Sugano, K., Semine, H., Takagi, A., Ajioka, Y., and Takizawa, T. 2002, Aliment. Pharmacol. Ther., 16(Suppl. 2), 187.
65. Kim, N., Lim, S.H., Lee, K.H., Choi, S.E., Jung, H.C., Song, I.S., and Kim, C.Y. 2000, Dig. Dis. Sci., 45, 1754.
66. Komoto, M., Tominaga, K., Nakata, B., Takashima, T., Inoue, T., Hirakawa, K. 2006, J. Exp. Clin. Cancer. Res., 25, 283.
67. Frendrick, A.M., Chernew, M.E., Hirt, R.A., Bloom, B.S., Bandekar, R.R., and Scheiman J.M. 1999, Arch. Intern. Med., 159, 142.
68. Mason, J., Axon, A.T., Forman, D., Duffett, S., Drummond, M., Crocombe, W., Feltbower, R., Mason, S., Brown, J., and Moayyedi, P. 2002, Aliment. Pharmacol. Ther., 16, 559.
69. Ikeda, S., Tamamuro, T., Hamashima, C., and Asaka, M. 2001, Aliment. Pharmacol. Ther., 15, 1777.
70. Moayyedi, P., Soo, S., Deeks, J., Delaney, B., Harris, A., Innes, M., Oakes, R., Wilson, S., Roalfe, A., Bennett, C., and Forman, D. 2006, Cochrane Database Syst. Rev., 2, CD002096.
71. Blaser, M.J. 1999, Ann. Intern. Med., 130, 695.
72. Blaser, M.J. 1999, J. Infect. Dis. 179, 1523.
73. Koike, T., Ohara, S., Sekine, H., Iijima, K., Abe, Y., Kato, K., Toyota, T., and Shimosegawa, T. 2001, Gut, 49, 330.
74. Jones, A.D., Bacon, K.D., Jobe, B.A., Sheppard, B.C., Deveney, C.W., and Rutten, M.J. 2003, J. Gastrointest. Surg., 7, 68.
75. Moayyedi, P., Bardhan, C., Young, L., Dixon, M.F., Brown, L., and Axon, A.T. 2001, Gastroenterology, 121, 1120.
76. Schwizer, W., Thumshirn, M., Dent, J., Guldenschuh, I., Menne, D., Cathomas, G., and Fried, M. 2001, Lancet, 357, 1738.
77. Malfertheiner, P., Dent, J., Zeijlon, L., Sipponen, P., Veldhuyzen Van Zanten, S.J., Burman, C.F., Lind, T., Wrangstadh, M., Bayerdorffer, E., and Lonovics, J. 2002, Aliment. Pharmacol. Ther., 16, 1431.
78. Nordenstedt, H., Nilsson, M., Johnsen, R., Lagergren, J., and Hveem, K. 2007, Helicobacter, 12, 16.
79. Wu, A.H., Crabtree, J.E., Bernstein, L., Hawtin, P., Cockbur,n M., Tseng, C.C., and Forman, D. 2003, Int. J. Cancer., 103, 815.
80. Eaton, K.A., Morgan, D.R., and Krakowka, S. 1992, J. Med. Microbiol., 37, 123.
81. Testerman, T.L., McGee, D.J., and Mobley, H.L.T. 2001, H.L.T. Mobley, G.L. Mendz, S.L. Hazell (Ed.), ASM Press, Washington DC, USA, 381.
82. Ernst, P.B., Peura, D.A., Crowe, S.E. 2006, Gastroenterol. 130, 188.
83. Screiber, S., Konradt, M., Groll, C., Scheid, P., Hanauer, G., Werling, H-O., Josenhans, C., Suerbaum, S. 2004, PNAS, 101, 5024.
84. Eaton, K.A., Brooks, C.L., Morgan, D.R., and Krakowka, S. 1991, Infect. Immun., 59, 2470.
85. Harris, P.R., Mobley, H.L., Perez-Perez, G.I., Blaser, M.J., and Smith, P.D. 1996, Gastroenterology, 111, 419.

86. Gobert, A.P., Mersey, B.D., Cheng, Y., Blumberg, D.R., Newton, J.C., and Wilson, K.T. 2002, J. Immunol., 168, 6002.
87. Tonello, F., Dundon, W.G., Satin, B., Molinari, M., Tognon, G., Grandi, G., Del Giudice, G., Rappuoli, R., and Montecucco, C. 1999, Mol. Microbiol., 34, 238.
88. Montecucco, C., and De Bernard, M. 2003, Microbes Infect., 5, 715.
89. Satin, B., Del Giudice, G., Della Bianca, V., Dusi, S., Laudanna, C., Tonello, F., Kelleher, D., Rappuoli, R., Montecucco, C., and Rossi, F. 2000, J. Exp. Med., 191, 1467.
90. Polenghi, A., Bossi, F., Fischetti, F., Durigutto, P., Cabrelle, A., Tamassia, N., Cassatela, M.A., Montecucco, C., Tedesco, F., and De Bernard, M. 2007, J. Immunol. 178, 1312.
91. Montemurro, P., Nishioka, H., Dundon, W.G., De Bernard, M., Del Giudice, G., Rappuoli, R., and Montecucco, C. 2002, Eur. J. Immunol., 32, 671.
92. Hacker, J., and Carniel, E. 2001, EMBO Rep., 2, 376.
93. Stein, M., Rappuoli, R., and Covacci, A. 2000, PNAS, 97, 1263.
94. Stein, M., Bagnoli, F., Halenbeck, R., Rappuoli, R., Fantl, W.J., and Covacci, A. 2002, Mol. Microbiol., 43, 971.
95. Bagnoli, F., Buti, L., Tompkins, L., Covacci, A., and Amieva, M.R. 2005, PNAS, 102, 16339.
96. Yamazaki, S., Yamakawa, A., Ito, Y., Ohtani, M., Higashi, H., Hatakeyama, M., and Azuma, T. 2003, J. Infect. Dis., 187, 334.
97. Akanuma, M., Maeda, S., Ogura, K., Mitsuno, Y., Hirata, Y., Ikenoue, T., Otsuka, M., Watanabe, T., Yamaji, Y., Yoshida, H., Kawabe, T., Shiratori, Y., and Omata, M. 2002, J. Infect. Dis., 185, 341.
98. Lupetti, P., Heuser, J.E., Manetti, R., Massari, P., Lanzavecchia, S., Bellon, P.L., Dallai, R., Rappuoli, R., and Telford, J.L., 1996, J. Cell. Biol., 133, 801.
99. Reyrat, J.M., Pelicic, V., Papini, E., Montecucco, C., Rappuoli, R., and Telford, J.L. 1999, Mol. Microbiol., 34, 197.
100. De Bellis, L., Papini, E., Caroppo, R., Montecucco, C., and Curci, S. 2001, Am. J. Physiol. Gastrointest. Liver Physiol., 281, G1440.
101. Tombola, F., Morbiato, L., Del Giudice, G., Rappuoli, R., Zoratti, M., Papini, E. 2001. J. Clin. Invest., 108, 929.
102. Telford, J.L., Ghiara, P., Dell'Orco, M., Comanducci, M., Burroni, D., Bugnoli, M., Tecce, M.F., Censini, S., Covacci, A., Xiang, Z., Papini, E., Montecucco, C., Parente, L., and Rappuoli, R. 1994, J. Exp. Med. 179, 1653.
103. Molinari, M., Salio, M., Galli, C., Norais, N., Rappuoli, R., Lanzavecchia, A., and Montecucco, C. 1998, J. Exp. Med., 187, 135.
104. Petersen, A.M., Sorensen, K., Blom, J., and Krogfelt, K.A., 2001, FEMS Immunol. Med. Microbiol., 30, 103.
105. Gebert, B., Fischer, W., Weiss, E., Hoffmann, R., Haas, R. 2003. Science, 301, 1099.
106. Boncristiano, M., Paccani, S.R., Barone, S., Ulivieri, C., Patrussi, L., Ilver, D., Amedei, A., D'Elios, M.M., Telford, J.L., Baldari, C.T. 2003. J. Exp. Med., 198, 1887.
107. Supajatura, V., Ushio, H., Wada, A., Yahiro, K., Okumura, K., Ogawa, H., Hirayama, T., Ra, C. 2002, J. Immunol., 168, 2603.
108. De Bernard, M., Cappon, A., Pancotto, L., Ruggiero, P., Rivera, J., Del Giudice, G., Montecucco, C. 2005, Cell Microbiol. 7, 191.
109. Salama, N.R., Otto, G., Tompkins, L., and Falkow, S. 2001, Infect. Immun., 69, 730.
110. Yamaoka, Y., Kwon, D.H., and Graham, D.Y. 2000, PNAS, 97, 7533.
111. Dossumbekova, A., Prinz, C., Mages, J., Lang, R., Kusters, J.G., Van Vliet, A.H., Reindl, W., Backert, S., Saur, D., Schmid, R.M., and Rad, R. 2006, J Infect Dis., 194, 1346.
112. Aspholm-Hurtig, M., Dailide, G., Lahmann, M., Kalia, A., Ilver, D., Roche, N., Vikstrom, S., Sjostrom, R., Linden, S., Backstrom, A., Lundberg, C., Arnqvist, A., Mahdavi, J., Nilsson, U.J., Velapatino, B., Gilman, R.H., Gerhard, M., Alarcon, T., Lopez-Brea, M., Nakazawa, T., Fox, J.G., Correa, P., Dominguez-Bello, M.G., Perez-Perez, G.I., Blaser, M.J., Normark, S., Carlstedt, I., Oscarson, S., Teneberg, S., Berg, D.E., and Boren, T. 2004, Science, 305, 519.

113. Carlsohn, E., Nystrom, J., Bolin, I., Nilsson, C.L., and Svennerholm, A.M. 2006, Infect. Immun., 74, 920.
114. Malfertheiner, P., Megraud, F., O'Morain, C., Hungin, A.P., Jones, R., Axon, A., Graham, D.Y., and Tytgat, G. 2002, Aliment. Pharmacol. Ther., 16, 167.
115. Gerrits. M,M,, Van Vliet, A.H., Kuipers, E.J., and Kusters, J.G., Lancet. Infect. Dis. 2006, 6, 699.
116. Gisbert, J.P. 2005, Am. J. Gastroenterol., 100, 2083.
117. Kamiji, M.M., and de Oliveira, R.B. 2005, Eur. J. Gastroenterol. Hepatol. 17. 973.
118. De Bortoli, N., Leonardi, G., Ciancia, E., Merlo, A., Bellini, M., Costa, F., Mumolo, M.G., Ricchiuti, A., Cristiani, F., Santi, S., Rossi, and M., Marchi, S. 2007, Am. J. Gastroenterol., 102, 1.
119. Rupnow, M.F., Owens, D.K., Shachter, R., and Parsonnet, J. 1999, Helicobacter, 4, 272.
120. Rupnow, M.F.T., Shachter, R.D., Douglas, K.O., and Parsonnet, J. 2002, Vaccine, 20, 879.
121. Sullivan, P.B., Thomas, J.E., Wight, D.G., Neale, G., Eastham, E.J., Corrah, T., Lloyd-Evans, N., and Greenwood, B.M. 1990, Arch. Dis. Child., 65, 189.
122. Dale, A., Thomas, J.E., Darboe, M.K., Coward, W.A., Harding, M., and Weaver, L.T. 1998, J. Pedriatr. Gastroenterol. Nutr., 26, 393.
123. Passaro, D.J., Taylor, D.N., Meza, R., Cabrera, L., Gilman, R.H., and Parsonnet, J. 2001, Pediatrics, 108, E87.
124. Shahinian, M.L., Passaro, D.J., Swerdlow, D.L., Mintz, E.D., Rodriguez, M., and Parsonnet, J. 2000, Lancet, 355, 377.
125. Bhan, M.K., Bahl, R., Sazawal, S., Sinha, A., Kumar, R., Mahalanabis, D., and Clemens, J.D. 2002, J. Infect. Dis., 186, 1857.
126. Marshall, B.J., Amstrong, J.A., McGechie, D.B., and Glancy, R.J. 1985, Med. J. Austr., 142, 436.
127. Morris, A., and Nicholson, G. 1987, Am. J. Gastroenterol., 82, 192.
128. Sobala, G.M., Crabtree, J.E., Dixon, M.F., Schorah, C.J., and Taylor, J.D. 1991, Gut, 32, 1415.
129. Graham, D.Y., Opekun, A.R., Osato, M.S., El-Zimaity, H.M., Lee, C.K., Yamaoka, Y., Qureshi, W.A., Cadoz, M., and Monath, T.P. 2004, Gut. 53, 1235.
130. Karita, M., Kouchiyama, T., Okita, K., and Nakazawa, T. 1991, Am. J. Gastroenterol., 86, 1596.
131. Marchetti, M., Aricò, B., Burroni, D., Figura, N., Rappuoli, R., and Ghiara, P., Science, 1995, 267, 1655.
132. Lee, A., O'Rourke, J., De Unghia, M.C., Robertson, B., Daskalopoulos, G., and Dixon, M.F. 1997, Gastroenterology, 112, 1386.
133. Li, H., Kalies, I., Mellgard, B., and Helander, H.F. 1998, Scand. J. Gastroenterol., 33, 370.
134. Shomer, N.H., Dangler, C.A., Whary, M.T., and Fox, J.G. 1998, Infect. Immun., 66, 2614.
135. Hirayama, F., Takagi, S., Kusuhara, H., Iwao, E., Yokoyama, Y., and Ikeda, Y. 1996, J. Gastroenterol., 31, 755.
136. Lee, C.K. 2001, Scand. J. Immunol., 53, 437.
137. Radin, M.J., Eaton, K.A., Krakowka, S., Morgan, D.R., and Lee, A. 1990, Infect. Immun., 58, 2606.
138. Rossi, G., Rossi, M., Vitali, C.G., Fortuna, D., Burroni, D., Pancotto, L., Capecchi, S., Sozzi, S., Tenzoni, G., Braca, G., Del Giudice, G., Rappuoli, R., Ghiara, P., and Taccini, E. 1999, Infect. Immun., 67, 3112.
139. Del Giudice, G., Covacci, A., Telford, J.L., Montecucco, C., and Rappuoli, R. 2001, Annu. Rev. Immunol., 19, 523.
140. Svennerholm, A.M. 2003, Vaccine, 21, 347.
141. Londono-Arcila, P., Freeman, D., Kleanthous, H., O'Dowd, A.M., Lewis, S., Turner, A.K., Rees, E.L., Tibbitts, T.J., Greenwood, J., Monath, T.P., and Darsley, M.J. 2002, Infect. Immun., 70, 5096.
142. Vajdy, M., Singh, M., Ugozzoli, M., Briones, M., Soenawan, E., Cuadra, L., Kazzaz, J., Ruggiero, P., Peppoloni, S., Norelli, F., Del Giudice, G., and O'Hagan, D. 2003, Immunology, 110, 86.

143. Taylor, J.M., Ziman, M.E., Fong, J., Solnick, J.V., Vajdy, M. 2007, Infect. Immun., 75, 3462.
144. Pizza, M., Giuliani, M.M., Fontana, M.R., Monaci, E., Douce, G., Dougan, G., Mills, K.H., Rappuoli, R., and Del Giudice, G. 2001, Vaccine, 19, 2534.
145. Bumann, D., Holland, P., Siejak, F., Koesling, J., Sabarth, N., Lamer, S., Zimny-Arndt, U., Jungblut, P.R., and Meyer, T.F. 2002, Infect. Immun., 70, 6494.
146. Nystrom, J., and Svennerholm, A.M. 2007, Vaccine, 25, 2591.
147. Kabir, S. 2007, Helicobacter, 12, 89.
148. Bumann, D., Aksu, S., Wendland, M., Janek, K., Zimny-Arndt, U., Sabarth, N., Meyer, T.F., and Jungblut, P.R. 2002, Infect. Immun., 70, 3396.
149. Sabarth, N., Hurwitz, R., Meyer, T.F., and Bumann, D. 2002, Infect. Immun., 70, 6499.
150. Baldwin D.N., Shepherd B., Kraemer P., Hall M.K., Sycuro L.K., Pinto-Santini D.M., and Salama N.R. 2007, Infect. Immun., 75, 1005.
151. Lee, C.K., Weltzin, R., Thomas, W.D. Jr., Kleanthous, H., and Ermak, T.H. 1995, J. Infect. Dis., 172, 161.
152. Ferrero, R.L., Thieberge, J.M., and Labigne, A. 1997, Gastroenterology, 113, 185.
153. Tosi, M.F., and Czinn, S.J. 1990, J. Infect. Dis., 162, 156.
154. Cover, T.L., Cao, P., Murphy, U.K., Sipple, M.S., and Blaser, M.J. 1992, J. Clin. Invest., 90, 913.
155. Bogstedt, A.K., Nava, S., Wadstrom, T., and Hammarstrom, L. 1996, Clin. Exp. Immunol., 105, 202.
156. Ermak, T.H., Giannasca, P.J., Nichols, R., Myers, G.A., Nedrud, J., Weltzin, R., Lee, C.K., Kleanthous, H., and Monath, T.P. 1998, J. Exp. Med., 188, 2277.
157. Blanchard, T.G., Czinn, S.J., Redline, R.W., Sigmund, N., Harriman, G., and Nedrud, J.G. 1999, Cell. Immunol., 191, 74.
158. Pappo, J., Torrey, D., Castriotta, L., Savinainen, A., Kabok, Z., and Ibraghimov, A. 1999, Infect. Immun., 67, 337.
159. Mohammadi, M., Czinn, S., Redline, R., and Nedrud, J. 1996, J. Immunol.. 156, 4729.
160. Rossi, G., Fortuna, D., Pancotto, L., Tenzoni, G., Taccini, E., Ghiara, P., Rappuoli, R., and Del Giudice, G. 2000, Infect Immun., 68, 4769.
161. Mattapallil, J.J., Dandekar, S., Canfield, D.R., and Solnick, J.V. 2000, Gastroenterology. 118, 307.
162. Wen, S., Felley, C.P., Bouzourene, H., Reimers, M., Michetti, P., and Pan-Hammarstrom, Q. 2004, J. Immunol., 172, 2595.
163. Windle, H.J., Ang, Y.S., Athie-Morales, V,, McManus, R., and Kelleher, D. 2005, Gut. 54, 25.
164. Amedei, A., Cappon, A., Codolo, G., Cabrelle, A., Polenghi, A., Benagiano, M., Tasca, E., Azzurri, A., D'Elios, M.M., Del Prete, G., and De Bernard, M. 2006, J. Clin. Invest., 116, 1092.
165. Taylor, J.M., Ziman, M.E., Huff, J.L., Morosi, N.M., Vajdy, M., and Solnick, J.V. 2006, Vaccine. 24, 4987.
166. Rathinavelu, S., Kao, J.Y., Zavros, Y., and Merchant, J.L. 2005, Helicobacter. 10, 424.
167. D'Elios, M.M., Amedei, A., and Del Prete, G. 2003, Microbes. Infect., 5, 723.
168. Mohammadi, M., Nedrud, J., Redline, R., Lycke, N., and Czinn, S.J. 1997, Gastroenterology, 113, 1848.
169. Radcliff, F.J., Ramsay, A.J., and Lee, A. 1996, Gastroenterology, 110, A-997.
170. Saldinger, P.F., Porta, N., Launois, P., Louis, J.A., Waanders, G.A., Bouzourene, H., Michetti, P., Blum, A.L., and Corthesy-Theulaz, I.E. 1998, Gastroenterology, 115, 891.
171. Gottwein, J.M., Blanchard, T.G., Targoni, O.S., Eisenberg, J.C., Zagorski, B.M., Redline, R.W., Nedrud, J.G., Tary-Lehmann, M., Lehmann, P.V., and Czinn, S.J. 2001, J. Infect. Dis., 184, 308.
172. Fox, J.G., Beck, P., Dangler, C.A., Whary, M.T., Wang, T.C., Shi, H.N., and Nagler-Anderson, C. 2000, Nat. Med., 6, 536.
173. Akhiani, A.A., Pappo, J., Kabok, Z., Schoen, K., Gao, W., Franzén, L.E., and Lycke, N. 2002, J. Immunol., 169, 6977.

174. Garhart, C.A., Heinzel, F.P., Czinn, S.J., and Nedrud, J.G. 2003, Infect. Immun., 71, 910.
175. Maeda, K., Yamashiro, T., Minoura, T., Fujioka, T., Nasu, M., and Nishizono, A. 2002, Microbiol. Immunol., 46, 613.
176. Akhiani, A.A., Schon, K., and Lycke, N. 2004, J. Immunol. 173, 3348.
177. Yamamoto, T., Kita, M., Ohno, T., Iwakura, Y., Sekikawa, K., and Imanishi, J. 2004, Microbiol. Immunol., 48, 647.
178. Lehmann, F.S., Terracciano, L., Carena, I., Baeriswyl, C., Drewe, J., Tornillo, L., De Libero, G., and Beglinger, C. 2002, Am. J. Physiol. Gastrointest. Liver. Physiol., 283, G481.
179. D'Elios, M.M., Amedei, A., Benagiano, M., Azzurri, A., and Del Prete, G. 2005, FEMS Immunol. Med. Microbiol., 44, 113.
180. Garhart, C.A., Redline, R.W., Nedrud, J.G., and Czinn, S.J. 2002, Infect. Immun., 70, 3529.
181. Raghavan, S., Svennerholm, A.M., and Holmgren, J. 2002, Infect. Immun., 70, 4621.
182. Ramarao, N., and Meyer, T.F. 2001, Infect. Immun., 69, 2604.
183. Allen, L.A., Schlesinger, L.S., and Kang, B. 2000, J. Exp. Med., 191, 115.
184. Peppoloni, S., Mancianti, S., Volpini, G., Nuti, S., Ruggiero, P., Rappuoli, R., Blasi, E., and Del Giudice, G. 2002, Eur. J. Immunol., 32, 2721.
185. Amieva, M.R., Salama, N.R., Tompkins, L.S., and Falkow, S. 2002, Cell. Microbiol., 4, 677.
186. Lundgren, A., Stromberg, E., Sjoling, A., Lindholm, C., Enarsson, K., Edebo, A., Johnsson, E., Suri-Payer, E., Larsson, P., Rudin, A., Svennerholm, A.M., and Lundin, B.S. 2005, Infect Immun.. 73, 523.
187. Kao, J.Y., Rathinavelu, S., Eaton, K.A., Bai, L., Zavros, Y., Takami, M., Pierzchala, A., and Merchant, J.L. 2006, Am. J. Physiol. Gastrointest. Liver Physiol.. 291, G73.
188. Guiney, D.G., Hasegawa, P., and Cole, S.P. 2003, Infect. Immun., 71, 4163.
189. Mitchell, P., Germain, C., Fiori, P.L., Khamri, W., Foster, G.R., Ghosh, S., Lechler, R.I., Bamford, K.B., and Lombardi, G. 2007, Infect. Immun., 75, 810.
190. Bergman, M.P., Engering, A., Smits, H.H., Van Vliet, S.J., Van Bodegraven, A.A., Wirth, H.P., Kapsenberg, M.L., Vandenbroucke-Grauls, C.M., Van Kooyk, Y., and Appelmelk, B.J. 2004, J. Exp. Med. 200, 979.
191. Yun, C.H., Lundgren, A., Azem, J., Sjoling, A., Holmgren, J., Svennerholm, A.M., and Lundin, B.S. 2005, Infect. Immun., 73, 1482.
192. Betten, A., Bylund, J., Christophe, T., Boulay, F., Romero, A., Hellstrand, K., and Dahlgren, C. 2001, J. Clin. Invest., 108, 1221.
193. Velin, D., and Michetti, P. 2006, Digestion, 73, 116.
194. Meyer-ter-Vehn, T., Covacci, A., Kist, M., and Pahl, H.L. 2000, J. Biol. Chem. 275, 16064.
195. Keates, S., Hitti, Y.S., Upton, M., and Kelly, C.P. 1997, Gastroenterology, 113, 1099.
196. Smith, M.F. Jr., Mitchell, A., Li, G., Ding, S., Fitzmaurice, A.M., Ryan, K., Crowe, S., and Goldberg, J.B. 2003, J. Biol. Chem., 278, 32552.
197. Torok, A.M., Bouton, A.H., and Goldberg, J.B. 2005, Infect. Immun., 73, 1523.
198. Stromberg, E., Edebo, A., Lundin, B.S., Bergin, P., Brisslert, M., Svennerholm, A.M., and Lindholm, C. 2005, Clin. Exp. Immunol., 140, 117.
199. Svennerholm, A.M., and Quiding-Järbrink, M. 2003, Microbes Infect., 5, 731.
200. Rahn, W., Redline, R.W., and Blanchard, T.G. 2004, Vaccine, 23, 807.
201. Raghavan, S., Hjulstrom, M., Holmgren, J., and Svennerholm, A.M. 2002, Infect. Immun., 70, 6383.
202. Nedrud, J.G., Blanchard, S.S., and Czinn, S.J. 2002, Helicobacter, 7(Suppl. 1), 24.
203. Nagai, S., Mimuro, H., Yamada, T., Baba, Y., Moro, K., Nochi, T., Kimono, H., Suzuki, T., Sasakawa, C., Koyasu, S. 2007, PNAS, 104. 8971.
204. Kotloff, K.L., Sztein, M.B., Wasserman, S.S., Losonsky, G.A., Di Lorenzo, S.C., and Walker, R.I. 2001, Infect. Immun., 69, 3581.
205. Losonsky, G., Kotloff, K.L., and Walker, R.I. 2003, Vaccine, 21, 562.

206. Michetti, P., Kreiss, C., Kotloff, K., Porta, N., Blanco, J.L., Bachmann, D., Herranz, M., Saldinger, P.F., Corthésy-Theulaz, I., Losonsky, G., Nichols, R., Simon, J., Stolte, M., Ackerman, S., Monath, T.P., and Blum, A.L. 1999, Gastroenterology, 116, 804.
207. Banerjee, S., Medina-Fatimi, A., Nichols, R., Tendler, D., Michetti,M., Simon, J., Kelly, C.P., Monath, T.P., and Michetti, P. 2002, Gut, 51, 634.
208. Kleanthous, H., Myers, G.A., Georgakopoulos, K.M., Tibbitts, T.J., Ingrassia, J.W., Gray, H.L., Ding, R., Zhang, Z.Z., Lei, W., Nichols, R., Lee, C.K., Ermak, T.H., and Monath, T.P. 1998, Infect. Immun., 66, 2879.
209. Sougioultzis, S., Lee, C.K., Alsahli, M., Banerjee, S., Cadoz, M., Schrader, R., Guy, B., Bedford P., T.P. Monath, C.P. Kelly, and Michetti, P. 2002, Vaccine, 21, 194.
210. Di Petrillo, M.D., Tibbetts, T., Kleanthous, H., Killeen, K.P., and Hohmann, E.L. 2000, Vaccine, 18, 449.
211. Angelakopoulos, H., and Hohmann, E.L. 2000, Infect. Immun., 68, 2135.
212. Bumann, D., Metzger, W.G., Mansouri, E., Palme, O., Wendland, M., and Hurwitz, R. 2001, Vaccine, 20, 845.
213. Metzger, W.G., Mansouri, E., Kronawitter, M., Diescher, S., Soerensen, M., Hurwitz, R., Bumann, D., Aebischer, T., Von Specht, B.U., and Meyer, T.F. 2004, Vaccine, 22, 2273.
214. Aebischer, T., Bumann, D., Epple, H. J., Graham, D.Y., Metzger, W., Schneider, T., Stolte, M., Zeitz, M., and Meyer, T.F. 2005, Helicobacter, 10, 547.
215. Xu, C., Li, Z.S., Du, Y.Q., Gong, Y.F., Yang, H., Sun, B.,and Jin, J. 2007, World J. Gastroenterol., 13, 939.
216. Rappuoli, R. 1997, Nat. Med., 3, 374.
217. Marchetti, M., Rossi, M., Giannelli, V., Giuliani, M.M., Pizza, M., Censini, S., Covacci, A., Massari, P., Pagliaccia, C., Manetti, R., Telford, J.L., Douce, G., Dougan, G., Rappuoli, R., and Ghiara, P. 1998, Vaccine, 16, 33.
218. Rossi, G., Ruggiero, P., Peppoloni, S., Pancotto, L., Fortuna, D., Lauretti, L., Volpini, G., Mancianti, S., Corazza, M., Taccini, E., Di Pisa, F., Rappuoli, R., and Del Giudice, G. 2004, Infect. Immun., 72, 3252.
219. Malfertheiner, P., Schultze, V., Del Giudice, G., Rosenkranz, B., Kaufmann, S.H.E., Winau, F., Ulrichs, T., Theophil, E., Jue, C.P., Novicki, D., Norelli, F., Contorni, M., Berti, D., Lin, J.S., Schwenke, C., Goldman, M., Tornese, D., Ganju, J., Palla, E., Rappuoli, R., and Scharschmidt, B. 2002, Gastroenterology, 122(Suppl. 1), A585.

Chapter 9
Immunology of Norovirus Infection

Juan S. Leon, Menira Souza, Qiuhong Wang, Emily R. Smith, Linda J. Saif and Christine L. Moe

Abstract Noroviruses are the leading cause of epidemic non-bacterial gastroenteritis worldwide. Despite their discovery over three decades ago, little is known about the host immune response to norovirus infection. The purpose of this chapter is to review the field of norovirus immunology and discuss the contributions of outbreak investigations, human and animal challenge studies and population-based studies. This chapter will survey both humoral and cellular immunity as well as recent advances in norovirus vaccine development.

9.1 Introduction

9.1.1 Norovirus Disease Etiology

Noroviruses (NoV) are the major cause of epidemic gastroenteritis in the United States and a significant cause of severe diarrhea in young children in developing countries [1, 2]. NoV is also the most frequent cause of acute gastroenteritis after ingestion of raw shellfish [3, 4, 5]. NoV symptomatic infection causes vomiting, watery diarrhea, nausea, abdominal cramps, fever and general malaise. Gastroenteritis induced by NoV is self-limiting and rarely fatal. Fatality in children and the elderly is usually caused by severe dehydration after NoV infection [2, 6, 7].

9.1.2 Classification

NoV belong to the family *Caliciviridae*, genus *norovirus*, and are currently divided into five distinct genetic classifications called genogroups (GI-V). Genogroups are

C.L. Moe
Hubert Department of Global Health, Rollins School of Public Health, Emory University, Atlanta GA, USA
e-mail: clmoe@sph.emory.edu.

M. Vajdy (ed.), *Immunity Against Mucosal Pathogens*,

further subdivided into clusters, each categorized with a number and the name of the prototype strain. Each cluster is comprised of individual strains identified from various outbreaks and human and animal infections. GI, GII, and GIV affect humans, and currently there are 8 clusters for GI, and 17 clusters for GII [8]. It is not possible to determine the exact number of strains for each cluster and genogroup because new strains continue to be added.

9.1.3 Transmission

Transmission of NoV may occur via ingestion of fecal-contaminated food or water, exposure to contaminated fomites or aerosolized vomitus, and direct person-to-person contact [9, 10, 11, 12, 13, 14, 15, 16, 17]. In rare cases, transmission can occur through organ transplantation [16, 17]. A low infectious dose of less than 5 genomic copies (viral particles) could be enough to infect a healthy adult (Moe, C.L. et al., unpublished data). In certain symptomatic and asymptomatic individuals, virus can be shed for more than 3 weeks post-challenge or exposure [18, 19, 20]. In immunocompromised individuals, such as transplant recipients, NoV has been detected in stool specimens for up to two years after initial infection [16, 17, 21, 22, 23]. These individuals may be asymptomatic carriers of NoV and a possible reservoir for human NoV in a population.

9.1.4 Epidemiology

NoVs are the second most important cause of severe gastroenteritis in young children [24, 25], and may cause about 20% of endemic gastroenteritis in families [26]. Each year in the U.S., the public health impact of NoV is evidenced by the estimated 23 million infections that result in an estimated 50,000 hospitalizations and 310 fatal cases [27]. This number is probably a severe underestimate of the true burden of disease. In 2004, in the U.S., NoV was responsible for 48% of outbreaks among all reported gastroenteritis outbreaks and 79% of outbreaks among reported non-bacterial gastroenteritis outbreaks [28]. GI and GII strains cause the majority of human outbreaks. Analyses, based on published outbreak reports and national surveillance systems, suggests most outbreaks are associated with GII strains [29]. Within GII, cluster 4 (GII.4), "Bristol", has been currently associated with most of the published outbreak reports among all the GI and GII clusters.

NoV are classified by the Centers for Disease Control and Prevention (CDC) and the National Institute of Allergy and Infectious Diseases (NIAID) as a Bioterrorism Category B Priority Pathogen based on their high transmissibility, low infectious dose, and serious public health and economic impact. No vaccine is currently available.

9.1.5 Overview

The purpose of this chapter is to review current advances in the field of NoV immunology. The reader is encouraged to read recent reviews on NoV immunology [30, 31, 32]. This chapter incorporates a synthesis of previous reviews and our recent understanding of NoV pathogenesis based on the development of new models of NoV disease.

At the onset, it is important to define key terms used throughout this chapter. NoV may be internalized, usually through the mucosal route, after active administration of NoV inocula to volunteers (*challenge*) or animals (*inoculation*) or passive contact of humans or animals with NoV, usually in outbreaks (*exposure*). Challenge, inoculation or exposure with NoV may or may not lead to infection. *Infection* with NoV is defined as replication of NoV in the host. NoV infection can be determined by detection of viral particles or viral RNA in the host or by the presence of viral nonstructural proteins in host cells. *Viral shedding* is defined as the release of quantifiable virus in host specimens such as feces and vomitus. NoV may be detected in these specimens by molecular detection techniques, such as various forms of reverse transcription PCR (RT-PCR: conventional PCR, real time PCR, quantitative real time PCR, etc.), antigen capture immunoassays (e.g. ELISA), or electron microscopy. Infection may or may not lead to *illness*. The terms *illness, disease*, and *symptomatic infection* will be used interchangeably. NoV immunology refers to the host immune response induced after challenge, inoculation, or exposure to NoV.

9.2 How to Study NoV Immunology

9.2.1 Population Studies

Until recently, there were few non-human models of NoV disease, and the majority of our understanding of NoV immunology came from human challenge studies and studies of human populations. Prospective, retrospective, and case-control studies were common study designs and often took advantage of existing community and clinical specimens collected from studies [33, 34, 35]. In addition, clinical specimens (e.g. saliva, sera, stool) taken from outbreak investigations continue to provide a wealth of information to our understanding of NoV immunology. The advantages of these approaches are that the results are more representative of *real* conditions than laboratory studies and are directly applicable to the human condition. Disadvantages include the difficulty in obtaining clinical specimens, limited control of study design and low or inadequate study power.

9.2.2 Human Challenge Studies

From the 1970s to the present, there have been over a dozen human challenge studies with NoV. In these studies, volunteers were enrolled in a clinical trial (after all the necessary regulatory approvals were met) and studied under controlled clinical conditions, usually in a hospital setting. Volunteers were challenged with various preparations of NoV, clinical specimens collected, and their responses recorded. The advantages of human challenge studies include: direct applicability of results to the human condition, control of study design and study power, and ease of obtaining clinical specimens. The disadvantages include: difficulty in obtaining all the relevant regulatory and ethical reviews and approvals, logistical issues involved with human volunteers, and difficulty in obtaining adequate sample sizes (study power). Another disadvantage is that human volunteers often have a range of pre-challenge exposures that will affect their response to the challenge. The difficulty is two-fold: (1) volunteers have diverse pre-challenge exposure history and (2) we do not have adequate methods of measuring and characterizing their pre-challenge exposure history. Lastly, it is difficult to create approved NoV inocula for use in human challenge studies.

Four NoV clusters, with possibly varying strains, have been used in human challenge studies since the 1970s: Norwalk virus (GI.1), Montgomery County (GI.5) [36], Snow Mountain virus (GII.2) and Hawaii virus (GII.1) (classification determined from [8]). Other human challenge studies, prior to 1970, have used uncharacterized challenge agents and therefore it is unclear whether these challenge agents were only NoV or included other viruses (reviewed in [37, 38, 39]). Currently (i.e. 2007), two groups perform human challenge studies; these are Dr. Mary Estes' group at Baylor College of Medicine, in Houston, Texas, and Dr. Christine Moe's group at Emory University, in Atlanta, Georgia.

9.2.3 *Non-Human Models*

During the past decade, several new promising animal models have been developed that provide more tools for studying NoV pathogenesis and immunology. These models include the gnotobiotic (Gn) pig, Gn calves, non-human primates, mice, as well as other invertebrate models. All models shed virus and seroconverted to NoV antigens. Currently, both the Gn pig and murine models hold the most promise for a greater understanding of NoV immunology.

9.2.3.1 Gnotobiotic Pigs

In 2006, Dr. Linda Saif's lab developed a new model of human NoV disease that promises to advance the field of NoV immunology. This model utilizes gnotobiotic (Gn) pigs as an experimental animal model [40, 41, 42]. In this model, the investigators were able to infect the Gn pigs with a GII.4 human NoV after oral inoculation. Gn pigs have long been used to study human rotavirus pathogenesis and host immune responses and have the advantages of similar gastrointestinal physiology, mucosal immune responses and milk diet to that of infants [43, 44, 45, 46]. Other advantages of this model include susceptibility to human NoV infection, and similarity of symptoms with human disease, such as diarrhea and pathological changes in the gastrointestinal system [40]. Additionally, Gn pigs have histo-blood group antigens, like humans, that influence susceptibility to NoV. The incubation period for GII.4 NoV in Gn pigs (12 to 48 hours) was similar to that observed in humans experimentally infected with Snow Mountain virus (SMV) (19 to 41 hours) [47], and the diarrhea was mild and of short duration (1 to 4 days in most pigs). However, there are some differences compared to human disease. Multiple passages of NoV through various pigs (up to 3 serial passages) seemed to reduce the duration of diarrhea, reduce the prevalence of Gn pigs that shed virus, and reduced the duration of viral shedding although the virus was still infectious; therefore, multiple passages in the Gn pigs may be required for more complete adaptation of the GII.4 NoV to this animal model. The human NoVs infect the proximal small intestine in both humans and in Gn pigs. However, infected humans reportedly exhibit inflammation in the duodenum and jejunum [48, 49, 50] whereas Gn pigs exhibit mild pathological changes in the duodenum, but not jejunum and ileum. Furthermore, only 1 of 7 inoculated pigs exhibited pathological changes [40].

One interesting observation was that intravenous administration of human NoV also consistently infected Gn pigs, and NoV was transiently detected in serum of orally inoculated pigs [40]. In humans, NoV viremia, i.e. systemic spread of the virus, has not yet been carefully examined; although it is hypothesized that NoV cannot be transmitted through blood products. However, based on the finding in pigs, re-examination of human NoV viremia is warranted.

9.2.3.2 Gnotobiotic Calves

In Dr. Linda Saif's lab, the Gn calf has also been used as an animal model to study pathogenesis and immune responses to viruses such as a human enteric coronavirus [51], rotaviruses [52] and NoV [53, 54]. Because Gn calves are raised on a milk diet and are free of microbes, their rumen does not develop and their gut physiology and mucosal immune responses (dominance of secretory IgA) remain similar to that of other monogastrics, including human infants. They provide an alternative animal model for the study of gastrointestinal viruses, such as the fastidious human NoV. Of importance, the occurrence of GIII NoVs in cattle also permits comparative studies of host-specific versus NoV adaptive host strains. In one model, the bovine GIII.2 NoV CV186-OH and the unassigned calicivirus genus, NB strain, infected the villous epithelial cells of the small intestine of Gn calves, especially the duodenum, causing cellular destruction and resulting in mild to severe diarrhea [55].

The Gn calves, like the Gn pigs, also serve as important animal models for the study of primary immune responses to NoV that are difficult to assess in adult volunteers due to the presence of pre-challenge antibodies to human NoVs. The immunogenicity of bovine NoV-like particle (VLP) and human VLP vaccines and the protection induced in Gn calves upon homologous viral inoculation have been evaluated [53]) and are further discussed in Sections 9.5.2 and 9.5.6 of this chapter.

9.2.3.3 Non-Human Primates

Non-human primates provided one of the earliest models of NoV disease. Several groups reported infection of non-human primates with human NoV [56, 57]. Similar to responses observed in adult human volunteers, newborn pigtail macaques reportedly began viral shedding after 24 hours, and it was reported that some shed virus in excess of 21 days [56]. Interestingly, the mother of one infected macaque that resided in the same cage also became ill with the same strain of human NoV. This finding suggested that horizontal transmission, probably by the fecal-oral route, had occurred, similar to NoV transmission in humans. Other investigators showed that inoculation of Rhesus macaques induced several weeks of viral shedding in one asymptomatic macaque [58]. Different non-human primate species exhibited varying clinical disease and shedding. Neither marmosets, cotton top tamarins, nor cynomolgus macaques developed diarrhea or clinical symptoms [58]. Both marmosets and tamarins shed virus for between 3 and 4 days post-inoculation, while cynomolgus macaques did not shed virus. Some non-human primates exhibited NoV-specific IgM and IgG (rapid rise by 14 days post inoculation); no NoV-specific IgA was detected in the saliva of any animal. The reported detection of IgG cross-reactive with human NoV in mangabey, pigtail, Rhesus, and chimpanzee species suggests that non-human primates may be naturally infected by NoV [59].

Advantages of this model are the physiologic similarity of non-human primates to humans, the ability to use human NoV to inoculate non-human primates, and the ability to control the immunologic exposure of the non-human primate if necessary. Disadvantages include finding access to a well-maintained primate facility, the high cost of the study and the effort to assure compliance with regulatory guidelines.

9.2.3.4 Mice

The murine NoV model is a recent exciting discovery by Dr. Herbert Virgin's group [60]. In this model, either immunocompromised or wild type mice can be infected with a mouse-specific NoV. This mouse-specific NoV was named murine NoV (MNV-1) and was classified as a Genogroup V virus. To date, MNV is known to infect only mice and is a common pathogen in mouse colonies [61]. Immunocompromised STAT 1 knockout mice, when inoculated with MNV-1, developed massive viremia and inflammation in all organs and died within 10 days post-inoculation (d.p.i.) [60]. In contrast, wild type adult (8 week) 129 mice, when inoculated with MNV-1, exhibited viremia, and virus in a few organs including liver, spleen, and proximal intestine. These mice were able to clear MNV-1 by 3 d.p.i. and did not die. Of five mice inoculated, all seroconverted to NoV antigen between 14 and 21 d.p.i. [60]. Infection by other wild type mouse strains, MNV-2, MNV-3, MNV-4, in juvenile (4 week) CD1 mice resulted in viral shedding in feces and the presence of virus in mesenteric lymph node, spleen, and jejunum up to 8 weeks post-inoculation ([62] and discussed in [63]).

The advantages of this model are that MNV-1 infected mice have some characteristics similar to infected humans. Mice can acquire the virus through the fecal oral route, similar to humans. Wild-type mice housed with persistently-infected immunocompromised mice (RAG knockout mice) develop MNV-specific antibodies within 3 to 4 weeks of exposure (discussed in [63]). Interestingly, mice can also acquire the virus through the respiratory route. Immunocompromised mice, but not wild type 129 mice, develop gastric bloating and diarrhea among other diverse clinical signs. Other clear advantages are the wealth of immunologic tools and mouse strains available to study the immunology of infection in mice and the ease of maintaining and setting up experiments. One disadvantage of this model is that wild type mice are asymptomatic and do not show lesions in the small intestine upon macroscopic inspection as is seen in human disease. Also, mice do not have the emetic physiologic response, so this variable and route of transmission cannot be assessed. Additionally, whether mice have the corresponding histo-blood group antigens like humans and pigs that influence susceptibility to NoV is unclear and precludes the use of mouse models for such investigations. In addition, infection of some immunocompromised mice with MNV-1 is lethal and leads to disseminated viremia, unlike human hosts. For additional discussion of the strengths and weaknesses of the MNV model, please consult [63].

Murine models of MNV disease may not be limited to 129, CD1 and immunocompromised mice. As discussed, MNV is a prevalent pathogen of outbred mouse colonies and induces immune responses in these mice, as measured by seroconversion to NoV antigen [61]. Therefore, it is likely that improved MNV models will be used in the future for the study of NoV immunology.

9.2.3.5 Cell Culture

Since 2006, two novel cell culture systems for NoV infection have been developed. The first *in vitro* NoV cell system is based on MNV infection of tissue culture macrophages (RAW lineage) or tissue-derived cells such as dendritic cells or bone marrow derived cells [64]. This model will help investigators dissect the host cell-NoV interactions at the level of immune cells. The second model is based on human NoV infection (GI and GII viruses) of a three dimensional, organoid model of human small intestinal epithelium [65]. Both infection and at least limited replication (discussed in [66]) were demonstrated in this model. Whether this model can be adapted to other cells, such as immune cells of the macrophage or dendritic lineage is yet to be determined. These two models are likely only the beginning of in vitro models of NoV infection that will help our understanding of the relationship between the host immune system and NoV.

9.3 Population

9.3.1 Seroprevalence

Before discussing the various seroprevalence studies, it is important to point out that different studies used different sources and types of antigens. For example, older studies (prior to the '90s) used virus antigen purified from stool samples while more recent studies (in the '90s to present) used recombinant VLP as antigen. In addition, different strains of NoV were used as antigen in these studies. These different antigens may have different cross-reactivity and binding affinities to human immunoglobulin in serum specimens. Therefore, while general trends on seroprevalence and epidemiology may be gleaned from these studies, the exact seroprevalence against NoV among the various studies are not comparable to each other.

Examination of age-specific antibody seroprevalence in a variety of populations provides another key to understanding the body's immunologic response to NoV. Pediatric seroprevalence is particularly insightful as it illustrates the role of maternal antibodies and varied rates of antibody acquisition in different environments. The seroprevalence of NoV antibody in infants less than 4 months of age is similar to the number of seropositive women of childbearing age in the same community suggesting that these infants possess maternal antibodies [67, 68, 69, 70, 71, 72]. In general, these maternal antibodies persist for several months before a decrease in seroprevalence occurs. This decrease is usually observed between 3 and 6 months of age, although Parker et al. suggests that maternal antibody may be detectable in infants up to 8 months of age [73]. Figure 9.1, collected from a representative set of studies, illustrates this trend. However, after 4 months of age, pediatric NoV seroprevalence shows variation across demographic lines. The contrasting rates of antibody acquisition in developed and developing countries is well documented and best understood within the context of pediatric seroprevalence. In several developing countries, including Bangladesh [74], Kuwait [68], rural Thailand [69], South Africa [33] and Mexico [75] children appear to acquire "primary infection" during or prior to their second year of life. In Bangladesh, 100% of four-year-old children were seropositive for NoV antibody and, in general, childhood seroprevalence reaches

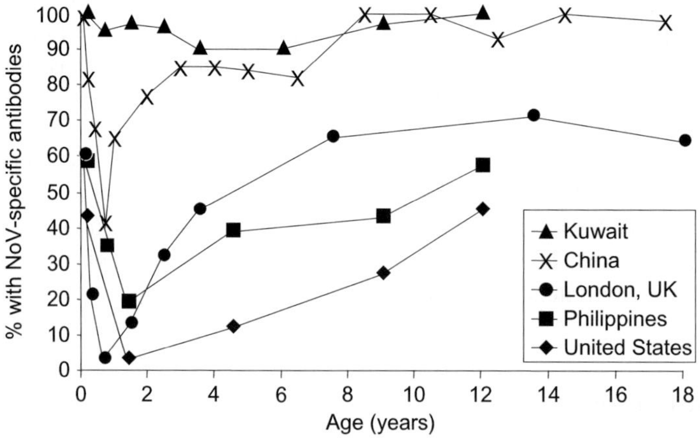

Fig. 9.1 Pediatric seroprevalence around the world. Data on seroprevalence in children ages 1 month to 18 years was collected from representative studies [67, 68, 70, 71]. The presence of maternal antibodies at birth (0 years) and subsequent decrease in seroprevalence between 3 months and 2 years is illustrated. Each data marker indicates the median age in years for the age range specified in the original report and seroprevalence value given for each age group in a population. Please see Section 9.3.1 for additional information regarding this figure

the maximum population-specific seroprevalence before age 5 in developing countries. In contrast, in developed countries, maternal antibodies decrease to low levels after 3 to 4 months, and the majority of children do not demonstrate subsequent serologic NoV response as a result of presumed exposure until age 12 [67, 76]. In Japan, NoV-specific seroprevalence is only 11% for children less than 6 years old while about 70% of adults are seropositive [76], and, in the United States, adult levels of NoV-specific seroprevalence are not reached until adolescence [67]. Seroprevalence data from Beijing, China, Hokkaido, Japan and Southeast Asia associate steep increases of childhood acquisition of antibody against NoV with daycare and elementary school attendance [70, 76, 77]. Ultimately, the varied patterns of age-specific antibody acquisition may provide insight into the environment and behavior associated with exposure to NoV.

Although the pattern of pediatric exposure to NoV is consistent, the pattern of adult seroprevalence of NoV appears to vary by populations. Greenberg et al. found that among 7 rural, urban, developed and developing countries there was no significant difference among adult seroprevalence [78]. However, examination of multiple seroprevalence studies indicates that there is variation among adult levels of seroprevalence. For example, an isolated Indian tribe in Ecuador, the Gabaro, was found to have no NoV antibodies in the late 1970s, while a Massachusetts cohort had a 50% NoV-specific seroprevalence in adulthood [79], and Kuwaiti adults had a 100% NoV-specific seroprevalence [68]. Table 9.1 illustrates adult seroprevalence from a representative number of international studies. While study design and method of

Table 9.1 Percent of adults* positive for NoV-specific antibody from representative regions

Region	Year of study	Percentage of Positive for NoV-specific antibody†	Reference
Beijing, China	1996–1997	100‡	[70]
Papua New Guinea	1979	100	[76]
Kuwait	1997**	100‡	[68]
Australian aborigines, Australia	1977	94	[71]
South Africa	1997**	93‡	[33]
Indonesia	1975–1976	90	[76]
Japan (*Hokkaido, Miyagi-ken, Saitama-ken, Kyoyo-fu, Fukuoka-ken*)	1984–1990	81	[76]
Dijon, France	2000–2001	78‡	[192]
Netherlands	1998–2001	74	[84]
United States	1974–1977	66	[78]
Singapore	1975	64	[76]
London, UK	1992	63	[71]
Massachusetts, USA	1975–1979	50	[79]
Ecuador (Gabaro)	1974–1977	0	[78]

* Adults are defined as individuals over 20 years, unless otherwise defined within the publication.

** Year of study was not available so the date the study was published was used instead.

† NoV-specific antibody for adults is defined as any isotype (IgG, IgA or IgM) seroprevalence for adults.

‡ Because no general adult prevalence value was given, the median adult prevalence value was calculated and rounded from published data.

antibody detection may account for some differences in the seroprevalence between populations, these findings also likely point to the roles of behavior and environment associated with exposure to NoV in a given population.

In general, certain behavioral and environmental characteristics can be linked to antibody acquisition against NoV in developing countries. First, lower socioeconomic status, poor standards of hygiene and limited access to potable water are found to be significantly associated with a higher NoV seroprevalence [35, 80, 81]. In agreement with these associations, a sero-epidemiological study of Chinese medical students found that students from rural towns had higher NoV-specific seroprevalence than students from urban settings [82]. Hygiene standards and socioeconomic status are often lower in rural settings. Other socio-behavioral factors associated with a higher NoV-specific seroprevalence include crowding within the home [35, 70, 80] and lower levels of maternal education [80]. In developed countries, the most common behavioral association with increased NoV-specific seroprevalence is childhood attendance at childcare centers or enrollment in elementary school [70, 76, 77, 80]. It is also notable that large outbreak situations are the most common pattern of infection in developed countries, indicating that high population density and common-source food or water contamination are important risk factors for infection [70, 83]. Additionally, there are a number of other factors associated with NoV-specific seroprevalence that are not specific to developed or developing countries including recreational water contact [80], consumption of sea food [80, 83], eating raw vegetables [80], blood type [82], and agricultural work [35, 84]. Ultimately, understanding the global patterns of seroprevalence and the associated risk behaviors will provide insight into the human immune response to NoV.

9.3.2 Immunology Learned from Outbreaks

Outbreaks have provided a wealth of information regarding the immune response of the host after NoV exposure and immunity against NoV. This information is supported and refined with data from human challenge studies that provide more controlled conditions (discussed in subsequent sections). For example, in outbreaks, several NoV-specific antibody isotypes have been identified including serum IgA [85], IgM [86], and IgG [87] as well as antibodies in saliva [88]. These NoV-specific antibodies are cross-reactive and bind to multiple NoV strains within various clusters within and across genogroups [87, 89]. In general, higher cross-reactivity is observed within clusters [89] and genogroups than across genogroups [6, 89]. As will be later discussed, pre-challenge serum antibodies, determined from acute serum specimens, did not provide protection from infection [90]. Lastly, outbreaks also show us that infection with one NoV strain does not protect from re-infection with a different strain [90].

Serology has been used in outbreaks to both diagnose and classify NoV strains. Multiple tests have been developed and used, including immune electron microscopy [91, 92], radioimmunoassay tests [93, 94, 95], isotype specific tests (e.g.

IgG, IgM, IgA) [85, 88, 96], tests using serum and saliva specimens [88], western blots [97], enzyme immunoassays (e.g. ELISAs) using single [96] and multiple antigens [6, 98], and "blocking" antibody tests [93, 94, 99, 100]. In general, a fourfold rise in NoV-specific antibody levels (from samples collected at least after 2 weeks post challenge) compared to baseline (from samples collected before or temporally close to NoV challenge) suggests infection. Unfortunately, serology is not as specific as genomic methods (e.g. real time PCR, reverse transcriptase PCR, genomic based phylogeny and classification) to diagnose and classify NoV strains because of the presence of pre-challenge antibodies and cross-reactivity across strains. Serology may also not be as sensitive as genomic methods at diagnosing infections. For example, high levels of pre-challenge antibodies may mask a fourfold rise in antibody levels after exposure. Genomic methods should be used as the main diagnostic and classification tool for NoV strains, and serology should be used to complement existing results or as an aid in outbreak investigations when stool specimens or genomic methods are not available.

9.4 Individuals

9.4.1 Clinical Features of NoV Gastroenteritis

A subset of NoV exposed individuals may become infected. Within this subset, a further subset of infected individuals may experience symptoms of gastroenteritis. In volunteer studies, the time to first symptom, or incubation period, ranged from 10 to 51 hours for Norwalk virus (NV) challenge (reviewed in [38]) and 19 to 41 hours for Snow Mountain virus challenge [47]. Symptoms may last from a few hours to several days. In our experience, NV-infected volunteers usually exhibit symptoms from 1 to 2 days while Snow Mountain virus infected volunteers usually exhibit symptoms from 1 to 5 days (Leon, J.S. and Moe, C.L. unpublished data). Symptoms include nausea, vomiting, diarrhea, abdominal cramps, headache, fever, chills and myalgia. Diarrhea stools are non-bloody, watery, and do not contain mucus or fecal leukocytes [37]. In agreement with these manifestations, fecal lactoferrin, a sensitive marker for intestinal polymorphonuclear leukocyte inflammation, is generally not detected [101]. In rare cases, NoV has been associated with encephalopathy (NoV was also detected in stool, cerebrospinal fluid, and serum) [102], neck stiffness, disseminated intravascular coagulation, photophobia, reduced alertness [103] and sore throat [104]. Dehydration is often the most serious complicating factor and may induce death in young children and elderly individuals. There is a relationship between symptoms and NoV-specific antibodies and this will be discussed in Section 9.4.4.4. In an immunocompetent host, symptoms usually resolve after a few days. The severity of illness and manifestation of specific symptoms seems to depend on both the NoV strain and the individual.

NoV shedding varies with each host. In our challenge studies, we have observed that NoV shedding often continues beyond the cessation of symptoms. In one extreme case, an immunocompetent individual exhibited symptoms until 4 days post-challenge and shed virus in their stool for beyond one month post-challenge (Leon, J.S. and Moe, C.L. unpublished data). Levels of viral shedding in stool may be as high as 10^{10} genomic copies per gram of stool (Leon, J.S. and Moe, C.L. unpublished data). These observations suggest that infected individuals could be asymptomatic NoV carriers and shed virus for many days beyond the cessation of symptoms. In contrast, we have observed NV-challenged individuals who seroconvert and do not shed detectable levels of NV in their stool, as evaluated by RT-PCR (Leon, J.S. and Moe, C.L. unpublished data).

Gastrointestinal function is altered after NV challenge. Challenged volunteers exhibited transient malabsorption of lactose, D-xylose, and fat [37]. Levels of small intestinal brush border enzymes, trehalase and alkaline phosphatase, were decreased 48 hours after NV challenge compared to baseline values [105]. Among Norwalk and Hawaii virus-challenged volunteers, gastric emptying was significantly delayed in those who exhibited illness compared those who did not develop illness [106]. Rates of gastric emptying returned to baseline levels usually once symptoms had resolved. Adenylate cyclase activity in the jejunum [107], gastric secretion of pepsin, hydrochloric acid, and intrinsic factor were not altered after NV challenge [106].

Human and animal models of disease suggest that the inflammation associated with NoV infection occurs in the small intestine. Jejunum biopsies of challenged volunteers exhibited histopathological lesions at the time of illness [49, 50, 105, 108]. The jejunum exhibited histological abnormalities five and six days after Hawaii and NV challenge and two days after symptom clearance [50, 108]. No abnormalities were observed two weeks [49, 105] and six to eight weeks after challenge [108]. Interestingly, these jejunum lesions were also observed 48 hours after challenge with Norwalk or Hawaii virus in some asymptomatic volunteers [50, 105, 108]. However, other asymptomatic volunteers exhibited no jejunum lesions 48 hours after NV challenge [48]. It is unlikely that these asymptomatic volunteers had no lesions because they were genetically resistant to NV infection, as some of these asymptomatic volunteers had previously been challenged and became ill after NV infection. Histological lesions were not observed in the gastric fundus, antrum, or rectal mucosa of volunteers challenged with NV [105, 109]. Based on this data, it is worth speculating that jejunum lesions are associated with NoV infection in the presence or absence of clinical illness.

The inflammation and lesion observed in the infected host had particular characteristics. The villi of the proximal small intestine exhibited blunting and broadening while the mucosa was histologically intact [49, 50, 105, 108]. Increased numbers of mononuclear cells and some polymorphonuclear leukocytes (neutrophil granulocytes) infiltrated the intercellular spaces between epithelial cells of the lamina propia [50, 108]. In the pig model, higher levels of apoptosis were observed among enterocytes in the small intestine of human NoV-inoculated pigs compared to mock-inoculated pigs [40]. In the mouse model, apoptosis of cells of the small intestine and spleen was only observed in STAT1 knockout mice but not in 129 mice infected with MNV [110]. Apoptosis has not yet been assayed in human biopsies.

There is much speculation about which cells are infected by NoV in humans. NoV was not detected by histology in any human biopsy of the small intestine taken after NoV challenge [48, 50, 105, 108]. However, animal models have provided several clues as to the identity of the target cell. In Gn pigs inoculated with human NoV, the NoV capsid protein was detected, via confocal immunofluorescent microscopy, in enterocytes in discrete areas of the villi of the duodenum and jejunum and in some cells in the ileum [40]. Positive enterocytes of these inoculated pigs were mainly located at the sides or tips of each villus. The viral capsid was also observed in deep areas of the duodenum. It was not clear whether the capsid protein was located in Brunner glands or crypts. To identify areas of viral replication, the nonstructural N-terminal viral protein was identified in some, but not all, enterocytes that expressed the capsid protein. The N-terminal protein was found in the apical end of enterocytes. By transmission electron microscopy, calicivirus-like particles of 25–40 nm in diameter were observed in cytoplasmic vesicles in enterocytes.

Other results were reported in studies in wild type mice and STAT1 knockout mice inoculated with MNV. To identify sites of MNV replication, the non-structural ProPol processing intermediate was detected in cells of the lamina propia in the villi within 24 hours post-inoculation with MNV in wild type 129 mice [110]. In STAT1 knockout mice, the ProPol protein was mainly detected in epithelial cells lining the villi, both the basolateral and apical regions by 12 hours post-inoculation. A few lamina propia cells exhibited the ProPol protein by 12 hours post-inoculation, but by 48 hours post-inoculation, the majority of cells expressing ProPol were in cells of the lamina propia and in the Peyer's patches. In contrast, very few epithelial cells expressed ProPol by 48 hours post-inoculation. The authors hypothesized that MNV initially seeds and replicates in epithelial cells in STAT1 knockout mice, but later replicates in the lamina propia and Peyer's patches.

In wild type 129 mice, infectious virus is first detected in the proximal small intestine 3 hours post-inoculation. MNV later spreads to peripheral organs including the spleen, lymph nodes, liver and lungs after 1 d.p.i. By 7 d.p.i., MNV levels were low or undetectable in all organs. In mice, the intestine may be the site of primary NoV seeding that later facilitates systemic spread [110]. Because MNV has been shown to replicate efficiently in wild type murine macrophages and dendritic cells [64], Wobus et al. hypothesized that infection of transepithelial dendritic cells in the lumen of the intestine may provide a pathway for initial NoV infection of the intestine [63]. In addition, we speculate that NoV-infected dendritic cells and macrophages could also be potential carriers of NoV infection to other peripheral organs including the spleen and lymph nodes.

9.4.2 Protective Immunity Against NoV Infection

There are two questions in the field of NoV immunology that are essential to the success of a NoV vaccine. The first question is whether short-term or long-term immunity can be induced and protect from future NoV infection. An effective vaccine should be able to induce host immunity that protects from future NoV infection. The second question is whether protective homologous and heterologous immunity can be induced. An effective vaccine should induce protective immunity against antigens in the vaccine (homologous) and, ideally, against related antigens not present in the vaccine (heterologous). These questions have been incompletely addressed in three separate human NoV challenge-re-challenge studies [48, 111, 112]. These studies were unable to use the current definition of infection (detection of viral shedding) because molecular assays were not available at the time these studies were conducted. Instead, clinical symptoms were used as a marker of NoV infection. Because infected individuals do not always exhibit clinical symptoms, the results of these studies should be interpreted with caution. In addition, it is difficult to distinguish whether a positive immune response in a volunteer, especially to a cross-reactive antigen, may be the result of either a current or past norovirus infection.

Two general hypotheses can be derived from these studies (reviewed [30, 38, 113, 114]). The first hypothesis is that challenge or exposure to NoV induces "short-term immunity" (6 months or less) but not "long-term immunity" (beyond 2 years) that protects from subsequent re-challenge or re-exposure to the same strain. This is based on the observation that challenged volunteers who became ill initially were re-challenged, within 6–15 weeks with the *same* virus and did not become ill [112]. Illness after re-challenge was 0/5 volunteers for NV (GI.1), 0/6 volunteers for Montgomery County virus (GI.5), and 0/3 for Hawaii virus (GII.1). The inability of each strain to induce illness in the re-challenged host was not due to a genetic factor because each strain was able to induce illness in the initial challenge. Therefore, these results suggest that previous challenge or exposure to NoV may induce protective immunity (6–15 weeks) that protects from subsequent challenge or exposure to the same strain. In agreement with this hypothesis, challenge and re-challenge with the same strain of bovine NoV within 30 days induced protection from viral shedding and clinical signs in Gn calves (discussed in 5.6 and in [53]). A separate NV challenge study demonstrated that re-challenge 6 months after the initial challenge did not result in illness in some volunteers who had become ill in the initial challenge [111]. A third NV challenge study demonstrated that re-challenge after 27 and 42 months induced illness in all volunteers that initially developed illness upon first challenge [48]. Collectively, these observations suggest that challenge with NoV provides protection against subsequent illness if volunteers are re-challenged with NoV within 6 months, but not after 2 years. Whether re-challenge, between 6 months and 2 years, protects individuals from human illness has not yet been addressed.

The second hypothesis is that challenge with NoV can induce both protective homologous immunity against re-challenge with the same strain, shown above, and protective heterologous immunity against re-challenge with *certain* other strains. Heterologous immunity is thought to depend on immune reactivity to shared epitopes in the virus capsid between different NoV strains (shared epitopes will be discussed in Sections 9.4.4.2 and 9.4.4.3). This hypothesis is based on the observation that challenged volunteers who became ill and were re-challenged, within

6–13 weeks with a *different* virus, did not become ill [112]. Illness after Montgomery County virus (GI.5) re-challenge occurred in 0/8 volunteers initially challenged with NV (GI.1) and in 0/4 volunteers initially challenged with Hawaii virus (GII.1). Interestingly, illness after NV re-challenge occurred in 1/3 volunteers initially challenged with Montgomery County, suggesting that *complete* protection may have been in one direction, but that *incomplete* protection may occur in the other direction. Within GI NoVs, one Montgomery County challenged individual developed a significant rise of NV-virus-specific antibodies, suggesting antigenic relatedness between these viruses [112]. However, challenge and re-challenge between genogroups did not provide *complete* protection in either direction. For example, challenge with NV (GI) and re-challenge with Hawaii virus (GII) resulted in 3/6 ill volunteers while challenge with Hawaii virus and re-challenge with NV resulted in 2/3 ill volunteers. This result suggests that Norwalk and Hawaii virus are antigenically different which is confirmed by their classification into distinct genogroups [8] and the absence of antibody cross-reactivity in serologic studies [112, 115, 116]. This result is also in agreement with a study of two outbreaks involving some overlapping cases. One case who became ill upon infection with a Norwalk-like virus (GI) later became ill upon infection with a Hawaii-like virus (GII) 6 months later in another outbreak (classification of Norwalk and Hawaii virus was based on the reactivity of the individual's sera to NV and Hawaii antigen) [90]. Based on the challenge-re-challenge study [112], because some volunteers did not become ill upon re-challenge (NV-Hawaii, Hawaii-NV, and Montgomery County-NV challenge-re-challenge), it is still possible that heterologous immunity protected these individuals from illness, albeit to a lesser degree. It is also possible that while these volunteers were genetically susceptible to the initial challenge strain and all became ill, some volunteers could have been genetically resistant to the different re-challenge strain and therefore did not become ill.

In one challenge-re-challenge study, two observations contradict the aforementioned hypotheses. The first observation was that NV challenge induced illness in two individuals but did not induce illness upon NV re-challenge *three years* later [111]. This observation suggests that long-term immunity (beyond two years) can be induced after one NoV challenge. These two individuals developed seroconversion to NV after re-challenge. It could be argued that these individuals were not protected because they seroconverted which may indicate viral shedding. However, seroconversion does not always indicate viral shedding and seroconversion can still occur in a protected host. Therefore, "long-term" immunity may have protected these two challenged-re-challenged human volunteers from clinical disease. In comparison, initial inoculation of calves with bovine NoV induced diarrhea, viral shedding, and seroconversion, and subsequent challenge 30 days later with bovine NoV failed to significantly increase IgA or IgG titer consistent with lack of diarrhea or detectable viral shedding in these "protected" calves (discussed in Section 9.5.6) [51].

The second observation that contradicts the current hypotheses regarding NoV immunity was that one individual who had undetected pre-challenge NV-specific IgG titers did not become ill, seroconvert, or increase NV-specific IgG after three separate challenges (spaced 6 months to 3 years apart) but seroconverted, without illness, *after the fourth challenge* [111]. Genetic resistance to NV in this individual would explain the results after the first three challenges. However, the current hypothesis is that genetically resistant individuals do not respond immunologically to NoV [117]. If this hypothesis is valid, and the individual was genetically resistant, then they could not have responded immunologically by seroconversion after the fourth challenge. Therefore, either the individual was genetically resistant and this hypothesis is invalid or the individual was genetically susceptible and this hypothesis is still valid. A second explanation for this observation could be that resistance to infection in this individual was mediated by a non-genetic, non-serum immune mechanism, such as a local mucosal immune mechanism, and that after four repeated immunological insults, the serum immune mechanism became activated and the individual seroconverted. Because no additional data is available on this individual, it is difficult to determine a definitive mechanism for this observation.

Lastly, it is important to acknowledge the contributions of early investigators like Parrino et al., who proposed several hypotheses that have been confirmed two and three decades later. For example, Parrino et al. [48] stated, "Our data suggests that serum antibody by immune electron microscopy reflects infection in susceptible persons but does not appear to play a uniformly protective part in Norwalk-agent illness. ... One hypothesis that could explain our findings is that local antibody rather than circulating antibody may determine clinical response after challenge with Norwalk agent." This hypothesis is based on the observation that levels of pre-challenge NV-specific titers did not correlate with illness in challenged-re-challenged volunteers with NV. We and many other groups have also observed this finding (discussed in Section 9.4.4.4). In addition, our group has found that "local antibody", in the form of mucosal IgA, was associated with protection to infection [117]. A similar association between induction of fecal IgA antibodies by wild type NoV (one inoculation) or NoV VLPs (2 or 3 immunizations), using LT(192G) as adjuvant, and complete or partial protection, respectively against homologous GIII NoV inoculation at post-inoculation day 30 was reported for Gn calves by Han et al. [53]. Parrino et al. [48] also stated that, "A second hypothesis that could explain our findings is based upon a genetic control of susceptibility to Norwalk-agent infection." This hypothesis is based on the observation that some individuals remained illness-free upon challenge and re-challenge, after 31–24 months, with NV. We and others have also identified genetic factors mediating resistance (discussed in 4.3) to Norwalk virus infection.

9.4.3 Genetics and Immunity

Though the focus of this chapter is on NoV immunology, a brief mention will be made of general trends in our understanding of how host genetic factors influence NoV infection and therefore impact NoV-specific immune responses. For a more in depth review of the association between genetics and NoV infection, the reader is encouraged to read the following excellent reviews [31, 114, 118, 119].

The current model of susceptibility and resistance to NoV infection states that, within a human population, there will be individuals who are genetically resistant to infection with *specific* NoV strains [117]. The mechanism of this resistance is thought to be due to the absence of a NoV receptor or binding molecule in genetically resistant individuals. These genetically resistant individuals, when exposed to NoV, do not provide any binding or internalization site for NoV, their cells do not become infected, and they do not develop illness. This mechanism is similar to that postulated for the CCR5 receptor in HIV infection [120, 121]. Presumably, because these individuals do not become infected, their immune system does not "see" the virus, and they do not mount an immune response to this specific NoV strain. The absence of immune response in genetically resistant individuals has not been rigorously tested but is supported by current evidence from human challenge and outbreak studies [117, 122, 123].

In contrast, it is thought that individuals who are genetically susceptible can become infected with the virus because they have a NoV receptor or binding site [117]. More recently, we have found that not all of these "genetically susceptible" individuals become infected upon challenge with NoV. We propose that not all of these "genetically susceptible" individuals become infected because they possess a "protective" immune response that protects them from infection. As discussed throughout this chapter, the existence of a protective NoV immune response is uncertain.

Host genetic factors thought to be important for susceptibility and resistance include the histo blood group antigens (HBGA). HBGAs are complex carbohydrates present on the surfaces of mucosal epithelium and red blood cells of the genitourinary, respiratory, and digestive tracts and are also present as free oligosaccharides in saliva, intestinal contents, breast milk and blood. Three families of HBGAs associated with NoV infection are thought to be the Lewis, secretor and ABO families (reviewed in [119]). Secretor status, one of the HBGA families, is associated with resistance to certain GI and GII NoV infections in challenge, endemic, and outbreak studies [117, 122, 123, 124], but seems to be NoV strain specific [125]. Because secretor status seems to affect susceptibility to NoV infection, it also appears to affect the levels and prevalence of NoV-specific antibodies. In 105 plasma specimens from Swedish blood donors, secretor negative individuals and Le^{a+b-} individuals had significantly lower GII.4 VLP-specific antibody titers and were more often antibody negative than secretor positive and Le^{a-b+} individuals [122]. Secretor negative individuals did not develop significant NV VLP-specific antibody response after NV challenge [117].

The ABO blood antigens, also members of the HBGA family, are also associated with resistance to infection. In general, blood type O individuals seem to be at increased risk of NV infection while blood type B individuals seem to be at decreased risk of infection. In agreement with this hypothesis, in two separate NV challenge studies, blood type O individuals were at a significantly increased risk of NV infection compared to other blood types [117, 126]. In one of these NV challenge studies, individuals who had the B antigen (B and AB blood phenotypes) were at a significantly decreased risk of infection and illness [126] while in the other NV challenge studies, investigators found a decreased risk of infection in blood type B individuals but this risk did not reach statistical significance [117]. In addition, in an outbreak study, blood type B was also associated with a significant decreased risk of infection from another GI virus [127].

By influencing susceptibility to infection, ABO status may also influence NoV-specific antibody levels. Though not statistically significant, one report found a trend of higher GII.4 VLP-specific antibody titers in individuals with blood group O and lower antibody titers in individuals with blood group B in 105 plasma specimens from Swedish blood donors [122]. In a study of a GI virus outbreak, individuals with blood group B were significantly less likely to acquire NV VLP-specific IgG and individuals with blood group A were significantly more likely to acquire NV VLP-specific IgG [127]. Taken together, these results suggest that an individual's ABO status affects their risk of infection and therefore their development of NoV-specific antibodies.

Care must be taken when interpreting the results of these studies and others like them. Recent models suggest that each NoV strain or cluster has a different pattern of HBGA binding than others and the ability of a NoV strain to bind to a host HBGA may affect its infectivity and ability to induce strain-specific antibodies. NoV strain binding patterns must take the ABO, secretor, and Lewis families into account [119, 128, 129]. One model suggests that most of the binding patterns can be sorted into two groups: the A/B and the Lewis (non-secretor) binding groups [119, 128]. In agreement with this model, the strain-specific antibody titers and prevalence of individuals from a Chinese military medical university were associated with the specific strain binding patterns [82]. Of note, the patterns of NoV-HBGA binding do not clearly match the genetic classification of NoVs although strains with the same or closely related binding patterns seem to be clustered [119, 128]. Overall, these findings indicate that a single strain cannot bind to all human HBGA variants although, collectively, NoV can almost cover the diversity of human HBGAs [118]. Interestingly, certain pig HBGAs (A, H) that correspond to the human counterparts can also bind human NoV VLPs and are associated with infection, viral shedding and seroconversion to NoV VLPs in pigs [41].

9.4.4 Humoral Immunology

Given our recent models of host genetics and NoV infection, the subsequent humoral and cellular immunology sections must be read with the understanding that the role of host genetics was not known at the time the majority of these studies were published.

There are two current questions about the role of the humoral immune response after NoV challenge or exposure: (1) How and what antibodies are generated after NoV challenge or exposure; (2) What role do NoV-specific antibodies play in infection and disease? These questions are addressed in the subsequent sections.

9.4.4.1 Antibody Isotypes and IgG Subclasses

Humans

Lindesmith et al. suggested that infection, *and not just exposure* (or challenge), with NoV is essential for activation of a NoV-specific antibody response even in volunteers with pre-challenge NoV-specific antibodies [117, 125]. This hypothesis is supported by data from various challenge studies where individuals with pre-challenge antibodies, when challenged with NoV, remained uninfected and did not seroconvert to NoV antigens [19, 111, 117, 125]. The exception to this hypothesis is in the case of NoV vaccination, where oral administration of high doses of NoV antigen, without infection, is sufficient to induce NoV-specific antibodies (reviewed in [32]). Therefore this hypothesis can be modified to state that, infection with NoV or vaccination with high doses of NoV antigen is necessary for activation of a NoV-specific antibody response.

In general, increases in NoV-specific antibody isotypes are associated with infection (viral replication in the host), but not necessarily challenge, inoculation or exposure (internalization of the virus) because some hosts may not become infected upon challenge, inoculation or exposure. Because increases in NoV-specific antibodies are usually induced after infection, they can be used as markers of infection. This is observed across various strains [117, 124], antibody isotypes, IgA, IgG, and IgM [88, 116, 130], in both volunteer studies [19, 117] and outbreak situations [6, 87, 131]. NoV-specific IgD and IgE levels have not been assessed. In animal models, seroconversion to antibody isotypes is also associated with infection, which may or may not be associated with illness. For example, wild type mice infected with MNV seroconvert to MNV antigen but do not exhibit illness [60, 63]. Interestingly, in Gn pigs inoculated with human NoV, seroconversion had a stronger correlation with illness (diarrhea) than with viral shedding [40].

NoV-specific IgG subclasses are generated after NoV challenge or exposure. One study of 132 Swedish serum specimens found that all IgG subclasses against NV were detected in varying degrees [132]. IgG1 predominated in all age groups, followed by IgG4, IgG3 and IgG2 had the lowest prevalence (3%). Similar findings were observed in a Snow Mountain virus challenge of volunteers: Snow Mountain virus-specific IgG1 was common while IgG2 was rarely detected [125]. Other antibody isotype subclasses (e.g. IgA1, IgA2) have not yet been assessed.

In general, a NoV-specific antibody rise should not be relied on as a diagnostic tool because of the high incidence of false positives perhaps due to NoV cross-reactivity [87]. In addition, not all NoV-infected individuals (defined as infected because they shed virus) exhibit a rise in NoV-specific antibodies (Leon, J.S. and Moe, C.L., unpublished data). However, serology data may be used to complement an outbreak investigation that uses RT-PCR (the gold standard) as a diagnostic test [133].

There seems to be a general temporal order of the appearance of NoV-specific antibody isotypes, IgG, IgM, and IgA. This order and the relative levels vary depending on the individual and may be confounded by pre-challenge NoV-specific antibodies due to previous NoV infection. The synthesis presented below is based on data from a few human challenge studies and has not been rigorously addressed. General trends and ranges of these NoV-specific isotypes are illustrated in Fig. 9.2. In general, in studies of human volunteers challenged with NoV, NoV-specific serum IgG, IgM and IgA increases are first observed after 5 days [86, 130, 134, 135]. In general, NoV-specific IgG levels reach their maximum at 3 weeks post-challenge, although some data suggests it may take up to 7 weeks for NoV-specific IgG levels to peak. Peak NoV-specific IgA and IgM levels occur earlier than IgG, generally during the second week of observation [86, 134, 135]. NoV-specific IgG levels tend to be higher than NoV-specific IgM levels that tend to be higher than NoV-specific IgA levels [86]. Furthermore, NoV-specific serum IgG levels tend to last longer at substantially elevated levels compared to NoV-specific IgM and IgA levels and maintain a plateau near their peak value for up to 100 days or more and decline thereafter. NoV-specific serum IgM levels revert to pre-challenge levels between 2 and 4 months [86, 130, 134], while NoV-specific serum IgA levels seem to persist for 1–2 months and then revert back to pre-challenge levels by 3 months [86, 130]. In some individuals, NoV-specific IgA may persist for 15 months after challenge [86]. Notably, several individuals failed to mount an IgG, IgA or an IgM response

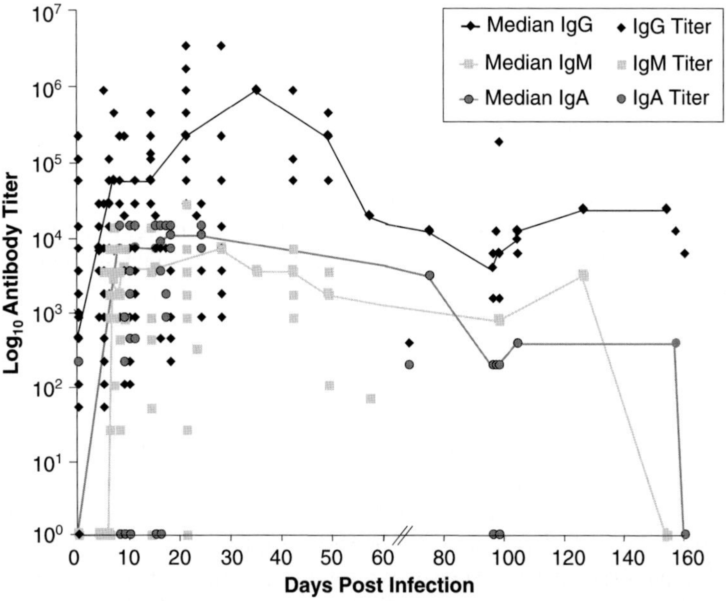

Fig. 9.2 Trends of antibody isotypes after infection. Data on titers was collected from various human challenge studies and outbreaks [86, 130, 134, 135]. The range of antibody isotypes can be observed from the individual symbols, as indicated in the figure legend, that correspond to antibody isotype titers of individuals. Symbols with lines indicate median titers for the range of titers at each specific day post-challenge. The two parallel lines in the X axis represent a break in scale. Please see Section 9.4.4.1 for additional information regarding this figure

after challenge [130, 135], perhaps because they were genetically resistant and not infected. Because of these trends, NoV-specific IgM and IgA seroconversion could be *cautiously* used as a marker of recent infection. In general, NoV-specific salivary IgA and IgG are induced one week after challenge and peak 2 weeks after initial challenge [117]. Some individuals may exhibit a rapid rise in NoV-specific salivary IgA prior to 5 days post-challenge. NoV-specific salivary IgG and IgA may persist for up to 40 days post-challenge (Leon, J.S. and Moe, C. L. unpublished data) but longer time-points post-challenge have not been reported. NoV-specific fecal IgA antibodies can be detected by two weeks post-challenge and may persist for up to two months [136]. Repeated infections may prolong the duration and magnitude of the antibody response [134].

The degree of correlation among the antibody isotypes has not been adequately addressed. In general the correlation between levels of serum IgG, IgA, and IgM seem to be high, but the degree of correlation has not yet been statistically analyzed [86, 116, 130]. We found that volunteers challenged with NV ($n = 77$) had NV-specific serum and saliva IgA and IgG levels that were significantly correlated with each other and had Pearson correlation coefficients between 0.5 and 0.8 (Leon J.S. and Moe. C. L. unpublished data). Because animal studies can use naïve hosts for inoculation, the duration of antibodies and degree of correlation between antibody isotypes obtained from animal studies will be clearer than the data collected from human studies.

Gnotobiotic Pigs

In the human NoV (HuNoV) HS66-inoculated pigs, although IgM, IgA and IgG antibodies were elicited at low titers, 65% of the HuNoV-HS66-inoculated pigs seroconverted, and 64% coproconverted (fourfold rise in HS66-specific fecal antibodies) with either IgA or IgG antibodies [42]. Positive associations were found between disease severity (diarrhea score) and serum and intestinal IgA and IgG antibody titers, suggesting that disease severity may reflect the intensity of intestinal stimulation, inducing increased titers of both local and systemic antibodies.

Low numbers of both IgA and IgG antibody-secreting cells (ASC) were elicited in the HuNoV-HS66-inoculated pigs. Similar numbers of IgA and IgG ASC were detected both locally (intestine) and systemically (spleen and blood) at 21 or 28 d.p.i., and the local ASC responses strongly correlated with systemic ASC responses after HuNoV-HS66 infection [42].

When we compared these results to that of Gn pigs infected with another human enteric virus, the virulent-Wa human rotavirus (HRV) [44], we observed that the HRV elicited higher numbers (10-fold) of IgA and IgG ASC in the intestine of the Gn pigs at 21 d.p.i. than the HuNoV-HS66 at 28 d.p.i.; however, similar numbers of IgA and IgG ASC were induced in spleen and blood in both HRV and HuNoV-infected pigs at 21 or 28 d.p.i. This higher level of intestinal ASC responses in the HRV-infected pigs is probably a consequence of the higher, and longer lasting, intestinal viral replication and diarrhea observed, and also of the more extensive intestinal lesions caused by the HRV in the Gn pigs [45].

9.4.4.2 Mapping of NoV antigenic sites

Because the main NoV target of the host immune response is the capsid protein, multiple groups have attempted various mapping strategies including genomic [17, 137], proteomic [138], monoclonal antibodies [139, 140], and crystallographic studies [141, 142] to identify antigenic, receptor binding, and cell binding regions. A single capsid protein has two main domains, the shell domain (S, aa residues 1–225) and the protruding domain (P) that is subdivided into two subdomains, P1 (aa residues 226–278 and 406–530) and P2 (aa residues 279–405) [143]. The distally located P2 subdomain is a large insertion in the P1 subdomain and protrudes more than the P1 subdomain. A comparison of capsid proteins from various calicivirus suggests that the S domain is well conserved, the P1 subdomain is moderately conserved, and the P2 subdomain is highly variable [142]. Chronic human infection with NoV induces the least number of mutations in the S domain, more in the P1 domain, and the majority in the P2 domain, suggesting that the P2 domain has important functions in host-pathogen interaction and pathogen or viral evasion [17]. Conserved regions in both the S domain and P1 domain may contribute to antibody cross-reactivity across clusters and genogroups [137]. For example, epitopes in the S domain from GI and GIII capsid proteins [144] or from GIII and GII VLPs [145] could bind the same monoclonal antibody derived from immunization of mice with NoV VLPs. The majority of monoclonal antibodies induced after immunization with NoV capsid protein targeted the S domain and cross-reacted with epitopes on GI and GII NoV capsid proteins [146]. One epitope in the P1 domain (aa 457–530) is common to GI viruses [139, 147]. The hypervariable P2 domain seems to contain multiple determinants for antigenicity, cell binding, and host specificity [138, 140, 148, 149, 150, 151, 152, 153] (reviewed in [114, 119]). With the development of in vitro NoV-specific antibody blocking assays, additional epitopes important for NoV neutralization and blocking will be identified.

9.4.4.3 NoV-Specific Antibody Cross-Reactivity

In general, infection with NoV or immunization with NoV antigens will induce a cross-reactive antibody response, most likely due to the conserved epitopes discussed in the preceding Section 9.4.4.2. These NoV-specific antibodies are cross-reactive and bind to multiple NoV strains within various clusters and across genogroups [87, 89]. In general, higher cross-reactivity is observed within clusters [89] and genogroups than across genogroups [6, 89]. In addition, the prevalence of cross-reactive antibodies after challenge, inoculation, exposure or immunization is usually lower than 100% in human and animal models. Because NoV infection induces cross-reactive antibodies, serologic assays are not useful for classification of NoV strains. In addition, the ability of serologic assays to detect cross-reactivity also depends on the quality and nature of the antigens used in the detection assay (e.g. fully formed VLPs versus recombinant capsid proteins that do not form VLPs). The cross-reactive nature of the immune response against NoV can be utilized to design effective vaccines that protect against antigens not present in the vaccine (discussed in Section 9.5.4) and may eventually be helpful in the design of broad serologic assays for the detection of NoV-specific antibodies in populations.

9.4.4.4 Contribution of NoV-Specific Antibodies to Infection and Illness

The exact contribution of antibodies to clinical NoV infection and disease is currently unknown. One of the main questions is whether antibodies are associated with protection from infection and disease or whether they are merely a marker of infection and disease. Various reports contradict each other in stating that pre-challenge NoV-specific antibodies are associated with a *decreased* risk of infection or illness [74, 154, 155, 156], are associated with an *increased* risk of infection or illness [19, 79, 116, 117, 122, 157], or are not associated with infection or illness [90, 115]. This contradiction may be due to differences in host populations, infectious NoV strains, varied study definitions of "infection", or the presence of host genetic confounders.

The majority of papers examining this question have focused on pre-challenge serum NoV-specific IgG antibodies. However after adjusting for secretor status in volunteers challenged with one virus strain, NV, (NV, n = 77), the presence of pre-challenge serum NV-specific IgG antibodies were associated with an *increased risk* of infection [117]. This association was also observed for all volunteers without adjustment for any confounders. Pre-challenge NV-specific saliva IgA was not associated with infection. In agreement with this finding, infected volunteers from another NV challenge study (n = 50) exhibited significantly higher pre-challenge NV-specific IgG than uninfected volunteers [19]. In another NV challenge study using similar volunteers (n = 38), a higher pre-challenge NV-specific fecal IgA antibody geometric mean titer was found in ill compared to well volunteers and in infected versus uninfected volunteers [157]. In these two studies, secretor status was unknown and therefore the effect of secretor status on the probability of becoming infected and having an antibody response was not taken into account. In another study, we challenged volunteers with a NoV genotype II (GII) virus, Snow Mountain (SMV) (n = 15) [125]. We did not observe any association between the presence of pre-challenge Snow Mountain-specific IgG or IgA and infection.

Interestingly, in our NV challenge study, 74% of secretor negative volunteers (genetically resistant) had pre-challenge NV-specific serum IgG. These secretor negative volunteers, although resistant to NV, may have been susceptible to other NoV strains and previously infected with other strains of NoV. Infection with these other strains may have induced cross-reactive IgG to NV antigen in these secretor negative volunteers.

Our results indicate that while pre-challenge NoV-specific serum IgG may not be associated with protection, rapidly induced NoV-specific salivary IgA may be associated with protection and may indicate a NoV-specific memory response. In our NV human challenge study, a portion of genetically susceptible individuals (secretor positive) was resistant to NV infection, suggesting that a memory immune response or other mechanism also protected them from NV infection [117]. The majority of the uninfected group (60%) exhibited a rapid rise in NV-specific salivary IgA titer *before* 5 days post-challenge. In contrast, the majority of the infected group (83%) exhibited a rise in NV-specific salivary IgA titer only *after* 5 days post-challenge. The rapid immunological response to NV challenge in secretor-positive, uninfected individuals suggests that acquired immunity may explain the difference in those secretor-positive volunteers who developed infection after challenge and those who did not. Therefore, the timing of the NV-specific mucosal IgA may be a good predictor of the risk of infection among genetically susceptible volunteers. At the same time, 6 out of the 15 uninfected volunteers did not develop a NV-specific salivary IgA response suggesting that other unidentified factors may also be associated with protection. In another one of our studies, where volunteers were challenged with Snow Mountain virus ($n = 15$), volunteers were infected regardless of their secretor status or blood group type [125]. Because we could not adjust for a genetic component (if there was one), we could not assess the effect of immunity independently and examine if the timing of salivary antibodies affected infection. Instead, we found Snow Mountain virus (SMV) infection increased SMV-specific IgG levels in serum and SMV-specific IgA levels in saliva after 14 or 21 days.

In summary, the majority of the reports indicate that presence of pre-challenge serum (and fecal) antibodies suggest prior infection and therefore susceptibility to future infection. Individuals who are genetically resistant to infection do not mount an immune response to NoV and therefore have low or no NoV-specific antibodies. As an additional confounder, there may be individuals who are genetically resistant to infection with some NoV strains, for example, strain A, but not others, such as strain B. These individuals, in general, will have lower NoV A-specific antibody levels, but once infected with strain B, may have higher NoV A and B specific antibody levels due to heterologous and homologous cross-reactivity. The timing of other antibodies, such as NoV-specific salivary IgA, may be associated with protection. However, there still may be other types of unidentified antibodies (e.g. mucosal, isotypes, IgG4) that may be associated with protection.

NoV-specific antibodies have also been shown to be associated with clinical illness. As discussed above, the presence of NoV-specific pre-challenge antibodies has been associated with increased risk of infection. Because infection may lead to symptoms, it is not surprising that the presence of pre-challenge antibodies (serum IgG, IgA, IgM, fecal IgA) has been associated with increased risk of developing symptoms [79, 116] and viral excretion [19]. Similarly, because infection may lead to both symptoms and rise of NoV-specific antibodies, the development of NoV-specific antibodies (serum IgG, IgA, IgM), as expected, was also associated with symptoms [79, 115, 130] and viral excretion [19]. In the NV challenge study reported by Cukor et al. [134], the association between NV-specific serum IgM and symptoms was significant ($RR = 3.8$, 95% CI [1.1–12.5], $p < 0.01$) (Leon, J.S. and Moe, C.L. unpublished data). Few reports have examined the association between NoV-specific antibodies and certain symptoms. One NV challenge study found a significant association between rise of the NV-specific titers and vomiting and nausea, headache or body ache, vomiting, and vomiting and diarrhea [19] while an epidemiologic study of Brazilian children found an association between NV-specific IgG seroconversion and diarrhea or vomiting [34].

9.4.4.5 NoV "Neutralizing" or "Blocking" Antibodies

To determine whether antibodies contribute to protection from NoV infection, it is necessary to determine whether NoV-specific antibodies can prevent infection. One direct way to assess their contribution in vivo is through passive transfer studies where serum or purified NoV-specific antibodies are transferred to recipients that will be challenged or exposed to NoV and then monitor them for infection and illness. In humans, epidemiological studies may address this by examining the association between transfer of maternal NoV-specific antibodies to children and children's risk of infection. In animal models, NoV-specific antibodies may be passively transferred to recipients to assess whether these antibodies prevent infection or diminish the duration or severity of infection in the recipient. Currently, both the pig and murine models are best suited for this approach. In unpublished observations, passive transfer of MNV1-reactive polyclonal serum was able to delay MNV1-induced lethality in immunocompromised mice (RAG2/STAT1) after intraperitoneal transfer suggesting that NoV-specific antibodies may protect the host from NoV infection and disease (discussed in [63]).

An indirect way to assess the ability of NoV-specific antibodies to protect from infection is to measure their ability to "block" NoV infection in vitro. Because, until recently, there was no in vitro model of NoV infection it was not possible to detect these "blocking" or "neutralization" antibodies. Instead, over the past 3 decades, groups have attempted to detect antibodies that can competitively prevent NoV binding to other antibodies [79, 96, 158, 159], other cells [153], or putative NoV receptors like the HBGA family [160, 161]. These assays have been adapted for the detection of various NoV antibody isotypes [134] and NoV specificities [86, 162, 163]. Antibodies that block binding of NoV to other antibodies have been associated with increased risk of infection and illness, as discussed in Section 9.4.4.4. Antibodies that block binding of NoV VLPs to tissue culture cells were obtained from a monoclonal antibody raised against NV VLP, but not sera from NV-infected individuals [153]. These antibodies identified site(s) of the P2 domain on VP1 important for binding of tissue culture cells to VLPs [140]. Antibodies that block binding of NoV VLPs to HBGA structures are present in infected individuals [100, 161], mice [160, 161], and pigs [41] and block HBGA binding in a strain-specific manner. These assays have not yet been used in human challenges studies to investigate the associations between these antibodies and infection, symptoms, viral shedding and other clinical parameters.

In 2004, NoV-specific antibodies that effectively block NoV from infecting cells have been detected in an in vitro tissue culture assay [64]. This in vitro assay takes advantage of the ability of MNV1 infection to form plaques in tissue culture murine macrophage cell monolayers. Through this assay, Thackray et al. demonstrated that MNV-specific antibodies induced after infection with 5 genetically distinct MNV strains were all equally effective at blocking MNV1 infection of RAW cells in a dose dependent manner. These results suggest that murine MNVs collected from research mouse colonies probably comprise only 1 serogroup [164]. In addition, these two reports are a few examples of the protective role of NoV-specific antibodies [64, 164]. In the coming years, other similar in vitro assays, perhaps based on the Gn pig model, will increase our understanding of the protective role of antibodies in human NoV infection.

9.4.5 Cellular Immunology

Little is known about the immune cells, cytokines, and chemokines associated with NoV infection and disease. Limited data on immune cells and serum immune factors and their role in NoV infection and disease comes from human challenge studies and murine, Gn pig and calf models. The recent introduction of these animal models will exponentially advance our understanding of the cellular immune response against NoV.

9.4.5.1 Innate Immunity

Immunity against NoV can be divided into innate and adaptive immunity. Innate immune protection seems to be mediated by downstream mediators of the STAT1 response, including the family of interferons, α, β, and γ. When the STAT-JNK pathway was deleted in mice, intracranial murine NoV (MNV) inoculation killed all STAT1 knockout mice [60]. Intracranial MNV inoculation of IFN-αβγ receptor knockout mice also killed all mice suggesting that deletion of *all* the interferon receptors renders a mouse susceptible to death from NoV inoculation. Neither, IFN-αβ receptor or IFN-γ receptor knockout mice died after MNV inoculation suggesting that *any* receptor is sufficient for survival. However, these mice had similar and persistent levels and distribution of NoV as their inoculated STAT1 counterparts suggesting that *other* factors are important for viral clearance. Knocking out other important IFN anti-viral mediators, such as inducible nitric oxide synthase (iNOS) and protein kinase RNA-activated (PKR), in mice also did not result in lethality. Interestingly, while intracranial inoculation killed all susceptible mice, intranasal and perioral inoculations were less lethal than intracranial inoculation. Other innate mucosal factors may mediate protection in these mice after inoculation. Taken together, these results suggest that innate immunity, specifically IFN or STAT1-dependent immune responses may be responsible for the rapid control of NoV infection in humans but not viral clearance [63].

STAT1 and the interferons seem to be necessary for the control of virus replication. Inoculated wild type mice that have STAT1-dependent immune responses have lower levels of virus in their intestine and peripheral organs than inoculated STAT1 knockouts or infected IFN-αβγ receptor knockout mice [110]. This low viremia in wild type mice was accompanied by detectable levels of IFN-α in sera and intestinal homogenates. In addition, bone marrow derived macrophages or dendritic cells from wild type mice exhibited lower MNV replication rates than STAT1, IFN-αβγ receptor knockout, and IFN-αβ receptor knockout mice [64]. This lower replication rate was accompanied by secretion of IFN-α in MNV-infected wild type bone marrow derived macrophages and dendritic cells [64]. Interestingly, infected macrophage or dendritic cells from IFN-γ receptor knockout, iNOS knockout, and PKR knockout mice had similar MNV replication rates as wild type suggesting that IFN-γ iNOS, and PKR do not affect MNV replication rates [64]. These results suggest that STAT1, the combined IFN-αβγ receptors, and the combined IFN-αβ receptors are important in the control of viral replication. Therefore, it is likely that both IFN-α and β are also important in the control of viral replication, but direct experiments to demonstrate this have yet to be done. In contrast, the IFN-γ receptor, PKR, and iNOS have no significant effect, compared to wild type, in the control of viral replication in macrophages and dendritic cells. Based on these results, Mumphrey et al. proposed two mechanisms for the role of STAT1 and IFNs: (1) IFN-induced host factors directly inhibit viral replication at the primary site of entry and (2) IFN responses limit NoV dissemination and replication in secondary tissues [110]. Interestingly, human challenge with NV or Hawaii virus did not induce detectable levels of interferon in sera, jejunal aspirates, or jejunal biopsy specimens taken 48–96 hours post-challenge [165].

9.4.5.2 Adaptive Immunity

Mice

If the STAT1 pathway, through the IFNs, inhibits viral replication and dissemination but not clearance, then other factors must be responsible for viral clearance. Experiments by Karts et al. showed that MNV inoculated RAG1 and RAG2 knockout mice, mice that lacked B and T cells, did not die [60]. However, these mice exhibited a persistent viremia throughout all their organs up to 90 d.p.i. Based on these results, the authors hypothesize that the adaptive immune response, in contrast to the STAT1-dependent innate immune response, was not required for protection against lethal MNV infection. Instead, the components of the adaptive immune response, specifically B and/or T cells, were required to contain and clear murine NoV infection [63]. These studies also showed that NoV can continuously replicate in tissues for long periods of time without causing severe morbidity or lethality. These results may explain the persistent viral shedding observed in immunocompromised patients, which may last up to two years in some cases [16, 17, 21, 22, 23]. These immunocompromised patients have impaired B and T cell responses that may not be able to clear replicating virus. One group found that reducing immunosuppression in transplant recipients was essential for clearing human calicivirus infection [22].

Humans

In humans, information on the adaptive immune system after human NoV challenge comes from our human challenge study performed with Snow Mountain virus

(SMV) [125] and one case-control study of American students with diarrhea after visiting Mexico [101]. In our human challenge study, we demonstrated that peripheral blood mononuclear cells (PBMCs) from challenged volunteers respond to Snow Mountain virus-like particles (VLPs) in vitro by secreting cytokines 8–21 days post-challenge [125]. Among all secreted cytokines assayed, IFN-γ and IL-2 exhibited the highest levels and prevalence among all specimens. Post-challenge PBMCs also secreted TNF-α, IL-4, IL-5 and IL-10 cytokines (at least fourfold level above baseline) in a few volunteers. There was no significant difference in pre-challenge and post-challenge PBMCs stimulated with SMV VLP for any cytokine in either uninfected or infected volunteers, perhaps due to small sample size. Interestingly, IFN-γ was detected in 91% of pre-challenge PBMCs after SMV VLP stimulation suggesting previous NoV exposure and immunological memory in these volunteers. Sera was also assayed for cytokine levels, and infected volunteers exhibited a significant rise in IFN-γ and IL-2 levels at 2 days post-challenge; no difference was observed for IL-6 and IL-10. There was no difference in IFN-γ, IL-2, IL-6 or IL-10 from post-challenge serum specimens compared to pre-challenge serum specimens for uninfected volunteers. These results suggest that: (1) challenged individuals respond to NoV antigen with secretion of anti-inflammatory and inflammatory cytokines, (2) serum cytokines can be quickly induced by 2 days post-challenge and (3) challenged individuals probably had pre-challenge NoV exposure.

Results similar to the human challenge study were found in the case-control study of American students with diarrhea [111]. Students that shed NoV ($n = 7$, 6 GI and 1 GII strain), in the absence of other detectable pathogens, had significantly higher levels of inflammatory cytokines, IFN-γ and IL-2, in their diarrhea specimens than students who had no detectable pathogens ($n = 19$). Interestingly, a group of students that had a NoV and enterotoxigenic *E. coli* (ETEC) co-infection ($n = 10$, 8 GI and 2 GII strains), exhibited the aforementioned cytokines and a significant and large increase in secreted IL-10 in their diarrhea specimens compared to students who had no detectable pathogens. It is important to note that cytokine secretion in this study was not antigen-specific. One advantage of this study is that fecal cytokines probably better reflect the gut mucosal immune response during NoV infection than serum cytokines. These results suggest that symptomatic NoV infection is associated with an inflammatory gut response, while symptomatic co-infection with ETEC is associated with both an inflammatory and anti-inflammatory gut response.

In our human challenge study, we also assessed whether cellular cross-reactive responses occurred after Snow Mountain virus (GII) challenge [125]. PBMCs from SMV infected volunteers were assayed for their cytokine response to Hawaii (GII) and NV (GI) antigen in vitro. Stimulation with Hawaii VLPs induced significantly higher levels and prevalence of cytokines in PBMCs compared to stimulation with NV VLPs. Hawaii VLP stimulation of PBMCs also induced significantly higher levels of IFN-γ and IL-2 compared to NV VLP stimulation of PBMCs. IL-10 (fourfold levels above baseline) was also detected in both Hawaii and NV VLP stimulation of PBMCs while TNF-α was only detected in Hawaii VLP stimulation of PBMCs.

We also investigated the relative roles of $CD4^+$ and $CD8^+$ cells on IFN-γ secretion after SMV VLP stimulation. PBMCs from 5 volunteers (both infected and uninfected) were depleted of either $CD4^+$ or $CD8^+$ cells and then stimulated with SMV VLPs. $CD4^+$ depleted PBMCs exhibited significantly lower IFN-γ levels (82–97% lower) compared to undepleted or $CD8^+$ depleted PBMCs after SMV VLP stimulation. $CD8^+$ depleted PBMCs from three infected volunteers exhibited no significant difference in IFN-γ levels compared to undepleted PBMCs. Interestingly, $CD8^+$ depleted PBMCs from two infected volunteers exhibited *higher* IFN-γ levels compared to undepleted PBMCs. In summary, we hypothesized that in PBMCs, $CD4^+$ cells are the primary source of SMV-specific IFN-γ secretion and that $CD8^+$ cells may regulate IFN-γ secretion in $CD4^+$ cells.

Gnotobiotic Pigs

Recently, the intestinal and systemic antibody titers, ASC responses, innate Type-I IFN (IFN-α), pro-inflammatory (IL-6), Th1 (IL-12 and IFN-γ), Th2 (IL-4) and Th2/T-reg (IL-10) cytokines and cytokine-secreting cell (CSC) profiles were evaluated in Gn pigs after oral inoculation with GII.4 HuNoV (HS66 strain) and compared to mock-inoculated controls [42].

In serum of the HuNoV-HS66-inoculated Gn pigs, delayed (10 and 12 d.p.i.) (2.2-fold over controls) innate (IFN-α) and low and early (4 d.p.i.) pro-inflammatory (IL-6) cytokine responses were observed. A balanced serum Th1/Th2 response was also observed with persistently higher levels of IL-12 at most d.p.i., and a transient, but significantly higher peak of IFN-γ was seen at 2 d.p.i. (2.5-fold over controls). This early peak of IFN-γ was similar to that observed in the serum of human volunteers orally challenged with a GII HuNoV, Snow Mountain virus (SMV) [125]. The Th2 (IL-4) and Th2/T-reg (IL-10) cytokines were only detected in the serum of the HuNoV-HS66-inoculated pigs at low concentrations and were significantly elevated above controls earlier in infection (2 to 8 d.p.i.).

In intestinal contents (IC) of the HuNoV-HS66-inoculated pigs, only innate (IFN-α) (early at 2 and 8 d.p.i. and later at 21 d.p.i.) and Th1 (IL-12) (later at 28 d.p.i.) responses were significantly higher compared to controls. Failure to detect significantly elevated concentrations of the other cytokines in IC compared to those found in serum may merely reflect the instability of these cytokines in IC. The local and systemic cytokine-secreting cell (CSC) responses were characterized by low numbers of pro-inflammatory (IL-6) CSC detected early in the gut and later systemically (in blood), and a generally more biased Th1 (IL-12 and IFN-γ) CSC response, both locally and systemically in the HuNoV-HS66-inoculated pigs.

In the HuNoV-HS66-infected Gn pigs, the low pro-inflammatory (IL-6) responses to viral infection, low levels of the anti-inflammatory IL-10 in serum and lack of significantly elevated numbers of IL-10 CSC could be due to the low levels of viral replication, mild pathology and low inflammation in the gut [40].

9.4.5.3 Other Immune Cells Involved in NoV Infection

Infection is also accompanied by inflammation in the intestine in mice and humans. In mice, granulocytes were observed 24 hours after infection [110]. In humans, duodenojejunal biopsies taken 48 hours after challenge showed monocyte and neutrophil infiltration of the lamina propia and epithelial cells in several volunteers 48 hours after infection [48, 50]. Murine NoV infection also increases the number of B cells and macrophages in the spleen after 72 hours [110].

Murine NoV can infect several immune cells in vitro including bone marrow derived murine macrophages and dendritic cells [64]. Murine NoV also seems to infect macrophages in vitro in immunodeficient mice as several cells co-stained with ProPol and F4/800, a macrophage lineage marker [166]. In STAT1 knockout mice, murine NoV seeds and replicates in intestinal epithelial cells but replicates predominantly in the lamina propia and Peyer's patch cells at later times post-infection [110]. Little to no viral replication was observed in wild type mice; it is unclear whether epithelial cell infection was prevented in wild type hosts. Wobus et al. hypothesized that infection of transepithelial dendritic cells in the lumen of the intestine may provide a pathway for NoV infection [63].

9.4.6 Maternal Factors Involved in NoV Protection

Human milk can protect children from infectious diseases through a number of mechanisms. To date, human milk is thought to protect the infant from NoV infection through two mechanisms (reviewed in [118, 119]). One mechanism is through transfer of NoV-specific antibodies from mother to child. Epidemiologic studies have suggested the presence of high titers of maternal NoV-specific antibodies in infants up to 8 months of age [67, 68, 69, 70, 71, 72, 79]. The ability of maternal milk to block NoV binding in a radioimmunoassay in the late 1970s indicated the presence of NoV-specific antibody. However, it was not until recently that NoV-specific IgA has been identified in human maternal milk [167, 168]. About 13% of human milk specimens, from mothers in Chiba City, Japan, had NoV-specific IgA that bound to different and multiple NoV clusters, including those within GI and GII. The prevalence of NoV-specific IgA to certain clusters varied depending on the clusters (e.g. antibodies to GII.6 had a prevalence of 13.6% while antibodies to GI.8 had a prevalence of 0.8%) [167]. In addition, the prevalence of NoV-specific IgA to certain clusters also depended on the mothers' HBGA makeup, specifically her secretor status [168].

A second mechanism of NoV protection in neonates is through decoy receptors. These decoy receptors in human milk are thought to bind NoV, prevent NoV from binding to host gastrointestinal cells, and therefore prevent NoV infection in infants. These decoy receptors are glycoproteins that correspond to the bile salt stimulated lipases and a milk fraction including the mucins MUC1 and MUC4 [169]. The ability of these glycoproteins to protect from NoV infection may depend on the mother's HBGA makeup, the infant's HBGA makeup and the HBGA binding pattern of the NoV strain [118]. In agreement with this model, two groups found a concordance between the mother's secretor status and Lewis phenotype and the ability of their milk to inhibit the binding of different NoV strains [168, 169]. Interestingly, not all of the mother's HBGAs were expressed in their milk specimens [168]. It is likely that, in the future, additional protective factors against NoV will be identified in mother's milk.

9.5 Vaccine Efforts

A NoV vaccine will be superior to other public health interventions at reducing morbidity and mortality associated with NoV illness. Clean water, hygiene and sanitation are effective public health interventions at reducing diarrhea morbidity and mortality associated with most enteric pathogens [170, 171]. However, developed countries, like the U.S. and Japan, that generally have clean water, good sanitation and hygiene practices, still have a high rate of outbreaks associated with NoV. This suggests that, for enteric pathogens that are highly infectious, these measures are not sufficient to prevent transmission. Therefore, an effective NoV vaccine is a necessary intervention to reduce the public health burden of this disease, especially in developed countries. To date, there is no commercially available NoV vaccine, but much progress has been made in the past decade to develop a vaccine. This section summarizes these advances, and the reader is encouraged to read other reviews on this topic [30, 32, 172, 173, 174, 175].

9.5.1 NoV VLPs are Immunogenic

Initial studies in mice have shown that oral administration or subcutaneous immunization with virus-like particles (VLPs) alone induce NoV-specific immune responses in mice. Both CD1 and Balb/c mice were fed recombinant NV VLPs produced by a bacculovirus expression vector. Increasing doses of NV VLPs induced higher NoV-specific serum IgG and fecal IgA and a higher prevalence of seroconversion in both strains of mice [176]. Among all NV-specific IgG isotypes, NV-specific IgG2b levels were the highest levels observed after VLP administration. In agreement with this study, oral administration of NV VLPs to Balb/c mice induced high levels of NV-specific serum, fecal, and vaginal antibodies, NV-specific T cell proliferation, IFN-γ and IL-4 secretion, and NV-specific $CD4^+$ and $CD8^+$ proliferation [177, 178]. In a separate study, Balb/c mice were fed NV VLPs from a Venezuelan equine encephalitis (VEE) replicon expression system [179]. In contrast to the Ball et al study [176], NV-specific IgM and IgG levels were low or non-existent [179]. However, footpad inoculation of Venezuelan equine encephalitis virus replicating particles (VEE-VRP) expressing NV antigen induced robust levels of NV-specific serum IgG, IgM and fecal IgA.

In humans, oral administration of NV VLPs, produced in a bacculovirus expression system, induced NV-specific immune responses. Various doses of VLPs induced serum NV-specific IgA antibody secreting cells, seroconversion and NV-specific IgA and IgG [180, 181]. NV-specific IgA was detected in saliva, stool, semen, and vaginal washes. Three doses of VLPs, 250 μg, 500 μg and 2000 μg, were administered to different groups of volunteers, and no significant differences were observed in either levels of antibodies, seroconversion rates, or antibody secreting cells among the various volunteer groups. PBMCs harvested from the groups receiving 250 μg and 500 μg of VLPs at day 21 post-immunization, proliferated and secreted IFN-γ in response to NV VLP antigens. IFN-γ secretion was not detected in PBMCS collected 56 days post-immunization. Interestingly, PBMCs harvested from the group receiving 2000 μg did not develop any cell-mediated response at either day 21 or 56 post-immunization. The authors could not explain this finding, but it is possible that a high VLP antigen dose could have induced oral tolerance to this antigen, similar to that observed in other systems (reviewed in [182]). No IL-4 was detected from any of the stimulated PBMCs. Ingestion of VLPs was safe in humans and did not induce any symptoms different in severity or prevalence than that seen in placebos (e.g. headache, malaise, nausea).

9.5.2 Adjuvants Enhance NoV VLPs' Immunogenicity

In general, adjuvants administered with NoV VLPs enhanced their immunogenicity in mice. Administration of cholera toxin together with oral feeding of NV VLPs to Balb/c mice and CD1 mice induced higher levels of NV-specific antibodies, a higher response at lower VLP immunization doses, and a higher rate of seroconversion [176]. In a similar study, a modified cholera toxin administered to mice that received oral NV VLPs also boosted the levels of NV-specific IgA present in various organs including the lungs, trachea, small intestine, and Peyer's patches [177]. The authors observed higher NV-specific IgG, IgA, and IgA antibody secreting cells, higher T cell proliferation and IFN-γ and IL-4 secretion and higher $CD4^+$ and $CD8^+$ stimulation in spleens and Peyer's patches. Interestingly, in this study, administration of bicarbonate also had a mild "adjuvant" effect. Another mucosal adjuvant, mucosal *Escherichia coli* heat-labile toxin, LT and its nontoxic mutant, LT(R192G), also enhanced the magnitude and duration of NV-specific IgG and IgA antibody responses after oral and intranasal administration of NV VLPs to Balb/c mice [178, 183]. Mice inoculated with these two adjuvants developed NV-specific IFN-γ, IL-2, IL-4, and IL-5 secretion in splenocytes, lymph node cells, and Peyer's patches [183]. Lastly, oral administration of a raw material from a yeast cell lysate containing a GII.4 VLP induced NoV-specific serum IgG, fecal IgA and blocking antibodies to histo-blood group putative receptors. The raw yeast material could have served as an adjuvant although this was not tested [184]. Purified GII.4 VLP intramuscular administration with Ribi adjuvant also induced NoV-specific immunity [184].

In addition to increasing the magnitude of NoV-specific immune responses, adjuvants may also affect the type of the immune response. For example, NV-specific IgG2b was the highest IgG subtype induced in mice orally immunized with NV VLPs. In contrast, NV-specific IgG1 was the highest IgG subtype induced in mice orally immunized with NV VLP and cholera toxin [176]. Similarly, levels of NV-specific secretion of IFN-γ were higher than IL-4 in mice orally immunized with VLPs. In contrast, levels of NV-specific secretion of IL-4 were higher than IFN-γ in mice orally immunized with VLPs and modified cholera toxin [177].

Gn calves were immunized with bovine GIII NoV VLPs in the presence of various adjuvants [53]. Intramuscular immunization with the oil adjuvant ISA50V induced the highest VLP-specific serum IgG antibody titers compared to intranasal VLP immunization with ISCOM or LT (R192G) or oral immunization with LT (R192G), but failed to prevent diarrhea or viral shedding post-inoculation in the Gn calves. However, only the LT (R192G) co-administered intranasally with VLPs induced fecal IgA antibodies to bovine NoV (BoNoV) in the calves and partial protection (delayed and shortened diarrhea and shedding of 1–2 days) compared to controls (8–9 days diarrhea). In contrast to the murine studies described above, LT (R192G) given orally with VLPs induced low or non-existent NoV-specific antibodies in the serum of immunized calves. However, the LT (R192G) administered orally with VLPs induced BoNoV-specific fecal IgA antibodies in Gn calves.

9.5.3 Immunization Route Affects NoV VLPs' Immunogenicity

In general, in mice, intranasal administration of VLPs seems to induce a stronger and longer immune response against NoV than oral feeding of VLPs. Balb/c mice that were immunized by the intranasal route with NV VLPs had higher levels of NV-specific serum, fecal (mucosal SIgA was not assayed so IgA could have been serum derived), and vaginal IgG and IgA than oral immunization [178, 183]. Levels induced by intranasal immunization persisted longer than levels induced by oral immunization. It is possible that the use of appropriate adjuvants for oral immunization would enhance the responses seen after this route of immunization. In mice, other investigators used intramuscular and intranasal immunization routes but did not directly compare them with each other or with oral routes [177, 184]. In a separate study in mice, oral immunization was superior to intraperitoneal or subcutaneous immunization (without adjuvant) at generating NoV-specific hybridomas, suggesting that oral immunization induced a higher number of antibody-producing cells than the other tested routes 2[185]. Lastly, one group successfully infected mice with Venezuelan equine encephalitis virus replicating particles (VEE-VRP) expressing various NoV antigens in the footpad [160, 179]. Mice mounted high levels of serum and gut NoV-specific IgA and IgG after infection. Differing immunization routes also induced different types of humoral and cellular immune responses [183, 185].

9.5.4 Cross-Protection After Immunization with NoV VLPs

Because of the diversity of NoV strains, an effective NoV vaccine needs to protect against NoV strains that were not present in the vaccine. Cross-protection can be *indirectly* assayed by determining whether cross-reactive humoral and cellular responses are present against various antigens or *directly* assayed by immunizing a host against one antigen or infecting a host with a NoV strain and later determining whether challenge with a different strain results in decreased or no disease/infection. It is important to note that only direct assays of cross-protection, and not indirect assays of cross-protection, will indicate whether immunization or infection with an antigen/NoV strain will protect from future challenge with a different NoV strain. Cross-protection was indirectly observed in humans and animals after the induction of cross-reactive antibodies and cellular responses after immunization with NoV VLPs and after infection with NoV, as previously discussed in Sections 9.4.2, 9.4.4, and 9.4.5. Cross-protection was directly observed in early human challenge studies where volunteers were initially challenged with NV (GI.1) or Hawaii virus (GII.1) and did not become ill when later challenged with Montgomery County virus (GI.5) [112] (classification of Montgomery County virus determined from [36]), as discussed in 4.2. In general, it seems as though cross-reactivity and protection is preferentially genogroup specific. For example, challenge with GII antigens induced stronger immune responses against other GII antigens than GI antigens [115, 116, 125].

The ability of vaccine VLPs to induce NoV cross-reactive antibodies was investigated in one study where Balb/c mice were inoculated with infectious Venezuelan equine encephalitis virus (VEE) replicon particles (VRPs) expressing Norwalk, Hawaii, Snow Mountain, or Lordsdale antigen. These mice developed homotypic antibodies to the specific antigens they received as well as heterotypic antibodies to other antigens (Norwalk, Hawaii, Snow Mountain, and Lordsdale). As previously discussed, the levels of the cross-reactive antibodies were highest among genogroups than across genogroups [160]. Interestingly, naïve Balb/c mice immunized with a trivalent VEE-VRP cocktail containing Norwalk, Hawaii, and Snow Mountain antigen and *lacking* Lordsdale antigen, generated levels of Lordsdale-specific IgG of similar magnitude as Balb/mice immunized with VEE-VRP cocktails *containing* Lordsdale antigen. In addition, the levels of the heterotypic Lordsdale-specific serum IgG after trivalent immunization were significantly higher than combining each level of the heterotypic Lordsdale-specific serum IgG after monovalent (Norwalk, Hawaii, Snow Mountain antigen) immunization. This exciting finding suggests that NoV cocktail vaccines may protect against NoV strains not in the original immunization and therefore may be useful against new NoV variants. The new murine and Gn pig models will facilitate the testing of these multivalent vaccines against both homotypic and heterotypic infections.

9.5.5 VLP Vaccine Delivery Vehicles

In addition to evaluating the vaccination route and adjuvants used, it is also important to evaluate the delivery vehicles used to administer the vaccine to humans. To date, three general types of delivery vehicles have been proposed: purified VLPs,

VLPs in unicellular organisms, and VLPs in plants. High yields of VLPs can be purified from baculovirus [180], yeast [184], and Venezuelan equine encephalitis vectors [160]. VLPs could be administered as part of a unicellular organism like in yeast and *Lactobacillus casei*, but these VLP-expressing organisms have not been tested in human volunteers [184, 186]. VLPs have also been expressed in tomatoes [187], potatoes [188, 189], and tobacco [189]. All three NV VLP-expressing plants induced NV-specific antibodies when fed to mice [187, 189]. In CD1 mice, inoculation with tobacco expressing NV VLPs induced higher levels of NV-specific serum IgG and fecal IgA than inoculation with potato expressing VLPs [189]. Additionally, tomatoes expressing NV VLP still retained their ability to induce NV-specific antibodies after they were freeze-dried [187]. Of the three plants, only potatoes expressing NV VLP have been tested in human volunteers. Feeding of raw potatoes expressing NV VLP to volunteers was safe and induced no symptoms other than nausea in volunteers receiving wild type potatoes or potatoes expressing NV VLP [188]. Potatoes expressing NV VLP induced NV-specific serum IgG, IgA, and IgM, fecal IgA and seroconversion. The rise in NV-specific IgG titers persisted until day 61, the latest time point measured, while NV-specific IgM levels decreased to pre-immunization levels by day 28 [188]. For additional reviews on plants that express NoV VLPs, please consult the following articles [32, 174, 175, 190].

Purified VLPs, VLPs in unicellular organisms and VLPs in plants have advantages and disadvantages. They are all easy to administer by the oral route, but purified VLPs have the highest likelihood of approval for immunization via non-oral routes. VLPs in unicellular organisms and plants may have higher acceptance by consumers, than purified VLPs, because they can be eaten and administered in foods that are commonly consumed (e.g. yogurts, beer, raw fruits and vegetables). In contrast, purified VLPs may be easier to standardize and assay for efficacy because there are less external factors that may affect the host immune system (e.g. other plant or unicellular proteins). Ultimately, there are several promising vaccine delivery vehicles available once a NoV vaccine has been shown to protect human volunteers from infection.

9.5.6 VLP Vaccines can Partially Protect from NoV Infection

Currently, there is only one study that demonstrated *partial* protection from NoV challenge after VLP immunization. This study addressed whether immunization with bovine NoV VLPs would protect Gn calves from infection with virulent bovine NoV (BoNoV) using BoNoV strain CV186-OH for both the VLP vaccine and challenge. This BoNoV is classified as GIII.2 NoV and has 45–50% amino acid sequence identity to GI NoV and 43–46% amino acid identity to GII NoV [53]. BoNoV VLPs were used to immunize 4- to 5-day-old Gn calves via the intramuscular, oral or intranasal routes in two or three doses. BoNoV VLPs were inoculated with different adjuvants including immunostimulating complexes (ISCOM), *E. coli* LT toxin (mLT, R192G) or oil (ISA50V). Immunized calves were then challenged with virulent BoNoV 20 days after immunization. As a positive control, a separate group of Gn calves were orally inoculated with one dose of virulent BoNoV (9×10^5 RT-PCR detectable units/calf) and orally re-inoculated with the same dose of virulent BoNoV 20 days after initial inoculation.

In this positive control group of Gn calves, initial inoculation with virulent BoNoV induced diarrhea and viral shedding. Protection was evident (no diarrhea or viral shedding) upon subsequent re-inoculation with virulent BoNoV, suggesting that one dose of virulent BoNoV is sufficient to induce protective immunity from BoNoV infection and disease. Examination of the immune response of these calves revealed that both BoNoV-specific serum and fecal IgA and IgG antibody responses were induced after initial inoculation. Low titers of serum BoNoV-specific IgA antibodies were detected at 10 and 20 days after initial inoculation and at 10 days post-re-inoculation (30 days after the initial inoculation). The highest fecal NoV-specific IgA antibody GMT (1,280) was detected at 10 days post inoculation. Titers were maintained through 20 days post-inoculation and did not increase post-re-inoculation. Serum BoNoV-specific IgG antibodies reached a geometric mean titer (GMT) of 400 and 2,560 at 10 and 20 days after inoculation, respectively. Fecal BoNoV-specific IgG antibody titers decreased from a GMT of 240 to 30 between 10 and 20 days after inoculation. Fecal BoNoV-specific IgG antibody titers then increased 6-fold after re-inoculation.

In contrast, Gn calves that were intranasally immunized with two or three doses of bovine VLPs administered with LT (R192G) and then were orally challenged with virulent BoNoV were *partially* protected showing a delayed onset of diarrhea and a shorter period of diarrhea and viral shedding. This immunization regime induced lower BoNoV-specific serum IgA and IgG antibody titers and no BoNoV-specific fecal IgG compared to inoculation with virulent BoNoV. These immunized calves were the only vaccinated group that developed NoV-specific fecal IgA antibody responses. Gn calves immunized intramuscularly with two doses of the BoNoV VLPs (250 μg/dose, 10 days apart) emulsified with oil adjuvant were not protected from diarrhea or infection after challenge, although they developed higher or similar serum IgA and IgG antibody titers, compared to calves that received virulent BoNoV. NoV-specific fecal IgG was also induced after the second dose of this vaccine (GMT 80). However, no NoV-specific fecal IgA was detected in these calves.

These data indicate that BoNoV-specific mucosal (fecal) IgA antibody responses are associated with protection of Gn calves against BoNoV infection and disease, but BoNoV-specific fecal and serum IgG antibody responses are not. Other factors, such as innate immunity and cellular immune responses may also contribute to protection. Similarly, in human volunteers orally challenged with NV, mucosal (salivary) IgA antibody responses were associated with protection against NV infection [117]. However, some individuals who were susceptible to NV and who did not have strong salivary IgA antibody responses were also resistant to infection, suggesting that other factors may be necessary to confer protection against NV infection.

The Gn calf study also indicated that virus-specific, short-term protective immunity exists after BoNoV infection, similar to human NoV infection [48, 112, 191].

Whether long-term protection can be induced by BoNoV infection has yet to be determined. Based on similarities shared between BoNoV and HuNoV infections and immune responses, investigation of innate and cell-mediated immune responses in Gn calves could provide additional information essential to understand immunity against NoV infections and for future vaccine development against NoV.

9.6 Conclusions

This is an exciting time for NoV immunology research. The advent of new animal models of disease and in vitro assays promises to provide new breakthroughs in our understanding of pathogenesis and immunity. Seroepidemiology studies are identifying risk factors and differences in immunity across diverse populations worldwide. The current clinical trials of NoV vaccines may lead to the development of commercially available vaccines against NoV that will reduce the substantial morbidity and mortality associated with these infections.

Acknowledgments Dr. Juan Leon was supported by a postdoctoral fellowship in immunology (Human Immunology Award) from the Irvington. Institute for Immunological Research and the Dana Foundation. The calicivirus research from Dr. Linda Saif's lab, and discussed in this chapter, was supported by grant #R01-AI49742 from the NIAID, National Institutes of Health. The authors are grateful to Mr. Owen Herzegh for manuscript assistance and Ms. Melissa Dowd for statistical assistance. We are also grateful to the Emory General Clinical Research Center (GCRC) and their NIH National Center for Research Resources grant number M01 RR00039 for their support of our Emory norovirus human challenge study.

References

1. Fankhauser, R.L., Noel, J.S., Monroe, S.S., Ando, T., and Glass, R.I. 1998. J. Infect. Dis. 178, 1571.
2. Parks, C.G., Moe, C.L., Rhodes, D., Lima, A., Barrett, L., Tseng, F., Baric, R., Talal, A., and Guerrant, R. 1999. J. Med. Virol. 58, 426.
3. Shieh, Y., Monroe, S.S., Fankhauser, R.L., Langlois, G.W., Burkhardt, I.W., and Baric, R.S. 2000. J. Infect. Dis. 181, S360.
4. Shieh, Y.C., Calci, K.R., and Baric, R.S. 1999. Appl. Environ. Microbiol. 65, 4709.
5. Rippey, S.R. 1994. Clin. Microbiol. Rev. 7, 419.
6. Green, K.Y., Belliot, G., Taylor, J.L., Valdesuso, J., Lew, J.F., Kapikian, A.Z., and Lin, F.Y. 2002. J. Infect. Dis. 185, 133.
7. Ward, J., Neill, A., McCall, B., Stafford, R., Smith, G., and Davison, R. 2000. Commun. Dis. Intell. 24, 229.
8. Zheng, D.P., Ando, T., Fankhauser, R.L., Beard, R.S., Glass, R.I., and Monroe, S.S. 2006. Virology 346, 312.
9. Becker, K.M., Moe, C.L., Southwick, K.L., and MacCormack, J.N. 2000. New Engl. J. Med. 343, 1223.
10. Berg, D.E., Kohn, M.A., Farley, T.A., and McFarland, L.M. 2000. J. Infect. Dis. 181, S381.

11. Kuritsky, J.N., Osterholm, M.T., Greenberg, H.B., Korlath, J.A., Godes, J.R., Hedberg, C.W., Forfang, J.C., Kapikian, A.Z., McCullough, J.C., and White, K.E. 1984. Ann. Intern. Med. 100, 519.
12. Long, S.M., Adak, G.K., O'Brien, S.J., and Gillespie, I.A. 2002. Commun. Dis. Public Health 5, 101.
13. Lawson, H.W., Braun, M.M., Glass, R.I., Stine, S.E., Monroe, S.S., Atrash, H.K., Lee, L.E., and Englender, S.J. 1991. Lancet 337, 1200.
14. Baron, R.C., Murphy, F.D., Greenberg, H.B., Davis, C.E., Bregman, D.J., Gary, G.W., Hughes, J.M., and Schonberger, L.B. 1982. Am. J. Epidemiol. 115, 163.
15. Marks, P.J., Vipond, I.B., Carlisle, D., Deakin, D., Fey, R.E., and Caul, E.O. 2000. Epidemiol. Infect. 124, 481.
16. Kaufman, S.S., Chatterjee, N.K., Fuschino, M.E., Magid, M.S., Gordon, R.E., Morse, D.L., Herold, B.C., LeLeiko, N.S., Tschernia, A., Florman, S.S., Gondolesi, G.E., and Fishbein, T.M. 2003. Am. J. Transplantation 3, 764.
17. Nilsson, M., Hedlund, K.O., Thorhagen, M., Larson, G., Johansen, K., Ekspong, A., and Svensson, L. 2003. J. Virol. 77, 13117.
18. Rockx, B., De Wit, M., Vennema, H., Vinje, J., De Bruin, E., Van Duynhoven, Y., and Koopmans, M. 2002. Clin. Infect. Dis. 35, 246.
19. Graham, D.Y., Jiang, X., Tanaka, T., Opekun, A.R., Madore, H.P., and Estes, M.K. 1994. J. Infect. Dis. 170, 34.
20. Patterson, T., Hutchings, P., and Palmer, S. 1993. Epidemiol. Infect. 111, 157.
21. Gallimore, C.I., Lewis, D., Taylor, C., Cant, A., Gennery, A., and Gray, J.J. 2004. J. Clin. Virol. 30, 196.
22. Kaufman, S.S., Chatterjee, N.K., Fuschino, M.E., Morse, D.L., Morotti, R.A., Magid, M.S., Gondolesi, G.E., Florman, S.S., and Fishbein, T.M. 2005. J. Pediatr. Gastroenterol. Nutr. 40, 328.
23. Morotti, R.A., Kaufman, S.S., Fishbein, T.M., Chatterjee, N.K., Fuschino, M.E., Morse, D.L., and Magid, M.S. 2004. Hum. Pathol. 35, 1236.
24. Glass, R.I., Noel, J., Ando, T., Fankhauser, R., Belliot, G., Mounts, A., Parashar, U.D., Bresee, J.S., and Monroe, S.S. 2000. J. Infect. Dis. 181, S254.
25. Pang, X.L., Honma, S., Nakata, S., and Vesikari, T. 2000. J. Infect. Dis. 181, S288.
26. Koopmans, M., Vinje, J., de Wit, M., Leenen, I., van Der Poel, W., and van Duynhoven, Y. 2000. J. Infect. Dis. 181, S262.
27. Mead, P.S., Slutsker, L., Dietz, V., McCaig, L.F., Bresee, J.S., Shapiro, C., Griffin, P.M., and Tauxe, R.V. 1999. Emerging Infect. Dis. 5, 607.
28. Centers for Disease Control and Prevention 2004, posting date. Foodborne Outbreak Response and Surveillance Unit. Centers for Disease Control and Prevention. [Online.]
29. Honorat, E. 2007. Epidemiology of norovirus outbreaks. MPH. Emory University, Atlanta.
30. Matsui, S.M., and Greenberg, H.B. 2000. J. Infect. Dis. 181, S331.
31. Hutson, A.M., Atmar, R.L., and Estes, M.K. 2004. Trends Microbiol. 12, 279.
32. Estes, M.K., Ball, J.M., Guerrero, R.A., Opekun, A.R., Gilger, M.A., Pacheco, S.S., and Graham, D.Y. 2000. J. Infect. Dis. 181 Suppl 2, S367.
33. Smit, T.K., Bos, P., Peenze, I., Jiang, X., Estes, M.K., and Steele, A.D. 1999. J. Med. Virol. 59, 227.
34. Talal, A.H., Moe, C.L., Lima, A.A., Weigle, K.A., Barrett, L., Bangdiwala, S.I., Estes, M.K., and Guerrant, R.L. 2000. J. Med. Virol. 61, 117.
35. Peasey, A.E., Ruiz-Palacios, G.M., Quigley, M., Newsholme, W., Martinez, J., Rosales, G., Jiang, X., and Blumenthal, U.J. 2004. J. Infect. Dis. 189, 2027.
36. Bok, K., Abente, E., Santos, N., Utagawa, E., Sosnovtsev, S.V., Kapikian, A.Z., and Green, K.Y. 2007. Presented at the American Society of Virology 2007, Corvallis, Oregon.
37. Blacklow, N.R., Dolin, R., Fedson, D.S., DuPont, H., Northrup, R.S., Hornick, R.B., and Chanock, R.M. 1972. Ann Intern Med 76, 993.

38. Kapikian, A.Z., Estes, M.K., and Chanock, R.M. (ed.). 1996. Norwalk group of viruses., 3rd ed. Lippincott-Ravon, Philadelphia.
39. Kapikian, A.Z. 2000. J. Infect. Dis. 181, S295.
40. Cheetham, S., Souza, M., Meulia, T., Grimes, S., Han, M.G., and Saif, L.J. 2006. J. Virol. 80, 10372.
41. Cheetham, S., Souza, M., McGregor, R., Meulia, T., Wang, Q., and Saif, L.J. 2007. J. Virol. 81, 3535.
42. Souza, M., Cheetham, S.M., Azevedo, M.S., Costantini, V., and Saif, L.J. 2007. Cytokine and antibody responses in gnotobiotic pigs after infection with human norovirus genogroup II. 4 (HS66 strain). J Virol. 81 (17), 9183–92.
43. Azevedo, M.S., Yuan, L., Jeong, K.I., Gonzalez, A., Nguyen, T.V., Pouly, S., Gochnauer, M., Zhang, W., Azevedo, A., and Saif, L.J. 2005. J. Virol. 79, 5428.
44. Yuan, L., Ward, L.A., Rosen, B.I., To, T.L., and Saif, L.J. 1996. J. Virol. 70, 3075.
45. Ward, L.A., Rosen, B.I., Yuan, L., and Saif, L.J. 1996. J. Gen. Virol. 77 (Pt 7), 1431.
46. Saif, L.J., Ward, L.A., Yuan, L., Rosen, B.I., and To, T.L. 1996. Arch. Virol. Suppl 12, 153.
47. Dolin, R., Reichman, R.C., Roessner, K.D., Tralka, T.S., Schooley, R.T., Gary, W., and Morens, D. 1982. J. Infect. Dis. 146, 184.
48. Parrino, T.A., Schreiber, D.S., Trier, J.S., Kapikian, A.Z., and Blacklow, N.R. 1977. New Engl. J. Med. 297, 86.
49. Dolin, R., Levy, A.G., Wyatt, R.G., Thornhill, T.S., and Gardner, J.D. 1975. Am. J. Med. 59, 761.
50. Schreiber, D.S., Blacklow, N.R., and Trier, J.S. 1974. J. Infect. Dis. 129, 705.
51. Han, M.G., Cheon, D.S., Zhang, X., and Saif, L.J. 2006. J. Virol. 80, 12350.
52. Saif, L.J., Redman, D.R., Smith, K.L., and Theil, K.W. 1983. Infect. Immun. 41, 1118.
53. Han, M.G., Cheetham, S., Azevedo, M., Thomas, C., and Saif, L.J. 2006. Vaccine 24, 317.
54. Souza, M., Azevedo, M.S., Jung, K., Cheetham, S., and Saif, L.J. 2008. Pathogenesis and immune responses in gnotobiotic calves after infection with the genogroup II. 4–HS66 strain of human norovirus. J. Virol. 82 (4), 1777–86.
55. Smiley, J.R., Chang, K.O., Hayes, J., Vinje, J., and Saif, L.J. 2002. J. Virol. 76, 10089.
56. Subekti, D.S., Tjaniadi, P., Lesmana, M., McArdle, J., Iskandriati, D., Budiarsa, I.N., Walujo, P., Suparto, I.H., Winoto, I., Campbell, J.R., Porter, K.R., Sajuthi, D., Ansari, A.A., and Oyofo, B.A. 2002. J. Med. Virol. 66, 400.
57. Wyatt, R.G., Greenberg, H.B., Dalgard, D.W., Allen, W.P., Sly, D.L., Thornhill, T.S., Chanock, R.M., and Kapikian, A.Z. 1978. J. Med. Virol. 2, 89.
58. Rockx, B.H., Bogers, W.M., Heeney, J.L., van Amerongen, G., and Koopmans, M.P. 2005. J. Med. Virol. 75, 313.
59. Jiang, B., McClure, H.M., Fankhauser, R.L., Monroe, S.S., and Glass, R.I. 2004. J. Med. Primatol. 33, 30.
60. Karst, S.M., Wobus, C.E., Lay, M., Davidson, J., and Virgin, H.W.T. 2003. Science 299, 1575.
61. Hsu, C.C., Wobus, C.E., Steffen, E.K., Riley, L.K., and Livingston, R.S. 2005. Clin. Diagn. Lab. Immunol. 12, 1145.
62. Hsu, C.C., Riley, L.K., Wills, H.M., and Livingston, R.S. 2006. Comp. Med. 56, 247.
63. Wobus, C.E., Thackray, L.B., and Virgin, H.W.T. 2006. J. Virol. 80, 5104.
64. Wobus, C.E., Karst, S.M., Thackray, L.B., Chang, K.O., Sosnovtsev, S.V., Belliot, G., Krug, A., Mackenzie, J.M., Green, K.Y., and Virgin, H.W.T. 2004. PLoS Biol. 2, e432.
65. Straub, T.M., Honer zu Bentrup, K., Orosz-Coghlan, P., Dohnalkova, A., Mayer, B.K., Bartholomew, R.A., Valdez, C.O., Bruckner-Lea, C.J., Gerba, C.P., Abbaszadegan, M., and Nickerson, C.A. 2007. Emerging Infect. Dis. 13, 396.
66. Chan, M., Wong, Y., and Leung, W. 2007. Emerging Infect. Dis. 13, 1117.
67. Cukor, G., Blacklow, N.R., Echeverria, P., Bedigian, M.K., Puruggan, H., and Basaca-Sevilla, V. 1980. Infect. Immun. 29, 822.

68. Dimitrov, D.H., Dashti, S.A., Ball, J.M., Bishbishi, E., Alsaeid, K., Jiang, X., and Estes, M.K. 1997. J. Med. Virol. 51, 115.
69. Echeverria, P., Burke, D.S., Blacklow, N.R., Cukor, G., Charoenkul, C., and Yanggratoke, S. 1983. J. Clin. Microbiol. 17, 923.
70. Jing, Y., Qian, Y., Huo, Y., Wang, L.P., and Jiang, X. 2000. J. Med. Virol. 60, 97.
71. Parker, S.P., Cubitt, W.D., Jiang, X.J., and Estes, M.K. 1994. J. Med. Virol. 42, 146.
72. Smit, T., Steele, A., Peenze, I., Jiang, X., and Estes, M. 1997. J. Clin. Microbiol. 35, 2381.
73. Parker, S.P., Cubitt, W.D., and Jiang, X. 1995. J. Med. Virol. 46, 194.
74. Black, R.E., Greenberg, H.B., Kapikian, A.Z., Brown, K.H., and Becker, S. 1982. J. Infect. Dis. 145, 483.
75. Farkas, T., Deng, X., Ruiz-Palacios, G., Morrow, A., and Jiang, X. 2006. J. Clin. Microbiol. 44, 3674.
76. Numata, K., Nakata, S., Jiang, X., Estes, M.K., and Chiba, S. 1994. J. Clin. Microbiol. 32, 121.
77. Honma, S., Nakata, S., Numata, K., Kogawa, K., Yamashita, T., Oseto, M., Jiang, X., and Chiba, S. 1998. J. Clin. Microbiol. 36, 2481.
78. Greenberg, H.B., Valdesuso, J., Kapikian, A.Z., Chanock, R.M., Wyatt, R.G., Szmuness, W., Larrick, J., Kaplan, J., Gilman, R.H., and Sack, D.A. 1979. Infect. Immun. 26, 270.
79. Blacklow, N.R., Cukor, G., Bedigian, M.K., Echeverria, P., Greenberg, H.B., Schreiber, D.S., and Trier, J.S. 1979. J. Clin. Microbiol. 10, 903.
80. O'Ryan, M.L., Vial, P.A., Mamani, N., Jiang, X., Estes, M.K., Ferrecio, C., Lakkis, H., and Matson, D.O. 1998. Clin. Infect. Dis. 27, 789.
81. Gurwith, M., Wenman, W., Gurwith, D., Brunton, J., Feltham, S., and Greenberg, H. 1983. J. Infect. Dis. 147, 685.
82. Dai, Y.C., Nie, J., Zhang, X.F., Li, Z.F., Bai, Y., Zeng, Z.R., Yu, S.Y., Farkas, T., and Jiang, X. 2004. J. Clin. Microbiol. 42, 4615.
83. Kapikian, Z.Z., Estes, M.K., and Chanock, R.M. 1996. Norwalk group of viruses. Lippincott-Raven, Philadelphia.
84. Widdowson, M.A., Rockx, B., Schepp, R., van der Poel, W.H., Vinje, J., van Duynhoven, Y.T., and Koopmans, M.P. 2005. J. Med. Virol. 76, 119.
85. Anderson, A.D., Garrett, V.D., Sobel, J., Monroe, S.S., Fankhauser, R.L., Schwab, K.J., Bresee, J.S., Mead, P.S., Higgins, C., Campana, J., and Glass, R.I. 2001. Am. J. Epidemiol. 154, 1013.
86. Erdman, D.D., Gary, G.W., and Anderson, L.J. 1989. J. Clin. Microbiol. 27, 1417.
87. Tseng, F.C., Leon, J.S., MacCormack, J.N., Maillard, J.M., and Moe, C.L. 2007. J. Med. Virol. 79, 84.
88. Moe, C.L., Sair, A., Lindesmith, L., Estes, M.K., and Jaykus, L.A. 2004. Clin. Diagn. Lab. Immunol. 11, 1028.
89. Noel, J.S., Ando, T., Leite, J.P., Green, K.Y., Dingle, K.E., Estes, M.K., Seto, Y., Monroe, S.S., and Glass, R.I. 1997. J. Med. Virol. 53, 372.
90. Taylor, M.B., Schildhauer, C.I., Parker, S., Grabow, W.O., Xi, J., Estes, M.K., and Cubitt, W.D. 1993. J. Med. Virol. 41, 18.
91. Okada, S., Sekine, S., Ando, T., Hayashi, Y., Murao, M., Yabuuchi, K., Miki, T., and Ohashi, M. 1990. J. Clin. Microbiol. 28, 1244.
92. Lewis, D. 1991. J. Infect. 23, 220.
93. Gordon, S.M., Oshiro, L.S., Jarvis, W.R., Donenfeld, D., Ho, M.S., Taylor, F., Greenberg, H.B., Glass, R., Madore, H.P., Dolin, R., et al. 1990. Am. J. Epidemiol. 131, 702.
94. Fleissner, M.L., Herrmann, J.E., Booth, J.W., Blacklow, N.R., and Nowak, N.A. 1989. Am. J. Epidemiol. 129, 165.
95. Cubitt, W.D., Blacklow, N.R., Herrmann, J.E., Nowak, N.A., Nakata, S., and Chiba, S. 1987. J. Infect. Dis. 156, 806.
96. Erdman, D.D., Gary, G.W., and Anderson, L.J. 1989. J. Virol. Methods 24, 57.

97. Hayashi, Y., Ando, T., Utagawa, E., Sekine, S., Okada, S., Yabuuchi, K., Miki, T., and Ohashi, M. 1989. J. Clin. Microbiol. 27, 1728.
98. Cubitt, W.D., Green, K.Y., and Payment, P. 1998. J. Med. Virol. 54, 135.
99. Guest, C., Spitalny, K.C., Madore, H.P., Pray, K., Dolin, R., Herrmann, J.E., and Blacklow, N.R. 1987. Pediatrics 79, 559.
100. Rockx, B., Baric, R.S., de Grijs, I., Duizer, E., and Koopmans, M.P. 2005. J. Med. Virol. 77, 439.
101. Ko, G., Jiang, Z.D., Okhuysen, P.C., and DuPont, H.L. 2006. J. Med. Virol. 78, 825.
102. Ito, S., Takeshita, S., Nezu, A., Aihara, Y., Usuku, S., Noguchi, Y., and Yokota, S. 2006. Pediatr. Infect. Dis. J. 25, 651.
103. Centers for Disease Control and Prevention. 2002. JAMA 287, 3203.
104. Kaplan, J.E., Gary, G.W., Baron, R.C., Singh, N., Schonberger, L.B., Feldman, R., and Greenberg, H.B. 1982. Ann. Intern. Med. 96, 756.
105. Agus, S.G., Dolin, R., Wyatt, R.G., Tousimis, A.J., and Northrup, R.S. 1973. Ann. Intern. Med. 79, 18.
106. Meeroff, J.C., Schreiber, D.S., Trier, J.S., and Blacklow, N.R. 1980. Ann. Intern. Med. 92, 370.
107. Levy, A.G., Widerlite, L., Schwartz, C.J., Dolin, R., Blacklow, N.R., Gardner, J.D., Kimberg, D.V., and Trier, J.S. 1976. Gastroenterology 70, 321.
108. Schreiber, D.S., Blacklow, N.R., and Trier, J.S. 1973. New Engl. J. Med. 288, 1318.
109. Widerlite, L., Trier, J.S., Blacklow, N.R., and Schreiber, D.S. 1975. Gastroenterology 68, 425.
110. Mumphrey, S.M., Changotra, H., Moore, T.N., Heimann-Nichols, E.R., Wobus, C.E., Reilly, M.J., Moghadamfalahi, M., Shukla, D., and Karst, S.M. 2007. J. Virol. 81, 3251.
111. Johnson, P.C., Mathewson, J.J., DuPont, H.L., and Greenberg, H.B. 1990. J. Infect. Dis. 161, 18.
112. Wyatt, R.G., Dolin, R., Blacklow, N.R., DuPont, H.L., Buscho, R.F., Thornhill, T.S., Kapikian, A.Z., and Chanock, R.M. 1974. J. Infect. Dis. 129, 709.
113. Blacklow, N.R., Herrmann, J.E., and Cubitt, W.D. 1987. Ciba Found. Symp. 128, 144.
114. Tan, M., and Jiang, X. 2007. Expert Rev. Mol. Med. 9, 1.
115. Madore, H.P., Treanor, J.J., Buja, R., and Dolin, R. 1990. J. Med. Virol. 32, 96.
116. Treanor, J.J., Jiang, X., Madore, H.P., and Estes, M.K. 1993. J. Clin. Microbiol. 31, 1630.
117. Lindesmith, L., Moe, C., Marionneau, S., Ruvoen, N., Jiang, X., Lindblad, L., Stewart, P., LePendu, J., and Baric, R. 2003. Nat. Med. 9, 548.
118. Le Pendu, J., Ruvoen-Clouet, N., Kindberg, E., and Svensson, L. 2006. Semin. Immunol. 18, 375.
119. Tan, M., and Jiang, X. 2005. Trends Microbiol. 13, 285.
120. Samson, M., Libert, F., Doranz, B.J., Rucker, J., Liesnard, C., Farber, C.M., Saragosti, S., Lapoumeroulie, C., Cognaux, J., Forceille, C., Muyldermans, G., Verhofstede, C., Burtonboy, G., Georges, M., Imai, T., Rana, S., Yi, Y., Smyth, R.J., Collman, R.G., Doms, R.W., Vassart, G., and Parmentier, M. 1996. Nature 382, 722.
121. Liu, R., Paxton, W.A., Choe, S., Ceradini, D., Martin, S.R., Horuk, R., MacDonald, M.E., Stuhlmann, H., Koup, R.A., and Landau, N.R. 1996. Cell 86, 367.
122. Larsson, M.M., Rydell, G.E., Grahn, A., Rodriguez-Diaz, J., Akerlind, B., Hutson, A.M., Estes, M.K., Larson, G., and Svensson, L. 2006. J. Infect. Dis. 194, 1422.
123. Hutson, A.M., Airaud, F., LePendu, J., Estes, M.K., and Atmar, R.L. 2005. J. Med. Virol. 77, 116.
124. Thorven, M., Grahn, A., Hedlund, K.O., Johansson, H., Wahlfrid, C., Larson, G., and Svensson, L. 2005. J. Virol. 79, 15351.
125. Lindesmith, L., Moe, C., Lependu, J., Frelinger, J.A., Treanor, J., and Baric, R.S. 2005. J. Virol. 79, 2900.
126. Hutson, A.M., Atmar, R.L., Graham, D.Y., and Estes, M.K. 2002. J. Infect. Dis. 185, 1335.
127. Rockx, B.H., Vennema, H., Hoebe, C.J., Duizer, E., and Koopmans, M.P. 2005. J. Infect. Dis. 191, 749.

128. Huang, P., Farkas, T., Zhong, W., Tan, M., Thornton, S., Morrow, A.L., and Jiang, X. 2005. J. Virol. 79, 6714.
129. Harrington, P.R., Vinje, J., Moe, C.L., and Baric, R.S. 2004. J. Virol. 78, 3035.
130. Gray, J.J., Cunliffe, C., Ball, J., Graham, D.Y., Desselberger, U., and Estes, M.K. 1994. J. Clin. Microbiol. 32, 3059.
131. Tsugawa, T., Numata-Kinoshita, K., Honma, S., Nakata, S., Tatsumi, M., Sakai, Y., Natori, K., Takeda, N., Kobayashi, S., and Tsutsumi, H. 2006. J. Clin. Microbiol. 44, 177.
132. Hinkula, J., Ball, J.M., Lofgren, S., Estes, M.K., and Svensson, L. 1995. J. Med. Virol. 47, 52.
133. Gary, G.W., Anderson, L.J., Keswick, B.H., Johnson, P.C., DuPont, H.L., Stine, S.E., and Bartlett, A.V. 1987. J. Clin. Microbiol. 25, 2001.
134. Cukor, G., Nowak, N.A., and Blacklow, N.R. 1982. Infect. Immun. 37, 463.
135. Brinker, J.P., Blacklow, N.R., Estes, M.K., Moe, C.L., Schwab, K.J., and Herrmann, J.E. 1998. J. Clin. Microbiol. 36, 1064.
136. Iritani, N., Seto, T., Hattori, H., Natori, K., Takeda, N., Kubo, H., Yamano, T., Ayata, M., Ogura, H., and Seto, Y. 2007. J. Med. Virol. 79, 1187.
137. Hansman, G.S., Natori, K., Shirato-Horikoshi, H., Ogawa, S., Oka, T., Katayama, K., Tanaka, T., Miyoshi, T., Sakae, K., Kobayashi, S., Shinohara, M., Uchida, K., Sakurai, N., Shinozaki, K., Okada, M., Seto, Y., Kamata, K., Nagata, N., Tanaka, K., Miyamura, T., and Takeda, N. 2006. J. Gen. Virol. 87, 909.
138. Chakravarty, S., Hutson, A.M., Estes, M.K., and Prasad, B.V. 2005. J. Virol. 79, 554.
139. Hardy, M.E., Tanaka, T.N., Kitamoto, N., White, L.J., Ball, J.M., Jiang, X., and Estes, M.K. 1996. Virology 217, 252.
140. Lochridge, V.P., Jutila, K.L., Graff, J.W., and Hardy, M.E. 2005. J. Virol. 86, 2799.
141. Chen, R., Neill, J.D., Estes, M.K., and Prasad, B.V. 2006. Proc Natl Acad Sci USA 103, 8048.
142. Chen, R., Neill, J.D., Noel, J.S., Hutson, A.M., Glass, R.I., Estes, M.K., and Prasad, B.V. 2004. J. Virol. 78, 6469.
143. Prasad, B.V., Hardy, M.E., Dokland, T., Bella, J., Rossmann, M.G., and Estes, M.K. 1999. Science 286, 287.
144. Batten, C.A., Clarke, I.N., Kempster, S.L., Oliver, S.L., Bridger, J.C., and Lambden, P.R. 2006. Virology 356, 179.
145. Oliver, S.L., Batten, C.A., Deng, Y., Elschner, M., Otto, P., Charpilienne, A., Clarke, I.N., Bridger, J.C., and Lambden, P.R. 2006. J. Clin. Microbiol. 44, 992.
146. Yoda, T., Terano, Y., Suzuki, Y., Yamazaki, K., Oishi, I., Kuzuguchi, T., Kawamoto, H., Utagawa, E., Takino, K., Oda, H., and Shibata, T. 2001. BMC Microbiol. 1, 24.
147. Hale, A.D., Tanaka, T.N., Kitamoto, N., Ciarlet, M., Jiang, X., Takeda, N., Brown, D.W., and Estes, M.K. 2000. J. Clin. Microbiol. 38, 1656.
148. Tohya, Y., Masuoka, K., Takahashi, E., and Mikami, T. 1991. Arch. Virol. 117, 173.
149. Thouvenin, E., Laurent, S., Madelaine, M.F., Rasschaert, D., Vautherot, J.F., and Hewat, E.A. 1997. J. Mol. Biol. 270, 238.
150. Matsuura, Y., Tohya, Y., Mochizuki, M., Takase, K., and Sugimura, T. 2001. J. Gen. Virol. 82, 1695.
151. Hutson, A.M., Atmar, R.L., Marcus, D.M., and Estes, M.K. 2003. J. Virol. 77, 405.
152. Guo, M., Chang, K.O., Hardy, M.E., Zhang, Q., Parwani, A.V., and Saif, L.J. 1999. J Virol 73, 9625.
153. White, L.J., Ball, J.M., Hardy, M.E., Tanaka, T.N., Kitamoto, N., and Estes, M.K. 1996. J. Virol. 70, 6589.
154. Ryder, R.W., Singh, N., Reeves, W.C., Kapikian, A.Z., Greenberg, H.B., and Sack, R.B. 1985. J. Infect. Dis. 151, 99.
155. Lew, J.F., Valdesuso, J., Vesikari, T., Kapikian, A.Z., Jiang, X., Estes, M.K., and Green, K.Y. 1994. J. Infect. Dis. 169, 1364.
156. Nakata, S., Chiba, S., Terashima, H., Yokoyama, T., and Nakao, T. 1985. J. Infect. Dis. 152, 274.

157. Okhuysen, P.C., Jiang, X., Ye, L., Johnson, P.C., and Estes, M.K. 1995. J. Infect. Dis. 171, 566.
158. Greenberg, H.B., Wyatt, R.G., Valdesuso, J., Kalica, A.R., London, W.T., Chanock, R.M., and Kapikian, A.Z. 1978. J. Med. Virol. 2, 97.
159. Madore, H.P., Treanor, J.J., Pray, K.A., and Dolin, R. 1986. J. Clin. Microbiol. 24, 456.
160. LoBue, A.D., Lindesmith, L., Yount, B., Harrington, P.R., Thompson, J.M., Johnston, R.E., Moe, C.L., and Baric, R.S. 2006. Vaccine 24, 5220.
161. Harrington, P.R., Lindesmith, L., Yount, B., Moe, C.L., and Baric, R.S. 2002. J. Virol. 76, 12335.
162. Dolin, R., Roessner, K.D., Treanor, J.J., Reichman, R.C., Phillips, M., and Madore, H.P. 1986. J. Med. Virol. 19, 11.
163. Treanor, J.J., Madore, H.P., and Dolin, R. 1988. J. Virol. Methods 22, 207.
164. Thackray, L.B., Wobus, C.E., Chachu, K.A., Liu, B., Alegre, E.R., Henderson, K.S., Kelley, S.T., and Virgin, H.W. 2007. 4th Murine noroviruses comprising a single genogroup exhibit biological diversity deposite limited sequence divergence. J Virol. 81 (19), 10460–73.
165. Dolin, R., and Baron, S. 1975. Proc. Soc. Exp. Biol. Med. 150, 337.
166. Ward, J.M., Wobus, C.E., Thackray, L.B., Erexson, C.R., Faucette, L.J., Belliot, G., Barron, E.L., Sosnovtsev, S.V., and Green, K.Y. 2006. Toxicol. Pathol. 34, 708.
167. Makita, K., Hayakawa, Y., Okame, M., Homma, K., Phan, T.G., Okitsu, S., and Ushijima, H. 2007. Clin Lab 53, 125.
168. Jiang, X., Huang, P., Zhong, W., Morrow, A.L., Ruiz-Palacios, G.M., and Pickering, L.K. 2004. Adv. Exp. Med. Biol. 554, 447.
169. Ruvoen-Clouet, N., Mas, E., Marionneau, S., Guillon, P., Lombardo, D., and Le Pendu, J. 2006. Biochem. J. 393, 627.
170. Fewtrell, L., Kaufmann, R.B., Kay, D., Enanoria, W., Haller, L., and Colford, J.M., Jr. 2005. Lancet Infect. Dis. 5, 42.
171. Esrey, S.A., Potash, J.B., Roberts, L., and Shiff, C. 1991. Bull. W.H.O. 69, 609.
172. Estes, M.K., Ball, J.M., Crawford, S.E., O'Neal, C., Opekun, A.A., Graham, D.Y., and Conner, M.E. 1997. Adv. Exp. Med. Biol. 412, 387.
173. Ball, J.M., Estes, M.K., Hardy, M.E., Conner, M.E., Opekun, A.R., and Graham, D.Y. 1996. Arch Virol Suppl 12, 243.
174. Santi, L., Huang, Z., and Mason, H. 2006. Methods 40, 66.
175. Tacket, C.O. 2005. Vaccine 23, 1866.
176. Ball, J.M., Hardy, M.E., Atmar, R.L., Conner, M.E., and Estes, M.K. 1998. J. Virol. 72, 1345.
177. Periwal, S.B., Kourie, K.R., Ramachandaran, N., Blakeney, S.J., DeBruin, S., Zhu, D., Zamb, T.J., Smith, L., Udem, S., Eldridge, J.H., Shroff, K.E., and Reilly, P.A. 2003. Vaccine 21, 376.
178. Guerrero, R.A., Ball, J.M., Krater, S.S., Pacheco, S.E., Clements, J.D., and Estes, M.K. 2001. J. Virol. 75, 9713.
179. Harrington, P.R., Yount, B., Johnston, R.E., Davis, N., Moe, C., and Baric, R.S. 2002. J. Virol. 76, 730.
180. Tacket, C.O., Sztein, M.B., Losonsky, G.A., Wasserman, S.S., and Estes, M.K. 2003. Clin. Immunol. 108, 241.
181. Ball, J.M., Graham, D.Y., Opekun, A.R., Gilger, M.A., Guerrero, R.A., and Estes, M.K. 1999. Gastroenterology 117, 40.
182. Faria, A.M., and Weiner, H.L. 2005. Immunol. Rev. 206, 232.
183. Nicollier-Jamot, B., Ogier, A., Piroth, L., Pothier, P., and Kohli, E. 2004. Vaccine 22, 1079.
184. Xia, M., Farkas, T., and Jiang, X. 2007. J. Med. Virol. 79, 74.
185. Tanaka, T., Kitamoto, N., Jiang, X., and Estes, M.K. 2006. Microbiol. Immunol. 50, 883.
186. Martin, M.C., Fernandez, M., Martin-Alonso, J.M., Parra, F., Boga, J.A., and Alvarez, M.A. 2004. FEMS Microbiol. Lett. 237, 385.
187. Huang, Z., Elkin, G., Maloney, B.J., Beuhner, N., Arntzen, C.J., Thanavala, Y., and Mason, H.S. 2005. Vaccine 23, 1851.

188. Tacket, C.O., Mason, H.S., Losonsky, G., Estes, M.K., Levine, M.M., and Arntzen, C.J. 2000. J. Infect. Dis. 182, 302.
189. Mason, H.S., Ball, J.M., Shi, J.J., Jiang, X., Estes, M.K., and Arntzen, C.J. 1996. Proc. Natl. Acad. Sci. USA 93, 5335.
190. Tacket, C.O. 2004. Expert Opin. Biol. Ther. 4, 719.
191. Dolin, R., Blacklow, N.R., DuPont, H., Buscho, R.F., Wyatt, R.G., Kasel, J.A., Hornick, R., and Chanock, R.M. 1972. Proc. Soc. Exp. Biol. Med. 140, 578.
192. Nicollier-Jamot, B., Pico, V., Pothier, P., and Kohli, E. 2003. J. Clin. Microbiol. 41, 3901.

Chapter 10
Intestinal and Systemic Immunity to Rotavirus in Animal Models and Humans

Ana María González, Marli S.P. Azevedo and Linda J. Saif

Abstract Rotavirus (RV) is an enteric virus causing severe diarrhea in young infants and children worldwide. After RV infection, mainly enteric, but also systemic immune responses are elicited. Recently, extraintestinal RV antigen was observed in various animal models and humans, explaining the induction of systemic immunity after infection. This chapter will summarize the B and T cell responses and innate immunity to RV after infection in animal models and humans. Additionally, current human rotavirus (HRV) vaccines and correlates of protection after vaccination will be discussed.

10.1 Introduction

Rotaviruses (RV) are non-enveloped, icosahedral, double-stranded RNA viruses that belong to the *Reoviridae* family. The virus capsid is composed of three concentric protein layers: a core composed of VP1, VP2, and VP3, an inner layer composed of VP6 and an outermost layer composed of VP4 and VP7 that contain the viral neutralizing antigens. Based on VP6 structural characteristics, RV are classified in to groups A to G [59]. Humans are mainly infected by group A RV (mainly children under the age of 5), which also infects many newborn animals including pigs [18, 159, 160], mice [163], rats [53] and calves [37]. Group A human RV (HRV) induces diarrhea and/or shedding in newborn animals such as gnotobiotic pigs [193], calves [146], lambs [156], newborn cynomolgus monkeys [109], sub-adult macaques [79], puppies [161] and mice [17].

Mature enterocytes in the small intestine are the main target of infectious RV. Entry of RV into the cells is a multistep process involving the two outermost capsid proteins VP7 and VP4 and various potential cell receptors such as integrins ($\alpha2\beta1$, $\alpha v\beta3$, $\alpha x\beta2$) and the heat shock protein 70 [105]. The mechanism of RV extraintestinal dissemination it is not completely understood. Children with and

---

L.J. Saif
Food Animal Health Research Program, Ohio Agricultural Research and Development Center, 1680 Madison Ave, The Ohio State University, Wooster, Ohio 44691, USA
e-mail: saif.2@osu.edu

M. Vajdy (ed.), *Immunity Against Mucosal Pathogens*,

without diarrhea developed RV antigenemia suggesting that enterocyte destruction may not be necessary for systemic dissemination to occur [24].

Transmission of RV is mainly fecal-oral but aerosol dissemination possibly via vomitus has also been reported [68]. Diarrhea induced by RV is multifactorial. Osmotic diarrhea can be induced by malabsorption secondary to enterocyte disruption [54] and reduction of mature absorptive enterocytes [132]. Furthermore secretory diarrhea is also observed in neonatal mice due to the production of the viral enterotoxin NSP4 [8] and the stimulation of the enteric nervous system [106]. Consequently severe diarrhea and secondary dehydration are often described in RV infected children [68].

The neonatal gnotobiotic pig model is one of the most useful models to study homologous and heterologous RV [e.g. HRV] infection and disease because of its prolonged susceptibility to RV induced diarrhea (at least 8 weeks) and lack of maternal antibodies transferred transplacentally. Their gnotobiotic status assures an environment free of wild type (WT) RV or other extraneous infections [148]. Furthermore, and most importantly, pigs are optimal models of HRV infection because of their similarities to humans in gastrointestinal physiology, their immune system and their susceptibility to HRV diarrhea [145, 147, 193].

The mouse model has been widely used to study RV infection. The major advantage of the mouse model is the availability of "knockouts" that can be genetically targeted to create immunodeficient mice that lack a specific population of immune cells, or molecules involved in immune functions. Gene targeted immunodeficient mice allow studies of the role of RV-specific B and T cells in protection after primary or secondary RV infection. Also, antibodies to mouse CD surface molecules are widely available permitting detailed characterization of RV specific B and T cells as well as innate immune responses. Important disadvantages of the mouse model include the distinct differences in the immune system compared to that of humans, e.g. unlike humans they acquire long-term, if not life-long immunity to RV [95] and the commonly used adult mouse model is not susceptible to RV diarrhea.

10.2 Studies of RV Immunity in the Mouse Model

10.2.1 Study of Acquired RV Immunity in the Mouse Model

The mouse model has been widely used to characterize B and T cell responses after inoculation with virulent homologous or heterologous RV. Adult mice inoculated by gastric gavage with homologous RV develop a peak of intestinal RV antigen at post-inoculation day (PID) 1-2 whereas neonatal mice present two peaks at PID2 and PID4. However both adult and neonatal mice shed RV antigen for 4–6 days post-inoculation (PI) [32]. The kinetics of RV antigen and infectious RV shedding are usually similar [163]. Furthermore RV antigen is associated with the induction of intestinal lesions [157] suggesting RV antigen is an adequate measure of infectious RV. Intestinal B lymphocytes are mainly located in lymphoid follicles of the organized gut associated lymphoid tissues (oGALT) represented mainly by the Peyer's patches (PP), appendix and the isolated lymphoid follicles (ILF) located throughout the small intestinal wall, as well as within the intestinal lamina propria (LP) [28]. Primary lymphoid follicles located within the oGALT structures are rich in naïve B cells (IgD^+IgM^+). After antigen encounter, germinal centers (GC) are generated and primary follicles are transformed to secondary follicles [28]. The B cells can be intrafollicular or extrafollicular. Intrafollicular B cells typically are T-cell dependent, they undergo somatic hypermutation and they generate memory B cells. Extrafollicular B cells, on the other hand, develop low level immunoglobulin hypermutation, and can be either T cell-dependent or independent and they undergo *in-situ* apoptosis early after activation [108]. Characterization of B cell responses to RV was performed in adult C57BL/6 mice after primary homologous RV (EC strain, G3P[16]) infection [186]. Activated B cells that secrete antibodies or plasma cells were studied by the ELISPOT assay and RV-specific antibody secreting cell (ASC) responses were characterized. Additionally, various surface markers expressed by RV-specific B cells were studied by flow cytometry. The RV-specific IgM ASC predominated in the PP, the intestinal LP and the mesenteric lymph nodes (MLN) and were first detected at PID 4 (soon after the peak of intestinal RV antigen), peaked at PID 4–14 and were low or absent at PID 28. The RV-specific IgA ASC were present in all tissues studied (PP, MLN, intestinal LP and spleen). They were first detected at PID 7, peaked at PID 14–28 in the MLN, intestinal LP and spleen, but continued to increase in the PP and the bone marrow (BM) up to at least 9 months PI. The IgG ASC predominated in the PP, intestinal LP, MLN and spleen, peaking at PID 14–28 and diminished in all tissues except the BM at 9 months PI.

Lymphocyte trafficking to the GALT is mediated by the interaction of the integrin $\alpha4\beta7$ expressed on intestinally primed lymphocytes and its ligand the mucosal addressin cell adhesion molecule-1 (MadCAM-1) [184]. Large and small IgD^-B220^{low} B cells were first observed in PP and MLN at PID 4. These cells are characteristically extrafollicular cells [class switched non-GC B cells] and have been used to identify antigen-specific ASC that first appear following immunization [113, 155]. Large extrafollicular cells peaked in the PP and MLN at PID 7 and 20% and > 50% of the large extrafollicular cells in the PP and MLN, respectively, expressed the intestinal homing marker $\alpha4\beta7$. By PID 14, > 50% of the large extrafollicular cells were $\alpha4\beta7^+$ in the PP and MLN, and by 4 months, these cells diminished to baseline levels in both tissues. Most RV-specific large extrafollicular cells secreted antibodies. However, the kinetics of large extrafollicular cells differed from that of RV-specific IgA ASC located in the PP. The RV-specific IgA ASC were observed in the PP at 9 months PI, whereas large extrafollicular cells were absent in PP after 4 months, suggesting that long-lived IgA ASC in the PP and the large extrafollicular cells are different cell populations. Characteristic GC RV-specific B cells (large IgD^-B220^{high}) peaked at PID 14 mainly in the MLN and the PP (but were also present in the spleen) and persisted up to 4 months. Memory cells (small IgD^-B220^{high} cells) [184] peaked at PID 14 in PP, MLN and spleen, but they remained high in the PP (up to 9 months) whereas they diminished in MLN and spleen.

The kinetics of memory B cells in the PP corresponded to the kinetics of RV-specific IgA ASC in the PP suggesting that both populations could be overlapping. Possibly, RV-specific IgA ASC detected in the PP were mainly long-lived "memory" plasma cells. The fact that IgG and IgA ASC were also detected in the BM at 9 months PI supports this possibility since various studies have described that the main location of long-lived plasma cells is in the BM of mice after systemic infection and antigen inoculation (OVA) [114, 152]. Also, at 9 months PI, $> 60\%$ of RV-specific memory B cells (small, IgD^-B220^{high} cells) in the PP were $\alpha4\beta7^+$.

In summary, the first RV specific B cells to be activated after inoculation were IgM ASC and large extrafollicular $B220^{low}$ cells in the MLN (also observed in lower numbers in the PP and the spleen). These cells returned to baseline levels at PID 28 and 4 months PI, respectively. Large and small GC IgD^- $B220^{high}$ B cells peaked at day 14 in the MLN and PP, but large cells disappeared by 4 months PI. On the other hand, memory B cells (small GC B cells) and RV-specific IgA ASC (possibly long-lived plasma cells) remained elevated in the PP for at least 9 months. It is known that protection against reovirus reinfection is dependent on intestinal IgA which prevents viral entry into the PP [81]. However to prevent RV reinfection, long-lived memory IgA ASC possibly play a role but this requires further investigation.

Responses by $CD4^+$ T cells are classified into Th1 and Th2 types: IFNγ, and IL-2 cytokine production characterize Th1 responses, and IL-12 is a Th1 inducer whereas Th2 responses are typically characterized by IL-4, IL-5 and IL-13. The Th1 responses are important for induction of cytotoxic cellular immunity whereas Th2 responses are important for humoral immunity, particularly in immunity against parasites [70]. After RV infection, T cells were characterized in adult mice and pups [84] and responses were studied in MLN, intestinal [PP and intraepithelial lymphocytes (IEL)], and systemic tissues (spleen, liver), lung and blood. Inoculation of adult C57BL/6 mice with homologous RV (EC) induced Th1 IFNγ secretion by both $CD4^+$ and $CD8^+$ T cells. The $CD8^+$ IFNγ secreting T cell frequencies were significantly higher than those of $CD4^+$ T cells and peaked at 5–7 days in the gut and the liver followed by a rapid decline. Responses to $VP6_{357-366}$ and $VP7_{33-42}$ peptides were similar and located mainly in the gut and liver, with lower frequencies detected in the MLN and spleen. Neonatal C57BL/6 mice inoculated with a homologous murine RV differed from adults by developing two peaks of hepatic $CD8^+$ T cell responses to VP6 at PID 7 and at PID 14. In comparison, adults had only one peak at PID 5–7 in all tissues inspected. These findings agree with previous observations demonstrating a biphasic RV shedding pattern in mouse pups and a monophasic pattern in adult mice [32]. These observations suggest that fecal RV antigen shedding (representative of intestinal replicating virus) is associated with subsequent induction of RV-specific T cells in intestinal, respiratory and systemic organs, with the latter possibly due to concomitant extraintestinal viral dissemination. The frequencies of the $CD8^+$ T cell responses to VP6 were similar between adults and pups (except at PID 14 when pups had the second peak). However, for unexplained reasons $CD8^+$ T cell responses to VP7 were significantly higher in adults in the liver and spleen compared to those in the neonatal mice.

Interestingly, adult mice infected with homologous (EC) or heterologous rhesus RV (RRV) had distinct responses. At PID 5–7 homologous RV elicited VP6 and VP7 specific T cell responses in the gut and liver whereas responses to heterologous RV were predominantly against VP7 in the gut, liver and spleen. The heterologous strain RRV has been associated with the development of biliary atresia in neonatal mice which is an obstruction of the extrahepatic and intrahepatic bile ducts due to chronic inflammation and fibrosis [153]. Furthermore RRV causes viremia in adult mice [23]. These observations may explain in part the marked systemic responses detected after RRV infection. However, the cause of the predominant VP7 response observed after RRV infection is unknown.

Two different types of memory T cells have been described: effector memory and central memory T cells. Effector memory T cells are rapidly activated after antigen encounter and express low to no L-selectin. Central memory T cells express L-selectin that facilitates homing to secondary lymph nodes, and they lack immediate effector function after antigen encounter [149]. The phenotype of RV-specific $CD8^+$ T cells was that of an effector memory T cell at PID 7 ($CD27^{int}$, $CD44^{high}$, L-selectin$^-$) and independent of the tissue of origin, they expressed $\alpha4\beta7$. These observations agree with previous studies that describe effector memory $CD8^+$ T cell migration to different nonlymphoid tissues regardless of the original site of activation [110].

In summary, both $CD4^+$ and $CD8^+$ T cells are activated by 5–7 days after RV infection and they secrete the Th1 cytokine IFNγ. The peak for $CD8^+$ T cell frequencies varied between mouse pups and adults and generally $CD8^+$ T cell responses were higher than $CD4^+$ T cell responses. The RV-specific $CD8^+$ T cells were widely distributed throughout the body shortly (PID 2) after infection. These cells were characteristically effector memory $CD8^+$ T cells and expressed $\alpha4\beta7$, suggesting that they were primed in the gut and not in systemic tissues. Studies of knockout mice are more descriptive to define the role of each lymphocyte subset in protection against RV reinfection and will be addressed in more detail in the following section.

10.2.2 Studies of RV Immunity in Knockout Mice

Knockout mice are a useful tool to delineate the role of a specific arm of immunity in infection induced by certain microorganisms. The role of B and T cells in protective immunity against RV shedding was studied in the adult mouse model using J_HD (which lack B cells) and β2-microglobulin (lack CD8 T cells) knockout mice, respectively [65]. The J_HD knockout mice developed similar virus shedding patterns to WT mice and cleared the virus after primary infection, but after a secondary challenge, the J_HD knockout mice were re-infected whereas WT mice were protected against reinfection. On the other hand, β2-microglobulin knockout mice had a 2-day delay in viral clearance after primary infection compared to WT mice. However, β2-microglobulin knockout mice were resistant or almost resistant to reinfection after a secondary infection. It was also observed that J_HD knockout mice depleted of $CD8^+$ T cells by CD8-specific monoclonal antibodies, developed chronic virus shedding. From these studies it was concluded that B cells are necessary for complete long-term protection and that $CD8^+$ T cells together with B cells play a complementary role in clearance after primary infection. Also, $CD8^+$ T cells induce timely resolution of primary RV infection and partial short-term protection against reinfection [65]. To further investigate the mechanism whereby $CD8^+$ T cells conferred timely partial short-term protection, perforin, fas, type I IFN receptor, stat1 and IFNγ knockout mice were infected with RV, but lack of these effector molecules did not affect RV clearance [2, 65, 67, 97, 162].

To delineate the role of antibodies in protective immunity to RV, IgA knockout mice in a C57BL/6 background (lacking IgA in serum and the intestine) challenged with homologous RV, cleared RV after primary infection and were protected from RV reinfection similarly to WT mice [125]. On the other hand, mice lacking intestinal IgA in a BALB/c background (J-chain -/-) and challenged with homologous RV had a 2-day delay in viral clearance compared to WT mice [151]. The IgA knockout mouse study suggests that possibly intestinal IgA can be substituted by other isotypes (IgM and IgG) and the study on the J-chain-/- mice suggests that other immune mechanisms (T cells, innate cells) may confer protection against RV infection. Also, it is known that protection against RV infection is mediated by T dependent B cells (B2) and not B1 cells according to studies of mice with severe combined immunodeficiency (SCID) that develop chronic RV shedding after inoculation and the observation that adoptive transfer of total B cells, but not purified B1 cells cleared RV shedding [101].

The role of $CD4^+$ T cells was determined in WT adult BALB/c mice [115] and C.B-17 SCID (T and B cell deficient) mice [101]. With the aid of monoclonal antibodies against CD4, infected mice were depleted of $CD4^+$ T cells and a chronic low level of RV antigen shedding was described. When CD4 depleted mice were further depleted of $CD8^+$ T cells, the amount of RV antigen shed increased. The CD4 depleted mice did not develop RV-specific antibodies providing further evidence of the T cell dependent B2 cell type nature of RV-specific B cell responses. In this study, it was concluded that CD4, CD8 and B cells each play a role in protection. However, depletion of $CD4^+$ T cells in C57BL/6 mice did not affect RV clearance and it has been suggested that this effect is strain dependent and secondary to an uncharacterized innate immune mechanism [66]. Additionally, the individual role of $CD4^+$ T cells was later confirmed in studies of SCID mice adoptively transferred with RV-specific $CD4^+$ T cells, which did not produce detectable intestinal RV specific IgA but were able to resolve RV shedding. This study demonstrated the direct role of $CD4^+$ T cells in the induction of protection against RV infection [101].

To determine if intestinally committed B cells are necessary to confer protection against RV infection, RAG-2 knockout mice (lacking B and T cells) were infected with RV [184]. It is known that lymphocyte trafficking to the gut is not only controlled by the expression of the integrin $\alpha 4\beta 7$ and its ligand MadCAM-1, but also the chemokine receptor CCR9 and its ligand CCL25 (TECK) and the chemokine receptor CCR10 and its ligand CCL28 (MEC) [99, 100, 133, 184]. The chemokine CCL25 (TECK) is widely expressed on crypt epithelial and endothelial cells of the small intestine [27]. On the other hand, the chemokine CCL28 is expressed in the colon, but also in other tissues such as tonsils, salivary glands, mammary glands, trachea, small intestine and appendix [133]. Only mice that received passive transfer of $\alpha 4\beta 7^{+high}$ IgD^{-} (memory) B cells produced intestinal RV-specific IgA, cleared RV and were protected from RV reinfection. On the other hand, transfer of $\alpha 4\beta 7^{-}$ IgD^{-} or $\alpha 4\beta 7^{-}$ IgD^{+} did not protect the mice and virus shedding persisted for at least 60 days after infection. Furthermore, blocking of both CCL25 and CCL28 with antibodies prevented RV specific B cells from localizing in the LP [63]. When either CCL25 or CCL28 were blocked, entry of RV specific cells to LP was still observed suggesting a redundant role of both chemokines for intestinal lymphocyte homing. Also, RV infected CCR9 knockout mice had significant accumulation of RV-specific IgA^{+} plasmablasts in the LP and only when CCL28 was also blocked with antibodies, was accumulation of IgA^{+} plasmablasts prevented. These studies suggest that the localization of B lymphocytes to the LP is a crucial event for the induction of RV immunity and possibly the subsequent protection from reinfection. Also, the phenotype of RV specific B cells is characterized by the expression of $\alpha 4\beta 7$, CCR9 and/or CCR10.

On the other hand, expression of homing markers on the surface of RV specific T cells is not necessary for the cells to confer protective immunity. Mice deficient in $\beta 7$ integrin and devoid of $\alpha 4\beta 7$ expression on the surface of T lymphocytes, clear RV with the same kinetics as WT mice. Furthermore, by transferring B7-/- $CD8^{+}$ T cells into chronically infected Rag-2 mice, RV infection is efficiently resolved. These studies suggest that the expression of $\alpha 4\beta 7$ on $CD8^{+}$ T cells is not essential for their role in partial short term protection against RV reinfection [98]. However, the expression of $\alpha 4\beta 7$ on $CD8^{+}$ T cells depends on the site of RV inoculation. Oral inoculation induced mainly RV specific $CD8^{+}\alpha 4\beta 7^{high}$ T cells, but subcutaneous inoculation induced $CD8^{+}\alpha 4\beta 7^{high}$ and $CD8^{+}\alpha 4\beta 7^{low}$ [98]. Primary activated $CD8^{+}$ T cells and most memory $CD8^{+}$ T cells, regardless of the site of antigen priming, can migrate to many uninfected tissues [110]. The mechanism of $CD8^{+}$ T cell "body patrol" is unclear but it is thought to be an important event in the control and elimination of viral infections. Similarly, $CD4^{+}$ T cells develop a specific phenotype depending on the tissue of origin. The $CD4^{+}$ T cells activated in the skin express L-selectin and downregulate $\alpha 4\beta 7$, whereas $CD4^{+}$ T cells activated in the gut, downregulate L-selectin, upregulate $\alpha 4\beta 7$ and are highly reactive to TECK (CCL25), suggesting their expression of CCR9 [34]. Small intestinal LP $CD4^{+}$ T cells expressing $\alpha 4\beta 7$ were mostly $CCR9^{+}$ but in studies of CCR9 knockout mice, a population of $CD4^{+}$ T cells expressing $\alpha 4\beta 7$ was observed in the gut suggesting a CCR9 independent mechanism for $CD4^{+}$ T cell intestinal homing, probably by the expression of CCR10 [63, 158]. *In-vitro* studies showed that $CD4^{+}\alpha 4\beta 7^{high}$ T cells are much more reactive to RV than $CD4^{+}\alpha 4\beta 7^{-}$ T cells suggesting an enrichment of RV specific $CD4^{+}$ T cells in the $\alpha 4\beta 7^{high}$ population [144]. No transfer studies of RV-specific $CD4^{+}$ T cells expressing $\alpha 4\beta 7^{high}$ or $\alpha 4\beta 7^{low}$ have been performed, so it is unclear if gut homing marker expression is necessary for RV specific $CD4^+$ T cells to localize in the gut and induce protection from RV reinfection.

In summary, RV-specific B and $CD4^+$ T cells in adult mice play an important role in complete long-term protection against RV. Both subsets show selective intestinal compartmentalization by the expression of $\alpha4\beta7$, CCR9 and CCR10. Intestinal B cell localization (not yet studied for $CD4^+$ T cells) is pivotal for protection from RV reinfection. On the other hand, $CD8^+$ T cells expressing intestinal homing markers are dispensable for their role in partial short-term protection and independent of their expression of $\alpha4\beta7$, $CD8^+$ T cells can exert their anti-viral role throughout the body.

10.2.3 Mouse Model for the Study of Innate Immunity after RV Infection

Approximately 40% of SCID mice in a C57BL/6 background are able to clear a primary RV infection suggesting an innate immune mechanism as a possible mediator of RV clearance [66]. Furthermore, when challenged with RV, STAT knockout adult mice that lack many IFN mediated responses, shed 100 times more virus than immunocompetent mice from 2–6 days after infection, but were able to resolve the infection with similar kinetics as the infected WT mice [162]. This observation also suggests that the initial control of virus replication might be achieved by innate immune mechanisms. Intestinal antigen presenting cells (APCs) are probably important players in this initial RV control and their importance in the generation of RV immunity has been addressed in mice. Using CCR6 knockout mice, intestinal innate immunity after RV infection was studied. The chemokine receptor CCR6 is expressed on DCs and its ligand is MIP3α or LARC. The CCR6 ligand is expressed in the epithelium of the PP aiding in the migration of immature DCs to the underlying sub-epithelial dome (SED) [48]. The CCR6-/- mice had significantly less fecal RV specific IgA antibody compared to WT mice, whereas the amount of systemic RV specific IgG antibody was similar between knockout and WT mice. The CCR6 knockout mice lacked $CD11c^+CD11b^+$ DCs at the SED of the PP [48] suggesting a role for this particular type of DC in the induction of intestinal adaptive RV-specific B cell immunity (and possibly to other intestinal pathogens). However, it is likely that other DCs might also be involved in the induction of intestinal responses to RV.

Macrophages, B cells and DCs uptake and process antigens for presentation via class I or II major histocompatibility complexes (MHC), acting as APCs and inducing adaptive immune responses. The observation that adult mice inoculated intramuscularly (IM) with live RV develop intestinal B cell responses, suggests that systemic APCs presenting RV antigens, home to the gut and induce local B cell responses [46]. Injection of non-dividing cells (mitomycin C treated) or untreated cells purified from the draining lymph nodes of RV inoculated mice induced intestinal RV specific B cell responses in naïve recipient mice [45]. Furthermore, if only B cells from a specific mouse strain were transferred, intestinal IgA secreting B cells from the recipient mice, as determined by Ig allotype analysis, were induced. These studies suggest that systemic B cells acting as APCs and possibly presenting RV antigens, home to the gut and induce intestinal RV-specific responses. The RV non-structural protein NSP4 has been detected in $B220^+$ cells [marker of B cells and plasmacytoid DC(pDC)], $CD11c^+$ cells (a DC marker) and $CD11b^+$ cells (a marker of macrophages, natural killer cells, granulocytes and activated T cells) in the MLN of RV inoculated mice suggesting RV was engulfed and possibly replicated inside these cells [61]. Alternatively, these APC may have engulfed RV-infected cells in which NSP4 was expressed. A different study detected virus proteins mainly inside macrophages in the PP, MLN and inguinal lymph nodes [31]. Infectious RV, positive and negative RV RNA strands were measured by plaque-assay and quantitative real time PCR, respectively in MLN derived cells [61]. High infectious RV titers and amounts of RNA were observed suggesting RV replication inside cells located in the MLN. However, NSP4 has been suggested to be a structural protein and thus it is not only exclusively synthesized during virus replication [52]. Further studies are needed to confirm RV replication inside APCs, but evidence of RV proteins inside macrophages, B cells and DCs supports RV antigen presentation.

Early after RV infection, massive B cell activation takes place [25]. This B cell proliferation was detected in the MLN and PP at PID 1 to 6. At PID 3–4 fragmented cultures derived from MLN and PP were positive for RV specific IgM, but not IgA antibodies. Furthermore, the B cell proliferation was dependent on the presence of VP7. These findings suggest that B1 cells (T independent B cells) were stimulated soon after RV infection because lymph node hyperplasia and B cell proliferation were also observed in TCR knockout mice that lack T lymphocytes [22]. Although it is known that B1 cells alone do not resolve RV infection [101], RV is able to stimulate B1 cells soon after infection. However, the *in-vivo* relevance of the activation of B1 cells after RV infection needs to be elucidated.

Neonatal mice are particularly susceptible to biliary atresia after heterologous RRV infection [62, 97, 138]. Mouse pups deficient in type I and type II IFN receptors were infected with RRV and only IFN type I receptor deficient mice were susceptible to biliary atresia [91]. Treatment with IFNα prevented the mortality due to cholestasis (or accumulation of toxic bile by-products in the blood due to the biliary duct obstruction) suggesting that after RV infection, type I IFN secreting cells, possibly pDC, are stimulated and potentially regulate inflammatory responses. Similar virus titers were observed in the livers and brains derived from WT and IFN type I receptor knockout RV infected mice suggesting that type I IFNs are important in modulation of the induction of RV immune responses, but they do not directly control RV replication. The lack of effect of type I IFNs on RV replication was further proven when treatment of mice with oral or intraperitoneal injections of IFNα did not prevent RV replication or diarrhea in mouse pups [2]. Also, in the mouse model of biliary atresia, treatment with 10,000 IU of IFNα at 6 hrs after RV infection resulted in identical RV titers in the liver and brain [97]. Although there are reports of Th1 induction by type I IFNs [38, 96], there is also increasing evidence of type I IFNs mediating partial antagonistic effects on Th1 responses by induction of IL-10. This in turn lowers inflammatory responses by the induction of regulatory T cells [128], playing a critical role in the generation of tolerance [10, 21]. Probably, biliary atresia induced by the lack of type I IFNs is not mediated by a larger RV load but by the higher induction of RV-specific adaptive T cell immunity. Proof of this concept derives from studies of IFNγ knockout mice after being exposed to RRV. The IFNγ knockout mice developed a suppressed phenotype of biliary atresia, were able to resolve jaundice and had a prolonged long-term survival compared to wild-type mice, whereas the RV titers remained similar between knockout and WT mice [153]. The IFNγ -/- mice treated with IFNγ developed bile duct obstruction after RV infection. A different study tested the development of biliary atresia by adoptive transfer of RRV-specific T cells into naïve syngeneic SCID mice. Bile duct pathology was induced after T lymphocyte transfer in the absence of detectable infectious RV [107]. These studies suggest that adaptive immunity to RV (RV-specific IFNγ secreting T cells) mediated the biliary duct inflammation and secondary atresia and type I IFNs possibly regulate the induction of adaptive RV-specific T cells after infection.

Interestingly, mice inoculated with RV dsRNA alone or the synthetic dsRNA poly I:C orally (a known TLR3 ligand) developed severe small intestinal pathology via a TLR3 mediated mechanism [196]. Moreover, after inoculation, intestinal epithelial cells secreted elevated amounts of IL-15 induced by TLR3 signaling that secondarily increased the influx and cytotoxic activity of intraepithelial $NK1.1^{+}$ $CD3^{+}$ lymphocytes. These observations suggest that dsRNA and intraepithelial lymphocytes play a crucial role in RV pathogenesis.

In summary, early in RV infection, $CD11c^{+}CD11b^{+}$ DCs located at the SED are responsible for the generation of local acquired B and T cell immune responses to RV. Furthermore, when RV is located in systemic compartments (after natural extraintestinal dissemination or after systemic injection), different types of APCs such as B cells, DCs and macrophages uptake the virus and home from systemic tissues to the gut to induce local adaptive immunity to RV. Possibly pDC are also stimulated after RV infection *in-vivo*, but further studies are needed to confirm this. The role of type I IFN producing cells *in-vivo* may be to modulate the induction of adaptive immune T cell responses after infection, but not to prevent RV replication. Also, RV VP7 stimulates B1 cells that apparently alone do not play a crucial role in protection against RV. Furthermore, inoculation with RV dsRNA, which activates TLR3, induced small intestinal pathology and it was associated with secretion of IL-15 by intestinal epithelial cells that increases the influx and cytotoxicity of intraepithelial lymphocytes suggesting their role in RV pathogenicity. Studies characterizing the different types of APCs activated after virulent RV infection are lacking but are necessary to better understand the induction of RV-specific adaptive immune responses after infection and to improve protective immunity induced by RV vaccines.

10.3 Studies of RV Immunity in the Gnotobiotic Pig Model

10.3.1 Acquired RV-Specific Immune Responses after RV Infection of Pigs

Neonatal pigs are susceptible to porcine RV and HRV strains and are an ideal model for the study of RV immunity and testing of HRV vaccines. Gnotobiotic piglets have been widely used for the study of HRV induced immunity because of their prolonged susceptibility (at least 8 weeks of age), their similarity to humans in gastrointestinal physiology, and mucosal immunity development and lack of contamination with WT RV or other pathogens [148]. Importantly, HRV inoculated gnotobiotic pigs developed similar titers of fecal RV, diarrhea scores and B cell responses compared to HRV inoculated pigs colonized with *lactobacillus* species suggesting that lack of a commensal flora does not interfere with RV pathogenesis or its immunogenicity [195]. Studies of HRV in gnotobiotic piglets were performed with the group A RV P1AG1 that is the most common G and P serotype associated with RV gastroenteritis in infants and children worldwide. Comparative B cell immune responses after porcine [OSU P7G5, [40]] and HRV [Wa P1AG1, [193]] inoculation were initially performed to delineate the similarities and differences induced by the homologous porcine RV versus the heterologous HRV [145].

10.3.1.1 Virulent Porcine RV Versus HRV-Specific ASC Responses in Pigs

Gnotobiotic pigs inoculated with virulent porcine RV (OSU) developed virus-specific IgM ASC as early as PID3 that were first observed in the MLN. At PID7, IgM responses were predominant in the spleen and MLN but lower in the intestinal LP. Similarly, after HRV inoculation, from PID 8–13 the first responses detected were IgM ASC; however, they were predominant in the MLN and the intestinal LP [193]. After porcine RV inoculation, IgA and IgG ASC to RV were first detected at PID 7 in all tissues tested (MLN, spleen, intestine) reaching a peak at PID 14 in the MLN and spleen and at PID 21 in the gut [145]. However, after HRV inoculation, the peak of IgA and IgG ASC for all tissues tested (intestinal LP, MLN, spleen and blood) was later at PID 21 with the highest responses in the gut and MLN [193]. Nevertheless, the numbers of systemic IgA and IgG ASC and intestinal LP RV specific IgA ASC were similar after porcine or human virulent RV inoculation at PID 21, but the HRV strain induced higher IgG ASC in the gut [145]. Both viruses induced villous atrophy and diarrhea by 24–72 hours PI [40] demonstrating that pigs were susceptible to both the homologous and heterologous virus strains. In conclusion, even though there were some differences in the early B cell responses elicited by porcine vs. HRV, later after infection (PID21) both viruses induced similar numbers of intestinal IgA ASC and the HRV induced higher IgG ASC in the gut suggesting that human strains were as immunogenic as porcine strains in the piglets.

Using the ELISPOT assay, protein-specific (VP4, VP6, VP7, NSP4, NSP3) ASC were studied in gnotobiotic pigs inoculated with virulent HRV [39]. At PID28, the main responses were of intestinal IgA and against the VP6 and VP4 structural viral proteins. Low numbers of ASC against VP7 and the non-structural proteins NSP3 and NSP4 were detected. These results agree with previous studies reporting VP6 as one of the most immunogenic RV proteins in humans [47]. However, in humans after natural infection, VP7 responses have been shown to be predominant in some studies [71], but not in others [181]. Perhaps the degree and type of protein-specific responses depends on the virus strain, but this requires further investigation. Additionally, in the assays used to detect VP7 specific responses based on use of baculovirus recombinant VP7 in ELISPOT or VP7 gene reassortants in RV neutralization, the recombinant or reassortant expressed VP7 may differ from the original protein in the native RV particle and influence the VP7-specific responses obtained.

Additionally, memory B cell responses were evaluated after HRV inoculation using a memory ELISPOT assay to detect ASC after a 5-day incubation with RV antigen *in-vitro* [187]. Plasma cells had a short half-life and vanished after the 5-days of incubation whereas memory B cells activated by RV antigen added to the culture medium survived and were enumerated as ASC. At PID28 after virulent Wa HRV inoculation, the highest memory responses were IgG, but IgA memory B cell responses were also present, and both were located mainly in the ileum. However, at PID83, IgG and IgA memory B cell responses were much lower in ileum, but still present, although they then mainly resided in the spleen, and not in bone marrow (BM). Both intestinal IgG and IgA effector and short-term (PID28) intestinal memory IgA and IgG ASC responses were correlated with protection against RV reinfection. Contrary to mice that develop long-lived plasma and memory B cells in the gut and BM [186], by PID83, piglets developed no or low memory B cell responses in BM and low longer-term memory B cell responses in the gut and spleen suggesting that in piglets, RV induces a weaker memory B cell response, perhaps explaining in part their higher susceptibility to RV disease at older ages compared to mice. It is possible that by disrupting the natural environment where long-lived plasma cells reside and by the prolonged incubation time that the assay requires, these cells died and were not detected. For this reason specific markers of long-lived plasma cells and memory B cells need to be used to confirm these results.

In summary, virulent HRV was as immunogenic as a porcine RV strain in gnotobiotic piglets after infection. Adaptive B cell responses in gnotobiotic piglets inoculated with virulent HRV were mainly intestinal IgA and IgG with the majority of the ASC directed against VP6 and VP4 structural proteins. Longer-term (PID83) memory B cell responses were mainly associated with IgG and IgA and located in the spleen. Importantly in the gnotobiotic pig model, effector B cells and memory intestinal IgG and IgA ASC were correlated with protection against RV reinfection.

10.3.1.2 HRV-Specific Antibody Responses in Pigs

Serum and intestinal HRV specific IgM, IgA or IgG antibodies were measured in gnotobiotic pigs after inoculation with virulent Wa HRV [160]. A positive correlation was demonstrated between protection and serum or intestinal IgA and intestinal IgG antibody titers. Serum neutralizing antibodies were first detected between PID 7–14, whereas intestinal neutralizing antibodies were only detected at PID 21. However, no correlation was observed between neutralizing antibodies and protection in this study. The correlation between protection and high titers of RV-specific IgA and IgG antibodies in serum and intestinal contents corresponded with the correlation observed between intestinal RV-specific IgA and IgG ASC and protection [193]. These findings suggest that IgA antibodies in blood and intestinal contents mirror intestinal IgA ASC responses and that IgA antibodies in serum or feces and transient IgA ASC in blood may be sufficient as markers of protection.

10.3.2 Immune Responses to Virulent HRV (Which Mimic Natural Infection) Versus Attenuated HRV (Which Mimic Live Vaccine) in Gnotobiotic Piglets

10.3.2.1 HRV-Specific ASC Responses after Virulent Versus Attenuated HRV

Attenuated HRV (AttHRV, Wa strain P1AG1) replicates poorly in the gut of piglets and it is not pathogenic but because it induces intestinal and systemic immune responses, it was initially tested as an optimal vaccine candidate for piglets and possibly humans [172, 189]. Because virulent HRV inoculated pigs are protected against reinfection after PID21–35 [187], immune B cell responses measured at PID28 induced by both virulent and attenuated HRV strains were compared to determine the B cell correlates of protection after vaccination. Gnotobiotic piglets inoculated with virulent HRV developed significantly higher numbers of IgA ASC in the intestinal LP and peripheral blood compared to pigs inoculated with AttHRV [191, 192, 193]. The numbers of IgA ASC correlated with the higher rates of protection against diarrhea in the virulent HRV group (89%) compared to piglets inoculated with 2–3 doses of attenuated HRV that had partial protection against diarrhea (33–44%) and lower numbers of intestinal LP IgA ASC [191, 193].

In all RV inoculated pigs (attenuated vs. virulent), regardless of tissue and isotype, the highest numbers of ASC were directed against VP6, followed by VP4, and to lesser degree against VP7, NSP3 and NSP4 [39]. The total HRV-specific ASC or VP6 and VP4-specific IgA and IgG ASC in intestinal LP and MLN were significantly higher in the group receiving virulent compared to AttHRV [39]. On the other hand, in the systemic compartment (blood and spleen), responses generally did not differ between the 2 groups. It is known that nasal priming induces lymphocytes that are capable of homing to the systemic compartment (express L-selectin and CCR7) [88]. A higher percentage of piglets inoculated orally with AttHRV developed nasal virus shedding (79–95%) compared to intestinal shedding (5–17%) whereas 100% of virulent HRV inoculated piglets developed intestinal and nasal shedding [4]. Therefore, these observations possibly explain the marked intestinal B cell response observed after virulent HRV compared to AttHRV inoculation and the similar systemic responses observed after inoculation with both types of HRV.

In conclusion, in piglets, the AttHRV induced moderate protection rates and lower intestinal IgA ASC compared to the virulent HRV that induced higher protection rates against reinfection and higher numbers of intestinal IgA ASC. Theses findings suggest that an effective RV vaccine must induce intestinal IgA B cells that serve as a correlate of protection against RV disease in neonates.

10.3.2.2 HRV-Specific Antibody Responses after Virulent HRV Versus AttHRV

In the virulent HRV versus AttHRV inoculated piglets, a positive correlation was observed among serum IgA antibody titers, intestinal IgA antibody titers, intestinal IgG antibody titers and protection rates, suggesting that serum IgA antibodies to HRV reflect the presence of intestinal IgA antibodies after rotavirus infection with the latter antibodies mediating local protection against reinfection with virulent HRV [160].

Antibody responses to the Wa HRV nonstructural protein NSP4 also were evaluated in neonatal gnotobiotic pigs inoculated orally with one dose of virulent or three doses of AttHRV. Passive antibodies to NSP4 have been shown to reduce diarrhea in neonatal mice; hence, it is possible that antibodies to NSP4 may play a role in controlling rotavirus diarrhea [9]. Subsets of the pigs were challenged with virulent HRV. Before challenge, both virulent and attenuated groups had similar moderate titers of serum IgG antibodies to NSP4. However, only the virulent HRV inoculated pigs developed detectable serum and intestinal IgA antibody titers to NSP4 before challenge. [82]. Moreover, pigs inoculated with virulent rotavirus and challenged with a virulent strain bearing the same NSP4 genotype but different VP4 and VP7 serotype, were not protected from reinfection suggesting that NSP4 antibodies in piglets do not play a singular role in protection [188]. In summary, intestinal IgA and IgG ASC and antibodies in intestinal contents and serum IgA antibodies were the best correlates of protection after primary virulent HRV infection. Most of the antibody specificities were against VP6 and VP4 structural proteins. Thus, an effective vaccine should elicit these types of local B cell responses.

10.3.2.3 HRV-Specific T Cell Responses after Virulent HRV Versus AttHRV

The T cell responses were initially studied using a lymphocyte proliferation assay [173] that is a measure of mainly $CD4^+$ T cell responses. However, the proliferation assay is not a specific measure of T cell responses because other non-T cells (e.g. B cells) can also proliferate upon RV exposure *in-vitro* in a T cell-independent manner [25]. Virulent HRV-inoculated pigs developed proliferative responses mainly in the MLN, with lower responses observed in the intestinal LP, spleen and blood at PID 21.

Recently, T cell responses were more accurately measured with the use of an ELISPOT assay that detects cytokine secreting cells (CSC) [5]. After incubating total mononuclear cells (MNC) with RV antigen (AttHRV) for 3 days, memory T cell responses were determined in systemic and intestinal tissues. One dose of virulent

Wa HRV (87% and 100% protection against diarrhea and shedding, respectively) was compared with one dose of AttHRV Wa (0 and 14% protection against diarrhea and shedding, respectively) in inoculated piglets at various PIDs [83, 193]. The Th1 and Th1 inducer responses represented by IFNγ and IL-12 CSC, respectively induce cell-mediated immunity whereas Th2 responses represented by IL-4 induce humoral immunity. The T immunoregulatory type 1 (Tr1) response represented by IL-10, is commonly induced after infection to downregulate immune responses and to prevent immune-mediated pathology (allergy, uncontrolled inflammation and autoimmunity) [116]. After virulent HRV inoculation, T cell responses were mainly Th1 (IFNγ) or Th1 inducer (IL-12) and Tr1 (IL-10) with lower Th2 (IL-4) CSC responses. The Th1 (IFNγ) CSC responses were highest at PID 28 and were located mainly in the ileum LP and spleen. Pigs inoculated with AttHRV developed lower Th1 responses in the gut and spleen at PID28. After virulent HRV inoculation, memory IL-12 CSC were lower than IFNγ CSC numbers and were detected as early as PID 5 in the gut and spleen and at PID 21 in the spleen and blood. Similar responses were detected in AttHRV inoculated pigs. After infection with virulent HRV, the Tr1 IL-10 CSC were as high or higher than IFNγ CSC numbers. They peaked at PID 28 in the ileum LP and at PID 14 in the spleen, whereas they remained high in the blood from PID 14 to 28. Furthermore, AttHRV inoculated pigs developed even higher IL-10 CSC in the ileum and spleen at later PIDs (21 and 28) compared to the virulent HRV inoculated pigs. The Th2 CSC responses after virulent RV inoculation were observed mainly in the ileum LP, spleen and blood from PID 5–21 corresponding with the previously described initiation and peak of RV-specific B cell responses. However, AttHRV inoculated pigs developed higher IL-4 CSC in the gut at PID 21. In summary, these results show that after virulent HRV inoculation, higher Th1 and lower Th2 CSC responses were elicited. After AttHRV inoculation, lower Th1 and higher Th2 CSC responses were elicited reflecting the lower pathogenicity of the attenuated strain but also suggesting that Th1 type immunity, specifically IFNγ responses may be necessary for protection against reinfection. Both HRV types induced Tr1 responses, but they were higher after AttHRV infection at later PIDs and may reflect an immunoregulatory mechanism to prevent an excessive immune response. Studies by flow cytometry showed that after virulent HRV infection, the effector memory T cells ($CD3^+$) that secreted IFNγ expressed CD4 or CD8, resided mainly in the gut and their higher frequencies correlated with protection [194].

Cytokines in serum may reflect the activation of acute and possibly innate immune mechanisms and were quantitated after virulent and AttHRV infection of Gn pigs [5, 6]. Serum Th1 (IFN-γ, IL-12, that may reflect later Th1 responses), Th2 (IL-4, that may reflect later Th2 responses), Tr1 (IL-10), pro-inflammatory (IL-6, TNFα) and innate (IFNα) cytokines were induced as early as 1 day PI with virulent HRV or AttHRV. However, higher and earlier IFNα, IFN-γ and proinflammatory cytokine (IL-6, TNF-α) responses were induced by the virulent HRV. Interestingly, serum cytokines responses measured in RV infected children corresponded with those observed in gnotobiotic piglets [7, 55, 185].

In serum, TNF-α levels peaked earlier and remained elevated in the virulent group, but peaked later in the AttHRV group. The higher pro-inflammatory TNF-α responses coincided with the viremia and diarrhea induced after virulent HRV infection with similar observations described in children after natural RV infection [87]. The IL-4 responses peaked at PID10 and were similar between both HRV groups. Only low and transient IFN-γ responses occurred in serum and intestinal contents of the AttHRV-infected pigs, compared to significantly higher and prolonged IFN-γ responses in the virulent HRV-infected pigs [5]. Serum IFN-γ responses peaked shortly after the peak of diarrhea and viremia induced by virulent HRV mimicking the early appearance of elevated IFN-γ in children with severe RV diarrhea [7, 87]. Thus, serum responses, similar to the CSC responses, also reflect the higher Th1-IFNγ induction by the virulent HRV strain compared to the AttHRV strain.

In conclusion, pigs inoculated with virulent HRV that are highly protected from RV reinfection developed a balanced Th1/Th2 response. On the other hand, one dose of AttHRV that induced a low protection rate [83], induced a low Th1 (IFNγ) response and high Th2 (IL-4) and Tr1 (IL-10) responses. Therefore, a balanced Th1 and Th2 cytokine response may play an important role in protection against RV reinfection. However, a more detailed characterization of $CD4^+$ and $CD8^+$ effector and central memory T cells including detection of other cytokines and expression of homing molecules after RV infection in pigs is important.

10.3.2.4 Impact of Passive Antibodies on Virulent HRV Infection and Responses in Pigs

Cesarean-derived gnotobiotic pigs are optimal for the study of passively acquired maternal antibodies because the diffuse epitheliochorial type of placenta of the sow prevents the passage of maternal antibodies during gestation such that colostrum-deprived newborn piglets are deficient in maternal antibodies. Evidence for reduced post-vaccinal seroconversion in infants with high titers-of passive (maternally acquired) serum antibodies has been derived from many reports of clinical vaccine trials [78]. The effects of passive antibodies on protection and active immune responses to HRV in gnotobiotic piglets were assessed by Parreno et al. [135] and Hodgins et al. [78]. Pigs were injected at birth with HRV hyperimmune serum (mimic non-breast fed infants with circulating maternal antibodies) or injected with serum and fed with colostrum from HRV-immunized sows (mimic breast fed infants with local and circulating maternal antibodies). They were inoculated with virulent HRV to assess passive protection and then challenged 21 days latter to assess active protection induced with or without maternal antibodies. Pigs receiving immune serum with or without immune colostrum showed partial protection against human virus shedding and diarrhea after primary oral inoculation with virulent HRV, but showed lower protection rates after challenge compared to control pigs. Before challenge with virulent HRV, pigs that received oral RV hyperimmune colostrum and/or intraperitoneal serum, developed lower intestinal IgA and IgG ASC compared to uninoculated pigs and these differences persisted after challenge. Thus, high titers of circulating with or without colostral maternal RV antibodies provide partial passive protection after HRV inoculation but suppress active mucosal

RV antibody responses and the pigs consequently developed lower protection rates against reinfection. Furthermore, a vaccine regime protective in the absence of maternal antibodies (one oral dose of AttHRV+ two intranasal doses of 2/6 virus-like particles), was less efficient in pigs with circulating maternal antibodies to HRV [123, 124]. These observations support the previous data showing a correlation between protection rates against RV reinfection and intestinal RV-specific-IgA and IgG ASC numbers (and IgA and IgG antibody titers in intestinal contents or IgA antibody titers in serum) and demonstrate that when B cell responses to RV are impaired, lower protection rates are achieved.

10.4 Acquired RV-Specific Responses after Natural Infection in Children and Adults

10.4.1 Humoral Antibody Responses

Humoral immune responses are characterized the most thoroughly after RV infection given the widespread access standardized antibody assays and the availability of serum and fecal samples from patients. Children with RV infection characteristically show increased RV-specific IgM antibody titers in serum, feces and saliva typical of a primary RV infection [74]. Serum IgM antibodies to RV are detected as early as 2 days after diarrhea onset, but more commonly they occur 6 days after the onset of symptoms. A four-fold increase in serum IgM antibody titers or seroconversion occurs in 100% of RV infected children, whereas fecal IgM antibody increases (or coproconversion) occur in only 61% of infected children [74]. Also, children infected with RV developed IgA and IgG antibodies in serum (68% and 91% seroconversion rates, respectively) as well as IgA antibodies in feces (77% IgA coproconversion), but did not develop fecal IgG antibodies [3, 74]. Fecal RV-specific IgM and IgA antibodies can be detected as early as 3 days after RV induced diarrhea and maximum titers appeared at 11–15 days, but by 5 weeks after the onset of diarrhea, the titers faded. However, serum IgA and IgG antibody titers to RV were more persisting than fecal antibodies, reaching a maximum peak at approximately 33.5 days post-diarrhea (1:400 and 1:6400 for IgA and IgG, respectively) and remaining elevated for at least 4 months [74]. Neutralizing antibodies that by definition block RV cell entry, were present in stools and peaked at 5–8 days post-diarrhea, but only 70% of infected children had detectable neutralizing antibodies in feces, whereas conversion rates for serum neutralizing antibodies reached 100% [49]. Adults after acute symptomatic or asymptomatic infection developed mainly RV-specific IgA antibodies in serum, jejunal fluids, stools and saliva and IgG antibodies in serum with low levels of IgM antibodies characteristic of secondary RV infections [140, 178].

Many groups have attempted to determine the correlation of humoral immune responses to RV with protection against RV reinfection in humans. Neutralizing antibodies were studied as correlates of RV immunity in infants from 1–24 months of age infected with RV [41]. Children with higher serum neutralizing antibody titers to RV were protected against RV diarrhea. A serum neutralizing antibody titer of 1:128

or greater against serotype G3 was associated with short-term protection against the G3 RV reinfection (homotypic protection), but the children were reinfected multiple times suggesting that immunity was short-lived [41]. These observations were confirmed in studies conducted in developing and developed countries [127, 141]. Furthermore, upon primary RV infection, children developed predominantly homotypic neutralizing antibodies whereas the subsequent secondary RV infections induced broader responses (heterotypic responses) and a wider range of protection against disease and infection to different RV serotypes [126]. This finding agrees with the observation that after a primary RV infection, some degree of protection against subsequent disease is evident, but after two RV infections almost complete protection against subsequent moderate to severe disease is achieved [164]. Moreover, heterotypic and not homotypic neutralizing antibody titers were correlated with protection in a case control study conducted in children from Bangladesh [179], contradicting previous findings suggesting that homotypic neutralizing antibodies play a role in protection against reinfection. Nevertheless, there is more evidence that support the correlation of higher neutralizing antibody titers with protection against reinfection in children.

Neutralizing antibodies and their correlation with protection against reinfection were also studied in adult volunteers, but often with conflicting results among the different studies. In one study, 18 individuals were inoculated orally with RV serotype G1P1A (D strain). Only 5 of the 18 volunteers shed RV in feces and 4 of the 5 developed diarrhea demonstrating that adults are often protected against reinfection. The pre-challenge virus neutralizing antibody titers ranged from < 20 to > 5120. Homotypic and heterotypic (G3) serum neutralizing antibody titers at $> 1{:}20$ against an immunogenic VP7 epitope were correlated with resistance against infection and disease; however, not all protected volunteers were positive for serum neutralizing antibodies. A similar correlation with protection was observed with VP4 epitopes [73]. Furthermore, neutralizing antibody titers in intestinal fluids were measured, but a correlation was not observed. However, volunteers with $> 1{:}100$ intestinal virus neutralizing antibody titers tended to develop less diarrhea than volunteers with $< 1{:}100$ antibody titers but these differences were not statistically significant [90]. Other investigators showed that neutralizing antibodies in serum and jejunal fluids correlated with protection, but also reported that not all protected volunteers were positive for serum or intestinal neutralizing antibodies suggesting that other immune parameters must also be involved in protection against disease and illness. In addition, serum RV specific IgG antibodies best predicted protection against infection, whereas RV neutralizing antibody titers in jejunal fluids best predicted protection against illness [176]. These observations were later confirmed [175]; however, a lack of correlation with neutralizing antibodies in jejunal fluids or serum and protection against RV infection and disease in adults has been described as well [178]. In conclusion, in adults the role of neutralizing antibodies in protection against reinfection is not as clear as it appears to be in children.

Studies of RV-specific IgA in serum and feces or serum IgG antibodies as correlates of protection against RV reinfection have also been performed. The RV specific IgA and IgG antibody titers in serum and IgA antibodies to RV in feces of adult volunteers receiving two doses of virulent HRV were examined and a correlation with protection against disease and illness was observed with all antibodies except fecal IgA. Many adult volunteers that had RV-specific IgA antibodies in stools were not protected from reinfection. Only serum RV-specific IgG antibodies were significantly correlated with protection [176]. Other studies performed using adults did not find a correlation with protection and serum antibodies [178]. Thus in adults the role of serum and fecal antibodies to RV in protection against RV infection and disease is unclear.

On the other hand, most of the studies of IgA and IgG antibodies induced after RV infection in children support a role of RV-specific IgA in protection against reinfection. In a cohort study in which children from a developing country were observed from birth up to 2 years of age, serum and stool sampling were performed every month. Children with serum RV-specific IgA antibody titers $> 1{:}800$ had a lower risk of RV reinfection and were protected against moderate-severe diarrhea. In addition, serum IgG antibody titers to RV $> 1{:}6400$ were correlated with protection against infection but not disease. Complete protection against severe-moderate diarrhea and serum IgA seroconversion was achieved after two consecutive symptomatic or asymptomatic infections [164, 165]. These studies suggest that serum IgA to RV correlates with protection against infection and disease. These observations were confirmed in a study performed in a developed country (USA) whereby a titer of serum IgA antibodies of $> 1{:}200$ and of IgG antibodies of $> 1{:}800$ to RV were correlated with protection against infection and disease [127]. Conversely in a study of Danish children with RV gastroenteritis, higher preexisting serum IgA antibody titers to RV were correlated with milder symptoms (e.g. less vomiting), but serum IgG antibody titers were not correlated [77]. The authors also reported that serum IgA antibody titers were correlated with the presence of serum immunoglobulin bound to secretory component suggesting that at least some of the detected serum IgA antibodies to RV were probably of intestinal origin, lasting about 4 months after infection. These findings also confirm that serum RV-specific IgA correlates with protection against infection and disease in children. Other studies have reported a correlation between lower serum RV-specific IgG, but not IgA antibodies, with the development of RV-induced diarrhea [44].

In addition, stool IgA antibodies to RV have been previously correlated with protection against RV infection in children. Two cohort studies demonstrated that fecal RV-specific IgA antibodies correlated with subsequent protection against RV reinfection [50, 51]. The most sensitive method for diagnosis of acute RV infection was RV-specific IgA coproconversion in stools because 92% of RV infected children coproconverted [51]. Nevertheless, a substantial number of children that developed disease had elevated pre-infection IgA antibody titers to RV in stools [50, 111]. Possibly the fecal IgA to RV could be of maternal origin and perhaps the passively acquired IgA antibodies did not prevent virus infection because they lack secretory component and can not mediate virus expulsion [64]. Moreover, fecal IgA antibodies, as previously mentioned, only last for about 5 weeks after the onset of diarrhea. Most children after infection do not maintain fecal IgA antibody titers and only frequent reinfections induce long-term RV-specific fecal IgA antibodies [50].

For these reasons fecal IgA antibodies are not considered an adequate long-term marker of protection against RV infection. Because some of the serum RV specific IgA probably derives from the intestine and it persists approximately 4 months after infection, many studies have observed a correlation between serum IgA antibodies to RV and protection from RV reinfection. Thus serum IgA antibodies have been suggested to be an adequate marker of protection against RV infection and disease [64]. However, others suggested that serum antibodies in general correlate with protection and perhaps, regardless of the isotype, VP4 and VP7 specific intestinal antibodies must be present to prevent RV infection [86]. In general, there is more evidence that supports the role of humoral immunity in protection against reinfection in children compared to adults. Possibly, distinct and more mature immune mechanisms (T cells and non-neutralizing antibody producing B cells) active in adults, but not yet developed in children, also play important roles in protection.

10.4.2 B Cell Responses after Natural RV Infection in Humans

Antibodies are secreted by plasma cells, which are mature antigen activated B cells that do not undergo cell division. Plasma cells are generally short lived, but there is a subpopulation of plasma cells that are long-lived residing primarily in the BM [33]. Characterization of RV-induced B cell responses has been performed in children after acute infection and in adults infected or exposed to RV [69, 89]. Children with acute RV diarrhea were bled early after onset of symptoms and blood MNC were isolated and subjected to an ELISPOT assay for detection of RV-specific ASC. All RV positive children developed circulating IgM ASC to RV (82 RV specific IgM ASC$/10^6$ MNC), 93% developed lower IgG ASC responses (3–4 IgG ASC to RV$/10^6$ MNC), but only 64% of the children developed low IgA ASC responses to RV (0.02–0.08 IgA ASC to RV$/10^6$ MNC). These responses reflect a typical primary immune response in agreement with the previously described humoral immune responses in acutely RV infected children. A different study showed that by magnetic isolation of CD38$^+$ (a marker of ASC) B cells, children apparently not recently infected had persisting CD38$^+$ RV specific ASC in the LP probably representing long-lived plasma cells. Additionally, the number of circulating CD38$^+$ IgA ASC to RV correlated with that of CD38$^+$ RV specific IgA ASC in the small intestinal LP of the non-infected children suggesting their constant transit through blood to the gut [30]. Other studies support the observations that circulating ASC to RV after primary infection are mainly of IgM isotype and can be detected as early as 1 day of onset of symptoms [69]. Furthermore, RV infected symptomatic and asymptomatic adults developed 10–100 times more circulating IgG and IgA ASC to RV than children and 63% and 83% of the adults developed IgG and IgA ASC respectively, reaching a peak at 5–9 days post-infection [69]. These are typical secondary responses and also agree with the previously described humoral immune responses observed in RV infected adults.

Children and adults recently infected with RV mainly had VP6 specific $CD19^+$ large B cells (presumably activated) that expressed the intestinal homing marker $\alpha4\beta7$, but the ratio between $\alpha4\beta7+/\alpha4\beta7-$ VP6-specific B cells was higher in adults (21.6) than in children (3.0) suggesting that children might have more systemic B cell responses to RV than adults, possibly secondary to higher viremia due to the lack of intestinal immunity [69]. A more detailed characterization was later performed in children after acute RV infection and it was described that circulating RV-specific B cells were mainly large $CD38^+$, $CD27^+$ (memory cell marker), $CD138^{+/-}$ (plasmablast marker), $CCR6^-$ (chemokine receptor expressed in naïve and memory cells), $\alpha4\beta7^+$ (intestinal homing marker), $CCR9^+$ and $CCR10^+$ (B cell homing chemokine receptor markers), cutaneous lymphocyte antigen $(CLA)^-$ (lymphocyte marker for skin migration), L-selectin$^{+/-}$ (peripheral lymph node homing marker), IgM^+ $IgA^{+/-}$ and IgG^- likely representing intestinally primed plasma cells and plasmablasts [85]. However, at convalescence circulating RV-specific B cells were small and large $CD38^{+/-}$, $CD27^{+/-}$, $CCR6^+$, $\alpha4\beta7^{+/-}$, $CCR9^{+/-}$, $CCR10^-$ possibly representing both local and systemic B cells with memory phenotype [85]. The CD27 was previously described to be a memory marker, but more recently it has been described that high expression of CD27 can be observed in ASC [94, 121, 129]. Adults demonstrated a phenotype similar to that of children at convalescence suggesting that the observed cells represented mature memory responses and are present in both children and adults after RV infection [85]. Importantly, the source of the protective RV-specific IgA antibodies detected in serum needs to be addressed. Possibly, intestinal long-lived plasma cells are a major source for the RV-specific IgA and may also correlate with protection against RV reinfection.

Furthermore, in adults, infants and neonates, similar frequencies of naïve B cells ($CD27^-$, $CD19^+$) recognize RV VP6, but not VP7 [134]. This phenomenon has been described in mice [151] and goats [134] suggesting that this is common among mammals, but its significance *in-vivo* is unknown. Probably, this will aid in the generation of stronger VP6 responses that agrees with the fact that the most immunogenic RV protein is VP6. Also children with acute RV infection develop B cells with few somatic hypermutations that are mainly RV-specific $CD5^+$ B cells (marker of B1a cells) that characteristically secrete natural polyreactive and low affinity antibodies [183]. Between 60–87% of circulating B cells in children were $CD5^+$ compared to 22% in adults. It is known that children are more frequently reinfected with RV than adults. Possibly, low affinity antibodies secreted by $CD5^+$ B cells may not protect from RV reinfection and this may explain why children have increased susceptibility to RV infection but this remains to be studied. Furthermore, whether the naïve B cells observed to bind VP6 in adults, neonates and children express CD5 and represent B1 cells, needs to be further investigated.

In conclusion, RV activates B cells to become short or long-lived plasma cells and memory B cells that can be detected during their circulation in blood. Children developed circulating RV-specific B cells, that based on homing markers, possibly derive from systemic tissues and the gut, whereas in adults, RV-specific B cells are mainly of intestinal origin. A detailed characterization of circulating RV-specific B cells induced after infection in children showed that plasma cells expressed mainly intestinal homing markers, whereas memory B cells expressed both intestinal and systemic homing markers. Nevertheless, the origins of the RV-specific plasma cells and memory B cells and their role in protection against reinfection need to be further investigated.

10.4.3 T Cell Responses after Natural RV Infection of Humans

Cellular immunity to RV has been studied in children after acute RV infection and in symptomatic and asymptomatic adults. Initially, cellular immunity was studied by proliferation assays that presumably measure mainly CD4 specific T cell responses and are not a specific measure of all T cell responses. In a study characterizing circulating cellular immunity to RV, investigators noted that 7 of 8 children with acute RV infection had a positive homotypic and heterotypic proliferative response at convalescence (2–8 weeks, 6 of 8 children) and at late convalescence (3–5 months post-infection, 4 of 4 children) [131]. Additionally, cytokine responses have been characterized in the supernatants of circulating MNC in children after acute infection. The supernatants were tested for IFNγ, IL-10, IL-8 and TNFα, but only IFNγ was significantly elevated in cell supernatants from RV infected children, suggesting that RV infection induced a Th1 response [7].

The characterization of RV specific T cells was then performed by intracellular cytokine staining of CD4$^+$ and CD8$^+$ T cells by flow cytometry after acute RV infection in children (4–84 months of age) and symptomatic, asymptomatic RV infected and non-infected adults [84]. Circulating MNC were isolated after acute infection from RV positive and negative children and adults and intracellular IFNγ (Th1) and IL-13 (Th2) were measured in activated (CD69$^+$) CD4$^+$ or CD8$^+$ T cells. Symptomatic adults had significantly higher activated CD4$^+$ and CD8$^+$ IFNγ secreting T cells than healthy adults and children with acute RV diarrhea. No IL-13 responses were detected in infected or healthy adults. Children after acute RV infection had low frequencies of CD4$^+$ and CD8$^+$ IFNγ secreting T cells and, contrary to infected adults, RV infected children developed low frequencies of IL-13 CD4$^+$ T cells. Also there was a tendency for slightly higher CD8$^+$ than CD4$^+$ T cell responses in both children and frequently exposed non-infected adults (laboratory workers), whereas infrequently exposed healthy adults had low and balanced CD4$^+$ and CD8$^+$ T cell responses. Using an ELISPOT assay for detection of RV-specific CD4$^+$ and CD8$^+$ IFNγ and IL-4 CSC in blood from RV infected children and non-infected healthy adults, mainly IFNγ CSC were detected, whereas IL-4 (Th2) CSC were not observed [142]. In accordance with the findings observed after intracellular cytokine staining detected by flow cytometry [84], magnetic isolation of CD4$^+$ and CD8$^+$ T cells and incubation with CD14$^+$ monocyte APCs plus RV antigen, children again demonstrated a slightly higher CD8$^+$ T cell response, whereas infrequently exposed healthy adults had a balanced CD4$^+$ and CD8$^+$ T cell response. However, other investigators reported circulating RV-specific IFNγ CSC by ELISPOT in healthy adults, and after depleting CD4$^+$ but not CD8$^+$ T cells, most CSC responses were abolished, suggesting that RV-specific T cell immunity in healthy adults is mediated by $CD4^+$ T cells [92]. This observation can also be attributed to the lack of CD4 help when depleting $CD4^+$ T cells and not to a predominant RV-specific $CD4^+$ T cell population induced after RV infection. Additionally, $CD4^+CD69^+IFN\gamma^+$ T cells to RV from adult volunteers predominantly expressed the intestinal homing marker $\alpha4\beta7$ and were also enriched in the L-selectin$^-$ population which is typical of cells with an intestinal homing pattern. This finding was contrary to CMV-specific $CD4^+CD69^+IFN\gamma^+$ T cells in adults that were mainly L-selectin$^+$ $\alpha4\beta7^-$ typical of cells with a systemic homing pattern [142]. These results confirmed that RV induces a mixed Th1-Th2 response in children, whereas in infected adults, RV induced a higher and predominant Th1 response and intestinally committed $CD4^+$ T lymphocytes. Previously, it was addressed that children develop humoral immunity that correlates with protection, whereas the evidence for humoral immunity in protection against reinfection in adults is less clear. Perhaps children that develop lower immune Th1 and Th2 responses rely more on humoral immune responses for protection against reinfection compared to adults. By inducing stronger Th1 responses, adults possibly rely more on cellular immunity for protection against RV reinfection, but this needs to be investigated further.

10.4.4 Innate Immunity after RV Infection in Humans

Cytokines in serum detected after acute primary RV infection reflect to some extent the innate immune responses induced by the virus. However because RV-specific T cell responses are detected as early as 1 day after the onset of diarrhea in a primary infection [84], which cells are the source of the observed early serum cytokines (innate cells vs. acquired T cells) has not been defined. Some studies of children with acute RV infection demonstrated that serum IL-12, IFNγ, IL-6 and IL-10 increased significantly [87, 185] and the type I IFNα reached a peak in serum 2 days after the onset of symptoms [55, 58]. The protein MxA, induced by IFNα, also increased in children with various acute viral infections [RV, adenovirus, CMV and respiratory syncytial virus (RSV)] compared to the levels in peripheral blood lysates from children with bacterial gastroenteritis [42, 75]. Type I IFN secretion is probably derived from pDC, but other DCs and epithelial cells [150] secrete these cytokines and the main source of IFNα secretion after RV infection is unclear. The importance of cytokine secretion in the control of viral infections has been studied *in-vitro* using human intestinal cell lines infected with RV. Pre-treatment of human colon carcinoma cells (Caco) with IFNγ, IL-1 and IFNα prevented RV replication. However, only treatment with IFNα and not IFNγ or IL-1, diminished lipofected RV replication in a dose dependent manner whereas viral entry was significantly inhibited by the treatment with IFNγ and IL-1, but not IFNα [16]. These investigators suggested that production of these cytokines may prevent or block to some extent infection or RV replication. Additionally, it was described recently that RV NSP1 protein inhibits the synthesis of type I IFNs by proteosome degradation of the

IRF3, IRF7 and IRF5 as a possible mechanism to counteract the inhibition of RV replication mediated by IFNα [13, 14]. However, it was recently described that adult peripheral blood mononuclear cells exposed to RV *in-vitro* secrete IFNα and pDCs were the main source of this type I IFN [117]. Because after RV infection, IFNα is detected in serum, it remains to be determined if IFNα inhibits RV replication *in-vivo* or if RV replicates and inhibits IFNα production by the main type I IFN producers, the pDCs and other types of DCs.

The epithelial cell responses after RV infection have been mainly studied *in-vitro* using human intestinal colonic cell lines. Studies of HT29 and Caco cells demonstrated that after RV infection, IL-8, RANTES, the growth related peptide alpha, IP-10 (a C-X-C chemokine), granulocyte-monocyte colony stimulating factor (GM-CSF) and IFNα mRNAs were up-regulated [35, 143]. The peak of the epithelial response occurred between 3–16 hours post-infection and lasted for more than 24 hrs. However, it was observed that 2/4/6/7VLPs also induced IL-8 secretion by HT29 cells, whereas incubation with 2/6/7VLPs induced lower quantities of IL-8 production suggesting a role of VP4 in the activation of epithelial cells after infection [143].

The mechanism of generation of cytokine responses by RV infected epithelial cells is unclear, but it was recently shown that RV dsRNA mediated TNFα, IFNβ and IL-6 mRNA production *in-vitro* as well as apoptosis and a delay in epithelial cell migration to repair the disrupted monolayer caused by RV infected dead cells [150]. The TLR3, that specifically recognizes dsRNA, was detected at the mRNA and protein level in various intestinal epithelial cell lines (IEC-6, HT29 and Caco). The abnormal monolayer repair was in part TLR3 dependent, because after the addition of TLR3 antibodies, cell migration was promoted. These observations suggest that TLR3 induction promotes apoptosis and prevents cell migration by mechanisms still unclear that probably also aid in containment of the viral infection. However, it is possible that these effects are related to RV pathogenesis as has been recently observed in mice [196]. A study of the expression of various TLRs in blood MNC from children with acute RV infection confirmed the up-regulation of TLR3 *in-vivo* as well as TLR8 (that recognizes ssRNA together with TLR7) during the first 3 days of illness [185]. It is known that via TLR3 signaling, NF-κβ and the IFNβ promoters are activated [112] and both mRNAs are known to be up-regulated after RV infection of human cell lines *in-vitro* [143]. However, addition of TLR3-specific antibodies does not prevent secretion of IL-12, IFNα or IFNβ by immature DCs exposed to RV dsRNA [112] suggesting that besides TLR3, other molecules may be involved in the induction of cytokine secretion by immature DCs. Other dsRNA recognition molecules that induce NF-κβ like RIG-1, mda-5 and PKR might be involved in the induction of cytokine responses by innate immune cells and epithelial cells after RV infection [57]. Rotavirus induced activation of these molecules and their pathways have not been explored.

Human DC activation after RV exposure has been studied in immature and mature myeloid DC (mDC) *in-vitro* from adult volunteers [122]. The DC maturation was not altered after RV exposure and immature DCs secreted mainly IL-6, whereas mature DCs secreted IL-6 and low amounts of IL-12 and IL-10. Also, intracellular NSP4 was observed in 4–46% of mature DCs, but rarely in immature DCs, suggesting that virus replication occurred and was promoted only in mature DCs for unknown reasons. However, with the recent evidence that NSP4 is a structural protein [52], virus replication in DCs needs to be studied further [122]. The majority of viruses that replicate in DCs do so in immature rather than mature DCs and it was suggested that because immature DCs have a higher endocytic capacity this facilitates virus cell entry and subsequent replication [104]. However, some viruses like RSV [15] and certain strains of CMV [76] infect mature DCs more efficiently as was also suggested for RV.

Paired serum specimens from children with RV diarrhea were studied to measure the levels of IL-1β, IL-2, IL-4, IL-6, IL-8, IL-10, IL-12 and IFNγ [87] as a possible reflection of innate immune responses. Children with acute RV infection had elevated median levels of IL-6, IL-10 and IFNγ compared to children without RV infection. In different studies, children also presented a peak of IFNα in serum at day 2 after the initiation of symptoms [55, 58]. Furthermore, children that developed fever and more episodes of diarrhea had higher levels of TNFα compared to children without fever or fewer episodes of diarrhea and children that presented vomiting had lower levels of IFNγ suggesting that innate cytokines induced after acute infection may play a role in promoting or ameliorating RV pathogenesis.

The contribution of innate immunity in the initial control of RV replication and in the induction of RV pathogenesis is unclear and requires additional characterization. Also, important differences in the innate immune responses to RV in neonates vs. adults might aid in the design of more efficient mucosal vaccines able to induce stronger immune responses by the immature neonatal immune system or overcome the suppressive effects of maternal antibodies.

10.5 Candidate HRV Vaccines

Because RV causes significant morbidity and mortality impacting the healthcare system and because of the fact that public health interventions are unlikely to decrease the incidence of disease, vaccines are the main strategy for disease prevention [29].

10.5.1 Monovalent Jennerian Vaccines

The initial candidate vaccines were developed from animal RV strains of bovine or non-human primate origin because human and animal rotaviruses share a common group antigen (VP6) and heterologous protection against infection has been demonstrated using animal models [197]. A "Jennerian approach," was tested in which a live RV of non-human origin is used to vaccinate human volunteers, to elicit cross-protective immunity. One vaccine was an attenuated bovine RV strain (Nebraska calf diarrhea virus [NCDV]). Three monovalent non-HRV vaccine candidates underwent field trials: two bovine strains (RIT4238, P6[1]G6 and WC3 P7[5]G6) and one simian strain (RRV MMU 18006, P5B[3]G3). The efficacy of these candidate vaccines to prevent severe HRV disease varied in clinical trials [102, 171]. A lack of heterologous protection was one reason for vaccine failure, mainly because of the differences between developed and developing countries, where in the latter, increased strain diversity, more mixed infections, a larger exposure inoculum and higher titers of maternal antibodies are present. Protection induced by the monovalent RRV vaccine was apparently not serotype specific and did not correlate with serum neutralizing antibodies or serum IgA antibodies [103]. The bovine RIT 4237 strain that in early studies induced high protection rates and heterotypic protection, induced RV-specific IgG antibody titers in serum (serum IgA antibodies not measured) that correlated with clinical protection [167]. However, later studies failed to provide evidence of protection by the bovine strain and it was discontinued. Similarly, the WC3 bovine strain induced variable protection rates, and induction of neutralizing antibodies or serum IgA antibodies to RV did not correlate with protection [64].

10.5.2 Multivalent Jennerian Vaccines

Because monovalent live non-HRV vaccines failed to induce significant protection in children, which was attributed to their lack of induction of heterotypic protective antibody to multiple rotavirus serotypes, it was thought that a successful RV vaccine would need to induce neutralizing antibodies to the major serotypes of RV (G1-G4) commonly seen in children [136]. Gorrell & Bishop [71], corroborating other studies, suggested that for RV vaccines to be effective they should be multidose to elicit high titers of cross-reactive VP7 neutralizing antibodies. In an attempt to induce heterotypic immune responses to multiple HRV serotypes, animal-HRV reassortants were included in a second-generation vaccine [36]. Several reassortant RV vaccines were constructed by incorporating HRV genes that code for the surface proteins VP7 and/or VP4 into a bovine or simian RV genome background.

10.5.2.1 Rotashield

A tetravalent Rhesus-HRV reassortant (RRV-TV) vaccine (Rotashield®, Wyeth Laboratories, Inc) was designed to protect against the four most important HRV serotypes (G1-G4). This vaccine showed about an 80% protection rate against severe diarrhea in some, but not all clinical trials [60, 72, 166]. Post-licensure evaluation of the Rotashield® vaccine identified intussusception (intestinal blockage or obstruction) as a serious adverse event potentially associated with this vaccine and prompted its withdrawal [120]. The Rotashield® RRV-TV vaccine was licensed in 1998 and withdrawn in 1999 after the uncommon but serious adverse event intussusception occurred [137]. Whether this association will occur with other candidate live RV vaccines strains at similar risk in other populations is unclear. Rhesus-human reassortant RV replicates in the human small intestine and causes diarrhea in several species other than monkeys. Offit et al. [130] suggested that intussusception might be related to high virus dose, the RV strain or due to the extent of virus replication. To overcome the highest period of susceptibility of children to develop intussusception, the use of live attenuated RV vaccines at 0–4 weeks of age for the first dose and 4–8 weeks of age for the second dose was recommended [154]. With this strategy, the risk of intussusception should be reduced, since with Rotashield®, no intussusception was observed in infants vaccinated younger than 60 days of age [26]. Moser et al. [119] studied the capacity of simian-human and bovine-human reassortant RV to cause lymphoid hypertrophy and hyperplasia of PP of adult BALB/c mice. They found neither hypertrophy nor hyperplasia in the mouse PP after oral inoculation with the reassortant RV. However, infectious virus was detected in PP and MLN after oral inoculation with the simian, but not the bovine reassortant RV. Moreover, vaccination with the RRV-TV strain and the titers of RV-specific IgA in serum or neutralizing antibodies induced were not correlated with protection [182].

10.5.2.2 RotaTeq

More recently, a human-bovine reassortant RV vaccine, RotaTeq® (Merck & Co), which contains five serotypes: G1, G2, G3, G4 and P1, has been licensed in the U.S since 2006 [1]. This vaccine is also in phase III clinical trials in other countries. It has shown high rates of seroconversion even after only the first dose [177]. In early studies, a quadrivalent vaccine human (WI79) and bovine (WC3) reassortant precursor of the current RotaTeq®, was evaluated for its efficacy, safety and immunogenicity in healthy infants [43]. The vaccine showed 74.6% efficacy against RV acute diarrhea and 100% efficacy against severe diarrhea. The quadrivalent vaccine was well tolerated, with no significant differences between vaccine and placebo groups regarding side-effects [1].

10.5.3 Non-Jennerian Vaccines

The first non-Jennerian vaccine approach used a HRV strain from an infected newborn baby. Evidence shows that human strains that infect these newborns are usually naturally attenuated [20, 80]. Such human strains have a theoretical advantage over reassortant vaccines, in that they should induce a broad antibody response and reduce the need for a complex mixture of strains [11]. The first vaccine tested using this approach was M37, carrying a P[6]G1 serotype; but a field trial showed no protection against RV gastroenteritis. Serological studies showed that the M37 vaccine induced very limited antibody responses [190] and the neutralizing antibodies induced were primarily vaccine-strain-specific, not serotype-specific [118]. Another neonatal strain tested was RV3 (serotype G3), characterized as a P[6]G3 strain. Neonates naturally exposed to this strain developed neutralizing antibodies to G1, G3 and G4, and were protected against severe diarrhea until 3 years of life [20]. A phase III clinical trial with 3 doses of RV3 showed 46% of vaccinees developed immune responses and in the ones that seroconverted, the efficacy was 54% [12].

10.5.3.1 Rotarix

An attenuated G1P[8] HRV strain, RIX4414 vaccine strain (Rotarix® – Glaxo-SmithKline) was cloned from the parental strain 89–12. In clinical trials this vaccine has shown 85 to 96% efficacy against severe RV gastroenteritis and 72% efficacy against any RV gastroenteritis [56, 168, 169]. The RIX4414 vaccine reduced gastroenteritis-related hospitalizations by more than 40% in Latin America and by 75% in Europe [139]. It proved to be immunogenic and to elicit non-G1 neutralizing antibodies as well as crossprotection [180]. Rotarix® was introduced in Mexico and licensed at the end of 2004 and after that in over 70 countries [93].

10.5.4 Monovalent vs Pentavalent RV Vaccines

These two new vaccines were licensed concomitantly in different countries and are being administered to children worldwide. Castello et al. [36] reviewed strains prevalent in Latin America, compared the genotypes in circulation during a 9-year period and calculated the probable efficacy of the two new candidate vaccines: RotaTeq® and Rotarix®. Their survey indicated that RotaTeq® would target 56% of all single infections by both VP7 and VP4 antigens, 42% would be targeted by only one antigen and only 1% of the uncommon strains would not be targeted by a single outer capsid protein. Whether the protection would be equal for strains bearing 1 or 2 antigens in common with the vaccine has yet to be determined. Alternatively Rotarix® that is composed of a Wa genogroup, has shown a wide crossreactivity and protection against rare G types (e.g G9 strains) with similar genogroups. Therefore, the greatest challenge for Rotarix® will be to protect against strains from different genogroups (DS-1 or AU-1) that can reach a ~30% prevalence rate. Ward et al. [174] suggested that the quality of responses induced by a HRV vaccine against a human challenge should be significantly better due to the conserved B and T cell epitopes within strains and in common with the circulating HRV. Correlates of protection after vaccination are still under study and a strict surveillance of intussusception induced after vaccination is in progress. No higher association with intussusception has yet been observed in the vaccinated group compared to placebo group for any of the two RV vaccines [19, 170]. However both vaccines remain to be tested in large clinical trials in the poorest of developing countries, particularly in Africa and Asia, where the need is greatest.

References

1. American Academy of Pediatrics Committee on Infectious, D. 2007. Pediatrics 119, 171.
2. Angel, J., Franco, M. A., Greenberg, H. B., and Bass, D. 1999. J Interferon Cytokine Res 19, 655.
3. Arias, C. F., Lopez, S., Mascarenhas, J. D., Romero, P., Cano, P., Gabbay, Y. B., de Freitas, R. B., and Linhares, A. C. 1994. Clin Diagn Lab Immunol 1, 89.

4. Azevedo, M. S., Yuan, L., Jeong, K. I., Gonzalez, A., Nguyen, T. V., Pouly, S., Gochnauer, M., Zhang, W., Azevedo, A., and Saif, L. J. 2005. J Virol 79, 5428.
5. Azevedo, M. S., Yuan, L., Pouly, S., Gonzales, A. M., Jeong, K. I., Nguyen, T. V., and Saif, L. J. 2006. J Virol 80, 372.
6. Azevedo, M. S. P., Gonzalez, A. M., Yuan, L., Zhang, W., Wen, K., Lovgren-Bengtsson, K., Morein, B., and Saif, L. J. 2007. In "The 4th International Conference on Vaccine for Enteric Diseases", Pestana Palace Hotel, Lisbon, Postugal.
7. Azim, T., Zaki, M. H., Podder, G., Sultana, N., Salam, M. A., Rahman, S. M., Sefat e, K., and Sack, D. A. 2003. J Med Virol 69, 286.
8. Ball, J. M., Mitchell, D. M., Gibbons, T. F., and Parr, R. D. 2005. Viral Immunol 18, 27.
9. Ball, J. M., Tian, P., Zeng, C. Q., Morris, A. P., and Estes, M. K. 1996. Science 272, 101.
10. Banchereau, J., Briere, F., Caux, C., Davoust, J., Lebecque, S., Liu, Y. J., Pulendran, B., and Palucka, K. 2000. Annu Rev Immunol 18, 767.
11. Barnes, G. 2004. Expert Rev Vaccines 3, 359.
12. Barnes, G. L., Lund, J. S., Mitchell, S. V., De Bruyn, L., Piggford, L., Smith, A. L., Furmedge, J., Masendycz, P. J., Bugg, H. C., Bogdanovic-Sakran, N., Carlin, J. B., and Bishop, R. F. 2002. Vaccine 20, 2950.
13. Barro, M., and Patton, J. T. 2005. Proc Natl Acad Sci U S A 102, 4114.
14. Barro, M., and Patton, J. T. 2007. J Virol 81, 4473.
15. Bartz, H., Turkel, O., Hoffjan, S., Rothoeft, T., Gonschorek, A., and Schauer, U. 2003. Immunology 109, 49.
16. Bass, D. M. 1997. Gastroenterology 113, 81.
17. Bell, L. M., Clark, H. F., O'Brien, E. A., Kornstein, M. J., Plotkin, S. A., and Offit, P. A. 1987. Proc Soc Exp Biol Med 184, 127.
18. Bernard, S., and Jestin, A. 1985. Zentralbl Veterinarmed B 32, 306.
19. Bernstein, D. I. 2006. Semin Pediatr Infect Dis 17, 188.
20. Bishop, R. F. 1993. Vaccine 11, 247.
21. Blanco, P., Palucka, A. K., Gill, M., Pascual, V., and Banchereau, J. 2001. Science 294, 1540.
22. Blutt, S. E., Crawford, S. E., Warfield, K. L., Lewis, D. E., Estes, M. K., and Conner, M. E. 2004. J Virol 78, 6974.
23. Blutt, S. E., Fenaux, M., Warfield, K. L., Greenberg, H. B., and Conner, M. E. 2006. J Virol 80, 6702.
24. Blutt, S. E., Matson, D. O., Crawford, S. E., Staat, M. A., Azimi, P., Bennett, B. L., Piedra, P. A., and Conner, M. E. 2007. PLoS Med 4, e121.
25. Blutt, S. E., Warfield, K. L., Lewis, D. E., and Conner, M. E. 2002. J Immunol 168, 5716.
26. Bonn, D. 2004. Lancet Infect Dis 4, 658.
27. Bowman, E. P., Kuklin, N. A., Youngman, K. R., Lazarus, N. H., Kunkel, E. J., Pan, J., Greenberg, H. B., and Butcher, E. C. 2002. J Exp Med 195, 269.
28. Brandtzaeg, P., and Johansen, F. E. 2005. Immunol Rev 206, 32.
29. Bresee, J. S., Glass, R. I., Ivanoff, B., and Gentsch, J. R. 1999. Vaccine 17, 2207.
30. Brown, K. A., Kriss, J. A., Moser, C. A., Wenner, W. J., and Offit, P. A. 2000. J Infect Dis 182, 1039.
31. Brown, K. A., and Offit, P. A. 1998. Microb Pathog 24, 327.
32. Burns, J. W., Krishnaney, A. A., Vo, P. T., Rouse, R. V., Anderson, L. J., and Greenberg, H. B. 1995. Virology 207, 143.
33. Calame, K. L. 2001. Nat Immunol 2, 1103.
34. Campbell, D. J., and Butcher, E. C. 2002. J Exp Med 195, 135.
35. Casola, A., Estes, M. K., Crawford, S. E., Ogra, P. L., Ernst, P. B., Garofalo, R. P., and Crowe, S. E. 1998. Gastroenterology 114, 947.
36. Castello, A. A., Arvay, M. L., Glass, R. I., and Gentsch, J. 2004. Pediatr Infect Dis J 23, S168.
37. Castrucci, G., Ferrari, M., Frigeri, F., Traldi, V., and Angelillo, V. 1994. Comp Immunol Microbiol Infect Dis 17, 321.
38. Cella, M., Facchetti, F., Lanzavecchia, A., and Colonna, M. 2000. Nat Immunol 1, 305.

39. Chang, K. O., Vandal, O. H., Yuan, L., Hodgins, D. C., and Saif, L. J. 2001. J Clin Microbiol 39, 2807.
40. Chen, W. K., Campbell, T., VanCott, J., and Saif, L. J. 1995. Vet Immunol Immunopathol 45, 265.
41. Chiba, S., Yokoyama, T., Nakata, S., Morita, Y., Urasawa, T., Taniguchi, K., Urasawa, S., and Nakao, T. 1986. Lancet 2, 417.
42. Chieux, V., Hober, D., Harvey, J., Lion, G., Lucidarme, D., Forzy, G., Duhamel, M., Cousin, J., Ducoulombier, H., and Wattre, P. 1998. J Virol Methods 70, 183.
43. Clark, H. F., Bernstein, D. I., Dennehy, P. H., Offit, P., Pichichero, M., Treanor, J., Ward, R. L., Krah, D. L., Shaw, A., Dallas, M. J., Laura, D., Eiden, J. J., Ivanoff, N., Kaplan, K. M., and Heaton, P. 2004. J Pediatr 144, 184.
44. Clemens, J. D., Ward, R. L., Rao, M. R., Sack, D. A., Knowlton, D. R., van Loon, F. P., Huda, S., McNeal, M., Ahmed, F., and Schiff, G. 1992. J Infect Dis 165, 161.
45. Coffin, S. E., Clark, S. L., Bos, N. A., Brubaker, J. O., and Offit, P. A. 1999. J Immunol 163, 3064.
46. Coffin, S. E., and Offit, P. A. 1998. J Virol 72, 3479.
47. Colomina, J., Gil, M. T., Codoner, P., and Buesa, J. 1998. J Med Virol 56, 58.
48. Cook, D. N., Prosser, D. M., Forster, R., Zhang, J., Kuklin, N. A., Abbondanzo, S. J., Niu, X. D., Chen, S. C., Manfra, D. J., Wiekowski, M. T., Sullivan, L. M., Smith, S. R., Greenberg, H. B., Narula, S. K., Lipp, M., and Lira, S. A. 2000. Immunity 12, 495.
49. Coulson, B. S. 1998. Clin Diagn Lab Immunol 5, 897.
50. Coulson, B. S., Grimwood, K., Hudson, I. L., Barnes, G. L., and Bishop, R. F. 1992. J Clin Microbiol 30, 1678.
51. Coulson, B. S., Grimwood, K., Masendycz, P. J., Lund, J. S., Mermelstein, N., Bishop, R. F., and Barnes, G. L. 1990. J Clin Microbiol 28, 1367.
52. Crawford, S., Hyser J.,Cheng E.,Zeng C., and Estes M. K 2006. In "9th dsRNA virus symposium", Cape Town, South Africa.
53. Crawford, S. E., Patel, D. G., Cheng, E., Berkova, Z., Hyser, J. M., Ciarlet, M., Finegold, M. J., Conner, M. E., and Estes, M. K. 2006. J Virol 80, 4820.
54. Davidson, G. P., Gall, D. G., Petric, M., Butler, D. G., and Hamilton, J. R. 1977. J Clin Invest 60, 1402.
55. De Boissieu, D., Lebon, P., Badoual, J., Bompard, Y., and Dupont, C. 1993. J Pediatr Gastroenterol Nutr 16, 29.
56. De Vos, B., Vesikari, T., Linhares, A. C., Salinas, B., Perez-Schael, I., Ruiz-Palacios, G. M., Guerrero Mde, L., Phua, K. B., Delem, A., and Hardt, K. 2004. Pediatr Infect Dis J 23, S179.
57. Didierlaurent, A., Sirard, J. C., Kraehenbuhl, J. P., and Neutra, M. R. 2002. Cell Microbiol 4, 61.
58. Dubos, F., Lorrot, M., Soulier, M., Rozenberg, F., Lebon, P., and Gendrel, D. 2004. Med Mal Infect 34, 561.
59. Estes, M. 2001. "Rotavirus and their replication" 4th ed., Lippincott Williams & Wilkins, Philadelphia.
60. Estes, M. K. 1996. J Infect Dis 174 Suppl 1, S37.
61. Fenaux, M., Cuadras, M. A., Feng, N., Jaimes, M., and Greenberg, H. B. 2006. J Virol 80, 5219.
62. Feng, J., Li, M., Cai, T., Tang, H., and Gu, W. 2005. J Pediatr Surg 40, 630.
63. Feng, N., Jaimes, M. C., Lazarus, N. H., Monak, D., Zhang, C., Butcher, E. C., and Greenberg, H. B. 2006. J Immunol 176, 5749.
64. Franco, M. A., Angel, J., and Greenberg, H. B. 2006. Vaccine 24, 2718.
65. Franco, M. A., and Greenberg, H. B. 1995. J Virol 69, 7800.
66. Franco, M. A., and Greenberg, H. B. 1999. J Infect Dis 179 Suppl 3, S466.
67. Franco, M. A., Tin, C., Rott, L. S., VanCott, J. L., McGhee, J. R., and Greenberg, H. B. 1997. J Virol 71, 479.

68. Glass, R. I., Parashar, U. D., Bresee, J. S., Turcios, R., Fischer, T. K., Widdowson, M. A., Jiang, B., and Gentsch, J. R. 2006. Lancet 368, 323.
69. Gonzalez, A. M., Jaimes, M. C., Cajiao, I., Rojas, O. L., Cohen, J., Pothier, P., Kohli, E., Butcher, E. C., Greenberg, H. B., Angel, J., and Franco, M. A. 2003. Virology 305, 93.
70. Gor, D. O., Rose, N. R., and Greenspan, N. S. 2003. Nat Immunol 4, 503.
71. Gorrell, R. J., and Bishop, R. F. 1999. J Med Virol 57, 204.
72. Gouvea, V., Glass, R. I., Woods, P., Taniguchi, K., Clark, H. F., Forrester, B., and Fang, Z. Y. 1990. J. Clin. Microbiol. 28, 276.
73. Green, K. Y., and Kapikian, A. Z. 1992. J Virol 66, 548.
74. Grimwood, K., Lund, J. C., Coulson, B. S., Hudson, I. L., Bishop, R. F., and Barnes, G. L. 1988. J Clin Microbiol 26, 732.
75. Halminen, M., Ilonen, J., Julkunen, I., Ruuskanen, O., Simell, O., and Makela, M. J. 1997. Pediatr Res 41, 647.
76. Hertel, L., Lacaille, V. G., Strobl, H., Mellins, E. D., and Mocarski, E. S. 2003. J Virol 77, 7563.
77. Hjelt, K., Grauballe, P. C., Paerregaard, A., Nielsen, O. H., and Krasilnikoff, P. A. 1987. J Med Virol 21, 39.
78. Hodgins, D. C., Kang, S. Y., deArriba, L., Parreno, V., Ward, L. A., Yuan, L., To, T., and Saif, L. J. 1999. J Virol 73, 186.
79. Hoshino, Y., Honma, S., Jones, R. W., Santos, N., Nakagomi, O., Nakagomi, T., Kapikian, A. Z., and Thouless, M. E. 2006. Virology 345, 1.
80. Hoshino, Y., and Kapikian, A. Z. 1994. Trends Microbiol 2, 242.
81. Hutchings, A. B., Helander, A., Silvey, K. J., Chandran, K., Lucas, W. T., Nibert, M. L., and Neutra, M. R. 2004. J Virol 78, 947.
82. Iosef, C., Chang, K. O., Azevedo, M. S., and Saif, L. J. 2002. J Med Virol 68, 119.
83. Iosef, C., Van Nguyen, T., Jeong, K., Bengtsson, K., Morein, B., Kim, Y., Chang, K. O., Azevedo, M. S., Yuan, L., Nielsen, P., and Saif, L. J. 2002. Vaccine 20, 1741.
84. Jaimes, M. C., Rojas, O. L., Gonzalez, A. M., Cajiao, I., Charpilienne, A., Pothier, P., Kohli, E., Greenberg, H. B., Franco, M. A., and Angel, J. 2002. J Virol 76, 4741.
85. Jaimes, M. C., Rojas, O. L., Kunkel, E. J., Lazarus, N. H., Soler, D., Butcher, E. C., Bass, D., Angel, J., Franco, M. A., and Greenberg, H. B. 2004. J Virol 78, 10967.
86. Jiang, B., Gentsch, J. R., and Glass, R. I. 2002. Clin Infect Dis 34, 1351.
87. Jiang, B., Snipes-Magaldi, L., Dennehy, P., Keyserling, H., Holman, R. C., Bresee, J., Gentsch, J., and Glass, R. I. 2003. Clin Diagn Lab Immunol 10, 995.
88. Johansen, F. E., Baekkevold, E. S., Carlsen, H. S., Farstad, I. N., Soler, D., and Brandtzaeg, P. 2005. Blood 106, 593.
89. Kaila, M., Isolauri, E., Virtanen, E., and Arvilommi, H. 1992. Gut 33, 639.
90. Kapikian, A. Z., Wyatt, R. G., Levine, M. M., Yolken, R. H., VanKirk, D. H., Dolin, R., Greenberg, H. B., and Chanock, R. M. 1983. J Infect Dis 147, 95.
91. Katze, M. G., He, Y., and Gale, M. 2002. Nat Rev Immunol 2, 675.
92. Kaufhold, R. M., Field, J. A., Caulfield, M. J., Wang, S., Joseph, H., Wooters, M. A., Green, T., Clark, H. F., Krah, D., and Smith, J. G. 2005. J Virol 79, 5684.
93. Keating, G. M. 2006. Paediatr Drugs 8, 389.
94. Klein, U., Kuppers, R., and Rajewsky, K. 1997. Blood 89, 1288.
95. Kordasti, S., Istrate, C., Banasaz, M., Rottenberg, M., Sjovall, H., Lundgren, O., and Svensson, L. 2006. J Virol 80, 11355.
96. Krug, A., Veeraswamy, R., Pekosz, A., Kanagawa, O., Unanue, E. R., Colonna, M., and Cella, M. 2003. J Exp Med 197, 899.
97. Kuebler, J. F., Czech-Schmidt, G., Leonhardt, J., Ure, B. M., and Petersen, C. 2006. Pediatr Res 59, 790.
98. Kuklin, N. A., Rott, L., Darling, J., Campbell, J. J., Franco, M., Feng, N., Muller, W., Wagner, N., Altman, J., Butcher, E. C., and Greenberg, H. B. 2000. J Clin Invest 106, 1541.

99. Kunkel, E. J., Campbell, J. J., Haraldsen, G., Pan, J., Boisvert, J., Roberts, A. I., Ebert, E. C., Vierra, M. A., Goodman, S. B., Genovese, M. C., Wardlaw, A. J., Greenberg, H. B., Parker, C. M., Butcher, E. C., Andrew, D. P., and Agace, W. W. 2000. J Exp Med 192, 761.
100. Kunkel, E. J., Kim, C. H., Lazarus, N. H., Vierra, M. A., Soler, D., Bowman, E. P., and Butcher, E. C. 2003. J Clin Invest 111, 1001.
101. Kushnir, N., Bos, N. A., Zuercher, A. W., Coffin, S. E., Moser, C. A., Offit, P. A., and Cebra, J. J. 2001. J Virol 75, 5482.
102. Lanata, C. F., Black, R. E., del Aguila, R., Gil, A., Verastegui, H., Gerna, G., Flores, J., Kapikian, A. Z., and Andre, F. E. 1989. J Infect Dis 159, 452.
103. Lanata, C. F., Black, R. E., Flores, J., Lazo, F., Butron, B., Linares, A., Huapaya, A., Ventura, G., Gil, A., and Kapikian, A. Z. 1996. Vaccine 14, 237.
104. Larsson, M., Beignon, A. S., and Bhardwaj, N. 2004. Semin Immunol 16, 147.
105. Lopez, S., and Arias, C. F. 2006. Curr Top Microbiol Immunol 309, 39.
106. Lundgren, O., Peregrin, A. T., Persson, K., Kordasti, S., Uhnoo, I., and Svensson, L. 2000. Science 287, 491.
107. Mack, C. L., Tucker, R. M., Lu, B. R., Sokol, R. J., Fontenot, A. P., Ueno, Y., and Gill, R. G. 2006. Hepatology 44, 1231.
108. MacLennan, I. C., Toellner, K. M., Cunningham, A. F., Serre, K., Sze, D. M., Zuniga, E., Cook, M. C., and Vinuesa, C. G. 2003. Immunol Rev 194, 8.
109. Majer, M., Behrens, F., Weinmann, E., Mauler, R., Maass, G., Baumeister, H. G., and Luthardt, T. 1978. Infection 6, 71.
110. Masopust, D., Vezys, V., Usherwood, E. J., Cauley, L. S., Olson, S., Marzo, A. L., Ward, R. L., Woodland, D. L., and Lefrancois, L. 2004. J Immunol 172, 4875.
111. Matson, D. O., O'Ryan, M. L., Herrera, I., Pickering, L. K., and Estes, M. K. 1993. J Infect Dis 167, 577.
112. Matsumoto, M., Funami, K., Tanabe, M., Oshiumi, H., Shingai, M., Seto, Y., Yamamoto, A., and Seya, T. 2003. J Immunol 171, 3154.
113. McHeyzer-Williams, M., McLean, M., Lalor, P., and Nossal, G. 1993. J. Exp. Med. 178, 295.
114. McHeyzer-Williams, M. G., and Ahmed, R. 1999. Curr Opin Immunol 11, 172.
115. McNeal, M. M., Rae, M. N., and Ward, R. L. 1997. J Virol 71, 8735.
116. Mege, J. L., Meghari, S., Honstettre, A., Capo, C., and Raoult, D. 2006. Lancet Infect Dis 6, 557.
117. Mesa, M. C., Rodriguez, L. S., Franco, M. A., and Angel, J. 2007. Virology.
118. Midthun, K., Halsey, N. A., Jett-Goheen, M., Clements, M. L., Steinhoff, M., King, J. C., Karron, R., Wilson, M., Burns, B., and Perkis, V. 1991. J Infect Dis 164, 792.
119. Moser, C. A., Dolfi, D. V., Di Vietro, M. L., Heaton, P. A., Offit, P. A., and Clark, H. F. 2001. J. Infect. Dis. 183, 1108.
120. Murphy, T. V., Gargiullo, P. M., Massoudi, M. S., Nelson, D. B., Jumaan, A. O., Okoro, C. A., Zanardi, L. R., Setia, S., Fair, E., LeBaron, C. W., Wharton, M., Livengood, J. R., Livingood, J. R., and Rotavirus Intussusception Investigation, T. 2001. N Engl J Med 344, 564.
121. Nakayama, T., Hieshima, K., Izawa, D., Tatsumi, Y., Kanamaru, A., and Yoshie, O. 2003. J Immunol 170, 1136.
122. Narvaez, C. F., Angel, J., and Franco, M. A. 2005. J Virol 79, 14526.
123. Nguyen, T. V., Yuan, L., Azevedo, M. S., Jeong, K. I., Gonzalez, A. M., Iosef, C., Lovgren-Bengtsson, K., Morein, B., Lewis, P., and Saif, L. J. 2006. Clin Vaccine Immunol 13, 475.
124. Nguyen, T. V., Yuan, L., MS, P. A., Jeong, K. I., Gonzalez, A. M., Iosef, C., Lovgren-Bengtsson, K., Morein, B., Lewis, P., and Saif, L. J. 2006. Vaccine 24, 2302.
125. O'Neal, C. M., Harriman, G. R., and Conner, M. E. 2000. J Virol 74, 4102.
126. O'Ryan, M. L., Matson, D. O., Estes, M. K., and Pickering, L. K. 1994. Pediatr Infect Dis J 13, 890.
127. O'Ryan, M. L., Matson, D. O., Estes, M. K., and Pickering, L. K. 1994. J Infect Dis 169, 504.

128. Ochando, J. C., Homma, C., Yang, Y., Hidalgo, A., Garin, A., Tacke, F., Angeli, V., Li, Y., Boros, P., Ding, Y., Jessberger, R., Trinchieri, G., Lira, S. A., Randolph, G. J., and Bromberg, J. S. 2006. Nat Immunol 7, 652.
129. Odendahl, M., Jacobi, A., Hansen, A., Feist, E., Hiepe, F., Burmester, G. R., Lipsky, P. E., Radbruch, A., and Dorner, T. 2000. J Immunol 165, 5970.
130. Offit, P. A. 2002. Semin Pediatr Infect Dis 13, 190.
131. Offit, P. A., Hoffenberg, E. J., Santos, N., and Gouvea, V. 1993. J Infect Dis 167, 1436.
132. Osborne, M. P., Haddon, S. J., Spencer, A. J., Collins, J., Starkey, W. G., Wallis, T. S., Clarke, G. J., Worton, K. J., Candy, D. C., and Stephen, J. 1988. J Pediatr Gastroenterol Nutr 7, 236.
133. Pan, J., Kunkel, E. J., Gosslar, U., Lazarus, N., Langdon, P., Broadwell, K., Vierra, M. A., Genovese, M. C., Butcher, E. C., and Soler, D. 2000. J Immunol 165, 2943.
134. Parez, N., Garbarg-Chenon, A., Fourgeux, C., Le Deist, F., Servant-Delmas, A., Charpilienne, A., Cohen, J., and Schwartz-Cornil, I. 2004. J Virol 78, 12489.
135. Parreno, V., Hodgins, D. C., de Arriba, L., Kang, S. Y., Yuan, L., Ward, L. A., To, T. L., and Saif, L. J. 1999. J. Gen. Virol. 80, 1417.
136. Perez-Schael, I., Garcia, D., Gonzalez, M., Gonzalez, R., Daoud, N., Perez, M., Cunto, W., Kapikian, A. Z., and Flores, J. 1990. J Med Virol 30, 219.
137. Peter, G., Myers, M. G., National Vaccine Advisory, C., and National Vaccine Program, O. 2002. Pediatrics 110, e67.
138. Petersen, C., Grasshoff, S., and Luciano, L. 1998. J Hepatol 28, 603.
139. Rayani, A., Bode, U., Habas, E., Fleischhack, G., Engelhart, S., Exner, M., Schildgen, O., Bierbaum, G., Maria Eis-Hubinger, A., and Simon, A. 2007. Scand J Gastroenterol 42, 81.
140. Rodriguez, W. J., Kim, H. W., Brandt, C. D., Schwartz, R. H., Gardner, M. K., Jeffries, B., Parrott, R. H., Kaslow, R. A., Smith, J. I., and Kapikian, A. Z. 1987. Pediatr Infect Dis J 6, 170.
141. Rojas, A. M., Boher, Y., Guntinas, M. J., and Perez-Schael, I. 1995. J Med Virol 47, 404.
142. Rojas, O. L., Gonzalez, A. M., Gonzalez, R., Perez-Schael, I., Greenberg, H. B., Franco, M. A., and Angel, J. 2003. Virology 314, 671.
143. Rollo, E. E., Kumar, K. P., Reich, N. C., Cohen, J., Angel, J., Greenberg, H. B., Sheth, R., Anderson, J., Oh, B., Hempson, S. J., Mackow, E. R., and Shaw, R. D. 1999. J Immunol 163, 4442.
144. Rott, L. S., Rose, J. R., Bass, D., Williams, M. B., Greenberg, H. B., and Butcher, E. C. 1997. J Clin Invest 100, 1204.
145. Saif, L., Yuan, L., Ward, L., and To, T. 1997. Adv Exp Med Biol 412, 397.
146. Saif, L. J., and Fernandez, F. M. 1996. J Infect Dis 174 Suppl 1, S98.
147. Saif, L. J., H.G Greenberg 1997. In "Pathology of Infectious Diseases" (Chandler, F. W., D.H Connor, ed.), p. 297. Appleton and Lange, Stamford.
148. Saif, L. J., Ward, L. A., Yuan, L., Rosen, B. I., and To, T. L. 1996. Arch Virol Suppl 12, 153.
149. Sallusto, F., Geginat, J., and Lanzavecchia, A. 2004. Annu Rev Immunol 22, 745.
150. Sato, A., Iizuka, M., Nakagomi, O., Suzuki, M., Horie, Y., Konno, S., Hirasawa, F., Sasaki, K., Shindo, K., and Watanabe, S. 2006. J Gastroenterol Hepatol 21, 521.
151. Schwartz-Cornil, I., Benureau, Y., Greenberg, H., Hendrickson, B. A., and Cohen, J. 2002. J Virol 76, 8110.
152. hapiro-Shelef, M., Lin, K.-I., Savitsky, D., Liao, J., and Calame, K. 2005. J. Exp. Med. 202, 1471.
153. Shivakumar, P., Campbell, K. M., Sabla, G. E., Miethke, A., Tiao, G., McNeal, M. M., Ward, R. L., and Bezerra, J. A. 2004. J Clin Invest 114, 322.
154. Simonsen, L., Viboud, C., Elixhauser, A., Taylor, R. J., and Kapikian, A. Z. 2005. J Infect Dis 192 Suppl 1, S36.
155. Smith, K. G., Hewitson, T. D., Nossal, G. J., and Tarlinton, D. M. 1996. Eur J Immunol 26, 444.

156. Snodgrass, D. R., Madeley, C. R., Wells, P. W., and Angus, K. W. 1977. Infect Immun 16, 268.
157. Starkey, W. G., Collins, J., Wallis, T. S., Clarke, G. J., Spencer, A. J., Haddon, S. J., Osborne, M. P., Candy, D. C., and Stephen, J. 1986. J Gen Virol 67 (Pt 12), 2625.
158. Stenstad, H., Ericsson, A., Johansson-Lindbom, B., Svensson, M., Marsal, J., Mack, M., Picarella, D., Soler, D., Marquez, G., Briskin, M., and Agace, W. W. 2006. Blood 107, 3447.
159. Theil, K. W., Bohl, E. H., Cross, R. F., Kohler, E. M., and Agnes, A. G. 1978. Am J Vet Res 39, 213.
160. To, T. L., Ward, L. A., Yuan, L., and Saif, L. J. 1998. J Gen Virol 79 (Pt 11), 2661.
161. Tzipori, S. 1976. Med J Aust 2, 922.
162. Vancott, J. L., McNeal, M. M., Choi, A. H., and Ward, R. L. 2003. J Interferon Cytokine Res 23, 163.
163. VanCott, J. L., Prada, A. E., McNeal, M. M., Stone, S. C., Basu, M., Huffer, B., Jr., Smiley, K. L., Shao, M., Bean, J. A., Clements, J. D., Choi, A. H., and Ward, R. L. 2006. J Virol 80, 4949.
164. Velazquez, F. R., Matson, D. O., Calva, J. J., Guerrero, L., Morrow, A. L., Carter-Campbell, S., Glass, R. I., Estes, M. K., Pickering, L. K., and Ruiz-Palacios, G. M. 1996. N Engl J Med 335, 1022.
165. Velazquez, F. R., Matson, D. O., Guerrero, M. L., Shults, J., Calva, J. J., Morrow, A. L., Glass, R. I., Pickering, L. K., and Ruiz-Palacios, G. M. 2000. J Infect Dis 182, 1602.
166. Vesikari, T. 1997. Lancet 350, 1538.
167. Vesikari, T., Isolauri, E., Delem, A., d'Hondt, E., Andre, F. E., Beards, G. M., and Flewett, T. H. 1985. J Pediatr 107, 189.
168. Vesikari, T., Karvonen, A., Korhonen, T., Espo, M., Lebacq, E., Forster, J., Zepp, F., Delem, A., and De Vos, B. 2004. Vaccine 22, 2836.
169. Vesikari, T., Karvonen, A., Puustinen, L., Zeng, S. Q., Szakal, E. D., Delem, A., and De Vos, B. 2004. Pediatr Infect Dis J 23, 937.
170. Vesikari, T., Karvonen, A. V., Majuri, J., Zeng, S. Q., Pang, X. L., Kohberger, R., Forrest, B. D., Hoshino, Y., Chanock, R. M., and Kapikian, A. Z. 2006. J Infect Dis 194, 370.
171. Vesikari, T., Ruuska, T., Bogaerts, H., Delem, A., and Andre, F. 1985. Pediatr Infect Dis 4, 622.
172. Ward, L. A., Rosen, B. I., Yuan, L., and Saif, L. J. 1996. J Gen Virol 77 (Pt 7), 1431.
173. Ward, L. A., Yuan, L., Rosen, B. I., To, T. L., and Saif, L. J. 1996. Clin Diagn Lab Immunol 3, 342.
174. Ward, R. L. 2003. Viral Immunol 16, 17.
175. Ward, R. L., Bernstein, D. I., Shukla, R., McNeal, M. M., Sherwood, J. R., Young, E. C., and Schiff, G. M. 1990. J Infect Dis 161, 440.
176. Ward, R. L., Bernstein, D. I., Shukla, R., Young, E. C., Sherwood, J. R., McNeal, M. M., Walker, M. C., and Schiff, G. M. 1989. J Infect Dis 159, 79.
177. Ward, R. L., Bernstein, D. I., Smith, V. E., Sander, D. S., Shaw, A., Eiden, J. J., Heaton, P., Offit, P. A., and Clark, H. F. 2004. J Infect Dis 189, 2290.
178. Ward, R. L., Bernstein, D. I., Young, E. C., Sherwood, J. R., Knowlton, D. R., and Schiff, G. M. 1986. J Infect Dis 154, 871.
179. Ward, R. L., Clemens, J. D., Knowlton, D. R., Rao, M. R., van Loon, F. P., Huda, N., Ahmed, F., Schiff, G. M., and Sack, D. A. 1992. J Infect Dis 166, 1251.
180. Ward, R. L., Kirkwood, C. D., Sander, D. S., Smith, V. E., Shao, M., Bean, J. A., Sack, D. A., and Bernstein, D. I. 2006. J Infect Dis 194, 1729.
181. Ward, R. L., Knowlton, D. R., Schiff, G. M., Hoshino, Y., and Greenberg, H. B. 1988. J Virol 62, 1543.
182. Ward, R. L., Knowlton, D. R., Zito, E. T., Davidson, B. L., Rappaport, R., and Mack, M. E. 1997. J Infect Dis 176, 570.
183. Weitkamp, J. H., Lafleur, B. J., and Crowe, J. E., Jr. 2006. Hum Immunol 67, 33.
184. Williams, M. B., Rose, J. R., Rott, L. S., Franco, M. A., Greenberg, H. B., and Butcher, E. C. 1998. J Immunol 161, 4227.

185. Xu, J., Yang, Y., Sun, J., Ding, Y., Su, L., Shao, C., and Jiang, B. 2006. Clin Exp Immunol 144, 376.
186. Youngman, K. R., Franco, M. A., Kuklin, N. A., Rott, L. S., Butcher, E. C., and Greenberg, H. B. 2002. J Immunol 168, 2173.
187. Yuan, L., Geyer, A., and Saif, L. J. 2001. Immunology 103, 188.
188. Yuan, L., Honma, S., Ishida, S., Yan, X. Y., Kapikian, A. Z., and Hoshino, Y. 2004. Virology 330, 92.
189. Yuan, L., Iosef, C., Azevedo, M. S., Kim, Y., Qian, Y., Geyer, A., Nguyen, T. V., Chang, K. O., and Saif, L. J. 2001. J Virol 75, 9229.
190. Yuan, L., Ishida, S., Honma, S., Patton, J. T., Hodgins, D. C., Kapikian, A. Z., and Hoshino, Y. 2004. J Infect Dis 189, 1833.
191. Yuan, L., Kang, S. Y., Ward, L. A., To, T. L., and Saif, L. J. 1998. J Virol 72, 330.
192. Yuan, L., and Saif, L. J. 2002. Vet Immunol Immunopathol 87, 147.
193. Yuan, L., Ward, L. A., Rosen, B. I., To, T. L., and Saif, L. J. 1996. J Virol 70, 3075.
194. Yuan, L., G. A. M., Azevedo, M. S. P., Wen, K., Zhang, W., Lovgren-Bengtsson, K.,Morein, B., and Saif, L.J 2006. In "9th Ds RNA virus symposium", Cape Town, South Africa.
195. Zhang, W., A. M. S. P., Gonzalez, A. M., Saif, L. J., Nguyen, T. V., Wen, K., Yousef, A., and Yuan, L. 2006. In "25th Annual Meeting of American Society of Virology", Madison, Wisconsin.
196. Zhou, R., Wei, H., Sun, R., and Tian, Z. 2007. J Immunol 178, 4548.
197. Zissis, G., Marissens, D., Lambert, J. P., and Marbehant, P. 1983. Dev Biol Stand 53, 201.

Part IV
Respiratory Pathogens

Chapter 11
Mucosal Control of *Streptococcus pneumoniae* Infections

Jacinta E. Cooper and Edward N. Janoff

Abstract Infections due to *Streptococcus pneumoniae* are associated with up to a million deaths per year, especially among young children and older adults. Mucosal sites, particularly the nasopharynx and upper and lower respiratory tracts, serve as the initial site of exposure and the subsequent morbidity and morality associated with this invasive bacterial pathogen. Thus, control of the pneumococcus at mucosal sites, where infection begins, is a critical goal in preventing disease. The mechanisms of mucosal control are not well characterized. These host strategies may support both antibody- and complement-dependent bacterial control by phagocytes. Newer data suggest that antibody-independent but CD4+ T cell-dependent processes may play an important role in bacterial clearance at these sites of entry. The bacteria employ a multifaceted range of pathogenic and immune-evasive activities to subvert the host response. Vaccine development should continue to focus on exploiting both the shared and the serotype-specific characteristics and vulnerabilities of the organism to enhance mucosal defense against this common and virulent pathogen.

11.1 Impact of Pneumococcal Mucosal Disease

Streptococcus pneumoniae is among the most prominent causes of significant morbidity and mortality worldwide. Young children, the elderly, and African-Americans are disproportionately affected by pneumococcal disease [1]. The incidence of invasive pneumococcal infections, a specific but insensitive marker of all pneumococcal infections, is highest in very young children and then begins to increase again with age in adults (Fig. 11.1; left y-axis lacks label). Despite the

E.N. Janoff
Tim Gill Professor of Medicine and Microbiology, Mucosal and Vaccine Research Program Colorado (MAVRC), Division of Infectious Diseases, University of Colorado Denver, 4200 E. Ninth Ave.; Room 1813, Denver, CO 80220, USA
e-mail: Edward.Janoff@uchsc.edu

M. Vajdy (ed.), *Immunity Against Mucosal Pathogens*,

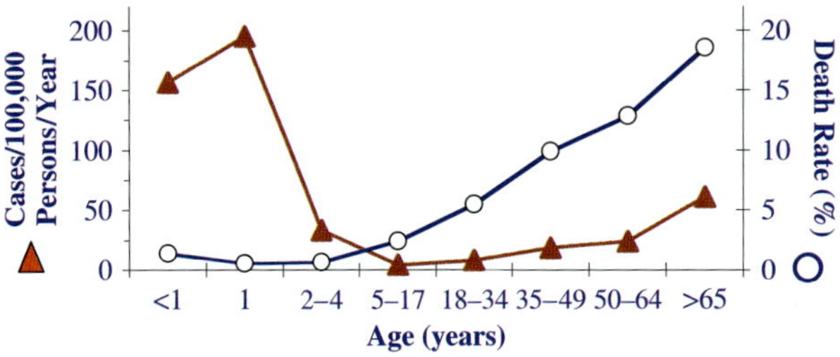

Fig. 11.1 Rates of Invasive Pneumococcal Infections 5 U.S. Cities - 1997–99. (Centers for Disease Control; Active Bacterial Core Surveillance)

increased incidence, bacteremia-related mortality is lowest among young children in the U.S. and increases with age.

The mucosal manifestations of pneumococcal disease, especially acute otitis media and pneumonia, are far more frequent and are associated with more morbidity and deaths than invasive systemic disease (e.g., bacteremia and meningitis). In the U.S., the estimated 3,000 cases of meningitis and 50,000 cases of bacteremia are far exceeded by the 125,000 cases of pneumonia requiring hospitalizations and the one quarter of the 7 million cases of otitis media caused by *S. pneumoniae* yearly [2, 3] In children, acute otitis media is the most common clinical manifestation of pneumococcal infection and outpatient diagnosis leading to an antibiotic prescription [3].

With 150 million cases in resource-limited countries yearly, pneumonia kills more children than AIDS, malaria, and measles combined [4]. Three quarters of childhood pneumonias occur in just 15 countries, particularly in Asia and Africa. Worldwide, 30,000 children under 5 years of age die each day; approximately 25% of these deaths (2.7 million/year) are due to pneumonia. About a third of pneumonia deaths are due to *S. pneumoniae* (up to a million deaths/year). Among adults in the U.S., community-acquired pneumonia (CAP) is the sixth leading cause of death, fifth among persons >65 years of age. *S. pneumoniae* is consistently the leading cause of bacterial CAP in children and adults, comprising a quarter to a third of all adult cases, as well as two thirds of all bacteremic pneumonias [5, 6]. In adults, the incidence of pneumococcal CAP continues to increase with age. In a prospective study in Spain, the rate of pneumococcal CAP was 14 per 10,000 inhabitants for those 75 years and older; 13 times higher than persons aged 15–44 and 2.7 fold higher than those 64–74 years [7]. Mortality with pneumococcal pneumonia increases from ≤5% for those treated as out-patients, to 12% for hospitalized cases, and ≥25% in patients with severe disease, including bacteremia [8]. Mortality also increases with age (Fig. 11.1). In an older Veteran population in Minneapolis, the mortality rate with pneumococcal bacteremia (n = 215; 1981–1994) at 3 weeks was 33% [9]. The vast majority of cases were due to pneumonia and most deaths occurred within the first week (28/33% = 85%), often within the first day (15/33% = 45%). The high proportion of all mortality that occurs early in the course of infection has persisted for over 35 years [10], suggesting that, despite the availability of potent antimicrobials and intensive care support, prevention must be the keystone to effective control of pneumococcal disease. Prevention has been focused to date on development and use of effective vaccines that generate antibodies to the bacterial capsule.

S. pneumoniae-specific antibodies likely provide the most effective mechanism of control and intervention [11, 12]. Antibodies that mediate opsonization, phagocytosis and killing of the organism are most consistently directed to the polysaccharide capsule. Challenges in vaccine development have been related in part to the large number of different capsular polysaccharide serotypes (>90), which are clustered into serogroups, but which otherwise have little cross-reactivity. Among young children, serogroups 1, 6, 14, 19, 23 are the most prominent serogroups around the world associated with disease. Serotypes 1, 5, and 46 are rarely found in carriage samples but cause a higher proportion of invasive pneumococcal disease [13]. A small number of serogroups are responsible for most of the cases of each disease syndrome. Serogroup 1 and 14 are more often isolated from blood; 6, 10, and 23 from cerebrospinal fluid; 3, 19, and 23 from middle ear fluid [14].

11.2 Colonization

Colonization is the critical first step in developing pneumonia for two reasons. First, pneumonia is thought to begin with inhalation of organisms into the lung (aspiration), after establishment of recently-acquired nasopharyngeal strains of *S. pneumoniae* [15]. Second, colonization, rather than pneumonia, is most consistently implicated in transmission. Blocking colonization should block pneumonia. Indeed, protein-conjugate vaccines decrease rates of colonization with the vaccine serotypes in children [16]. A striking correlation exists between the effect of the protein-conjugate vaccine on carriage of vaccine-related strains in the nasopharynx of children and its efficacy in preventing invasive pneumococcal disease [17, 18]. At a second level of impact, vaccination of young children with the protein-conjugate decreases rates of invasive disease in children, as well as among adults from 20−>65 years by 20–40% [19] (Fig. 11.2). This effect is related to 7-valent (capsules from 7 serotypes) conjugate vaccine rather than use of the adult 23-valent pneumococcal polysaccharide vaccine because the decline in adults was only among the 7 serotypes included in the pediatric conjugate vaccine whereas rates of disease due to the remaining serotypes in the adult vaccine did not change in this period. This decrement in adults is likely mediated by the effect of the vaccine on nasopharyngeal carriage in children, an apparently important reservoir for invasive strains in adults. Declines in carriage of vaccine-related strains in young children are associated with decreased carriage of vaccine, but not non-vaccine strains among adults in the same community [17, 18].

Colonization is a complex process. That pneumococcal colonization clears within 12 weeks in healthy adults suggests the effects of acquired immunity [20]. A cross-sectional study of adults and children in a childcare setting in the United States found that 15% of adults and 65% of children were colonized with the

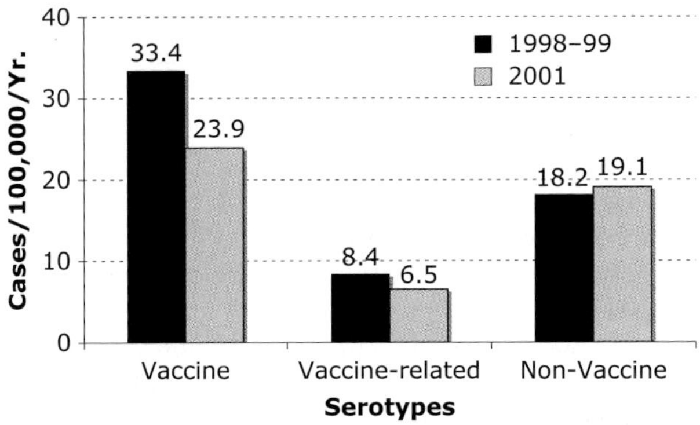

Fig. 11.2 Vaccination reduces the prevalence of vaccine and non-vaccine related serotypes Adapted from: Whitney C, et. al. N Engl J Med 2003; 348: 1737

pneumococcus [3]. Indeed, colonization with *S. pneumoniae* begins very early in life [21, 22, 23] and elicits antibodies to multiple bacterial antigens [15, 24], including pneumococcal capsule [25, 26] and the many proteins that intercalate through to the surface. Capsule may inhibit mucus-mediated clearance [27], and capsule-specific antibodies are associated with decreased rates of type-specific acquisition and/or clearance of the organism [28]. Two other bacterial surface proteins implicated in epithelial cell binding, pneumococcal surface antigen A (PsaA) and pneumococcal surface/choline-binding protein A (PspA; CbpA), elicit antibodies following colonization [15, 29] (reviewed in [24]). Antibodies to the former reduce carriage in mice [30]and to the latter are associated with protection against experimental colonization in humans [29].

In addition to specific antibodies, the inflammatory response elicited by colonizing organisms may contribute to clearance. Proinflammatory constituents of *S. pneumoniae* include choline-binding protein A (CbpA) [31], pneumolysin (the major pneumococcal protein toxin) [32, 33, 34], and the pneumococcal cell wall component phosphoryl choline (ChoP). Cell wall ChoP shares homology, and cross-reactive antibodies, with both the lipoteichoic acid of *S. pneumoniae* and lipopolysaccharide of *H. influenzae* [35]. Such epithelial cell activation facilitates binding of bacterial ChoP to its primary target, platelet activating factor receptor (PAFr), a prominent receptor for *S. pneumoniae* on epithelial cells [36, 37].

Despite the substantial body of accumulated evidence that capsular conjugate vaccine and antibodies to a range of pneumococcal antigens prevent colonization, as described above, several provocative recent studies from 2 independent groups provide persuasive evidence that protection against and particularly clearance of colonization in mice can be elicited with whole organisms in a CD4+ T cell-dependent, but antibody-independent manner. The latter point was confirmed in antibody-deficient mice, so the role of CD4+ T cells was not restricted to regulating B cell responses [32, 38]. These T cell responses may involve the cell wall polysaccharide via toll-like receptor 2 (TLR2) or the bacterial toxin pneumolysin via TLR 4, but the effector mechanisms of protection have not been proposed. A limitation of these data is the focus on mice, which are not natural hosts and do not become colonized or infected in the wild. However, potentially intriguing human correlates are the observations that although the frequencies of nasopharyngeal pneumococcal colonization among HIV-infected and -uninfected persons [39, 40, 41] are typically similar, patients with lower CD4 cell counts may have somewhat higher nasopharyngeal carriage [41, 42]. In the one longitudinal study, patients with HIV infection were more likely to acquire new strains of *S. pneumoniae* and less likely to lose the carrier state. As a result, carriage was more persistent among HIV-infected adults than among controls [41]. These data are consistent with a putative role for CD4+ T cells in resolution of colonization, although patients with HIV also exhibit B cell and other mucosal and systemic immunity defects.

Thus, we do not currently know with certainty whether antibodies to capsular polysaccharides, antibodies to surface protein antigens, antibody-independent CD4+ T cell-dependent mechanisms, or some combination underlie the susceptibility to colonization and the ability to clear the organism effectively. Nor do we know how these immunologic constituents mediate protection by enhancing opsonization and phagocytosis, preventing adherence to epithelial cells, or perhaps by CD4-dependent modulation of epithelial cell receptivity to attachment. We do know that protein-conjugated polysaccharide vaccines do decrease rates of colonization and disease, and specific antibody, rather than cellular responses are the most well documented vaccine outcomes.

As opposed to the specific acquired humoral and cellular mechanisms invoked above, innate or non-specific mechanisms are also active at the mucosal surface. Mechanical barriers of the respiratory tract provide the first line of defense against pathogens. Particles may be trapped in the nasal hairs, nasopharynx, or lower respiratory tract from the glottis to bronchioles. Particles can also become trapped in the mucociliary blanket that lines the lower respiratory tract. Mucin, a glycoprotein in the blanket, and its contents, can be microbiocidal. Organisms trapped and killed in the mucociliary blanket are removed by the ciliary escalator. The increased risk of pneumococcal disease associated with both active smoking, and passive exposure to tobacco smoke, may relate in part to the effects of smoke on inhibiting mucociliary activity. However, bacteria may overcome these mechanical defenses and reach the terminal airways, requiring other mechanisms to maintain sterility in the alveoli [43].

11.3 Role of Complement in Protection Against *S. pneumoniae*

Complement has been considered critical to protection against invasive pneumococcal disease. *S. pneumoniae* is the most common cause of invasive bacterial infection among patients with congenital or acquired deficiencies of early components of the classical pathway (C1, C4, and C2) and the alternative pathway (factor I, factor

H, and Factor B); C3 is common to both pathways. Indeed, Factor C2 deficiency, associated with an increased incidence of invasive *S. pneumoniae* infection, is the most common complement deficiency among persons of European origin [44]. Although less common, C3 deficiency is associated with increased rates of pyogenic infections with encapsulated organisms such as *S. pneumoniae*, and with recurrent pneumococcal infections [45, 46].

Complement activation may occur via the classical, alternative, or lectin pathways (reviewed in [47, 48, 49]), each of which converges at C3. Complement may facilitate clearance of the organism by the three potential mechanisms. First, although complement activation by any pathway may initiate the C3-activated membrane attack complex by terminal complement components ($C5 \rightarrow 9$), the thick-walled pneumococcus is not susceptible to such direct lysis. Second, complement can bind to *S. pneumoniae* and mediate phagocytosis by both antibody-dependent and -independent mechanisms [50]. The pneumococcal capsule, rather than other protein virulence factors, is the primary determinant and inhibitor of complement deposition [51]. However, several bacterial virulence factors also serve to inhibit complement deposition or activation. These surface proteins include pneumococcal surface proteins A (PspA) [52] and C (PspC); [PspC is also known as choline binding protein A (CbpA) or secretory immunoglobulin A (IgA) binding protein (SpsA)], pneumococcal surface adhesin A (PsaA), and pneumolysin [33, 53] (reviewed in [54]). Surface binding and activation of C3 and its products are most important in engaging complement receptors for phagocytosis and killing. The ability of mucosal IgA to fix and activate complement is not well characterized under physiologic conditions. IgG, however, can serve this function, and IgG comprises about half of the antibody present in the lower lung. In vitro, phagocytes will kill *S. pneumoniae* in the presence of both capsule-specific IgA and whole hemolytic complement, but not in the presence of either one alone at limiting concentrations (Fig. 11.3: [Anti-T14 IgA- Type 14 specific anti-IgA, C-baby rabbit complement, Phagocytes-human neutrophils]). Finally, the cleavage products of C5 and C3, C5a, and, to a lesser extent, C3a, serve as anaphylotoxins to recruit and activate phagocytes to and at the site of infection. C5-deficient mice also experience compromised

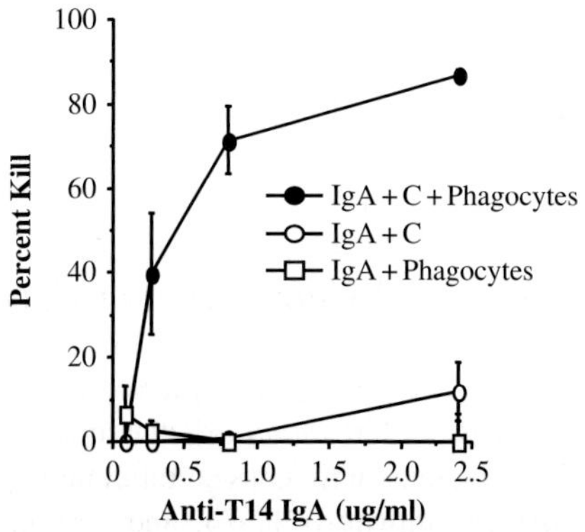

Fig. 11.3 Killing of *S. pneumoniae* with Type 14 specific IgA, baby rabbit complement and human neutrophils. Adapted from Janoff et al., [73]

ability to control respiratory pneumococcal infection [33, 55]. Indeed, C5a and its receptor may facilitate both mucosal and systemic bacterial clearance [33, 55].

In the absence of antibody, very high concentrations of complement alone, such as those found in serum, may promote killing of some *S. pneumoniae* by polymorphonuclear cells (PMNs) [50, 56, 57]. However, conditions at mucosal sites are quite different. Levels of complement at mucosal sites, including upper and lower respiratory tract, appear to be extremely low [58]. Individual complement components may be produced locally by activated monocytes or alveolar macrophages (AM) (C3, C5, C9) [59, 60, 61, 62, 63, 64, 65], recruited granulocytes (C3) [66], and alveolar epithelial cells (C2, C4, Factor B, C3 and C5 and others) [67, 68, 69, 70]. Whether these isolated components are sufficient to support antibody-dependent killing of *S. pneumoniae* by phagocytes in not known, and whether such complement activity would need to be targeted to the organism or the local inflammatory milieu in the mucosal environment, is also not known but would be quite relevant to investigate.

Thus, whether the role of complement is to mediate binding of the organism to complement receptors to enhance phagocytosis and oxidative killing or whether e.g., soluble C5a is a sufficient stimulus for phagocyte function in the presence of specific antibody and other proinflammatory cytokines, such as TNFα, is unclear. Classic studies suggest that complement receptors may serve varying functions in an activated compared with a resting state; the former supporting binding, the latter phagocytosis, but neither eliciting reactive oxygen species [71]. In contrast, immunoglobulin Fc receptors can support phagocytosis and reactive oxygen species production. The activation of neutrophils and AM may require local inflammation, which may abrogate the need for high levels of complement to affect IgA- and IgG-mediated killing of mucosal pathogens. Classically, Fc receptors mediate phagocytosis by resting macrophages [72], but, as we have shown with neutrophils, they neither phagocytose or kill pneumococci without activation [73]. Complement receptors may mediate binding to macrophages, but not internalization without additional stimuli. Complement activation by antibody-opsonized *S. pneumoniae* may also generate sufficient degradation products (particularly C5a) in the fluid phase to provide such phagocyte activation.

11.4 Phagocytosis

S. pneumoniae may be susceptible to several local innate immune factors, but typically require phagocytosis for effective clearance after opsonization by antibodies. Phagocytosis seems essential because pneumococci have the ability to evade extracellular killing by the recently described DNA "nets" and associated innate proteins produced by neutrophils [74]. Gordon, et al. found that *S. pneumoniae* bind to human AM and are subsequently internalized and sent to compartments containing lysosome-associated membrane protein-1 (LAMP-1), a transmembrane glycoprotein localized in lysosomes and late endosomes [75]. Opsonization increases binding, rates of internalization, and maturation of phagosomes. Internalization with complement as an opsonin occurs at similar rates as when IgG is the opsonin. AM can phagocytose and kill opsonized Type 1 pneumococci, and internalization is necessary for some of the bactericidal activity of these cells [75]. Hoidal, et al. also found that human AM can phagocytose *S. aureus* and *E. coli*. However, PMN ingested the bacteria faster and more efficiently generated a respiratory burst than AM. Over time, AM are able to take up more bacteria than PMN, but PMN are more effective at killing the ingested bacteria [76].

Pneumococci are opsonized by both antibody and complement. The deposition of complement on the surface of the organism is dependent on the capsule (serotype dependent) [50] and surface proteins [52] such as PspA. Complement deposition may also involve interaction between IgA (or IgG) and complement, although IgA has no C1q or other obvious mechanism of complement activation. The binding of antibody to the organism is dependent on the organism's strain and serotype and the avidity of the antibody. The opsonization of the organism by complement and IgA may be independent, but parallel events, both required but independent. Indeed, pneumococci alone, in the absence of antibody, can activate and bind complement. Opsonization of the pneumococcus with immune serum enhances uptake by AM and PMN [77]. Nevertheless, the opsonization by both complement and IgA allows for the engagement of both the complement receptors and the FcαRI of the phagocyte. The cross-linking and aggregation of FcαRI receptors initiates protein tyrosine kinase (syk) and Bruton tyrosine kinase (Btk) signaling through immunoreceptor tyrosine-based activation motifs (ITAM) on the FcR$_\gamma$ chain [78, 79, 80]. Ligation of complement receptors induces tyrosine kinase activity [81]. We observed dose-dependent phagocytosis when both class of receptors are engaged, but phagocytosis diminishes to virtually nil with unstimulated phagocytes when either IgA or complement alone is used to opsonize pneumococci (Fig. 11.3). Killing is reduced if either Fc$_\alpha$ or complement receptors are blocked (Fig. 11.4: [CR1-complement receptor 1, CR3-complement receptor 3]) [73].

Complement-independent phagocytosis is also under consideration. The ability of rats without an intact complement system to clear *S. pneumoniae* is somewhat

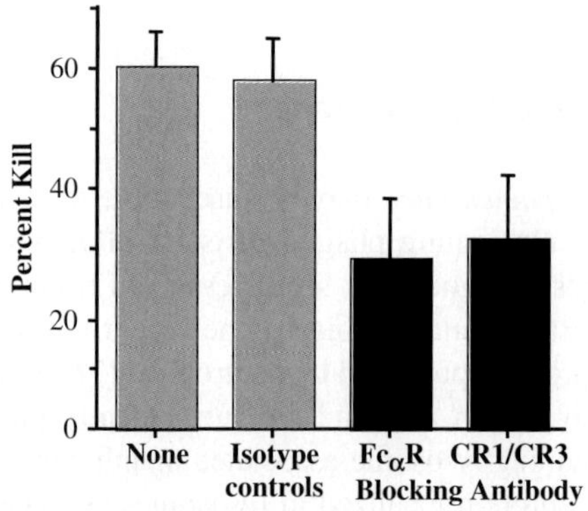

Fig. 11.4 Blocking Fc$_\alpha$ and complement receptors inhibits IgA mediated killing of the pneumococcus

preserved. These animals are able to clear pneumococci at rates that exceed that predicted for a mucociliary mechanism. Experiments with decomplemented animals lead Rehm, et al., to conclude that the bacteria are killed in situ, possibly by phagocytosis [82]. Complement depletion by heat inactivation may not decrease the ability of immune serum to act as an opsonin in vitro. Hof, et al., propose that in the low complement environment of the lung in a non-immune host, AM would not efficiently take up *S. pneumoniae* unless there are adequate opsonins present that do not require complement. [77]. Uptake of *S. pneumoniae* by AM required large concentrations of serum, whereas only small amounts of serum were required as an oposonin for efficient uptake of *S. aureus* by these cells. Levels of specific antibody to each organism were not compared. AM also phagocytosed *S. aureus* in the presence of albumin, suggesting that either albumin or a bound protein can function as an opsonin, or AM do not need an opsonin to phagocytose *S. aureus*. Thus, the lack of appropriate opsonins in the lungs may result in the ineffective uptake of certain bacteria, particularly *S. pneumoniae*, by AM.

Local infection and inflammation may serve as a stimulus for influx of PMN and additional opsonins, primarily complement, from the blood [83]. That PMN's appear more dependent on opsonins than alveolar macrophages, but mediate bactericidal activity more effectively in their presence, suggests AM may function better in the very early stages of bacterial exposure when there are few opsonins available, such as in the normal healthy lung [76, 84]. When PMN's become activated, the requirement for complement for phagocyte uptake and killing of pneumococci is diminished. This cellular activation could facilitate complement-independent phagocytosis by lowering requirements for the FcαR mediated pathway by changing the number and/or affinity of FcαRs [85]. Granulocyte and granulocyte-macrophage colony stimulating factors (G-CSF and GM-CSF, respectively) can both induce a transition from low to high-affinity neutrophil Fcα receptors. Saturation binding and Scatchard analysis demonstrate an increase in affinity, but a decrease in number of IgA receptors in response to these factors [86]. FcαR expression is upregulated 3–4-fold on blood neutrophils after exposure to formyl-MET-LEU-PHE (fMLP) and zymosan by flow cytometry, which may enhance the ability of these cells to support IgA-mediated phagocytosis. This increased expression was not inhibited by cyclohexamide or puromycin, suggesting release from intracellular storage pools or by amplifying FcαR signaling [78]. The aggregation of Fcα receptors on U937 cell lines and neutrophils result in tyrosine phosphorylation of FcR gamma chain, p72syk, and Bruton tyrosine kinase. Such events could bypass the complement receptor ligation requirement and provide signals for full phagocytosis and organism killing.

Cellular activation could also affect changes in the complement receptors, allowing for conformational changes, which may permit binding of the lectin-like domain of CD11b on macrophages directly to pneumococci [87]. The CD11b subunit of complement receptor 3 (CR3) has a lectin-like domain (LLD) that recognizes mannose and β-glucans. IgA recognizing the capsular glucuronoxylomannan of *Crytococcus neoformans* promotes complement-independent phagocytosis by macrophages. This uptake is proportional to CR3 expression and inhibited by antibodies to CR3 and CR4, by the presence of soluble glucuronoxylomannan and may involve cross-linking (auto-cross-linking) of the complement receptors by some other mechanism.

Work by Coonrod, et al. demonstrated the killing of pneumococci in the lung varies by species. In the rat lung, killing is mainly extracellular, whereas in the rabbit lung, killing occurs in the macrophage. The inability of rats to bind complement-coated particles may explain why rat macrophages do not readily uptake pneumococcus, but it is not clear whether complement is required in vivo to phagocytose inhaled pneumococci. The levels of human C3, in the alveolar lining material is very low, and may not be sufficient to opsonize pneumococci. In the lung, alveolar macrophages may be able to phagocytose pneumococci with smaller amounts of complement and antibody than are needed in vitro. In animals in which pneumococcal clearance is mediated by AM phagocytosis, opsonsins such as surfactants (e.g., SP-A) may be very important [88].

Neutrophils and other phagocytes are necessary for killing of pneumococci. Neutrophil activation can occur in a range of biological circumstances. First, activation can be initiated in the presence of cytokines and chemokines. Neutrophil chemotaxis, in the presence of suboptimal concentrations of IL-8, is enhanced by MCP-1, MCP-2, MCP-3, and SDF-1α [89]. IL-8, often derived from activated epithelial cells and macrophages, increases surface expression of CD11b/CD18 on neutrophils and increases the attachment index of erythrocytes coated with iC3b to neutrophils [90]. In addition, TNFα was shown to increase the surface density of CD11b [91]. The cytokine effects were not modified by cyclohexamide treatment, suggesting that their increased expression was not due to synthesis de novo. The inference is that CD11b/CD18 and CR1 in PMN's are stored in intracellular pools and transported to the cell surface upon activation. Another cytokine, IL-15, enhances neutrophil phagocytosis [92], and the Syk inhibitor, piceatannol, significantly inhibited the ability of IL-15 to enhance phagocytosis. Moreover, IL-15 increased the ability of PMNs to phagocytose heat-killed *Candida albicans* in a dose-dependent manner, without opsonization by antibody or complement related products. [93].

A second mechanism of neutrophil activation may result from recognition by antibody. In this context, FcαR intracellular signals via gamma chain increase the avidity of $\alpha_M\beta_2$, considered "inside out signaling" [94]. Indeed, PI3-K has a role in activation of $\alpha_M\beta_2$ (CR3) by immune complexes, and the FcγR-initiated pathway is dependent on PI3-K activity (inhibited by wortmannin). As a third mechanism of activation, PMN's may directly recognize microbial products through pathogen recognition receptors or TLR [95]. Peptidoglycan and PAM, TLR2 ligands, activated Akt in neutrophils and was inhibited by a PI3-K blockade (wortmannin). The PI3-K and Akt axis occupies a central role in TLR2-induced activation of neutrophils [96]. In this context, TREM-1 ligation synergizes with the activation by the TLR ligands LPS, PAM, and R-848 (resiquimod), increasing CD11b surface expression by almost two-fold [97]. LPS exposure increased surface expression of CD11b/CD18 on human neutrophils, which again, was not interrupted by protein synthesis inhibitors, but was reduced by treatment of neutrophils with

anti-CD14 monoclonal antibody indicating surface receptor involvement in activation.
Ultimately, neutrophils must ingest and kill bacteria by serially attaching and internalizing organisms and releasing antimicrobial factors (e.g, lysozyme, proteases, phospolipase A2, defensins, cathelicidins, B/P-IP) and reactive oxygen species into phagocytic vacuoles [98].

These data are consistent with earlier observations that serial exposures to two soluble, inflammatory stimuli elicit a greater and often more prolonged response by neutrophils [99] (reviewed in [100]). The first stimulus "primes" the response to the second. We are exploring whether such sequential exposures and subsequent activation may provide a viable sequence in which neutrophils and, perhaps, alveolar macrophages may support phagocytosis and killing of *S. pneumoniae* in the alveolus, in the presence of no or low complement levels. Killing of *S. pneumoniae* by AM in our lab shows a dose-dependent response to complement [101]. Activation requirements for complement may be lower if the phagocytes have been primed by chemotactic stimuli, such as IL-8 and other local factors, such as C-reactive protein (CRP). Epithelial cells may offer both cognate and soluble priming and activating signals, as may activated alveolar macrophages, thus providing a relevant model for understanding mucosal defense against local infections. The ability of epithelium-derived chemokines to recruit and stimulate neutrophils and monocytes (e.g., by C-X-C [IL-8] and C-C [MIP-1α, MIP-1β, RANTES] chemokines, respectively) and of direct interactions between phagocytes and epithelium may each facilitate the efficiency of antibody-mediated killing.

Once neutrophils migrate to the site of infection and bind to the pathogen, the organisms are phagocytized into the cell [102]. Antibodies or complement (esp. C3bi) increase the efficiency of this process [103]. The mechanisms of phagocytosis are not well defined but likely involve shifts in the mobility of β2 integrins (CD11/CD18). These adhesion molecules are essential for neutrophil adherence, migration and activation. Neutrophil contact between C3bi during initiation of phagocytosis leads to aggregation of β2 integrins and release of intracellular calcium.

11.5 Antibody Responses to Pneumococcal Capsular Polysaccharides

Antibodies likely provide a prominent mechanism of protection against colonization with *S. pneumoniae*, but antibodies are essential for bacterial clearance from the lung and blood. The polysaccharide capsule of *S. pneumoniae* is its primary virulence factor. Differences among the >90 capsular types account for approximately 60 fold differences in the invasiveness of the organism [104, 105, 106, 107]. The capsule confers the ability to resist recognition and uptake by phagocytes. Antibodies to the capsule opsonize the organism and provide the target-effector bridge between the bacteria and the phagocytes to mediate killing by phagocytes, classically

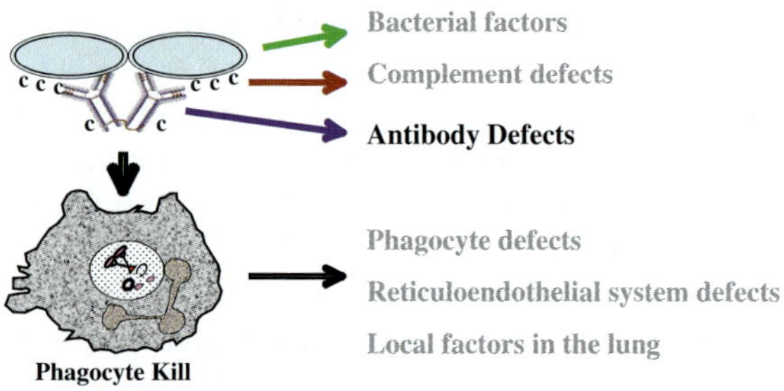

Fig. 11.5 Obstacles to Killing of *S. pneumoniae*

in the presence of complement. Failure to produce antibodies, e.g., in young children, adults with hypogammaglobulinemia, chronic lymphocytic leukemia, multiple myeloma, or polysaccharide-specific defects, result in increased rates of invasive disease [108, 109, 110] (reviewed in [111, 112]). Each group will also be deficient in mucosal antibodies as well. The ability of both 23-valent polysaccharide and protein-polysaccharide vaccines to provide protection is based on their ability to induce capsule-specific antibodies. Indeed, of the various obstacles posed by the bacteria and the host to effectively killing *S. pneumoniae* (Fig. 11.5), only enhancement of capsule-specific antibody production is amenable to intervention. Antibodies in the blood are important in preventing invasive disease, but cases of pneumonia outnumber those of detectable bacteremia by at least 5–10 fold, so the burden of disease is rooted in the mucosa, particularly the lung, where mucosal antibodies, both IgA and IgG, may be most relevant.

11.6 IgA Responses to *S. pneumoniae*

The 23-valent PPS vaccine, designed to induce antibodies of sufficient levels and function to prevent disease, consistently elicits rapid and high concentrations of capsule-specific serum IgA [113, 114, 115, 116]. However, the vaccine does not reliably elicit increased antibodies to the capsule at mucosal sites where infection and disease begin (nasopharynx and lung). Colonization with *S. pneumoniae* does elicit local IgA to the organism [29, 117, 118, 119]. Capsule-specific IgA is elicited by parenteral immunization in blood and breast milk. Although IgA is the most prominent antibody class at the mucosa, and *S. pneumoniae* is an invasive mucosal pathogen, only a subset of people with selective IgA deficiency show increased rates of mucosal infections. Those at risk exhibit concomitant IgG2 deficiency, those with mucosal irritation (e.g., allergies), and those unable to produce IgA *in vitro*, even with exogenous stimuli. This discrepancy may relate to the ability of most IgA-deficient persons to produce compensatory increases in mucosal IgM, which is also transported by the polymeric Ig receptor [pIgR]/secretory component across epithelial cells to the moist mucosal lumen in contact with the environment.

11.7 Monomeric, Polymeric and Secretory Capsule-Specific IgA

IgA differs from IgG by its prominence in mucosal sites and by its ability to form multimers (polymeric IgA [pIgA] with 2 or more IgA molecules bound by a J chain). In addition, pIgA can associate with the polymeric Ig receptor (pIgR) at the basal epithelial surface to allow transport from plasma cells in mucosal tissues across the epithelium for release into the lumen of airway, lung, intestine, etc. In this epithelial transport process, pIgA is bound by pIgR at the basal surface and released as secretory IgA [(S-IgA) pIgA bound to secretory component, a cleavage product of pIgR]. We characterized the biologic impact of the molecular forms of hMAb capsule-specific IgA (monomeric [mIgA], polymeric IgA [pIgA], and secretory IgA [SIgA]) on phagocytosis and susceptibility to cleavage by IgA1 protease [120]. The efficiency of S-IgA to support phagocytosis of *S. pneumoniae* was comparable to that of pIgA and both forms exceeded that of mIgA by 5-fold. This structure-function relationship was associated with three factors. First, the avidities, or functional affinities, of both pIgA and SIgA for pneumococcal capsules exceeded that of mIgA. Second, both pIgA and S-IgA required less complement to achieve similar levels of bacterial phagocytosis than did mIgA, indicating that secretory component does not hinder the effect of complement. Third, both pIgA and SIgA mediated agglutination of the organism, whereas mIgA did not. All three forms of capsule-specific IgA showed comparable susceptibilities to cleavage and functional inhibition by bacterial IgA1 protease, demonstrating that secretory component does not prevent the proteolytic degradation of IgA1 by pneumococcal IgA1 protease. IgA1 cleavage results in formation of identical Fab fragments for each of the molecular forms, thereby abolishing the contribution of multivalence of pIgA and S-IgA. In summary, the polymeric forms of IgA (both pIgA and SIgA) provide a substantial advantage in binding, agglutination and phagocytosis of the organism.

Studies on the functional interaction of human pneumococcal capsule-specific IgA and the organism have involved both polyclonal [73, 113, 121] and monoclonal antibodies [37, 120, 122, 123, 124, 125]. The latter reveal that variable regions of human monoclonal antibodies to the polysaccharide capsular polysaccharides show isotype class switching, high rates of mutation, and high percent replacement mutations, consistent with an antigen-derived process and immunologic memory, and comparable to the antibodies generated in response to pathogenic proteins. The availability of the specific human monoclonal antibodies and their related functional activity will facilitate characterization of the impact of a prominent and intriguing pneumococcal virulence factor, IgA1 protease, on protection against the organism.

11.8 IgA1 Protease (IgA1P) and Killing of *S. pneumoniae*

Bacterial IgA1 proteases may sabotage the protective effects of IgA by cleaving the Fcα1 heavy chain antigen-binding Fab portion from the phagocyte-binding Fc effector component. Because these enzymes are specific for IgA1 from humans and higher primates, the availability of pneumococcal capsule-specific human monoclonal antibodies permitted the characterization of the functional consequences of the enzyme using IgA1 and IgA2 recognizing and with functional activity against the same pneumococcal capsular serotype. This work showed 1) striking effects on phagocytosis of organisms in vitro and mouse protection against lethal infection in vivo [120, 123], as well as 2) enhancement rather than inhibition of adherence to epithelial cell in vitro by novel mechanisms [37]. This enhancement was dependent on cleavage of the IgA1 by IgA1 protease. The adherence-promoting properties of cleaved antibodies correlated with the cationic characteristics of their variable segments (pI of V_H predicted from nucleotide sequence) suggesting that bound Fab fragments may neutralize the inhibitory effect of negatively charged capsules on adhesive interaction with host cells. Coating of pneumococci with anti-capsular antibody and their subsequent cleavage by IgA1 protease unmasked the bacterial phosphorylcholine (ChoP) ligand, typically in large part buried within and below the capsule, allowing for increased adherence mediated by binding to the platelet activating factor receptor on epithelial cells. Enhancement was blocked by competitive inhibition with C reactive protein binding to ChoP, and by antibody to platelet activating factor (PAF) receptor and by PAF receptor antagonist. Thus, bacterial IgA1 proteases serve novel functions. In addition to inhibiting antibody-mediated killing of the organism, these enzymes may enable pathogens to subvert the antigen-specificity of the humoral immune response to facilitate adhesive interactions and persistence on the mucosal surface. In the context of the many mechanisms of defense and counter-offense displayed by the host and organism, the real question is whether there is relevant clinical evidence for mucosal defense against *S. pneumoniae* infection in vivo.

11.9 Breast Milk and Pneumococcal Infection

In resource-poor countries, breastfeeding is associated with protection against pneumonia, diarrhea and death [126, 127, 128] but the mechanisms of protection have not been well defined. In the last 40 years, only 30 or so articles have been published about the interaction between *S. pneumoniae* and breastfeeding. No published study has evaluated the effects of maternal or infant immunization on rates of colonization in breast-fed infants. Kaleida, et al. showed no significant differences in rates of colonization with 1 or more respiratory pathogens among breast-fed vs. formula fed infants at 2 months of age (34.8 vs 35.1%) [129]. Rosen, et al., attempted to correlate levels of capsule-specific antibodies to 3 of the 6 common serotypes in milk of 310 mothers with the frequency of colonization in their infants at 2, 6 and 10 months postpartum [130]. Anti-capsular antibodies were detected in $<10\%$ of samples although antibodies to the cell wall polysaccharide and its phosphorylcholine constituent (ChoP) were present in $>80\%$, and neither correlated with colonization. However, fewer than two milk samples were available from each mother, and correlations were not made between acquisition of specific types and specific antibodies over specific intervals.

Immunization of women in the Gambia in the 3rd trimester with the 23 valent vaccine was associated with higher cord blood levels of specific IgG, but levels fell to control values by 1–2 months [131]. This rapid decline in infant sera was comparable to that in children of immunized mothers in Bangladesh [132]. In contrast, levels of specific IgA varied substantially by capsule type both in magnitude and duration in milk over time. This variation highlights the need to correlate levels of specific antibody with the specific colonizing organisms and specific interval of acquisition of colonization to realistically determine this interaction. Immunization has been shown to elicit opsonizing specific IgA [133]. Most recently, breast milk from 8 women immunized with the 23 valent vaccine in late pregnancy showed an increase in levels of capsule-specific IgA and in the ability to inhibit adherence to a pharyngeal epithelial cell line (Detroit 562) of type 14 pneumococci, but not type 6B, compared with milk from 8 control women given meningococcal vaccine. Significant inhibition was also seen with milk from both groups vs. with media (HBSS), suggesting the presence of other antibodies, innate immune or non-immune factors [134].

Non-immune milk components likely play an important role in limiting adherence and colonization. A primary target of pneumococcal adherence is the sialyated GlcNAc-($\beta1 \rightarrow 3$)Gal disaccharide lectin on pharyngeal or type II alveolar epithelial cells [135]. Oligosaccharides, including those released from milk kappa-casein, bind this lectin and block epithelial cell adhesion [136, 137]. Pneumococcal surface protein A (PspA) and pneumococcal surface protein adhesin A (PsaA), which may contaminate the current 2-valent polysaccharide vaccine, are the likely targets of this competitive inhibition. Antibodies to PsaA have correlated inversely with the frequency of colonization [138].

Finn, et al. demonstrated that capsule-specific secretory IgA (S-IgA) in breast milk could enhance functional antibodies in infants [133]. Following immunization of 3 lactating mothers with 23 valent vaccine, specific S-IgA, but not IgG, increased by $\geq$2 fold in milk of 1–3 women for 6 of 7 serotypes. The S-IgA was predominantly IgA1, in secretory form, and highly specific, with avidity comparable to serum IgA and IgG. Milk whey from 2 immunized women tested supported dose-dependent killing of *S. pneumoniae* serotypes 19F and 14 with human neutrophils as did purified S-IgA to serotype 19F. Consistent with our earlier data [73], killing was complement-dependent and increased with increasing complement. Thus, capsule-specific human S-IgA in breast milk supported killing of *S. pneumoniae*, providing for the first time proof of concept that vaccine-induced human mucosal S-IgA can support functional bactericidal activity. These data provided an impetus for generating such specific antibodies at local sites with mucosal vaccines.

In a related study, based on the concept that breastfeeding reduces infant diarrheal and respiratory disease morbidity and mortality in the developing world [126, 128], Shapiro, et al., determined rates of pneumonia by 24 months among breastfed infants of 588 HIV-infected and 137 HIV-uninfected women in Botswana [84]. In this high-risk population, neither innate immune factors, nor antibodies to the pneumococcal capsular polysaccharides, nor *H. influenzae* correlated with risk or protection vs. pneumonia. Stopping breastfeeding was the single greatest risk of developing pneumonia and diarrhea (Cases 40% vs. Controls 7%; Odd Ratio 21 [95% CI

2.8–156; $p < .001$]). These data highlight the importance of breastfeeding in protecting against mucosal pathogens, but the antigen-specificity of the test used was limited (23-valent vaccine) and no microbiologic diagnoses were established.

11.10 Vaccines

The current 23-valent capsular polysaccharide (PPS) vaccine provides reasonable protection against invasive pneumococcal disease in adults (56–81%) for the 23 common serotypes included (of over 90 total serotypes) [139], as does the 7- or 9-valent PPS-protein conjugate vaccine in young children (≥90% protection) [140, 141, 142]. In stark contrast, substantial data and meta-analyses do not support the efficacy of the current 23 valent PPS vaccine against mucosal infections in adults, esp. pneumococcal pneumonia in older adults [143, 144, 145]. Similarly, the conjugate vaccine has more modest effects on presumed pneumococcal pneumonia and otitis media in children than against invasive disease [140, 141, 142, 146]. However, the conjugate vaccine appears substantially more effective against pneumonia in children than does the 23 valent polysaccharide vaccine in adults.

In summary, the burden of pneumococcal disease in the U.S. and abroad is substantial and occurs primarily at mucosal sites in children and adults, particularly from pneumonia. Unfortunately, current vaccines are least effective at these most common sites, particularly in adults, although substantial progress has been made in children with polysaccharide-protein conjugate vaccines. Vaccine efficacy likely involves both the vaccine formulation, the age and immune competence of the host, and the risk of environmental exposure. Clearly, greater focus and discovery must be directed to understanding the mechanisms of mucosal pathogenesis of *S. pneumoniae* infections, as well as to the immunologic mechanisms of local host protection.

References

1. Robinson, K. A., Baughman, W., Rothrock, G., Barrett, N. L., Pass, M., Lexau, C., Damaske, B., Stefonek, K., Barnes, B., Patterson, J., Zell, E. R., Schuchat, A., and Whitney, C. G. 2001, JAMA, 285, 1729.
2. Hausdorff, W. P. 2002, Eur J Pediatr, 161(Suppl 2), S135.
3. Hawley, L. A., Walker, F., and Whitner, C. G. 2002 Manual for the Surveillance of Vaccine Preventable Disease. Centers for Disease Control and Prevention, Atlanta.
4. Wardlaw, T., Salama, P., Johansson, E. W., and Mason, E. 2006, Lancet, 368, 1048.
5. Farr, B. M. and Mandell, G. L. 1988, Respiratory Infections: Diagnosis and Management, J.E. Pennington (ed.), Raven Press, New York, 298.
6. Mundy, L. M., Auwaerter, P. G., Oldach, D., Warner, M. L., Burton, A., Vance, E., Gaydos, C. A., Joseph, J. M., Gopalan, R., Moore, R. D., Quinn, T. C., Charache, P., and Bartlett, J. G. 1995, Am J Respir Crit Care Med, 152, 1309.
7. Gutierrez, F., Masia, M., Mirete, C., Soldan, B., Rodriguez, J. C., Padilla, S., Hernandez, I., Royo, G., and Martin-Hidalgo, A. 2006, J Infect, 53, 166.

8. Fine, M. J., Smith, M. A., Carson, C. A., Mutha, S. S., Sankey, S. S., Weissfeld, L. A., and Kapoor, W. N. 1996, JAMA, 275, 134.
9. Rubins, J. B., Boulware, D., and Janoff, E. N. 2007 Pneumococcal Vaccines, Siber, G., K, K., and Makela, H. (eds.), ASM Press, Washington, DC, in press.
10. Austrian, R. and Gold, J. 1964, Ann Intern Med, 60, 759.
11. Bruyn, G. A. W. 1992, Lancet, 340, 1418.
12. Gillespie, S. H. 1989, J Med Microbiol, 28, 237.
13. Hausdorff, W. P., Feikin, D. R., and Klugman, K. P. 2005, Lancet Infect Dis, 5, 83.
14. Hausdorff, W. P., Bryant, J., Kloek, C., Paradiso, P. R., and Siber, G. R. 2000, Clin Infect Dis, 30, 122.
15. Adrian, P. V., Bogaert, D., Oprins, M., Rapola, S., Lahdenkari, M., Kilpi, T., de Groot, R., Kayhty, H., and Hermans, P. W. 2004, Vaccine, 22, 2737.
16. Dagan, R., Givon-Lavi, N., Zamir, O., Sikuler-Cohen, M., Guy, L., Janco, J., Yagupsky, P., and Fraser, D. 2002, J Infect Dis, 185, 927.
17. Black, S., Shinefield, H., Fireman, B., Lewis, E., Ray, P., Hansen, J. R., Elvin, L., Ensor, K. M., Hackell, J., Siber, G., Malinoski, F., Madore, D., Chang, I., Kohberger, R., Watson, W., Austrian, R., and Edwards, K. 2000, Pediatr Infect Dis J, 19, 187.
18. Hammitt, L. L., Bruden, D. L., Butler, J. C., Baggett, H. C., Hurlburt, D. A., Reasonover, A., and Hennessy, T. W. 2006, J Infect Dis, 193, 1487.
19. Whitney, C. G., Farley, M. M., Hadler, J., Harrison, L. H., Bennett, N. M., Lynfield, R., Reingold, A., Cieslak, P. R., Pilishvili, T., Jackson, D., Facklam, R. R., Jorgensen, J. H., Schuchat, A., and for the Active Bacterial Core Surveillance of the Emerging Infections Program Network 2003, N Engl J Med, 348, 1737.
20. Ekdahl, K., Braconier, J. H., and Svanborg, C. 1997, Clin Infect Dis, 25, 654.
21. Granat, S. M., Mia, Z., Ollgren, J., Herva, E., Das, M., Piirainen, L., Auranen, K., and Makela, P. H. 2007, Pediatr Infect Dis J, 26, 319.
22. Kononen, E., Jousimies-Somer, H., Bryk, A., Kilp, T., and Kilian, M. 2002, J Med Microbiol, 51, 723.
23. Leino, T., Auranen, K., Jokinen, J., Leinonen, M., Tervonen, P., and Takala, A. K. 2001, Pediatr Infect Dis J, 20, 1022.
24. Bogaert, D., De Groot, R., and Hermans, P. W. 2004, Lancet Infect Dis, 4, 144.
25. Goldblatt, D., Hussain, M., Andrews, N., Ashton, L., Virta, C., Melegaro, A., Pebody, R., George, R., Soininen, A., Edmunds, J., Gay, N., Kayhty, H., and Miller, E. 2005, J Infect Dis, 192, 387.
26. Musher, D. M., Groover, J. E., Watson, D. A., Pandey, J. P., Rodriguez-Barradas, M. C., Baughn, R. E., Pollack, M. S., Graviss, E. A., de Andrade, M., and Amos, C. I. 1997, J Investig Med, 45, 57.
27. Nelson, A. L., Roche, A. M., Gould, J. M., Chim, K., Ratner, A. J., and Weiser, J. N. 2007, Infect Immun, 75, 83.
28. Dagan, R., Givon-Lavi, N., Fraser, D., Lipsitch, M., Siber, G. R., and Kohberger, R. 2005, J Infect Dis, 192, 367.
29. McCool, T. L., Cate, T. R., Moy, G., and Weiser, J. N. 2002, J Exp Med, 195, 359.
30. Johnson, S. E., Dykes, J. K., Jue, D. L., Sampson, J. S., Carlone, G. M., and Ades, E. W. 2002, J Infect Dis, 185, 489.
31. Murdoch, C., Read, R. C., Zhang, Q., and Finn, A. 2002, J Infect Dis, 186, 1253.
32. Malley, R., Henneke, P., Morse, S. C., Cieslewicz, M. J., Lipsitch, M., Thompson, C. M., Kurt-Jones, E., Paton, J. C., Wessels, M. R., and Golenbock, D. T. 2003, Proc Natl Acad Sci USA, 100, 1966.
33. Rubins, J. B., Charboneau, D., Paton, J. C., Mitchell, T. J., Andrew, P. W., and Janoff, E. N. 1995, J Clin Invest, 95, 142.
34. Rubins, J. B., Paddock, A. H., Charboneau, D., Berry, A. M., Paton, J. C., and Janoff, E. N. 1998, Microb Pathog, 25, 337.
35. Goldenberg, H. B., McCool, T. L., and Weiser, J. N. 2004, J Infect Dis, 190, 1254.
36. Cundell, D., Masure, R., and Tuomanen, E. I. 1995, Clin Infect Dis, 21(Suppl 3), S204.

37. Weiser, J. N., Bae, D., Fasching, C., Scamurra, R. W., Ratner, A. J., and Janoff, E. N. 2003, Proc Natl Acad Sci USA, 100, 4215.
38. van Rossum, A. M., Lysenko, E. S., and Weiser, J. N. 2005, Infect Immun, 73, 7718.
39. Janoff, E. N. 1993, J Infect Dis, 167, 49.
40. Polack, F. P., Flayhart, D. C., Zahurak, M. L., Dick, J. D., and Willoughby, R. E. 2000, Pediatr Infect Dis J, 19, 608.
41. Rodriguez-Barradas, M. C., Tharapel, R. A., Groover, J. E., Giron, K. P., Lacke, C. E., Houston, E. D., Hamill, R. J., Steinhoff, M. C., and Musher, D. M. 1997, J Infect Dis, 175, 590.
42. Blossom, D. B., Namayanja-Kaye, G., Nankya-Mutyoba, J., Mukasa, J. B., Bakka, H., Rwambuya, S., Windau, A., Bajaksouzian, S., Walker, C. J., Joloba, M. L., Kityo, C., Mugyenyi, P., Whalen, C. C., Jacobs, M. R., and Salata, R. A. 2006, Int J Infect Dis, 10, 458.
43. Tosi, M. F. 2005, J Allergy Clin Immunol, 116, 241.
44. Johnson, C. A., Densen, P., Wetsel, R. A., Cole, F. S., Goeken, N. E., and Colten, H. R. 1992, N Engl J Med, 326, 871.
45. Figueroa, J. E., and Densen, P. 1991, Clin Microbiol Rev, 4, 359.
46. Winkelstein, J. A. 1984, Crit Rev Microbiol, 11, 187.
47. Hostetter, M. K. 2004 The Pneumococcus, Tuomanen, E. I. (ed). ASM Press, Washington, D.C., 201.
48. Walport, M. J. 2001, N Engl J Med, 344, 1140.
49. Walport, M. J. 2001, N Engl J Med, 344, 1058.
50. Hostetter, M. K. 1986, J Infect Dis, 153, 682.
51. Abeyta, M., Hardy, G. G., and Yother, J. 2003, Infect Immun, 71, 218.
52. Ren, B., Szalai, A. J., Hollingshead, S. K., and Briles, D. E. 2004, Infect Immun, 72, 114.
53. Rubins, J. B., Charboneau, D., Fasching, C., Berry, A. M., Paton, J. C., Alexander, J. E., Andrew, P. W., Mitchell, T. J., and Janoff, E. N. 1996, Am J Respir Crit Care Med, 153, 1339.
54. Bergmann, S., and Hammerschmidt, S. 2006, Microbiology, 152, 295.
55. Hopken, U. E., Lu, B., Gerard, N. P., and Gerard, C. 1996, Nature, 383, 86.
56. Gordon, D. L., Johnson, G. M., and Hostetter, M. K. 1986, J Infect Dis, 154, 619.
57. Hostetter, M. K., Krueger, R. A., and Schmeling, D. J. 1984, J Infect Dis, 150, 653.
58. Reynolds, H. Y. 1987, Am Rev Respir Dis, 135, 250.
59. Cole, F. S., Matthews, W. J., Jr., Marino, J. T., Gash, D. J., and Colten, H. R. 1980, J Immunol, 125, 1120.
60. Colten, H. R., Ooi, Y. M., and Edelson, P. J. 1979, Ann NY Acad Sci, 332, 482.
61. Einstein, L. P., Schneeberger, E. E., and Colten, H. R. 1976, J Exp Med, 143, 114.
62. Hetland, G., Johnson, E., Falk, R. J., and Eskeland, T. 1986, Scand J Immunol, 24, 421.
63. Lappin, D. F., Birnie, G. D., and Whaley, K. 1990, Eur J Biochem, 194, 177.
64. Pettersen, H. B., Johnson, E., Mollnes, T. E., Garred P. 1998, Scand J Immunol, 28, 431.
65. Whaley, K. 1980, J Exp Med, 151, 501.
66. Botto, M., Lissandrini, D., Sorio, C., and Walport, M. J. 1992, J Immunol, 149, 1348.
67. Andoh, A., Fujiyama, Y., Bamba, T., and Hosoda, S. 1993, J Immunol, 151, 4239.
68. Rothman, B. L., Merrow, M., Despins, A., Kennedy, T., Kreutzer, D. L. 1989, J Immun, 143, 196.
69. Rothman, B. L., Despins, A. W., and Kreutzer, D. L. 1990, J Immunol, 145, 592.
70. Strunk, R. C., Eidlen, D. M., Mason, R. J. 1988, J Clin Invest, 81, 1419.
71. Wright, S. D., and Silverstein, S. C. 1983, J Exp Med, 158, 2016.
72. Aderem, A., and Underhill, D. M. 1999, Annu Rev Immunol, 17, 593.
73. Janoff, E. N., Fasching, C., Orenstein, J. M., Rubins, J. B., Opstad, N. L., and Dalmasso, A. P. 1999, J Clin Invest, 104, 1139.
74. Beiter, K., Wartha, F., Albiger, B., Normark, S., Zychlinsky, A., and Henriques-Normark, B. 2006, Curr Biol, 16, 401.
75. Gordon, S. B., Irving, G. R., Lawson, R. A., Lee, M. E., and Read, R. C. 2000, Infect Immun, 68, 2286.
76. Hoidal, J. R., Schmeling, D., and Peterson, P. K. 1981, J Infect Dis, 144, 61.

77. Hof, D. G., Repine, J. E., Giebink, G. S., and Hoidal, J. R. 1981, Am Rev Respir Dis, 124, 193.
78. Launay, P., Lehuen, A., Kawakami, T., Blank, U., and Monteiro, R. C. 1998, J Leukoc Biol, 63, 636.
79. Morton, H. C., and Brandtzaeg, P. 2001, Arch Immunol Ther Exp (Warsz), 49, 217.
80. Morton, H. C., van den Herik-Oudijk, I. E., Vossebeld, P., Snijders, A., Verhoeven, A. J., Capel, P. J. A., and van de Winkel, J. G. J. 1995, J Biol Chem, 270, 29781.
81. Weineisen, M., Sjobring, U., Fallman, M., and Andersson, T. 2004, J Immunol, 172, 3798.
82. Rehm, S. R., and Coonrod, J. D. 1982, Infect Immun, 36, 24.
83. Hof, D. G., Repine, J. E., Peterson, P. K., and Hoidal, J. R. 1980, Am Rev Respir Dis, 121, 65.
84. Shapiro, R. L., Lockman, S., Kim, S., Smeaton, L., Rahkola, J. T., Thior, I., Wester, C., Moffat, C., Arimi, P., Ndase, P., Asmelash, A., Stevens, L., Montano, M., Makhema, J., Essex, M., and Janoff, E. N. 2007, J Infect Dis, 196, 562.
85. Weisbart, R. H., Kacena, A., Schuh, A., and Golde, D. W. 1988, Nature, 332, 647.
86. Hostoffer, R. W., Krukovets, I., and Berger, M. 1993, J Immunol, 150, 4532.
87. Taborda, C. P., and Casadevall, A. 2002, Immunity, 16, 791.
88. Coonrod, J. D., Varble, R., and Jarrells, M. D. 1990, J Lab Clin Med, 116, 354.
89. Gouwy, M., Struyf, S., Catusse, J., Proost, P., and Van Damme, J. 2004, J Leukoc Biol, 76, 185.
90. Detmers, P. A., Lo, S. K., Olsen-Egbert, E., Walz, A., Baggiolini, M., and Cohn, Z. A. 1990, J Exp Med, 171, 1155.
91. Limb, G. A., Hamblin, A. S., Wolstencroft, R. A., and Dumonde, D. C. 1991, Immunology, 74, 696.
92. Ratthe, C. and Girard, D. 2004, J Leukoc Biol, 76, 162.
93. Musso, T., Calosso, L., Zucca, M., Millesimo, M., Puliti, M., Bulfone-Paus, S., Merlino, C., Savoia, D., Cavallo, R., Ponzi, A. N., and Badolato, R. 1998, Infect Immun, 66, 2640.
94. Jones, S. L., Knaus, U. G., Bokoch, G. M., and Brown, E. J. 1998, J Biol Chem, 273, 10556.
95. Strassheim, D., Asehnoune, K., Park, J. S., Kim, J. Y., He, Q., Richter, D., Kuhn, K., Mitra, S., and Abraham, E. 2004, J Immunol, 172, 5727.
96. Radsak, M. P., Salih, H. R., Rammensee, H. G., and Schild, H. 2004, J Immunol, 172, 4956.
97. Lynn, W. A., Raetz, C. R., Qureshi, N., and Golenbock, D. T. 1991, J Immunol, 147, 3072.
98. Zychlinsky, A., Weinrauch, Y., and Weiss, J. 2003, Microbes Infect, 5, 1289.
99. DeLeo, F. R., Renee, J., McCormick, S., Nakamura, M., Apicella, M., Weiss, J. P., and Nauseef, W. M. 1998, J Clin Invest, 101, 455.
100. Nathan, C. 2006, Nat Rev Immunol, 6, 173.
101. Janoff, E. N., Myers, N., Palaia, J. M., Rahkola, J. T., Fasching, C., Charboneau, D., and Rubins, J. B. 2006 Compartmentalization of humoral responses to pneumococcal polysaccharide vaccine in smoking and non-smoking adults. In. *44th Annual Infectious Disease Society of America Meeting*, Toronto, Ontario.
102. Moraes, T. J., and Downey, G. P. 2003, Microbes Infect, 5, 1293.
103. Cunnion, K. M., Zhang, H. M., and Frank, M. M. 2003, Infect Immun, 71, 656.
104. Briles, D. E., Crain, M. J., Gray, B. M., Forman, C., and Yother, J. 1992, Infect Immunol, 60, 111.
105. Brueggemann, A. B., Griffiths, D. T., Meats, E., Peto, T., Crook, D. W., and Spratt, B. G. 2003, J Infect Dis, 187, 1424.
106. Crook, D. W. 2006, Clin Infect Dis, 42, 460.
107. Sjostrom, K., Spindler, C., Ortqvist, A., Kalin, M., Sandgren, A., Kuhlmann-Berenzon, S., and Henriques-Normark, B. 2006, Clin Infect Dis, 42, 451.
108. Ambrosino, D. M., Schiffman, G., Gotschlich, E. C., Schur, P. H., Rosenberg, G. A., DeLange, G. G., vanLoghem, E., and Siber, G. R. 1985, J Clin Invest, 75, 1935.
109. Janoff, E. N., Fasching, C., Ojoo, J. C., O'Brien, J., and Gilks, C. F. 1997, J Infect Dis, 175, 975.

110. Siber, G. R., Schur, P. H., Aisenberg, A. C., Weitzman, S. A., and Schiffman, G. 1980, N Engl J Med, 303, 178.
111. Janoff, E. N., and Rubins, J. B. 2004 The Pneumococcus, Tuomanen, E. I., Mitchell, T. J., Morrison, D. A., and Spratt, B. G. (eds.), ASM Press, Washington, DC, 252.
112. Picard, C., Puel, A., Bonnet, M., Ku, C. L., Bustamante, J., Yang, K., Soudais, C., Dupuis, S., Feinberg, J., Fieschi, C., Elbim, C., Hitchcock, R., Lammas, D., Davies, G., Al-Ghonaium, A., Al-Rayes, H., Al-Jumaah, S., Al-Hajjar, S., Al-Mohsen, I. Z., Frayha, H. H., Rucker, R., Hawn, T. R., Aderem, A., Tufenkeji, H., Haraguchi, S., Day, N. K., Good, R. A., Gougerot-Pocidalo, M. A., Ozinsky, A., and Casanova, J. L. 2003, Science, 299, 2076.
113. Carson, P. J., Schut, R. L., Simpson, M. L., O'Brien, J., and Janoff, E. N. 1995, J Infect Dis, 172, 340.
114. Lue, C., Tarkowski, A., and Mestecky, J. 1988, J Immunol, 140, 3793.
115. Opstad, N. L., Daley, C. L., Thurn, J. R., Rubins, J. B., Merrifield, C., Hopewell, P. C., and Janoff, E. N. 1995, J Infect Dis, 172, 566.
116. Tarkowski, A., Lue, C., Moldoveanu, Z., Kiyono, H., McGhee, J. R., and Mestecky, J. 1990, J Immunol, 144, 3770.
117. Simell, B., Kilpi, T., and Kayhty, H. 2006, Clin Exp Immunol, 143, 543.
118. Simell, B., Kilpi, T. M., and Kayhty, H. 2002, J Infect Dis, 186, 1106.
119. Simell, B., Korkeila, M., Pursiainen, H., Kilpi, T. M., and Kayhty, H. 2001, J Infect Dis, 183, 887.
120. Fasching, C. E., Grossman, T., Corthesy, B., Plaut, A. G., Weiser, J. N., and Janoff, E. N. 2007, Infec Immun, 75, 1801.
121. Johnson, S., Opstad, N. L., Douglas, J. M., Jr., and Janoff, E. N. 1996, Infect Immun, 64, 4339.
122. Baxendale, H. E., Davis, Z., White, H. N., Spellerberg, M. B., Stevenson, F. K., and Goldblatt, D. 2000, Eur J Immunol, 30, 1214.
123. Janoff, E. N., Fasching, C., Scamurra, R. W., Kravis, B., Asrani, A. C., and Rubins, J. B. 2007, In preparation.
124. King, S. J., Hippe, K. R., Gould, J. M., Bae, D., Peterson, S., Cline, R. T., Fasching, C., Janoff, E. N., and Weiser, J. N. 2004, Mol Microbiol, 54, 159.
125. Steinitz, M., Tamir, S., Ferne, M., and Goldfarb, A. 1986, Eur J Immunol, 16, 187.
126. WHO. 2000, Lancet, 355, 451.
127. Molbak, K., Gottschau, A., Aaby, P., Hojlyng, N., Ingholt, L., and da Silva, A. P. 1994, Bmj, 308, 1403.
128. Victora, C. G., Vaughan, J. P., Lombardi, C., Fuchs, S. M. C., Gigante, L. P., Smith, P. G., Nobre, L. C., Teixeira, A. M. B., Moreira, L. B., and Barros, F. C. 1987, Lancet, ii, 319.
129. Kaleida, P. H., Nativio, D. G., Chao, H. P., and Cowden, S. N. 1993, J Clin Microbiol, 31, 2674.
130. Rosen, F. S., and Janeway, C. A. 1966, N Engl J Med, 275, 709.
131. O'Dempsey, T. J., McArdle, T., Ceesay, S. J., Banya, W. A., Demba, E., Secka, O., Leinonen, M., Kayhty, H., Francis, N., and Greenwood, B. M. 1996, Vaccine, 14, 963.
132. Shahid, N. S., Steinhoff, M. C., Hoque, S. S., Begum, T., Thompson, C., and Siber, G. R. 1995, Lancet, 346, 1252.
133. Finn, A., Booy, R., Moxon, R., Sharland, M., and Heath, P. 2002, Arch Dis Child, 87, 18.
134. Deubzer, H. E., Obaro, S. K., Newman, V. O., Adegbola, R. A., Greenwood, B. M., and Henderson, D. C. 2004, J Infect Dis, 190, 1758.
135. Andersson, B., Eriksson, B., Falsen, E., Fogh, A., Hanson, L. A., Nylén, O., Peterson, H., and Edén, C. S. 1981, Infect Immunity, 32, 311.
136. Andersson, B., and Svanborg-Eden, C. 1989, Respiration, 55(Suppl 1), 49.
137. Aniansson, G., Andersson, B., Lindstedt, R., and Svanborg, C. 1990, Microb Pathog, 8, 315.
138. Obaro, S. K., Adegbola, R. A., Tharpe, J. A., Ades, E. W., McAdam, K. P., Carlone, G., and Sampson, J. S. 2000, Vaccine, 19, 411.

139. Zheng, B., Han, S., Takahashi, Y., and Kelsoe, G. 1997, Immunol Rev, 160, 63.
140. Black, S., Shinefield, H., Ray, P., Lewis, E. M., Fireman, B., The Kaiser Permanente Vaccine Study Group, Austrian, R., Siber, G., Hackell, J., Kohberger, R., and Chang, I. 1999 Efficacy of heptavalent conjugate pneumococcal vaccine (Wyeth Lederle) in 37,000 infants and children: impact on pneumocia, otitis media, and an update on invasive disease – Results of the Northern California Kaiser Permanente Efficicy Trial. In. *39th Interscience Conference on Antimicrobial Agents and Chemotherapy*, San Francisco, CA.
141. Cutts, F. T., Zaman, S. M., Enwere, G., Jaffar, S., Levine, O. S., Okoko, J. B., Oluwalana, C., Vaughan, A., Obaro, S. K., Leach, A., McAdam, K. P., Biney, E., Saaka, M., Onwuchekwa, U., Yallop, F., Pierce, N. F., Greenwood, B. M., and Adegbola, R. A. 2005, Lancet, 365, 1139.
142. Klugman, K. P., Madhi, S. A., Huebner, R. E., Kohberger, R., Mbelle, N., Pierce, N., and for the Vaccine Trialists Group. 2003, N Engl J Med, 349, 1341.
143. Dear, K., Holden, J., Andrews, R., and Tatham, D. 2003, Cochrane Database Syst Rev, CD000422.
144. Hirschmann, J. V. and Lipsky, B. A. 1994, Arch Intern Med, 154, 373.
145. Jackson, L. A., Neuzil, K. M., Yu, O., Benson, P., Barlow, W. E., Adams, A. L., Hanson, C. A., Mahoney, L. D., Shay, D. K., Thompson, W. W., and for the Vaccine Safety Datalink. 2003, N Engl J Med, 348, 1747.
146. Fireman, B., Black, S. B., Shinefield, H. R., Lee, J., Lewis, E., and Ray, P. 2003, Pediatr Infect Dis J, 22, 10.

Chapter 12
Neisseria meningitidis

Barbara Baudner and Rino Rappuoli

Abstract *Neisseria meningitidis* is a Gram-negative encapsulated, aerobic diplococcus, which colonizes the nasopharynx of humans and has no other known environmental niche. It is one of the major causes of bacterial meningitis and sepsis, and is the only form of bacterial meningitis responsible for epidemic levels of meningitis infection worldwide. The bacteria are commonly carried asymptomatically in the upper respiratory tract. In a small number of persons, *Neisseria meningitidis* penetrates the mucosa and gains access to the bloodstream, causing systemic disease. In most persons, however, carriage is an immunizing process, resulting in a systemic protective antibody response. Nasopharyngeal colonization with Neisseria species initially generates mucosal effector T cells and secretory IgA in the absence of protective systemic immunity. With increasing age and repeated contact, local immunity in the nasopharyngeal mucosa and immunological memory increase and systemic effector T cells and memory responses develop. Natural immunological programming results in a mucosal response that is dominated by a pro-inflammatory interferon-γ (IFN-γ) phenotype. The T-cell phenotype of the *Neisseria meningitidis* antigen responding population in the circulation seems to have a more balanced Th1–Th2 profile. The cause of progression from carriage to invasive disease is dependent on both host and infecting organism factors. A lack of sufficient biologically active antibodies against *Neisseria meningitidis* is thought to be one of the major factors conferring susceptibility to infection.

Clinical syndromes caused by *Neisseria meningitidis* include meningitis, with or without meningococcemia, relatively mild bacteremia, fulminant meningococcemia, meningoencephalitis, pneumonia, and septic arthritis, as well as other presentations. Even when the disease is diagnosed early and adequate therapy is instituted, 5–10% of patients die. The rapid onset of invasive meningococcal disease, the high incidence in childhood, the severity of the sequellae and the high mortality rate clearly indicate the importance of vaccine development against *Neisseria meningitidis*. An ideal meningococcal vaccine should be able to protect against all disease-causing meningococci, developing protective immunity at the mucosal

---

B. Baudner
Novartis Vaccines and Diagnostics S.r.l, Via Fiorentina 1, 53100 Siena, Italy
e-mail: barbara.baudner@novartis.com

M. Vajdy (ed.), *Immunity Against Mucosal Pathogens*,

surface and in the circulation. Several polysaccharide vaccines, which have been available for over 30 years, exist against serogroups A, C, Y, W135 in various combinations. A major advance in the prevention of meningococcal disease has been the development and introduction of meningococcal polysaccharide-protein conjugate vaccines and the development of broadly effective serogroup B vaccines could eliminate the organism as a major threat to human health within the next decades.

12.1 Introduction (History)

Bacterial meningitis is a serious threat to global health, accounting for an estimated 170,000 death worldwide per year. Up to 25% of survivors develop permanent sequelae, such as epilepsy, mental retardation, or sensorineural deafness [1]. The disease remains one of the most feared infections due to its rapid progression and tendency to cause outbreaks and epidemics. Mortality from meningococcal septicaemia may be as high as 7–15%. Together with *Haemophilus influenzae* and *Streptococcus pneumoniae, Neisseria meningitidis* is one of the major causes of bacterial meningitis and sepsis, causing at least 500,000 cases and more than 50,000 deaths every year [2]. The bacteria are commonly carried asymptomatically in the upper respiratory tract of between 8–20% of healthy individuals [3, 4, 5, 6] and more frequently in adolescents and other populations with close contact. Only occasionally *Neisseria meningitidis* invades the bloodstream and meninges to cause disease. The disease incidence varies in human populations from rare to over 1000/100 000 population/year: Meningococcal disease occurs as sporadic cases, incidence <1/100 000 population/year, hypersporadic disease 1–10/100 000 population/year, localized outbreaks and case clusters, and epidemic and pandemic disease with rates >10 to >1000/100 000 population/year [7, 8, 9, 10, 11, 12, 13].

Meningococcal meningitis has been recognized as a serious problem for almost 200 years, and clinical symptoms of meningitis were first documented during the Bubonic Plague in the Middle Ages. During an outbreak of the disease in 1805 in Geneva, Switzerland meningitis was first identified and described by Vieusseux [7]. One year after its first description in Geneva the first definitive recognition of epidemic meningitis in the United States occurred in Medfield, Massachusetts, in 1806, by Danielson and Mann [8, 14]. In 1886, Hirsch described "epidemic cerebrospinal meningitis" throughout Europe, Africa, Asia, and the Americas, as well as some of the epidemiologic features of meningococcal disease [14, 15]. However, it wasn't until 1887 that the causative organism, *Neisseria meningitis*, was first isolated from cerebrospinal fluid of patients with meningitis and named "*Diplococcus intracellularis meningitidis*" by the Austrian pathologist and bacteriologist Anton Weichselbaum [16]. Asymptomatic carriage was first described in 1896 in Europe and subsequently correlated with meningococcal incidence [17, 18]. Also in the sub-Saharan Africa epidemics have been recognized for more than 100 years. First documented on the West Coast of Africa in 1909, meningitis has been a huge burden on the African continent in the last century in the form of cyclical epidemics. The "meningitis belt", a region of Africa hit hardest by the disease, stretches from Senegal and Gambia in the East to Ethiopia and Western Eritrea in the West. This area is home to a population of over 300 million, and suffers massive epidemics about every 8–12 years.

In the early part of the 20th century, the American physician Simon Flexner reported on the use of antimeningococcal serum therapy, first in an animal model and subsequently in humans, demonstrating that intrathecal administration reduced the mortality of meningococcal meningitis [19, 20, 21, 22]. In 1915, a serogrouping scheme was proposed that separated meningococcal isolates into types I to IV, later changed to the letter-based nomenclature that is used today [23, 24]. During the 20th century, meningitis was prevalent during both World Wars, due in large part to the poor living conditions of soldiers. In 1944, the infection was successfully treated for the first time with intravenous and intrathecal penicillin, and by 1950 clinical trials established high doses of intravenous penicillin as the most effective treatment for meningitis [25]. Due to its dramatic clinical presentation, there has been huge public pressure to find a definitive vaccine that will eradicate the disease. Ideally, such a vaccine must be immunogenic at all ages and provide cross-protection against all meningococci.

12.2 Meningococcal Bacteriology

Neisseria meningitidis is a Gram-negative encapsulated, oxidase positive, aerobic diplococcus [6, 26]. The bacterium readily colonizes the nasopharynx of humans and has no other known environmental niche [4, 27, 28]. It has inner and outer cell membranes, separated by a peptidoglycan cell wall. The outer membrane contains numerous different outer membrane proteins (OMPs), which enable the organism to interact with and adhere to host cells and act as transport proteins to control the intracellular environment. Additionally, it contains lipopolysaccharide, endotoxin), which helps confer resistance to bactericidal mechanisms and is involved in the pathogenesis of meningococcal disease [29]. A polysaccharide capsule, which covers the outer membrane of the bacteria, is a major virulence factor for meningococci and although capsular-deficient strains are commonly isolated, these are non-infectious [4]. Capsule mediates resistance to phagocytosis, complement-mediated lysis and protects against environmental insults (Fig. 12.1).

Thirteen serogroups of *Neisseria meningitidis*, based on different capsular polysaccharide structure, are known but only six serogroups (A, B, C, W-135, Y and recently X) are currently associated with significant pathogenic potential [30, 31]. The capsule of serogroup A is made of N-acetyl mannosamine-1-phosphate, while serogroups B, C, Y, and W-135 have capsules that consist of polysialic acid or sialic acid linked to glucose or galactose. For epidemiologic purposes, *Neisseria meningitidis* has been further classified into serotypes and serosubtypes, which are based on the antigenic differences of the major outer membrane proteins PorB

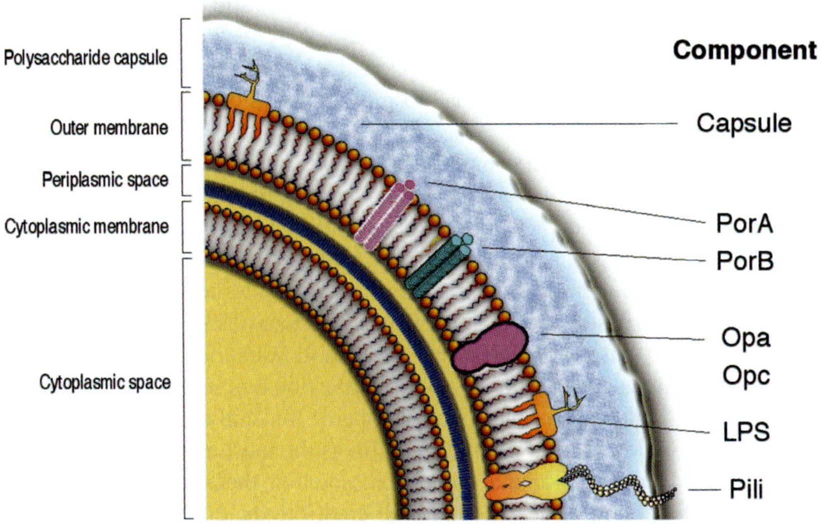

Fig. 12.1 Surface structures of *Neisseria meningitidis*. The cell envelope is composed of an outer membrane (OM) and an inner cytoplasmic membrane (IM), separated by a periplasmic space containing a peptidoglycan layer and proteins. The OM is an asymmetric bilayer with lipopolysaccharide (LPS) and phospholipids in the outer leaflet and phospholipids in the inner leaflet. In the OM, several integral outer membrane proteins (OMP) and the cell-surface exposed proteins are present. The OM is covered by a polysaccharide capsule. The IM is a symmetric bilayer of phospholipids with proteins. Pili, are complex outer membrane protein organelles that extend from the cell surface

(class 2 or 3 OMP/serotype) and PorA (class 1 OMP/serosubtype) [32, 33, 34, 35]. The lipopolysaccharide in the outer membrane is used to classify strains into immunotypes (Table 12.1). The nomenclature of *Neisseria meningitidis* is historically based on differences in monoclonal antibody recognition of the capsular polysaccharide, PorB, PorA, and lipopolysaccharide, and follows a classification scheme: serogroup:serotype:serosubtype:immunotype (eg, B:2a:P1.2:L3) [32, 34].

The complete nucleotide sequences for *Neisseria meningitidis* strains MC58 (serogroup B), and Z2491 (serogroup A) have been reported [36, 37]. Additionally *Neisseria meningitidis* strain FAM18 (serogroup C) and *Neisseria gonorrhoeae* strain FA1090 (GenBank AE004959) genome sequences are available.

Approaches such as multilocus enzyme electrophoresis (electrophoretic type-ET), now replaced by multilocus sequence typing (MLST) [38, 39], have been used to monitor the global epidemiology of meningococcal disease [40]. More than 5000 sequence types (ST) have been identified, forming 35 "ST complexes". Most of the strains isolated from invasive disease belong to ST-11, ST-32, ST-8, ST-7, ST-41, ST-1, ST-4 and ST-5. MLST and other genomic studies have revealed much about the meningococcal epidemiology, population structure, carriage and transmission dynamics and genetic requirements for invasive meningococcal disease [41].

Table 12.1 Classification of *Neisseria meningitidis*

Bases		Classification System	Function	N° of groups	Names
Capsule		Serogroup	Protects against complement dependent bacteriolysis and phagocytosis	13	A, B, C, E-29, H, I, K, L, M, W-135, X, Y, Z
Outer membrane proteins (porins)	PorB (Class 2/3 OMP)	Serotype	Create pores through which small hydrophilic solute pass, cationic-selective or anion-selective	10	1, 2a, 2b, ...21
	PorA (Class 1 OMP)	Serosubtype		20	P1.1, P1.2, P1.3 . . . .
Lipooligosaccharides		Immunotype	Has potent immunotoxic activity	13	L1, L2, ... L13
Pili			Promote initinal adherence to epithelial/endothelial cells and red blood cells	2	I, II

12.3 Meningococcal Epidemiology

Neisseria meningitidis is the only form of bacterial meningitis responsible for epidemic levels of meningitis infection worldwide. It is carried in the nasopharynx of about 10% of adults, around 25% of 15–19 year olds, but relatively few children: only 0.71% of children less than 4 years of age and 5.4% of teenagers carry *Neisseria meningitidis* [28]. However, analysis of tonsillectomy specimens using immunohistochemistry has suggested that carriage may be as high as 45% [42]. Acquisition rates vary, with the highest rates reported in military recruits and university/college students in dormitory accommodation or halls of residence [43, 44]. Meningococcal carriage is a natural immunizing process and results in the induction of protective antibody responses [45]. Despite the high rate of colonization, meningococcal disease is quite rare [46], and the mechanism by which bacteria invade the mucosal barrier and get access to the blood stream is not fully understood. Rates of meningococcal disease are highest in young children (related to waning of protective maternal antibody) and increase again in adolescents and young adults aged 14–24, probably related to increased transmission and acquisition [45]. There are significant geographic differences in the distribution of specific serogroups of meningococcal disease [30, 31] (Fig. 12.2).

Serogroup A isolates are predominantly found in sub-Saharan African countries in an area from Senegal to Ethiopia that is referred to as "the meningitis belt" and cause frequent epidemics [9, 10, 47, 48, 49]. Outbreaks occur during the dry season and diminish with the onset of the rainy season [9]. During the 1996 epidemic around 200,000 cases were reported with 20,000 deaths in areas of the belt (e.g. Burkina Faso, Mali, Niger, Nigeria and Chad) and these numbers likely reflect only part of the magnitude of the outbreaks [10, 12, 50]. Molecular epidemiology and genetic analysis suggest serogroup A meningococci causing these outbreaks are highly clonal, emerge and spread rapidly, such as ST-5 or ST-7 complexes, and may have descended from a common ancestor in the 19th century [51]. Group A epidemics also have occurred in other parts of Africa and in Asia, including China, India, Nepal and Mongolia [52]. In the US and Europe large outbreaks of serogroup A meningococcal disease occurred in the first part of the 20th century,

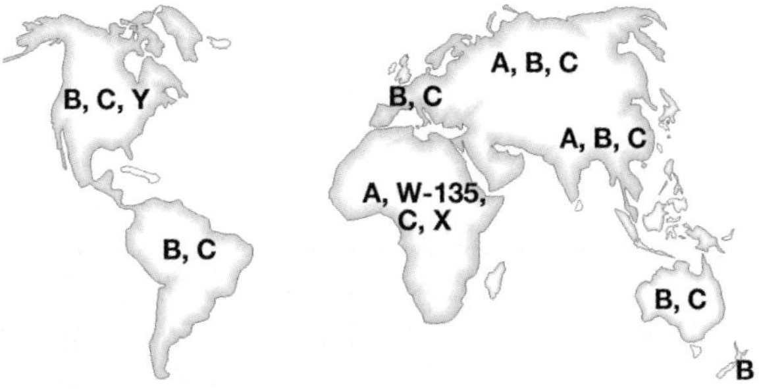

Fig. 12.2 Global meningococcal serogroup distibution

but since World War II serogroup A disease has virtually disappeared from the US and Western Europe.

Since 2000, the emergence of serogroup W-135 (also ET-37/ST-11 complexes) as the cause of outbreaks (Saudi Arabia, 2000 and 2001; Burkina Faso, 2002–2003) has added complexity to the epidemiological situation in the region [53, 54]. Diseases caused by serogroup W135 have been associated with the Haj pilgrimage in Saudi Arabia and significant disease due to this serogroup is seen in parts of the African meningitis belt [13, 49, 55].

Serogroup B is a major cause of sporadic or endemic meningitis in industrialized countries, accounting for up to 80% in certain European countries, with most of the remaining cases being caused by serogroup C strains. In the USA, the serogroup distribution differs, with 32% of the cases caused by serogroup B, 35% by serogroup C and 26% by serogroup Y [11]. The proportion of group B strains is especially high in Norway, The Netherlands, Germany and Denmark, while increasing proportions of group C strains have been reported from Slovak and Czech Republics, Greece, Republic of Ireland, Spain [56], Canada and the United Kingdom [57]. In all countries, the incidence of group B and C disease is highest in winter in infants less than 1 year-old and for the elderly. Group B meningococcus can also cause severe, persistent epidemics, such as those that occurred in Latin American countries (Cuba, Colombia, Brazil and Chile) or in Norway. Since 1991, New Zealand has had an epidemic of group B meningococcal disease with incidence rates of up to ten times the background incidence, and with much higher age-specific, area-specific, and ethnic-specific rates [12, 58, 59, 60]. Outer membrane protein types and subtype and specific genotypes ET-5 or ST-32 and ST41/44 complexes distinguish these outbreaks, but these types also cause endemic disease [61].

In contrast to serogroup A and C epidemics, which usually resolve in 1–3 years, serogroup B epidemics begin slowly but may persist for 10 years or longer, as seen in Cuba, Norway, areas of Chile and in New Zealand. Serogroup B global incidence has been estimated between 20,000 and 80,000 cases per year, accounting for 2000–8000 deaths annually.

Serogroup C (especially ET-37/ST-11 complexes) has caused major epidemic outbreaks in sub-Saharan Africa, Brazil, case clusters and local outbreaks in the US, Canada and Western Europe, especially among adolescents and young adults [11, 62, 63].

Since the mid-1990s, serogroup Y strains (ET-501/ST-23 and related STs) have caused increased rates of disease in the US and Israel, and serogroup X has been responsible for localized recent outbreaks in parts of sub-Saharan Africa such as Niger [64, 65, 66].

Interestingly, some parts of the world, for reasons not well understood, persistently have very low rates of meningococcal disease (Mexico, Japan and other parts of Asia and areas of South and Central America) and few or no outbreaks. Population carriage and invasive disease susceptibility factors as well as not fully understood environmental factors may be explanations.

12.4 Meningococcal Pathogenesis

Five to 10 percent of adults are asymptomatic nasopharyngeal carriers of strains of *Neisseria meningitidis* [3, 4], most of which are not pathogenic. In a small number of persons, *Neisseria meningitidis* penetrates the mucosa and gains access to the bloodstream, causing systemic disease [67]. In most persons, however, carriage is an immunizing process, resulting in a systemic protective antibody response [5].

The nasopharynx is the site of mucosal colonization by meningococci and assumed to be the primary site of invasion prior to the development of systemic infection. There is likely to be a number of factors that contribute to the integrity of the mucosal barrier and prevent both colonization and invasion. The surface charge and hydrophobicity of the nasal mucosa has a bearing on bacterial adhesion [68] and changes in charge, and thus adhesion, may result in an increased risk of invasive disease [69]. Various interactions between bacterial and mucosal cell surface structures are involved in adherence and invasion of meningococci [70] and polymorphisms in these molecules, both human and bacterial, may be associated with successful colonization or invasion following exposure to the organism.

12.4.1 Meningococcal Virulence Factors

Meningococci possess a number of virulence factors such as the major outer membrane components (capsular polysaccharide, outer membrane proteins, and lipooligosaccharide [endotoxin]) that contribute to immune evasion and pathogenesis [12]. Additionally, the potential for genetic change and surface structural variability of the meningococcus together with the ability to switch serogroups is quite remarkable and another means of immune evasion [71].

Other virulence-related mechanisms include rapid doubling time and release of outer membrane vesicles, so-called blebs, molecular mimicry, and the less defined possible release of toxins.

12.4.1.1 Capsule

The capsule of *Neisseria meningitidis* is an important virulence determinant required for survival in the blood, and furthermore helps with transmission and colonization, and protects the meningococcus from desiccation, phagocytic killing, opsonisation, and complement-mediated bactericidal killing [72, 73]. Antibodies directed versus the capsule play a major part in protection against meningococcal disease. On the other hand capsule is involved in inhibiting cellular interactions mediated by meningococcal outer membrane adhesins. However, there is some evidence that target receptor density on host cells may determine whether or not capsulated bacteria can adhere via outer membrane proteins. Thus, capsule may not be an adequate barrier for cellular interactions as far as up-regulation of carcinoembryonic antigen-related cellular adhesion molecule (CEACAMs), which can occur in response to inflammatory cytokines, could lead to translocation of a small number of fully capsulated bacteria across mucosal epithelium into the bloodstream sufficient to cause a rapid onset of disseminated disease. Capsule switching is the result of transformation and horizontal DNA exchange of portions of the capsule biosynthetic operon, and a mechanism of escape from vaccine-induced or natural protective immunity [60]. This virulence mechanism is shown by various encapsulated bacterial pathogens.

12.4.1.2 Pili

The pili are complex outer membrane protein organelles that extend several thousand nm from the cell surface, ease initial attachment of meningococci to host cells, and are associated with so-called twitching motility, which is important for passage through the mucus layer and movement over epithelial surfaces [74].

12.4.1.3 Outer Membrane Proteins (OMP)

Meningococci also express variable proteins on their surface, which are important for bacteria-host interactions [12]:

Outer membrane porins are involved in host-cell interactions and serve as targets for bactericidal antibodies [12]. PorB, which is the major outer membrane porin, inserts in membranes, induces Ca^{++} influx, and activates Toll-like receptor 2 (TLR2) and cell apoptosis [75]. PorA is a second important porin and a major target of outer membrane vesicle vaccines. Porins can stimulate B cell proliferation and Ig secretion in vitro and up-regulate expression of B7-2 on B cells [76, 77, 78]. In vivo, meningococcal porins can convert T-cell-independent antibody responses to be T-cell-dependent [79]. Thus, they may be necessary to facilitate immunological processing of meningococcal polysaccharides from strains carried in the nasopharynx and for development of "natural protein–polysaccharide conjugate vaccines".

Neisserial Opa proteins are important in adherence to various cell types. A single meningococcus contains genes encoding 3–4 different Opa variants. The individual Opa variants are antigenically distinct and can have different binding properties, typically binding to either heparan sulfate proteoglycan (HSPG) receptors or to one or more receptors of the CEACAM family of proteins [80].

Iron-acquiring proteins are necessary for colonization and infection. Throughout the process of meningococcal infection, the bacteria encounter many low-iron environments and have therefore developed ways to procure iron from the intracellular and extracellular environment [12]. Meningococci possess receptors that bind to hemoglobin within cells as well as human transferrin and lactoferrin in the extracellular space. Over 80 genes are regulated by the iron responsive repressor Fur [81]. Meningococcal iron-acquiring proteins include HmbR (haemoglobin), TbpA and TbpA-B (transferrin), HbpA and HbpA-B (lactoferrin), HpnA and HpnA-B (haemoglobin-haptoglobin complex), and possible siderophore homologues. New putative iron acquisition proteins, adhesion or invasion proteins, and toxin proteins have been identified in genome searches.

12.4.1.4 Endotoxin, or Lipooligosaccharide

This is a major component of the outer membrane, differs in structure from enteric endotoxins [82], and is crucial in inflammatory signaling via Toll-like receptor 4 (TLR4) [83]. Lipooligosaccharide is also important in adherence and colonization, as well as in the pathogenesis of fulminant sepsis and meningitis. The α-chain structures of meningococcal lipooligosaccharide are identical to the human I and i antigens, an example of host molecular mimicry as an immune escape mechanism [84].

12.4.1.5 Secreted Factors

IgA1 proteases Meningococci also secrete an IgA protease and produce factors that inhibit ciliary activity in order to escape host mucosal defenses [85]. IgA1 protease cleaves human IgA1 in the hinge region and is possibly an important bacterial virulence factor [86, 87]. IgA1 protease may cause masking of bacterial epitopes through generation of F(ab)2 IgA fragments that prevent IgM or IgG binding with the bacterial surface, thus inhibiting complement-mediated bactericidal activity or antibody-mediated opsonophagocytosis [88]. IgA1 also cleaves LAMP1, a major integral membrane glycoprotein of late endosomes and lysosomes, and promotes intracellular survival of pathogenic Neisseria spp. [89].

12.5 Mucosal Colonization of *Neisseria meningitidis*

Generally, the pathway of infection for any etiological agent of meningitis begins with the establishment of localized transmission in the host [90]. *Neisseria meningitidis* like most meningeal pathogens is transmitted via the respiratory route. The pathogen must then evade host immune responses and enter the submucosa (Fig. 12.3). Finally, the infectious agent must cross the blood-brain barrier and access the cerebrospinal fluid (CSF). This is most often accomplished by invasion of the bloodstream and subsequent seeding of the central nervous system (CNS) [90].

Meningococci must overcome host mucosal defenses and attach to the nonciliated columnar epithelial cells of the nasopharynx. Primary attachment of *Neisseria* to the apical surface of mucosal epithelial cells is mediated by the bacterial pilus. Pili, found on the surface of meningococci, bind to the cell surface receptor CD46 [12, 91]. This filamentous appendage then retracts [90], pulling the bacterium down onto the host cell membrane. Subsequent binding of the outer membrane proteins, Opa and Opc to CD66 (carcinoembryonic antigen) and heparin sulfate proteoglycan receptors, respectively, lead to engulfment of the meningococci by epithelial cells [29, 92, 93]. Afterwards, the bacteria may pass through the epithelial layer via

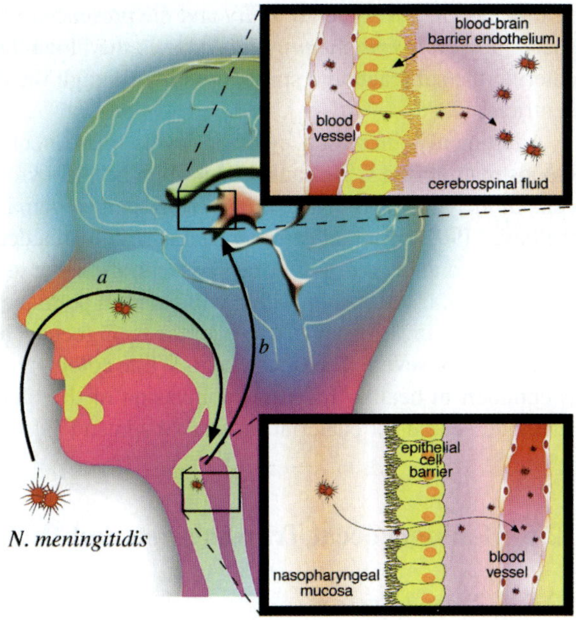

Fig. 12.3 Mucosal colonization of *Neisseria meningitidis* in the nasopharynx and entry into bloodstream and cerebrospinal fluid

phagocytic vacuoles [94] and exit into the subepithelial compartment. IgA1 protease cleaves the lysosome-associated membrane protein and may promote the survival of *Neisseria meningitidis* in epithelial cells, whereas PorB crosses the cell membrane and arrests the maturation of the phagosome [95]. The bacteria likely tend to remain within this niche, presumably being in contact with cells of the immune system but somehow avoiding triggering a specific immune response. Disseminated meningococcal infections can occur if the bacteria penetrate across the endothelium and spread through the bloodstream (Fig. 12.3).

Meningococci that have entered the bloodstream either seed the CSF if multiplication proceeds slowly, or cause meningococcemia and shock, if multiplication is rapid [73]. The morbidity and mortality of meningococcal meningitis has been directly correlated to the amount of circulating lipooligosaccharide (or endotoxin) [73]. The presence of lipooligosaccharide contributes to the inflammatory cascade of cytokines that is characteristic of severe disease [85]. Meningococci are also prone to frequent blebing of the outer membrane. Thus, in the bloodstream, the organisms release endotoxin in the form of blebs (vesicular outer membrane structures) that contain 50% lipooligosaccharide and 50% outer membrane proteins, phospholipids, and capsular polysaccharide that further enhance the inflammatory cascade. Exposure of cells to bacterial secretions results in the production of cytokines such as TNF-α and IL-11. These inflammatory mediators are believed to enhance the permeability of the blood-brain barrier and facilitate bacterial invasion of the CSF [12]. Secondary mediators such as IL-6, IL-8, nitric oxide, prostaglandins, and platelet activating factor amplify the effect, resulting in further vascular endothelial injury and the presence of bacteria in the subarachnoid space [90]. *Neisseria meningitidis* crosses the blood–brain barrier endothelium by entering the subarachnoid space, possibly through the choroid plexus of the lateral ventricles [73].

Additionally, increased cytokine levels cause neutrophils to migrate from the bloodstream to the CSF, causing neutrophilic pleocytosis (an increased number of cells in the CSF) and contributing to vasogenic edema (increased water content in the brain) [90]. Eventual cerebral edema results in a decreased cerebral blood flow. This leads to an increase in lactate concentration due to anaerobic metabolism [90]. This process, if left uncontrolled, will result in permanent neuronal dysfunction.

Therefore, invasion and survival of the menigococci in the bloodstream are essential factors for severe disease to occur. While colonization of the respiratory mucosa is common in healthy individuals, invasion of the bloodstream is very infrequent and not essential for bacterial survival or spread [73].

12.6 Meningococcal Transmission and Risk Factors

12.6.1 Transmission

The bacteria are transmitted from person to person through droplets of respiratory or throat secretions. Close and prolonged contact (e.g. kissing, sneezing and coughing on someone, living in close quarters or dormitories (military recruits, students), sharing eating or drinking utensils facilitate the spread of the pathogen. The average incubation period is four days, ranging between two and ten days. Naturally acquired immunity as well as the low virulence of many carrier strains is partly responsible for the disparity between carriage and disease rates. Furthermore, the disease incidence is influenced by host and environmental factors, which affect the human reservoir of meningococci and the host susceptibility to meningococcal disease. Understanding how the meningococcus can be both a common commensal and a devastating human pathogen has been a major quest in the biology and in the design of prevention strategies for *Neisseria meningitidis*.

12.6.2 Host Risk Factors

Several specific host factors that increase the risk of invasive meningococcal infection have been identified. Among males the risk seems higher until around age 45 years, at which time the risk becomes higher among women [11]. Lower socioeconomic status and genetic background have been associated with an increased risk, but whether race per se actually increases the risk is doubtful [11, 96, 97]. Increased risk to young children due to the lack of protective serum bactericidal antibodies has been known for decades to be one of the key host factors associated with an amplified risk [98, 99, 100]. Opsonization and phagocytic function appear to be further important host defenses as demonstrated by disease reduction after polysaccharide vaccination in individuals with complement deficiencies.

In addition, genetic deficiencies determine human susceptibility and disease outcome [101]. Defects in the complement system may lead to rapid fatal meningococcemia in patients who lack properidin in the alternative complement pathway whereas recurrent meningococcal infections are seen in patients with defects in the terminal pathway (C5–C9) [102, 103]. A relative deficiency of late complement components observed in infancy [104], may be adding to the susceptibility to meningococcal infection in this age group. Deficiency of mannan-binding lectin is associated with an increased susceptibility to infection including meningococcal infection [45, 98, 105, 106], although not all studies have been able to demonstrate this [107]. Variation in the plasminogen-activator inhibitor 1 (PAI-1) gene has been associated with increased susceptibility to meningococcal septic shock, but not an overall risk of meningococcal disease [108]. Moreover, polymorphisms in genes coding for the Fcγ-receptor II (CD32), Fcγ-receptor III (CD16), and Toll-like receptor 4 (TLR4) are associated with increased risk or severity of meningococcal sepsis [109, 110, 111, 112, 113, 114]. Fc receptor expression on the surface of phagocytes may be of particular importance in host defense against *Neisseria meningitidis* as it is required for the phagocytosis of IgG-opsonized bacteria [115]. Fc receptors with poor IgG2 binding are more common in children who develop meningococcal infection [116].

Meningococcal disease is also linked to immune suppression such as that seen in the nephritic syndrome, hypogammaglobulinemia, and HIV disease (but not serogroup A epidemics)

In addition, an increased risk of invasive disease after splenectomy was found [117, 118], and it is presumed that this susceptibility arises from a combination of the loss of splenic filtering of encapsulated bacteria from the circulation, reduced production of opsonins, and reduced anatomic juxtaposition of immune cells [117]. In children, functional or anatomic asplenia has been associated with very poor humoral responses to polysaccharide vaccines, suggesting that the spleen may also have a critical role in the generation of humoral immune responses to polysaccharides in infancy [101, 112, 119, 120, 121, 122, 123, 124, 125, 126, 127].

12.6.3 Environmental and Behavioral Risk Factors

An intact mucosal barrier is an important prerequisite to prevent meningococcal invasion. Therefore, factors leading to the disruption of the respiratory epithelium subsequently facilitate meningococcal invasion. Clearly, climatic conditions influence the risk of infection. In sub-Saharan Africa, epidemic meningococcal infection typically starts during the dry season, when it typically is hot, dusty and arid, and ends with the onset of the rainy season [128]. In the United States, meningitis cases occur mostly in the winter, peaking at late winter/early spring, as a result of the necessity of remaining indoors for many at risk populations. Infection of the upper respiratory tract and co-infections such as mycoplasma, influenza and other respiratory viral infections have been associated with an increase risk of carriage and invasive disease in some settings [99, 100, 129, 130, 131, 132]. Exposure to tobacco smoke by active and passive smoking has been associated with meningococcal disease and meningococcal carriage [133, 134, 135, 136, 137], possibly by increased transmission due to coughing.

Furthermore, closed populations such as military recruits and Haj pilgrims have been shown to have increased risk of meningococcal infection. Studies of the risk of meningococcal disease in college students yielded similar results [138, 139] showing furthermore that students living on campus had an over threefold increased risk relative to the off-campus population [140]. Transmission of serogroup C meningococcal disease during a small outbreak in Maryland appears to have occurred during a party [141], and bar or discotheque patronage has been associated with the risk of infection during a university outbreak [136]. In addition, there have been numerous reports of laboratory-acquired meningococcal infection, mostly among microbiologists manipulating the organism on an open laboratory bench [142, 143, 144, 145].

12.7 Meningococcal Disease

12.7.1 Symptoms

The cause of progression from carriage to invasive disease is dependent on both host and infecting organism factors. A lack of sufficient biologically active antibodies against *Neisseria meningitidis* is thought to be one of the major factors conferring susceptibility to infection. The peak incidence of disease is between 6 months and 2 years of age. It then falls throughout childhood only to rise again transiently in adolescence.

Meningococcal disease is characterized by a variety of clinical symptoms [146, 147, 148] and signs of the early phase may be indistinguishable from other bacterial or viral infections. The mild symptoms can change rapidly (within hours) in a life-threatening disease. The most common symptoms are stiff neck, high fever, sensitivity to light, confusion, headaches and vomiting. Even when the disease is diagnosed early and adequate therapy instituted, 5–10% of patients die, typically within 24–48 hours of onset of symptoms. Clinical syndromes caused by *Neisseria meningitidis* include meningitis, with or without meningococcemia, relatively mild bacteremia, fulminant meningococcemia, meningoencephalitis, pneumonia, and septic arthritis, as well as other presentations [26].

The three traditional symptoms of meningitis include fever, severe headache, and stiffness of the neck [149, 150, 151, 152]. Due to increased intracranial pressure resulting from the vigorous inflammatory immune response within the CSF, altered mental status is another extremely common symptom. Bacterial meningitis may result in brain damage, hearing loss, or learning disability in 10–20% of survivors. Meningococcemia without meningitis is the most deadly of the meningococcal syndromes [96]. Meningococcemia is characterized by an abrupt onset of fever and a petechial or purpuric rash, which may progress to purpura fulminans, and is often associated with the rapid onset of hypotension, acute adrenal hemorrhage (the Waterhouse–Friderichsen syndrome), and multiorgan failure [153].

12.7.2 Diagnosis

The diagnosis of meningococcal meningitis is suspected by the clinical presentation and a lumbar puncture showing a purulent spinal fluid; sometimes the bacteria can be seen in microscopic examinations of the spinal fluid. Meningococcemia can be confirmed with blood culture, lumbar puncture, and a Gram stain of lesional skin biopsy or aspirate specimens. Definitive diagnosis requires the isolation of *Neisseria meningitidis* from a normally sterile body fluid or the detection of meningococcal DNA by PCR [154, 155, 156, 157, 158].

The classic laboratory diagnosis of meningococcal disease has relied on bacteriologic culture. However, the sensitivity of culture may be reduced when performed after the initiation of antibiotic treatment [159]. Gram's staining of cerebrospinal fluid is still considered an important method for rapid and accurate identification of *Neisseria meningitidis* [160]. Nonculture methods, such as the use of commercially available kits to detect polysaccharide antigen in cerebrospinal fluid, have been used to enhance the laboratory diagnosis. These methods are rapid and specific and can provide a serogroup-specific diagnosis, but false negative results are common, especially in cases of serogroup B disease [161]. Polymerase-chain-reaction (PCR) analysis offers the advantages of detecting serogroup-specific *Neisseria meningitidis* DNA and of not requiring live organisms for a positive result. In addition, newer molecular-based subtyping techniques may allow further characterization of *Neisseria meningitidis* from PCR-derived products.

12.7.3 Treatment

Meningococcal disease is potentially fatal and should always be viewed as a medical emergency. Admission to a hospital or health centre is necessary [148]. Antimicrobial therapy must be commenced as soon as possible after the lumbar puncture has been carried out and should not be delayed whilst waiting for the results of laboratory investigations. In approximately 10% of the cases of meningitis and 20% of sepsis, the disease is fatal despite antibiotic treatment [162]. Without antibiotics, death rates increase up to 80% [163]. Of the surviving patients, 5–20% experiences permanent damage, such as amputation of necrotic extremities, mental retardation or deafness [163, 164]. Rarely, invasive disease may take the form of arthritis, pneumonia or endophthalmitis.

A range of antibiotics may be used for treatment. For confirmed infection, intravenous aqueous penicillin is generally considered to be the therapy of first choice, although treatment with penicillin has reportedly failed in a few patients with strains of *Neisseria meningitidis* that have intermediate resistance to the drug [165, 166].

Appropriate broader treatments include cefotaxime or ceftriaxone given by the intravenous or intramuscular route. The duration of antibiotic treatment is 10–14 days for most patients; however, persistent parameningeal infections will require therapy of longer duration. Additionally, should a patient suffer bacteriologic or symptomatic relapse-specifically fever, drowsiness, or hemiparesis—after the discontinuation of treatment, antibiotic therapy should begin again immediately [149, 152]. Under epidemic conditions in Africa, oily chloramphenicol is the drug of choice in areas with limited health facilities because a single dose of this long-acting formulation has been shown to be effective.

Unfortunately, many patients with severe meningococcemia respond poorly to treatment with antimicrobial agents, steroids, or vasopressor agents and death may occur within hours of onset. Nevertheless, more commonly therapeutic interventions with antibiotics and supportive treatments are successful and have dramatically reduced mortality due to meningococcal disease [165, 166, 167].

12.8 Meningococcal Immunity

Natural immunity plays an important role in the maintenance of the commensal state for *Neisseria meningitidis* [150]. Protection against disease depends on the ability of the human host to generate long-lived, antigen-specific memory T and B cells that can be rapidly mobilized to mediate microbial clearance. Protective immunity consists of mucosal immunity, which limits colonization and prevents early invasion and systemic immunity, which blocks multiplication and systemic dissemination of meningococci that breach the mucosal barrier [98, 168, 169, 170].

12.8.1 Innate Immunity

In infancy and early childhood, non-specific defense mechanisms are the first responses against invading pathogens, and may be the most important defense against meningococcal infection. Mechanical mechanisms serve the purpose of preventing infection and for controlling nasopharyngeal colonization to prevent possible invasion of the intravascular space and CNS. Laminar airflow across mucous layers, that filter inspired air, the guttural reflex, laryngeal disclosure, and the cough reflex are non-specific, physical mechanisms for preventing infection of the human nasopharynx [171]. Additionally, within the respiratory tract, mucosal epithelial cells produce and secrete hydrogen peroxide, a general bactericide. Interestingly, the middle ear and Eustachian tube are usually immune to bacterial colonization despite their direct connection to the nasopharynx and it appears that secretions of lysozyme, lactoferrin and β-defensins -1 and -2 play a role [171, 172].

Furthermore, macrophages and neutrophils within the mucosal membrane of the nasopharynx, will clear bacteria following their opsonization by complement [173]. This is especially important in the non-immune individual, particularly in infancy when specific antibody is absent. Complement activation seems primarily to be mediated by C-reactive proteins that have bound themselves to phosphatidylcholine found on the surface of bacterial cell walls [174].

For efficient immune clearance *Neisseria meningitidis* is capable of eliciting innate immune responses via several mechanisms: the mannose binding lectin pathway is instrumental in activating complement-mediated phagocytosis of the bacteria. The recognition of capsular lipooligosaccharide triggers the TLR4 pathway, which initiates a signal transduction process that attracts phagocytes, activates macrophages and dendritic cells, and ultimately leads to the phagocytosis and destruction of the offending bacteria [171]. Furthermore, lipooligosaccharide and other lipopolysaccharide structures trigger the alternative pathway by binding C1q, and like in the mannose binding lectin pathway, the alternative pathway leads to bacterial cell death via the membrane attack complex (C5–C9) [150, 175]. Maturational defects in innate immunity (low complement levels and decreased phagocytosis) in combination with low levels of antigen-specific immunity in early childhood are responsible for the age-dependent differences in disease incidence.

12.8.2 Adaptive Immunity

12.8.2.1 The Acquisition of Natural Immunity to *Neisseria meningitidis*

In addition to mechanical and innate immune mechanisms, during the first few months of life, most infants are protected from invasive meningococcal disease by the presence of maternal IgG directed against capsular polysaccharides and outer membrane antigens, and passively immunized by IgA from maternal breast-milk. The age of greatest vulnerability to infection is when the level of maternally derived antibodies falls through infancy and there is little or no specific humoral immunity, phagocyte function is poor and the terminal complement components are at relatively low levels.

Following waning of maternal antibodies, children begin to develop immune responses through exposure to cross-reacting species in the nasopharynx and gastrointestinal tract [176, 177, 178, 179, 180]. For example, the *Escherichia coli* K1 capsule cross-reacts antigenically with serogroup B meningococci [181, 182] and the K92 *Escherichia coli* capsular polysaccharide with serogroup C polysaccharide [183]. In addition, natural carriage of *Neisseria meningitidis* and related bacteria such as non-pathogenic *Neisseria lactamica* [184], that commonly colonizes the nasopharynx during childhood and that is highly cross reactive with *Neisseria meningitidis*, leads to increasing levels of specific antibodies through childhood and results in the development of protective immunity by programming appropriate components of an immune response at both the mucosal surface and in the circulation.

The carriage dynamics of *Neisseria* species are complex and can be long-lived [46, 185, 186, 187]. In teenage and early adulthood single or multiple *Neisseria meningitidis* colonization events occur, whereas carriage of *Neisseria lactamica* occurs earlier in childhood.

Initially, nasopharyngeal colonization with *Neisseria* species generates mucosal effector T cells and secretory IgA in the absence of protective systemic immunity. With increasing age and repeated contact, local immunity in the nasopharyngeal mucosa and immunological memory increase and systemic effector T cells and memory responses develop [188, 189, 190]. Natural immunological programming results in a mucosal response that is dominated by a pro-inflammatory interferon-γ (IFN-γ) phenotype. The T-cell phenotype of the *Neisseria meningitidis* antigen responding population in the circulation seems to have a more balanced Th1–Th2 profile (Fig. 12.4).

The T-cell memory in both compartments is broadly cross-reactive across a range of antigens [191, 192] and by adulthood, even if carriage is quite common, most individuals have developed sufficient immunity to have protection against invasive disease. Both the local and systemic arms are essential to natural immunity and interrupt the evolution of disease at two crucial points.

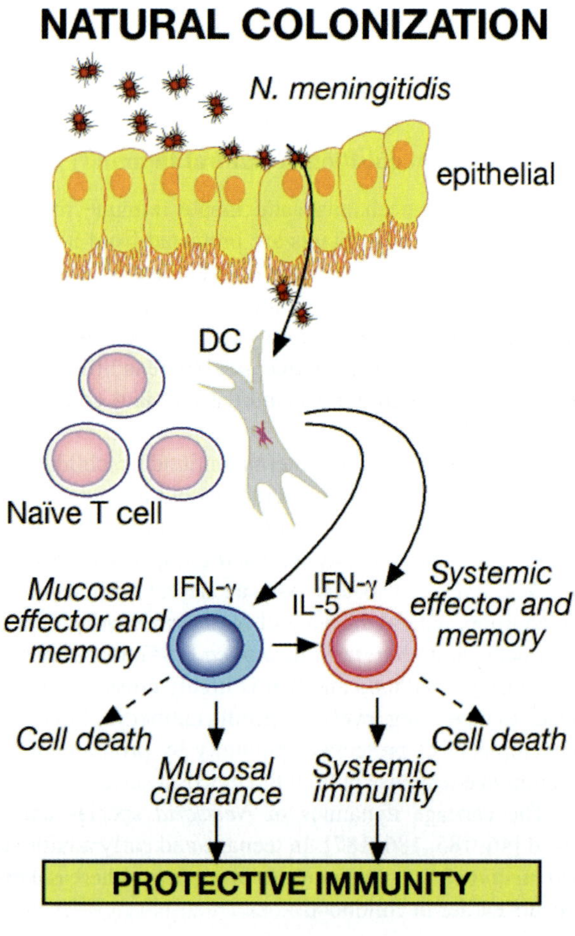

Fig. 12.4 Development of natural immunity to *Neisseria meningitidis*

The carriage of non-pathogenic organisms that induce immunity to meningococci appears not only to be essential for the acquisition of natural immunity, but it seems also plausible that population differences in carriage of such cross-reacting organisms may be partially responsible for the variation in incidence of meningococcal disease around the world and the variation in dominant serogroups and strains [184].

12.8.2.2 Mucosal Immunity

Mucosal immunity limits colonization and prevents early invasion. Capsular polysaccharides as the outermost organelle of meningococcal bacteria seem to be the "prime targets" for mucosal immunity [169] and are the most important antigens for mounting immune responses against serogroups A and C. Unfortunately, the polysaccharide capsule of the serogroup B meningococcus is chemically and antigenically related to human brain and fetal antigens and is therefore poorly immunogenic in humans [169]. Instead, for these serogroup antibody responses are mounted against other antigens including PorA, PorB, and lipopolysaccharide class 5 outer membrane proteins [169].

It has been suggested that in mucosal secretions, IgA antibodies against protein antigens are predominantly IgA1, whereas those directed against polysaccharides are almost equally distributed between the two subclasses, IgA1 and IgA2 [193, 194, 195]. The commonly observed close correlation between an individual's concentration of antigen specific secretory component and specific IgA antibodies in saliva suggests that these mucosal IgA antibodies are secretory IgA and that their release is locally regulated [196, 197]. The IgA2 subclass of IgA antibodies may provide some functional advantage in specific mucosal immune responses over IgA1 as a result of structural differences that make IgA2 relatively resistant to IgA1 protease activity [198, 199]. IgA1 protease produced by all strains of *Neisseria meningitidis* are potentially capable to cleave IgA1 to Fab and Fc fragments, and can therefore eliminate the Fc mediated functions of IgA1 [199, 200]. However, IgA1 proteases can be inactivated by "an abundance of antibodies" carried by the majority of humans [169]. Therefore, despite the presence of bacterial proteases, IgA may still provide an effective response. IgA can also inhibit immune lysis by out-competing or displacing IgM and IgG for antigen epitopes, thus blocking their stimulation of lytic activity. Conversely, as illogical a function as it may seem, IgA may have an important homeostatic function in down-regulating complement activation and the subsequent, potentially detrimental, inflammatory response [171].

12.8.2.3 Systemic Immunity

Systemic immunity blocks multiplication and systemic dissemination of meningococci that breach the mucosal barrier. It consists of complement-fixing IgG antibodies that are high in bactericidal activity [98, 184, 189]. Bactericidal activity is acquired or boosted through nasopharyngeal carriage of *Neisseria meningitidis* in adults [45, 201], which induces some cross-protection between strains [201]. Serum bactericidal activity has been the primary determinant of establishing a commensal relationship between the host and colonies of *Neisseria meningitidis*, and is primarily mediated via the lysis of meningococcal bacteria, which in turn is primarily mediated by the formation of membrane attack complex [169, 202]. Intravascular protection against gram-negative bacteria is believed to be mediated by antibody and complement recognition of the bacterial surface. C3b mediates opsonophagocytosis, and membrane attack complex mediates cell lysis, and other mechanisms such as non-complement fixing opsonophagocytosis and antibody-dependent cellular cytotoxicity [169, 184, 203]. Cell death limits the size of the responding cell populations, as is the case in any immune response, and occurs by antigenic neglect and activation-induced cell death.

The pathway of complement activation is considered to be central to protection against *Neisseria meningitidis* [204]. Moreover, the alternative complement pathway, which does not use antibody, is crucial in protection from meningococcal disease. Mannan-binding lectin binds to carbohydrate surface structures [205, 206] on bacteria, yeasts, parasitic protozoa and viruses [207] and is structurally similar to C1q and is able to activate complement via associated serine proteases (MASP-1 and MASP-2) [208, 209]. Killing is mediated by the terminal lytic complement pathway [210] or by increased phagocytosis [211]. However, some data suggest that the mannan-binding lectin is not important in phagocytosis of serogroup A meningococci [212].

Most IgG to meningococcal polysaccharides are of the IgG2 subclass, and deficiencies in IgG2 are often associated with more severe disease. The polysaccharide capsule prevents classical pathway activation by IgG that bind subcapsular antigens. Especially for serogroup B there are few antibodies against the polysaccharide capsule due to humans' high tolerance of the sialylated capsule that mimics the protective sialylation of host cells primarily from complement-mediated destruction.

Other antibodies that are naturally acquired during childhood include antibodies to non-sialylated lipooligosaccharide after repeated colonization by *Neisseria meningitidis*. This antibody-mediated immunity should last through adulthood, although later in life, as the efficacy of antibody responses decline, susceptibility increases. All lipooligosaccharides bear an epitope for IgM, however, sialiation prevents galactosaminylation, consequently preventing opsonization by IgM.

12.8.2.4 Cellular Immunity

Cellular immune mechanisms can be clearly demonstrated after meningococcal infection. T cell responses are likely to be central to acquired protective immunity through appropriate regulation and stimulation of antibody production. Furthermore, T-cells play an important role in the establishment of immunologic memory. In particular, the pattern of dendritic cell (DC) activation and antigen processing [213, 214, 215] and the resulting T-cell profile are vital for antibody and phagocytic control of both colonization and disease. T-helper (Th)-1 cells secrete IFN-γ and provide help for pro-inflammatory immune responses that involve macrophage and neutrophil activation and complement-fixing antibody production. Th2 cells secrete cytokines including interleukin (IL)-4 and IL-5, which promote B-cell activation, non-complement-fixing antibody production and secretory IgA. Distinct $CD4^+$ T-regulatory cells (Treg) suppress inflammatory responses and include a thymically derived $CD25^+$ subset that is crucial in preventing autoimmunity, and inducible subsets that probably arise in the periphery in response to antigen exposure [216, 217, 218]. Inducible Treg are generated locally in the mucosae as part of the normal response to the microbial flora [218] and probably arise as part of the natural response to *Neisseria meningitidis*, potentially suppressing the response to carriage and modulating immunity. The pattern of DC activation and the balance between Th1 versus Th2 and Treg is dependent on the nature, length, strength and site of *Neisseria meningitidis* antigenic exposure and the characteristics of the local microenvironment [213, 214, 215, 219, 220, 221].

A Th2 response is stimulated by meningococcal membrane protein antigens including PorA, PorB, Opa, and Opc [222]. Studies have demonstrated that the Th1 response is particularly prevalent among infant immune responses and become less prevalent with age as an individual switches from primarily a Th1 response to a Th2 response. Immunologic memory improves as the immune system matures past infancy and early childhood. Moreover, although infants produce the same amounts of IgG1 and IgG3, as older children, their immune response is still poor, suggesting differences in Ab specificity or affinity [169, 202, 223].

Cellular immune responses to OMVs following meningococcal infection have been examined [169], and showed that proliferative responses to the OMVs were not strain specific or dependent on age. In contrast, cytokine production was age-dependent with more IL-10 produced by peripheral blood mononuclear cells from older children and more interferon-gamma produced by infants [169]. These may reflect important differences in specific immune responses that could affect the quality or specificity of antibody and explain the lack of bactericidal responses in infants, noted above.

Following infection, there is a broad range of opsonins cross-reacting with different serogroups of meningococci, which are not directed against capsule, suggesting that these antibodies may recognize conserved regions of surface structures [180]. Fc receptor expression on the surface of phagocytes may be of particular importance in host defense against *Neisseria meningitidis* as it is required for the phagocytosis of IgG-opsonized bacteria [115]. Fc receptors with poor IgG2 binding are more common in children who develop meningococcal infection [116].

Neonatal polymorphonuclear cells have a reduced capability for adherence, migration, phagocytosis and intracellular killing when compared with adult cells [224]. It is not clear at which age adult activity is attained, but it is possible that the reduced phagocytic activity of these cells may be directly related to the increased incidence of meningococcal disease in early childhood, particularly since phagocytes may be important in defense against serogroup B organisms, which are common causes of disease at this age.

Non-opsonic phagocytosis by neutrophils, via interactions with bacterial Opa and Opc proteins, may also have a role in defense against meningococci although activity is reduced by sialylation of the polysaccharide capsule and lipopolysaccharide [225, 226]. The neutrophil surface receptor involved in these interactions with Opa is CD66 [227]. These bacterial proteins are involved in adherence and invasion of epithelial cells and are an essential virulence mechanism of the bacteria. Non-opsonic phagocytosis may also be mediated via phagocyte complement receptor 3 (CR3; CD18/CD11b) [228] or interactions of CD14 with lipopolysaccharide.

12.9 Meningococcal Control, Prevention and Vaccinaton

12.9.1 Control

Persons in close contact with patients who have meningococcal disease are at elevated risk for contracting the disease, including household members, contacts at day care centers, and anyone else directly exposed to an infected patient's oral and nasal secretions [29]. Because the risk of secondary disease among close contacts is highest during the first few days after the onset of disease in the index patient, chemoprophylaxis should be administered preferably within 24 hours of identification of the index patient [229]. Rifampin, ciprofloxacin, and ceftriaxone have been demonstrated to effectively eliminate nasopharyngeal carriage of *Neisseria meningitidis*. Notwithstanding, mass chemoprophylaxis to control large meningococcal epidemics is not recommended [230] and increases the likelihood of emergence of antimicrobial resistance. However, it might be considered for smaller outbreaks in well-defined populations, such as school outbreaks [231, 232].

12.9.2 Prevention

The rapid onset of invasive meningococcal disease, the high incidence in childhood, the severity of the sequellae and the high mortality rate clearly indicate the importance of vaccine development against *Neisseria meningitidis*. An ideal meningococcal vaccine should be able to protect against all disease-causing meningococci, developing protective immunity at the mucosal surface and in the circulation.

Several polysaccharide vaccines, which have been available for over 30 years, exist against serogroups A, C, Y, W135 in various combinations (Fig. 12.5). A major advance in the prevention of meningococcal disease has been the development and introduction of meningococcal polysaccharide-protein conjugate vaccines [27, 230, 233, 234, 235, 236, 237, 238, 239, 240]. These vaccines are immunogenic, particularly for children under 2 years of age whereas polysaccharide vaccines are not. All these vaccines have been proven to be safe and effective with infrequent and mild side effects. The development of vaccines for serogroup B *Neisseria meningitidis* remains a challenge [241]. The serogroup B capsule has an identical structure to polysialic structures expressed in fetal neural tissue, and does not induce a protective IgG response, therefore a different class of antigens has to be identified for an efficient vaccine formulation.

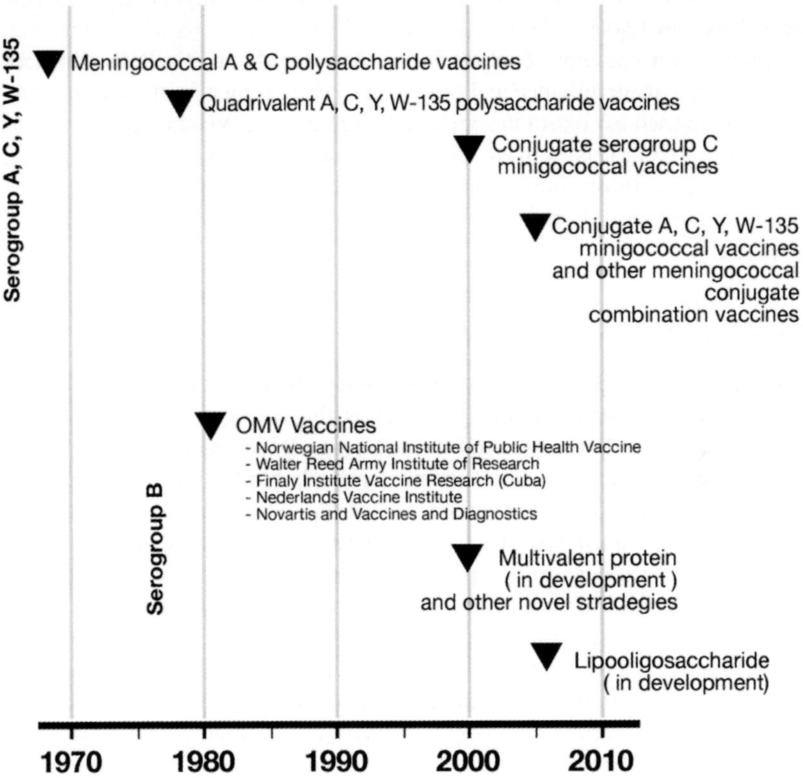

Fig. 12.5 Timeline of meningococcal vaccine development

12.9.3 Vaccination

Vaccination against meningococcal disease is used in the following circumstances:

I. Routine vaccination: Routine preventive mass vaccination has been attempted and its effect has been extensively debated. Saudi Arabia, for example, offers routine immunization of its entire population. Sudan and other countries routinely vaccinate school children) [242].

II. Preventive vaccination can be used to protect individuals at risk (e.g. travellers, military recruits, pilgrims, patients with complement or other immune deficiencies, microbiologists who are routinely exposed to isolates of *Neisseria meningitidis*) [243].

III. Protection of close contacts: When a sporadic case occurs, the close contacts need to be protected by a vaccine and chemoprophylaxis with antibiotics to cover the delay between vaccination and protection (the vaccines may not provide adequate protection for 10–14 days following injection). Antibiotics used for chemoprophylaxis are rifampicin, minocycline, spiramycin, ciprofloxacin and ceftriaxone [244].

IV. Vaccination for epidemic control: In the African Meningitis Belt, enhanced epidemiological surveillance and prompt case management with oily chloramphenicol are used to control the epidemics. The current WHO recommendation for outbreak control is to mass vaccinate every district that is in an epidemic phase, as well as those contiguous districts that are in alert phase. It is estimated that a mass immunization campaign, promptly implemented, can avoid 70% of cases [245].

12.9.3.1 Serum Bactericidal Activity as a Correlate of Protection

The classical studies performed by Goldschneider and colleagues [45, 98] established serum bactericidal antibody (SBA) as the surrogate of protection for meningococcal disease. Serum bactericidal activity measures bacterial killing mediated by complement. Meningococcal-specific antibody binds to the target cell surface via meningococcal-specific protein or carbohydrate moieties. The C1q subunit of C1 binds to the Fc portion of the surface-bound Ig. The binding of C1q–Ig activates the classical pathway of complement, which results ultimately in death of the target cell. The SBA titer for each unknown serum is expressed as the reciprocal serum dilution yielding at least 50% killing or more as compared with the number of target cells present before incubation with serum and complement. It is thought to be a more clinically relevant measure of the immune response than IgG titer. Total serum IgG correlates poorly with SBA, but a modified assay detecting high avidity IgG has been proposed [246]. The complement source used in the SBA assay is of importance. The Goldschneider studies used human sera as the source of exogenous complement in the SBA assay (hSBA) and was an individual surrogate following natural infection. A hSBA titer of four or more was shown to be protective. The World Health Organization (WHO)-recommended procedure uses baby rabbit complement as the source and recommends that an efficacious meningococcal vaccine should induce at least a 4-fold rise in SBA in more than 90% of vaccinees [247].

12.9.3.2 Vaccines Against *Neisseria meningitidis* Serogroup A, C, Y and W135

Polysaccharide Vaccines

Capsular, polysaccharide vaccines to decrease A, C, Y, and W-135 meningococcal disease were introduced in the 1970s and 1980s on the basis of Gotschlich, Gold, Goldschneider, and Artenstein's classic studies (Fig. 12.5) [45, 72, 98, 105, 239, 240, 248, 249]. They are made of extracted and purified forms of the bacterial outer polysaccharide capsule. Capsular polysaccharide vaccines against *Neisseria meningitidis* are licensed in bivalent and tetravalent formulations. Groups A and C polysaccharide vaccines (Mengivac^{TM} and AC VAX^{TM}) were developed in the 1960s in response to meningitis epidemics among military recruits in the USA and were widely tested in Europe, Latin America and Africa [72, 250, 251]. Meningococcal polysaccharide vaccines have been used extensively to control epidemics in countries of the African meningitis belt. Furthermore, tetravalent polysaccharide vaccines against serogroups A, C, Y, and W-135 have been developed and are available worldwide (Menomune™ and ACWY VAX™) [252, 253, 254, 255] (Fig. 12.5). Plain serogroup B polysaccharide vaccine development has been unsuccessful for the reasons of poor immunogenicity of this polysaccharide.

Polysaccharide vaccines trigger the formation of circulating antibodies. These antibodies then coat the bacterial capsules with antibodies and/or complement, promoting phagocytosis and removal by cells of the innate immune system (through macrophages and neutrophils). Purified polysaccharides activate B cells in a thymus-independent type 2 (TI-2) manner, they cross-link antibodies on the B cell surface, which leads to the formation of plasma B cells and antibodies without the help of helper T cells. As a result of this T independent activation, IgM is the predominant isotype produced due to little class switching, no affinity maturation, and little development of memory cells. A consequence of the T-cell-independent immunity is the poor efficacy of polysaccharide based vaccines in young children and infants and no long-term immune memory [106, 256, 257, 258, 259, 261]. At all ages, immune responses to polysaccharide antigens appear to be relatively short-lived.

The ability of older children and adults to produce serum antibody responses to capsular polysaccharides is considered to be an indicator of intrinsic B cell maturation. However, there is growing evidence that acquisition of antibody responsiveness, at least in adults, also reflects natural priming and generation of anti-capsular B cell populations in the memory state.

Experience from several A/C polysaccharide vaccine studies in the Gambia and in Saudi Arabia has shown that repeated doses of the group C polysaccharide vaccine may induce immunologic hyporesponsiveness in children and adults [106, 262, 263, 264, 265, 267, 268] possibly by depleting the pool of antigen-specific memory B-cells [263, 269]. By contrast, the serogroup A meningococcal polysaccharide, unlike most other polysaccharides, induces immunological memory in young children [263, 270]. This implies fundamental differences in their respective mechanisms of immunogenicity that remain poorly understood. Furthermore, it is not clear whether this is an entirely maturational phenomenon related to a deficiency in the infant's immune response to T-cell-independent antigens [271, 272].

Consequently, vaccines composed solely of purified polysaccharides are only effective in older children and adults for an average of three to five years. They are not effective for infants and children under two years of age, who have immature immune systems. Therefore, conjugate vaccines against some forms of meningitis have been developed.

Conjugate Vaccines

By conjugating purified bacterial polysaccharides to purified protein carriers, T-cell-dependent responses can be induced, leading to the establishment of immunological memory [273]. Polysaccharide-protein conjugate vaccines are safe, immunogenic in young infants and induce long-term immune memory [239, 240, 255, 274].

Several protein carriers have been used effectively for conjugate vaccine strategies, including tetanus toxoid, diphtheria toxoid, and a non-toxic variant of diphtheria toxin (CRM197) [275]. The protein carrier of the conjugate vaccine leads to antigen presentation on B cells, or other antigen presenting cells. The interaction of the CD40 on B cells and CD40L on T helper cells allow class switching from IgM to IgG, and longer-lasting effector functions. Memory B cells are also induced to participate in a secondary response upon exposure to the bacterial pathogen [276].

Immunization with conjugate vaccines also decreases nasopharyngeal carriage and transmission of the pathogen through the establishment of herd immunity [27, 234]. Whereas only high molecular weight polysaccharide preparations are immunogenic, effective glycoconjugate vaccines can be prepared from low molecular weight oligosaccharides such as those present in the lipooligosaccharides of *Neisseria meningitidis* [277, 278].

Meningococcal group C conjugate (MenC) vaccines were developed (Fig. 12.5), consisting of saccharides derived from capsular polysaccharide and chemically conjugated to carrier proteins, using CRM197 as the protein carrier (Meningitec^{TM} and Menjugate^{TM}), or using tetanus toxoid as the carrier (NeisVac-C^{TM}). Since 1999 group C glycoconjugate vaccines have been adopted for infant immunization program at 2, 3 and 4 months of age in the UK [237]. Immunogenicity studies showed that infants vaccinated at 2–4 months of age with group C glycoconjugate vaccine produced high levels of high avidity bactericidal antibodies [246, 279, 280]. Good antibody responses with booster doses of plain group C polysaccharide in infants given a primary course of conjugate vaccine indicate that immunological memory can be induced even in the very young [279, 281]. The successful introduction of the vaccine into the UK infant immunization schedule, in combination with a catch-up campaign for individuals less than 18 years of age targeting all children >4 months and <18 years of age (individuals <1 year of age received 2 doses of vaccine, and those $>=1$ year of age received 1 dose) [253], has virtually eliminated all group C disease in childhood [282]. Importantly, the vaccine program has driven tremendous impact on the incidence of the disease, resulting in a more than 90% decrease in the number of deaths and clinical cases, and a 66% decrease in asymptomatic carriage [282, 283, 284, 285]. Interestingly, the vaccine also decreased the number of cases in non-vaccinated people by 70%, a substantial benefit due to herd immunity, likely related to the effects on colonization and decreased circulation of the bacteria in the population [27, 234]. Group C glycoconjugate vaccine has also been licensed in other countries around Europe and was found to be effective in a mass-immunization campaign in Quebec, Canada, that was implemented in response to a MenC epidemic that occurred there in 2001 [286]. Diverse countries implemented different immunization strategies for the control of serogroup C meningococcal disease (CMD). Surprisingly, children given lower doses of antigen and/or fewer numbers of doses mount higher memory antibody responses to a subsequent boost with unconjugated polysaccharide vaccine than those who received higher doses or more injections of the conjugated antigen. Thus, existing routine immunization schedules may not be optimal and should be designed to achieve the highest level of protection using the lowest number of doses. However, more studies are needed to determine the long-term efficacy of different routine immunization schedules and the level of herd immunity provided by different levels of coverage [287].

A tetravalent conjugate vaccine incorporating polysaccharide from groups A, C, Y and W135 covalently linked to diphtheria toxin (Menactra™) has recently been developed and has shown a 100% efficiency at raising bactericidal antibody titers in 11- to 18-year olds. The vaccine has been licensed in the USA for 11- to 55-year olds and an application has been filed with the FDA to obtain licensure for 2- to 10-year olds [288, 289, 290, 291].

In 2001, the Meningitis Vaccine Project (MVP) was created as a partnership between Program for Appropriate Technology in Health (PATH) and the World Health Organization (WHO) with the goal of eliminating meningococcal epidemics in Africa through the development, licensure, introduction, and widespread use of affordable (US$ 0.40 per dose) meningococcal A (Men A) conjugate vaccines [292, 293]. A phase 1 study of the vaccine in India has shown that the product is safe and immunogenic. Phase 2 studies have begun in Africa, and a large demonstration study of the conjugate vaccine is envisioned for 2008–2009. Together with African public health officials a vaccine introduction plan has been developed that includes introduction of the Men A conjugate vaccine into standard Expanded Program on Immunization (EPI) schedules and further emphasizes mass vaccination of 1–29 year olds to induce herd immunity. The MVP model is a clear example of the usefulness of a "push mechanism" to finance the development of a needed vaccine for the developing world. Licensure of the vaccine is expected in 2009 [294]. Additionally a combined pediatric vaccine that contains serogroup A and C meningococcal conjugates has also been tested in Africa [295].

The chemical conjugation of capsular polysaccharide to protein carrier molecules ensures that a T-cell-dependent immune response is induced [296]. This approach, however, does not hold for group B meningococci, for which the prospect of developing a vaccine still is remote. A number of alternative approaches to serogroup B meningococcal vaccine development, like the use of recombinant outer membrane proteins, lipopolysaccharides and vesicle vaccines, have been considered over the past 20 years. Additionally, information resulting from different genomic sequencing projects of meningococcus is having a major influence on vaccine research and development [297].

12.9.3.3 Vaccines Against *Neisseria meningitidis* Serogroup B

Polysaccharide and Polysaccharide- Conjugate Based MenB Vaccines

Group B, a major serotype causing meningococcal meningitis, continues to be a global health burden. The ideal serogroup B vaccine would provide cross-protection throughout the serogroup. By definition, the shared antigen for these bacteria is the serogroup B polysaccharide capsule. Unfortunately, vaccines based on the serogroup B capsular polysaccharide are reported to elicit predominantly IgM antibodies of low avidity, and thus are poorly immunogenic [298, 299, 300, 301]. This poor immunogenicity is probably due to immunological tolerance induced by cross-reactive polysialylated glycoproteins. The MenB capsular polysaccharide structure ([alpha]2-8-N-acetylneuraminic acid) is homologous to that of a nerve cell adhesion molecule that is present in developing neural tissue and in small amounts in adult tissues [302]. Their similarities raise the risk of including autoimmune phenomena following immunization with vaccines based on group B polysaccharides. Attempts have been made to make a polysaccharide-based serogroup B conjugate vaccine in which the polysaccharide has been chemically modified. One such candidate vaccine was recently shown to be poorly immunogenic in adults, suggesting that this may not be a fruitful approach [303]. Knowing the complications in developing safe and effective serogroup B capsular polysaccharide based vaccines alternative strategies focus on non-capsular surface protein antigens. Accordingly, serogroup B outer membrane vesicle (OMV) and recombinant protein based vaccines are under investigation.

12.9.3.4 Protein Based OMV MenB Vaccines

Outer membrane vesicles consist of intact outer membrane and contain outer membrane proteins (OMPs) and lipooligosaccharides [241, 297]. They readily can be prepared by detergent extraction from the bacteria. However, diversity of major outer membrane structures among meningococci has limited these approaches. For example, in a study of the variability of endemic serogroup B strains from a variety of sites throughout the United States, it was found that 20 PorA types would have to be included in a serogroup B outer membrane vesicle vaccine to cover 80% of strains that cause endemic disease [304]. Nevertheless, serogroup B OMV vaccines can be useful during epidemics, when disease is caused by a single circulating serosubtype and an appropriately matched MenB OMV vaccine can be developed. Because MenB epidemics can last for more than a decade, the development of a matched vaccine may be feasible in such settings.

To date, the immunogenicity of a variety of OMV-based serogroup B vaccines have been studied. Adult volunteers have lymphoproliferative responses to Opa, Opc and PorA proteins of the bacterial outer membrane [192] and proliferative responses to OMV have been documented after vaccination with Dutch and Norwegian OMV vaccines [305, 306] and infection [169]. The T cell epitopes of PorA have been identified in adult volunteers as being in conserved regions of the protein and the MHC restriction of these epitopes was documented [307, 308, 309, 310]. Human T cell responses to purified meningococcal OMPs were higher to class 5 OMPs (Opa) and Opc than to PorA with some epitopes more widely recognized by different human leukocyte antigen (HLA) types and some showing greater HLA restriction [192]. PorA and PorB T cell epitopes are in regions of OMPs, which are not only conserved between strains but also highly conserved amongst Neisserial porin proteins [360]. T cell epitopes have also been described for both PorA and PorB OMP in mice that are conserved between strains [311].

The Cuban VA-MENGOC-BC® is a bivalent vaccine of capsular polysaccharide of *Neisseria meningitidis* group C and outer membrane vesicles of serogroup B meningococcus that include PorA, PorB, Opa, Opc, Tbp, NspA, and other proteins [312]. The vaccine was licensed in 1989 and was successfully used for epidemic control in Cuba, Brazil, Colombia and Uruguay [313, 314, 315]. The Cuban bivalent vaccine showed to induce proper bactericidal and opsonophagocytic activity [316] and was included in the Cuban infant vaccination schedule in 1991. More than 55 million doses have been applied. In an evaluation of three doses of the Cuban VA-MENGOC-BC® (B:4:P1.19,15) vaccine and a Norwegian MenBvac (B:15:P1.7,16) vaccine in Chile, at least two-thirds or more of children and adults had at least a fourfold rise in SBA, as did at least 90% of infants [317, 318, 319]. Efficacy studies of these vaccines were conducted during epidemics that occurred in Norway, Cuba, Brazil, and Chile [317, 318, 320, 321, 322]. The vaccines were found to be immunogenic among adults and older children (70–80% efficacy within 1 year), but immunogenicity was markedly reduced among young children and infants [253]. This limitation is mainly due to the vaccine's high strain-specificity showing good protective immune responses in both infants and adults only for the homologous situation [323, 324]. Furthermore, the immune response was of short duration, and the vaccine did not seem to have an impact on nasopharyngeal carriage. A combination of the Norwegian OMV MenB vaccine with a conjugate MenC vaccine was studied in adult volunteers and shown to be immunogenic with regard to both serogroup B and C meningococci [325].

Additionally, an OMV strain-specific vaccine containing PorA and PorB as well as lipopolysaccharide from the New Zealand serogroup B strain was developed (MeNZB®) and has been given to 1 million young people (approximately 3 million doses) in New Zealand to control the 15-year epidemic of ST40/41 serotype-specific serogroup B meningococcal disease, with promising safety and efficacy results [326, 327, 328]. A three-dose regimen of immunization was found to elicit bactericidal antibodies in 70% of children 6–24 months of age as well as in 8–12 years old children. The vaccine elicited a 90% response in teenagers. Strain-specific anti-OMP immune responses also were generated in infants. Sequential, nationwide introduction of the vaccine in the under 20 years of age population is currently ongoing with intensified phase III/IV monitoring [329, 330, 331].

In an attempt to develop more broadly protective OMV vaccines, a hexavalent PorA and more recently a nonavalent PorA OMV vaccine were developed at the National Institute for Public Health and the Environment, The Netherlands [332]. Two (or three) vaccine strains expressing 3 different PorA proteins each were constructed through recombinant DNA technology, in order to cover the majority of circulating serosubtypes in The Netherlands and other countries in Europe [332, 333]. Phase I and II studies have demonstrated that hexavalent PorA OMV vaccine is safe and immunogenic [334, 335]. Furthermore, it is estimated that the nonavalent PorA OMV vaccine would cover approximately 75% of the globally circulating serogroup B strains. Additionally, OMV antigens of *Neisseria lactamica*, are also being explored as a possible meningococcal vaccine [336, 337, 338, 339].

Thus, various methods have been used for the development of an effective vaccine against MenB disease. Although some vaccines have proved to be effective in epidemic situations and others provide some inconsistent protection against MenB strains, it is clear that novel approaches are necessary for the development of a safe and efficacious vaccine. The sequencing of the genome of a serogroup MenB strain (MC58) allowed a unique, genome-based approach to vaccine development.

12.9.3.5 Genome Derived MenB Vaccine Development

The genome-based approach to vaccine development uses the genomic sequence of an organism to identify sequences of proteins that are candidate antigens. This approach has been called reverse vaccinology [340], and has the potential to identify broadly protective vaccine candidates with improved immunogenicity, because it can access targets that might have been missed with the conventional methods for purification of bacterial components.

Several genome-derived antigens have been further evaluated as potential vaccine candidates, including genome-derived Neisserial A (GNA) 2,132, GNA1870 and surface-expressed proteins such as adhesion proteins (like NadA) and lipoproteins [297, 341, 342, 343]. Recently, GNA1870 has been renamed Factor H-binding Protein (fHBP), in order to reflect its important role in binding a down-regulatory component of the human complement alternative pathway and enhancing serum resistance of *Neisseria meningitidis* [344]. A recombinant pentavalent protein vaccine, containing fHBP, NadA, GNA2132 and two other antigens discovered by reverse vaccinology induced bactericidal antibodies in mice against 78% of a panel of 85 pathogenic meningococcal strains representative of the global population diversity [345]. Currently, this vaccine is undergoing evaluation in clinical trials [346].

Genome driven vaccine candidates together with the discovery of novel group B capsule-specific epitopes [347] as well as progress with detoxified lipooligosaccharide and conserved inner-core lipooligosaccharide [348, 349, 350] suggest new opportunities for vaccine prevention of serogroup B meningococcal disease. Furthermore, monoclonal antibodies to one relatively conserved antigen among genetically diverse serogroup B strains (neisserial antigen 1870) were recently found to be protective in infant rats [203]. In preclinical studies promising results also have been found using OMV vaccines prepared from *Neisseria meningitidis* strains that have been engineered to over-express certain conserved, cross-reactive antigens [351, 352]. Other antigens that have been explored with various degrees of success include transferrin-binding protein B, and another outer membrane protein referred to as H0.8 [353, 354, 355, 356, 357, 358, 359].

In order to obtain broad coverage and reduce the risk of disease from multiple endemic strains future vaccines should consist of multiple different antigens and/or contain immunogenic epitopes from multiple different strains. The development of serogroup B vaccines will require standardized methodologies [360] and a better definition of the role of activity other than the serum bactericidal assays. Up to now, the focus has been mainly on the induction of bactericidal antibodies but protection may also require generation of opsonizing antibodies or augmentation of mucosal immune responses through mucosal immunization.

12.9.3.6 Capsular Switching and Serotype Replacement

There is some anxiety that widespread vaccination may exert evolutionary pressure on the meningococcus and allow hyperinvasive lineages with a different capsule (e.g., serogroup B) to become established [361]. Thus, multivalent vaccines may well be required in order to prevent capsular switching and serotype replacement. Capsular switching refers to the horizontal exchange of genetic information between two different meningococci. A phenomenon which has been shown to occur [362]. Such instances lead to concerns that the administration of for example MenC vaccine could cause the serogroup C to switch to another serogroup, including B. Given that there is no viable vaccine for serogroup B yet, this could have serious effects on global meningitis infection. Serotype replacement is another vaccine concern, as a reduction in a particular vaccine-targeted serotype could lead to an increase in another non-vaccine serotype.

However, a recent workshop on vaccine pressure and *Neisseria meningitidis* (Annecy, France, 9–11 March 2005) concluded that there is currently no evidence of meningococcal serogroup switching or replacement induced by vaccine pressure when serogroup C conjugate vaccines have been used [363]. In countries where meningococcal polysaccharide vaccines were used routinely prior to conjugate vaccines, significant changes in the bacterial populations have been reported. Nevertheless, it remains unknown whether those changes were caused by vaccine-induced switching. Hence, carriage studies are essential to assess the impact of vaccine pressure, and to improve our understanding of disease transmission and the impact of conjugate vaccines on the carriage state, as well as the role vaccines may play in the selection of dominant strains [364].

12.9.3.7 Mucosal Vaccines

It is hypothesized that vaccine strategies that mimic the natural immunization process would better-optimize vaccine-induced protective immunity. With the purpose of inducing defense against a mucosal pathogen, mucosal immunization before a systemic booster vaccination could provide a solution and reduce the necessity for multiple injections to achieve immunity.

Intranasal vaccination with outer membrane vesicles derived from meningococci has been proven safe in phase I studies and induces a local and systemic humoral immune response that may be of importance in prevention of meningococcal disease [365, 366]. Comparison of functional immune responses in humans after intranasal and intramuscular immunizations with outer membrane vesicle vaccines against group B meningococcal disease, showed that intranasal OMV immunization can be used to induce mucosal and systemic antibodies with antibacterial activity. However, to represent an alternative to the present parenteral vaccine, the nasal approach would need to be improved both regarding the vaccine formulation and immunization schedule [367]. More recently, the induction of protective serum meningococcal bactericidal and diphtheria-neutralizing antibodies and mucosal IgA in human volunteers, were reported after nasal insufflations of the *Neisseria meningitidis* serogroup C polysaccharide-CRM197 conjugate vaccine mixed with chitosan [368]. Therefore, an available, alum-free, polysaccharide conjugate vaccine could reliably induce levels of immunity associated with protection against both meningococcal C disease and diphtheria when delivered nasally as powder, comparable with the same vaccine delivered intramuscularly.

Thus, it may be important to include adjuvants in the vaccine formulations that direct the response in a favorable direction and help to reduce the antigen doses needed to elicit strong and functional immune responses after intranasal immunization. Furthermore, the possibility to combine mucosal and systemic immunization should be considered.

12.10 Conclusion

Widespread use of effective vaccines will be necessary in order to ultimately control diseases caused by *Neisseria meningitidis*. Worldwide surveillance, the expanded use of polysaccharide conjugate vaccines, and the development of broadly effective serogroup B vaccines could eliminate the organism as a major threat to human health in industrialized countries within the next decade. Nevertheless, strong, coordinated, and continuous support will be needed from the international community if meningococcal disease is to be controlled in Africa and other developing areas, where the infection poses the greatest persistent threat.

References

1. http: and www.who.int/vaccine_research/diseases/soa_bacterial/en/index2.html 2007.
2. Anonymous 1998, WHO practical guidelines. 2nd ed. World Health Organization.
3. Greenfield, S., Sheehe, P.R., and Feldman, H.A. 1971, J. Infect. Dis., 123, 67.
4. Caugant, D.A., Hoiby, E.A., Magnus, P., Scheel, O., Hoel, T., Bjune, G., Wedege, E., Eng, J., and Froholm, L.O. 1994, J. Clin. Microbiol., 32, 323.
5. Stephens, D.S. 1999, Lancet, 353, 941.
6. Janda, W.M. and Knapp, J.S. 2003, Manual of Clinical Microbiology, 1, (Murray PR, Baron EJ, Jorgensen JH, Pfaller MA & Yolken RH, eds), 585.
7. Vieusseux, M. 1806, Letter to Medical and Agricultural Registrar. Boston, MA. Danielson, L., and E. Mann., 11, 163.
8. Danielson L. and Mann E. 1806, Med. Agric. Reg. Boston, MA, 1, 65.
9. Greenwood, B.M., Bradley, A.K., and Wall, R.A. 1985, Lancet, 2, 829.
10. World Health Organization Working Group 1995, Edition Foundation Marcel Merieux, Lyon, France.
11. Rosenstein, N.E., Perkins, B.A., Stephens, D.S., Lefkowitz, L., Cartter, M.L., Danila, R., Cieslak, P., Shutt, K.A., Popovic, T., Schuchat, A., Harrison, L.H., and Reingold, A.L. 1999, J. Infect. Dis., 180, 1894.
12. Tzeng, Y.L. and Stephens, D.S. 2000, Microbes. Infect., 2, 687.

13. Raghunathan, P.L., Jones, J.D., Tiendrebeogo, S.R., Sanou, I., Sangare, L., Kouanda, S., Dabal, M., Lingani, C., Elie, C.M., Johnson, S., Ari, M., Martinez, J., Chatt, J., Sidibe, K., Schmink, S., Mayer, L.W., Konde, M.K., Djingarey, M.H., Popovic, T., Plikaytis, B.D., Carlone, G.M., Rosenstein, N., and Soriano-Gabarro, M. 2006, J. Infect. Dis., 193, 607.
14. Danielson, L. and Mann, E. 1983, Rev. Infect. Dis., 5, 969.
15. Hirsch, A. 1886, A. Hirsch (ed.), Handbook of geographical and historical pathology, vol. 3. The New Sydenham Society, London, England.
16. Weichselbaum, A. 1887, Fortschr. Med., 5, 573.
17. Kiefer, F. 1896, Berl. Klin. Wochenschr., 33, 628.
18. Bruns, H. and J.Hohn 1908, Klin. Jahrb., 18, 825.
19. Flexner, S. 1906, JAMA, 47, 560.
20. Flexner, S. 1913, J. Exp. Med., 17, 553.
21. Flexner, S. and Jobling J.W. 1908, J. Exp. Med., 10, 690.
22. Flexner, S. and Jobling J.W. 1908, J. Exp. Med., 10, 141.
23. Gordon, M.H. and Murray E.G.D. 1915, J. R. Army Med. Corps, 25, 411.
24. Vedros, N. 1987, N. A. Vedros (ed.), Evolution of meningococcal disease, vol. 2. CRC Press, Boca Raton, Fla, 2, 33.
25. http: and www.emedicine.com/NEURO/topic210.htm 2007.
26. Apicella, M.A. 2005, GL Mandell, JE Bennett and R Dolin, Editors, Principles and Practice of Infectious Diseases (6th edn.), Elsevier Churchill Livingstone Publishers, Philadelphia, 2498.
27. Maiden, M.C. and Stuart, J.M. 2002, Lancet, 359, 1829.
28. Gold, R., Goldschneider, I., Lepow, M.L., Draper, T.F., and Randolph, M. 1978, J. Infect. Dis., 137, 112.
29. Rosenstein, N.E., Perkins, B.A., Stephens, D.S., Popovic, T., and Hughes, J.M. 2001, N. Engl. J. Med., 344, 1378.
30. Stephens, D.S. 2007, FEMS Microbiol. Rev., 31, 3.
31. Stephens, D.S., Greenwood, B., and Brandtzaeg, P. 2007, Lancet, 369, 2196.
32. Frasch, C.E., Zollinger, W.D., and Poolman, J.T. 1985, Rev. Infect. Dis., 7, 504.
33. Aaberge I, Helland O, Oster P, and et al 2002, Caugant D, Wedege E, eds. Abstracts of the Thirteenth International Pathogenic Neisseria Conference, Oslo, Norway; September 1–6. Oslo, Norway: Norwegian Institute of Public Health, 62.
34. Abdillahi, H. and Poolman, J.T. 1988, FEMS Microbiol. Immunol., 1, 139.
35. Poolman JT, van der Ley P, and Tommassen J 1995, Cartwight K, ed. Meningococcal Disease. New York, NY: John Wiley and Sons, 21.
36. Tettelin, H., Saunders, N.J., Heidelberg, J., Jeffries, A.C., Nelson, K.E., Eisen, J.A., Ketchum, K.A., Hood, D.W., Peden, J.F., Dodson, R.J., Nelson, W.C., Gwinn, M.L., DeBoy, R., Peterson, J.D., Hickey, E.K., Haft, D.H., Salzberg, S.L., White, O., Fleischmann, R.D., Dougherty, B.A., Mason, T., Ciecko, A., Parksey, D.S., Blair, E., Cittone, H., Clark, E.B., Cotton, M.D., Utterback, T.R., Khouri, H., Qin, H., Vamathevan, J., Gill, J., Scarlato, V., Masignani, V., Pizza, M., Grandi, G., Sun, L., Smith, H.O., Fraser, C.M., Moxon, E.R., Rappuoli, R., and Venter, J.C. 2000, Science, 287, 1809.
37. Parkhill, J., Achtman, M., James, K.D., Bentley, S.D., Churcher, C., Klee, S.R., Morelli, G., Basham, D., Brown, D., Chillingworth, T., Davies, R.M., Davis, P., Devlin, K., Feltwell, T., Hamlin, N., Holroyd, S., Jagels, K., Leather, S., Moule, S., Mungall, K., Quail, M.A., Rajandream, M.A., Rutherford, K.M., Simmonds, M., Skelton, J., Whitehead, S., Spratt, B.G., and Barrell, B.G. 2000, Nature, 404, 502.
38. Maiden, M.C., Bygraves, J.A., Feil, E., Morelli, G., Russell, J.E., Urwin, R., Zhang, Q., Zhou, J., Zurth, K., Caugant, D.A., Feavers, I.M., Achtman, M., and Spratt, B.G. 1998, Proc. Natl. Acad. Sci. U. S. A, 95, 3140.
39. Maiden, M.C. 2006, Annu. Rev. Microbiol., 60, 561.
40. Caugant, D.A. 1998, APMIS, 106, 505.
41. Brehony, C., Jolley, K.A., and Maiden, M.C. 2007, FEMS Microbiol. Rev., 31, 15.

42. Sim, R.J., Harrison, M.M., Moxon, E.R., and Tang, C.M. 2000, Lancet, 356, 1653.
43. Riordan, T., Cartwright, K., Andrews, N., Stuart, J., Burris, A., Fox, A., Borrow, R., Douglas-Riley, T., Gabb, J., and Miller, A. 1998, Epidemiol. Infect., 121, 495.
44. Neal, K.R., Nguyen-Van-Tam, J.S., Jeffrey, N., Slack, R.C., Madeley, R.J., Ait-Tahar, K., Job, K., Wale, M.C., and Ala'Aldeen, D.A. 2000, BMJ, 320, 846.
45. Goldschneider, I., Gotschlich, E.C., and Artenstein, M.S. 1969, J. Exp. Med., 129, 1327.
46. Ala'Aldeen, D.A., Neal, K.R., Ait-Tahar, K., Nguyen-Van-Tam, J.S., English, A., Falla, T.J., Hawkey, P.M., and Slack, R.C. 2000, J. Clin. Microbiol., 38, 2311.
47. Greenwood, B.M., Bradley, A.K., Smith, A.W., and Wall, R.A. 1987, Trans. R. Soc. Trop. Med. Hyg., 81, 536.
48. Greenwood, B. 1999, Trans. R. Soc. Trop. Med. Hyg., 93, 341.
49. Outbreak News 2006, Wkly Epidemiol Rec, 81, 119.
50. Yaro S, Traore Y, Tarnagda Z, Sangare L, Njanpoplafourcade BM, Drabo A, Findlow H, Borrow R, Nicolas P, Gessner BD, and Mueller JE 2007, Vaccine, 7, [Epub ahead of print].
51. Morelli, G., Malorny, B., Muller, K., Seiler, A., Wang, J.F., del Valle, J., and Achtman, M. 1997, Mol. Microbiol., 25, 1047.
52. Wang, J.F., Caugant, D.A., Li, X., Hu, X., Poolman, J.T., Crowe, B.A., and Achtman, M. 1992, Infect. Immun., 60, 5267.
53. Wilder-Smith, A., Barkham, T.M., Earnest, A., and Paton, N.I. 2002, BMJ, 325, 365.
54. Mayer, L.W., Reeves, M.W., Al Hamdan, N., Sacchi, C.T., Taha, M.K., Ajello, G.W., Schmink, S.E., Noble, C.A., Tondella, M.L., Whitney, A.M., Al Mazrou, Y., Al Jefri, M., Mishkhis, A., Sabban, S., Caugant, D.A., Lingappa, J., Rosenstein, N.E., and Popovic, T. 2002, J. Infect. Dis., 185, 1596.
55. Aguilera, J.F., Perrocheau, A., Meffre, C., and Hahne, S. 2002, Emerg. Infect. Dis., 8, 761.
56. Berron, S., De La, F.L., Martin, E., and Vazquez, J.A. 1998, Eur. J. Clin. Microbiol. Infect. Dis., 17, 85.
57. Ramsay, M., Kaczmarski, E., Rush, M., Mallard, R., Farrington, P., and White, J. 1997, Commun. Dis. Rep. CDR Rev., 7, R49.
58. Martin, D.R., Walker, S.J., Baker, M.G., and Lennon, D.R. 1998, J. Infect. Dis., 177, 497.
59. Dyet, K., Devoy, A., McDowell, R., and Martin, D. 2005, Vaccine, 23, 2228.
60. Swartley, J.S., Marfin, A.A., Edupuganti, S., Liu, L.J., Cieslak, P., Perkins, B., Wenger, J.D., and Stephens, D.S. 1997, Proc. Natl. Acad. Sci. U. S. A, 94, 271.
61. Dyet, K.H. and Martin, D.R. 2006, Epidemiol. Infect., 134, 377.
62. Jackson, L.A., Schuchat, A., Reeves, M.W., and Wenger, J.D. 1995, JAMA, 273, 383.
63. Raymond, N.J., Reeves, M., Ajello, G., Baughman, W., Gheesling, L.L., Carlone, G.M., Wenger, J.D., and Stephens, D.S. 1997, J. Infect. Dis., 176, 1277.
64. Racoosin, J.A., Whitney, C.G., Conover, C.S., and Diaz, P.S. 1998, JAMA, 280, 2094.
65. Djibo, S., Nicolas, P., Alonso, J.M., Djibo, A., Couret, D., Riou, J.Y., and Chippaux, J.P. 2003, Trop. Med. Int. Health, 8, 1118.
66. Boisier, P., Nicolas, P., Djibo, S., Taha, M.K., Jeanne, I., Mainassara, H.B., Tenebray, B., Kairo, K.K., Giorgini, D., and Chanteau, S. 2007, Clin. Infect. Dis., 44, 657.
67. AYCOCK, W.L. and MUELLER, J.H. 1950, Bacteriol. Rev., 14, 115.
68. Criado, M.T., Ferreiros, C.M., and Sainz, V. 1985, Med. Microbiol. Immunol., 174, 151.
69. Stuart, J.M., Cartwright, K.A., Robinson, P.M., and Noah, N.D. 1989, Lancet, 2, 723.
70. Dehio, C., Gray-Owen, S.D., and Meyer, T.F. 2000, Subcell. Biochem., 33, 61.
71. Hilse, R., Hammerschmidt, S., Bautsch, W., and Frosch, M. 1996, J. Bacteriol., 178, 2527.
72. Gotschlich, E.C., Goldschneider, I., and Artenstein, M.S. 1969, J. Exp. Med., 129, 1367.
73. http: and www3.accessmedicine.com/popup.aspx?aID=70200& print=yes 2007.
74. Merz, A.J. and So, M. 2000, Annu. Rev. Cell Dev. Biol., 16, 423.
75. Massari, P., Ram, S., Macleod, H., and Wetzler, L.M. 2003, Trends Microbiol., 11, 87.
76. Vordermeier, H.M., Drexler, H., and Bessler, W.G. 1987, Immunol. Lett., 15, 121.
77. Vordermeier, H.M. and Bessler, W.G. 1987, Immunobiology, 175, 245.
78. Wetzler, L.M., Ho, Y., and Reiser, H. 1996, J. Exp. Med., 183, 1151.

79. Donnelly, J.J., Deck, R.R., and Liu, M.A. 1990, J. Immunol., 145, 3071.
80. Dehio, C., Gray-Owen, S.D., and Meyer, T.F. 1998, Trends Microbiol., 6, 489.
81. Delany, I., Grifantini, R., Bartolini, E., Rappuoli, R., and Scarlato, V. 2006, J. Bacteriol., 188, 2483.
82. Kahler, C.M. and Stephens, D.S. 1998, Crit Rev. Microbiol., 24, 281.
83. Zughaier, S.M., Tzeng, Y.L., Zimmer, S.M., Datta, A., Carlson, R.W., and Stephens, D.S. 2004, Infect. Immun., 72, 371.
84. Mandrell, R.E., Griffiss, J.M., and Macher, B.A. 1988, J. Exp. Med., 168, 107.
85. Pathan, N., Faust, S.N., and Levin, M. 2003, Arch. Dis. Child, 88, 601.
86. Vitovski, S., Read, R.C., and Sayers, J.R. 1999, FASEB J., 13, 331.
87. Reinholdt, J. and Kilian, M. 1997, Infect. Immun., 65, 4452.
88. Jarvis, G.A. and Griffiss, J.M. 1991, J. Immunol., 147, 1962.
89. Lin, L., Ayala, P., Larson, J., Mulks, M., Fukuda, M., Carlsson, S.R., Enns, C., and So, M. 1997, Mol. Microbiol., 24, 1083.
90. http: and www.emedicine.com/MED/topic2613.htm 2007.
91. Johansson, L., Rytkonen, A., Bergman, P., Albiger, B., Kallstrom, H., Hokfelt, T., Agerberth, B., Cattaneo, R., and Jonsson, A.B. 2003, Science, 301, 373.
92. de Vries, F.P., Cole, R., Dankert, J., Frosch, M., and van Putten, J.P. 1998, Mol. Microbiol., 27, 1203.
93. Virji, M., Evans, D., Hadfield, A., Grunert, F., Teixeira, A.M., and Watt, S.M. 1999, Mol. Microbiol., 34, 538.
94. Stephens, D.S., Hoffman, L.H., and McGee, Z.A. 1983, J. Infect. Dis., 148, 369.
95. Mosleh, I.M., Huber, L.A., Steinlein, P., Pasquali, C., Gunther, D., and Meyer, T.F. 1998, J. Biol. Chem., 273, 35332.
96. Harrison, L.H. and Broome, C.V. 1987, N. A. Vedros (ed.), Evolution of meningococcal disease, vol. 1. CRC Press, Boca Raton, Fla., 1, 27.
97. Jackson, L.A. and Wenger, J.D. 1993, MMWR CDC Surveill Summ., 42, 21.
98. Goldschneider, I., Gotschlich, E.C., and Artenstein, M.S. 1969, J. Exp. Med., 129, 1307.
99. Harrison, L.H., Armstrong, C.W., Jenkins, S.R., Harmon, M.W., Ajello, G.W., Miller, G.B., Jr., and Broome, C.V. 1991, Arch. Intern. Med., 151, 1005.
100. Krasinski, K., Nelson, J.D., Butler, S., Luby, J.P., and Kusmiesz, H. 1987, Am. J. Epidemiol., 125, 499.
101. Emonts, M., Hazelzet, J.A., de Groot, R., and Hermans, P.W. 2003, Lancet Infect. Dis., 3, 565.
102. Sjoholm, A.G., Braconier, J.H., and Soderstrom, C. 1982, Clin. Exp. Immunol., 50, 291.
103. Fijen, C.A., Kuijper, E.J., te Bulte, M.T., Daha, M.R., and Dankert, J. 1999, Clin. Infect. Dis., 28, 98.
104. Lassiter, H.A., Watson, S.W., Seifring, M.L., and Tanner, J.E. 1992, J. Infect. Dis., 166, 53.
105. Gold, R., Lepow, M.L., Goldschneider, I., Draper, T.L., and Gotschlich, E.C. 1975, J. Clin. Invest, 56, 1536.
106. Gold, R., Lepow, M.L., Goldschneider, I., and Gotschlich, E.C. 1977, J. Infect. Dis., 136 Suppl, S31.
107. Goldschneider, I., Lepow, M.L., Gotschlich, E.C., Mauck, F.T., Bachl, F., and Randolph, M. 1973, J. Infect. Dis., 128, 769.
108. Westendorp, R.G., Hottenga, J.J., and Slagboom, P.E. 1999, Lancet, 354, 561.
109. Hibberd, M.L., Sumiya, M., Summerfield, J.A., Booy, R., and Levin, M. 1999, Lancet, 353, 1049.
110. Fijen, C.A., Bredius, R.G., Kuijper, E.J., Out, T.A., De Haas, M., De Wit, A.P., Daha, M.R., and De Winkel, J.G. 2000, Clin. Exp. Immunol., 120, 338.
111. Read, R.C., Pullin, J., Gregory, S., Borrow, R., Kaczmarski, E.B., di Giovine, F.S., Dower, S.K., Cannings, C., and Wilson, A.G. 2001, J. Infect. Dis., 184, 640.
112. Smirnova, I., Mann, N., Dols, A., Derkx, H.H., Hibberd, M.L., Levin, M., and Beutler, B. 2003, Proc. Natl. Acad. Sci. U. S. A, 100, 6075.

113. Faber, J., Meyer, C.U., Gemmer, C., Russo, A., Finn, A., Murdoch, C., Zenz, W., Mannhalter, C., Zabel, B.U., Schmitt, H.J., Habermehl, P., Zepp, F., and Knuf, M. 2006, Pediatr. Infect. Dis. J., 25, 80.
114. Tully, J., Viner, R.M., Coen, P.G., Stuart, J.M., Zambon, M., Peckham, C., Booth, C., Klein, N., Kaczmarski, E., and Booy, R. 2006, BMJ, 332, 445.
115. van de Winkel, J.G. and Capel, P.J. 1993, Immunol. Today, 14, 215.
116. Bredius, R.G., Derkx, B.H., Fijen, C.A., de Wit, T.P., De Haas, M., Weening, R.S., van de Winkel, J.G., and Out, T.A. 1994, J. Infect. Dis., 170, 848.
117. Styrt, B. 1990, Am. J. Med., 88, 33N.
118. Ellison, E.C. and Fabri, P.J. 1983, Surg. Clin. North Am., 63, 1313.
119. Molrine, D.C., George, S., Tarbell, N., Mauch, P., Diller, L., Neuberg, D., Shamberger, R.C., Anderson, E.L., Phillips, N.R., Kinsella, K., and Ambrosino, D.M. 1995, Ann. Intern. Med., 123, 828.
120. Biggar, W.D., Ramirez, R.A., and Rose, V. 1981, Pediatrics, 67, 548.
121. Figueroa, J.E. and Densen, P. 1991, Clin. Microbiol. Rev., 4, 359.
122. Francke, E.L. and Neu, H.C. 1981, Surg. Clin. North Am., 61, 135.
123. Holmes, F.F., Weyandt, T., Glazier, J., Cuppage, F.E., Moral, L.A., and Lindsey, N.J. 1981, JAMA, 246, 1119.
124. Kuipers, S., Aerts, P.C., Cluysenaer, O.J., Bartelink, A.K., Ezekowitz, R.A., Bax, W.A., Salimans, M., and Vandyk, H. 2003, Adv. Exp. Med. Biol., 531, 351.
125. Platonov, A.E., Vershinina, I.V., Kuijper, E.J., Borrow, R., and Kayhty, H. 2003, Vaccine, 21, 4437.
126. Salimans, M.M., Bax, W.A., Stegeman, F., van Deuren, M., Bartelink, A.K., and van Dijk, H. 2004, Clin. Diagn. Lab Immunol., 11, 806.
127. Stephens, D.S., Hajjeh, R.A., Baughman, W.S., Harvey, R.C., Wenger, J.D., and Farley, M.M. 1995, Ann. Intern. Med., 123, 937.
128. Greenwood, B.M., Bradley, A.K., Cleland, P.G., Haggie, M.H., Hassan-King, M., Lewis, L.S., Macfarlane, J.T., Taqi, A., Whittle, H.C., Bradley-Moore, A.M., and Ansari, Q. 1979, Trans. R. Soc. Trop. Med. Hyg., 73, 557.
129. Cartwright, K.A., Jones, D.M., Smith, A.J., Stuart, J.M., Kaczmarski, E.B., and Palmer, S.R. 1991, Lancet, 338, 554.
130. Moore, P.S., Hierholzer, J., DeWitt, W., Gouan, K., Djore, D., Lippeveld, T., Plikaytis, B., and Broome, C.V. 1990, JAMA, 264, 1271.
131. Olcen, P., Kjellander, J., Danielsson, D., and Lindquist, B.L. 1981, Scand. J. Infect. Dis., 13, 105.
132. Young, L.S., LaForce, F.M., Head, J.J., Feeley, J.C., and Bennett, J.V. 1972, N. Engl. J. Med., 287, 5.
133. Cookson, S.T., Corrales, J.L., Lotero, J.O., Regueira, M., Binsztein, N., Reeves, M.W., Ajello, G., and Jarvis, W.R. 1998, J. Infect. Dis., 178, 266.
134. Fischer, M., Hedberg, K., Cardosi, P., Plikaytis, B.D., Hoesly, F.C., Steingart, K.R., Bell, T.A., Fleming, D.W., Wenger, J.D., and Perkins, B.A. 1997, Pediatr. Infect. Dis. J., 16, 979.
135. Imrey, P.B., Jackson, L.A., Ludwinski, P.H., England, A.C., III, Fella, G.A., Fox, B.C., Isdale, L.B., Reeves, M.W., and Wenger, J.D. 1995, J. Clin. Microbiol., 33, 3133.
136. Imrey, P.B., Jackson, L.A., Ludwinski, P.H., England, A.C., III, Fella, G.A., Fox, B.C., Isdale, L.B., Reeves, M.W., and Wenger, J.D. 1996, Am. J. Epidemiol., 143, 624.
137. Tappero, J.W., Reporter, R., Wenger, J.D., Ward, B.A., Reeves, M.W., Missbach, T.S., Plikaytis, B.D., Mascola, L., and Schuchat, A. 1996, N. Engl. J. Med., 335, 833.
138. Bruce, M.G., Rosenstein, N.E., Capparella, J.M., Shutt, K.A., Perkins, B.A., and Collins, M. 2001, JAMA, 286, 688.
139. Froeschle, J.E. 1999, Clin. Infect. Dis., 29, 215.
140. Harrison, L.H., Dwyer, D.M., Maples, C.T., and Billmann, L. 1999, JAMA, 281, 1906.
141. Finn, R., Groves, C., Coe, M., Pass, M., and Harrison, L.H. 2001, South. Med. J., 94, 1192.

142. Bhatti, A.R., DiNinno, V.L., Ashton, F.E., and White, L.A. 1982, J. Infect., 4, 247.
143. Centers for Disease Control and Prevention 2002, Morb. Mortal. Wkly. Rep., 51, 141.
144. Centers for Disease Control and Prevention 1991, Morb. Mortal. Wkly. Rep., 46, 55.
145. Paradis, J.F. and Grimard, D. 1994, Can. Commun. Dis. Rep., 20, 12.
146. Steven N and Wood M 1995, Cartwright K, editor. Meningococcal disease. Chichester, England, Willy and sons Ltd., 177.
147. van Deuren, M., Brandtzaeg, P., and van der Meer, J.W. 2000, Clin. Microbiol. Rev., 13, 144.
148. Nadel, S. and Kroll, J.S. 2007, FEMS Microbiol. Rev., 31, 71.
149. Harrison's Principles of Internal Medicine, S.E.M.-H. 2005, Chapter 360, Online.
150. Cartwright, K. 1995, John Wiley & Sons; New York,
151. Tyler, K. and Martin J B 1993, F. A. Davis Company; Philadelphia.
152. Victor M and Ropper A.H. 2001, Adams & Victor's Principles of Neurology, 7th Edition. McGraw-Hill, Chapter 32, Online.
153. Cartwright, K.A., Stuart, J.M., Jones, D.M., and Noah, N.D. 1987, Epidemiol. Infect., 99, 591.
154. Borrow, R., Claus, H., Chaudhry, U., Guiver, M., Kaczmarski, E.B., Frosch, M., and Fox, A.J. 1998, FEMS Microbiol. Lett., 159, 209.
155. Corless, C.E., Guiver, M., Borrow, R., Edwards-Jones, V., Fox, A.J., and Kaczmarski, E.B. 2001, J. Clin. Microbiol., 39, 1553.
156. Diggle, M.A. and Clarke, S.C. 2003, J. Med. Microbiol., 52, 51.
157. Parent, d.C., I, Traore, Y., Gessner, B.D., Antignac, A., Naccro, B., Njanpop-Lafourcade, B.M., Ouedraogo, M.S., Tiendrebeogo, S.R., Varon, E., and Taha, M.K. 2005, Clin. Infect. Dis., 40, 17.
158. Seward, R.J. and Towner, K.J. 2000, J. Med. Microbiol., 49, 451.
159. Wylie, P.A., Stevens, D., Drake, W., III, Stuart, J., and Cartwright, K. 1997, BMJ, 315, 774.
160. Dunbar, S.A., Eason, R.A., Musher, D.M., and Clarridge, J.E., III 1998, J. Clin. Microbiol., 36, 1617.
161. Zollinger WD and Boslego J 1997, In: Rose NR, Conway de Macario E, Folds JD, Lane HC, Nakamura RM, eds. Manual of clinical laboratory immunology. 5th ed. Washington, D. C.: ASM Press, 473.
162. Peltola, H. 1983, Rev. Infect. Dis., 5, 71.
163. Poolman, J.T. 1995, Infect. Agents Dis., 4, 13.
164. Diaz, R.J. and Outschoorn, I.M. 1994, Clin. Microbiol. Rev., 7, 559.
165. Turner, P.C., Southern, K.W., Spencer, N.J., and Pullen, H. 1990, Lancet, 335, 732.
166. Casado-Flores, J., Osona, B., Domingo, P., and Barquet, N. 1997, Clin. Infect. Dis., 25, 1479.
167. Perez-Trallero, E., Aldamiz-Echeverria, L., and Perez-Yarza, E.G. 1990, Lancet, 335, 1096.
168. Lehmann, A.K., Halstensen, A., Aaberge, I.S., Holst, J., Michaelsen, T.E., Sornes, S., Wetzler, L.M., and Guttormsen, H. 1999, Infect. Immun., 67, 2552.
169. Pollard, A.J., Galassini, R., Rouppe van der Voort EM, Hibberd, M., Booy, R., Langford, P., Nadel, S., Ison, C., Kroll, J.S., Poolman, J., and Levin, M. 1999, Infect. Immun., 67, 2452.
170. Heyderman, R.S., Davenport, V., and Williams, N.A. 2006, Trends Microbiol., 14, 120.
171. Kasper, D.e.al. and Harrison's Principles of Internal Medicine The McGraw Hill Companies, I.U.S.o.A. 2005, Sixteenth Edition,
172. Lee, H.Y., Andalibi, A., Webster, P., Moon, S.K., Teufert, K., Kang, S.H., Li, J.D., Nagura, M., Ganz, T., and Lim, D.J. 2004, BMC. Infect. Dis., 4, 12.
173. Meli, D.N., Christen, S., Leib, S.L., and Tauber, M.G. 2002, Curr. Opin. Infect. Dis., 15, 253.
174. Immunology, F.E.W.H.F.a.C.N.Y. 2003.
175. http: and www.emedicine.com/ped/topic910.html 2007.
176. Poolman, J.T., Hopman, C.T., and Zanen, H.C. 1983, Infect. Immun., 40, 398.
177. Hoff, G.E. and Hoiby, N. 1978, Acta Pathol. Microbiol. Scand. [C.], 86, 1.
178. Hoff, G.E. and Frasch, C.E. 1979, Infect. Immun., 25, 849.
179. Hoff, G.E. and Hoiby, N. 1978, Acta Pathol. Microbiol. Scand. [B], 86, 87.

180. Guttormsen, H.K., Bjerknes, R., Naess, A., Lehmann, V., Halstensen, A., Sornes, S., and Solberg, C.O. 1992, Infect. Immun., 60, 2777.
181. Grados, O. and Ewing, W.H. 1970, J. Infect. Dis., 122, 100.
182. Kasper, D.L., Winkelhake, J.L., Zollinger, W.D., Brandt, B.L., and Artenstein, M.S. 1973, J. Immunol., 110, 262.
183. Glode, M.P., Robbins, J.B., Liu, T.Y., Gotschlich, E.C., Orskov, I., and Orskov, F. 1977, J. Infect. Dis., 135, 94.
184. Pollard, A.J. and Frasch, C. 2001, Vaccine, 19, 1327.
185. Coen, P.G., Cartwright, K., and Stuart, J. 2000, Int. J. Epidemiol., 29, 180.
186. Bennett, J.S., Griffiths, D.T., McCarthy, N.D., Sleeman, K.L., Jolley, K.A., Crook, D.W., and Maiden, M.C. 2005, Infect. Immun., 73, 2424.
187. Yazdankhah, S.P. and Caugant, D.A. 2004, J. Med. Microbiol., 53, 821.
188. Robinson, K., Neal, K.R., Howard, C., Stockton, J., Atkinson, K., Scarth, E., Moran, J., Robins, A., Todd, I., Kaczmarski, E., Gray, S., Muscat, I., Slack, R., and Ala'Aldeen, D.A. 2002, Infect. Immun., 70, 1301.
189. Wedege, E., Kuipers, B., Bolstad, K., van Dijken, H., Froholm, L.O., Vermont, C., Caugant, D.A., and van den, D.G. 2003, Infect. Immun., 71, 3775.
190. Jones, G.R., Christodoulides, M., Brooks, J.L., Miller, A.R., Cartwright, K.A., and Heckels, J.E. 1998, J. Infect. Dis., 178, 451.
191. Davenport, V., Guthrie, T., Findlow, J., Borrow, R., Williams, N.A., and Heyderman, R.S. 2003, J. Immunol., 171, 4263.
192. Wiertz, E.J., Delvig, A., Donders, E.M., Brugghe, H.F., van Unen, L.M., Timmermans, H.A., Achtman, M., Hoogerhout, P., and Poolman, J.T. 1996, Infect. Immun., 64, 298.
193. Brown, T.A. and Mestecky, J. 1985, Infect. Immun., 49, 459.
194. Brown, T.A., Murphy, B.R., Radl, J., Haaijman, J.J., and Mestecky, J. 1985, J. Clin. Microbiol., 22, 259.
195. Ladjeva, I., Peterman, J.H., and Mestecky, J. 1989, Clin. Exp. Immunol., 78, 85.
196. Kauppi, M., Eskola, J., and Kayhty, H. 1995, Pediatr. Infect. Dis. J., 14, 286.
197. Zhang, Q., Choo, S., Everard, J., Jennings, R., and Finn, A. 2000, Infect. Immun., 68, 2692.
198. Mestecky, J. and McGhee, J.R. 1987, Adv. Immunol., 40, 153.
199. Kilian, M., Mestecky, J., and Russell, M.W. 1988, Microbiol. Rev., 52, 296.
200. Plaut, A.G. 1983, Annu. Rev. Microbiol., 37, 603.
201. Reller, L.B., MacGregor, R.R., and Beaty, H.N. 1973, J. Infect. Dis., 127, 56.
202. Williams, J.N., Jones, G.R., Christodoulides, M., and Heckels, J.E. 2003, J. Infect. Dis., 187, 1433.
203. Welsch, J.A., Rossi, R., Comanducci, M., and Granoff, D.M. 2004, J. Immunol., 172, 5606.
204. Kolmer J.A., Toyama I., and Matsunami T. 1918, J. Immunol, III, 157.
205. Kawasaki, T., Etoh, R., and Yamashina, I. 1978, Biochem. Biophys. Res. Commun., 81, 1018.
206. Ikeda, K., Sannoh, T., Kawasaki, N., Kawasaki, T., and Yamashina, I. 1987, J. Biol. Chem., 262, 7451.
207. Turner, M.W. 1996, Immunol. Today, 17, 532.
208. Thiel, S., Vorup-Jensen, T., Stover, C.M., Schwaeble, W., Laursen, S.B., Poulsen, K., Willis, A.C., Eggleton, P., Hansen, S., Holmskov, U., Reid, K.B., and Jensenius, J.C. 1997, Nature, 386, 506.
209. Matsushita, M. and Fujita, T. 1992, J. Exp. Med., 176, 1497.
210. Kawasaki, N., Kawasaki, T., and Yamashina, I. 1989, J. Biochem. (Tokyo), 106, 483.
211. Kuhlman, M., Joiner, K., and Ezekowitz, R.A. 1989, J. Exp. Med., 169, 1733.
212. Drogari-Apiranthitou, M., Fijen, C.A., Thiel, S., Platonov, A., Jensen, L., Dankert, J., and Kuijper, E.J. 1997, Immunopharmacology, 38, 93.
213. Dixon, G.L., Newton, P.J., Chain, B.M., Katz, D., Andersen, S.R., Wong, S., van der, L.P., Klein, N., and Callard, R.E. 2001, Infect. Immun., 69, 4351.
214. Al Bader, T., Christodoulides, M., Heckels, J.E., Holloway, J., Semper, A.E., and Friedmann, P.S. 2003, Infect. Immun., 71, 5590.

215. Singleton, T.E., Massari, P., and Wetzler, L.M. 2005, J. Immunol., 174, 3545.
216. Taams, L.S., Smith, J., Rustin, M.H., Salmon, M., Poulter, L.W., and Akbar, A.N. 2001, Eur. J. Immunol., 31, 1122.
217. Roncarolo, M.G., Levings, M.K., and Traversari, C. 2001, J. Exp. Med., 193, F5.
218. Cong, Y., Weaver, C.T., Lazenby, A., and Elson, C.O. 2002, J. Immunol., 169, 6112.
219. Guermonprez, P., Valladeau, J., Zitvogel, L., Thery, C., and Amigorena, S. 2002, Annu. Rev. Immunol., 20, 621.
220. Kelsall, B.L., Biron, C.A., Sharma, O., and Kaye, P.M. 2002, Nat. Immunol., 3, 699.
221. Davenport, V., Groves, E., Hobbs, C.G., Williams, N.A., and Heyderman, R.S. 2007, Cell Microbiol., 9, 1050.
222. Vermont, C. and van den, D.G. 2002, FEMS Immunol. Med. Microbiol., 34, 89.
223. Pollard, A.J., Galassini, R., van der Voort, E.M., Booy, R., Langford, P., Nadel, S., Ison, C., Kroll, J.S., Poolman, J., and Levin, M. 1999, Infect. Immun., 67, 2441.
224. Hill, H.R. 1987, Pediatr. Res., 22, 375.
225. Estabrook, M.M., Zhou, D., and Apicella, M.A. 1998, Infect. Immun., 66, 1028.
226. McNeil, G. and Virji, M. 1997, Microb. Pathog., 22, 295.
227. Virji, M., Watt, S.M., Barker, S., Makepeace, K., and Doyonnas, R. 1996, Mol. Microbiol., 22, 929.
228. Heyderman, R.S., Ison, C.A., Peakman, M., Levin, M., and Klein, N.J. 1999, J. Infect. Dis., 179, 1288.
229. Kimmel, S.R. 2005, Am. Fam. Physician, 72, 2049.
230. Bilukha, O.O. and Rosenstein, N. 2005, MMWR Recomm. Rep., 54, 1.
231. Jackson, L.A., Alexander, E.R., DeBolt, C.A., Swenson, P.D., Boase, J., McDowell, M.G., Reeves, M.W., and Wenger, J.D. 1996, Pediatr. Infect. Dis. J., 15, 992.
232. Zangwill, K.M., Schuchat, A., Riedo, F.X., Pinner, R.W., Koo, D.T., Reeves, M.W., and Wenger, J.D. 1997, JAMA, 277, 389.
233. Trotter, C.L., Andrews, N.J., Kaczmarski, E.B., Miller, E., and Ramsay, M.E. 2004, Lancet, 364, 365.
234. Ramsay, M.E., Andrews, N.J., Trotter, C.L., Kaczmarski, E.B., and Miller, E. 2003, BMJ, 326, 365.
235. Larrauri, A., Cano, R., Garcia, M., and Mateo, S. 2005, Vaccine, 23, 4097.
236. Trotter, C.L., Edmunds, W.J., Ramsay, M.E., and Miller E 2006, Hum Vaccin, 2, 68.
237. Snape, M.D. and Pollard, A.J. 2005, Lancet Infect. Dis., 5, 21.
238. Price, A.A. 2007, Curr. Pharm. Des, 13, 2009.
239. Danzig, L. 2004, Pediatr. Infect. Dis. J., 23, S285.
240. Girard, M.P., Preziosi, M.P., Aguado, M.T., and Kieny, M.P. 2006, Vaccine, 24, 4692.
241. Zimmer, S.M. and Stephens, D.S. 2006, Curr. Opin. Investig. Drugs, 7, 733.
242. Bovier, P.A., Wyss, K., and Au, H.J. 1999, Soc. Sci. Med., 48, 1205.
243. Wilder-Smith, A. and Memish, Z. 2003, Int. J. Antimicrob. Agents, 21, 102.
244. Prasad, K. and Karlupia, N. 2007, Respir. Med., 101, 2037.
245. http: and www.who.int/mediacentre/factsheets/2003/fs141/en/ 2007.
246. Maslanka, S.E., Gheesling, L.L., Libutti, D.E., Donaldson, K.B., Harakeh, H.S., Dykes, J.K., Arhin, F.F., Devi, S.J., Frasch, C.E., Huang, J.C., Kriz-Kuzemenska, P., Lemmon, R.D., Lorange, M., Peeters, C.C., Quataert, S., Tai, J.Y., and Carlone, G.M. 1997, Clin. Diagn. Lab Immunol., 4, 156.
247. Borrow, R., Andrews, N., Goldblatt, D., and Miller, E. 2001, Infect. Immun., 69, 1568.
248. Artenstein, M.S., Gold, R., Zimmerly, J.G., Wyle, F.A., Schneider, H., and Harkins, C. 1970, N. Engl. J. Med., 282, 417.
249. Committee on Infectious Diseases, M.d.p.a.c.s.f.f.p.A.r.f.c.s.P.1. 2000, 1500.
250. Gotschlich, E.C., Liu, T.Y., and Artenstein, M.S. 1969, J. Exp. Med., 129, 1349.
251. Kabat, E.A. and BEZER, A.E. 1958, Arch. Biochem. Biophys., 78, 306.
252. Rosenstein, N.E., Fischer, M., and Tappero, J.W. 2001, Infect. Dis. Clin. North Am., 15, 155.
253. Jodar, L., Feavers, I.M., Salisbury, D., and Granoff, D.M. 2002, Lancet, 359, 1499.

254. Lepow, M.L., Beeler, J., Randolph, M., Samuelson, J.S., and Hankins, W.A. 1986, J. Infect. Dis., 154, 1033.
255. Harrison, L.H. 2006, Clin. Microbiol. Rev., 19, 142.
256. Zangwill, K.M., Stout, R.W., Carlone, G.M., Pais, L., Harekeh, H., Mitchell, S., Wolfe, W.H., Blackwood, V., Plikaytis, B.D., and Wenger, J.D. 1994, J. Infect. Dis., 169, 847.
257. Goldschneider, I., Lepow, M.L., and Gotschlich, E.C. 1972, J. Infect. Dis., 125, 509.
258. Reingold, A.L., Broome, C.V., Hightower, A.W., Ajello, G.W., Bolan, G.A., Adamsbaum, C., Jones, E.E., Phillips, C., Tiendrebeogo, H., and Yada, A. 1985, Lancet, 2, 114.
259. Gold, R., Lepow, M.L., Goldschneider, I., Draper, T.F., and Gotshlich, E.C. 1979, J. Infect. Dis., 140, 690.
260. Lepow, M.L., Goldschneider, I., Gold, R., Randolph, M., and Gotschlich, E.C. 1977, Pediatrics, 60, 673.
261. Kayhty, H., Karanko, V., Peltola, H., Sarna, S., and Makela, P.H. 1980, J. Infect. Dis., 142, 861.
262. Leach, A., Twumasi, P.A., Kumah, S., Banya, W.S., Jaffar, S., Forrest, B.D., Granoff, D.M., Libutti, D.E., Carlone, G.M., Pais, L.B., Broome, C.V., and Greenwood, B.M. 1997, J. Infect. Dis., 175, 200.
263. Maclennan, J., Obaro, S., Deeks, J., Williams, D., Pais, L., Carlone, G., Moxon, R., and Greenwood, B. 1999, Vaccine, 17, 3086.
264. Scholten, R.J., Kuipers, B., Valkenburg, H.A., Dankert, J., Zollinger, W.D., and Poolman, J.T. 1994, J. Med. Microbiol., 41, 236.
265. Smith, J.M., Dowson, C.G., and Spratt, B.G. 1991, Nature, 349, 29.
266. Maiden, M.C.J. and Feavers, I.M. 1995, Baumberg S, Young JPW, Suanders JR, Welington EMH, (eds). Population genetics of bacteria. Cambridge: Cambridge University Press, 269.
267. Caugant, D.A., Mocca, L.F., Frasch, C.E., Froholm, L.O., Zollinger, W.D., and Selander, R.K. 1987, J. Bacteriol., 169, 2781.
268. Feavers, I.M., Heath, A.B., Bygraves, J.A., and Maiden, M.C. 1992, Mol. Microbiol., 6, 489.
269. Peltola, H., Makela, H., Kayhty, H., Jousimies, H., Herva, E., Hallstrom, K., Sivonen, A., Renkonen, O.V., Pettay, O., Karanko, V., Ahvonen, P., and Sarna, S. 1977, N. Engl. J. Med., 297, 686.
270. Gold, R. 1979, J. Rudbach and P. Baker, Editors, Immunology of bacterial polysaccharides, Elsevier, New York, 121.
271. Muller, E. and Apicella, M.A. 1988, Infect. Immun., 56, 259.
272. Taylor, C.E. and Bright, R. 1989, Infect. Immun., 57, 180.
273. Lesinski, G.B. and Westerink, M.A. 2001, J. Microbiol. Methods, 47, 135.
274. Wuorimaa, T., Dagan, R., Vakevainen, M., Bailleux, F., Haikala, R., Yaich, M., Eskola, J., and Kayhty, H. 2001, J. Infect. Dis., 184, 1211.
275. Bernatoniene, J. and Finn, A. 2005, Drugs, 65, 229.
276. Goldsby, R.A. and et al. 2003, 5th Ed. New York: W. H. Freeman & Co.,
277. Jones, C. 2005, An. Acad. Bras. Cienc., 77, 293.
278. Mieszala, M., Kogan, G., and Jennings, H.J. 2003, Carbohydr. Res., 338, 167.
279. MacLennan, J.M., Shackley, F., Heath, P.T., Deeks, J.J., Flamank, C., Herbert, M., Griffiths, H., Hatzmann, E., Goilav, C., and Moxon, E.R. 2000, JAMA, 283, 2795.
280. Choo, S., Zuckerman, J., Goilav, C., Hatzmann, E., Everard, J., and Finn, A. 2000, Vaccine, 18, 2686.
281. Cartwright, K., Noah, N., and Peltola, H. 2001, Vaccine, 19, 4347.
282. Ramsay, M.E., Andrews, N., Kaczmarski, E.B., and Miller, E. 2001, Lancet, 357, 195.
283. Miller, E., Salisbury, D., and Ramsay, M. 2001, Vaccine, 20 Suppl 1, S58.
284. Balmer, P., Borrow, R., and Miller, E. 2002, J. Med. Microbiol., 51, 717.
285. De Wals, P., Nguyen, V.H., Erickson, L.J., Guay, M., Drapeau, J., and St Laurent, J. 2004, Vaccine, 22, 1233.
286. De Wals P, Deceuninck G, Boulianne N, and Serres G. 2004, Int J Infect Dis., 8 (suppl 1), S190.

287. De Wals, P., Trottier, P., and Pepin, J. 2006, Vaccine, 24, 3500.
288. Pollard, A.J. 2004, Pediatr. Infect. Dis. J., 23, S274.
289. Pichichero, M., Casey, J., Blatter, M., Rothstein, E., Ryall, R., Bybel, M., Gilmet, G., and Papa, T. 2005, Pediatr. Infect. Dis. J., 24, 57.
290. Pichichero, M.E. 2005, Expert. Opin. Biol. Ther., 5, 1475.
291. Pichichero, M.E. 2005, Clin. Pediatr. (Phila), 44, 479.
292. LaForce FM, Konde K, Viviani S, and Preziosi MP 2007, Vaccine, 7, [Epub ahead of print].
293. Jodar, L., LaForce, F.M., Ceccarini, C., Aguado, T., and Granoff, D.M. 2003, Lancet, 361, 1902.
294. http: and www.meningvax.org/timeline.htm 2007.
295. Chippaux, J.P., Garba, A., Ethevenaux, C., Campagne, G., de Chabalier, F., Djibo, S., Nicolas, P., Ali, H., Charrondiere, M., Ryall, R., Bybel, M., and Schuchat, A. 2004, Vaccine, 22, 3303.
296. Robbins, J.B., Schneerson, R., Anderson, P., and Smith, D.H. 1996, JAMA, 276, 1181.
297. Pizza, M., Scarlato, V., Masignani, V., Giuliani, M.M., Arico, B., Comanducci, M., Jennings, G.T., Baldi, L., Bartolini, E., Capecchi, B., Galeotti, C.L., Luzzi, E., Manetti, R., Marchetti, E., Mora, M., Nuti, S., Ratti, G., Santini, L., Savino, S., Scarselli, M., Storni, E., Zuo, P., Broeker, M., Hundt, E., Knapp, B., Blair, E., Mason, T., Tettelin, H., Hood, D.W., Jeffries, A.C., Saunders, N.J., Granoff, D.M., Venter, J.C., Moxon, E.R., Grandi, G., and Rappuoli, R. 2000, Science, 287, 1816.
298. Andersen, J., Berthelsen, L., and Lind, I. 1997, Clin. Diagn. Lab Immunol., 4, 345.
299. Leinonen, M. and Frasch, C.E. 1982, Infect. Immun., 38, 1203.
300. Mandrell, R.E. and Zollinger, W.D. 1982, J. Immunol., 129, 2172.
301. Zollinger, W.D., Mandrell, R.E., Griffiss, J.M., Altieri, P., and Berman, S. 1979, J. Clin. Invest, 63, 836.
302. Finne, J., Leinonen, M., and Makela, P.H. 1983, Lancet, 2, 355.
303. Bruge, J., Bouveret-Le Cam, N., Danve, B., Rougon, G., and Schulz, D. 2004, Vaccine, 22, 1087.
304. Tondella, M.L., Popovic, T., Rosenstein, N.E., Lake, D.B., Carlone, G.M., Mayer, L.W., and Perkins, B.A. 2000, J. Clin. Microbiol., 38, 3323.
305. van der Voort, E.R., van Dijken, H., Kuipers, B., van der, B.J., van der, L.P., Meylis, J., Claassen, I., and Poolman, J. 1997, Infect. Immun., 65, 5184.
306. Naess, L.M., Oftung, F., Aase, A., Wetzler, L.M., Sandin, R., and Michaelsen, T.E. 1998, Infect. Immun., 66, 959.
307. McGuiness B, Barlow A.K., Clarcke I.N., Farley J.E., Anilionis A., and Poolman J.T. 1990, J. Exp. Med., 171, 1871.
308. Wiertz, E.J., van Gaans-van den Brink JA, Schreuder, G.M., Termijtelen, A.A., Hoogerhout, P., and Poolman, J.T. 1991, J. Immunol., 147, 2012.
309. Wiertz, E.J., van Gaans-van den Brink JA, Gausepohl, H., Prochnicka-Chalufour, A., Hoogerhout, P., and Poolman, J.T. 1992, J. Exp. Med., 176, 79.
310. Wiertz, E., van Gaans-van den Brink, Hoogerhout, P., and Poolman, J. 1993, Eur. J. Immunol., 23, 232.
311. Lifely, M.R., Rogers, M.V., Esdaile, J., Payne, M., and Tite, J.P. 1992, Vaccine, 10, 159.
312. Uli, L., Castellanos-Serra, L., Betancourt, L., Dominguez, F., Barbera, R., Sotolongo, F., Guillen, G., and Pajon, F.R. 2006, Proteomics., 6, 3389.
313. Azeredo E., do Amaral Cl., and Juarez E. 1994, Informe Epidemiológico do Sus 3, 2, 35.
314. Pirez M.C., Picon T., Galazka J., Rubio I., Montano A., and Ferrari A.M. 2004, Rev Med Uruguay, 20, 92.
315. Rodriguez A.P., Dickinson F., Baly A., and Martinez R. 1999, Mem Inst Oswaldo Cruz, 94, 433.
316. Perez, O., Lastre, M., Lapinet, J., Bracho, G., Diaz, M., Zayas, C., Taboada, C., and Sierra, G. 2001, Infect. Immun., 69, 4502.
317. Boslego, J., Garcia, J., Cruz, C., Zollinger, W., Brandt, B., Ruiz, S., Martinez, M., Arthur, J., Underwood, P., Silva, W., and. 1995, Vaccine, 13, 821.

318. Sierra, G.V., Campa, H.C., Varcacel, N.M., Garcia, I.L., Izquierdo, P.L., Sotolongo, P.F., Casanueva, G.V., Rico, C.O., Rodriguez, C.R., and Terry, M.H. 1991, NIPH Ann., 14, 195.
319. Bjune, G., Hoiby, E.A., Gronnesby, J.K., Arnesen, O., Fredriksen, J.H., Halstensen, A., Holten, E., Lindbak, A.K., Nokleby, H., Rosenqvist, E., and. 1991, Lancet, 338, 1093.
320. Zollinger, W.D., Boslego, J., Moran, E., Garcia, J., Cruz, C., Ruiz, S., Brandt, B., Martinez, M., Arthur, J., Underwood, P., and. 1991, NIPH Ann., 14, 211.
321. de Moraes, J.C., Perkins, B.A., Camargo, M.C., Hidalgo, N.T., Barbosa, H.A., Sacchi, C.T., Landgraf, I.M., Gattas, V.L., Vasconcelos, H.G., and. 1992, Lancet, 340, 1074.
322. Milagres, L.G., Ramos, S.R., Sacchi, C.T., Melles, C.E., Vieira, V.S., Sato, H., Brito, G.S., Moraes, J.C., and Frasch, C.E. 1994, Infect. Immun., 62, 4419.
323. Holst, J., Feiring, B., Naess, L.M., Norheim, G., Kristiansen, P., Hoiby, E.A., Bryn, K., Oster, P., Costantino, P., Taha, M.K., Alonso, J.M., Caugant, D.A., Wedege, E., Aaberge, I.S., Rappuoli, R., and Rosenqvist, E. 2005, Vaccine, 23, 2202.
324. Tappero, J.W., Lagos, R., Ballesteros, A.M., Plikaytis, B., Williams, D., Dykes, J., Gheesling, L.L., Carlone, G.M., Hoiby, E.A., Holst, J., Nokleby, H., Rosenqvist, E., Sierra, G., Campa, C., Sotolongo, F., Vega, J., Garcia, J., Herrera, P., Poolman, J.T., and Perkins, B.A. 1999, JAMA, 281, 1520.
325. Aaberge, I.S., Oster, P., Helland, O.S., Kristoffersen, A.C., Ypma, E., Hoiby, E.A., Feiring, B., and Nokleby, H. 2005, Clin. Diagn. Lab Immunol., 12, 599.
326. Oster, P., O'hallahan, J., Aaberge, I., Tilman, S., Ypma, E., and Martin, D. 2007, Vaccine, 25, 3075.
327. Nokleby, H., Aavitsland, P., O'hallahan, J., Feiring, B., Tilman, S., and Oster, P. 2007, Vaccine, 25, 3080.
328. Thornton, V., Lennon, D., Rasanathan, K., O'hallahan, J., Oster, P., Stewart, J., Tilman, S., Aaberge, I., Feiring, B., Nokleby, H., Rosenqvist, E., White, K., Reid, S., Mulholland, K., Wakefield, M.J., and Martin, D. 2006, Vaccine, 24, 1395.
329. Wedege, E., Bolstad, K., Aase, A., Herstad, T.K., McCallum, L., Rosenqvist, E., Oster, P., and Martin, D. 2007, Clin. Vaccine Immunol., 14, 830.
330. McNicholas, A., Galloway, Y., Stehr-Green, P., Reid, S., Radke, S., Sexton, K., Kieft, C., Macdonald, C., Neutze, J., Drake, R., Isaac, D., O'donnell, M., Tatley, M., Oster, P., and O'hallahan, J. 2007, Hum. Vaccin., 3,
331. Sandbu, S., Feiring, B., Oster, P., Helland, O.S., Bakke, H.S., Naess, L.M., Aase, A., Aaberge, I.S., Kristoffersen, A.C., Rydland, K.M., Tilman, S., Nokleby, H., and Rosenqvist, E. 2007, Clin. Vaccine Immunol., 14, 1062.
332. Peeters, C.C., Rumke, H.C., Sundermann, L.C., Rouppe van der Voort EM, Meulenbelt, J., Schuller, M., Kuipers, A.J., van der, L.P., and Poolman, J.T. 1996, Vaccine, 14, 1009.
333. Connolly, M. and Noah, N. 1999, Epidemiol. Infect., 122, 41.
334. Cartwright, K., Morris, R., Rumke, H., Fox, A., Borrow, R., Begg, N., Richmond, P., and Poolman, J. 1999, Vaccine, 17, 2612.
335. de Kleijn, E.D., de Groot, R., Labadie, J., Lafeber, A.B., van den, D.G., van Alphen, L., van Dijken, H., Kuipers, B., van Omme, G.W., Wala, M., Juttmann, R., and Rumke, H.C. 2000, Vaccine, 18, 1456.
336. Braun, J.M., Beuth, J., Blackwell, C.C., Giersen, S., Higgins, P.G., Tzanakaki, G., Unverhau, H., and Weir, D.M. 2004, Vaccine, 22, 898.
337. Litt, D.J., Savino, S., Beddek, A., Comanducci, M., Sandiford, C., Stevens, J., Levin, M., Ison, C., Pizza, M., Rappuoli, R., and Kroll, J.S. 2004, J. Infect. Dis., 190, 1488.
338. Mukhopadhyay, T.K., Halliwell, D., O'Dwyer, C., Shamlou, P.A., Levy, M.S., Allison, N., Gorringe, A., and Reddin, K.M. 2005, Biotechnol. Appl. Biochem., 41, 175.
339. Oliver, K.J., Reddin, K.M., Bracegirdle, P., Hudson, M.J., Borrow, R., Feavers, I.M., Robinson, A., Cartwright, K., and Gorringe, A.R. 2002, Infect. Immun., 70, 3621.
340. Rappuoli, R. 2001, Vaccine, 19, 2688.
341. Welsch, J.A., Moe, G.R., Rossi, R., Adu-Bobie, J., Rappuoli, R., and Granoff, D.M. 2003, J. Infect. Dis., 188, 1730.

342. Comanducci, M., Bambini, S., Caugant, D.A., Mora, M., Brunelli, B., Capecchi, B., Ciucchi, L., Rappuoli, R., and Pizza, M. 2004, Infect. Immun., 72, 4217.
343. Masignani, V., Comanducci, M., Giuliani, M.M., Bambini, S., Adu-Bobie, J., Arico, B., Brunelli, B., Pieri, A., Santini, L., Savino, S., Serruto, D., Litt, D., Kroll, S., Welsch, J.A., Granoff, D.M., Rappuoli, R., and Pizza, M. 2003, J. Exp. Med., 197, 789.
344. Madico, G., Welsch, J.A., Lewis, L.A., McNaughton, A., Perlman, D.H., Costello, C.E., Ngampasutadol, J., Vogel, U., Granoff, D.M., and Ram, S. 2006, J. Immunol., 177, 501.
345. Giuliani, M.M., Adu-Bobie, J., Comanducci, M., Arico, B., Savino, S., Santini, L., Brunelli, B., Bambini, S., Biolchi, A., Capecchi, B., Cartocci, E., Ciucchi, L., Di Marcello, F., Ferlicca, F., Galli, B., Luzzi, E., Masignani, V., Serruto, D., Veggi, D., Contorni, M., Morandi, M., Bartalesi, A., Cinotti, V., Mannucci, D., Titta, F., Ovidi, E., Welsch, J.A., Granoff, D., Rappuoli, R., and Pizza, M. 2006, Proc. Natl. Acad. Sci. U. S. A, 103, 10834.
346. Rappuoli R 2006, 15th International Pathogenic Neisseria conference. Cairns, Australia.
347. Moe, G.R., Dave, A., and Granoff, D.M. 2005, Infect. Immun., 73, 2123.
348. Plested, J.S., Harris, S.L., Wright, J.C., Coull, P.A., Makepeace, K., Gidney, M.A., Brisson, J.R., Richards, J.C., Granoff, D.M., and Moxon, E.R. 2003, J. Infect. Dis., 187, 1223.
349. Plested, J.S., Makepeace, K., Jennings, M.P., Gidney, M.A., Lacelle, S., Brisson, J., Cox, A.D., Martin, A., Bird, A.G., Tang, C.M., Mackinnon, F.M., Richards, J.C., and Moxon, E.R. 1999, Infect. Immun., 67, 5417.
350. Verheul, A.F., Snippe, H., and Poolman, J.T. 1993, Microbiol. Rev., 57, 34.
351. Koeberling, O., Welsch, J.A., and Granoff, D.M. 2007, Vaccine, 25, 1912.
352. Hou, V.C., Koeberling, O., Welsch, J.A., and Granoff, D.M. 2005, J. Infect. Dis., 192, 580.
353. Bhattacharjee, A.K., Moran, E.E., Ray, J.S., and Zollinger, W.D. 1988, Infect. Immun., 56, 773.
354. Bhattacharjee, A.K., Moran, E.E., and Zollinger, W.D. 1990, Can. J. Microbiol., 36, 117.
355. Cadieux, N., Plante, M., Rioux, C.R., Hamel, J., Brodeur, B.R., and Martin, D. 1999, Infect. Immun., 67, 4955.
356. Danve, B., Lissolo, L., Mignon, M., Dumas, P., Colombani, S., Schryvers, A.B., and Quentin-Millet, M.J. 1993, Vaccine, 11, 1214.
357. Martin, D., Cadieux, N., Hamel, J., and Brodeur, B.R. 1997, J. Exp. Med., 185, 1173.
358. Moe, G.R., Tan, S., and Granoff, D.M. 1999, Infect. Immun., 67, 5664.
359. West, D., Reddin, K., Matheson, M., Heath, R., Funnell, S., Hudson, M., Robinson, A., and Gorringe, A. 2001, Infect. Immun., 69, 1561.
360. Borrow, R., Carlone, G.M., Rosenstein, N., Blake, M., Feavers, I., Martin, D., Zollinger, W., Robbins, J., Aaberge, I., Granoff, D.M., Miller, E., Plikaytis, B., van Alphen, L., Poolman, J., Rappuoli, R., Danzig, L., Hackell, J., Danve, B., Caulfield, M., Lambert, S., and Stephens, D. 2006, Vaccine, 24, 5093.
361. De Wals, P., De Serres, G., and Niyonsenga, T. 2001, JAMA, 285, 177.
362. Stefanelli, P., Fazio, C., Neri, A., Sofia, T., and Mastrantonio, P. 2003, J. Clin. Microbiol., 41, 5783.
363. Alonso, J.M., Gilmet, G., Rouzic, E.M., Nassif, X., Plotkin, S.A., Ramsay, M., Siegrist, C.A., Stephens, D.S., Teyssou, R., and Vogel, U. 2007, Vaccine, 25, 4125.
364. Caugant, D.A., Tzanakaki, G., and Kriz, P. 2007, FEMS Microbiol. Rev., 31, 52.
365. Haneberg, B., Dalseg, R., Wedege, E., Hoiby, E.A., Haugen, I.L., Oftung, F., Andersen, S.R., Naess, L.M., Aase, A., Michaelsen, T.E., and Holst, J. 1998, Infect. Immun., 66, 1334.
366. Drabick, J.J., Brandt, B.L., Moran, E.E., Saunders, N.B., Shoemaker, D.R., and Zollinger, W.D. 1999, Vaccine, 18, 160.
367. Aase, A., Naess, L.M., Sandin, R.H., Herstad, T.K., Oftung, F., Holst, J., Haugen, I.L., Hoiby, E.A., and Michaelsen, T.E. 2003, Vaccine, 21, 2042.
368. Huo, Z., Sinha, R., McNeela, E.A., Borrow, R., Giemza, R., Cosgrove, C., Heath, P.T., Mills, K.H., Rappuoli, R., Griffin, G.E., and Lewis, D.J. 2005, Infect. Immun., 73, 8256.

Chapter 13
Mucosal Immunity Against Anthrax

Prosper N. Boyaka, Alexandra Duverger, Estelle Cormet-Boyaka and Jean-Nicolas Tournier

Abstract Anthrax has recently gained much attention due to the potential use of *Bacillus anthracis* spores as an agent of biowarfare. In the event of intentional dissemination of *Bacillus anthracis* spores, the most likely routes of infection are through the nasopharyngeal and gastrointestinal tracts. Since inhalational and gastrointestinal anthrax are the most severe and potentially lethal forms of the disease, appropriate vaccine strategies are needed to provide optimal protection against these mucosally initiated infections. This chapter will review the role of anthrax toxins in the pathogenesis of anthrax, with special emphasis on how *Bacillus anthracis* lethal toxin and edema toxin affect immune cells and alter innate and adaptive immunity. We also will summarize traditional strategies for vaccination against anthrax and discuss new approaches to inducing both mucosal and systemic immunity against anthrax toxins for enhanced protection against inhalational and gastrointestinal anthrax.

13.1 Introduction to *Bacillus Anthracis* and the Disease Anthrax

The Gram-positive, spore forming, rod shaped *Bacillus anthracis* produces a poly-γ-D-glutamic acid capsule attached to the peptidoglycan of the cell wall and a tripartite exotoxin for initiation of the clinical manifestations of anthrax [1, 2, 3]. The capsule enables the recently germinated anthrax *Bacillus* to resist phagocytic destruction [2]. It also provides a poorly immunogenic surface vital for full virulence. The three-components exotoxin produced is thought to contribute directly to the lethality of anthrax [1, 2, 3]. *Bacillus anthracis* infection leads to three forms of diseases depending on the cutaneous, gastrointestinal or inhalational route of exposure to *Bacillus anthracis* spores. Cutaneous anthrax traditionally has been reported in agricultural settings and this form of infection rarely evolves into a

P.N. Boyaka
Departments of Veterinary Biosciences, Internal Medicine, and Center for Microbial Interface Biology, The Ohio State University, Columbus, Ohio, USA
e-mail: boyaka.1@osu.edu

M. Vajdy (ed.), *Immunity Against Mucosal Pathogens*,

lethal disease. The gastrointestinal and inhalational forms of diseases traditionally have occurred among individuals in close contact with animal or animal products. These two mucosal routes of infection generally are associated with more severe symptoms than cutaneous infectious and often lead to death. Thus, biowarfare and bioterrorist development of anthrax focuses on inhalation and ingestion of anthrax spores.

13.2 Gastrointestinal Anthrax

Cases of gastrointestinal anthrax occur in areas endemic for anthrax, which exist in tropical and subtemperate regions. Contamination often results from consumption of undercooked meat or meat product. It also is believed that cases of gastrointestinal anthrax may have resulted from the consumption of anthrax-containing chocolate developed as biological weapons during World War II [4]. Despite the wide geographic distribution of anthrax endemicity, detailed epidemiologic data on gastrointestinal anthrax are rare. This mostly is due to lack of microbiologic, pathologic, or serologic testing in developing countries. Based on limited reports of gastrointestinal anthrax, it appears that the disease develops a few days after ingesting spore-contaminated meat. The spectrum of the resulting disease ranges from asymptomatic to nausea, vomiting, abdominal pain, bloody diarrhea and even death by shock or sepsis. Ingested anthrax spores can cause lesions in all segments of the gastrointestinal tract from the oral cavity to the coecum. Ulcerative lesions, usually multiple and superficial, may occur in the stomach, and possibly, in association with similar lesions of the esophagus and jejunum [5, 6]. These ulcerative lesions may bleed, resulting in massive and fatal hemorrhage in severe cases [5, 6]. Lesions also were reported farther down the gastrointestinal tract, in the mid-jejunum, terminal ileum, or cecum. These lesions of the lower gastrointestinal tract tend to develop around a single site or a few sites of ulceration and edema. In addition, these lesions may lead to hemorrhage, obstruction, perforation, or any combination of these symptoms [7]. Pathologic examination of gastrointestinal anthrax showed lesions on the mucosa, as well as in mesenteric lymph nodes which are enlarged and become hemorrhagic [7]. A bioluminescent *Bacillus anthracis* was recently constructed to examine the site of bacteria multiplication after ingestion of spores. These bacteria provide a unique experimental system where strongly luminescent bacteria can easily be discriminated from nonluminescent dormant spores. This system has helped demonstrate the presence of bacterial rods, numerous polymorphonuclear cells, necrosis, and hemorrhage in the Peyer's patches of mice orally administered *Bacillus anthracis* spores [8]. Thus, following ingestion of *Bacillus anthracis* spores, bacterial growth occurs primarily in Peyer's patches [8]. This study showed that spores are continuously shed for up to 48 h post-inoculation after a majority of the initial inoculum has been excreted. It also demonstrated that anthrax spores germinate in the gastrointestinal tract and that vegetative *Bacillus anthracis* transiently colonize the intestinal lumen [8].

13.3 Inhalational Anthrax

Inhalational anthrax was known in the 1800s and early 1900s as a disease related to the textile (woolsorters) and tanning industries. The largest outbreak of inhalational anthrax documented in the 20th century occurred in Sverdlovsk in the former Soviet Union, as a result of aerosol release of *Bacillus anthracis* spores from a military facility [9, 10]. Less than two dozen cases of inhalational anthrax were reported in the US before the intentional release of *Bacillus anthracis* spores in 2001 [11]. In contrast to the gastrointestinal form of disease, more epidemiological studies are available on inhalational anthrax. It is well established that inhalation anthrax is a biphasic illness characterized by a silent initial phase of 1 to 4 days followed by a fulminant phase. The initial silent phase is associated with flu-like symptoms including malaise, fatigue, fever and non-productive cough. Symptoms of the fulminant phase, which can be followed by death, include respiratory distress, bluish discoloration of the skin or mucous membranes caused by lack of oxygen in the blood, and excessive sweating. It is important to note that nausea and vomiting also were observed in the initial phase of disease among the recent case of inhalational anthrax in the US [11]. This finding could indicate an early involvement of the gastrointestinal tract after aerosol infection.

The current model of inhalational anthrax considers that anthrax spores are take-up by alveolar macrophages and transported to the draining mediastinal lymph nodes, where they germinate and establish infection within the lymphatics before the vegetative forms ultimately disseminate via the general circulation. Recent studies have confirmed the role of alveolar macrophages in the uptake of *Bacillus anthracis* spores. However, it now appears that dendritic cells also can endocytose *Bacillus anthracis* spores [12], and interstitial lung dendritic cells were identified as the cell population that transports spores into the thoracic lymph nodes [13]. Since spores were seen in lung dendritic cells up to 72 h after intranasal infection [13], the current view that inhaled spores of *Bacillus anthracis* do not germinate before they reach the draining lymph nodes is in question. In this regard, studies with recombinant bioluminescent *Bacillus anthracis* have shown that inhaled spores establish initial infection in nasal-associated lymphoid tissues [8]. Thus far, inhalational anthrax has been well-accepted as leading to an infection of the general bloodstream. It now appears that this form of infection includes a significant mucosal component, against which mucosal immunity likely is required for optimal protection of the host.

13.4 *Bacillus Anthracis* Exotoxins

Bacillus anthracis produces a tripartite exotoxin, which consists of the protective antigen (PA), the edema factor (EF) and the lethal factor (LF) [14]. The PA subunit targets cells via the anthrax toxin receptor 1 (ATR1), which resembles the tumor endothelial marker 8 (TEM8) [15], and the related ATR2, which is similar to the capillary morphogenesis gene 2 (CMG2) [16]. After the 83 kDa PA (PA83) binds to the ATR1/TEM8 or ATR2/CMG2, a protease furin cleaves a 20 kDa peptide (PA20) [17] which allows the formation of a PA63 heptamer in the host cell membrane [18]. This PA heptamer ring serves to anchor EF, LF or the two in combination [19]. The three exotoxins are not toxic separately, but only in combination. Thus, the combination of PA and EF is usually termed edema toxin (EdTx), while the combination of PA and LF is termed lethal toxin (LeTx). These names originate from the early observations that intravenous injection of PA plus LF is lethal [20] and that intradermal injection of PA plus EF induces edema in the skin [21]. It has been suggested that the low-density lipoprotein receptor-related protein 6 (LRP6) is a co-receptor with either ATR1/TEM8 or ATR2/CMG2, which promotes endocytosis of anthrax toxin into cells [22]. However, this role of LRP6 has been challenged recently [23]. The LF has been crystallized [24] and was shown to be a protease, which cleaves mitogen-activated protein kinase kinase (MAPKK) and thus inhibits intracellular signaling [25]. On the other hand, EF is an adenylate cyclase [26] whose structure has been solved [27]. After binding to calmodulin, EF undergoes a molecular rearrangement to acquire its adenylate cyclase activity and elevate intracellular cyclic adenosine monophosphate (cAMP) levels [27].

13.4.1 Anthrax Lethal Toxin Alters Innate and Adaptive Immune Functions

A number of studies have established that LeTx contributes to the pathogenesis of anthrax by altering both the innate and adaptive branches of host immunity. In fact, LeTx is known to inhibit the secretion of proinflammatory cytokines by macrophages [28, 29, 30]. In more recent studies, LeTx was shown also to inhibit the secretion of all cytokines by dendritic cells [31]. However, others observed more selective effects of LeTx on cytokine secretion by dendritic cells, which correlated with a selective inhibition of p38 and ERK signaling [32]. More specifically, this study has shown that LeTx inhibits dendritic cell production of IL-10 and TNF-α, but not IL-12p70 [32]. Phospholipase A2 type IIA expression was demonstrated to be crucial for killing of anthrax spores by pulmonary phagocytes [33]. LeTx also alters innate immunity by inhibiting the production of phospholipase A2 type IIA thus, inhibiting the bactericidal activity of phagocytes [33]. A recent study has shown that LeTx could inhibit the bactericidal activity of alveolar macrophages in non-human primate [30]. In addition, LeTx has been shown to prevent chemotaxis of polymorphonuclear cells. One potential mechanism for alteration of chemotaxis includes MKK-1-independent signaling and the inhibition of actin polymerization [34]. The two other mechanisms involve reduction or switch-off of chemokine production by dendritic cells [12], and inhibition of chemokine receptor expression on responding cells with subsequent blocking of chemotaxis [35].

In addition to affecting innate immune responses, LeTx can alter the development of adaptive immune responses. For example, LeTx was reported to impair the ability of antigen presentation by LPS-stimulated dendritic cells through inhibition of MHC and co-stimulatory molecule expression [31, 36]. Furthermore, LeTx has been shown to inhibit antigen-specific antibody responses [31]. It has become clear that the inhibitory effects of LeTx on adaptive immune responses occur via direct effects on both T- and B-lymphocytes. LeTx inhibits mouse lymphocyte activation in vivo [37] and blocks IL-2 production by $CD4^{+}$T cells [38]. LeTx also has been reported to block MKK-dependent B-cell proliferation and IgM production [39].

The best described role of LeTx is certainly its ability to induce the death of macrophages [40, 41, 42, 43, 44, 45] and dendritic cells [46]. Interestingly, several studies have shown that MKK cleavage occurs in both resistant and susceptible mouse cells [42, 46], suggesting that LeTx-induced cell death is independent of MKK cleavage. In this regard, one study suggested that LeTx induces apoptosis of activated macrophages via a p38a-dependent pathway [42, 47]. Another apoptosis pathway, involving the activation of PKR downstream from TLR4 by *Bacillus anthracis* motifs recently has been described [48]. Of interest, the susceptibility of mouse macrophages to LeTx recently has been linked to a polymorphism in the locus *Nalp1b*, which encodes NALP1 [49]. In fact, NALP1 activation only occurs in macrophages susceptible to LeTx [49]. The NALP1 is a key component of the inflammasome or the multiprotein complex responsible for the activation of caspases 1 and 5, leading to the processing and secretion of the pro-inflammatory cytokines IL-1β and/or IL-18 leading to a potent inflammatory response. Therefore, LeTx-induced death could involve the activation of caspases, alone or together with secretion of pro-inflammatory cytokines [50].

13.4.2 Effect of Anthrax Edema Toxin on Innate and Adaptive Immune Functions

Until recently, little was known about the role played by EdTx in the alteration of host immune functions during anthrax infection. EF, the enzymatic moiety of EdTx, binds/utilizes ATP and alters the functions of macrophages. In fact, it now is evident that EdTx alone or in combination with LeTx affects innate immunity. Thus, EdTx inhibits LPS-induced oxidative burst [51] and other functions of neutrophils [34, 52]. EdTx activates the expression of activator protein-1 (AP-1) and CAAAT/enhancer-binding protein-beta (C/EBP-beta) by macrophages, but also inhibits their phagocytic ability and production of TNF-α [53]. It has been shown that EdTx selectively inhibits the production of IL-12p70 by murine dendritic cells in vitro, but also, secretion of IL-12p70 and IFN-γ secretion induced by LPS, in vivo [32]. Both LeTx and EdTx inhibit TNF-α production by dendritic cells, but in contrast with LeTx, EdTx does not inhibit production of the anti-inflammatory cytokine IL-10 [32]. In summary, while anthrax spores stimulate proinflammatory cytokine responses by macrophages and dendritic cells [54, 55], each of the anthrax toxin targets either pro-inflammatory cytokines or anti-inflammatory cytokine responses to better counter host innate defenses.

The brisk increase of cAMP induced by EdTx can alter PKA-dependent intracellular signaling. Interestingly, PKA inhibits Mkk4/7, via its effects on Rho and Mek1/2 through alteration of Ras and Raf signaling. In this regard, the mammalian Rho family of GTPases including RhoA, Rac1, and Cdc42 play pivotal roles in controlling many cellular functions including cell polarity, motility, proliferation, apoptosis, and cytokinesis. MEK1 and MEK2, also called MAP kinase kinase and ERK activator kinase, are dual-specificity protein kinases that function in a mitogen activated protein kinase cascade controlling both cell growth and differentiation. Therefore, Mkk4/7 and MAPK/ERK kinase (Mek) are potential common targets of both LeTx and EdTx and it is not surprising that EdTx also alters the functions of cells that mediate adaptive immune responses. For example, sublethal doses of EdTx can suppress T cell activation through the inhibition of NFAT (nuclear factor of activated T cells) and AP-1, two transcription factors essential for cytokine secretion [56]. In addition, a recent study has shown that both LeTx and EdTx inhibit chemotaxis of T cells and macrophages by compromising signaling through CXC (by macrophage inflammatory protein 1 alpha or MIP1α) and CC (by Stromal Cell Derived 1 alpha SDF1α) chemokine receptors [35]. It is important to indicate that despite these analogies, EdTx differs from LeTx with regard to inhibition of T cell responses. Thus, T cells from mice intraperitoneally injected with LeTx exhibited lower proliferative and cytokine responses to in vitro stimulation via the TCR. Similar treatment with EdTx only transiently affected T cell responsiveness to TCR stimulation [37]. However, T cells that recovered responsiveness to TCR four days after EdTx produced lower levels of TNF-α, IFN-γ and IL-17 [37], which is consistent with other studies that showed EdTx preferentially inhibits production of inflammatory cytokines [32].

Mice injected with high doses of EdTx exhibited intestinal fluid accumulation, suggesting that EdTx also could affect immune homeostasis in the mucosal tissues of the gut [57]. Interestingly, while expression of ATR1 on the apical surface of intestinal epithelial cells has been reported, no fluid accumulation was seen after intragastric administration of EdTx (Cormet-Boyaka E, unpublished observation). This observation is consistent with the concept that symptoms of gastrointestinal anthrax are not due to toxins released in the intestinal lumen. It is more likely that toxins reach GALT tissues from the basolateral side after being released in the Peyer's patches by bacteria that germinate in this site after uptake of spores. Alternatively, toxins affect GALT tissues after being released into the general circulation.

While most studies have emphasized the inhibitory effect of EdTx on immune cells, evidence existsthat EdTx upregulates the expression of anthrax toxin receptor by macrophages [58] This mechanism could contribute to disease pathogenesis by increasing in the rate of toxin internalization and indicates that EdTx may provide a positive signal for enhanced rather than impaired functions by target cells. In this regard, the transcription factor cAMP response-element binding protein (CREB) induced by EdTx was found unexpectedly to protect macrophages from LeTx-induced death [42]. The EdTx also stimulates the expression of MHC and costimulatory molecules by antigen presenting cells [32, 59]. In contrast with LeTx which inhibits antibody responses, EdTx was shown to act as an adjuvant following administration through the subcutaneous [60], the nasal [59], or the transcutaneous routes (Duverger A, unpublished observation) and augment antibody responses.

13.5 New Approaches for Inducing Immunity to Anthrax

Live spores of avirulent strains of *Bacillus anthracis* are being used for vaccination of domestic animals worldwide [61, 62]. Until recently, live spore vaccines were used for dermal vaccination of humans in a number of countries [61, 62, 63]. The injection of anthrax toxin into primates reproduces the pathogenesis of infection with *Bacillus anthracis* and as a result, anthrax is considered an exotoxin-mediated disease [64]. Cell-free PA-based vaccines for humans are used both in the United States [65] and the United Kingdom. These vaccines are aluminum hydroxide anthrax vaccine adsorbed (AVA) supernatants from cultures of *B. anthracis* strain V770-NPI-R [61] or culture filtrates from the Sterne strain [65, 66]. The latter AVA currently is used for military personnel in the USA. These vaccines require a series of six injections over an 18-month period with annual boosters.

Several studies have unveiled a major role for anti-PA antibodies in the protection against anthrax [67, 68, 69]. Thus, passive transfer of polyclonal or monoclonal PA-specific IgG antibodies to mice, rabbits, guinea pigs and monkeys conferred protection against both cutaneous and inhalational forms of anthrax [64, 67]. In vitro, anti-PA antibodies can protect mouse macrophages from toxicity induced by the LeTx and thereby providing the basis for the in vitro macrophage toxicity assay used as a correlate of protection [70, 71]. In addition, other studies suggest that protection against anthrax involves immune responses to other anthrax toxin components. Thus, both anti-PA and anti-LF antibodies were detected in the sera of individuals with histories of clinical anthrax [72]. As we will be discussed below, effort is being dedicated to the development of new anthrax vaccines to be administered by non-invasive routes, including nasal and oral vaccines.

13.5.1 Approaches for the Development of New PA-Based Vaccines

Most experimental anthrax vaccines that either are or have been tested are PA-based and consist of recombinant PA given with various adjuvants [68, 73]. The first evidence that non-invasive nasal immunization may be useful for improving vaccination strategies against anthrax came from the observation that both intranasal and subcutaneous immunizations with PA induced high levels of anti-PA serum IgG antibodies, which could protect J774A.1 cells against LeTx, in vitro [74]. However, only nasally immunized mice contained high titers of serum IgA antibodies [74]. Another study showed that neutralizing anti-PA antibodies could be induced in mucosal secretions [73]. Thus, mice nasally immunized with recombinant PA and cholera toxin or CpG oligodeoxynucleotides, as mucosal adjuvants, developed high plasma PA-specific IgG responses and neutralizing anti-PA IgA antibodies in mucosal secretions [73]. The current BioThrax anthrax vaccine (i.e., AVA) also was tested as a nasal vaccine in mice. Nasal administration of BioThrax elicited robust anti-PA serum and mucosal antibody responses against PA [75]. More interestingly, two nasal doses of BioThrax were sufficient to completely protect the susceptible A/J mice against challenge with *Bacillus anthracis* Sterne spores [75].

In addition to adjuvants, new formulations and delivery systems are being tested for inducing both mucosal and systemic immunity by PA-based anthrax vaccines. Thus, a powder formulation administered intranasally to rabbits provided complete protection, whereas a liquid formulation provided only partial protection [76]. These authors also showed that intradermal administration of PA achieved higher seroconversion rates than intramusculular injection, and more efficiently protected rabbits against aerosol challenge with anthrax spores [76]. Water-in-oil nanoemulsions also were tested for nasal immunization of mice and guinea pigs. Intranasal immunization with recombinant PA mixed in a water-in-oil nanoemulsion as an adjuvant, was effective in inducing both serum anti-PA IgG and bronchial anti-PA IgA and IgG antibodies after one or two mucosal administrations [77]. During the last decade, the transcutaneous route has emerged as an alternative means to induce both mucosal and systemic immunity [78]. Transcutaneous immunization of mice with recombinant PA was shown to induce long-term neutralizing antibody titers, which were higher than those obtained with aluminum-adsorbed PA [79]. A powder formulation of PA given by the transcutaneous route also protected rabbits against aerosol challenge with spores, suggesting that protection against inhalational anthrax could be achieved by transcutaneous application of PA-based vaccines [76].

Besides the protein-based PA vaccines discussed above, plasmid DNA and several live recombinant vectors expressing PA were developed for immunization against anthrax. Almost a decade ago, a plasmid DNA encoding PA was injected intramuscularly into mice and was found to induce serum antibodies with protective activity [80]. One approach to improve the immunogenicity of antigens expressed by plasmid DNA vaccines consists in fusion with the VP22 protein of Herpes Simplex Virus type I. Despite the success of this approach with other plasmid DNA, the immune enhancing mechanisms of VP22 remain poorly understood. Therefore, it is rather difficult to explain the fact that immunization with a plasmid DNA expressing PA63 fused to VP22 failed to enhance anti-PA immunity and did not improve the protection of A/J mice against live spore challenge [81]. Recombinant *Salmonella* vectors are well-recognized as an effective system for delivery of heterologous transgenes by multiple routes, including the oral route. In an attempt to develop live recombinant anthrax vaccines, whole PA or portions of this molecule were expressed in *Salmonella* vectors [82, 83, 84]. Intravenous injection of mice with an auxotrophic mutant of *Salmonella typhimurium* expressing the PA gene as a fusion with the signal sequence of the hemolysin (Hly) export system, led to high levels of anti-PA antibodies, which could protect mice against intraperitoneal challenge with lethal doses of *Bacillus anthracis* [83]. Another export system for the expression of heterologous antigens as secreted molecules, the cytolysin A hemolysin of *Salmonella enterica serovar typhi* (ClyA), was used to enhance the immunogenicity of the PA domain 4 expressed in the *Salmonella typhi* vaccine strain CVD 908-htrA [85]. More than 70% of mice immunized intranasally with the *Salmonella* vaccine secreting PA domain 4 developed high serum anti-PA antibody responses, while no significant responses were seen in mice that received the live vector expressing PA domain 4 in the cytoplasm [85]. However, the PA antigen expressed by *the Salmonella* vector seems be crucial for the efficacy of these recombinant vectors as oral vaccines. In this regard, *Salmonella typhimurium* vectors, which expressed full-length PA, PA domains 1 and 4, or PA domain 4 fused to the Salmonella typhi ClyA, were produced. Oral immunization of mice with these constructs showed that only the *Salmonella* expressing full-length PA protected mice against aerosol challenge with *Bacillus anthracis* STI spores [84]. The recombinant *Salmonella* expressing PA domains 1 and 4 protected 25% of mice, while the vector expressing PA domain 4 alone afforded no protection [84].

Replication-deficient adenoviruses are proven vectors for nasal and oral delivery of foreign DNA. One such vector, the human adenovirus serotype 5 (Ade5), was engineered to express PA. A single intramuscular injection of mice with the Ade5-PA evoked higher levels of anti-PA antibodies and enhanced protection against *Bacillus anthracis* lethal toxin challenge than the AVA [86]. However, mucosal application of this Ade5-PA has not been reported. An attenuated nontoxinogenic nonencapsulated *Bacillus anthracis* spore vaccine expressing high levels of PA also has been developed and tested in oral immunization studies in animal models. This vaccine induced anti-PA secretory IgA antibodies into the intestinal lumen of orally immunized guinea pigs, indicating that spores of attenuated *Bacillus* strains could be a viable oral vaccine delivery system for induction of mucosal immunity [87].

The use of attenuated bacteria or viruses as vaccine vectors has raised safety concerns. *Lactobacilli* are commensals of the gut and genitourinary tract and thus, generally are regarded as safer than attenuated bacteria for use as delivery vectors for vaccine antigens and other regulatory molecules. Attempts were made to express PA as an intracellular, surface-anchored or extracellular antigen in *Lactobacilli casei* and these recombinant vectors were tested for oral immunization [88]. Plants also could represent a "safer" vaccine vector alternative to attenuated bacteria and viruses. In this regard, PA has been successfully expressed in transgenic tobacco chloroplasts [89, 90]. Subcutaneous immunization of mice with partially purified chloroplast-derived PA yielded anti-PA antibody levels, which protected mice against intraperitoneal challenge with lethal toxin [90]. After the development of a rice-based oral vaccine expressing cholera toxin B subunit [91], one would expect that same strategy could aid the development of a plant-based cold-chain-free oral anthrax vaccine.

13.5.2 Experimental Anthrax Vaccines for Induction of Immunity Against Anthrax Toxin Components and Spores

Immunization with plasmid DNA encoding the LF antigen alone was reported to provide complete protection in mice [92]. This study also showed that co-immunization with PA and LF DNA promoted four to five times greater titers of anti-PA and anti-LF antibodies than those induced by the individual DNA plasmid given alone. The PA binding sites of anthrax lethal factor and edema factor are highly homologous, and are both located in the N-terminal fragment of these molecules. An attenuated adenovirus encoding the N-terminal region 1–254 amino acids of EF was constructed [93]. After three intramuscular immunizations, this vaccine could protect approximately 50% of mice from subcutaneous challenge with 100 × LD50 of *Bacillus anthracis* spores [93]. The study also demonstrated that anti-EF antibodies cross-react with anthrax lethal factor and can neutralize the activity of both LeTx and EdTx [93].

Other investigators addressed the question of whether or not protection against *Bacillus anthracis* infection can be improved by inducing antibodies that target anthrax toxin components and spores or the capsule. In fact, the addition of formaldehyde-inactivated spores of *Bacillus anthracis* to PA could elicit total protection against challenge with virulent *Bacillus anthracis* strains in mice and guinea pigs [94]. Other studies now have demonstrated that the *Bacillus* collagen-like surface protein of anthracis (BclA), which is located on the surface of exosporium could substitute whole spores in vaccine formulations. For example, mice immunized with the combination of PA- and BclA-encoding plasmid DNA better survived challenge with *Bacillus anthracis* Ames spores than those immunized with PA- or BclA-encoding plasmids alone [95]. In another study of mice primed two weeks earlier with suboptimal doses of PA, injection of a non-glycosylated recombinant BclA PA protected mice against lethal challenge with *Bacillus anthracis* Stern spores [96]. Components of the capsule also were reported to improve anthrax immunity. Thus, nasal immunization of rabbits with a vaccine consisting of PA conjugated to a 10-mer peptide of the poly-D-glutamic acid capsule of *Bacillus anthracis* was found to provide superior protection against *Bacillus anthracis* infection than PA alone [97].

The combination of PA plus the N-terminal 1–254 fragment of LF (LFn^{1-254}) has been used as a "molecular syringe" to introduce foreign antigens coupled to LFn^{1-254} into target cells. This strategy successfully delivered short peptides [98, 99] as well as proteins [100, 101]. Mechanistic studies with PA mutants attenuated in self-assembly or translocation [98, 102] or cells with disrupted furin genes [103] showed that delivery of Ag involves the same toxin self-assembly and translocation steps that occurr during intoxication. Most interestingly, systemic injection of mice with PA in association with EF was found to enhance the levels of serum anti-PA antibody responses above those achieved by injection of PA alone [60]. The adjuvant activity of EdTX for anti-PA immunity was confirmed after nasal immunization with a mutant of EdTx consisting of a mutant EF with reduced adenylyl cyclase activity [59]. It is important to note that EdTx as a nasal adjuvant also induced neutralizing anti-PA antibodies in mucosal secretions [59]. Furthermore, high anti-EF antibodies were induced after nasal immunization with EdTx (Duverger A, unpublished observation). Both native and mutant EdTx were shown to upregulate the expression of costimulatory molecules by antigen presenting cells [32, 59], and this certainly accounts in the adjuvant activity of these molecules.

13.6 Summary and Perspectives

The recent cases of inhalation anthrax in the US, as well as the threat of other deliberate dissemination of anthrax spores as agents of bio-warfare has spurred increased anthrax research since 2001. A major result of this burst is a greater appreciation of mucosal tissues of the gastrointestinal and respiratory tract as both portals of entry of *Bacillus anthracis* spores and initial sites of interaction of host immune cells with *Bacillus anthracis* pathogenesis factors. Consistent with this new understanding, strategies are being developed for potential induction of both mucosal and systemic immunity to *Bacillus anthracis* toxins and other spore-related pathogenesis factors (Table 13.1). The recent identification of EdTx derivatives as adjuvants for induction of high levels of immunity against PA and EF, as well as co-administered unrelated antigens [59], could provide a unique platform for the development of multivalent mucosal anthrax vaccines.

Table 13.1 Summary of new approaches investigated for induction of protective immunity against inhalational and gastrointestinal anthrax[a]

Antigens	Formulations	Delivery Routes
PA (Recombinant protein)	In saline with CpG or cholera toxin derivatives as adjuvants	Nasal
	With alum	Nasal
	Powder	Nasal
	Water-in-oil nanoemulsion	Nasal
	Powder	Transcutaneous
PA (DNA)	Plasmid DNA	Injection sc or im
	Salmonella vector	Nasal
	Salmonella vector	Oral
	Adenoviral vector	Injection im
	Lactobacilli vector	Oral
	Transgenic plants	Injection sc
EF (DNA)	Plasmid DNA	Injection
PA (DNA) and EF (DNA)	Plasmid DNA	Injection
PA (DNA) and BclA (DNA)	Plasmid DNA	Injection
		Injection
Native PA plus native or mutant EF (Recombinant proteins)	In saline	Nasal
		Transcutaneous
		Oral
Recombinant PA conjugated to a 10-mer peptide of the poly-D-glutamic acid capsule	In saline	Nasal

[a]While most current vaccines are given by intramuscular or subcutaneous injection, the new approaches recently investigated for induction of anti-anthrax immunity involved (or could be adapted for) delivery by routes (i.e., oral, nasal or transcutaneous) were more likely to induce both mucosal and systemic immunity. Sc: subcutaneous; im: intra-muscular.

Acknowledgments The authors thank Dr. Kathleen Hayes-Ozello for editorial assistance. This work was supported in part by NIH grant AI 43197.

References

1. Dixon, T.C., M. Meselson, J. Guillemin, and P.C. Hanna, 1999. N Engl J Med. 341, 815.
2. Mock, M. and A. Fouet, 2001. Annu Rev Microbiol. 55, 647.
3. Pezard, C., P. Berche, and M. Mock, 1991. Infect Immun. 59, 3472.
4. Harris, S., 1992. Ann N Y Acad Sci. 666, 21.
5. Kunanusont, C., K. Limpakarnjanarat, and H.M. Foy, 1990. Ann Trop Med Parasitol. 84, 507.
6. Viratchai, C., 1974. J Med Assoc Thai. 57, 147.
7. Sirisanthana, T. and A.E. Brown, 2002. Emerg Infect Dis. 8, 649.
8. Glomski, I.J., A. Piris-Gimenez, M. Huerre, M. Mock, and P.L. Goossens, 2007. PLoS Pathog. 3, e76.
9. Grinberg, L.M., F.A. Abramova, O.V. Yampolskaya, D.H. Walker, and J.H. Smith, 2001. Mod Pathol. 14, 482.
10. Meselson, M., J. Guillemin, M. Hugh-Jones, A. Langmuir, I. Popova, A. Shelokov, and O. Yampolskaya, 1994. Science. 266, 1202.
11. Jernigan, J.A., D.S. Stephens, D.A. Ashford, C. Omenaca, M.S. Topiel, M. Galbraith, M. Tapper, T.L. Fisk, S. Zaki, T. Popovic, R.F. Meyer, C.P. Quinn, S.A. Harper, S.K. Fridkin, J.J. Sejvar, C.W. Shepard, M. McConnell, J. Guarner, W.J. Shieh, J.M. Malecki, J.L. Gerberding, J.M. Hughes, and B.A. Perkins, 2001. Emerg Infect Dis. 7, 933.
12. Brittingham, K.C., G. Ruthel, R.G. Panchal, C.L. Fuller, W.J. Ribot, T.A. Hoover, H.A. Young, A.O. Anderson, and S. Bavari, 2005. J Immunol. 174, 5545.
13. Cleret, A., A. Quesnel-Hellmann, A. Vallon-Eberhard, B. Verrier, S. Jung, D. Vidal, J. Mathieu, and J.N. Tournier, 2007. J Immunol. 178, 7994.
14. Collier, R.J. and J.A. Young, 2003. Annu Rev Cell Dev Biol. 19, 45.
15. Bradley, K.A., J. Mogridge, M. Mourez, R.J. Collier, and J.A. Young, 2001. Nature. 414, 225.
16. Scobie, H.M., G.J. Rainey, K.A. Bradley, and J.A. Young, 2003. Proc Natl Acad Sci USA. 100, 5170.
17. Molloy, S.S., P.A. Bresnahan, S.H. Leppla, K.R. Klimpel, and G. Thomas, 1992. J Biol Chem. 267, 16396.
18. Milne, J.C., D. Furlong, P.C. Hanna, J.S. Wall, and R.J. Collier, 1994. J Biol Chem. 269, 20607.
19. Elliott, J.L., J. Mogridge, and R.J. Collier, 2000. Biochemistry. 39, 6706.
20. Beall, F.A., M.J. Taylor, and C.B. Thorne, 1962. J Bacteriol. 83, 1274.
21. Stanley, J.L. and H. Smith, 1961. J Gen Microbiol. 26, 49.
22. Wei, W., Q. Lu, G.J. Chaudry, S.H. Leppla, and S.N. Cohen, 2006. Cell. 124, 1141.
23. Young, J.J., J.L. Bromberg-White, C. Zylstra, J.T. Church, E. Boguslawski, J.H. Resau, B.O. Williams, and N.S. Duesbery, 2007. PLoS Pathog. 3, e27.
24. Pannifer, A.D., T.Y. Wong, R. Schwarzenbacher, M. Renatus, C. Petosa, J. Bienkowska, D.B. Lacy, R.J. Collier, S. Park, S.H. Leppla, P. Hanna, and R.C. Liddington, 2001. Nature. 414, 229.
25. Duesbery, N.S., C.P. Webb, S.H. Leppla, V.M. Gordon, K.R. Klimpel, T.D. Copeland, N.G. Ahn, M.K. Oskarsson, K. Fukasawa, K.D. Paull, and G.F. Vande Woude, 1998. Science. 280, 734.
26. Leppla, S.H., 1982. Proc Natl Acad Sci USA. 79, 3162.
27. Drum, C.L., S.Z. Yan, J. Bard, Y.Q. Shen, D. Lu, S. Soelaiman, Z. Grabarek, A. Bohm, and W.J. Tang, 2002. Nature. 415, 396.
28. Bergman, N.H., K.D. Passalacqua, R. Gaspard, L.M. Shetron-Rama, J. Quackenbush, and P.C. Hanna, 2005. Infect Immun. 73, 1069.

29. Erwin, J.L., L.M. DaSilva, S. Bavari, S.F. Little, A.M. Friedlander, and T.C. Chanh, 2001. Infect Immun. 69, 1175.
30. Ribot, W.J., R.G. Panchal, K.C. Brittingham, G. Ruthel, T.A. Kenny, D. Lane, B. Curry, T.A. Hoover, A.M. Friedlander, and S. Bavari, 2006. Infect Immun. 74, 5029.
31. Agrawal, A., J. Lingappa, S.H. Leppla, S. Agrawal, A. Jabbar, C. Quinn, and B. Pulendran, 2003. Nature. 424, 329.
32. Tournier, J.N., A. Quesnel-Hellmann, J. Mathieu, C. Montecucco, W.J. Tang, M. Mock, D.R. Vidal, and P.L. Goossens, 2005. J Immunol. 174, 4934.
33. Gimenez, A.P., Y.Z. Wu, M. Paya, C. Delclaux, L. Touqui, and P.L. Goossens, 2004. J Immunol. 173, 521.
34. During, R.L., W. Li, B. Hao, J.M. Koenig, D.S. Stephens, C.P. Quinn, and F.S. Southwick, 2005. J Infect Dis. 192, 837.
35. Paccani, S.R., F. Tonello, L. Patrussi, N. Capitani, M. Simonato, C. Montecucco, and C.T. Baldari, 2007. Cell Microbiol. 9, 924.
36. Cleret, A., A. Quesnel-Hellmann, J. Mathieu, D. Vidal, and J.N. Tournier, 2006. J Infect Dis. 194, 86.
37. Comer, J.E., A.K. Chopra, J.W. Peterson, and R. Konig, 2005. Infect Immun. 73, 8275.
38. Fang, H., R. Cordoba-Rodriguez, C.S. Lankford, and D.M. Frucht, 2005. J Immunol. 174, 4966.
39. Fang, H., L. Xu, T.Y. Chen, J.M. Cyr, and D.M. Frucht, 2006. J Immunol. 176, 6155.
40. Friedlander, A.M., 1986. J Biol Chem. 261, 7123.
41. Lin, C.G., Y.T. Kao, W.T. Liu, H.H. Huang, K.C. Chen, T.M. Wang, and H.C. Lin, 1996. Curr Microbiol. 33, 224.
42. Park, J.M., F.R. Greten, Z.W. Li, and M. Karin, 2002. Science. 297, 2048.
43. Popov, S.G., R. Villasmil, J. Bernardi, E. Grene, J. Cardwell, A. Wu, D. Alibek, C. Bailey, and K. Alibek, 2002. Biochem Biophys Res Commun. 293, 349.
44. Shin, S., Y.B. Kim, and G.H. Hur, 1999. Cell Biol Toxicol. 15, 19.
45. Kassam, A., S.D. Der, and J. Mogridge, 2005. Cell Microbiol. 7, 281.
46. Alileche, A., E.R. Serfass, S.M. Muehlbauer, S.A. Porcelli, and J. Brojatsch, 2005. PLoS Pathog. 1, e19.
47. Park, J.M., F.R. Greten, A. Wong, R.J. Westrick, J.S. Arthur, K. Otsu, A. Hoffmann, M. Montminy, and M. Karin, 2005. Immunity. 23, 319.
48. Hsu, L.C., J.M. Park, K. Zhang, J.L. Luo, S. Maeda, R.J. Kaufman, L. Eckmann, D.G. Guiney, and M. Karin, 2004. Nature. 428, 341.
49. Boyden, E.D. and W.F. Dietrich, 2006. Nat Genet. 38, 240.
50. Tschopp, J., F. Martinon, and K. Burns, 2003. Nat Rev Mol Cell Biol. 4, 95.
51. Wright, G.G. and G.L. Mandell, 1986. J Exp Med. 164, 1700.
52. O'Brien, J., A. Friedlander, T. Dreier, J. Ezzell, and S. Leppla, 1985. Infect Immun. 47, 306.
53. Comer, J.E., C.L. Galindo, F. Zhang, A.M. Wenglikowski, K.L. Bush, H.R. Garner, J.W. Peterson, and A.K. Chopra, 2006. Microb Pathog. 41, 96.
54. Pickering, A.K., M. Osorio, G.M. Lee, V.K. Grippe, M. Bray, and T.J. Merkel, 2004. Infect Immun. 72, 6382.
55. Glomski, I.J., J.H. Fritz, S.J. Keppler, V. Balloy, M. Chignard, M. Mock, and P.L. Goossens, 2007. Cell Microbiol. 9, 502.
56. Paccani, S.R., F. Tonello, R. Ghittoni, M. Natale, L. Muraro, M.M. D'Elios, W.J. Tang, C. Montecucco, and C.T. Baldari, 2005. J Exp Med. 201, 325.
57. Firoved, A.M., G.F. Miller, M. Moayeri, R. Kakkar, Y. Shen, J.F. Wiggins, E.M. McNally, W.J. Tang, and S.H. Leppla, 2005. Am J Pathol. 167, 1309.
58. Maldonado-Arocho, F.J., J.A. Fulcher, B. Lee, and K.A. Bradley, 2006. Mol Microbiol. 61, 324.
59. Duverger, A., R.J. Jackson, F.W. van Ginkel, R. Fischer, A. Tafaro, S.H. Leppla, K. Fujihashi, H. Kiyono, J.R. McGhee, and P.N. Boyaka, 2006. J Immunol. 176, 1776.
60. Quesnel-Hellmann, A., A. Cleret, D.R. Vidal, and J.N. Tournier, 2006. Vaccine. 24, 699.

61. Turnbull, P.C., 1991. Vaccine. 9, 533.
62. Sternbach, G., 2003. J Emerg Med. 24, 463.
63. Shlyakhov, E.N. and E. Rubinstein, 1994. Vaccine. 12, 727.
64. Leppla, S.H., J.B. Robbins, R. Schneerson, and J. Shiloach, 2002. J Clin Invest. 110, 141.
65. Grabenstein, J.D., 2003. Immunol Allergy Clin North Am. 23, 713.
66. Baillie, L., R. Hebdon, H. Flick-Smith, and D. Williamson, 2003. FEMS Immunol Med Microbiol. 36, 83.
67. Little, S.F., B.E. Ivins, P.F. Fellows, and A.M. Friedlander, 1997. Infect Immun. 65, 5171.
68. Ivins, B.E., M.L. Pitt, P.F. Fellows, J.W. Farchaus, G.E. Benner, D.M. Waag, S.F. Little, G.W. Anderson, Jr., P.H. Gibbs, and A.M. Friedlander, 1998. Vaccine. 16, 1141.
69. Singh, Y., B.E. Ivins, and S.H. Leppla, 1998. Infect Immun. 66, 3447.
70. Pitt, M.L., S. Little, B.E. Ivins, P. Fellows, J. Boles, J. Barth, J. Hewetson, and A.M. Friedlander, 1999. J Appl Microbiol. 87, 304.
71. Hering, D., W. Thompson, J. Hewetson, S. Little, S. Norris, and J. Pace-Templeton, 2004. Biologicals. 32, 17.
72. Turnbull, P.C., M.G. Broster, J.A. Carman, R.J. Manchee, and J. Melling, 1986. Infect Immun. 52, 356.
73. Boyaka, P.N., A. Tafaro, R. Fischer, S.H. Leppla, K. Fujihashi, and J.R. McGhee, 2003. J Immunol. 170, 5636.
74. Gaur, R., P.K. Gupta, A.C. Banerjea, and Y. Singh, 2002. Vaccine. 20, 2836.
75. Zeng, M., Q. Xu, and M.E. Pichichero, 2007. Vaccine. 25, 5388.
76. Mikszta, J.A., V.J. Sullivan, C. Dean, A.M. Waterston, J.B. Alarcon, J.P. Dekker, 3rd, J.M. Brittingham, J. Huang, C.R. Hwang, M. Ferriter, G. Jiang, K. Mar, K.U. Saikh, B.G. Stiles, C.J. Roy, R.G. Ulrich, and N.G. Harvey, 2005. J Infect Dis. 191, 278.
77. Bielinska, A.U., K.W. Janczak, J.J. Landers, P. Makidon, L.E. Sower, J.W. Peterson, and J.R. Baker, Jr., 2007. Infect Immun. 75, 4020.
78. Boyaka, P.N., J.R. McGhee, C. Czerkinsky, and J. Mestecky, 2005. in *Mucosal Immunology*, Mestecky, J.E.A., Editor., Elsevier Academic Press: San Diego, CA, 855.
79. Matyas, G.R., A.M. Friedlander, G.M. Glenn, S. Little, J. Yu, and C.R. Alving, 2004. Infect Immun. 72, 1181.
80. Gu, M.L., S.H. Leppla, and D.M. Klinman, 1999. Vaccine. 17, 340.
81. Perkins, S.D., H.C. Flick-Smith, H.S. Garmory, A.E. Essex-Lopresti, F.K. Stevenson, and R.J. Phillpotts, 2005. Genet Vaccines Ther. 3, 3.
82. Coulson, N.M., M. Fulop, and R.W. Titball, 1994. Vaccine. 12, 1395.
83. Garmory, H.S., R.W. Titball, K.F. Griffin, U. Hahn, R. Bohm, and W. Beyer, 2003. Infect Immun. 71, 3831.
84. Stokes, M.G., R.W. Titball, B.N. Neeson, J.E. Galen, N.J. Walker, A.J. Stagg, D.C. Jenner, J.E. Thwaite, J.P. Nataro, L.W. Baillie, and H.S. Atkins, 2007. Infect Immun. 75, 1827.
85. Galen, J.E., L. Zhao, M. Chinchilla, J.Y. Wang, M.F. Pasetti, J. Green, and M.M. Levine, 2004. Infect Immun. 72, 7096.
86. Tan, Y., N.R. Hackett, J.L. Boyer, and R.G. Crystal, 2003. Hum Gene Ther. 14, 1673.
87. Aloni-Grinstein, R., O. Gat, Z. Altboum, B. Velan, S. Cohen, and A. Shafferman, 2005. Infect Immun. 73, 4043.
88. Zegers, N.D., E. Kluter, H. van Der Stap, E. van Dura, P. van Dalen, M. Shaw, and L. Baillie, 1999. J Appl Microbiol. 87, 309.
89. Aziz, M.A., D. Sikriwal, S. Singh, S. Jarugula, P.A. Kumar, and R. Bhatnagar, 2005. Faseb J. 19, 1501.
90. Koya, V., M. Moayeri, S.H. Leppla, and H. Daniell, 2005. Infect Immun. 73, 8266.
91. Nochi, T., H. Takagi, Y. Yuki, L. Yang, T. Masumura, M. Mejima, U. Nakanishi, A. Matsumura, A. Uozumi, T. Hiroi, S. Morita, K. Tanaka, F. Takaiwa, and H. Kiyono, 2007. Proc Natl Acad Sci USA. 104, 10986.
92. Price, B.M., A.L. Liner, S. Park, S.H. Leppla, A. Mateczun, and D.R. Galloway, 2001. Infect Immun. 69, 4509.
93. Zeng, M., Q. Xu, E.D. Hesek, and M.E. Pichichero, 2006. Vaccine. 24, 662.

94. Brossier, F., M. Levy, and M. Mock, 2002. Infect Immun. 70, 661.
95. Hahn, U.K., R. Boehm, and W. Beyer, 2006. Vaccine. 24, 4569.
96. Brahmbhatt, T.N., S.C. Darnell, H.M. Carvalho, P. Sanz, T.J. Kang, R.L. Bull, S.B. Rasmussen, A.S. Cross, and D. O'Brien A, 2007. Infect Immun. 75, 5240.
97. Wimer-Mackin, S., M. Hinchcliffe, C.R. Petrie, S.J. Warwood, W.T. Tino, M.S. Williams, J.P. Stenz, A. Cheff, and C. Richardson, 2006. Vaccine. 24, 3953.
98. Ballard, J.D., R.J. Collier, and M.N. Starnbach, 1996. Proc Natl Acad Sci USA. 93, 12531.
99. Doling, A.M., J.D. Ballard, H. Shen, K.M. Krishna, R. Ahmed, R.J. Collier, and M.N. Starnbach, 1999. Infect Immun. 67, 3290.
100. Goletz, T.J., K.R. Klimpel, N. Arora, S.H. Leppla, J.M. Keith, and J.A. Berzofsky, 1997. Proc Natl Acad Sci USA. 94, 12059.
101. Lu, Y., R. Friedman, N. Kushner, A. Doling, L. Thomas, N. Touzjian, M. Starnbach, and J. Lieberman, 2000. Proc Natl Acad Sci USA. 97, 8027.
102. Goletz, T.J., K.R. Klimpel, S.H. Leppla, J.M. Keith, and J.A. Berzofsky, 1997. Hum Immunol. 54, 129.
103. Zhang, Y., Y. Kida, K. Kuwano, Y. Misumi, Y. Ikehara, and S. Arai, 2001. Microbiol Immunol. 45, 119.

Chapter 14
Structure, Immunopathogenesis and Vaccines Against SARS Coronavirus

Indresh K. Srivastava, Elaine Kan, Isha N. Srivastava, Jimna Cisto and Zohar Biron

Abstract A new disease, severe atypical respiratory syndrome (SARS), emerged in China in late 2002 and developed into the first epidemic of the 21st century. The disease was caused by an unknown animal coronavirus (CoV) that had crossed the species barrier through close contact of humans with infected animals, and was identified as the etiological agent for SARS. This new CoV not only became readily transmissible between humans but also was also more pathogenic. The disease spread across the world rapidly due to the air travel, and infected 8096 people and caused 774 deaths in 26 countries on 5 continents. The disease is characterized by flu-like symptoms, including high fever, malaise, cough, diarrhea, and infiltrates visible on chest radiography. The overall mortality was about 10%, but varied profoundly with age; the course of disease seemed to be milder in the pediatric age group and resulted rarely in a fatal outcome, but the mortality in the elderly was as high as 50%. Aggressive quarantine measures taken by the health authorities have successfully contained and terminated the disease transmission. As a result there are no SARS cases recorded recently. Nevertheless there is a possibility that the disease may emerge in the population with high vigor. Significant progress has been made in understanding the disease biology, pathogenesis, development of animal models, and design and evaluation of different vaccines, and these are the focus of this chapter.

14.1 Introduction

A new infectious disease, known as severe acute respiratory syndrome (SARS), appeared in the Guangdong province of southern China in 2002. It was characterized mainly by flu-like symptoms, including high fevers, dry nonproductive dyspnea, and infiltrates visible on chest radiography. In about a third of all cases, the resulting

I.K. Srivastava
Novartis Vaccines and Diagnostics, Inc., 4560 Horton Street, Emeryville, CA 94608, USA
e-mail: Indresh.Srivastava@novartis.com

M. Vajdy (ed.), *Immunity Against Mucosal Pathogens*,

pneumonia led to acute breathing problems requiring artificial respirators [1]. The overall mortality for SARS was about 10%, but varied greatly with age with a mortality rate in the elderly as high as 50% [2, 3, 4]. A previously unidentified coronavirus was isolated from Vero and FRhK-4 cells that were inoculated with clinical specimens (nasopharyngeal, oropharyngeal and sputum) from SARS patients [5, 6, 7]. The association of the virus with the disease was confirmed when monkeys that were inoculated with the virus developed symptoms similar to those observed in human cases of SARS [8]. Although accurate information about the onset of the SARS epidemic is not available, the Chinese Ministry of Health reported an outbreak of unexplained pneumonia to the World Health Organization (WHO) in February 2003. The SARS associated coronavirus (SARS-CoV) is believed to have jumped from an animal host to humans in rural areas of the Guangdong province, and then spread rapidly throughout the world via air travel. During the period from November 2002 to July 2003, the epidemic of SARS spread to 29 countries, affected approximately 8,000 people, resulted in about 800 deaths and severely crippled the Asian economy. The overall cost of the outbreak was estimated to approach $100 billion, mostly as a result of cancelled travel and decreased investment in the affected region [9]. However, aggressive quarantine measures successfully controlled the emergence of SARS in 2003, yet in January 2004 two new confirmed cases of community acquired SARS has been reported in China. This suggests that this infectious disease has not been completely eliminated, and may dramatically re-emerge in the human population [10].

Based on the phylogenetic analysis of the replicase genes of Coronaviruses (CoVs), they are divided into three main sero groups: ***group I CoVs***, including transmissible gastroenteritis virus and human CoV 229E; ***group II CoVs***, including mouse hepatitis virus and bovine CoV; and ***group III***, including infectious bronchitis virus (Fig. 14.1) [11]. Phylogenetic analyses of the complete genome sequence of the SARS-CoV suggests that it is not closely related to any of the three previously identified coronavirus groups, nor does it seem to be a reassortant of known coronaviruses [12]. Its unique sequence suggests that the virus has evolved independently from the other members of the family for a long period of time. The search for a possible natural reservoir of the SARS-CoV is ongoing, since it could serve as the launch pad for another SARS outbreak. To date, these efforts have limited success because a virus with very close sequence homology was isolated from palm civets, raccoons, dogs, and the Chinese ferret badger, indicating that the virus may have jumped recently from these mammals to humans [13]. Cats may be infected with the virus and can spread it, but do not show clinical signs of infection [14]. Moreover, the virus has been detected on the body-surface and gut contents of cockroaches by PCR, but their organs were negative, so they might act as a mechanical vector of virus transmission [15]. The pandemic potential and pathogenicity of SARS-CoV, as well as the absence of effective licensed drugs, highlights the need for aggressive efforts directed toward the development of a safe and effective vaccine. The availability of a prophylactic vaccine would be a particularly desirable solution, since it would not only prevent disease in vaccinated people, but it would also reduce overall spread of the virus. While the development of coronavirus vaccines generally has

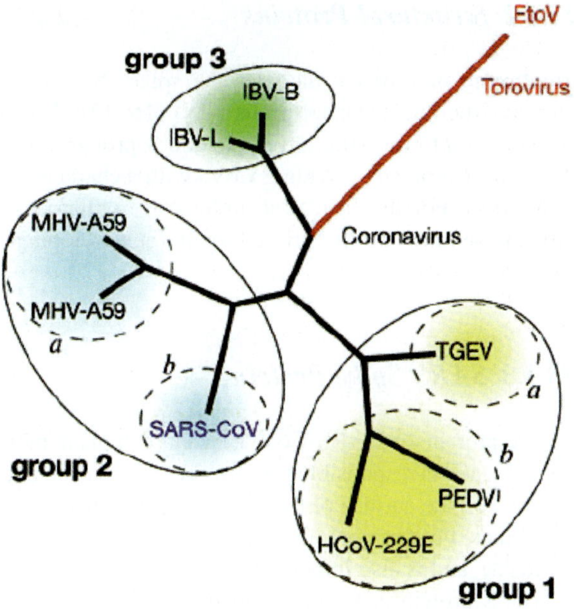

Fig. 14.1 Phylogenetic analysis of coronavirus replicase genes. SARS-CoV linage is derived from group 2 viruses. IBV, infectious bronchitis virus; TGEV, porcine transmissible gastroenteritis virus; EtoV, equine torovirus; MHV-A59, murine hepatitis virus A59; PEDV, porcine epidemic diarrhea virus; HCoV-229E, human coronavirus 229E (Figure is adapted from [11, 148])

been challenging, there are several encouraging factors which point towards the feasibility of developing a SARS-CoV vaccine: (1) the evidence that SARS-CoV is inducing an acute infection and disease (it is generally more complicated to develop a vaccine against a microorganism that induces a chronic infection); (2) the infection mounts a strong humoral response; (3) passive transfer of sera obtained from convalescent patients to SARS patients resulted in the reduction of the viral load that saved the lives of the patients; (4) the relative ease with which the virus can be propagated in vitro; and (5) as a proof of concept there are some effective licensed veterinary coronavirus vaccines based on inactivated or live attenuated virus, including those against a canine coronavirus and avian infectious bronchitis virus.

In this chapter, we will focus on: (a) Genes and proteins of the SARS-CoV; (b) correlates of protection; (c) animal models; (d) application of different technologies for developing SARS vaccine; (e) use of adjuvant and delivery systems for enhancing the potency of the vaccine, and (f) the potential issue of disease enhancement due to vaccines.

14.1.1 Gene Organization of SARS Coronavirus

Similar to other coronaviruses, SARS-CoV is an enveloped positive-strand RNA virus, featuring a large viral genome encoding for the three different types of proteins known as (i) structural, (ii) non-structural and (iii) accessory proteins. We will discuss these different proteins one by one.

14.1.2 Structural Proteins

This group of proteins includes the spike (S), envelope (E), matrix (M) glycoproteins, the nucleocapsid protein (N) [16, 17]. The S, M and E proteins are incorporated into the viral envelope, and S-protein dimers or trimers protrude from the viral membrane, providing CoVs with a characteristic corona-resembling shape (Fig. 14.2). For the structural proteins we will focus on Spike and nucleocapsid proteins, as they are the prime targets for developing an effective and efficacious vaccine.

14.1.3 SARS Spike Protein

The S protein of SARS-CoV is a large transmembrane glycoprotein of coronaviruses, and is responsible for virus binding, fusion and entry. Since the S protein is exposed on the surface of the virion, it is the major target for inducing neutralizing antibodies. Furthermore, the S protein plays critical roles in viral pathogenesis and virulence, and is also important for viral functions and antigenicity [18].

The S protein is a type I transmembrane glycoprotein with 1255 amino acids. All CoV S-proteins contain an N-terminal signal peptide, which facilitates transport into the endoplasmic reticulum, where the proteins are extensively glycosylated. Notably, SARS-CoV S has 23 consensus sites for N-linked glycosylation [16, 17, 19, 20, 21]. Comparison of SARS S protein with the S proteins of other coronaviruses, revealed that the sequence of the SARS-CoV S protein may have the following hypothetical features (i) a 13-amino-acid cleavable secretory signal [16], (ii) a putative S1 globular domain (residues 15–680) with a potential receptor binding

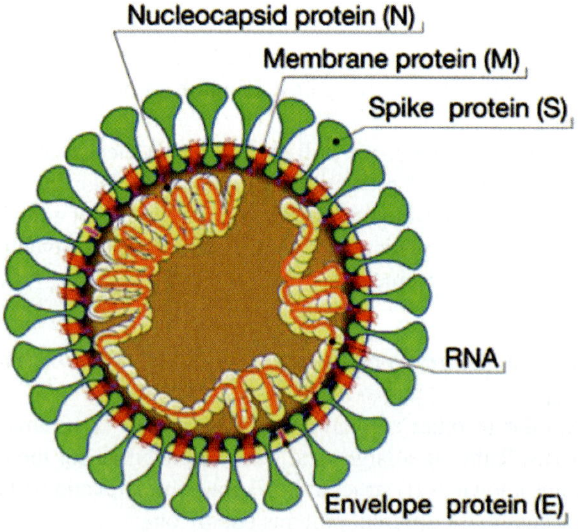

Fig. 14.2 Morphology of the SARS-CoV. A Schematic presentation of the virus. A lipid bilayer comprising the spike protein, the membrane protein and the envelope protein cloaks the helical nucleocapsid, which consist of the nucleocapsid protein that is associated with the viral RNA (adapted from [148])

site [7, 21, 22, 23], (iii) a putative S2 stalk domain (residues 681–1255) with a fusion peptide and heptad repeats, (iv) a hydrophobic transmembrane (TM) domain near the C terminus that could be responsible for anchoring the S protein to the virion lipid envelope [24], and (v) a cysteine-rich (Cy) domain immediately following the membrane anchor region, a feature common to all other coronaviruses which may be involved in stabilizing protein-lipid interactions.

The S1 domain is responsible for virus binding to the receptor on the target cells. It has been demonstrated that angiotensin-converting enzyme 2 (ACE2) is a functional receptor for SARS-CoV [7, 22, 25, 26]. Investigators have mapped an approximately 200 amino acid region of S1 domain (318–510 amino acids) that is responsible for interacting with ACE2, its receptor. This 200 amino acid domain is known as the receptor-binding domain (RBD) [27, 28, 29]. The S2 domain contains a putative fusion peptide and two heptad repeat (HR1 and HR2) regions. Upon binding of RBD on the viral S protein to ACE2 on target cells, S2 changes conformation by interaction between the HR1 and HR2 regions to form fusogenic core and bring viral and target cell membrane into close proximity, resulting in virus fusion and entry [30]. The HR1 and HR2 regions can associate to form a six-helix bundle structure [30, 31], and a peptide derived from the HR2 region of SARS-CoV S protein had inhibitory activity on SARS-CoV infection [30]. This indicates that the fragments containing the functional domains on the S protein may be used as antigens for inducing antibodies to block virus binding or fusion. Notably, glycoproteins (GPs) of highly divergent viruses, including human immunodeficiency virus (HIV) [32], and mouse hepatitis coronavirus (MHV) [24, 33] exhibit a similar architecture. These GPs, referred to as class I fusion proteins, use similar mechanisms to promote membrane fusion, which has important implications for therapeutic intervention. However, despite the similarities in domain organization, the SARS-CoV S-protein does not exhibit significant sequence identity with S-protein of any other CoVs; the highest sequence conservation is found in heptad repeats (HRs) located within the S2 regions underlining their important function.

The S polypeptide is N-glycosylated co-translationally in the endoplasmic reticulum (ER) and further processed in the Golgi apparatus [24]. The S glycoproteins of human coronavirus 229E, transmissible gastroenteritis virus (TGEV), porcine respiratory coronavirus, feline infectious peritonitis virus, and canine coronavirus remain as single glycoproteins, those of mouse hepatitis virus, bovine coronaviruses, and human coronavirus OC43 are proteolytically cleaved into two subunits, S1 and S2, during the cellular transport process. Earlier studies of TGEV indicate that the S protein is an oligomer composed of three copies of the monomeric S glycoprotein [34]. Such a quaternary structure has been reported for other enveloped RNA viruses and has been demonstrated to be important for eliciting neutralizing antibodies against hemagglutinin A (HA) of influenza virus [35], the gp120-gp41 heterodimer of human immunodeficiency virus [36], and the G protein of vesicular stomatitis virus [37].

It is believed that S-protein is present as a trimer on the surface of SARS CoV. Li and colleagues [38] have characterized four sequential states of a purified recombinant S ectodomain (S-e) comprising S1 and the ectodomain of S2. They are S-e monomers, uncleaved S-e trimers, cleaved S-e trimers, and dissociated S1 monomers and S2 trimer rosettes. Lowered pH induces an irreversible transition from flexible, L-shaped S-e monomers to clove-shaped trimers. Protease cleavage of the trimer occurs at the S1-S2 boundary; an ensuing S1 dissociation leads to a major rearrangement of the trimeric S2 and to formation of rosettes likely to represent clusters of elongated, post-fusion trimers of S2 associated through their fusion peptides. The states and transitions of S suggest conformational changes that mediate viral entry into cells. However, S-protein in different conformations is yet to be evaluated for its efficacy in inducing potent neutralizing antibody responses.

14.2 The Interaction of S-protein with Dendritic Cells

Binding of viral glycoproteins to cellular factors other than the receptor(s) does not enable entry but can enhance viral infection. Therefore, various pathogens, including HIV, are thought to interact with factors on dendritic cells (DCs) to promote their spread within infected individuals [39]. Binding of HIV to DCs facilitates infection of nearby susceptible cells through a mechanism that is not completely understood. The lectin DC-SIGN (dendritic cell-specific ICAM-grabbing non-integrin) or related molecules might be instrumental to this process because DC-SIGN expressed on cell lines binds to the GP of HIV and catalyzes infection of adjacent receptor-positive cells. DC-SIGN also interacts with SARS-CoV S-protein and augments infection with retroviral particles bearing the S-protein on their envelope [40, 41]. This observation is reflected by efficient DC-mediated SARS-CoV transmission to target cells [40, 41]. Furthermore, additional factors other than DC-SIGN are clearly involved in viral transfer [40]. Notably, DCs are not permissive to virus infection, indicating that productive infection is not required for transmission [40, 41]. However, the interaction of SARS-CoV with DCs could contribute to SARS pathogenesis. Attachment of SARS-CoV to dermal DCs might facilitate viral spread in the skin, whereas DC-SIGN-positive alveolar macrophages could promote SARS-CoV replication in the lung. Moreover, internalization of SARS-CoV by DCs might provide the virus with means of immune escape.

14.3 Receptor for SARS-CoV

Efforts from several groups were focused on the identification of the cellular receptor for SARS-CoV. Li et al. [7] reported that the metallopeptidase angiotensin-converting enzyme 2 (ACE2) is a receptor for SARS-CoV. They had used a soluble S1-immunoglobulin (Ig) fusion protein for immunoprecipitation experiments with lysates from Vero E6 cells, the cell type used for the isolation of SARS-CoV. Subsequent proteomic analysis revealed ACE2 to be a high-affinity binding partner of S1. Inhibition of SARS-CoV infection of susceptible cells using antibodies against

ACE2, in conjunction with the observation that ACE2 expression in a resistant cell line for SARS CoV makes it highly susceptible for SARS infection, indicated that the interaction of S1 protein to ACE2 facilitated SARS-CoV infection [7]. Wong et al. also identified ACE2 as a SARS-CoV receptor by using a different approach [28]. Binding studies using soluble fragments of SARS-CoV S-protein revealed residues 318 to 510 to be the minimal receptor-binding domain [21, 27, 28]. An initial search for S-protein residues that are important for ACE2 binding pointed to E452 and D454, with the latter being crucial for association with ACE2 [28]. However, the exact regions of ACE2, which are in contact with the S-protein, remain to be identified. Recently, the structure of the ACE2 ectodomain has been resolved, revealing two ridges flanking the catalytic site [42]. Molecular modeling suggests that these ridges might interact with the S-protein [26]. Interestingly, binding of ACE2 to an inhibitor and probably also to substrate induces structural changes within these ridges [42] and might interfere with binding to the S-protein. Detailed analysis will be required to identify residues within the S-protein and ACE2 that are critical for their interaction; and therefore represent attractive targets for developing inhibitors. It is also conceivable that inducing potent antibodies against the receptor-binding domain could be an attractive strategy for developing a vaccine against the SARS.

These studies might also have important implications for the development of small animal models. Thus, ACE2 from African green monkeys enables efficient entry of SARS-CoV [43], and infection of some macaque species reproduces aspects of SARS in humans [8, 44], while viral replication in mice is less robust and does not induce disease [45]. It will, therefore, be important to determine if a potentially reduced interaction of SARS-CoV S-protein with murine ACE2 limits viral spread in these animals. If that were the case, then the generation of transgenic animals would be of greater significance [7]. ACE2, a carboxypeptidase that cleaves polypeptides from the renal-angiotensin system [46], is essential for cardiac function [47] and is expressed in various tissues and organs [48]. Importantly, major target cells of SARS-CoV, such as pneumocytes [1, 49, 50, 51], express ACE2 [48], and expression in cell lines correlates with permissiveness to SARS-CoV S-driven infection [43], indicating that ACE2 plays a central role in SARS-CoV replication.

14.4 Co-receptor or Alternative Receptors for SARS CoV

In contrast to HIV, the evidence for the requirement of a co-receptor or the existence of alternative receptors for SARS-CoV to entry into certain tissues has not yet been demonstrated. However, ACE2-dependent infection of organs other than the lung might contribute to SARS pathogenesis. For example, small intestinal enterocytes express ACE2 [48] and are permissive for SARS-CoV [49, 51, 52]. Moreover, the efficient infection of renal epithelial cells of different species [41] and isolation of the virus from kidney tissue of a SARS patient [6] suggest that SARS-CoV infection of kidney cells might contribute to acute renal failure observed in some SARS patients [41, 53]. Infection of intestinal enterocytes and kidney cells could facilitate viral transmission via the fecal-oral route. Viral RNA has been detected in stool samples from SARS patients [1, 52], however, it is unclear if transmission via feces promoted viral spread during the 2003 outbreak. Hepatocytes from SARS patients were also infected [49], and some ACE2-expressing hepatoma cell lines are highly permissive to replication of SARS-CoV [41, 43, 54]. Infection of hepatocytes might therefore partially account for the altered levels of liver-specific enzymes commonly observed in SARS patients [1, 54]. Because liver tissue was found to be largely negative for ACE2 protein expression [48], it will be interesting to examine whether viral entry is facilitated by low levels of ACE2 expression or other factors.

14.5 SARS-CoV Spike Protein Triggering: Low pH Versus Receptor Engagement

Receptor engagement can activate the fusion machinery of viral glycoproteins in two ways. First, binding to receptor can directly activate the fusion process [25, 55], which is the case for HIV and murine leukemia virus (MLV) glycoproteins. Alternatively, receptor engagement can trigger the internalization of viral particles into endosomes where protonation activates glycoprotein-driven membrane fusion [25, 55]. Influenza hemagglutinin and the vesicular stomatitis virus G-protein (VSV-G) are activated by low pH. In the case of SARS CoV it has been demonstrated that inhibitors of vacuolar acidification also block infection by S-bearing pseudotypes, suggesting potential triggering of the fusion activity of SARS-CoV S-protein by low pH [41, 54, 56]. However, SARS-CoV S-driven cell-to-cell fusion can occur in the absence of low pH [21, 25]. Therefore, the S-protein of SARS-CoV might be able to mediate membrane fusion in a pH-dependent and independent fashion, and several parameters might control which stimulus is required under what conditions. One such parameter could be the association between the S1 and S2 subunits of the SARS-CoV S-protein. Many class I fusion proteins are cleaved into an outer and a transmembrane subunit by cellular proteases. In addition, the cleavage is essential for the functionality of the glycoprotein such as the case for HIV envelope. By contrast, S-proteins of group I CoVs are not cleaved at all, and cleavage of the S-protein of MHV, a group II CoV, appears to be cell-type dependent and not required for its functionality [24, 57]. Based on the limited set of data, it seems there are no obvious consensus sites for cellular proteases present in SARS-CoV S protein [16, 17], and efficient cleavage of the protein has not been reported [21, 24, 30, 58]. Notably, protease treatment of cells expressing S-protein resulted in an increased cell-to-cell fusion activity [56], indicating that cleavage of S protein might enable pH-independent, receptor-dependent triggering of the fusion activity. Further work is needed to clearly demonstrate the role of cellular proteases in triggering the S-protein driven membrane fusion of the SARS CoV.

14.6 Membrane Fusion

Two functional elements located in the transmembrane domain of CoV S-proteins are key to the membrane fusion: a putative fusion peptide and two heptads repeat (HR). The function of these elements has been elucidated in the context of prototype class I fusion proteins, such as HIV gp160 [59]. Cleavage of gp160 produces the fusion active form of the transmembrane subunit gp41, which is oriented perpendicular to the viral membrane and contains a fusion peptide (a stretch of hydrophobic amino acids) at its N-terminus. Two HRs (HR1 and HR2) are located between the fusion peptide and the transmembrane domain. During the fusion process, the fusion peptide inserts into the target cell membrane, HR2 folds back onto HR1, resulting in the formation of a six-helix bundle structure (trimer of dimers). In this conformation, the HRs are oriented in an anti-parallel fashion, thereby bringing the fusion peptide (inserted into the target cell membrane) and the transmembrane domain (inserted into the viral membrane) into close contact, which ultimately facilitates the membrane fusion [59].

Bosch et al. [24] demonstrated that fusion driven by the S-protein of MHV follows similar principles; however, a major difference compared to HIV was observed. The fact that the cleavage of many CoV S-proteins, including that of SARS-CoV, is not needed for exerting their function suggests that these proteins must have an internal fusion peptide similar to the G-protein of VSV. A computer-based analysis has predicted a potential fusion peptide at the N-terminus of HR1 in SARS-CoV S [59]. In light of these observations, the model for membrane fusion illustrated previously must be revised for the SARS-CoV S-protein. Thus, upon exposure to low pH it is possible that an internal fusion peptide, which is covalently associated with both the S1 and S2 subunits, inserts into the target cell membrane, and the membrane fusion is driven by the formation of the six-helix bundle between HR1 and HR2. In case of influenza HA, a low pH environment triggers irreversible conformational changes associated with membrane fusion, whereas exposure of VSV-G to low pH induces a reversible transition into the fusion active state [59]. The nature of pH-induced conformational changes in SARS-CoV S-protein remains to be determined. If the model proposed above accurately describes SARS-CoV S-driven membrane fusion, one would expect that peptides mimicking HR1 or HR2 should assemble into a six-helix bundle and that such peptides would inhibit SARS-CoV S-mediated membrane fusion. The latter speculation is based on evidence obtained with several viral class I fusion proteins, including MHV S-protein [24], for which HR-derived peptides were shown to inhibit fusion by preventing the formation of the six-helix bundle [59]. The peptide T20, which potently inhibits HIV gp160-driven membrane fusion when present in the low-nanomolar range, has been approved for use in patients, representing the first member of entry inhibitors, a new class of therapeutics [60]. Therefore, blocking the six-bundle formation either by chemical means such as T-20 is a proven therapeutic agent or whereas directing antibodies to the critical elements may represent an attractive target for developing a vaccine against SARS-CoV.

14.7 SARS Nucleocapsid Protein

The N protein, which binds to the genomic RNA via a leader sequence, recognizes a stretch of RNA that serves as a packaging signal and leads to the formation of the helical ribonucleoprotein (RNP) complex during assembly. The structure of the RNA-binding domain of the SARS-CoV N protein was determined by NMR spectroscopy in 2004 [61]. It consists of a five-stranded β sheet whose folding is unrelated to that of other RNA-binding proteins. The authors identified a binding site for single-stranded RNA (ssRNA), using NMR to determine the resonance of residues perturbed by the addition of RNA, and revealed a similar mode of interaction to RNA-binding proteins such as U1A RNP. They also identified small molecules from an NMR-based screen that bind to the RNA-binding domain and might impair its function. Antigenic peptides of the coronavirus N protein can be recognized by T cells on the surface of infected cells [62, 63]. The structure of the MHC-I molecule HLAA* 1101 in complex with such a peptide derived from the SARS-CoV N protein, a nonamer with a SARS-specific sequence, has recently been determined to 1.45 A° resolution [61]. It is similar to other MHC-I molecules and shows a similar peptide-binding mode, and thus this structure adds to the growing library of MHC-I structures and could be used as a template for peptide-based vaccine design.

14.8 Non-structural Proteins

The SARS-CoV replicase gene encodes for 16 non-structural proteins (nsp), with multiple enzymatic functions. These are known or predicted to include types of enzymes that are common components of the replication machinery of plus-strand RNA viruses: an RNA-dependent RNA polymerase activity (RdRp, nsp12); a 3C-like serine protease activity (Mpro or 3CLpro, nsp5); a papain-like protease 2 activity (PL2pro, nsp3); and a superfamily-1 helicase activity (HEL1, nsp13) [16, 64, 65]. In addition, the replicase gene encodes proteins that are indicative of 30–50 exoribonuclease activity (ExoN homologue, nsp14), endoribonuclease activity (XendoU homologue, nsp15), adenosine diphosphate-ribose 10-phosphatase activity (ADRP, nsp3) and ribose 20-O-methyltransferase activity (20-O-MT, nsp16). These enzymes are less common in plus-strand RNA viruses, and may therefore be related to the unique properties of coronavirus replication and transcription. Finally, the replicase gene encodes another nine proteins, of which little is known about their structure or function. The nsp 4, 10 and 16 have been implicated by genetic analysis in the assembly of a functional replicase–transcriptase complex.

14.9 Accessory Proteins

The genomic sequences of numerous SARS-CoV isolates have been determined. The 'conserved' open reading frames (ORFs) of the SARS-CoV genome occur in the same order as and are of similar size to those found in other coronaviruses. However, in addition to the conserved genes, the SARS-CoV genome contains eight novel

ORFs at the 3′ end (ORFs 3a, 3b, 6, 7a, 7b, 8a, 8b and 9b) [11]. To date, the functions of these genes remain largely unknown, although their absence from other genomes suggests unique functions that might be advantageous to SARS-CoV replication, assembly or virulence [66]. Only one of these so-called accessory proteins has a known structure and further studies are required to elucidate their precise functions.

14.10 Correlate of Protection for SARS

SARS CoV infection of humans results in the development of acute respiratory syndrome in about 10% of the patients, which usually results in mortality. Therefore, one of the major questions in SARS is how the immune system manages to control the infection in the majority of patients. The S glycoprotein is a major structural protein of SARS-CoV and a potential target for SARS-specific humoral immunity and/or cell-mediated immune responses. Recent studies in animal models have demonstrated that vaccines based on the S protein of SARS-CoV seem to induce a considerable neutralizing antibody response [58, 67, 68, 69] and provide protection via inhibition of viral replication after challenge with live SARS-CoV [68]. In addition, SARS-CoV specific IgG can be detected at week 3 after the onset of syndromes in SARS patients and persist for a long period of time [70, 71]. Polyclonal immune sera from convalescent SARS patients were passively transferred to treat SARS patients during the outbreak in 2003 [72]. These findings revealed that humoral immune responses, therefore, play an important role in controlling and clearing SARS-CoV infection in humans and mice. Moreover, the S protein in its native conformation might be a suitable candidate for vaccine approaches. Indeed, immunization of macaques with adenoviruses coding for S, M, and N triggered the production of neutralizing antibodies, providing at least some indirect proof that S-based vaccines hold promise. To clearly demonstrate that antibodies are really involved in the control of SARS CoV infection, investigators resorted to the use of passive transfer experiments, a practice of administering polyclonal immunoglobulin isolated from hyperimmune sera of animal or human origin, introduced by Von Behring and Kitasato. It has been used extensively in prophylactic as well as in therapeutic settings [73]. However, the risks related to the use of human blood products make them problematic as a standard therapy. Human monoclonal antibodies may be a solution to some of the problems. In addition the use of mAbs has the added advantage of using higher tittered and higher avidity antibody directed against the protective epitope(s). Furthermore, it may also alleviate the concern of disease enhancement, since generally low avidity antibodies are considered to cause disease enhancement.

14.11 Passive Transfer of Monoclonal Antibodies

The first human monoclonal antibody (mAb) resulted from a screen of a human non-immune single-chain variable region fragment (scFv) phage library constructed from B cells against the S1 domain of the SARS-CoV spike protein. The antibody had a high neutralization activity in vitro and blocked syncytia formation of 293T cells expressing the S protein on their surface [74]. When the antibody was given prophylactically to mice at doses therapeutically achievable in humans, viral replication was reduced by more than 4 orders of magnitude and was no longer detectable [74]. Also employing phage display technology to screen a naïve antibody library for antibodies reactive against the S protein, ter Meulen and colleagues isolated a human mAb belonging to the IgG1 subclass [75]. Prophylactic administration of this antibody to ferrets, an animal model of SARS-CoV infection and disease, has reduced replication of SARS-CoV in the lungs of infected ferrets by 3.3 logs ($p < 0.001$), completely prevented the development of SARS-CoV induced lung pathology ($p < 0.013$), and abolished shedding of the virus in pharyngeal secretions. Traggiai and coworkers [76] chose B cells from a convalescent person to identify human mAbs recognizing SARS-CoV. Through a combination of magnetic and fluorescence-activated cell sorting they isolated memory B-lymphocytes, which were subsequently immortalized with Epstein-Barr virus (EBV). B lymphocyte clones were selected according to their ability to recognize the SARS-CoV proteins and/or to neutralize the virus. Surprisingly, only a small fraction of memory B cells specific for SARS-CoV antigens were directed against neutralizing epitopes present in the spike protein. One of the human mAbs was tested in a mouse model for its in vivo neutralizing activity [45]. It protected the lungs completely and the upper respiratory tract partially from virus replication when given two days prior to challenge with SARS-CoV. This approach has the advantage that it is fast, as it can be completed within 3 months, and efficient. It generates large numbers of antibodies that can immediately be screened for the most favorable affinity and epitope specificity. Passive immunization certainly has its merits especially when it comes to the prophylactic protection of high-risk groups like health care workers but might also be of therapeutic benefit when given early after onset of the disease.

14.12 The Role of Cell Mediated Immunity in SARS-CoV Infection

The role of cell-mediated immunity in the resolution of SARS-CoV infection in humans is still not well understood. In an study by Yang and colleagues, the investigators observed that memory T-cell responses against the S protein were persistent for more than 1 year after SARS-CoV infection by detecting the production of IFN-γ using ELISA and ELISpot assays [77]. Flow cytometric analysis demonstrated that both CD4+ and CD8+ T cells were involved in cellular responses against SARS-CoV infection. Interestingly, most of SARS-CoV S-specific memory CD4+ T cells were central memory cells with the CD45RO+ CCR7+ CD62L phenotype. However, the majority of memory CD8+ T cells revealed effector memory phenotype, CD45RO-CCR7-CD62L-. Thus, the study provides evidence that SARS-CoV infection in humans can induce cellular immune responses that are persistent for a long period of time. These data may argue for the design a vaccine that may be effective in inducing not only potent neutralizing antibody responses but also cellular responses and T-helper cell responses as well. Nevertheless, further work would be needed to establish a direct correlation between strong cellular responses and reduced viral load.

14.13 Animal Models

An authentic animal model that represents infections of humans with SARS-CoV is critical for a better understanding of the disease. In addition, animal models are key for pre-clinical evaluations of the most effective vaccines leading to clinical evaluations. Efforts have been made by different groups to develop rhesus, mouse, and ferret challenge models for SARS-CoV infection. SARS-CoV can infect cynomolgus macaques (*Macaca fascicularis*) following intratracheal inoculation [8, 44, 78, 79]. The histopathologic pattern resembled that seen in humans dying of SARS when the animals were analyzed 4 or 6 days after infection with a Hong Kong SARS-CoV isolate [8, 44]. A different study using the Tor2 isolate found only mild, self limited, respiratory symptoms in some animals and low-level virus replication [78]. Experimental infection of the three species of Old World monkeys (African Green, rhesus, and cynomolgus monkeys) with SARS-CoV strain Urbani *via* the respiratory route revealed a quantitative difference in virus replication in the upper and lower respiratory tract of the three species; the virus replicated to higher titers and for a longer time in the respiratory tract of African Green monkeys (AGM) compared to cynomolgus or rhesus monkeys. The titer of serum neutralizing antibodies induced in these animals correlated with the level of viral replication in the respiratory tract. Histopathologic examination of African green monkey lungs were consistent with those reported by Kuiken et al. and Fouchier et al. although they found more evidence for pneumonitis at earlier time-points post infection (2 days vs. 4 or 6 days, respectively) [79]. Ferrets, cats, mice, and Golden Syrian hamsters have also been successfully infected with SARS-CoV. All of these animal models support viral replication in the upper and lower respiratory tract although no clinical signs were seen in SARS-CoV inoculated cats. Infected ferrets on the other hand became lethargic from day 2–4 post infection, and developed a lung pathology similar to but milder than those described for infected macaques. Both infected cats and ferrets were able to efficiently transmit the virus to other animals living in close proximity [14]. Following intranasal administration of SARS-CoV strain Urbani, the virus replicates in the respiratory tract of BALB/c mice [45] and Golden Syrian hamsters [80]. The kinetics of virus replication in both species resembles each other, peaking at day 2, post infection and the virus clears after 5–7 days. In hamsters, however, the virus reaches higher titers especially in the upper respiratory tract, and the animals are shedding the virus for a longer time. In contrast to mice, hamsters showed pathology in the upper and lower respiratory tract, as well as viremia and extrapulmonary spread of the virus to liver and spleen. Neither Golden Syrian hamster nor BALB/c mice showed any clinical signs of disease with the exception of aged mice. Twelve to fourteen months old BALB/c mice infected with SARS-CoV demonstrated signs of clinical illness as characterized by weight loss, hunching, ruffled fur, and slight dehydration, which were resolved by day 7-post infection. Compared to young BALB/c mice, SARS-CoV replication was enhanced and prolonged in the aged mice; virus titers were higher in the lungs and accompanied by alveolar damage and interstitial pneumonitis [80]. All animal models described above differ from the human disease in two important aspects: (a) the period between infection and the peak in viral load is shorter; and (b) pathology is shortened and the disease rarely progresses to a fatal outcome. But despite these differences, these animal models are important and useful for vaccine evaluation.

14.14 Development of SARS Vaccine using Different Platform Technologies

Vaccine efficacy is generally measured by the ability of the antigen to raise a protective immunologic response from B and/or T cells after exposure to the pathogen. Ideally, by creating antigen-specific memory within the immune system, individuals will be protected from infection for decades. Several veterinary coronavirus vaccines are currently available, but their efficacy is variable. The vaccine for prevention of infectious bronchitis virus (IBV), which infects chickens, is effective [81], but the canine and porcine vaccines are only partially effective [82]. The feline infectious peritonitis (FIP) vaccine is actually deleterious to the health of the animal.

There are several independent and parallel vaccine approaches being evaluated against the SARS-CoV: (i) live attenuated or inactivated virus, (ii) DNA Vaccines; (iii) DNA prime and boost approach; iv) use of viral vectors for delivering the vaccine; (v) recombinant subunit vaccine, and (vi) use of virus like particles (VLP). All these approaches have potential advantages and disadvantages, and one needs to weigh the pros and cons for each technology, and select the technology, which is most potent in inducing protective immune responses.

14.14.1 Attenuated and Inactivated Whole Virus Based Vaccines

The choice of an inactivated vaccine is, without question, the most expeditious route that can be pursued to reach the clinical evaluation stage of a potential SARS-CoV vaccine. It has a safety record established by immunizing people with hundreds of million doses and they are generally easy to manufacture. Moreover, they are able to induce a broad immune response against all antigenic determinants of the virus. In fact, many groups from academic institutions and industry are working on the development of an inactivated vaccine against the SARS-CoV and in December 2004 Sinovac Biotech announced that all 36 subjects participating in a phase I human clinical trial testing an inactivated SARS-CoV vaccine had received their second and last vaccination (http://www.news-medical.net/?id=4560). So far, no adverse reactions have been reported and all participants are in good health. A recent study by Zhou and colleagues [83] describes a formaldehyde inactivated whole virus vaccine, which was tested for its immunogenicity, safety, and protective efficacy in rhesus monkeys. The animals were immunized twice intramuscularly (i.m) at one-week interval with 0.5, 5, or 50 μg of non-adjuvanted vaccine and challenged 2 weeks after the second immunization. None of the vaccinated animals developed any clinical symptoms upon virus challenge. However, the control group showed only minor clinical signs. Two weeks after the second immunization SARS-CoV specific IgG and neutralization titers were highest in the 50 μg group and those animals were protected from virus challenge, as no replicating virus could be detected in the lungs 15 days post challenge. However, animals receiving lower dosages were only partially protected. The detection of increased IFN-γ concentrations and almost constant IL-4 concentrations in vaccinated animals indicated a Th-1 driven immune response. He and colleagues (2004) demonstrated that SARS-CoV inactivated by β-propiolactone (BPL) elicited high titers of antibodies in the immunized mice and rabbits that recognize the S protein, especially the receptor-binding domain (RBD) in the S1 region [84]. The antisera from the immunized animals efficiently bound to the RBD and blocked binding of RBD to angiotensin-converting enzyme 2, the functional receptor on the susceptible cells for SARS-CoV. With a sensitive and quantitative single-cycle infection assay, using pseudovirus bearing the SARS-CoV S protein, the investigators demonstrated that mouse and rabbit antisera significantly inhibited S protein-mediated virus entry with mean 50% inhibitory titers of 1:7393 and 1:2060, respectively. These data suggest that the RBD of S protein is a major neutralization determinant in the inactivated SARS vaccine, which can induce potent neutralizing antibodies to block SARS-CoV entry [84]. In another study, a BPL inactivated SARS-CoV vaccine was developed and tested for its immunogenicity and efficacy in BALB/c mice in collaboration with the NIH [85]. Animals were immunized at 0, 2, and 4 weeks with 5 μg of inactivated virus with or without the adjuvant MF59, an oil squalene-in-water emulsion, approved for human use in Europe for an influenza vaccine [86]. After three doses, the MF59 adjuvanted BPL-inactivated SARS virus vaccine induced a ten fold higher neutralizing titer (1:645) than the non-adjuvanted vaccine (1:64). IgG subclass determination indicated predominant Th2-type immune response to the adjuvanted vaccine. Two weeks after the last vaccine dose, mice were challenged intranasally and nasal turbinates and lung tissues were analyzed for infectious virus two days later. Complete protection from virus replication was observed in mice that received the MF59 adjuvanted vaccine; neither the nasal turbinates nor the lungs of these mice contained recoverable virus. Immunization with the non-adjuvanted vaccine resulted in complete protection of the upper respiratory tract and a significant ($p < 0.00001$) reduction of viral titers in the lower respiratory tract of 30,000-fold compared to the control groups. A number of other investigators have also evaluated whole virus vaccines using UV, BPL or formaldehyde to inactive the virus in mice and rabbits and combined them with different types of adjuvant [84, 87, 88, 89]. Regardless of the method or combination used, all vaccines elicited strong immune responses underscoring the potential of this approach. One major drawback inherent to all vaccines using inactivated viruses is the high risk of accidents during their production. High amounts of infectious virus have to be cultivated and purified raising the probability of incidents. Using live attenuated virus strains instead of wild type virus could minimize this risk.

In summary, vaccines can be produced by inactivation of the virus, by using an attenuated or weak form of the virus, or by using recombinant forms of viral components. Inactivated virus vaccines are relatively safe because they cannot revert back to the live form. They are also relatively stable and may not even require refrigeration. This is important in developing countries and for ease in mobilization during outbreak or emergency situations. However, there are limitations to their use. Inactivated vaccines usually require several doses and some are weakly effective at stimulating an immune response. The vaccine to prevent hepatitis A is an example of an inactivated viral vaccine [90]. Furthermore, these vaccines are less characterized, may require special laboratory for the development therefore it may be even harder to get the FDA approval for clinical evaluation. An unfortunate example of this general lack of characterization is the inactivated respiratory syncytial virus, which caused two deaths and many hospitalizations due to disease enhancement in vaccinated infants.

14.14.2 DNA Vaccines

DNA vaccination is quite a powerful strategy, and receiving considerable attention due to its ability to induce both humoral and cellular immune responses. It induces immune responses to, and in some cases even leads to, the protection against various types of infections, such as influenza, malaria, and SARS [91, 92, 93]. A common feature of DNA vaccination is that the synthesis of the antigen occurs in intracellular compartments, allowing the processed antigen to enter the MHC class I pathway, in turn to generate CD8+ cytotoxic T-lymphocyte (CTL) activities. High CTL and antibody responses were observed after mice were injected three times with a recombinant plasmid vector expressing the N protein [94]. Mice immunized with a plasmid containing the S protein produced anti-SARS-CoV IgG and developed neutralizing antibodies and a T-cell mediated response resulting in a six-fold reduction in viral titers in the lungs. Plasmids encoding either the S1 or S2 regions of the spike protein elicited antibody production in mice. Neither the S1 or S2 antibodies alone were capable of neutralizing the virus; however, cooperatively they enabled neutralization of the virus, suggesting that both regions of the spike protein are important for host-cell viral entry.

Jin and colleagues have demonstrated that both humoral- and cellular-mediated specific responses could be induced by DNA vaccines for N, M, and E antigens [95]. All three DNA constructs induced SARS-CoV-specific antibodies in mice, however the highest antibodies were induced against N protein, followed by M and E proteins. In addition, T cell proliferation and DTH responses were also successfully induced in mice after vaccinations with these constructs. These results suggested that the DNA vaccine was effective to prime a specific anti-SARS-CoV response and apparently generate a broad range of both T-helper and B-cell memory responses during the priming. This was also consistent with previous observations of the protective immunity [96, 97]. Furthermore, the DNA administrations generated a lower level of IL-10, suggesting that DNA vaccines may help polarizing Th-1 type of responses.

Recently, Zhu et al. [98] have demonstrated that a DNA construct based on the pcD3d vector could successfully induce SARS-CoV N protein-specific antibody titers and CTL responses. Kim [99] has demonstrated that vaccination with N DNA vaccine could successfully induce a SARS-CoV antigen-specific CD8+ T cell response and distinctly reduce the titer of recombinant vaccinia virus expressing SARS-CoV N protein after the challenge and that the co-expression of calreticulin (CRT), a 46 kDa Ca2+-binding protein, with N gene could enhance its ability to protect against viral challenges. Therefore, it is consistent that the N protein construct could induce the highest SARS-specific IgG, T cell proliferation, and in vivo CTL response were once again induced for N protein, followed by M and E. Furthermore, the highest Th-1 type responses based on IFN-γ and IL-2 production were also induced against the N protein. It is difficult to answer if N is truly more immunogenic compared to M and E, or the increased level of immune responses may be due to more epitopes in the N nucleocapsid protein since it contains 422 amino acid residues, while the M and E have 220 and 76 amino acids respectively. In addition, the differential concentrations of each antigen presented in the killed SARS-CoV preparation could have also contributed to the differences.

It has been demonstrated in several studies that there is an interaction between the N and M proteins. Shi and colleagues [100] have tried to answer two questions: (a) Can N or M membrane proteins be expressed in a DNA vaccine? and (b) Can the expression of a membrane protein (M) affect the immune responses induced by N protein in the context of a DNA vaccine? The animals were injected with 20 μg of the mixture of DNA vaccines encoding for M and N. The ELISA analysis using the N antigen or inactivated SARS-CoV particles as capture antigen showed that co-injection of SARS-M could enhance the antibody responses against N, especially of the IgG2a subclass. After lymphocytes were stimulated with 10 μg/ml purified N antigen, the CD4+ and CD8+ T cells of N and M plus N group were increased compared with those of control groups. Cytokine ELISA analysis revealed that co-injection of M could enhance the levels of IFN-γ, and IL-2 production induced by the N antigen. Virus challenge test was conducted in BSL3 bio safety laboratory with Brandt's vole SARS-CoV model, and the results indicated that co-immunization of M and N antigens could reduce the mortality and pathological changes in the lungs from virus infection in the mixed vaccine immunized group compared to the single vaccine groups.

In another study, Huang et al. [101] have also demonstrated that immunization of mice with SARS-CoV spike DNA vaccine induced antigen-specific cellular and humoral immune responses. The cellular immune responses were mediated by both CD4+ and CD8+ T cells [101].

14.14.3 DNA Prime Vector or Protein Boost Vaccines

Different forms of SARS coronavirus spike protein-based vaccines were evaluated for the generation of neutralizing antibody responses against SARS-CoV in a mouse model [102]. In this study, they compared six combinations: (a) intra-peritoneal (i.p.) immunization with recombinant spike polypeptide produced in *Escherichia coli* (S-peptide), (b) mice primed with tPA-optimized DNA vaccine (tPA-SDNA) and boosted with S-peptide i.p.; (c) mice primed with CTLA4HingeS ARS800 DNA vaccine (CTLA4-S-DNA) and boosted with S-peptide i.p.; (d) mice primed with oral live-attenuated *Salmonella typhimurium* (*Salmonella*-S-DNA-control) and boosted with. S-peptide i.p, (e) mice primed with oral live-attenuated *S. typhimurium* that contained tPA-optimize800 DNA vaccine (*Salmonella*-tPA-S-DNA) and boosted with. S-peptide i.p; and (f) mice primed with oral live-attenuated *S. typhimurium* that contained CTLA4 Hinge SARS800 DNA vaccine (*Salmonella* tPA-S-DNA) and boosted with. S-peptide i.p. There was no statistically significant difference among the Th-1/Th-2 profile among these six groups of mice with had high antigen-specific IgG levels. Sera of all six mice immunized i.p. with S-peptide, i.m. with DNA vaccine control and oral *Salmonella*-SDNA- control showed no neutralizing antibody against SARS-CoV. Sera of the mice immunized with i.m. tPA-S-DNA, i.m. CTLA4-S-DNA, oral *Salmonella*-S-DNA-control boosted with i.p. S-peptide, oral *Salmonella*-tPA-S-DNA, oral *Salmonella*-tPA-S-DNA boosted with i.p S peptide, oral *Salmonella*-CTLA4-S-DNA and oral *Salmonella*-CTLA4-S-DNA boosted with i.p. S-peptide showed neutralizing antibody titers of <1:20–1:160. Sera of all the mice immunized with i.m. tPA-S-DNA boosted with i.p. S-peptide and i.m. CTLA4-S-DNA boosted with i.p. S-peptide showed neutralizing antibody titers of $\geq$ 1:1280. The present observation may have major practical value, such as immunization of civet cats, since production of recombinant proteins from *E. coli* is far less expensive than production of recombinant proteins using eukaryotic systems.

Among all the combinations of vaccines examined in this study, mice primed with SARS-CoV human codon usage optimized spike polypeptide DNA vaccines and boosted with S-peptide produced by *E. coli* generated the highest titers of neutralizing antibody against SARS-CoV. It has been demonstrated that the S-peptide produced by *E. coli* did not induce neutralizing antibody against SARS-CoV infection. On the other hand, recombinant spike polypeptide generated by eukaryotic systems such as transfection of COS7 and BHK21 cells or DNA vaccine was able to elicit high neutralizing antibody titers against SARS-CoV infection [67, 103, 104]. This was probably because the S-peptide produced in *E. coli* did not have the same structure and conformation compared to the S-protein produced in the mammalian expression system. In this study, we documented that although recombinant S-peptide produced by *E. coli* itself was not able to generate neutralizing antibody against SARS-CoV infection, mice primed with spike polypeptide DNA vaccine and boosted with S-peptide from *E. coli* were able to generate high titers of neutralizing antibody against SARS-CoV. This indicates that the type of vaccine used for priming is crucial in determining the type of immune response developed after the boost.

Furthermore, it has also been demonstrated that the humoral immune response developed in mice primed with spike DNA vaccine and boosted with S-peptide from *E. coli* did not develop a Th1 type immune response. However, mice immunized with S-peptide from *E. coli* alone developed a Th1 type response. This indicates that a Th1 type immune response may not be essential for the generation of neutralizing antibodies against SARS-CoV. Although our results suggest that priming with DNA vaccines and boosting with S-peptide produced by *E. coli* was successful in the generation of neutralizing antibody against SARS-CoV, further experiments using infection models to evaluate its protective immunity are warranted, since anti-spike antibodies have been shown to enhance the infectivity of coronaviruses in some cell culture systems, as occurred with SARS-CoV and feline infectious peritonitis virus [105, 106].

Zakhartchouk and colleagues [107] evaluated the efficacy of DNA prime and whole killed SARS-CoV vaccines in combination vs. both the vaccines alone. They have clearly demonstrated that a combination of the vaccines is more immunogenic in mice than the DNA vaccine alone. Higher antibody responses (as compared to DNA vaccine and the whole killed virus vaccine alone) as well as higher cell-mediated responses (as compared to DNA vaccine alone) were elicited. Their finding also suggests that the S protein is expressed in 293 transfected cells as a single, uncleaved polypeptide, but in two differentially glycosylated forms. A combination of the vaccines and the DNA vaccine induced Th1-dominated immune response, while two injections of the whole killed vaccine induced Th2-biased response. It has been shown previously, that aluminum adjuvants skewed the immune response towards a Th2 response and a DNA vaccine enhanced T-cell immune responses [108, 109]. Immunity associated with a Th1-type immune response is thought to be essential for the control intracellular pathogens; therefore, changing the bias of the immune response may be an attractive feature of a vaccine combination strategy.

14.14.4 Use of Viral Vectors for Delivering the Vaccine

Compared to the DNA vaccines, whole killed virus, and live attenuated virus, delivering the vaccines using viral vectors is more effective for the induction of functional immune responses. Various viral vectors such as recombinant adeno associated virus (rAAV), rhabdovirus, adenovirus, and MVA have been used for the delivering the gene or genes of interest for developing SARS vaccines. In this section, we will briefly review various viral vectors.

14.14.4.1 rAAV for Vaccine Delivery

Du and colleagues have used rAAV for delivering the RBD in mice [110]. The investigators have demonstrated: (1) a single dose of RBD-rAAV vaccination could induce sufficient neutralizing antibody against SARS-CoV infection; (2) two more repeated doses of the vaccination boosted the neutralizing antibody to about 5 times of the level achieved by a single dose of the immunization; and (3) the level of the antibody continued to increase for the entire duration of the experiment (5.5 months). It was very interesting to see that neutralizing activity of these antibodies continued to increase with the number of immunizations. These data suggest that the RBD-rAAV vaccination can deliver a prolonged immune response. This may be due to the fact that the gene expression of the recombinant AAV goes through a slow onset initially, taking a course of a few days or weeks, followed by persistent gene expression for many months, which is supported by reports that the AAV may express foreign genes long-term in vivo in different organisms without resulting in significant toxicity [111, 112, 113]. However, the limitation of this study is that it did not provide the information if the antibody response has reached its highest level at the time the study was ended, and also the longevity of the antibody responses. Nevertheless, it is quite clear that the rAAV has the potential to be used as a delivery vector. In addition, it has been generally observed that some of the viral delivery systems, i.e., adenovirus and vaccinia virus may not be used for repeated immunizations either due to the pre-existing antibodies against the vector or the induction of vector specific antibodies during the primary immunization. The vector specific antibodies limit the efficacy of subsequent immunizations for enhancing the immune response for the target gene by repeated immunizations [114, 115]. However, these results clearly demonstrate that this is not the case at least for rAAV. This may be due to the lower antigenicity of the AAV delivery vector used in this study.

14.14.4.2 Use of Rhabdovirus for Vaccine Delivery

Rhabdovirus (RV) has been introduced as a vaccine vector [116, 117, 118, 119, 120, 121, 122, 123] that could also be used for the expression of relevant SARS virus antigens. There are several advantages of RV that suggest its suitability as an expression vector for SARS virus proteins: (i) the modular genome of RV is organized with short transcription stop/start sequences flanking the genes making it readily amenable to manipulation [116]; (ii) the RV genome is RNA and the life cycle of RV is exclusively cytoplasmic so no DNA recombination, reversion or integration is observed [119, 124]; (iii) stable incorporation of large and multiple foreign genes of up to 6.5 kb offers advantages over plus stranded RNA virus vectors [119]; (iv) RV is non-cytopathic in infected cells and expresses high levels of foreign proteins over extended periods of time [118, 119]; (v) RV can induce a protective immune response in a variety of animals (e.g. dog and mongoose) following immunization by the oral route and attenuated RV can target cells in the tonsils and buccal mucosa [125]; (vi) multiple mutations introduced into the RV genome that completely abolish the pathogenicity of RV render the RV vector extremely safe [119] and replication-defective RVs can be produced that are safe for even completely immunocompromised individuals [126, 127]; and (vii) since RV contains a nucleocapsid protein that has the properties of a superantigen [128], the RV vector is a unique vaccine delivery vehicle. They have evaluated the ability of RV for delivering nucleocapsid protein or envelope spike protein genes in mice. A single inoculation with the RV-based vaccine expressing SARS-CoV S protein induced a strong SARS-CoV-neutralizing antibody response. The ability of the RV-SARS-CoV S vector to induce strong immune responses after a single inoculation makes this a promising candidate for further evaluation in larger animals as well as in challenge studies to determine the protective efficacy of the immune responses.

14.14.4.3 Use of Adenovirus for Vaccine Delivery

Replication-deficient human adenovirus type 5 (AdH5) is yet another promising vector that can induce strong transgene product-specific cellular and humoral responses. However, one of the major limitations associated with this vector is the presence of neutralizing antibodies (NAbs) against AdH5. Therefore, Zhi and colleagues [129] developed a chimpanzee adenovirus C7 (AdC7) vector to circumvent interference by pre-existing immunity to AdH5, and to evaluate the impact of pre-existing immunity to human adenovirus on the efficacy of adenovirus-based vaccines against SARS CoV. Efficacy was assessed after intramuscular injection of the vector into mice and was measured as the frequency of SARS-CoV-specific T cells and neutralizing antibodies (Nab) against SARS-CoV. Immunogenicity of the AdH5-based vaccine was significantly attenuated or completely abolished when the pre-existing anti-AdH5 NAb titer was higher than 40. In contrast, preexisting anti-AdH5 NAbs have a minimal effect on the potency of the AdC7-based genetic vaccine. Taken together, these results warrant further development of AdC7 as a vaccine vector for human trials.

14.14.4.4 Use of MVA for Vaccine Delivery

As early as in 2004, Bisht and colleagues have determined the immunogenicity of S protein delivered by modified vaccinia Ankara (MVA) in a BALB/c mouse challenge model following the intranasal or intramuscular route of immunization [58]. Irrespective of the route of immunization, immunized mice induced serum antibodies that recognized the SARS S-protein in ELISA, and also neutralized SARS-CoV in vitro. Moreover, MVA_S administered by either route elicited protective immunity, as shown by reduced titers of SARS-CoV in the upper and lower respiratory tracts of mice after challenge. Passive transfer of serum from mice immunized with MVA_S to naive mice also reduced the replication of SARS-CoV in the respiratory tract after challenge, demonstrating a role for anti-S antibodies in protection. The attenuated nature of MVA and the ability of MVA_S to induce neutralizing antibody that protects mice support further development of a candidate vaccine.

More recently, Ba and colleagues have performed a head to head comparison of different delivery technologies such as DNA, MVA, and Ad5 for the full-length SARS-CoV S gene either alone or in combination [130]. They have reported that Ad5-S elicited the highest level of Nabs against SARS-pseudovirus, MVA-S induces about tenfold (IC50) lower levels of Nabs than Ad5-S, with DNA-S being the lowest. Therefore, the live vector Ad5 may offer some advantages for inducing the highest level of NAb response after one immunization. After the boost, DNA primed/DNA boosted animals induced the lowest level of NAb activity. On the other hand, priming with MVA and boosting with Ad-5 induced the highest levels of antibody responses against the S protein. It was interesting to observe that mice primed with MVA-S or DNA-S and boosted with MVA-S induced relatively lower levels of antibodies compared to animal that were primed with MVA and boosted with Ad-5. Surprisingly, mice primed with Ad5-S and boosted with either Ad5-S or MVA-S (AA, AM) induced antibodies that were higher than what was observed following the DNA primed and DNA boosted regimen, but lower than the other combinations. This finding indicates the importance of heterologous MVA-S prime and Ad5-S boost regimen for inducing the substantial level of NAb response. The use of this regimen may offer an alternative approach to overcome the problems associated with the limitations of using live viral vectors for multiple immunizations due to the pre-existing immunity against the vectors.

Ishii et al. [131] constructed a series of recombinant Dis (rDIs), a highly attenuated vaccinia strain (highly restricted host range mutant of vaccinia virus isolated by successive 1-day egg passage of the DIE vaccinia strain), expressing a gene encoding four structural proteins (E, M, N and S) of SARS-CoV [131, 132]. These rDIs elicited SARS-CoV-specific serum IgG antibody and T-cell responses in vaccinated mice following intranasal or subcutaneous administration. Mice that were subcutaneously vaccinated with rDIs expressing S protein with or without other structural proteins induced a high level of serum neutralizing IgG antibodies and demonstrated marked protective immunity against SARS-CoV challenge in the absence of a mucosal IgA response. These results indicate that the potent immune response elicited by subcutaneous injection of rDIs containing S is able to control mucosal infection by SARS-CoV. Thus, replication-deficient DIs constructs hold promise for the development of a safe and potent SARS vaccine.

14.14.4.5 Recombinant Subunit Vaccines

To sidestep the problems linked to the risks of inactivated SARS-CoV vaccine production, a **recombinant subunit vaccine** based on the spike protein but also different SARS-CoV proteins could be designed. The S protein already has been shown to be the major antigenic site in the virus and antibodies directed against this protein efficiently block SARS-CoV infection in vitro and in vivo [76, 133, 134]. The protein can be expressed and purified in its full-length form as a cell membrane anchored trimer, or as a truncated protein lacking its transmembrane region, from the supernatant of transfected cells [104]. Using the baculovirus expressed S protein, He and colleagues characterized the antigenic structure of the S protein against a panel of 38 monoclonal antibodies (MAbs) isolated from the immunized mice [135]. The epitopes of most anti-S MAbs (32 of 38) were localized within the S1 domain, and those of the remaining 6 MAbs were mapped to the S2 domain. Among the anti-S1 MAbs, 17 MAbs targeted the N-terminal region (amino acids [aa] 12 to 327), 9 MAbs recognized the receptor-binding domain (RBD; aa 318 to 510), and 6 MAbs reacted with the C-terminal region of S1 domain that contains the major immunodominant site (aa 528 to 635). Strikingly, all of the RBD-specific MAbs had potent neutralizing activity, 6 of which efficiently blocked the receptor binding, confirming that the RBD contains the main neutralizing epitopes and that blockage of the receptor association is the major mechanism of SARS-CoV neutralization. Five MAbs specific for the S1 N-terminal region exhibited moderate neutralizing activity, but none of the MAbs reacting with the S2 domain and the major immunodominant site in S1 showed neutralizing activity. All of the neutralizing MAbs recognized conformational epitopes. This panel of anti-S MAbs can be used as tools for studying the structure and function of the SARS-CoV S protein.

Bisht and colleagues [136] expressed a truncated version of the S protein (amino acid residue 14-762) in the baculovirus system, and purified the S protein to homogeneity, and evaluated in mice its ability to induce protective antibody responses. The truncated form of the S-protein elicited higher levels of neutralizing antibodies in mice and protected them against a viral challenge. In another study, He and colleagues have expressed full-length S protein (FL-S) or its extracellular domain (EC-S) in baculoviruses, and compared their immunogenicity in mice [137]. The immunized mice developed high titers of anti-S antibodies with potent neutralizing activities against SARS pseudoviruses constructed with the S proteins of diverse heterologous isolates of Tor2, GD03T13, and nSZ3, the representative strains of 2002 to 2003 and 2003 to 2004 human SARS-CoV and palm civet SARS-CoV, respectively. These data suggest that the recombinant baculovirus-expressed S protein vaccines possess important conformational neutralizing epitopes, which are important for protection against the challenge infection. They have further demonstrated that a recombinant S-protein containing193-amino acid receptor-binding domain (residues 318–510) fused in frame to human IgG1 Fc fragment induced highly potent antibody responses in the immunized rabbits. The antibodies recognized RBD on S1 domain and completely inhibited SARS-CoV infection at a serum dilution of 1:10,240. Rabbit anti-sera effectively blocked binding of S1, which contains RBD, to ACE2. This suggests that RBD can induce highly potent neutralizing antibody responses and has potential to be developed as an effective and safe subunit vaccine for prevention of SARS.

To map the precise epitope in the RBD He and colleagues have immunized mice with the RBD-IgG1 FC recombinant protein. The RBD-IgG1 FC induced high titer of RBD-specific Abs in the immunized mice, and these sera effectively neutralized infection by both SARS-CoV and SARS pseudoviruses [137]. They have developed a series of mAbs (27) and to characterize the neutralization determinants on the RBD of S protein. Six groups of conformation-dependent epitopes, designated as Conf I–VI, and two adjacent linear epitopes were identified by ELISA and binding competition assays. The Conf IV and Conf V mAbs significantly blocked RBD-Fc binding to angiotensin-converting enzyme 2, suggesting that their epitopes overlaps with the receptor-binding sites in the S protein. Most of the mAbs (23 of 25) that recognized the conformational epitopes possessed potent neutralizing activities against SARS pseudovirus with 50% neutralizing activity in doses ranging from 0.005 to 6.569 ug/ml. Therefore, the RBD of SARS S protein contains multiple conformational epitopes capable of inducing potent NAb responses, and is an important target site for developing vaccines and immunotherapeutics.

It was very interesting to note that S protein produced in baculo virus induced potent binding and neutralizing antibodies in mice and rabbits [137]. However, it has been shown by Bai and colleagues that the S protein expressed in E. coli failed to induce NAb in mice [138]. It is not clear what the reason could be. One potential difference between the S protein produced in E. coli and baculovirus is that the baculovirus produced protein is glycosylated, and the E. coli produced protein is non-glycosylated. However, further work is needed to define that it is the sugars that are directly involved in creating immunodominant epitopes, or sugars are involved in producing a correctly folded S-protein. In either situations, perhaps the expression of S-protein in mammalian expression system will be more desirable. In a proof of concept study, Chang et al. studied the effect of intron and exon splicing enhancers to improve the expression of STR2 (88 kDa), carrying three S fragments (S74–253, S294–739, and S1129–1255) in mammalian cells [139]. The investigators demonstrated that the addition of an 138 base-pair intron increased the expression of STR2 protein by 1.9, 2.5, and 4.1-fold in Vero E6, 293A cells, and CHO cells respectively. Furthermore, exon-splicing enhancers also increased the STR2 expression 1.7-2.8 fold. However, the combination of intron and exon splicing enhancer resulted in the suppression of STR2 expression. These results can provide an optimal strategy to enhance SARS-CoV S protein expression in mammalian cells and may contribute to the development of SARS-CoV subunit vaccine.

Based on the data reviewed here, it is quite evident that that full length, truncated, and also the receptor binding domain of S protein induced strong binding and neutralizing antibody responses against the S-protein in mice and rabbits. Furthermore, upon challenge with pathogenic SARS virus, these animals were protected. However, it is not clear if the S protein alone will be able to induce a long lasting immunity against the virus, or other proteins such as M and N will need to be included in an effective vaccine against SARS.

14.14.4.6 Use of VLPs for Developing a Vaccine

Subunit vaccines based on recombinant proteins can suffer from poor immunogenicity owing to incorrect folding of the target protein or poor presentation to the immune system. Virus-like particles (VLPs) represent a specific class of recombinant vaccines that mimic the outer structure of authentic virus particles but do not cause a productive infection since they lack the viral genome. Yet, they are recognized readily by the immune system and present viral antigens in their authentic or near-authentic conformation. Co-expression of mouse hepatitis virus (MHV) M and E proteins resulted in the assembly of particles. The S protein was dispensable but was incorporated when present. The resulting secreted VLPs are indistinguishable from authentic virions in size and shape. The N protein was neither required nor packaged into the particles when present [140]. Similar observations were made when SARS-CoV M, E, and S protein were co-expressed in insect cells using a baculovirus expression system [141, 142]. Whether those VLPs are able to elicit a protective immune response remains to be determined. In contrast, formation of SARS-Co VLPs in mammalian cells seems not to be dependent on the expression of the E but rather the N protein. Therefore, SARS-Co VLP assembly in human 293 renal epithelial cells relies on the expression of the S, M, and N protein [143]. Whether this discrepancy is caused by the different expression systems needs to be studied further.

14.15 Use of Adjuvant for Enhancing Functional Immune Responses

The vaccination modality may play a crucial role in the type of antibodies or T cell responses induced [144]. Therefore, it is possible that different vaccination modalities such as inactivated virus, live virus vector delivery, DNA vaccine, protein subunit vaccines and VLP will elicit antibodies that are qualitatively different even if the immunogen is similar. Additionally, the formulation of immunogens with different adjuvants may affect the maturation of antibody responses differently. This is suggested by the qualitative differences recorded in antibody responses generated by a given antigen formulated in different adjuvants in case of HIV vaccines [145]. This observation suggests that the structure of specific epitopes within the immunogen may be presented differently in combinations with different adjuvants. Therefore, even if the overall structure of the immunogen remains stable, the exposure of specific epitopes may be modified. In a proof of concept study Hu and colleagues have demonstrated that co-inoculation of DNA vaccine and IL2 had induced a significantly higher immune response compared to the spike DNA vaccine alone [146]. In the same study, they evaluated the route of immunization upon the type and magnitude of immune responses and found that oral vaccination evoked a vigorous T-cell response and a weak IgG2a antibody responses; intramuscular immunization evoked a vigorous antibody response and a weak T-cell response, and vaccination by electroporation evoked a vigorous response with a predominant subclass IgG1 antibody response and a moderate T-cell response.

14.16 Potential of Disease Enhancement Due to Candidate SARS Vaccines

Enhanced disease in previously immunized individuals is a concern for the development of any vaccine. This may be particularly true for SARS-CoV vaccines since adverse effects have already been reported for one coronavirus vaccine, feline infectious peritonitis virus [147]. In fact, it has been demonstrated that some S variants were not only resistant to antibody neutralization [133], but also showed enhanced entry in the presence of certain antibodies in vitro [106]. The S protein from different SARS-CoV strains isolated during the outbreak in 2002/2003 and 2003/2004 but also S protein of virus isolated from civet cats was used to generate pseudo typed viruses. Each pseudovirus was incubated independently with immune IgG purified from mice vaccinated with S protein from strain Urbani. Inhibition of entry was demonstrated for the prototype strain (Urbani) but also for pseudoviruses from 2002/2003 human isolates. However, one S (GD03T0013) pseudovirus was markedly resistant to antibody inhibition by the polyclonal IgG and, unexpectedly, entry of two pseudoviruses from palm civet S glycoproteins was markedly enhanced. A similar effect could be detected when human mAbs derived from a recovered SARS patient [76] were used instead of the polyclonal IgG. Antibodies with insufficient cross-neutralization capacity may enhance rather than protect from virus infection. Epitope mapping of human mAbs using recombinant S protein fragments demonstrated a significant reduction in reactivity with an S fragment (residues 318–510) containing a N479S substitution [134]. Sui et al. obtained similar results [133] when they tested the neutralization ability of another human mAb. The antibody could protect mice from a SARS-CoV (strain Urbani) challenge when given 1 day before inoculation. However, when tested in vitro on its ability to neutralize various S protein-pseudo typed viruses, this antibody did not bind to pseudo typed virus containing a D480G substitution in the spike. D480G is a naturally occurring variation in the S protein found in a SARS-CoV isolate (GD03T0013). Although this virus, which was isolated in December 2003 from a SARS patient and viruses isolated from civet cats seem to be only weakly pathogenic in humans, other mutations in the viral genome might occur that impact viral tropism and virulence. Variations in the viral genome will likely continue to occur in the animal reservoirs due to the high mutation rate of RNA viruses and especially coronaviruses. Additionally, antibodies mediating virus entry, could function as a facilitating portal for viruses to gain entrance into the human population. SARS-CoV-like strains which normally do not infect or hardly infect humans could get the chance to replicate in and thus also to adapt to the new host. Therefore, any antiviral strategy based on neutralizing antibodies, whether passive immunotherapy or active immunization, has to be carefully evaluated in appropriate animal models that closely resemble the human disease.

Acknowledgments We would like to thank Ms Nelle Cronen for her expert editorial and administrative help, and the members of the Vaccines Research for their contribution. The SARS project work is supported by an NIH grant.

References

1. Peiris, J.S., Lai, S.T., Poon, L.L., Guan, Y., Yam, L.Y., Lim, W., Nicholls, J., Yee, W.K., Yan, W.W., Cheung, M.T., Cheng, V.C., Chan, K.H., Tsang, D.N., Yung, R.W., Ng, T.K., and Yuen, K.Y. 2003, Lancet, 361, 1319.
2. Donnelly, C.A., Ghani, A.C., Leung, G.M., Hedley, A.J., Fraser, C., Riley, S., Abu-Raddad, L.J., Ho, L.M., Thach, T.Q., Chau, P., Chan, K.P., Lam, T.H., Tse, L.Y., Tsang, T., Liu, S.H., Kong, J.H., Lau, E.M., Ferguson, N.M., and Anderson, R.M. 2003, Lancet, 361, 1761.
3. Hon, K.L., Leung, C.W., Cheng, W.T., Chan, P.K., Chu, W.C., Kwan, Y.W., Li, A.M., Fong, N.C., Ng, P.C., Chiu, M.C., Li, C.K., Tam, J.S., and Fok, T.F. 2003, Lancet, 361, 1701.
4. Peiris, J.S., Chu, C.M., Cheng, V.C., Chan, K.S., Hung, I.F., Poon, L.L., Law, K.I., Tang, B.S., Hon, T.Y., Chan, C.S., Chan, K.H., Ng, J.S., Zheng, B.J., Ng, W.L., Lai, R.W., Guan, Y., and Yuen, K.Y. 2003, Lancet, 361, 1767.

5. Drosten, C., Gunther, S., Preiser, W., van der Werf, S., Brodt, H.R., Becker, S., Rabenau, H., Panning, M., Kolesnikova, L., Fouchier, R.A., Berger, A., Burguiere, A.M., Cinatl, J., Eickmann, M., Escriou, N., Grywna, K., Kramme, S., Manuguerra, J.C., Muller, S., Rickerts, V., Sturmer, M., Vieth, S., Klenk, H.D., Osterhaus, A.D., Schmitz, H., and Doerr, H.W. 2003, N. Engl. J. Med., 348, 1967.
6. Ksiazek, T.G., Erdman, D., Goldsmith, C.S., Zaki, S.R., Peret, T., Emery, S., Tong, S., Urbani, C., Comer, J.A., Lim, W., Rollin, P.E., Dowell, S.F., Ling, A.E., Humphrey, C.D., Shieh, W.J., Guarner, J., Paddock, C.D., Rota, P., Fields, B., DeRisi, J., Yang, J.Y., Cox, N., Hughes, J.M., LeDuc, J.W., Bellini, W.J., and Anderson, L.J. 2003, N. Engl. J. Med., 348, 1953.
7. Li, W., Moore, M.J., Vasilieva, N., Sui, J., Wong, S.K., Berne, M.A., Somasundaran, M., Sullivan, J.L., Luzuriaga, K., Greenough, T.C., Choe, H., and Farzan, M. 2003, Nature, 426, 450.
8. Fouchier, R.A., Kuiken, T., Schutten, M., Van Amerongen, G., Van Doornum, G.J., Van Den Hoogen, B.G., Peiris, M., Lim, W., Stohr, K., and Osterhaus, A.D. 2003, Nature, 423, 240.
9. Pearson, H., Clarke, T., Abbott, A., Knight, J., and Cyranoski, D. 2003, Nature, 424, 121.
10. WHO 2004, Update 3: Announcement of suspected SARS case in southern China; Investigation of source of infection for confirmed case begins tomorrow.
11. Snijder, E.J., Bredenbeek, P.J., Dobbe, J.C., Thiel, V., Ziebuhr, J., Poon, L.L., Guan, Y., Rozanov, M., Spaan, W.J., and Gorbalenya, A.E. 2003, J. Mol. Biol., 331, 991.
12. Ruan, Y.J., Wei, C.L., Ee, L.A., Vega, V.B., Thoreau, H., Yun, S.T.S., Chia, J.M., Ng, P., Chiu, K.P., Lim, L., Tao, Z., Peng, C.K., Ean, L.O.L.E., Lee, N.M., Sin, L.Y., Ng, L.F.P., Chee, R.e., Stanton, L.W., Long, P.M., and Liu, E.T. 2003, Lancet, 361, 1779.
13. Guan, Y., Zheng, B.J., He, Y.Q., Liu, X.L., Zhuang, Z.X., Cheung, C.L., Luo, S.W., Li, P.H., Zhang, L.J., Guan, Y.J., Butt, K.M., Wong, K.L., Chan, K.W., Lim, W., Shortridge, K.F., Yuen, K.Y., Peiris, J.S., and Poon, L.L. 2003, Science, 302, 276.
14. Martina, B.E., Haagmans, B.L., Kuiken, T., Fouchier, R.A., Rimmelzwaan, G.F., Van Amerongen, G., Peiris, J.S., Lim, W., and Osterhaus, A.D. 2003, Nature, 425, 915.
15. WHO. 2003, Wkly. Epidemiol. Rec., 78, 86.
16. Marra, M.A., Jones, S.J., Astell, C.R., Holt, R.A., Brooks-Wilson, A., Butterfield, Y.S., Khattra, J., Asano, J.K., Barber, S.A., Chan, S.Y., Cloutier, A., Coughlin, S.M., Freeman, D., Girn, N., Griffith, O.L., Leach, S.R., Mayo, M., McDonald, H., Montgomery, S.B., Pandoh, P.K., Petrescu, A.S., Robertson, A.G., Schein, J.E., Siddiqui, A., Smailus, D.E., Stott, J.M., Yang, G.S., Plummer, F., Andonov, A., Artsob, H., Bastien, N., Bernard, K., Booth, T.F., Bowness, D., Czub, M., Drebot, M., Fernando, L., Flick, R., Garbutt, M., Gray, M., Grolla, A., Jones, S., Feldmann, H., Meyers, A., Kabani, A., Li, Y., Normand, S., Stroher, U., Tipples, G.A., Tyler, S., Vogrig, R., Ward, D., Watson, B., Brunham, R.C., Krajden, M., Petric, M., Skowronski, D.M., Upton, C., and Roper, R.L. 2003, Science, 300, 1399.
17. Rota, P.A., Oberste, M.S., Monroe, S.S., Nix, W.A., Campagnoli, R., Icenogle, J.P., Penaranda, S., Bankamp, B., Maher, K., Chen, M.H., Tong, S., Tamin, A., Lowe, L., Frace, M., DeRisi, J.L., Chen, Q., Wang, D., Erdman, D.D., Peret, T.C., Burns, C., Ksiazek, T.G., Rollin, P.E., Sanchez, A., Liffick, S., Holloway, B., Limor, J., McCaustland, K., Olsen-Rasmussen, M., Fouchier, R., Gunther, S., Osterhaus, A.D., Drosten, C., Pallansch, M.A., Anderson, L.J., and Bellini, W.J. 2003, Science, 300, 1394.
18. Holmes, K.V. 2003, N. Engl. J. Med., 348, 1948.
19. Eickmann, M., Becker, S., Klenk, H.D., Doerr, H.W., Stadler, K., Censini, S., Guidotti, S., Masignani, V., Scarselli, M., Mora, M., Donati, C., Han, J., Song, H.C., Abrignani, S., Covacci, A., and Rappuoli, R. 2003, Science, 302, 1504.
20. Eickmann, M., Becker, S., Klenk, H.D., Doerr, H.W., Stadler, K., Censini, S., Guidotti, S., Masignani, V., Scarselli, M., Mora, M., Donati, C., Han, J.H., Song, H.C., Abrignani, S., Covacci, A., and Rappuoli, R. 2003, Science, 302, 1504.
21. Xiao, X., Chakraborti, S., Dimitrov, A.S., Gramatikoff, K., and Dimitrov, D.S. 2003, Biochem. Biophys. Res. Commun., 312, 1159.

22. Wang, P., Chen, J., Zheng, A., Nie, Y., Shi, X., Wang, W., Wang, G., Luo, M., Liu, H., Tan, L., Song, X., Wang, Z., Yin, X., Qu, X., Wang, X., Qing, T., Ding, M., and Deng, H. 2004, Biochem. Biophys. Res. Commun., 315, 439.
23. WHO 2003, Consultation on needs and opportunities for SARS vaccine research and development.
24. Bosch, B.J., van der Zee, R., de Haan, C.A., and Rottier, P.J. 2003, J. Virol., 77, 8801.
25. Dimitrov, D.S. 2003, Cell., 115, 652.
26. Prabakaran, P., Xiao, X., and Dimitrov, D.S. 2004, Biochem. Biophys. Res. Commun., 314, 235.
27. Babcock, G.J., Esshaki, D.J., Thomas, W.D., Jr., and Ambrosino, D.M. 2004, J. Virol., 78, 4552.
28. Wong, S.K., Li, W., Moore, M.J., Choe, H., and Farzan, M. 2004, J. Biol. Chem., 279, 3197.
29. Xiang, S.H., Wang, L., Abreu, M., Huang, C.C., Kwong, P.D., Rosenberg, E., Robinson, J.E., and Sodroski, J. 2003, Virology, 315, 124.
30. Liu, S., Xiao, G., Chen, Y., He, Y., Niu, J., Escalante, C.R., Xiong, H., Farmar, J., Debnath, A.K., Tien, P., and Jiang, S. 2004, Lancet, 363, 938.
31. Tripet, B., Howard, M.W., Jobling, M., Holmes, R.K., Holmes, K.V., and Hodges, R.S. 2004, J Biol Chem., 279, 20836.
32. Chan, D.C., Fass, D., Berger, J.M., and Kim, P.S. 1997, Cell, 89, 263.
33. Xu, Y., Liu, Y.J., and Yu, Q. 2004, J. Biol. Chem., 279, 41179.
34. Delmas, B., and Laude, H. 1990, J. Virol., 64, 5367.
35. Wiley, D.C., Skehel, J.J., and Waterfield, M. 1977, Virology, 79, 446.
36. Wyatt, R., Kwong, P.D., Desjardins, E., Sweet, R.W., Robinson, J., Hendrickson, W.A., and Sodroski, J.G. 1998, Nature, 393, 705.
37. Doms, R.W., Keller, D.S., Helenius, A., and Balch, W.E. 1987, J. Cell Biol., 105, 1957.
38. Li, F., Li, W., Farzan, M., and Harrison, S.C. 2006, Adv. Exp. Med. Biol., 581, 229.
39. van Kooyk, Y., and Geijtenbeek, T.B. 2003, Nat. Rev. Immunol., 3, 697.
40. Marzi, A., Gramberg, T., Simmons, G., Moller, P., Rennekamp, A.J., Krumbiegel, M., Geier, M., Eisemann, J., Turza, N., Saunier, B., Steinkasserer, A., Becker, S., Bates, P., Hofmann, H., and Pohlmann, S. 2004, J. Virol., 78, 12090.
41. Yang, Z.Y., Huang, Y., Ganesh, L., Leung, K., Kong, W.P., Schwartz, O., Subbarao, K., and Nabel, G.J. 2004, J. Virol., 78, 5642.
42. Towler, P., Staker, B., Prasad, S.G., Menon, S., Tang, J., Parsons, T., Ryan, D., Fisher, M., Williams, D., Dales, N.A., Patane, M.A., and Pantoliano, M.W. 2004, J. Biol. Chem., 279, 17996.
43. Hofmann, H., Geier, M., Marzi, A., Krumbiegel, M., Peipp, M., Fey, G.H., Gramberg, T., and Pohlmann, S. 2004, Biochem. Biophys. Res. Commun., 319, 1216.
44. Kuiken, T., Fouchier, R.A., Schutten, M., Rimmelzwaan, G.F., van Amerongen, G., van Riel, D., Laman, J.D., de Jong, T., van Doornum, G., Lim, W., Ling, A.E., Chan, P.K., Tam, J.S., Zambon, M.C., Gopal, R., Drosten, C., van der Werf, S., Escriou, N., Manuguerra, J.C., Stohr, K., Peiris, J.S., and Osterhaus, A.D. 2003, Lancet, 362, 263.
45. Subbarao, K., McAuliffe, J., Vogel, L., Fahle, G., Fischer, S., Tatti, K., Packard, M., Shieh, W.J., Zaki, S., and Murphy, B. 2004, J. Virol., 78, 3572.
46. Tipnis, S.R., Hooper, N.M., Hyde, R., Karran, E., Christie, G., and Turner, A.J. 2000, J. Biol. Chem., 275, 33238.
47. Crackower, M.A., Sarao, R., Oudit, G.Y., Yagil, C., Kozieradzki, I., Scanga, S.E., Oliveira-dos-Santos, A.J., da Costa, J., Zhang, L., Pei, Y., Scholey, J., Ferrario, C.M., Manoukian, A.S., Chappell, M.C., Backx, P.H., Yagil, Y., and Penninger, J.M. 2002, Nature, 417, 822.
48. Hamming, I., Timens, W., Bulthuis, M., Lely, A., Navis, G., and van Goor, H. 2004, J. Pathol., 203, 631.
49. Ding, Y., Wang, H., Shen, H., Li, Z., Geng, J., Han, H., Cai, J., Li, X., Kang, W., Weng, D., Lu, Y., Wu, D., He, L., and Yao, K. 2003, J. Pathol., 200, 282.

50. Nicholls, J.M., Poon, L.L., Lee, K.C., Ng, W.F., Lai, S.T., Leung, C.Y., Chu, C.M., Hui, P.K., Mak, K.L., Lim, W., Yan, K.W., Chan, K.H., Tsang, N.C., Guan, Y., Yuen, K.Y., and Peiris, J.S. 2003, Lancet, 361, 1773.
51. To, K.F., Tong, J.H., Chan, P.K., Au, F.W., Chim, S.S., Chan, K.C., Cheung, J.L., Liu, E.Y., Tse, G.M., Lo, A.W., Lo, Y.M., and Ng, H.K. 2004, J. Pathol., 202, 157.
52. Leung, W.K., To, K.F., Chan, P.K., Chan, H.L., Wu, A.K., Lee, N., Yuen, K.Y., and Sung, J.J. 2003, Gastroenterology, 125, 1011.
53. Lew, T.W., Kwek, T.K., Tai, D., Earnest, A., Loo, S., Singh, K., Kwan, K.M., Chan, Y., Yim, C.F., Bek, S.L., Kor, A.C., Yap, W.S., Chelliah, Y.R., Lai, Y.C., and Goh, S.K. 2003, JAMA, 290, 374.
54. Hofmann, H., Hattermann, K., Marzi, A., Gramberg, T., Geier, M., Krumbiegel, M., Kuate, S., Uberla, K., Niedrig, M., and Pohlmann, S. 2004, J. Virol., 78, 6134.
55. Smith, A.E., and Helenius, A. 2004, Science, 304, 237.
56. Simmons, G., Reeves, J.D., Rennekamp, A.J., Amberg, S.M., Piefer, A.J., and Bates, P. 2004, Proc. Natl. Acad. Sci. U. S. A., 101, 4240.
57. de Haan, C.A., Stadler, K., Godeke, G.J., Bosch, B.J., and Rottier, P.J. 2004, J. Virol., 78, 6048.
58. Bisht, H., Roberts, A., Vogel, L., Bukreyev, A., Collins, P.L., Murphy, B.R., Subbarao, K., and Moss, B. 2004, Proc. Natl. Acad. Sci. U. S. A., 101, 6641.
59. Eckert, D.M., and Kim, P.S. 2001, Annu. Rev. Biochem., 70, 777.
60. Kilby, J.M., and Eron, J.J. 2003, N. Engl. J. Med., 348, 2228.
61. Huang, Q., Yu, L., Petros, A.M., Gunasekera, A., Liu, Z., Xu, N., Hajduk, P., Mack, J., Fesik, S.W., and Olejniczak, E.T. 2004, Biochemistry (Mosc). 43, 6059.
62. Boots, A.M., Kusters, J.G., van Noort, J.M., Zwaagstra, K.A., Rijke, E., van der Zeijst, B.A., and Hensen, E.J. 1991, Immunology, 74, 8.
63. Bergmann, C., McMillan, M., and Stohlman, S. 1993, J. Virol., 67, 7041.
64. Thiel, V., Ivanov, K.A., Putics, A., Hertzig, T., Schelle, B., Bayer, S., Weissbrich, B., Snijder, E.J., Rabenau, H., Doerr, H.W., Gorbalenya, A.E., and Ziebuhr, J. 2003, J. Gen. Virol., 84, 2305.
65. Yang, Z.Y., Wyatt, L.S., Kong, W.P., Moodie, Z., Moss, B., and Nabel, G.J. 2003, J. Virol., 77, 799.
66. Ziebuhr, J. 2004, Curr. Opin. Microbiol., 7, 412.
67. Yang, Z.Y., Kong, W.P., Huang, Y., Roberts, A., Murphy, B.R., Subbarao, K., and Nabel, G.J. 2004, Nature, 428, 561.
68. Buchholz, U.J., Bukreyev, A., Yang, L., Lamirande, E.W., Murphy, B.R., Subbarao, K., and Collins, P.L. 2004, Proc. Natl. Acad. Sci. U. S. A., 101, 9804.
69. Gao, W., Tamin, A., Soloff, A., D'Aiuto, L., Nwanegbo, E., Robbins, P.D., Bellini, W.J., Barratt-Boyes, S., and Gambotto, A. 2003, Lancet, 362, 1895.
70. Li, G., Chen, X., and Xu, A. 2003, N. Engl. J. Med., 349, 508.
71. Wang, Y.D., Li, Y., Xu, G.B., Dong, X.Y., Yang, X.A., Feng, Z.R., Tian, C., and Chen, W.F. 2004, Clin. Immunol., 113, 145.
72. Soo, Y.O., Cheng, Y., Wong, R., Hui, D.S., Lee, C.K., Tsang, K.K., Ng, M.H., Chan, P., Cheng, G., and Sung, J.J. 2004, Clin. Microbiol. Infect., 10, 676.
73. Keller, M.A., and Stiehm, E.R. 2000, Clin. Microbiol. Rev., 13, 602.
74. Sui, J., Li, W., Murakami, A., Tamin, A., Matthews, L.J., Wong, S.K., Moore, M.J., Tallarico, A.S., Olurinde, M., Choe, H., Anderson, L.J., Bellini, W.J., Farzan, M., and Marasco, W.A. 2004, Proc. Natl. Acad. Sci. U. S. A., 101, 2536.
75. ter Meulen, J., Bakker, A.B., van den Brink, E.N., Weverling, G.J., Martina, B.E., Haagmans, B.L., Kuiken, T., de Kruif, J., Preiser, W., Spaan, W., Gelderblom, H.R., Goudsmit, J., and Osterhaus, A.D. 2004, Lancet, 363, 2139.
76. Traggiai, E., Becker, S., Subbarao, K., Kolesnikova, L., Uematsu, Y., Gismondo, M.R., Murphy, B.R., Rappuoli, R., and Lanzavecchia, A. 2004, Nat. Med., 10, 871.
77. Yang, L.T., Peng, H., Zhu, Z.L., Li, G., Huang, Z.T., Zhao, Z.X., Koup, R.A., Bailer, R.T., and Wu, C.Y. 2006, Clin. Immunol., 120, 171.

78. Rowe, T., Gao, G., Hogan, R.J., Crystal, R.G., Voss, T.G., Grant, R.L., Bell, P., Kobinger, G.P., Wivel, N.A., and Wilson, J.M. 2004, J Virol, 78, 11401.
79. McAuliffe, J., Vogel, L., Roberts, A., Fahle, G., Fischer, S., Shieh, W.J., Butler, E., Zaki, S., St Claire, M., Murphy, B., and Subbarao, K. 2004, Virology, 330, 8.
80. Roberts, A., Kretzschmar, E., Perkins, A.S., Forman, J., Price, R., Buonocore, L., Kawaoka, Y., and Rose, J.K. 1998, J. Virol., 72, 4704.
81. Ladman, B.S., Pope, C.R., Ziegler, A.F., Swieczkowski, T., Callahan, C.J., Davison, S., and Gelb, J., Jr. 2002, Avian Dis., 46, 938.
82. Pratelli, A., Tinelli, A., Decaro, N., Martella, V., Camero, M., Tempesta, M., Martini, M., Carmichael, L.E., and Buonavoglia, C. 2004, Vet. Microbiol., 99, 43.
83. Zhou, J., Wang, W., Zhong, Q., Hou, W., Yang, Z., Xiao, S.Y., Zhu, R., Tang, Z., Wang, Y., Xian, Q., Tang, H., and Wen, L. 2005, Vaccine, 23, 3202.
84. He, Y., Zhou, Y., Siddiqui, P., and Jiang, S. 2004, Biochem. Biophys. Res. Commun., 325, 445.
85. Stadler, K., Roberts, A., Becker, S., Vogel, L., Eickmann, M., Kolesnikova, L., Klenk, H.D., Murphy, B., Rappuoli, R., Abrignani, S., and Subbarao, K. 2005, Emerg. Infect. Dis., 11, 1312.
86. Podda, A., Del Giudice, G. 2004, New Generation Vaccines, M. Levine, Kaper, J.B., Rappuoli, R., Liu, M., Good, M.F. (Ed.), Marcel Dekker, New York, 225.
87. Takasuka, N., Fujii, H., Takahashi, Y., Kasai, M., Morikawa, S., Itamura, S., Ishii, K., Sakaguchi, M., Ohnishi, K., Ohshima, M., Hashimoto, S., Odagiri, T., Tashiro, M., Yoshikura, H., Takemori, T., and Tsunetsugu-Yokota, Y. 2004, Int. Immunol., 16, 1423.
88. Tang, L., Zhu, Q., Qin, E., Yu, M., Ding, Z., Shi, H., Cheng, X., Wang, C., Chang, G., Fang, F., Chang, H., Li, S., Zhang, X., Chen, X., Yu, J., Wang, J., and Chen, Z. 2004, DNA Cell Biol., 23, 391.
89. Zhang, C.H., Lu, J.H., Wang, Y.F., Zheng, H.Y., Xiong, S., Zhang, M.Y., Liu, X.J., Li, J.X., Wan, Z.Y., Yan, X.G., Qi, S.Y., Cui, Z., and Zhang, B. 2005, Vaccine, 23, 3196.
90. Purcell, R.H., D'Hondt, E., Bradbury, R., Emerson, S.U., Govindarajan, S., and Binn, L. 1992, Vaccine, 10 Suppl 1, S148.
91. Sauzet, J.P., Perlaza, B.L., Brahimi, K., Daubersies, P., and Druilhe, P. 2001, Infect. Immun., 69, 1202.
92. Sedegah, M., Weiss, W., Sacci, J.B., Jr., Charoenvit, Y., Hedstrom, R., Gowda, K., Majam, V.F., Tine, J., Kumar, S., Hobart, P., and Hoffman, S.L. 2000, J. Immunol., 164, 5905.
93. Ulmer, J.B., Donnelly, J.J., Parker, S.E., Rhodes, G.H., Felgner, P.L., Dwarki, V.J., Gromkowski, S.H., Deck, R.R., DeWitt, C.M., Friedman, A., et al. 1993, Science, 259, 1745.
94. Robinson, H.L., and Torres, C.A. 1997, Semin. Immunol., 9, 271.
95. Jin, H., Xiao, C., Chen, Z., Kang, Y., Ma, Y., Zhu, K., Xie, Q., Tu, Y., Yu, Y., and Wang, B. 2005, Biochem. Biophys. Res. Commun., 328, 979.
96. Dunachie, S.J., and Hill, A.V. 2003, J. Exp. Biol., 206, 3771.
97. McShane, H., Brookes, R., Gilbert, S.C., and Hill, A.V. 2001, Infect. Immun., 69, 681.
98. Zhu, M.S., Pan, Y., Chen, H.Q., Shen, Y., Wang, X.C., Sun, Y.J., and Tao, K.H. 2004, Immunol. Lett., 92, 237.
99. Kim, T.W., Lee, J.H., Hung, C.F., Peng, S., Roden, R., Wang, M.C., Viscidi, R., Tsai, Y.C., He, L., Chen, P.J., Boyd, D.A., and Wu, T.C. 2004, J. Virol., 78, 4638.
100. Shi, S.Q., Peng, J.P., Li, Y.C., Qin, C., Liang, G.D., Xu, L., Yang, Y., Wang, J.L., and Sun, Q.H. 2006, Mol. Immunol., 43, 1791.
101. Huang, J., Ma, R., and Wu, C.Y. 2006, Vaccine, 24, 4905.
102. Woo, P.C., Lau, S.K., Tsoi, H.W., Chen, Z.W., Wong, B.H., Zhang, L., Chan, J.K., Wong, L.P., He, W., Ma, C., Chan, K.H., Ho, D.D., and Yuen, K.Y. 2005, Vaccine, 23, 4959.
103. Zhang, M.Y., Xiao, X., Sidorov, I.A., Choudhry, V., Cham, F., Zhang, P.F., Bouma, P., Zwick, M., Choudhary, A., Montefiori, D.C., Broder, C.C., Burton, D.R., Quinnan, G.V., Jr., and Dimitrov, D.S. 2004, J. Virol., 78, 9233.

104. Song, H.C., Seo, M.Y., Stadler, K., Yoo, B.J., Choo, Q.L., Coates, S.R., Uematsu, Y., Harada, T., Greer, C.E., Polo, J.M., Pileri, P., Eickmann, M., Rappuoli, R., Abrignani, S., Houghton, M., and Han, J.H. 2004, J. Virol., 78, 10328.
105. Olsen, C.W., Corapi, W.V., Ngichabe, C.K., Baines, J.D., and Scott, F.W. 1992, J. Virol., 66, 956.
106. Yang, Z.Y., Werner, H.C., Kong, W.P., Leung, K., Traggiai, E., Lanzavecchia, A., and Nabel, G.J. 2005, Proc. Natl. Acad. Sci. U. S. A., 102, 797.
107. Zakhartchouk, A.N., Liu, Q., Petric, M., and Babiuk, L.A. 2005, Vaccine, 23, 4385.
108. Sin, J.I., Bagarazzi, M., Pachuk, C., and Weiner, D.B. 1999, DNA Cell Biol., 18, 771.
109. Ioannou, X.P., Gomis, S.M., Karvonen, B., Hecker, R., Babiuk, L.A., and van Drunen Littel-van den Hurk, S. 2002, Vaccine, 21, 127.
110. Du, L., He, Y., Wang, Y., Zhang, H., Ma, S., Wong, C.K., Wu, S.H., Ng, F., Huang, J.D., Yuen, K.Y., Jiang, S., Zhou, Y., and Zheng, B.J. 2006, Virology, 353, 6.
111. Monahan, P.E., and Samulski, R.J. 2000, Mol. Med. Today, 6, 433.
112. Rabinowitz, J.E., and Samulski, J. 1998, Curr. Opin. Biotechnol., 9, 470.
113. Smith-Arica, J.R., and Bartlett, J.S. 2001, Curr. Cardiol. Rep., 3, 43.
114. Muruve, D.A. 2004, Hum. Gene Ther., 15, 1157.
115. Nasz, I., and Adam, E. 2001, Acta Microbiol. Immunol. Hung., 48, 323.
116. Foley, H.D., McGettigan, J.P., Siler, C.A., Dietzschold, B., and Schnell, M.J. 2000, Proc. Natl. Acad. Sci. U. S. A., 97, 14680.
117. Foley, H.D., Otero, M., Orenstein, J.M., Pomerantz, R.J., and Schnell, M.J. 2002, J. Virol., 76, 19.
118. McGettigan, J.P., Sarma, S., Orenstein, J.M., Pomerantz, R.J., and Schnell, M.J. 2001, J. Virol., 75, 8724.
119. McGettigan, J.P., Naper, K., Orenstein, J., Koser, M., McKenna, P.M., and Schnell, M.J. 2003, J. Virol., 77, 10889.
120. Morimoto, K., Schnell, M.J., Pulmanausahakul, R., McGettigan, J.P., Foley, H.D., Faber, M., Hooper, D.C., and Dietzschold, B. 2001, J. Immunol. Methods, 252, 199.
121. Schnell, M.J., Foley, H.D., Siler, C.A., McGettigan, J.P., Dietzschold, B., and Pomerantz, R.J. 2000, Proc. Natl. Acad. Sci. U. S. A., 97, 3544.
122. Siler, C.A., McGettigan, J.P., Dietzschold, B., Herrine, S.K., Dubuisson, J., Pomerantz, R.J., and Schnell, M.J. 2002, Virology, 292, 24.
123. Faber, M., Lamirande, E.W., Roberts, A., Rice, A.B., Koprowski, H., Dietzschold, B., and Schnell, M.J. 2005, J. Gen. Virol., 86, 1435.
124. Schnell, M.J., Mebatsion, T., and Conzelmann, K.K. 1994, EMBO J., 13, 4195.
125. Orciari, L.A., Niezgoda, M., Hanlon, C.A., Shaddock, J.H., Sanderlin, D.W., Yager, P.A., and Rupprecht, C.E. 2001, Vaccine, 19, 4511.
126. Dietzschold, B., Faber, M., and Schnell, M.J. 2003, Expert. Rev. Vaccine, 2, 399.
127. Shoji, Y., Kobayashi, Y., Sato, G., Itou, T., Miura, Y., Mikami, T., Cunha, E.M., Samara, S.I., Carvalho, A.A., Nocitti, D.P., Ito, F.H., Kurane, I., and Sakai, T. 2004, J. Vet. Med. Sci., 66, 1271.
128. Lafon, M., Lafage, M., Martinez-Arends, A., Ramirez, R., Vuillier, F., Charron, D., Lotteau, V., and Scott-Algara, D. 1992, Nature, 358, 507.
129. Zhi, Y., Figueredo, J., Kobinger, G.P., Hagan, H., Calcedo, R., Miller, J.R., Gao, G., and Wilson, J.M. 2006, Hum. Gene Ther., 17, 500.
130. Ba, L., Yi, C.E., Zhang, L., Ho, D.D., and Chen, Z. 2007, Appl. Microbiol. Biotechnol., 76, 1131.
131. Ishii, K., Hasegawa, H., Nagata, N., Mizutani, T., Morikawa, S., Suzuki, T., Taguchi, F., Tashiro, M., Takemori, T., Miyamura, T., and Tsunetsugu-Yokota, Y. 2006, Virology, 351, 368.
132. Ishii, K., Hasegawa, H., Nagata, N., Mizutani, T., Morikawa, S., Tashiro, M., Suzuki, T., Taguchi, F., Takemori, T., Miyamura, T., and Tsunetsugu-Yokota, Y. 2006, Adv. Exp. Med. Biol., 581, 593.

133. Sui, J., Li, W., Roberts, A., Matthews, L.J., Murakami, A., Vogel, L., Wong, S.K., Subbarao, K., Farzan, M., and Marasco, W.A. 2005, J. Virol., 79, 5900.
134. van den Brink, E.N., Ter Meulen, J., Cox, F., Jongeneelen, M.A., Thijsse, A., Throsby, M., Marissen, W.E., Rood, P.M., Bakker, A.B., Gelderblom, H.R., Martina, B.E., Osterhaus, A.D., Preiser, W., Doerr, H.W., de Kruif, J., and Goudsmit, J. 2005, J. Virol., 79, 1635.
135. He, Y., Li, J., Du, L., Yan, X., Hu, G., Zhou, Y., and Jiang, S. 2006, Vaccine, 24, 5498.
136. Bisht, H., Roberts, A., Vogel, L., Subbarao, K., and Moss, B. 2005, Virology, 334, 160.
137. He, Y., Li, J., and Jiang, S. 2006, Biochem. Biophys. Res. Commun., 344, 106.
138. Bai, B., Lu, X., Meng, J., Hu, Q., Mao, P., Lu, B., Chen, Z., Yuan, Z., and Wang, H. 2008, Mol. Immunol., 45, 868.
139. Chang, C.Y., Hong, W.W., Chong, P., and Wu, S.C. 2006, Vaccine, 24, 1132.
140. Gorbalenya, A.E., Snijder, E.J., and Spaan, W.J. 2004, J. Virol., 78, 7863.
141. Ho, Y., Lin, P.H., Liu, C.Y., Lee, S.P., and Chao, Y.C. 2004, Biochem. Biophys. Res. Commun., 318, 833.
142. Mortola, E., and Roy, P. 2004, FEBS Lett., 576, 174.
143. Huang, Y., Yang, Z.Y., Kong, W.P., and Nabel, G.J. 2004, J. Virol., 78, 12557.
144. Richmond, J.F., Lu, S., Santoro, J.C., Weng, J., Hu, S.L., Montefiori, D.C., and Robinson, H.L. 1998, J. Virol., 72, p9092.
145. VanCott, T.C., Mascola, J.R., Kaminski, R.W., Kalyanaraman, V., Hallberg, P.L., Burnett, P.R., Ulrich, J.T., Rechtman, D.J., and Birx, D.L. 1997, J. Virol., 71, 4319.
146. Hu, H., Lu, X., Tao, L., Bai, B., Zhang, Z., Chen, Y., Zheng, F., Chen, J., Chen, Z., and Wang, H. 2007, Clin. Vaccine Immunol., 14, 894.
147. Vennema, H., de Groot, R.J., Harbour, D.A., Dalderup, M., Gruffydd-Jones, T., Horzinek, M.C., and Spaan, W.J. 1990, J. Virol., 64, 1407.
148. Stadler, K., and Rappuoli, R. 2005, Curr. Mol. Med., 5, 677.

Chapter 15
Influenza Virus Pathogenesis and Vaccines

Michael Vajdy

Abstract Infections with seasonally prevalent strains of influenza cause a substantial healthcare burden worldwide. Influenza gains access to the host through the mucosa of the upper respiratory tract, but exerts most of its pathologic effects in the lower respiratory tract. The threat of avian influenza strains to cause bird to human and then human to human transmission and hence pandemic influenza, is serious. Thus, while improved vaccines for seasonal influenza, particularly for infants and the elderly, are sought for, new vaccines that are effective against pandemic influenza strains are urgently needed.

15.1 Introduction

Infections with influenza virus cause considerable morbidity and mortality in the world [1]. Currently, in the USA alone, seasonal influenza infections are estimated to cost 16.3 billion dollars associated with death and illness [2]. In the USA, every year, 5–20% of the population get infected with influenza, resulting in over 200,000 hospitalizations and 36,000 deaths (WWW.CDC.gov). Influenza virus by definition belongs to the family of mucosally transmitted viruses (*Orthomyxoviridae*). While the virus enters the host primarily through the upper respiratory tract, it exerts many of its pathologic effects in the lower respiratory tract. Nonetheless, nasal secretions constitute an important pathway of transmission to and from others.

Influenza infections result in many deaths mainly in infant and the aged populations. The major symptoms include fever, headaches, malaise, dry cough, sore throat, runny nose, muscle aches, nausea, diarrhea and respiratory distress. The period of the year with the most influenza infections is mainly from November through March, with the highest number of cases seen in January and February. Infected persons can transmit the virus from 1 day before the onset of symptoms to 5 days after the onset of symptoms (WWW.CDC.gov).

M. Vajdy
University of California, Davis, School of Medicine, Department of Medical Microbiology and Immunology, Davis CA 95616, USA
e-mail: michaelvajdy@comcast.net

Immunity against influenza is generally believed to consist of both antibodies and cytotoxic T lymphocytes (CTL). However, for seasonal influenza a universally accepted correlate of protection is serum hemagglutination inhibition (HI) assay titer of over 40. Such antibody responses are targeted against the hemagglutinin (HA) portion of the virus envelope. Seasonal influenza strains from year to year possess considerable homology, albeit not sufficient to confer complete protection. Therefore, the world health organization (WHO) closely monitors incubating influenza strains and predicts each year the predominant strains (usually 3) for the following year. Equipped with such knowledge vaccine manufacturers design and develop the following year's vaccine.

While seasonal influenza causes considerable global mortality and morbidity, emergence of pandemic influenza strains are the cause of even greater concern. Pandemic influenza occurs when major segments of HA or neuraminidase (NA) of human and avian tropic influenza strains undergo major genetic reassortment through recombination in an intermediary host, which possesses receptors for both the human-and avian-tropic strains, such as swine. This renders the general population, who is completely naïve with regards to this new emergent strain, and with no pre-existing immunity, extremely susceptible to infection. The Spanish pandemic influenza caused over 50 million deaths worldwide in the beginning of the 20th century. The global impact of a pandemic influenza at the present time when travel and transportations are extremely widespread cannot be overemphasized.

Vaccinations against seasonal influenza currently consist of a single annual intramuscular (IM) injection of a cocktail of purified HA protein from egg-derived influenza viruses based on the 3 WHO recommended strains. More recently, live attenuated influenza strains for intra-nasal (IN) administration have been licensed for human use. While it is widely expected that one or two IM doses of pandemic influenza vaccines will confer protection, whether IM injections can be effective to protect against a pandemic influenza strain remains to be seen. Given the importance of IN immunizations to induce better herd immunity (local immune responses at the portal of virus entry) as well as cross-protection against heterologous strains (Secretory IgA), it may be expected that IN or combinations of IN and IM immunizations may provide optimal protection against pandemic influenza against which there is no pre-existing immunity.

15.2 Description of the Virus

Influenza (stemming from Italian/Latin *influentia* meaning epidemic) belongs to the family of *Orthomyxoviridae. Orthos means correct, and myxa means mucus in Greek*. As such, it may be argued that members of this family of viruses are essentially mucosal pathogens. *Orthomyxoviridae* are enveloped viruses, with shapes ranging from small spherical to long filamentous. The viruses contain a segmented single stranded (ss), negative strand, RNA. The negative strand serves as a template to mRNA synthesis as well as the antigenome positive strand. Unlike other members of the *Orthomyxoviridae*, influenza virus transcription and replication of its RNA occurs in the nucleus of the target cell [3].

Influenza A, B and C viruses (i.e. 3 different genera, A, B, C) can be distinguished based on their antigenic differences in the nucleocapsid (NP) and neuraminidase (NA) glycoproteins. Because the majority of respiratory diseases as well as the potential pandemic strains are caused by influenza A viruses, this chapter focuses on this particular virus. The virus is composed of a lipid bilayer covering the M1, matrix protein. Protruding from the lipid bilayer are the hemagglutinin (HA) and the neuraminidase (NA) spike glycoproteins. Subtypes of influenza are classified based on their expression of HA (H) and NA (N) genes. As such, H1N1, H2N2, H3N2, H5N1, H7N7 and H9N2 are the only subtypes isolated from humans thus far. The lipid bilayer and M1 are traversed by the M2 protein, which forms ion channels. Within the envelope are 8 segments of ssRNA contained in the form of helical ribonucleoproteins (RNP). Transcriptase and polymerase complexes are associated with the RNPs. The RNA segments encode the following gene products: Segment 1: PB (polymerase basic protein) 2; segment 2: PB1; segment 3: PA (polymerase acidic protein); segment 4: HA; segment 5: NP; segment 6: NA; segment 7: M1 and M2; segment 8: NS1 and NS2 (nuclear second; found only in infected cells).

Influenza A envelope contains a lipid envelope derived from the host's cell membrane during the budding process. The HA spikes are rod-shaped, whereas the NA spikes are mushroom-shaped. The HA binds to sialic acid receptors on respiratory epithelium, and mediates the fusion of the virus membrane to the endosomal membrane of the target cell in low pH. NA exists as a homo-tetramer and has enzymatic activity to cleave the α-ketosidic linkage between D-galactose or D-galactoseamine residues. While NA does not appear to be necessary for viral infectivity, it is believed to be important in releasing progeny virions from infected cells. Glycans ending in α-2,3-linked sialic acid gaalctose preferentially bind avian influenza strains and those ending in α-2,6-linked sialic acid galactose preferentially bind human influenza strains. A recent report further demonstrated that a two-amino acid change in the HA of the 1918 pandemic strains of influenza caused a switch in receptor binding preference from the human α-2,6 to the avian α-2,3 sialic acid residues [4]. Despite many such proofs, it was recently demonstrated that H5N1 influenza strains can bind ex vivo cultures of human nasopharyngeal, adenoid and tonsilar tissues, despite a lack of expression of α-2,3-sialic acid receptors [5]. Novel glycan micro-array technologies offer a new tool to study the ability of new emergent influenza strains to bind to host receptors and may shed light on such discrepancies [6].

15.3 Viral Entry and Pathogenesis

The first step in the viral entry is the binding of the hemagglutinin-neuraminidase (HN) to its sialic acid receptor on respiratory epithelial cells, causing the HN to undergo a conformational change. This in turn triggers a conformational change in the fusion (F) protein allowing a perturbation of the cell membrane, leading to the fusion of the host cell membrane with the viral lipid bilayer. NA may also remove sialic acid residues from SIgA and $\gamma\delta$TCRA+ T Cells in the respiratory mucosa, thus providing further pathogenesis [7].

Influenza virus replicates in both the upper and lower respiratory tracts. Virus replication peaks at 48 h after inoculation and declines slowly over the following 6 days. The peak titer of influenza is seen on the first day of symptomatic illness with an average of 10^4 tissue culture infectious dose $(TCID)_{50}/ml$ of nasal wash. While pathologic changes occur in both the upper and lower respiratory tracts, the latter are the most affected. Acute inflammation of the larynx, trachea and bronchi with mucosal inflammation and edema are common. Columnar ciliated epithelial cells become vacuolated, edematous and loose cilia before shedding off. Viral antigens are present in epithelial and mononuclear cells. In infected cells, protein synthesis is shut off and apoptosis (cell death) is induced. Necrosis of the alveoli and bronchioles ensues, complete resolution of which may take up to a month [8].

Influenza also causes death by various additional mechanisms including primary viral pneumonia, secondary bacterial pneumonia in virus-damaged lungs, and by exacerbating serious chronic diseases such as diabetes mellitus, renal diseases, and congestive heart failure.

15.4 Host Innate Responses

The respiratory epithelium contains a host of innate defense mechanisms comprising large proteins such as lysozyme and lactoferrin as well as cathelicidin and cationic defensin peptides (including e.g. the human betadefensin-1) [9]. Salivary agglutinin and lung scavenger receptor glycoprotein 340 display anti-infleunza activities [10]. Human neutrophil pepride 1, a member of the alpha-defensins which are cationic antimicrobial peptides produced by polymorphonuclear neutrophils, inhibits influenza virus infections in vitro [11]. The interferon-induced resistance factor Mx1 mediated protection against pandemic 1918 and H5N1 strains in mice [12]. A family of host defense lectins, called collectins, including surfactant-associated proteins SP-A and SP-D opsonize and enhance microbial (including influenza) killing [13]. Specifically, SP-A and SP-D have been shown to inhibit the neuraminidase activity of influenza [14]. Moreover, the mannan-binding lectin (MBL) is a C-type serum collectin that can inhibit infection of mammalian cells with influenza virus in vitro [15]. A role of complement components, e.g. CD59a, has also been suggested in defense against influenza in that lack of this protein was shown to exacerbate influenza-induced lung inflammation through complement dependent pathways [16].

TLR7/8 appear to play an important role in protection following IN challenge of rats, and this was dependent on IFNα and TNFα production [17]. TLR7 is recognized by influenza virus [18]. TLR2 does not appear to be involved in host response during post-influenza pneumococcal pneumonia [19]. Infection of cynomulgus macaques with the 1918 pandemic influenza strain induced strong serum levels of CCL2 (monocyte chemotactic protein-1), CCL5 (RANTES), IL-8 and

IL-6 by day 6 post infection [20]. Infection of murine alveolar epithelial cells with influenza virus induced the release of monocyte chemoattractants CCL2 and CCL5 followed by monocyte transepithelial migration, and this was dependent on expression of CCR2 but not CCR5 [21]. Exposure of human monocyte derived DC to influenza induced the expression of a number of genes measured by microarray analysis. These included IL-8, TNFα, IL-1β, IL-6, prostaglandin/leukotriene-associated genes, type I interferons, CD86, ICAM-1, IL-15R, IL-7R, IL-3R, CD83, and NFκB p50, p52 and p65 [22].

Neutrophils and macrophages accumulate in the lungs following influenza challenge and TLR4 may also play a role in viral pathogenesis [23]. Influenza infection results in rapid upregulation of TLR2 on neutrophils [24]. TLR3 has also been implicated in the inflammatory responses following influenza infection of lung epithelial cells [25]. As HA also binds to sialic acid receptors on NK cells, these cells are activated by influenza [26]. Important studies in mice demonstrated that in the absence of an NK receptor (Ncr1 in mice and NKpg46 in humans), influenza infection was lethal [27, 28]. In this regard, IL-18 may augment the NK-mediated cytotoxicity against influenza-infected cells [29]. The monocyte attractant protein 1 (MCP-1) has been found to be important for immunity against influenza in mice lacking this protein. While wild type mice have increased levels of MCP-1 in their lungs after influenza challenge, the MCP-1 deficient mice had enhanced weight-loss, elevated viral loads, reduced leukocyte (mainly macrophages and granulocytes) recruitment into lungs and reduced pulmonary IgA levels [30]. Activation of TLR3 on respiratory epithelial cells by influenza virus may mediate signaling through TRIF (TLR associated adaptor molecule) but not MyD88, leading to NfκB activation and ultimately secretion of IL-8, IL-6, RANTES and IFNβ as well as upregulation of ICAM-1 [31].

Anti-influenza humoral responses were shown to be induced by direct stimulation of TLR on B cells through production of IFNα and increased IgG2a/c switching [32]. Moreover, IFNα production by lung B cells occurred within 48 h post influenza infection, causing an arrest in clonal expansion of B cells in regional lymph nodes [33]. Because Th-deficient mice controlled influenza infection better than combined Th/B cell deficient mice, a Th-independent role of B cells in influenza virus clearance has been suggested [34]. Natural antibody, defined as IgM antibodies that bind to antigens that the host has not previously been exposed to, together with complement can mediate neutralization of influenza virus [35]. Such natural IgM producing B cells apparently do increase IgM production or undergo isotype switch to IgG2a upon influenza infection [36]. Type I IFN receptors on B cells stimulate early B cell responses in lungs through upregulation of the activation molecules CD69 and CD86 [37].

Recently, it was shown that IL-18, but not IL-12, is important for the development of anti-influenza CD8+ T cells [38]. Earlier reports attributed an important role for IL-12 to early (days 3 and 5 post infection) NK cell dependent IFNγ production but not late (day 7 post infection) T cell dependent IFNγ production [39]. Interestingly, it was demonstrated that early on following influenza re-infection NK-mediated IFNγ production was enhanced by pre-existing virus specific IL-2 producing T cells [40], thus suggesting a positive feed-back loop between memory type and innate immunity.

15.5 Correlates of Protection

Presence of serum HI activity against prevalent influenza viruses strongly correlate with protection from disease and an essential role of B cells in heterosubtypic cross protection against lethal Influenza A H5N1 virus infection has been reported [41]. Although in murine models, a controversial role of mucosal IgA has also been suggested [42], many murine and human studies support the importance of mucosal IgA responses in protection against influenza infection and disease. In this regard, of particular importance have been the findings in both murine and human studies that mucosal IgA induced by intra-nasal immunization, as opposed to serum IgG induced by systemic immunization, protected against multiple strains of influenza virus [41, 43, 44, 45, 46, 47, 48, 49, 50, 51, 52]. Therefore, recent efforts have focused on intra-nasal immunization strategies that induce both local IgA and systemic IgG responses [51, 52, 53].

15.6 T and B Cell Mediated Adaptive Immunity Against Influenza

The adaptive immune responses against influenza include antibodies produced by B cells, cytotoxic granules produced by both CD4+ and CD8+ T cells and helper cytokines for both B and T cells produced by CD4+ T cells. It is believed that CD8+ T cells or B cells can each independently control influenza infections. Thus, mice lacking either CD8+ T cells or B cells can survive influenza challenge, while mice lacking both CD8+ and B cells succumb to infection.

15.6.1 T Cell-Mediated Responses

15.6.1.1 CD4+ T Cell-Mediated Responses

Following intra-nasal challenge with live influenza virus, HA RNA was first detected in mediastinal lymph node (MLN) by 48 h post infection, and continued the expression until days 6 and 10 post infection in both CD4+ T cell-depleted and intact mice [54]. Influenza-specific CD4+ effector T cells protect against influenza by both a cytotoxic perforin-mediated and promotion of antibody-mediated responses [55]. The signaling lymphocytic activation molecule associated protein (SAP) appears to be required for T helper function in protection against influenza [56]. Influenza-specific IFNγ-expressing CD4+ T cells in the lungs lacked the expression of the chemokine receptor CCR7, and thus homing of effector CD4+ T cells towards chemokines that bind this receptor is not required [57].

Early on, the importance of CD4+ T cell help for cytotoxic CD8+ T cell responses was debated [58]. Later, it was demonstrated that primary and secondary influenza-virus specific CD8+ T cell responses were clearly decreased in lungs, and furthermore, delayed virus clearance and increased morbidity were observed in CD4-depleted mice [59, 60]. Collagen, which is abundant and accounts for about 15% of dry weight of lungs, expresses both of the collagen binding integrins $\alpha1\beta1$ and $\alpha2\beta1$. While CD4+ T cells express CD49b, CD8+ T cells express CD49a and thus these two T cell subsets differ in their binding to the interstitial environment of the lungs [61]. In the absence of MHCII-expressing lung epithelial cells, monocyte/macrophages and B cells can interact with CD4+ T Cells for virus clearance [62]. Although influenza neuraminidase-primed CD4+ T cells may traffic to the lungs, they do not proliferate in the absence of antigen [63]. In the absence of B cells, CD4+ T cells are unable to clear an influenza infection [64, 65]. Interestingly, CD4-independent influenza-specific IgG responses can occur during primary infection of CD40-deficient or MCHII-deficient mice [66]. Curiously, in MHCII-deficient or CD40-deficient mice, early during influenza infection, a CD4-dependent IgA response is generated which does not depend on cognate B cell-T cell interactions. However, in support of the previous findings, in these mice, influenza-specific IgM or IgG responses were not generated [67].

Antigen-presentation and antigen–antibody complexes on follicular Dendritic cells are not essential for persistence of CD4+ memory T cells [68]. Generation and maintenance of influenza-specific long-term CD4+ memory T cell responses depends on the antigen dose, duration of repeated interactions with antigen-presenting cells, and the micro-environment of inflammatory and growth cytokines around the CD4+ T cells [69]. Following influenza infection, mRNA expression of IL-4 and IL-10 occurs in CD4+ T cells first in MLN and later in lungs [70]. After primary challenge of rats CD4+ T cells expand in both MLN and bronchoalveolar lavage (BAL) and subsequent infection causes further expansion [71].

15.6.1.2 CD8+ T Cell-Mediated Responses

Following intra-nasal inoculation with influenza virus, peak virus titers were detected in the MLN at 2 days post infection, while influenza-specific CD8+ T cells were detected in MLN at 1 day post infection. Although Dendritic cells (DC), macrophages and B cells isolated from the infected mice all had the capacity to activate an influenza-specific T cell hybridoma in an MHCI-restricted manner, the DC were deemed to be responsible as the primary source of antigen-presentation [72].

CD8+ cytotoxic T cell responses against influenza are generally induced against conserved, internal proteins, such as polymerase and nucleoproteins [73]. IFNγ and TNFα mRNA are found in CD8+ T cells in MLN following influenza infection [70]. Both Fas and perforin-dependent cytotoxicity is required for clearance of influenza from the lungs [74]. Acute heterosubtypic CTL responses in spleen, cervical lymph nodes (CLN) and MLN were induced following pulmonary priming with nonpathogenic influenza strain Udorn (H3N2), and challenge with the mouse adapted pathogenic PR/8/34 strain (H1N1), while mucosal memory CTL responses were highly dependent on mucosal route of priming [75]. While presence of influenza-derived antigens is required to activate CD8+ T cells in the lungs, subsequent localization of influenza-specific cells in the lungs is not antigen-driven [76]. The magnitude of the recall response against influenza is strongly CD4+ T cell dependent [60]. However, the long-term (200 days) distribution of memory T cell clones in blood, spleen, regional LN, bone marrow and lungs appears to be random [77]. Adoptive transfer of influenza A nucleoprotein epitope-specific CD8+ T cells into naïve mice, and subsequent challenge with influenza B, indicated that recall responses against influenza may involve both naïve and memory CD8+ T cells [78]. However, the specificity of CD8+ T cell responses can substantially vary in primary vs. secondary influenza infections, a phenomenon that depends on the participation of both DC and non-DC cells in antigen-presentation during the primary response and mostly DC cells during the secondary response [79]. The size of the pre-existing antigen-specific CD8+ T cell pool as well as the amount of antigen and mode of antigen-presentation can all affect the size of the endogenous recall CD8+ T cell responses [80].

Prior infection with a persistent virus, i.e. gammaherpesvirus68, reduced the numbers of influenza-specific CD8+ T cells by 50% [81]. Following primary influenza infection, treatment of mice with antibodies against CD62L, forced the homing of both CD4+ and CD8+ T cells from MLN to spleen, and further splenectomy of these mice induced their migration to bone marrow [82]. Virus-specific cells in lungs are more activated, are more cytolytic and secrete more cytokines compared to virus-specific cells in spleen. This is observed during influenza-induced pneumonia and is due to the higher antigenic load at this site [83]. Lung CD8+ T cells specific for influenza virus nucleocapsid and acid polymerase-derived epitopes produced IL-2, TNFα and IFNγ [84]. In vaccinated humans, higher numbers of IFNγ-producing NK and T cells were detected in peripheral blood following vaccination with live attenuated virus compared to inactivated influenza virus [85]. The collagen binding $\alpha 1\beta 1$ integrin, which is important for CD4+ T cell binding to lung interstitial microenvironment [61], also regulates CD8+ T cell responses against influenza [86].

15.6.2 B Cell-Mediated Responses

In the absence of antibodies and B cells, mice can clear influenza infection through the actions of CD4+ and CD8+ T cells [87]. B cell immunity against influenza includes antigen-presentation to CD4+ T cells, which then provide help for both B cells and CD8+ cytotoxic T cells, as well as antigen-presentation for CD8+ T cells. However, the main function of B cells in immunity against influenza is production of antibodies. Three different classes of antibodies participate in protection against influenza. These include IgM, IgG and IgA.

Surprisingly, unmutated IgM antibodies protected mice from lethal primary and secondary infection with influenza [88]. These were seemingly not stemming from natural IgM sources, as they were produced de novo after infection. Natural antibodies are produced in the complete absence of antigenic stimulation by B-1 cells, with selective antigenic specificities that have evolutionarily risen to respond to pathogens in the absence of an as yet undeveloped adaptive response [89].

While it is generally believed that influenza-specific IgG antibody responses are CD4 and CD40-dependent, mice lacking CD40 or CD4 developed anti-influenza IgG responses and recovered from primary influenza infection similar to intact mice. However, mice lacking B cells were not protected against primary influenza infection [66]. The majority of influenza HA-specific antibody secreting cells in draining LN of influenza challenged mice are CD19+IgD-CD23-CD24highCD38low germinal center B cells [90]. The first stimulatory signal the local B cells receive following influenza infection is through their type I IFNR [37]. Influenza infection of mice caused depletion of bone marrow B cell lineage cells due to apoptosis of early B cells through actions of TNFα and lymphotoxin α. Moreover, prior infection of mice with lymphocytic choriomeningitis virus, reduced influenza-specific serum antibody titers following a subsequent influenza infection [91].

The role of IgA in mucosal immunity against influenza, although favored, is controversial. IgA-deficient mice were protected against IN influenza challenge similar to wild type mice [92]. This is not surprising given the fact that even B cell-deficient mice survive an IN influenza challenge. Conversely, lack of SIgA in polymeric Ig receptor (pIgR) deficient mice resulted in reduced protection and cross-protection, eventhough serum IgG and IgA transudates were present in nasal secretions [49]. The role of SIgA in protection against influenza may be complementary and particularly important in memory type responses. Moreover, although it is thought that induction of local IgA through IN vaccinations may induce better cross-reactivity with heterologous influenza strains (i.e. those not included in the vaccine) [49], this remains to be conclusively demonstrated in humans.

15.7 Vaccines Against Influenza

Commercially available inactivated whole- and split-virus vaccines have been successful to prevent disease caused by influenza infection [93, 94], and a licensed IN cold adapted live attenuated influenza vaccine is currently in use [51, 53]. However, these vaccines have not convincingly shown efficacy in generating long-lasting immunity, particularly in the elderly, and may not be sufficiently cross-reactive to protect against antigenic variants [95, 96, 97]. Although the intra-muscularly injected vaccines are known to induce serum immunoglobulin G (IgG) antibodies, they are poor stimulators of secretory IgA at respiratory mucosal sites and show sporadic $CD8^+$ cytotoxic T-lymphocyte (CTL) activation [1, 98, 99]. Efforts are currently under way to develop influenza vaccines that generate significant secretory IgA, as well as maintain high serum IgG titers, by exploiting mucosal immunization [95, 100, 101, 102].

The role of intra-nasal vs. systemic immunization and induction of local IgA as opposed to serum IgG in protection against replication in the nose, the lung or protection from disease is well established [103, 104, 105, 106]. Therefore, IN immunizations alone or in combination with IM immunization may show superior local and systemic antibody responses. For optimal induction of immune responses through the IN route, effective and safe mucosal adjuvants are required. Mutants of the heat labile enterotoxin from enteropathogenic E. coli have been shown to be safe in animal and human studies [107]. LTK63 is an effective mucosal adjuvant with no detectable toxic ADP-ribosyltransferase activity [107], and holds promise as a mucosal adjuvant.

We recently reported that mucosal followed by parenteral immunizations with Helicobacter pylori derived protein antigens induced enhanced local and systemic responses compared to mucosal alone, parenteral alone, or parenteral followed by mucosal immunizations [108]. Moreover, data from a rhesus macaque study suggested that IN followed IM immunizations significantly enhanced serum and vaginal antibody responses against HIV-env [109]. Therefore, in a recent study we tested whether the combination of IN followed IM immunization with cell culture derived HA from two strains of influenza A and a strain of influenza B virus induced enhanced local and systemic immune responses compared to IM followed by IN, IM alone or IN alone immunizations. Local responses were measured in cervical lymph nodes (CLN), which are widely accepted to drain the nasal mucosa [110, 111, 112, 113], as well as in nasal washes. One intranasal (IN) followed by one intramuscular (IM) immunizations with a combination of cell culture produced hemagglutinin (HA) antigens derived from 3 different influenza strains induced significantly higher serum hemagglutination inhibition (HI) and serum IgG antibody titers as well as T cell responses, compared to 2 IM, 2 IN or 1 M followed by 1 IN immunizations. Moreover, while 2 IM immunizations did not induce any antibody responses in nasal secretions or cervical lymph nodes, which drain the nasal mucosa, IN immunizations alone or in combination with IM immunization induced mucosal and local responses. These data showed that the IN followed by IM immunization strategy holds promise to significantly raise serum and local antibody and T cell responses against seasonal influenza strains, and possibly pandemic influenza strains, for which no pre-existing immunity exists [114].

A variety of immunopotentiating adjuvants and delivery systems have been used to enhance the efficacy of mucosal and systemic experimental influenza vaccines against homologous as well as heterologous strains. These include liposomes, oil in water emulsions, immunostimulating complexes (ISCOMS), heat labile toxin from E. coli (LT) and its mutated derivatives, cholera toxin (CT) and its mutated derivatives, CpG oligonucleotides, plasmid DNA and RNA (e.g. alphavirus-based replicon particles). Of these the oil in water emulsion (MF59) has been shown to be safe and effective for IM immunizations.

A word of caution regarding the use of immunostimulating adjuvants for intranasal vaccinations is warranted. In 1997, a Swiss company (Berna Biotech) marketed an IN vaccine against influenza consisting of HA and NA in liposomes mixed with wild type LT. While this had proved efficacious in clinical trials, post marketing there was a close link between cases of Bell's palsy (Bp; temporary partial facial paralysis) and vaccinations with this vaccine [115]. Hence this product was withdrawn from the market. It has since been shown that CT (and possibly LT) can access the olfactory bulb and bind to neuronal cells, hence causing Bp either directly or through activation of dormant viruses such as herpes virus. While recent research indicates that for the neurotoxic effects of these toxins via IN administration both receptor binding (monoganglioside-1 (GM1)) and toxicity (ADP-ribosyltransferase activity) should occur, it remains to be seen how safe their nontoxic moieties prove to be in large-scale clinical trials.

The need to continuously monitor for changes in the HA membrane protein of seasonal influenza virus strains have lead to the idea that design of a vaccine based on the portions of the virus that do not undergo such high rates of mutations may be possible and obviously more advantageous. Several attempts have been made to design a universal vaccine against influenza based on the transmembrane viral M2 protein [116]. Another approach for universal influenza vaccine is through the design of plasmid DNA encoding NP and/or M1 [117].

15.8 Preparing for a Potential Pandemic Influenza

Epidemics of influenza occur when small mutations in the HA and NA genes creating new antigenic epitopes, called antigenic drift, render existing neutralizing antibodies against previous epitopes obsolete. Pandemic influenza occurs when major segments of HA or NA of human and avian tropic influenza strains undergo major genetic reassortment through recombination in an intermediary host, which possesses receptors for both the human-and avian-tropic strains, such as swine. All major pandemic influenza strains in the 20th century were derived from influenza A strains. The pandemic that caused the most death, up to 50 million, was the Spanish influenza in 1918 caused by an H1N1 influenza strain. The other major pandemics were the Asian (1957) and Hong Kong (1968–1969), each resulting in an estimated 1–2 million deaths. Currently, the H5N1 strain has resulted in the death of millions of domestic and wild birds. Moreover, this strain has resulted in bird to human transmissions and death of several hundred humans. It is suspected that once the virus can cause direct human to human transmission, the pandemic will be imminent. This is likely due to the considerable difference in the antigenicity of a mutated H5N1, and no pre-existing immunity in the general population, compared to the more common seasonal influenza strains, for which there is a level of pre-existing immunity. Therefore, there is an urgent need to establish vaccines and immunization protocols that can prevent an influenza pandemic.

Several companies have produced mock vaccines against putative pandemic strains of influenza. Vaccination against pandemic influenza is arguably the best means of protection against the disease, and animal studies suggest the vaccines could prove efficacious. The generally accepted protective serum HI titer value of 1:40 against seasonal influenza strains remains to be shown to hold true for pandemic influenza strains. Recent human studies suggest that IN immunizations provide better cross-protective responses against diverse strains of seasonal influenza, and it is plausible that the same may hold true for pandemic strains of influenza. However, the use of better adjuvants or delivery systems for systemic vaccinations may also improve cross-protective responses. There are also several antiviral drugs being stockpiled that can reduce the severity of disease caused by pandemic influenza. Major investments have been made by both governments and! large vaccine manufacturing companies alike to complete additional manufacturing facilities for vaccines against both seasonal and pandemic influenza. The current major focus of these efforts is to transform the egg-based to cell-based manufacturing technology, in order to save precious processing and development time once imminent pandemic strains are officially announced by the WHO.

References

1. Glezen, W.P. 1982, Epidemiol Rev, 4, 25.
2. Molinari, N.A., Ortega-Sanchez, I.R., Messonnier, M.L., Thompson, W.W., Wortley, P.M., Weintraub, E., and Bridges, C.B. 2007, Vaccine, 25, 5086.
3. Cheung, T.K., and Poon, L.L. 2007, Ann NY Acad Sci, 1102, 1.
4. Tumpey, T.M., Maines, T.R., Van Hoeven, N., Glaser, L., Solorzano, A., Pappas, C., Cox, N.J., Swayne, D.E., Palese, P., Katz, J.M., and Garcia-Sastre, A. 2007, Science, 315, 655.
5. Nicholls, J.M., Chan, M.C., Chan, W.Y., Wong, H.K., Cheung, C.Y., Kwong, D.L., Wong, M.P., Chui, W.H., Poon, L.L., Tsao, S.W., Guan, Y., and Peiris, J.S. 2007, Nat Med, 13, 147.
6. Stevens, J., Blixt, O., Paulson, J.C., and Wilson, I.A. 2006, Nat Rev Microbiol, 4, 857.
7. Bhatia, A., and Kast, R.E. 2007, Cell Mol Biol Lett, 12, 111.
8. Wright, P.F., and Webster, R.G. 2001, Fields virology, D.M. Knipe (Ed.), Lippincott Williams & Wilkins, Philadelphia, PA, 1533.
9. Laube, D.M., Yim, S., Ryan, L.K., Kisich, K.O., and Diamond, G. 2006, Curr Top Microbiol Immunol, 306, 153.
10. Hartshorn, K.L., Ligtenberg, A., White, M.R., Van Eijk, M., Hartshorn, M., Pemberton, L., Holmskov, U., and Crouch, E. 2006, Biochem J, 393, 545.
11. Salvatore, M., Garcia-Sastre, A., Ruchala, P., Lehrer, R.I., Chang, T., and Klotman, M.E. 2007, J Infect Dis, 196, 835.
12. Tumpey, T.M., Szretter, K.J., Van Hoeven, N., Katz, J.M., Kochs, G., Haller, O., Garcia-Sastre, A., and Staeheli, P. 2007, J Virol, 81, 10818.
13. Crouch, E., Hartshorn, K., and Ofek, I. 2000, Immunol Rev, 173, 52.
14. Tecle, T., White, M.R., Crouch, E.C., and Hartshorn, K.L. 2007, Arch Virol, 152, 1731.
15. Kase, T., Suzuki, Y., Kawai, T., Sakamoto, T., Ohtani, K., Eda, S., Maeda, A., Okuno, Y., Kurimura, T., and Wakamiya, N. 1999, Immunology, 97, 385.
16. Longhi, M.P., Williams, A., Wise, M., Morgan, B.P., and Gallimore, A. 2007, Eur J Immunol, 37, 1266.
17. Hammerbeck, D.M., Burleson, G.R., Schuller, C.J., Vasilakos, J.P., Tomai, M., Egging, E., Cochran, F.R., Woulfe, S., and Miller, R.L. 2007, Antiviral Res, 73, 1.
18. Lund, J.M., Alexopoulou, L., Sato, A., Karow, M., Adams, N.C., Gale, N.W., Iwasaki, A., and Flavell, R.A. 2004, Proc Natl Acad Sci USA, 101, 5598.
19. Dessing, M.C., van der Sluijs, K.F., Florquin, S., Akira, S., and van der Poll, T. 2007, Am J Respir Cell Mol Biol, 36, 609.
20. Kobasa, D., Jones, S.M., Shinya, K., Kash, J.C., Copps, J., Ebihara, H., Hatta, Y., Kim, J.H., Halfmann, P., Hatta, M., Feldmann, F., Alimonti, J.B., Fernando, L., Li, Y., Katze, M.G., Feldmann, H., and Kawaoka, Y. 2007, Nature, 445, 319.

21. Herold, S., von Wulffen, W., Steinmueller, M., Pleschka, S., Kuziel, W.A., Mack, M., Srivastava, M., Seeger, W., Maus, U.A., and Lohmeyer, J. 2006, J Immunol, 177, 1817.
22. Huang, Q., Liu, D., Majewski, P., Schulte, L.C., Korn, J.M., Young, R.A., Lander, E.S., and Hacohen, N. 2001, Science, 294, 870.
23. Hashimoto, Y., Moki, T., Takizawa, T., Shiratsuchi, A., and Nakanishi, Y. 2007, J Immunol, 178, 2448.
24. Lee, R.M., White, M.R., and Hartshorn, K.L. 2006, Scand J Immunol, 63, 81.
25. Le Goffic, R., Pothlichet, J., Vitour, D., Fujita, T., Meurs, E., Chignard, M., and Si-Tahar, M. 2007, J Immunol, 178, 3368.
26. Owen, R.E., Yamada, E., Thompson, C.I., Phillipson, L.J., Thompson, C., Taylor, E., Zambon, M., Osborn, H.M., Barclay, W.S., and Borrow, P. 2007, J Virol, 81, 11170.
27. Gazit, R., Gruda, R., Elboim, M., Arnon, T.I., Katz, G., Achdout, H., Hanna, J., Qimron, U., Landau, G., Greenbaum, E., Zakay-Rones, Z., Porgador, A., and Mandelboim, O. 2006, Nat Immunol, 7, 517.
28. Draghi, M., Pashine, A., Sanjanwala, B., Gendzekhadze, K., Cantoni, C., Cosman, D., Moretta, A., Valiante, N.M., and Parham, P. 2007, J Immunol, 178, 2688.
29. Liu, B., Mori, I., Hossain, M.J., Dong, L., Takeda, K., and Kimura, Y. 2004, J Gen Virol, 85, 423.
30. Dessing, M.C., van der Sluijs, K.F., Florquin, S., and van der Poll, T. 2007, Clin Immunol, 7, 7.
31. Guillot, L., Le Goffic, R., Bloch, S., Escriou, N., Akira, S., Chignard, M., and Si-Tahar, M. 2005, J Biol Chem, 280, 5571.
32. Heer, A.K., Shamshiev, A., Donda, A., Uematsu, S., Akira, S., Kopf, M., and Marsland, B.J. 2007, J Immunol, 178, 2182.
33. Chang, W.L., Coro, E.S., Rau, F.C., Xiao, Y., Erle, D.J., and Baumgarth, N. 2007, J Immunol, 178, 1457.
34. Mozdzanowska, K., Maiese, K., and Gerhard, W. 2000, J Immunol, 164, 2635.
35. Jayasekera, J.P., Moseman, E.A., and Carroll, M.C. 2007, J Virol, 81, 3487.
36. Baumgarth, N., Herman, O.C., Jager, G.C., Brown, L., and Herzenberg, L.A. 1999, Proc Natl Acad Sci USA, 96, 2250.
37. Coro, E.S., Chang, W.L., and Baumgarth, N. 2006, J Immunol, 176, 4343.
38. Denton, A.E., Doherty, P.C., Turner, S.J., and La Gruta, N.L. 2007, Eur J Immunol, 37, 368.
39. Monteiro, J.M., Harvey, C., and Trinchieri, G. 1998, J Virol, 72, 4825.
40. He, X.S., Draghi, M., Mahmood, K., Holmes, T.H., Kemble, G.W., Dekker, C.L., Arvin, A.M., Parham, P., and Greenberg, H.B. 2004, J Clin Invest, 114, 1812.
41. Tumpey, T.M., Renshaw, M., Clements, J.D., and Katz, J.M. 2001, J Virol, 75, 5141.
42. Zhang, Y., Pacheco, S., Acuna, C.L., Switzer, K.C., Wang, Y., Gilmore, X., Harriman, G.R., and Mbawuike, I.N. 2002, Immunology, 105, 286.
43. Asahi-Ozaki, Y., Yoshikawa, T., Iwakura, Y., Suzuki, Y., Tamura, S., Kurata, T., and Sata, T. 2004, J Med Virol, 74, 328.
44. Asahi, Y., Yoshikawa, T., Watanabe, I., Iwasaki, T., Hasegawa, H., Sato, Y., Shimada, S., Nanno, M., Matsuoka, Y., Ohwaki, M., Iwakura, Y., Suzuki, Y., Aizawa, C., Sata, T., Kurata, T., and Tamura, S. 2002, J Immunol, 168, 2930.
45. Tamura, S., Ito, Y., Asanuma, H., Hirabayashi, Y., Suzuki, Y., Nagamine, T., Aizawa, C., and Kurata, T. 1992, J Immunol, 149, 981.
46. Asanuma, H., Koide, F., Suzuki, Y., Nagamine, T., Aizawa, C., Kurata, T., and Tamura, S. 1995, Vaccine, 13, 3.
47. Tamura, S.I., Samegai, Y., Kurata, H., Kikuta, K., Nagamine, T., Aizawa, C., and Kurata, T. 1989, Vaccine, 7, 257.
48. Tamura, S.I., Asanuma, H., Ito, Y., Hirabayashi, Y., Suzuki, Y., Nagamine, T., Aizawa, C., Kurata, T., and Oya, A. 1992, Eur J Immunol, 22, 477.
49. Brandtzaeg, P. 2003, Dev Biol (Basel), 115, 39.
50. Tamura, S., Tanimoto, T., and Kurata, T. 2005, Jpn J Infect Dis, 58, 195.

51. Mendelman, P.M., Rappaport, R., Cho, I., Block, S., Gruber, W., August, M., Dawson, D., Cordova, J., Kemble, G., Mahmood, K., Palladino, G., Lee, M.S., Razmpour, A., Stoddard, J., and Forrest, B.D. 2004, Pediatr Infect Dis J, 23, 1053.
52. Lee, M.S., and Yang, C.F. 2003, J Infect Dis, 188, 1362.
53. Harper, S.A., Fukuda, K., Cox, N.J., and Bridges, C.B. 2003, MMWR Recomm Rep, 52, 1.
54. Eichelberger, M.C., Wang, M.L., Allan, W., Webster, R.G., and Doherty, P.C. 1991, J Gen Virol, 72, 1695.
55. Brown, D.M., Dilzer, A.M., Meents, D.L., and Swain, S.L. 2006, J Immunol, 177, 2888.
56. Kamperschroer, C., Dibble, J.P., Meents, D.L., Schwartzberg, P.L., and Swain, S.L. 2006, J Immunol, 177, 5317.
57. Debes, G.F., Bonhagen, K., Wolff, T., Kretschmer, U., Krautwald, S., Kamradt, T., and Hamann, A. 2004, J Virol, 78, 7528.
58. Allan, W., Tabi, Z., Cleary, A., and Doherty, P.C. 1990, J Immunol, 144, 3980.
59. Riberdy, J.M., Christensen, J.P., Branum, K., and Doherty, P.C. 2000, J Virol, 74, 9762.
60. Belz, G.T., Wodarz, D., Diaz, G., Nowak, M.A., and Doherty, P.C. 2002, J Virol, 76, 12388.
61. Richter, M., Ray, S.J., Chapman, T.J., Austin, S.J., Rebhahn, J., Mosmann, T.R., Gardner, H., Kotelianski, V., deFougerolles, A.R., and Topham, D.J. 2007, J Immunol, 178, 4506.
62. Topham, D.J., Tripp, R.A., Sarawar, S.R., Sangster, M.Y., and Doherty, P.C. 1996, J Virol, 70, 1288.
63. Chapman, T.J., Castrucci, M.R., Padrick, R.C., Bradley, L.M., and Topham, D.J. 2005, Virology, 340, 296.
64. Mozdzanowska, K., Furchner, M., Maiese, K., and Gerhard, W. 1997, Virology, 239, 217.
65. Topham, D.J., and Doherty, P.C. 1998, J Virol, 72, 882.
66. Lee, B.O., Rangel-Moreno, J., Moyron-Quiroz, J.E., Hartson, L., Makris, M., Sprague, F., Lund, F.E., and Randall, T.D. 2005, J Immunol, 175, 5827.
67. Sangster, M.Y., Riberdy, J.M., Gonzalez, M., Topham, D.J., Baumgarth, N., and Doherty, P.C. 2003, J Exp Med, 198, 1011.
68. Topham, D.J., Tripp, R.A., Hamilton-Easton, A.M., Sarawar, S.R., and Doherty, P.C. 1996, J Immunol, 157, 2947.
69. Swain, S.L., Agrewala, J.N., Brown, D.M., Jelley-Gibbs, D.M., Golech, S., Huston, G., Jones, S.C., Kamperschroer, C., Lee, W.H., McKinstry, K.K., Roman, E., Strutt, T., and Weng, N.P. 2006, Immunol Rev, 211, 8.
70. Carding, S.R., Allan, W., McMickle, A., and Doherty, P.C. 1993, J Exp Med, 177, 475.
71. Eichelberger, M.C., Bauchiero, S., Point, D., Richter, B.W., Prince, G.A., and Schuman, R. 2006, Cell Immunol, 243, 67.
72. Hamilton-Easton, A., and Eichelberger, M. 1995, J Virol, 69, 6359.
73. Webby, R.J., Andreansky, S., Stambas, J., Rehg, J.E., Webster, R.G., Doherty, P.C., and Turner, S.J. 2003, Proc Natl Acad Sci USA, 100, 7235.
74. Topham, D.J., Tripp, R.A., and Doherty, P.C. 1997, J Immunol, 159, 5197.
75. Nguyen, H.H., Moldoveanu, Z., Novak, M.J., van Ginkel, F.W., Ban, E., Kiyono, H., McGhee, J.R., and Mestecky, J. 1999, Virology, 254, 50.
76. Topham, D.J., Castrucci, M.R., Wingo, F.S., Belz, G.T., and Doherty, P.C. 2001, J Immunol, 167, 6983.
77. Turner, S.J., Diaz, G., Cross, R., and Doherty, P.C. 2003, Immunity, 18, 549.
78. Turner, S.J., Cross, R., Xie, W., and Doherty, P.C. 2001, J Immunol, 167, 2753.
79. Crowe, S.R., Turner, S.J., Miller, S.C., Roberts, A.D., Rappolo, R.A., Doherty, P.C., Ely, K.H., and Woodland, D.L. 2003, J Exp Med, 198, 399.
80. Jenkins, M.R., Webby, R., Doherty, P.C., and Turner, S.J. 2006, J Immunol, 177, 2917.
81. Liu, H., Andreansky, S., Diaz, G., Turner, S.J., Wodarz, D., and Doherty, P.C. 2003, J Virol, 77, 7756.
82. Tripp, R.A., Topham, D.J., Watson, S.R., and Doherty, P.C. 1997, J Immunol, 158, 3716.
83. Marshall, D.R., Olivas, E., Andreansky, S., La Gruta, N.L., Neale, G.A., Gutierrez, A., Wichlan, D.G., Wingo, S., Cheng, C., Doherty, P.C., and Turner, S.J. 2005, Proc Natl Acad Sci USA, 102, 6074.

84. La Gruta, N.L., Turner, S.J., and Doherty, P.C. 2004, J Immunol, 172, 5553.
85. He, X.S., Holmes, T.H., Zhang, C., Mahmood, K., Kemble, G.W., Lewis, D.B., Dekker, C.L., Greenberg, H.B., and Arvin, A.M. 2006, J Virol, 80, 11756.
86. Ray, S.J., Franki, S.N., Pierce, R.H., Dimitrova, S., Koteliansky, V., Sprague, A.G., Doherty, P.C., de Fougerolles, A.R., and Topham, D.J. 2004, Immunity, 20, 167.
87. Epstein, S.L., Lo, C.Y., Misplon, J.A., and Bennink, J.R. 1998, J Immunol, 160, 322.
88. Harada, Y., Muramatsu, M., Shibata, T., Honjo, T., and Kuroda, K. 2003, J Exp Med, 197, 1779.
89. Baumgarth, N., Tung, J.W., and Herzenberg, L.A. 2005, Springer Semin Immunopathol, 26, 347.
90. Doucett, V.P., Gerhard, W., Owler, K., Curry, D., Brown, L., and Baumgarth, N. 2005, J Immunol Methods, 303, 40.
91. Borrow, P., Hou, S., Gloster, S., Ashton, M., and Hyland, L. 2005, Eur J Immunol, 35, 524.
92. Mbawuike, I.N., Pacheco, S., Acuna, C.L., Switzer, K.C., Zhang, Y., and Harriman, G.R. 1999, J Immunol, 162, 2530.
93. Ghendon, Y. 1989, Adv Exp Med Biol, 257, 37.
94. Riddiough, M.A., Sisk, J.E., and Bell, J.C. 1983, JAMA, 249, 3189.
95. Clements, M.L., and Stephens, I. 1997, New Generation Vaccines, M.M. Levine, G.C. Woodrow, J.B. Kaper and G.S. Cobon (Eds.), Marcel Dekker, Inc., New York, 545.
96. Hoskins, T.W., Davies, J.R., Smith, A.J., Miller, C.L., and Allchin, A. 1979, Lancet, 1, 33.
97. Patriarca, P.A., Weber, J.A., Parker, R.A., Hall, W.N., Kendal, A.P., Bregman, D.J., and Schonberger, L.B. 1985, JAMA, 253, 1136.
98. Bender, B.S., Johnson, M.P., and Small, P.A. 1991, Immunology, 72, 514.
99. Renegar, K.B., and Small, P.A., Jr. 1991, J Virol, 65, 2146.
100. de Haan, A., Geerligs, H.J., Huchshorn, J.P., van Scharrenburg, G.J., Palache, A.M., and Wilschut, J. 1995, Vaccine, 13, 155.
101. Oh, Y., Ohta, K., Kuno-Sakai, H., Kim, R., and Kimura, M. 1992, Vaccine, 10, 506.
102. Waldman, R.H., Wood, S.H., Torres, E.J., and Small, P.A., Jr. 1970, Am J Epidemiol, 91, 574.
103. Bizanov, G., Janakova, L., Knapstad, S.E., Karlstad, T., Bakke, H., Haugen, I.L., Haugan, A., Samdal, H.H., and Haneberg, B. 2005, Scand J Immunol, 61, 503.
104. Plante, M., Jones, T., Allard, F., Torossian, K., Gauthier, J., St-Felix, N., White, G.L., Lowell, G.H., and Burt, D.S. 2001, Vaccine, 20, 218.
105. Verweij, W.R., de Haan, L., Holtrop, M., Agsteribbe, E., Brands, R., van Scharrenburg, G.J., and Wilschut, J. 1998, Vaccine, 16, 2069.
106. Haan, L., Verweij, W.R., Holtrop, M., Brands, R., van Scharrenburg, G.J., Palache, A.M., Agsteribbe, E., and Wilschut, J. 2001, Vaccine, 19, 2898.
107. Peppoloni, S., Ruggiero, P., Contorni, M., Morandi, M., Pizza, M., Rappuoli, R., Podda, A., and Del Giudice, G. 2003, Expert Review Vaccines, 2, 285.
108. Vajdy, M., Singh, M., Ugozzoli, M., Briones, M., Soenawan, E., Cuadra, L., Kazzaz, J., Ruggiero, P., Peppoloni, S., Norelli, F., del Giudice, G., and O'Hagan, D. 2003, Immunology, 110, 86.
109. Vajdy, M., Singh, M., Kazzaz, J., Soenawan, E., Ugozzoli, M., Zhou, F., Srivastava, I., Bin, Q., Barnett, S., Donnelly, J., Luciw, P., Lourdes, L., Montefiori, D., and O'Hagan, D. 2004, AIDS Res Hum Retroviruses, 20, 1269.
110. Oh, Y.K., Kim, J.P., Hwang, T.S., Ko, J.J., Kim, J.M., Yang, J.S., and Kim, C.K. 2001, Vaccine, 19, 4519.
111. Hagiwara, Y., McGhee, J.R., Fujihashi, K., Kobayashi, R., Yoshino, N., Kataoka, K., Etani, Y., Kweon, M.N., Tamura, S., Kurata, T., Takeda, Y., Kiyono, H., Walter, B.A., Valera, V.A., Takahashi, S., and Ushiki, T. 2003, J Immunol, 170, 1754.
112. Valosky, J., Hishiki, H., Zaoutis, T.E., and Coffin, S.E. 2005, Clin Diagn Lab Immunol, 12, 171.
113. Walter, B.A., Valera, V.A., Takahashi, S., and Ushiki, T. 2006, Neuropathol Appl Neurobiol, 32, 388.

114. Vajdy, M., Baudner, B., Del Giudice, G., and O'Hagan, D. 2007, Clin Immunol, 123, 166.
115. Couch, R.B. 2004, N Engl J Med, 350, 860.
116. Fiers, W., De Filette, M., Birkett, A., Neirynck, S., and Min Jou, W. 2004, Virus Res, 103, 173.
117. Ulmer, J.B. 2002, Vaccine, 20, S74.

Part V
Genital Pathogens

Chapter 16
Immunity Against *Chlamydia trachomatis*

Ellen Marks and Nils Lycke

Abstract *Chlamydia trachomatis*, the etiology of one of the most common human infections, is responsible for an increasing number of genital tract infections worldwide. In addition, this obligate intracellular pathogen infects epithelial cells of the eye, resulting in trachoma and, as a sequela, blindness. The genital tract infection often leads to severe damage of the reproductive tract, resulting in tubal factor infertility. Infection of the genital tract stimulates a complex array of host immune responses involving innate as well as adaptive immune responses. Cells of the innate immune system recognize and limit the spread of infection, and influence the outcome of infection through the modulation of the adaptive immune response. Protective immunity against *C. trachomatis* involves primarily Th1 $CD4^+$ T cells and IFN-γ production, while antibodies, and possibly $CD8^+$ T cells, can contribute to protection. However, the immune response to infection is complex and those components that convey protective immunity may contribute to the pathogenesis. Therefore, it is believed that a balance between Th1/Th2 effector and regulatory T cell populations is required to avoid harmful immunopathology and permanent sequelae. Advances in our understanding of the immunobiology of the genital tract and better knowledge about chlamydial infections are key issues for the development of effective vaccines. In this context, identifying protective chlamydial antigens is crucial as well as investigating optimal immunization routes and vaccine formulations, including choice of adjuvant, will probably be central to an effective anti-chlamydial vaccine. In the present chapter, we discuss the current knowledge of correlates of protection and vaccine strategies.

N. Lycke
Department of Microbiology and Immunology, Mucosal Immunobiology and Vaccine Research Center (MIVAC), Institute of Biomedicine, The Sahlgrenska Academy at Gothenburg University, Gothenburg, Sweden
e-mail: Nils.lycke@microbio.gu.se

M. Vajdy (ed.), *Immunity Against Mucosal Pathogens*,

16.1 Introduction

The female genital tract mucosa is challenged with two immunologically distinct and opposing functions; i.e. maintaining tolerance to allogenic spermatozoa and the developing fetus, on the one hand, and ability to mount a protective immune response against pathogenic microorganisms that use the genital tract as a portal of entry, on the other. This delicate balance between tolerance and immunity is a hallmark of mucosal barrier functions in the body and particularly well studied for the gut intestinal immune system (reviewed [1]). Much less known are mechanisms responsible for this balance in the genital tract. From a host protection point of view the ability to recognize invasive pathogens is paramount to the ability to develop an adaptive immune response in the genital tract. A better understanding of regulatory functions in the genital tract immune system is, therefore, much needed if effective vaccines are to be developed. This has become increasingly important because sexually transmitted diseases (STDs) caused by bacterial or viral infections are a major problem world-wide. For example, genital tract HIV and human papilloma viruses (HPV) are causative agents of severe viral diseases, however, bacterial STDs are also a growing problem. In this regard, infections caused by *Chlamydia trachomatis* are of particular interest. *Chlamydia trachomatis* infections can affect not only the genital tract, but also the respiratory tract and the eye. Thus, *Chlamydia* infections in humans are a major medical concern in both industrialized and developing countries, with approximately 90 million new infections reported annually [2], representing the most common bacterial STD and the number one cause of preventable blindness worldwide [3]. More than two-thirds of the *Chlamydia*-infected cases are found in developing countries, where resources for diagnostic- and treatment services are limited. It was recently estimated that sub-Saharan countries and southern Asia together have more than 60 million new cases every year [2]. In addition, studies have indicated that genital infection with *C. trachomatis* correlates with an increased HIV transmission rate and cervical neoplasia, which in turn is associated with infections with HPV [4, 5], thus making infections and co-infections a key concern for health workers and the launch of effective treatment plans, including prophylactic vaccine development, much needed.

C. trachomatis infects both men and women. In females the primary site for genital tract infection is the columnar epithelium of the endocervix, where the bacteria divide and infect neighboring cells. Characteristic symptoms caused by local inflammation include abnormal vaginal discharge, lower abdominal pain, mucopurulent cervicitis and post-coital bleeding. The infection is effectively treated with antibiotics, however, the number of asymptomatic infections is high, with approximately 70–90% of women and 30–50% of men remaining asymptomatic [6]. If untreated in females, the *C. trachomatis* infection may invade and ascend to the fallopian tubes where it can persist for several months. The local inflammation caused at the site of infection can be severe and disrupt tissue functions, leading to pelvic inflammatory disease (PID), tubal factor infertility and ectopic pregnancy. Although antimicrobial treatment is effective a majority of cases remain undetected. In fact, antimicrobial treatment has been suspected to promote transmission of infection and reduce natural immunity in the population [7]. Therefore, there is growing interest in prophylactic vaccines against chlamydial infections.

Also, high morbidity and economic costs of infection call for a vaccine against *C. trachomatis*, but several factors have delayed progress in this field. Firstly, we do not have a good understanding of which antigens are protective. It is believed that an effective vaccine should mimic the natural immune response to infection. However, the vaccine must not induce the severe inflammatory reactions sometimes associated with infection or, indeed, should not aggravate the inflammatory response when vaccinated individuals are naturally exposed to *C. trachomatis* bacteria, as has been seen (reviewed [8]). Secondly, the route for effective immunization of the genital tract is still much debated. Strong advocating arguments for local as well as well systemic vaccination in animal models can be found but few studies in humans have resolved the issue. Thirdly, the adjuvant formulation to be used for an effective genital tract vaccine is poorly defined as well as which host factors convey resistance to infection and the role of innate and adaptive immunity. It is a fact that, while a natural infection does induce some protection against reinfection, especially in animal models, this appears to be less true in humans, especially when reinfection is of a different serovar [9]. Notwithstanding this, our knowledge about vaccine candidates and immune responses to *C. trachomatis* is growing. Most studies in animal models, in particular using the mouse model of infection with *C. muridarium* or human isolates of *C. trachomatis* have given us a better understanding of immune protection against *Chlamydia* infections in the genital tract. Of note though, caution should be used when extrapolating results from *C. muridarium* and *C. trachomatis* in the mouse model to the human infection, because these species differ considerably with regard to susceptibility to IFN-γ as well as in the allelic variation in genes encoding protective antigens, such as major outer membrane protein (MOMP) [10, 11, 12]. In this chapter we will focus on what has been learned about genital tract *Chlamydia* infection referring mostly to the literature but also to our own work in gene-knockout mouse models.

16.1.1 Chlamydia trachomatis

The genus *Chlamydia* comprises obligate, gram-negative, aerobic, intracellular bacteria with a variety of tissue tropisms and host species. Human *Chlamydia* pathogens include *C. pneumoniae* and *C. trachomatis*, which infect the respiratory and occulogenital mucosa respectively. *C. trachomatis* consists of 15 identified serovariants which differ primarily in the properties of their major outer membrane proteins. Servovars A, B, Ba and C cause trachoma while serovars D to K are sexually transmitted, but may also infect the respiratory epithelium. Serovars L1, L2 and L3 infect the genital epithelium but disseminate to the lymphatics, leading to a systemic disease know as lymphogranuloma venereum (Table 16.1). Serovars D and E are the most prevalent infections in humans, while H and I represent the least number of cases reported [13].

Table 16.1 *Chlamydia trachomatis*, the causative agent of many diseases, the related groups of serovars and their incidence in different countries

Species	Disease	Riskgroup	Incidence*
C. trachomatis			
Serovar A, B, Ba, C	conjunctivitis trachoma	Children low socioeconomic regions	Up to 50% of children in endemic areas
Serovar D, Da, E, F, G, H, I, Ia, J, Ja, K	Urethritis cervicitis pelvic inflammatory disease ectopic pregnancy tubal factor infertility neonatal pneumonia neonatal conjunctivitis proctitis epididymitis vesiculitis	18–26 year olds	2.5% China 4.2% United States 2.1% Britain
L1, L2, L3	lymphogranuloma venerum proctitis	MSM Sex workers	Up to 8% clinic patients endemic areas

* Figures presented vary dramatically from country to country, and also with the diagnostic methods used, ages and socioeconomic groups of the populations that were studied [14, 15, 16, 17, 18]

The defining features common to chlamydiae are unique and illustrate a complex biphasic developmental cycle (Fig. 16.1). Small, infectious, but metabolically inert, elementary bodies (EBs) attach and bind to the host epithelium, where they induce endocytosis into a vacuole, termed an inclusion. In order to evade the host immune defense, the bacterium inhibits fusion of the inclusion with host cell lysosomes [19]. Within 2–6 hours after internalization the EBs differentiate into metabolically active, but non-infectious, reticulate bodies (RBs) that divide exponentially by binary fission before condensing back into EB form. These EBs are subsequently released from the cell by firstly disrupting the host cell plasma membrane, followed by disruption of the inclusion membrane, releasing the bacteria to infect neighboring cells [20].

16.2 *Chlamydia* Immunobiology

The female genital tract mucosa contains a full repertoire of immune competent cells belonging to both the innate and adaptive immune systems, which vary in number, type and distribution throughout the menstrual cycle. It is thought that the female reproductive tract lacks organized lymphoepithelial structures found at other

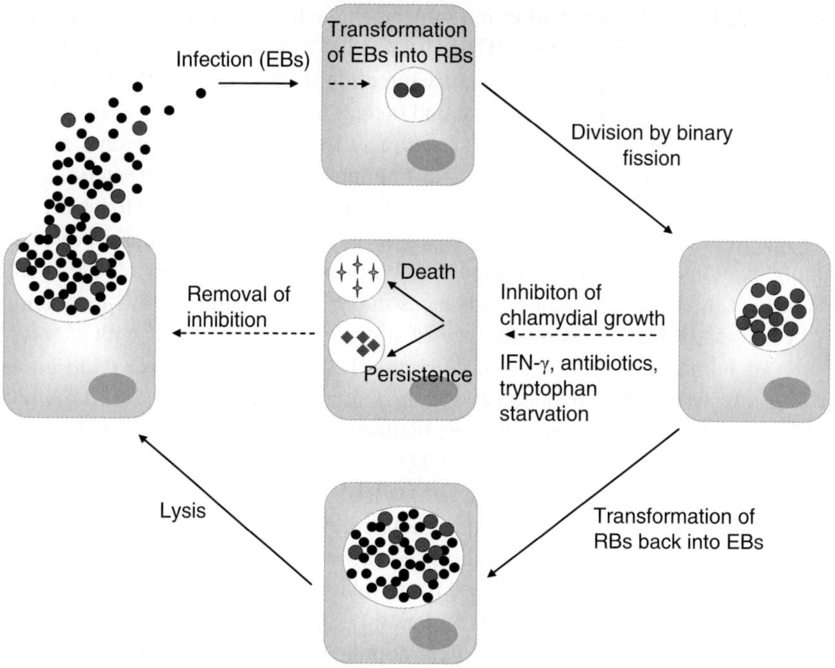

Fig. 16.1 The growth cycle of *Chlamydia*. The productive infection cycle involves transformation of EBs into RBs and the division of *Chlamydia* RBs inside inclusion bodies by binary fission. RBs then differentiate back into infectious EBs before lysis of the cell and release of infectious particles for reinfection. RBs may respond to inhibition of the growth cycle by acquiring a non-replicating, but still viable, aberrant form. The infection will become productive again once the inhibiting factors are removed and the persistent forms re-differentiate back into infectious EBs

mucosal tissues, such as the Peyer's patches or isolated lymphoid follicles in the small and large intestine. However, studies have shown the presence of lymphoid aggregates (LA) consisting of $CD8^+$ T cells with a core of B cells surrounded by an outer mantle of macrophages, which can be found in the human cycling uterus, but are absent at menses and during gestation [21]. In women suffering from *C. trachomatis* infection these LA develop into lymphoid follicles, presenting with germinal centers, the sites for expansion of antigen-triggered B cells [21, 22]. While these LA can act as local inductive sites, it is likely that T and B cell responses in the female genital tract may be induced both locally and through recruitment from distant inductive sites. However, because the genital tract mucosa of uninfected individuals contains relatively few lymphocytes, recruitment of circulating lymphocytes appears to be a critical part of the immune response. We know that chemokines and adhesion molecules act together to recruit lymphocytes from the circulation and distant secondary lymphoid tissues into the mucosa. These mechanisms appear to also occur in the female genital tract. Chemokines capable of directing distant lymphocytes to the genital tract, such as CCL5, CCL7 and CXCL10, have been shown to be expressed in murine epithelial cell lines in response to *C. muridarium* infection [23]. Expression of adhesion molecules on the endothelium is required for tissue emigration of leukocytes. VCAM1 and MADCAM1 are up-regulated in the fallopian tubes following *C. trachomatis* infection, indicating that the response to *Chlamydia* infection evokes both local and distant immune responses [24].

To mount an effective immune response a series of coordinated events activate the innate and adaptive immune systems. The intricacies of those interactions during *Chlamydia* genital tract infection are still under investigation. However, it is clear that the cell-mediated immune response plays an important role in protective immune responses. Given the limited feasibility of detailed studies of *Chlamydia* genital tract infections in humans, advances in our understanding of host immunity has been greatly facilitated by the use of murine models infected with *C. muridarium* or human serovars of *C. trachomatis*. Like humans, mice develop an ascending infection that spontaneously resolves after 2–4 weeks [25], during which time there is an influx of macrophages, neutrophils and lymphocytes, including B cells, $CD8^+$ T cells and, in particular, $CD4^+$ T cells [26]. Following resolution of infection, $CD4^+$ T cells remain in the genital tract. These $CD4^+$ T cells are thought to significantly contribute to protection against reinfection with the same serovar [9]. Noteworthy, mice that do become reinfected suffer only a transient infection of the lower genital tract with greatly reduced bacterial loads and milder inflammation [9]. Importantly, the mouse infection is remarkably similar to the human chlamydial infection in several regards. As in human populations, different strains of mice respond differently to infection and are differently susceptible to the development of post infection sequelae [27, 28].

16.2.1 Innate Immunity

The innate immune response is the first line of defense against invading pathogens and may reduce the initial load of the pathogen and promote the induction of an adaptive immune response capable of specifically combating the infection. Innate immunity encompasses mechanisms of defense such as the production of soluble molecules and anti-microbial substances, the detection of pathogen-specific molecular structures, and the actions of macrophages, neutrophils, dendritic cells, natural killer (NK) cells and NK T cells. These cells of the innate immune system may have profound implications for the eventual outcome and immunopathology of the genital tract infection. The pro-inflammatory response that is triggered by infection is characterized by production of cytokines and chemokines by epithelial and other cells in the mucosa. Of the cytokines, IL-1α, IL-6, TNFα, GM-CSF, IL-8, Type I IFNs, and IL-12 are strongly produced [23, 29, 30]. The cytokine microenvironment promotes the development of Th1-dominated $CD4^+$ T cell immunity (see below).

16.2.1.1 TLR Recognition of *Chlamydia*

Epithelial cells, neutrophils, NK and NKT cells, macrophages, and dendritic cells (DCs), are important in the innate immune response through their expression of toll-like receptors (TLRs). TLRs act as pattern recognition receptors (PRRs) that enable cells to recognize pathogen-associated molecular patterns (PAMPs), which are conserved molecules in bacteria and distinguishable from host molecules. TLRs 1–9 can be found throughout the female genital tract. However, TLR4 is weakly expressed or absent from the lower genital tract [31]. Upon entry into the genital tract, *Chlamydia* could be recognized by TLR4, which recognizes bacterial LPS and heat shock proteins (HSPs), or through TLR2, which binds bacterial lipoproteins. However, surprisingly, it was found that mice lacking expression of TLR2 or TLR4 were unimpaired in their clearance of *Chlamydia* bacteria from the genital tract. This occured despite reduced ability by macrophages in $TLR2^{-/-}$ mice to produce cytokines, resulting in less immunopathology compared to wild-type control mice [29]. It, thus, appears that the importance of TLRs still is insufficiently investigated to help explain host resistance against genital tract infections with *C. trachomatis*. Also, a role of TLR2 for the immunopathology in humans has been suggested [32, 33]. However, single gene nucleotide polymorphisms (SNPs) in the TLR4 or CD14, a co-receptor to TLR4-signalling, did not affect tubal pathology [34, 35, 36]. Future studies are warranted to dissect the precise role of TLRs for host resistance and the immunopathology in genital tract *Chlamydia* infections.

16.2.1.2 NOD Proteins

Another family of PRRs is the nucleotide-binding oligomerization domain (NOD) protein family. NOD1 and NOD2 are intracellular PRRs which recognize ligands including LPS and peptidoglycans [37, 38]. When activated they up-regulate the transcription factor NF-κB and c-Jun N-terminal kinase (JNK) in the host cell [39]. Although peptidoglycans have not been detected in *Chlamydia*, the gene encoding peptidoglycan is present in the *Chlamydia* genome and may be expressed [40]. Since the *Chlamydia* infection is strictly intracellular it is likely that NOD-receptors are important for host recognition. Studies in epithelial cell lines have shown that NOD1 recognition of *Chlamydia* can cause up-regulation of NF-kB expression and fibroblasts were shown to augment IL-6 and MIP-2 production, an effect strongly impaired in $NOD1^{-}/^{-}$ mice [41]. Thus, NOD-signaling may be involved in the host recognition of *Chlamydia* bacteria.

16.2.1.3 The Early Host Response to *Chlamydia*

Natural killer (NK) and NK T cells may be important factors in host resistance against genital tract infections not only due to their ability to lyse infected target cells, but also because they produce certain cytokines, such as IFN-γ, known to promote phagocytic activity of macrophages as well as driving T cells towards the Th1 lineage. NK cells were first proposed to play a role in host resistance against *Chlamydia* infection when it was observed that bacterial elimination from the genital tract was marked already 7 days into the primary infection and coincided with an increased number of IFN-γ producing cells, at a time when adaptive immunity was still developing [42]. Treatment *in vitro* with anti-CD4 antibody did not diminish this production, suggesting that a cell type other than the CD4$^+$ cells was responsible for the IFN-γ production. NK cells, therefore, are likely candidates, and in fact, NK cell activity in the genital tract can be detected as early as 12–24 hours after inoculation of bacteria [42]. Since IFN-γ-deficient mice do not show enhanced bacterial colonization at this early stage of infection it is assumed that the function of NK cell IFN-γ production is not acting by impairing bacterial growth [43]. Rather, it is thought that NK cells are critical to the overall resolution of infection, through polarization of the immune response towards Th1 and down-regulating the Th2 response.

Studies in humans have suggested that NK cell activity is defective due to anergy in these *Chlamydia* infected patients. Decreased production of IFN-γ and TNF-α was associated with a reduced ability for antibody-dependent cellular cytotoxicity (ADCC) [44]. Others have reported anergy to chlamydial antigens in adults with scarring sequelae of trachoma and peripheral blood T cell responses were suppressed in patients suffering from chronic disease [45, 46]. Alterations in the activity of NK cells could potentially change the outcome of infection and have important implications for the early stages of the immunobiology of *C. trachomatis* infections.

NK T cells, a subset of CD4$^+$ innate-like lymphocytes, were recently ascribed a role in linking innate and adaptive immune responses against various bacteria, viruses, and parasites [47, 48, 49]. In contrast to antigen presentation on MHC-II molecules, recognized by classical CD4$^+$ T cells, NK T cells recognize lipid antigen in the context of CD1d molecules for activation. Activated NK T cells act directly on infected cells by killing the CD1d-expressing cell or through production of IFN-γ, in turn activating NK cells. Most importantly though, is probably the downstream effect of NK T cells to modulate the early phase of the adaptive immune response, polarizing it into Th1/Th2 domination in the genital tract [50, 51]. In addition, it was demonstrated in the chlamydiae-respiratory tract infection model that bacteria evaded a protective immune response by down-regulating surface-expression of CD1d, thereby avoiding NK T cell activation and destruction of the infected cells [52]. Nevertheless, the role of NK T cells in response to genital tract *Chlamydia* infection has yet to be better explored.

16.2.2 Adaptive Immunity

16.2.2.1 Cell Mediated Immunity

Adaptive cell-mediated immunity is crucial for clearance of *C. trachomatis* from the genital tract. The uninfected genital tract hosts low numbers of CD4$^+$ T cells. Following infection, lymphocytes in the genital tract form immune inductive sites in which T and B cells are expanded [53]. In the mouse, these inductive lymphoid follicles consist mainly of clusters of CD4$^+$ T cells, while in human females, T cells, DCs as well as B cells are found [21]. However, it is thought that the majority of lymphocytes in the infected genital tract tissue represent cells that migrate in from other inductive sites, such as the paraaortic draining lymph nodes. This is supported by the documentation of homing receptors on the lymphocytes and adhesion molecules expressed locally in the genital tract mucosa. For example, the fallopian tubes in *Chlamydia*-infected mice express the adhesion molecules MADCAM-1 and VCAM-1, molecules responsible for recruitment of mucosal-$\alpha 4\beta 7$ and non-mucosal $\alpha 4\beta 1$ homing CD4$^+$ T cells, respectively [24].

An adaptive immune response that clears *Chlamydia* bacteria from the genital tract develops over a period of several weeks. The ensuing acquired immunity confers partial resistance to reinfection, especially when reinfection is caused by the same serovar as the primary infection [9]. It is now well established that clearance and protective immunity to *Chlamydia* is dependent primarily upon CD4$^+$ T cells [43, 54, 55, 56]. Nude mice, which lack T cells, were unable to control infection and mice that lack CD4$^+$ T cells but have CD8$^+$ T cells are unable to clear infection [25]. However, mice deficient in CD8$^+$ T cells but hosting CD4$^+$ T cells, such as the β2-microglobulin deficient mice, effectively eliminate a genital tract infection. Extended investigations have given that mice deficient in IFN-γ, MHC-II, or IL-12 or normal mice depleted of CD4$^+$ T cells, are severely impaired in clearing of infection [25, 43, 54, 57]. Whereas, Th1-dominated immunity is protective, Th2- cells have been associated with delayed clearance, as seen after transfer of *Chlamydia*-specific Th2 clones into nude mice or in mice immunized with EB pulsed-DCs, establishing a Th2-dominated immune response [58, 59].

The role of CD8$^+$ T cells for host resistance against genital tract *Chlamydia* infection is less obvious [55]. As aforementioned, CD8 or β2-microglobulin deficient mice clear infection well [25]. However, several studies have documented *Chlamydia*-specific CTL activity against infected target cells [60, 61]. In humans, it is possible that CD8$^+$ T cell-mediated recognition of peptides presented by epithelial cells, which constitutively express MHC-I, may be an important part of the immune response [62]. It is noteworthy that both CD8$^+$ and CD4$^+$ specific T cells are generated in response to a genital tract *Chlamydia* infection [63]. Notwithstanding this, CD4$^+$ T cell mediated immunity is thought to be superior to CD8$^+$-mediated CTL activity, as demonstrated with *C. muridarium*-specific CD4$^+$ splenic T cells or T-cell lines transferred into infected nude mice, rather than when corresponding specific CD8$^+$ T cells were transferred [55, 64].

16.2.2.2 Induction of CD4$^+$ T Cell Responses

Protective immunity is a complex coordinated activity involving effector as well as regulatory T cells. Upon encounter with a genital tract infection a series of adaptations of the T cell response would be expected in order to acquire strong local protection with minimal tissue destruction. However, these modifications require strict control of the different responses. Therefore, it is likely that naïve CD4$^+$ T cells differentiate into Th1,Th2, Th17 or T regulatory cell (Treg) subsets, which together convey resistance against chlamydial infection. According to the two-signal model for CD4$^+$ T cell activation, peptide-recognition by T cell receptor is followed by critical co-stimulation [65, 66]. Several factors impact on forming the appropriate T cell response. Of these, the type of APC, the expression of co-stimulatory molecules,

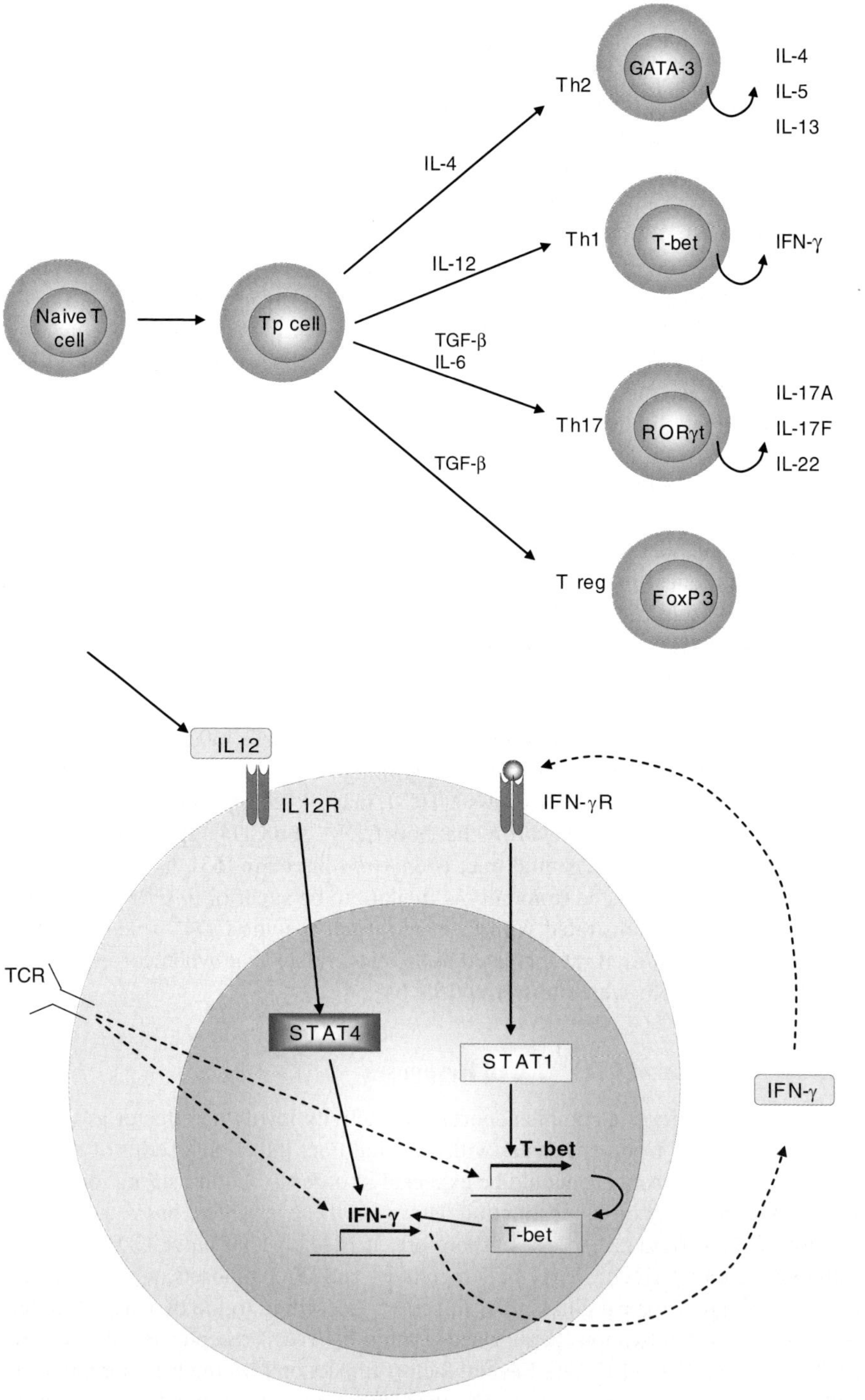

Fig. 16.2 (continued)

such as CD80 and CD86, and the cytokine environment are key components in this process [67, 68, 69]. The T cell expresses CD28 constitutively, but upon activation, ICOS and CTLA4 are also induced, which have key functions in modulating T cell differentiation into Th1, Th2 or Treg cells [70, 71]. Whereas CD80 and 86 are ligands for CD28 and CTLA4, ICOSL binds to ICOS on the activated T cell to promote expansion and differentiation of effector and regulatory $CD4^+$ T cells [72].

ICOS and CD28 have been shown to differentially regulate the polarization of an immune response towards Th1 or Th2 in a number of infectious disease models including *Leishmania major, Salmonella typhimurium* and *Listeria monocytogenes* [73, 74, 75]. The co-stimulation required to expand and differentiate the naïve T cells depend on receptor-ligand interactions and subsequent production of cytokines, which act in an autocrine and paracrine fashion to shape the T cell subset response (Fig. 16.2B). Therefore, the cytokine environment is vastly different depending on which costimulatory molecules are involved. ICOS-signaling has been associated with the production of IL-10, IL-4 and IFN-γ depending on the antigenic stimuli, whereas CD28-signaling is particularly important for IL-2 production [76, 77, 78]. ICOS has been shown to be more highly expressed on Th2 than Th1 cells and is involved in germinal center (GC) formation, somatic hypermutation and class-switch recombination of activated B cells. Moreover, Tregs are thought to develop as a consequence of signaling through co-stimulatory molecules [79]. Currently, 2 broad categories of Tregs have been described: (1) The natural $FOXP3^+CD4^+CD25^+$ Treg, which develop in the thymus, and (2) the inducible Tregs (Th3 and Tr1 cells), which develop in the periphery. The importance of CD28 in the development, maintenance and function of Tregs has been established. Interestingly, some subsets of Treg cells have been shown to depend on ICOS expression for *in vivo* and *in vitro* suppression through the production of IL-10.

Since immunity against *C. trachomatis* is based primarily on the contribution of Th1 $CD4^+$ cells, in a recent study, we characterized the generation of protective specific memory responses during infection in the absence of co-stimulatory molecules. We analysed the differential requirements of co-stimulatory signaling through CD28 and ICOS in the course of an adaptive immune response against *C. trachomatis serovar D*. We found that CD28 was required for the priming and expansion of Th1 effector cells, responsible for eradicating the genital tract infection and for the development of long-term memory against reinfection [80]. The

Fig. 16.2 Differentiation of T cells into specific subsets is mediated by cytokines. (**A**) $CD4^+$ naïve T cells generate T progenitor (Tp) cells following TCR-MHC-II interactions, which can then differentiate into subsets of Th1, Th2, Th17 or Treg cells following cytokine stimulus. Once differentiation is initiated, cytokines produced by the T cells act to enhance expansion and differentiation of the specific subsets of $CD4^+$ T cells. (**B**) Th1 differentiation showing the IFN-γ positive feedback loop. Differentiation into Th1 cells depends on IL-12-mediated activation of STAT4, which in turn supports IFN-γ production. A positive feedback loop follows, whereby IFN-γ induces STAT1 which activates T-bet, which increases expression of IFN-γ and the β2-subunit of the IL-12 receptor. This IFN-γ produced by Th1 cells acts to promote APCs' function, thereby additionally driving Th1 differentiation lack of Th1 priming was apparent in CD28-deficient mice and the poor IFN-γ production resulted in impaired protection and insufficient immunity to reinfection. These mice exhibited low expression of the Th1-specific transcription factor, T-bet, in their draining lymph nodes, but prominent expression of GATA-3, which is a transcription factor linked to Th2-development. By contrast, in the absence of ICOS Th1 development was greatly augmented and protection was enhanced with almost sterile immunity to a secondary challenge infection. Therefore, we concluded that ICOS signaling modulates the genital tract T cell response by dampening the Th1-domination while promoting Th2-development [80]. Also, Treg development was impaired in ICOS-deficient mice as demonstrated by lower expression of Foxp3 mRNA, the specific transcription factor for Treg-development, and lower levels of IL-10 in the genital tract lymphocytes than that found in normal mice infected with *C. trachomatis*. This resulted in better protection, but also more immunopathology in ICOS$^{-/-}$ mice compared to normal mice following *C. trachomatis* infection of the genital tract [80]. Thus, it appears that CD28-signaling is critical for immune protection against a genital tract chlamydial infection, but ICOS signalling is important for dampening Th1-development and driving Th2-immunity as well as for anti-inflammatory IL-10 production and promotion of Foxp3^{+} Treg populations in the genital tract. Since ICOS is required for GC-formation and specific antibody production, no antibodies were found in ICOS$^{-/-}$ mice, albeit immune protection against a genital tract chlamydial infection was enhanced. Therefore, and in agreement with several other mouse models, there was poor correlation between immune protection against chlamydial infection and presence of specific antibodies [55, 56, 81, 82].

16.2.2.3 Mechanisms of Protection: The Role of IFN-γ

There is general agreement that the anti-chlamydial action of CD4^{+} and CD8^{+} T-cells, as well as NK cells, could be mediated through production of IFN-γ [43]. IFN-$\gamma^{-/-}$ mice are unable to clear a genital tract infection with human strains of *C. trachomatis* [43]. IFN-γ is thought to directly affect survival of *Chlamydia* bacteria and not exclusively through its ability to activate macrophages to release factors that drive inflammation. During a *Chlamydia* infection in mice, IFN-γ is produced in a biphasic pattern, peaking at 1 week and 3 weeks after infection [42]. The early peak of IFN-γ has been attributed to NK and NKT cells, while the latter corresponds to the high levels of proliferation and influx of CD4^{+} and CD8^{+} T cells into the genital tract tissues [42]. It is known that high doses of IFN-γ may confer protection against *Chlamydia* infection by blocking chlamydial growth, albeit low doses of IFN-γ appears to rather promote persistent infection by increasing production of aberrant RBs [83]. Thus, the exact anti-chlamydial effect of IFN-γ *in vivo* is still incompletely understood. It may include many factors such as reactive nitrogen species through inducible nitric oxide synthase (iNOS), IDO-mediated tryptophan starvation, deprivation of iron (Fe) via down-regulation of transferrin receptors, stimulation of phagolyzosomal fusion, and the disruption of selective vesicular nutrient transport via p47/GTPase activation (reviewed in [84]).

Particularly, the growth-inhibiting effect has attracted much attention. Induction of the enzyme iNOS by activated phagocytes is key to this effect. Many different factors other than IFN-γ may promote the production of iNOS, such as other cytokines and bacterial components, including LPS and lipotechoic acid [85, 86]. iNOS catalyzes the production of reactive nitrogen species (RNS), including nitric oxide (NO).

$$\text{L-arginine} + O_2 + \text{NADPH} \xrightarrow[iNOS]{} \text{NO} + \text{L-citrulline} + \text{NADP}$$

NO is known to posses powerful immunoregulatory and antimicrobial properties. It can inactivate enzymes of the citric acid cycle and induce DNA damage [87, 88]. RNS have been shown to inhibit chlamydial growth in isolated fibroblasts [89], macrophages [90] and murine epithelial cells [91]. But, interestingly, mice lacking iNOS resisted infection well and were able to normally clear chlamydiae bacteria from the genital tract [92, 93]. Perhaps, it can be concluded from these latter studies that presence of iNOS-induced RNS can protect mice from more severe chronic infection, since the prevalence of hydrosalpinx in $iNOS^{-/-}$ mice was significantly increased compared to infected normal mice. Furthermore, induction of the enzyme indoleamine 2,3-dioxygenase (IDO) also plays a central role in IFN-γ-mediated growth inhibition of *Chlamydia* in human model systems [94]. Noteworthy, this has not been observed in murine models of genital tract *Chlamydia* infection [95]. IDO is responsible for the degradation of tryptophan to kynurenine, thereby causing inhibition of bacterial replication and growth by deprivation of the intracellular tryptophan pools (Fig. 16.3).

16.2.2.4 Humoral Immunity

Antibodies are critical for protection, especially against pathogenic microorganisms that are extracellular. *C. trachomatis* is an obligate intracellular pathogen and there-

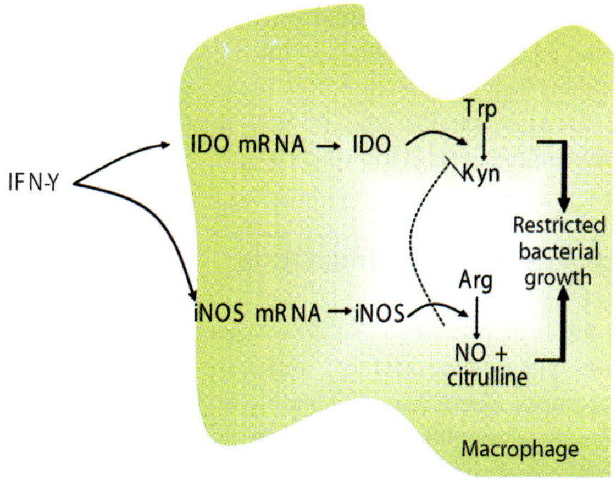

Fig. 16.3 The antimicrobial effects of IFN-γ, iNOS and IDO restrict *C. trachomatis* growth in different ways. Whereas IDO catalyses the conversion of tryptophan to kynurenine, which depletes tryptophan stores and ablates growth of the bacteria, iNOS catalyses the production of nitric oxide, thereby contributing to lyzosomal pathogen killing

fore there has been extensive debate as to the role of specific antibodies for host resistance against *Chlamydia* infections. IgG and IgA-secreting plasma cells are present in the lamina propria of the endocervix but scarce in the vagina. However, unlike other mucosal surfaces of the body, the dominant antibody isotype in the female genital tract is IgG, while IgA is present in significant amounts only in the cervical mucosa and fallopian tubes (reviewed [96]). Diffusion of serum-derived and locally produced IgG and IgA through epithelia is an important part of humoral immunity in the female genital tract. Since specific antibodies can protect the host in many ways it has been speculated as to what antibodies could do to eliminate a genital tract *Chlamydia* infection [56]. Of course, they could act as opsonins by binding to the surface of the bacteria or the infected cells and together with Fc-receptor carrying phagocytes or complement components causing lysis of bacteria and host cells. Activation of complement also contributes to the release of chemotactic factors that can attract inflammatory and phagocytic cells to the area of infection. Specific antibodies may also neutralize infection by preventing bacterial up-take or invasion into host cells. However, none of these mechanisms appear to be critical for host resistance against *C. trachomatis* genital tract infections in the mouse model. Mice deficient in B cells and antibodies demonstrate an intact ability to clear infection, exhibiting a pattern of bacterial clearance and long term acquired immunity very similar to that observed in normal mice [97, 98, 99, 100]. Furthermore, vaccine candidates that elicited only high titers of specific antibodies and no $CD4^+$ T cells were non-protective [43, 56, 101]. Nevertheless, recent studies have clearly shown that antibodies may contribute to resistance and dampening of the immunopathogenesis associated with infection [97, 98, 99, 100, 101]. Using an alternative approach, it was found that *Chlamydia*-immune mice, which were rendered susceptible to infection by deletion of T cells, were protected against a re-infection by passive transfer of immune serum [97]. Additionally, studies in $FcR^{-/-}$ mice showed reduced resistance against infection, suggesting a role for antibodies in host protection against reinfection of the genital tract [99]. The mechanisms for these protective effects are unknown, but it is noteworthy that passive transfer of immune serum does not protect against a primary infection, only against reinfection, a fact which argues against a neutralizing or complement activating effect [97]. To what extent these findings in mice bear any significance for the development of resistance against infection in humans is still an open question. Significantly elevated *Chlamydia*-specific IgG and IgA titers have been associated with chronic infection and pathogenesis [102, 103, 104].

16.3 Immunopathogenesis

It has been estimated that 20% of women with lower genital tract chlamydial infection will develop PID, 4% suffer from chronic pelvic pain, and 3% show signs of infertility. There is a considerable majority (50–70%) of women who remain asymptomatic throughout the infection, and both asymptomatic and symptomatic infections are at risk of developing adverse complications [105]. PID is an inflammation of the uterus, fallopian tubes and pelvic structures [106], and the risk of developing infertility after a single episode of PID is 10% and doubles for each consecutive episode [107]. Similarly, women suffering from PID are 7–10 fold more likely to develop ectopic pregnancy and 24–75% experience chronic pelvic pain (reviewed in [108]).

The pathogenesis of *C. trachomatis* genital tract infections is the result of an inflammation that is immunologically mediated, although surprisingly little in known about the inflammatory process that leads to fibrosis and damage of the genital tract. It is thought that persistent antigen synthesis and an ineffective immune response contribute to the chronic inflammation that ultimately lead to salpingitis and/or infertility [109]. Failure to effectively clear the pathogen from the genital tract may induce a state of aberrant *Chlamydia* inclusions [110]. Ultimately, the outcome of chlamydial infection depends on the balance between the Th1, Th2 and Treg subsets in the genital tract, which in turn is influenced by many different factors including genetics, bacterial strain, and the local microenvironment. In this context it is noteworthy that *C. trachomatis* infections have been associated with the induction of heat shock protein (HSP60), which belongs to a family of phylogenetically conserved proteins that are present in both bacteria and produced by host cells. *Chlamydia* HSP60 displays 48% homology to human HSP60 and is involved in the assembly of the outer membrane of EBs [111]. Its accumulation is thought to contribute to the pathogenesis by breaking tolerance to human HSP60 due to antigenic cross-reactivity and the induction of an autoimmune reaction [111, 112]. During chronic infection the accumulation of HSP60 provides a continuous antigenic stimulation, and thus prolongs an autoimmune response [98]. Alternatively, *Chlamydia*-HSP60 stimulates specific Tregs that produce IL-10, and when these cells are activated they promote a switch from Th1 domination to a predominant Th2 response, which delays resolution of the infection [112].

16.4 *Chlamydia* Infection of the Male Genital Tract

Urethritis, epididymitis, vesiculitis, and prostatitis in men can result from acute or chronic *C. trachomatis* infection of the genital tract [113]. Recent findings suggest that *Chlamydia* may cause male infertility by acting directly on sperm or seminal fluid. Studies have demonstrated lower pregnancy rates in couples where the male partner has *C. trachomatis*-specific serum IgG or IgA [114]. The semen of men with detectable levels of *Chlamydia*-specific serum IgA was subject to changes such as impaired motility, decreased sperm vitality, the presence of leucocytes and defective tails. Men positive for both *Chlamydia*-specific serum IgG and IgA displayed decreased sperm concentration and a lower percentage of progressive spermatozoa [100]. Negative effects on seminal quality following *Chlamydia* infection include a reduced motility of sperm [115], sperm DNA fragmentation [116], reduced survival of spermatozoa [117], lowered sperm concentration [100], a blunted acrosome reaction [118] and an increased amount of leucocytes [119]. While the mechanism behind these changes remains to be elucidated, there have been suggestions that the *Chlamydia* LPS is a potent inducer of apoptosis in sperm [120], because 500-fold more *E. coli* LPS is required to have the same apoptotic effect as *Chlamydia* LPS [117]. Interestingly, the relationship observed between HSP60-specific Abs and tubal factor infertility in women has not been demonstrated in men. In men, native host HSP60 is involved in normal spermatogenesis [121]. Therefore, it has been speculated that *Chlamydia* could negatively affect sperm development through an autoimmune cross-reaction with human HSP60 [100].

16.5 Vaccines Against *Chlamydia*

An important challenge when constructing an effective anti-chlamydial vaccine is to find an adequate mix of protective antigens and formulate these together with an appropriate adjuvant that stimulates genital tract protective immunity. Recent progress conveys optimism. A subunit vaccine will have to stimulate both systemic and local humoral as well as cell-mediated immunity. Studies in experimental models and human females using non-chlamydial antigens, such as the cholera B subunit, have clearly shown that intranasal or local intravaginal immunization are effective routes for induction of specific systemic and genital tract immunity [122, 123]. The fear that lack of detectable local immune inductive sites in the female genital tract would not allow for intravaginal immunizations, appear not to be limiting for a successful vaccination. However, the influence of sex hormones seems more difficult.

To avoid the influence of sex hormones upon vaccination, it is desirable to immunize at a remote site, eliciting an immune response in the inductive tissue and also stimulating immunity at the site of infection, i.e in the genital tract mucosa. In this respect, it has been demonstrated that immunization at mucosal sites, such as the nasal mucosa, with live chlamydiae correlates to a higher level of protective immunity, including strong Th1 responses and Ab titers, than when EBs are delivered parenterally [124]. In fact, following intranasal and intravaginal immunization, much higher levels of specific- IgA and IgG can be found in the genital tract secretions than immunization orally and rectally, although IgA and IgG are also induced in the genital tract via these latter routes [124]. This emphasizes the potential of the nasal route of delivery for immunization against genital tract pathogens.

Early attempts for vaccination against *Chlamydia* were focused on vaccination with inactivated whole EB preparations and while this approach was successful for vaccination against other pathogens, vaccination with whole *Chlamydia* EBs induced only short-lived immunity. But, a more severe finding was that hyper-reactivity to natural infection ensued as a negative side-effect in some vaccinated individuals, calling attention to potential deleterious components of the induced host immune response (reviewed [8]). This led to the development of subunit vaccines to avoid potential side effects of immunization. Subsequently efforts have focused on identifying the protective antigenic subunits as well as effective delivery systems, immunization routes and adjuvants to be included in an anti-chlamydial vaccine candidate.

Early experiments identified the major outer membrane protein (MOMP) to be a protective antigen [125]. MOMP accounts for 60% of the outer membrane of *C. trachomatis* and is composed of 4 variable regions, differing between serovars, and 5 constant regions. MOMP is a cysteine rich protein which is thought to act as a porin although its exact function is not yet clear [126]. Neutralizing sites have been identified as linear epitopes that reside within variable domains of MOMP [127]. However, while some vaccination attempts in animal models have yielded promising results, cross-serovar protection appears to be poor [9]. Although the different serotypes of *Chlamydia* share 84–97% homology of the MOMP, it is clear that the elements that are most immunogenic are also those which differ the most between the different serovars [12].

From animal studies we have learnt that an effective anti-chlamydial immunization must generate strong specific $CD4^+$ T cell immunity. The antigenic determinants that prime these cells most effectively are, therefore, likely to be critical components in a successful vaccine. Several MHC-II helper T cell-epitopes have been identified in MOMP, but few studies have successfully tested these MOMP-epitopes for stimulating protective immunity in mice [128]. This was probably due to a lack of an effective expression vector system or because of poor formulation of the epitopes for effective immunization. However, when whole MOMP was over-expressed in an inducible *E. coli* clone, significant progress was made towards immunizations that conveyed protective immunity [129]. Immunization with recombinant MOMP has produced a vast diversity of results, ranging from good to insufficient levels of protection, depending on the strain of mice, route of immunization, as well as the choice of adjuvant and delivery systems (Table 16.2) [130, 131, 132, 133, 134].

One of the most important recent developments within the field is the sequencing of the entire *Chlamydia* genome, which has allowed researchers to identify new virulence factors and membrane proteins that could be potential targets in a new vaccine. Foremost among newly identified proteins are several outer membrane proteins (Omp-1, Omp-2, and Omp-3) which are more highly conserved and contain both $CD4^+$ and $CD8^+$ T cell epitopes [135, 136]. Another set of promising molecules are polymorphic outer membrane proteins (pmp) [137], conserved PorB membrane proteins [138], an ADP/ATP translocase (Npt1) [139], a plasmid protein (pgp3) [140], the proteasome/protease-like activity factor (CPAF) [141], toxins [142] and members of the type III secretory machinery [143]. Recently, a study using the DNA expression library and gene gun immunization approach was published, in which the authors document a whole series of novel chlamydial vaccine antigens [144]. The particular strength with this technique is that it requires no prior knowledge of the antigenic targets and has the potential to screen the entire genome of the pathogen for host immune responses. Because the gene gun is used for vaccine delivery no assumptions regarding the best protective immune response need to be made and using a live challenge model the authors identified several strong protective *C. muridarium* antigens.

Many of these antigens have yet to be tested experimentally. It is clear that most immunizations with single subunits in various constellations with traditional

Table 16.2 Results of immunization of Balb/c mice with recombinant *C. Muridarium* MOMP when challenged was administered intravaginally/intrabursally

Adjuvant	Route of immunization	Booster	Outcome of Immunization		Protection level	Ref
			T cell responses	Antibody responses		
Fused to CT	i.n	3x i.n	ND	Strong IgG in serum and locally. IgA detected	~50% reduction in IFU after 3 weeks	Singh et al., (2004)
Fused to CT	i.vag	3x i.vag	ND	4-fold lower serum Ig, than when administered i.n. Local IgG present in low amounts. IgA absent	Not protective	Singh et al., (2004)
CT + CpG-ODN	t.c	4x t.c	90-fold increase in IFN-γ mRNA	Strong serum and local IgG2a/IgG1, moderate local IgA levels	~50% reduction after 6 days	Berry et al., (2003)
DspA from strain B31 of *B. burgdorferi*	i.n	NA	NA	Low levels of serum IgG and IgA. (IgG1 > IgG2a). No IgG detected locally	Not protective, no increase in fertility	Pal et al., (2003)
DspA from strain B31 of *B. burgdorferi*	i.n + i.p	NA	Moderate T cell proliferation, IFN-γ	High levels of serum IgG and IgA. (IgG1 > IgG2a). local IgG and IgA	30% reduction in IFU after 2 weeks, 16% increase in fertility	Pal et al., (2003)
DspA from strain B31 of *B. burgdorferi*	i.m + s.c	NA	Very strong T cell proliferation, IFN-γ	Moderate levels of serum IgG and IgA. (IgG1 > IgG2a). local IgG and IgA	50% reduction in IFU after 2 weeks, 35% increase in fertility	Pal et al., (2003)

i.n; intranasal, i.vag; intravaginal, t.c; transcutaneous, i.p; intraperitoneal, i.m; intramuscular, s.c; subcutaneous

adjuvants have not resulted in induction of strong protective immunity. However, because of a growing awareness that novel delivery systems and multiple subunit vaccines have proven superior to previous formulations we are optimistic about the possibility that an effective vaccine against genital tract chlamydial infection can be developed [145, 146]. Recently, mice immunized with DCs pulsed with whole EBs developed protection superior to that observed with DCs pulsed with MOMP alone, suggesting that there are complementary protective antigens in whole EBs, not present in MOMP alone [59]. Also, the delivery systems or vectors used appear to be crucial. Thus, mice immunized with bacterial ghosts from *Vibrio cholerae*, expressing both MOMP and OMP2 offered greater numbers of specific Th1 cells and resulted in significant protection against infection [147]. Moreover, the vector itself might be a key component of a vaccine, as seen when MOMP was expressed by an attenuated influenza A virus strain [148]. Furthermore, immunizations with MOMP in combinations with potent adjuvants, such as with Montanide ISA 720, ISCOMs, CpG plus cholera toxin, or as a fusion to the cholera holotoxin have also proven effective [131, 132, 149, 150, 151]. Thus, these results clearly hold promise as to our ability to develop an effective vaccine against *Chlamydia* genital tract infections in the future.

A fundamental limitation to any vaccine development and, in particular, mucosal vaccine development is the choice of adjuvant that is both safe and effective. Mucosal vaccines have been difficult to design and develop as a consequence of the few adjuvants that enhance mucosal immunizations. It should be remembered that ineffective stimulation of an immune response at mucosal sites may, in fact, increase the risk of development of antigen-specific immune tolerance rather than promote mucosal immunity [152, 153]. This would, of course, be deleterious to any chlamydial vaccine that is applied at a mucosal site, be it intranasally, rectally or through intravaginal application. Hence, effective adjuvants are required to stimulate strong protective immunity. Interestingly, the most effective substances that exert strong adjuvant function belong to the family of ADP-ribosylating bacterial enterotoxins, cholera toxin (CT) and the closely related *E.coli* heat-labile toxin (LT) [130, 132, 149]. These toxins are AB_5-complexes that bind to gangliosides on the cell membrane of most mammalian cells. In this way the toxins gain access to the cytoplasm and stimulate ADP-ribosylation, which leads to dramatic increases in cAMP. Unfortunately, due to the promiscious binding to all nucleated cells the holotoxins have been found to be too toxic to be used in clinical practice [154, 155]. As a means to avoid the toxicity problem some research groups have developed mutant toxins with less or no ADP-ribosylating activity [156, 157]. These molecules have been found quite effective as mucosal adjuvants, albeit the ADP-ribosylating enzyme has proven a key factor for a optimal adjuvant effect.

An alternative approach to circumvent the toxicity problem was to develop an adjuvant molecule, that combined the full enzymatic activity of CTA1 with a synthetic dimer of the D-fragment of *Staphylococcus aureus* protein A [158, 159]. The resulting CTA1-DD adjuvant, was devoid of the GM1-ganglioside binding B-subunit, and instead was targeted to antigen-presenting cells and B cells, in particular. The CTA1-DD was found to be completely non-toxic [158, 160]. Mice and monkeys were given doses of more than 200 μg of CTA1-DD without any apparent side effects or signs of reactogenicity, while similar doses of CT are known to be lethal. CTA1-DD was demonstrated to be safe and as potent as the CT holotoxin in augmenting immune responses after systemic as well as mucosal immunizations [158, 161, 162]. The adjuvant effect of CTA1-DD was dependent on an intact CTA1 enzymatic activity, because mutations that disrupted enzymatic activity also abolished the adjuvant effect [162]. The CTA1-DD adjuvant has recently been used together with MOMP in mice and it was found that this combination given intranasally stimulated strong protective immunity in the genital tract against a live challenge infection with *C. muridarium* (K.W. Beagley & N. Lycke, unpublished). This and other very promising results demonstrate that prospects have improved for the development of a mucosal vaccine against STDs in general and *Chlamydia* infections in particular.

16.6 Summary

C. trachomatis infection of the human genital tract is a major public health concern. Invasion of the female genital tract by this intracellular bacterium results in naturally limiting infection that is spontaneously cleared by the host immune response. This immune response involves components of both the innate and adaptive immune systems. Cells of the innate immune system sense the presence of the invading *Chlamydia* bacteria through TLRs including TLR4, which is the first step towards limiting the spread of infection and bacterial load. These cells also orchestrate the ensuing inflammatory and adaptive immune responses by release of cytokines and chemokines or expression of chemokine receptors and co-stimulatory molecules. In the early phase of infection NK and NK T cells may be involved through their production of IFN-γ, which has direct effects on the infection by reducing bacterial growth in addition to activation of protective cell-mediated cellular immunity. This protective immune response is thought to be dominated by Th1 $CD4^{+}$ T cells, although CTL as well as specific antibodies most likely contribute to protective immunity and host resistance against genital tract chlamydial-infection.

There is a considerable risk of tissue damage and permanent sequelae secondarily to a genital tract infection with *C. trachomatis*. The immunopathology of *Chlamydia* is complex and we are only beginning to understand some of the components that are responsible for this effect. The production of large quantities of IFN-γ in the microenvironment is the most important protective measure. However, IFN-γ can also promote the development of aberrant inclusions, evasion of adaptive immunity and persistent infection, leading ultimately to chronic infection. Further studies are necessary to unravel the mechanisms involved in the immunopathology of genital tract to chlamydial infections. An important question to resolve is the role of Tregs in preventing tissue damage.

Current vaccines against STDs are ineffective with few exceptions. Apart from the injectable anti-human papilloma virus (HPV) vaccine no anti-STD vaccine has been launched [163, 164]. There is general concensus that a *Chlamydia* vaccine must stimulate protective immunity based on $CD4^{+}$ T cells primarily, but also inducing specific CTL and antibody production. The choice of protective antigens is key to success. In this field, recent progress has been made [130, 132, 144, 149]. Most researchers have taken a subunit vaccine approach, incorporating epitopes from MOMP, OMP and other surface exposed antigens. Moreover, fundamental knowledge of antigen uptake and presentation in the genital tract will help develop local vaccines for intravaginal application. Other routes that are being explored are intranasal and transcutaneous vaccine administrations. The latter approach stimulated strong protective immunity in the genital tract in mice against a live challenge infection with *C. muridarium* [149]. Moreover, the use of innovative adjuvant and delivery systems, such as conjugation of subunit components to adjuvants, such as the CTA1-DD molecule, or viral expression vectors or the use of the combined CTA1-DD/ISCOMs vector, constitute novel approaches to the design of an effective mucosal anti-chlamydial vaccine. Importantly, knowledge of the immune response to *C. trachomatis* vaccines has relevance that extends beyond its field and can have implications for the development of vaccines also against other STDs such as *N. gonorrhoea* and HIV.

References

1. Iweala O.I., Nagler C.R. 2006, Immunol Rev, 213,82.
2. W.H.O. Global Prevalence and Incidence of Selected Curable Sexually Transmitted Infections: Overview and Estimates. World Health Organisation: Geneva, 2001.

3. Resnikoff S., Pascolini D., Etya'ale D., Kocur I., Pararajasegaram R., Pokharel G.P., Mariotti S.P. 2004, Bull World Health Organ, 82,844.
4. Anttila T., Saikku P., Koskela P., Bloigu A., Dillner J., Ikaheimo I., Jellum E., Lehtinen M., Lenner P., Hakulinen T., Narvanen A., Pukkala E., Thoresen S., Youngman L., Paavonen J. 2001, JAMA, 285,47.
5. Plummer F.A., Simonsen J.N., Cameron D.W., Ndinya-Achola J.O., Kreiss J.K., Gakinya M.N., Waiyaki P., Cheang M., Piot P., Ronald A.R., et al. 1991, J Infect Dis, 163,233.
6. Peipert J.F. 2003, N Engl J Med, 349,2424.
7. Yokoi S., Yasuda M., Ito S., Takahashi Y., Ishihara S., Deguchi T., Maeda S., Kubota Y., Tamaki M., Fukushi H. 2004, J Infect Chemother, 10,262.
8. Ward M.E. 1992, J Infect, 25 Suppl 1,11.
9. Lyons J.M., Morre S.A., Airo-Brown L.P., Pena A.S., Ito J.I. 2005, BMC Infect Dis, 5,105.
10. Nelson D.E., Virok D.P., Wood H., Roshick C., Johnson R.M., Whitmire W.M., Crane D.D., Steele-Mortimer O., Kari L., McClarty G., Caldwell H.D. 2005, Proc Natl Acad Sci USA, 102,10658.
11. Roshick C., Wood H., Caldwell H.D., McClarty G. 2006, Infect Immunol, 74,225.
12. Yuan Y., Zhang Y.X., Watkins N.G., Caldwell H.D. 1989, Infect Immunol, 57,1040.
13. Ito J.I., Jr., Lyons J.M., Airo-Brown L.P. 1990, Infect Immunol, 58,2021.
14. Behets F.M., Van Damme K., Rasamindrakotroka A., Hobbs M., McClamroch K., Rasolofomanana J.R., Raharimalala L., Dallabetta G., Andriamiadana J. 2005, Sex Health, 2,77.
15. Fenton K.A., Korovessis C., Johnson A.M., McCadden A., McManus S., Wellings K., Mercer C.H., Carder C., Copas A.J., Nanchahal K., Macdowall W., Ridgway G., Field J., Erens B. 2001, Lancet, 358,1851.
16. Miller W.C., Ford C.A., Morris M., Handcock M.S., Schmitz J.L., Hobbs M.M., Cohen M.S., Harris K.M., Udry J.R. 2004, Jama, 291,2229.
17. Parish W.L., Laumann E.O., Cohen M.S., Pan S., Zheng H., Hoffman I., Wang T., Ng K.H. 2003, Jama, 289,1265.
18. Schemann J.F., Sacko D., Malvy D., Momo G., Traore L., Bore O., Coulibaly S., Banou A. 2002, Int J Epidemiol, 31,194.
19. Scidmore M.A., Fischer E.R., Hackstadt T. 2003, Infect Immunol, 71,973.
20. Beatty W.L. 2007, Cell Microbiol, 9, 2147.
21. Yeaman G.R., Collins J.E., Fanger M.W., Wira C.R., Lydyard P.M. 2001, Immunology, 102,434.
22. Kiviat N.B., Wolner-Hanssen P., Eschenbach D.A., Wasserheit J.N., Paavonen J.A., Bell T.A., Critchlow C.W., Stamm W.E., Moore D.E., Holmes K.K. 1990, Am J Surg Pathol, 14,167.
23. Johnson R.M. 2004, Infect Immunol, 72,3951.
24. Kelly K.A., Natarajan S., Ruther P., Wisse A., Chang M.H., Ault K.A. 2001, J Infect Dis, 184,885.
25. Morrison R.P., Feilzer K., Tumas D.B. 1995, Infect Immunol, 63,4661.
26. Morrison S.G., Morrison R.P. 2000, Infect Immunol, 68,2870.
27. Darville T., Andrews C.W., Jr., Laffoon K.K., Shymasani W., Kishen L.R., Rank R.G. 1997, Infect Immunol, 65,3065.
28. den Hartog J.E., Ouburg S., Land J.A., Lyons J.M., Ito J.I., Pena A.S., Morre S.A. 2006, BMC Infect Dis, 6,122.
29. Darville T., O'Neill J.M., Andrews C.W., Jr., Nagarajan U.M., Stahl L., Ojcius D.M. 2003, J Immunol, 171,6187.
30. Maxion H.K., Kelly K.A. 2002, Infect Immunol, 70,1538.
31. Soboll G., Schaefer T.M., Wira C.R. 2006, Am J Reprod Immunol, 55,434.
32. Lorenz E., Mira J.P., Cornish K.L., Arbour N.C., Schwartz D.A. 2000, Infect Immunol, 68,6398.
33. Sutherland A.M., Walley K.R., Russell J.A. 2005, Crit Care Med, 33,638.
34. Erridge C., Stewart J., Poxton I.R. 2003, J Exp Med, 197,1787.

35. Morre S.A., Murillo L.S., Bruggeman C.A., Pena A.S. 2003, J Infect Dis, 187,341.
36. Ouburg S., Spaargaren J., den Hartog J.E., Land J.A., Fennema J.S., Pleijster J., Pena A.S., Morre S.A. 2005, BMC Infect Dis, 5,114.
37. Girardin S.E., Boneca I.G., Carneiro L.A., Antignac A., Jehanno M., Viala J., Tedin K., Taha M.K., Labigne A., Zahringer U., Coyle A.J., DiStefano P.S., Bertin J., Sansonetti P.J., Philpott D.J. 2003, Science, 300,1584.
38. Inohara N., Ogura Y., Chen F.F., Muto A., Nunez G. 2001, J Biol Chem, 276,2551.
39. Girardin S.E., Tournebize R., Mavris M., Page A.L., Li X., Stark G.R., Bertin J., DiStefano P.S., Yaniv M., Sansonetti P.J., Philpott D.J. 2001, EMBO Rep, 2,736.
40. Ghuysen J.M., Goffin C. 1999, Antimicrob Agents Chemother, 43,2339.
41. Welter-Stahl L., Ojcius D.M., Viala J., Girardin S., Liu W., Delarbre C., Philpott D., Kelly K.A., Darville T. 2006, Cell Microbiol, 8,1047.
42. Cain T.K., Rank R.G. 1995, Infect Immunol, 63,1784.
43. Johansson M., Schon K., Ward M., Lycke N. 1997, Infect Immunol, 65,1032.
44. Mavoungou E., Poaty-Mavoungou V., Toure F.S., Sall A., Delicat A., Yaba P., Mandeme Y., Nabias R., Lansoud-Soukate J. 1999, Trop Med Int Health, 4,719.
45. Bailey R.L., Holland M.J., Whittle H.C., Mabey D.C. 1995, Infect Immunol, 63,389.
46. Holland M.J., Bailey R.L., Hayes L.J., Whittle H.C., Mabey D.C. 1993, J Infect Dis, 168,1528.
47. Apostolou I., Takahama Y., Belmant C., Kawano T., Huerre M., Marchal G., Cui J., Taniguchi M., Nakauchi H., Fournie J.J., Kourilsky P., Gachelin G. 1999, Proc Natl Acad Sci U S A, 96,5141.
48. Exley M.A., Bigley N.J., Cheng O., Tahir S.M., Smiley S.T., Carter Q.L., Stills H.F., Grusby M.J., Koezuka Y., Taniguchi M., Balk S.P. 2001, J Leukoc Biol, 69,713.
49. Gonzalez-Aseguinolaza G., de Oliveira C., Tomaska M., Hong S., Bruna-Romero O., Nakayama T., Taniguchi M., Bendelac A., Van Kaer L., Koezuka Y., Tsuji M. 2000, Proc Natl Acad Sci U S A, 97,8461.
50. Leite-De-Moraes M.C., Hameg A., Pacilio M., Koezuka Y., Taniguchi M., Van Kaer L., Schneider E., Dy M., Herbelin A. 2001, J Immunol, 166,945.
51. Singh A.K., Wilson M.T., Hong S., Olivares-Villagomez D., Du C., Stanic A.K., Joyce S., Sriram S., Koezuka Y., Van Kaer L. 2001, J Exp Med, 194,1801.
52. Kawana K., Quayle A.J., Ficarra M., Ibana J.A., Shen L., Kawana Y., Yang H., Marrero L., Yavagal S., Greene S.J., Zhang Y.X., Pyles R.B., Blumberg R.S., Schust D.J. 2007, J Biol Chem, 282,7368.
53. Paavonen J., Teisala K., Heinonen P.K., Aine R., Laine S., Lehtinen M., Miettinen A., Punnonen R., Gronroos P. 1987, Br J Obstet Gynaecol, 94,454.
54. Perry L.L., Feilzer K., Caldwell H.D. 1997, J Immunol, 158,3344.
55. Su H., Caldwell H.D. 1995, Infect Immunol, 63,3302.
56. Johansson M., Ward M., Lycke N. 1997, Immunology, 92,422.
57. Wang S., Fan Y., Brunham R.C., Yang X. 1999, Eur J Immunol, 29,3782.
58. Hawkins R.A., Rank R.G., Kelly K.A. 2002, Infect Immunol, 70,5132.
59. Shaw J., Grund V., Durling L., Crane D., Caldwell H.D. 2002, Infect Immunol, 70,1097.
60. Beatty P.R., Stephens R.S. 1994, J Immunol, 153,4588.
61. Starnbach M.N., Bevan M.J., Lampe M.F. 1994, J Immunol, 153,5183.
62. Ljunggren G., Anderson D.J. 1998, J Reprod Immunol, 38,123.
63. Thoma-Uszynski S S.U., Marro R, Essig A. 1998, Microbiol Immunol 187,71.
64. Igietseme J.U., Magee D.M., Williams D.M., Rank R.G. 1994, Infect Immunol, 62,5195.
65. Harding F.A., McArthur J.G., Gross J.A., Raulet D.H., Allison J.P. 1992, Nature, 356,607.
66. June C.H., Ledbetter J.A., Linsley P.S., Thompson C.B. 1990, Immunol Today, 11,211.
67. Riley J.L., Mao M., Kobayashi S., Biery M., Burchard J., Cavet G., Gregson B.P., June C.H., Linsley P.S. 2002, Proc Natl Acad Sci U S A, 99,11790.
68. Scott P. 1991, J Immunol, 147,3149.
69. Sloan-Lancaster J., Steinberg T.H., Allen P.M. 1997, J Immunol, 159,1160.

70. McAdam A.J., Chang T.T., Lumelsky A.E., Greenfield E.A., Boussiotis V.A., Duke-Cohan J.S., Chernova T., Malenkovich N., Jabs C., Kuchroo V.K., Ling V., Collins M., Sharpe A.H., Freeman G.J. 2000, J Immunol, 165,5035.
71. Ubaldi V., Gatta L., Pace L., Doria G., Pioli C. 2003, Clin Dev Immunol, 10,13.
72. Gross J.A., Callas E., Allison J.P. 1992, J Immunol, 149,380.
73. Compton H.L., Farrell J.P. 2002, J Immunol, 168,1302.
74. Mittrucker H.W., Kohler A., Mak T.W., Kaufmann S.H. 1999, J Immunol, 163,6769.
75. Mittrucker H.W., Kursar M., Kohler A., Yanagihara D., Yoshinaga S.K., Kaufmann S.H. 2002, J Immunol, 169,5813.
76. Noel P.J., Boise L.H., Green J.M., Thompson C.B. 1996, J Immunol, 157,636.
77. Shahinian A., Pfeffer K., Lee K.P., Kundig T.M., Kishihara K., Wakeham A., Kawai K., Ohashi P.S., Thompson C.B., Mak T.W. 1993, Science, 261,609.
78. Tafuri A., Shahinian A., Bladt F., Yoshinaga S.K., Jordana M., Wakeham A., Boucher L.M., Bouchard D., Chan V.S., Duncan G., Odermatt B., Ho A., Itie A., Horan T., Whoriskey J.S., Pawson T., Penninger J.M., Ohashi P.S., Mak T.W. 2001, Nature, 409,105.
79. Tai X., Cowan M., Feigenbaum L., Singer A. 2005, Nat Immunol, 6,152.
80. Marks E., Verolin M., Stensson A., Lycke N. 2007, Infect Immunol.
81. Johansson M., Schon K., Ward M., Lycke N. 1997, Scand J Immunol, 46,546.
82. Ramsey K.H., Soderberg L.S., Rank R.G. 1988, Infect Immunol, 56,1320.
83. Beatty W.L., Belanger T.A., Desai A.A., Morrison R.P., Byrne G.I. 1994, Infect Immunol, 62,3705.
84. Rottenberg M.E., Gigliotti-Rothfuchs A., Wigzell H. 2002, Curr Opin Immunol, 14,444.
85. Hattor Y., Kasai K., Akimoto K., Thiemermann C. 1997, Biochem Biophys Res Commun, 233,375.
86. Xie Q.W., Cho H.J., Calaycay J., Mumford R.A., Swiderek K.M., Lee T.D., Ding A., Troso T., Nathan C. 1992, Science, 256,225.
87. Ouellet H., Ouellet Y., Richard C., Labarre M., Wittenberg B., Wittenberg J., Guertin M. 2002, Proc Natl Acad Sci U S A, 99,5902.
88. Wink D.A., Kasprzak K.S., Maragos C.M., Elespuru R.K., Misra M., Dunams T.M., Cebula T.A., Koch W.H., Andrews A.W., Allen J.S., et al. 1991, Science, 254,1001.
89. Mayer J., Woods M.L., Vavrin Z., Hibbs J.B., Jr. 1993, Infect Immunol, 61,491.
90. Chen B., Stout R., Campbell W.F. 1996, FEMS Immunol Med Microbiol, 14,109.
91. Igietseme J.U. 1996, Immunology, 87,1.
92. Ramsey K.H., Miranpuri G.S., Sigar I.M., Ouellette S., Byrne G.I. 2001, Infect Immunol, 69,5131.
93. Ramsey K.H., Sigar I.M., Rana S.V., Gupta J., Holland S.M., Byrne G.I. 2001, Infect Immunol, 69,7374.
94. Thomas S.M., Garrity L.F., Brandt C.R., Schobert C.S., Feng G.S., Taylor M.W., Carlin J.M., Byrne G.I. 1993, J Immunol, 150,5529.
95. Ramsey K.H., Miranpuri G.S., Poulsen C.E., Marthakis N.B., Braune L.M., Byrne G.I. 1998, Infect Immunol, 66,835.
96. Mestecky J., Russell M.W. 2000, FEMS Immunol Med Microbiol, 27,351.
97. Morrison S.G., Morrison R.P. 2005, J Immunol, 175,7536.
98. Hessel T., Dhital S.P., Plank R., Dean D. 2001, Infect Immunol, 69,4996.
99. Moore T., Ananaba G.A., Bolier J., Bowers S., Belay T., Eko F.O., Igietseme J.U. 2002, Immunology, 105,213.
100. Idahl A., Abramsson L., Kumlin U., Liljeqvist J.A., Olofsson J.I. 2007, Int J Androl, 30,99.
101. Johansson M., Lycke N. 2001, Immunology, 102,199.
102. Chaim W., Edelstein Z., Sarov B., Sarov I. 1992, Arch Gynecol Obstet, 251,159.
103. Numazaki K., Kusaka T., Chiba S. 1996, Clin Infect Dis, 23,208.
104. Sharara F.I., Queenan J.T., Jr. 1999, J Reprod Med, 44,581.
105. Patton D.L., Moore D.E., Spadoni L.R., Soules M.R., Halbert S.A., Wang S.P. 1989, Obstet Gynecol, 73,622.

106. McCormack W.M. 1994, N Engl J Med, 330,115.
107. Westrom L. 1995, Venereology, 8,219.
108. Westrom L. 1980, Am J Obstet Gynecol, 138,880.
109. Pal S., Hui W., Peterson E.M., de la Maza L.M. 1998, J Med Microbiol, 47,599.
110. Kunimoto D., Brunham R.C. 1985, Rev Infect Dis, 7,665.
111. Viale A.M., Arakaki A.K., Soncini F.C., Ferreyra R.G. 1994, Int J Syst Bacteriol, 44,527.
112. Yi Y., Yang X., Brunham R.C. 1997, Infect Immunol, 65,1669.
113. Wagenlehner F.M., Weidner W., Naber K.G. 2006, World J Urol, 24,4.
114. Idahl A., Boman J., Kumlin U., Olofsson J.I. 2004, Hum Reprod, 19,1121.
115. Gdoura R., Keskes-Ammar L., Bouzid F., Eb F., Hammami A., Orfila J. 2001, Eur J Contracept Reprod Health Care, 6,102.
116. Satta A., Stivala A., Garozzo A., Morello A., Perdichizzi A., Vicari E., Salmeri M., Calogero A.E. 2006, Hum Reprod, 21,134.
117. Hosseinzadeh S., Pacey A.A., Eley A. 2003, J Med Microbiol, 52,193.
118. Jungwirth A., Straberger A., Esterbauer B., Fink K., Schmeller N. 2003, Andrologia, 35,314.
119. Hosseinzadeh S., Eley A., Pacey A.A. 2004, J Androl, 25,104.
120. Eley A., Hosseinzadeh S., Hakimi H., Geary I., Pacey A.A. 2005, Hum Reprod, 20,2601.
121. Meinhardt A., Parvinen M., Bacher M., Aumuller G., Hakovirta H., Yagi A., Seitz J. 1995, Biol Reprod, 52,798.
122. Kozlowski P.A., Williams S.B., Lynch R.M., Flanigan T.P., Patterson R.R., Cu-Uvin S., Neutra M.R. 2002, J Immunol, 169,566.
123. Rudin A., Johansson E.L., Bergquist C., Holmgren J. 1998, Infect Immunol, 66,3390.
124. Kelly K.A., Robinson E.A., Rank R.G. 1996, Infect Immunol, 64,4976.
125. Maclean I.W., Peeling R.W., Brunham R.C. 1988, Can J Microbiol, 34,141.
126. Wyllie S., Ashley R.H., Longbottom D., Herring A.J. 1998, Infect Immunol, 66,5202.
127. Villeneuve A., Brossay L., Paradis G., Hebert J. 1994, Microbiology, 140 (Pt 9),2481.
128. Knight S.C., Iqball S., Woods C., Stagg A., Ward M.E., Tuffrey M. 1995, Immunology, 85,8.
129. Conlan J.W., Kajbaf M., Clarke I.N., Chantler S., Ward M.E. 1989, Mol Microbiol, 3,311.
130. Hickey D.K., Jones R.C., Bao S., Blake A.E., Skelding K.A., Berry L.J., Beagley K.W. 2004, Vaccine, 22,4306.
131. Pal S., Peterson E.M., de la Maza L.M. 2005, Infect Immunol, 73,8153.
132. Pal S., Peterson E.M., Rappuoli R., Ratti G., de la Maza L.M. 2006, Vaccine, 24,766.
133. Skelding K.A., Hickey D.K., Horvat J.C., Bao S., Roberts K.G., Finnie J.M., Hansbro P.M., Beagley K.W. 2006, Vaccine, 24,355.
134. Pal S., Luke C.J., Barbour A.G., Peterson E.M., de la Maza L.M. 2003, Vaccine, 21,1455.
135. Allen J.E., Stephens R.S. 1993, Eur J Immunol, 23,1169.
136. Freidank H.M., Herr A.S., Jacobs E. 1993, Eur J Clin Microbiol Infect Dis, 12,947.
137. Vandahl B.B., Birkelund S., Christiansen G. 2004, Proteomics, 4,2831.
138. Kawa D.E., Schachter J., Stephens R.S. 2004, Vaccine, 22,4282.
139. Murdin A.D., Dunn P., Sodoyer R., Wang J., Caterini J., Brunham R.C., Aujame L., Oomen R. 2000, J Infect Dis, 181 Suppl 3,S544.
140. Donati M., Sambri V., Comanducci M., Di Leo K., Storni E., Giacani L., Ratti G., Cevenini R. 2003, Vaccine, 21,1089.
141. Sharma J., Bosnic A.M., Piper J.M., Zhong G. 2004, Infect Immunol, 72,7164.
142. Belland R.J., Scidmore M.A., Crane D.D., Hogan D.M., Whitmire W., McClarty G., Caldwell H.D. 2001, Proc Natl Acad Sci U S A, 98,13984.
143. Slepenkin A., de la Maza L.M., Peterson E.M. 2005, J Bacteriol, 187,473.
144. McNeilly C.L., Beagley K.W., Moore R.J., Haring V., Timms P., Hafner L.M. 2007, Vaccine, 25,2643.
145. Dong-Ji Z., Yang X., Shen C., Lu H., Murdin A., Brunham R.C. 2000, Infect Immunol, 68,3074.
146. Eko F.O., He Q., Brown T., McMillan L., Ifere G.O., Ananaba G.A., Lyn D., Lubitz W., Kellar K.L., Black C.M., Igietseme J.U. 2004, J Immunol, 173,3375.

147. Eko F.O., Lubitz W., McMillan L., Ramey K., Moore T.T., Ananaba G.A., Lyn D., Black C.M., Igietseme J.U. 2003, Vaccine, 21,1694.
148. He Q., Martinez-Sobrido L., Eko F.O., Palese P., Garcia-Sastre A., Lyn D., Okenu D., Bandea C., Ananaba G.A., Black C.M., Igietseme J.U. 2007, Immunology.
149. Berry L.J., Hickey D.K., Skelding K.A., Bao S., Rendina A.M., Hansbro P.M., Gockel C.M., Beagley K.W. 2004, Infect Immunol, 72,1019.
150. Igietseme J.U., Murdin A. 2000, Infect Immunol, 68,6798.
151. Singh S.R., Hulett K., Pillai S.R., Dennis V.A., Oh M.K., Scissum-Gunn K. 2006, Vaccine, 24,1213.
152. Ngan J., Kind L.S. 1978, J Immunol, 120,861.
153. Walker L.S., Chodos A., Eggena M., Dooms H., Abbas A.K. 2003, J Exp Med, 198,249.
154. Northrup R.S., Chisari F.V. 1972, Prog Immunobiol Stand, 5,355.
155. Northrup R.S., Chisari F.V. 1972, J Infect Dis, 125,471.
156. Douce G., Turcotte C., Cropley I., Roberts M., Pizza M., Domenghini M., Rappuoli R., Dougan G. 1995, Proc Natl Acad Sci U S A, 92,1644.
157. Giuliani M.M., Del Giudice G., Giannelli V., Dougan G., Douce G., Rappuoli R., Pizza M. 1998, J Exp Med, 187,1123.
158. Agren L.C., Ekman L., Lowenadler B., Lycke N.Y. 1997, J Immunol, 158,3936.
159. Uhlen M., Guss B., Nilsson B., Gotz F., Lindberg M. 1984, J Bacteriol, 159,713.
160. Agren L., Lowenadler B., Lycke N. 1998, Immunol Cell Biol, 76,280.
161. Agren L., Sverremark E., Ekman L., Schon K., Lowenadler B., Fernandez C., Lycke N. 2000, J Immunol, 164,6276.
162. Agren L.C., Ekman L., Lowenadler B., Nedrud J.G., Lycke N.Y. 1999, J Immunol, 162,2432.
163. Billich A. 2003, Curr Opin Investig Drugs, 4,210.
164. Koutsky L.A., Ault K.A., Wheeler C.M., Brown D.R., Barr E., Alvarez F.B., Chiacchierini L.M., Jansen K.U. 2002, N Engl J Med, 347,1645.

Chapter 17
HIV and the Mucosa: No Safe Haven

Satya Dandekar, Sumathi Sankaran and Tiffany Glavan

Abstract Since its discovery in 1983, Human Immunodeficiency Virus-1 (HIV) infection has resulted in the death of several million people worldwide. Control of the spread of HIV infections is one of the major concerns of the scientific community today. This retrovirus houses a genome encoding three viral structural proteins: group-specific antigen, polymerase and envelope. The envelope gp120 binds to cell surface receptors on susceptible target cells in the host. Although HIV is known to infect various cell types, cells expressing the CD4 receptor support high levels of viral replication in vivo. Gastrointestinal complications are commonly seen in patients with HIV-1 infections progressing to AIDS. Recent studies have shown that HIV causes massive disruption of the Gut Associated Lymphoid Tissue (GALT) very early in infection. Intestinal dysfunction is often not resolved despite the success of HAART in improving the clinical status of these patients. GALT supports robust viral replication in mucosal memory CD4+ T cells during primary HIV infection, leading to a severe depletion of the CD4+ T cells which persists through all stages of HIV infection in the absence of therapeutic interventions. Functional genomic analysis of host responses to HIV shows that molecular processes regulating inflammation and immune activation are dominantly expressed in GALT of therapy-nave HIV-1 infected patients with a progressive clinical course. Heterosexual transmission is the primary mode for the spread of HIV infection and the virus gains entry through mucosal surfaces and infects susceptible target cells. Since a majority of new infections are acquired through the mucosal route, induction of mucosal immunity is an important factor in vaccine design.

17.1 Introduction

Since the discovery of Human Immunodeficiency Virus-1 [HIV] in 1983, more than 25 million people have died of the HIV-induced disease, acquired immunodeficiency syndrome [AIDS]. According to the UNAIDS 2007 report, an estimated

---

S. Dandekar
Department of Medical Microbiology and Immunology, University of California, Davis, CA, USA
e-mail: sdandekar@ucdavis.edu

M. Vajdy (ed.), *Immunity Against Mucosal Pathogens*,

33.2 million people around the world are living with HIV/AIDS, of whom 15.4 million are women and 2.5 million are children under the age of 15 [1]. In sub-Saharan Africa, more than 22.5 million people are reportedly infected with HIV, making this region the most affected area in the world [1]. A high incidence of new HIV infections has been reported among pregnant women and female sex workers in Botswana and Zimbabwe [2]. In the United States, approximately 1.2 million people are living with HIV, of whom 300,000 are women [3]. The development of innovative strategies for HIV prevention will be crucial to the control of new infections worldwide.

Globally, heterosexual transmission is the primary mode for the spread of HIV infection [3]. In China, India, and Eastern Europe, the increased prevalence of new infections has been associated with intravenous drug use, sexually transmitted diseases, and poorly monitored public health systems [4, 5, 6, 7, 8]. Young men who have sex with men remain the predominantly affected group in the United States. However, recent reports suggest an increased incidence of HIV infections among women and racial minorities, including African Americans and Hispanics, who represent the majority of new HIV cases in the United States irrespective of the mode of transmission. HIV transmission has been shown to occur through exposure to HIV-positive blood via blood transfusions, intravenous drug use, needle sticks and from mother-to-child during childbirth or through breast-feeding [9, 10, 11, 12, 13]. Although the availability of an array of anti-HIV drugs has significantly improved the management of HIV disease, eradication of HIV infection remains a major challenge. Novel approaches towards HIV prevention that are effective at mucosal sites of transmission will be required to prevent new HIV infections.

HIV has evolved over time into separate clades, possibly due to multiple transfers from animal reservoirs to humans. Three major classes of HIV have been identified: M [main], N [new], and O [outlier]. The M group viruses are subdivided into nine clades, designated by the letters A-D, F-H, J, and K [14]. The majority of HIV infections in the Unites States consist of HIV-1 clade B. Human immunodeficiency virus 2 [HIV-2] is closely related to HIV-1 with respect to viral structure and disease but the two species have distinct biologic and serologic features [15]. HIV-2 has been detected primarily in the African subcontinent but it was recently found in parts of Europe and India. The transmissibility and pathogenic potential of HIV-2 is lower than that of HIV-1 and people infected with HIV-2 survive for longer periods of time than those infected with HIV-1 [15]. These differences can be attributed to lower viral loads, highly effective virus neutralizing antibody levels and HIV-specific cellular immune responses, reduced cytopathicity, and a lack of modulation of CD4 expression on the target cell surface.

17.2 Structure of HIV

HIV is a lentivirus that belongs to the Retroviridae family of viruses. The viral particle has a diameter of approximately 100–120 nm [16]. The genome of HIV is approximately 9.8 kilobase in length. Alternative splicing events and post-transcriptional proteolytic cleavage allow for the production of multiple viral proteins that are involved in the formation of infectious viral particles and the regulation of viral and host gene expression. The HIV genome encodes three viral structural proteins, group specific antigen [Gag], polymerase [Pol] and envelope [Env]. In addition, it codes for six regulatory and accessory proteins that are involved in the modulation of host and viral gene expression [Tat, Rev, Vpu, Vpr, Vif, and Nef]. The coding sequences are flanked by long terminal repeat [LTR] sequences that harbor important molecular determinants for viral transcriptional regulation and integration. The enveloped virion contains two identical copies of the single stranded positive-sense RNA genome encapsulated in a non-covalently linked dimer and enclosed within a cone-shaped viral core (Fig. 17.1) [17]. The core also harbors the viral proteins reverse transcriptase, integrase, and protease that are crucial for the initial stage of viral replication. Several other viral proteins are packaged within or associated with the core of the virion, including Nef, Vif, Vpr [HIV-1 only], and Vpx [HIV-2 only]. These viral proteins lead to an increase in the replication and/or infectivity of the virus through multiple mechanisms. In addition, certain cytoskeletal proteins are also packaged within the virion, but their functions in viral infection have not been fully revealed [18]. A better understanding of the structure and functional domains of the viral antigens will be important for the development of effective anti-HIV therapy as well as vaccines.

The *gag* gene encoded Gag polyprotein is cleaved by the viral protease to produce proteins that form the viral capsid [p24], the nucleocapsid [p6 and p7], and the matrix [p17]. HIV nucleocapsid proteins are also found within the core and act as chaperones to protect the RNA from digestion by host cell nucleases. The viral RNA polymerase gene [Pol] codes for the p160 polyprotein that is proteolytically processed to release the protease, reverse transcriptase, and integrase enzymes. Reverse transcriptase is an RNA-dependent DNA polymerase necessary to convert the viral RNA genome into complementary DNA. Integrase trims the ends of the viral DNA and cleaves the host DNA so that the viral genome can be integrated

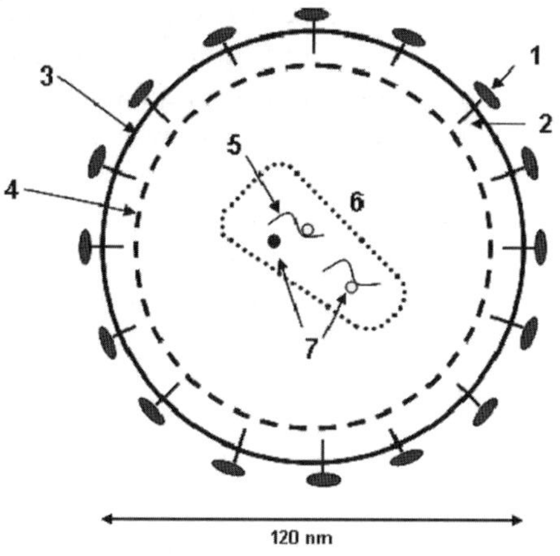

Fig. 17.1 HIV Structure. 1. Envelope surface protein (gp120) 2. Envelope transmembrane protein (gp41) 3. Cell-derived lipid bilayer 4. Matrix proteins (p17) 5. RNA genomes 6. Capsid (p24) 7. Various viral proteins, including: protease (p10), integrase (p32), reverse transcriptase (p66,p51), and the nucleocapsid protein (p7). Vif, Nef, and Vpr are also located inside the virion, but their exact locations are unknown

directly into the host chromosomal DNA. The protease enzyme is responsible for the post-transcriptional processing of viral polyproteins, converting them to their functional forms to generate infectious viral structures.

The viral envelope consists of two layers of phospholipids that are derived from the cell membrane during budding. The *envelope* gene codes for the Env polyprotein gp160, which is processed by cellular proteases to form surface envelope gp120 and transmembrane envelope gp41. These two proteins produce a structure consisting of a trimeric head that extends out of the viral envelope and a tripod-like stalk that anchors it into the bilayer [19]. The *envelope* gp120 binds to cell surface receptors on susceptible target cells in the host and gp41 facilitates fusion of the viral envelope with the host cell membrane. The matrix protein of HIV, p17, is inserted into the inner surface of the envelope layer and surrounds the inner core. The matrix confers stability to the envelope layer and protects the structural integrity of the virus particle. The matrix proteins may also participate in viral localization to lipid rafts [20].

Tat, the "trans-activator of transcription", plays a primary role in the regulation of HIV transcription. It binds to the transactivation response element [TAR] and activates transcription from the HIV-LTR, promoting elongation and production of full-length viral transcripts. The single-stranded viral RNA genome initially creates a hairpin that prohibits the polymerase enzyme from traveling along the entire genome. Tat phosphorylates the RNA and relieves the hairpin configuration. It has been shown that Tat may also directly alter the host cell gene expression, causing cytopathic effects [21]. Rev, the "Regulator of Virion", is crucial for the export of viral RNA from the nucleus to the cytoplasm. In the absence of Rev, host RNA splicing machinery processes the viral RNA. Rev binds to a cis-acting RNA loop structure, the Rev Response Element [RRE], promoting the transport of unspliced viral transcripts from the nucleus to the cytoplasm and facilitating production of viral structural proteins. Since Rev is produced in the early phase of the viral life cycle, its accumulation conducts an appropriately timed transition into late phase of the viral replication cycle.

Vif, a "viral infectivity factor", is an accessory protein that accumulates in the cytoplasm of infected cells. It has been demonstrated that infectivity of the HIV virus decreases up to 50-fold in the absence of Vif [22]. Recent studies have shown that Vif may function by degrading a cellular protein, APOBEC3G, through a ubiquitination-dependent pathway [23]. APOBEC3G catalyzes the deamination of deoxycytidine in the first DNA strand synthesized by reverse transcriptase. This introduced mutation may account for the lower infectivity of *vif* mutants and may represent an innate cellular defense against retroviruses. Nef, a "negative regulatory factor", is shown to be essential for the pathogenesis of HIV. It is translated from multiply spliced early transcripts. Nef has multiple functions and is shown to impact viral infection on various levels. The presence of Nef causes downregulation of CD4 and major histocompatibility complex class I surface expression [24]. This is one of the important mechanisms contributing to the immune evasion of HIV infected cells and their persistence. Vpr, or "viral protein R", regulates nuclear import and localization of the HIV preintegration complex. It also prevents the activation of the p34cdc2/cyclin B complex, causing cell cycle arrest in the G2 phase and subsequent apoptosis [25]. However, cell cycle arrest of the cells may also generate a pool of latently HIV infected cells. Vpx is found only in HIV-2 and carries out a role analogous to Vpr of HIV-1. It is thought that Vpx might have been formed from a duplication of Vpr. Another HIV accessory protein, Vpu [viral protein U], is a 16-kD membrane phosphoprotein that is unique to HIV-1 and plays an important role in the release of the virus during budding process. In the absence of Vpu, viral particles containing multiple cores are produced [26]. Vpu also plays a role in the degradation of CD4 by trapping it in the endoplasmic reticulum of the cell and triggering the ubiquitin-mediated degradation of CD4 molecules complexed with Env [27]. Since activation of the immune system is dependant on surface molecules displayed by the virion, comprehensive knowledge of HIV viral structural proteins is crucial to the development of successful vaccines.

The emergence of viral variants has been well documented in HIV infected patients due to the high mutation rate in the HIV genomes. The error-prone nature of reverse transcriptase is estimated to introduce a mutation into each viral replication cycle. Viral variants generated by new mutations could escape from detection by virus specific antibodies and cytotoxic T lymphocytes in the host, leading to viral persistence. Development of new mutations in HIV genomes have also contributed to resistance to anti-retroviral drugs and failure of therapy in HIV infected patients.

17.3 HIV Infection

Although HIV is known to infect various cell types, cells expressing the CD4 receptor seem to support high levels of viral replication *in vivo*. CD4 receptors are found on a variety of hematopoietic cells, including T lymphocytes, thymocytes, monocytes, macrophages, and dendritic cells. Some CD4-negative cells, such as follicular dendritic cells and epithelial cells, are also susceptible to HIV infection. The viral entry into epithelial cells may involve other molecules such as galactosyl ceramide [28]. Attachment of the viral envelope to the CD4 receptor in itself is not sufficient for entry into the cell. Two chemokine receptors, CCR5 and CXCR4 have been shown to serve as HIV co-receptors that enable fusion of the viral envelope with the cell membrane of the host cell. The CD4+ T cell tropic HIV variants have been shown to utilize the CXCR4 co-receptor for viral entry while CCR5 co-receptor utilizing viral variants primarily infect monocytes/macrophages. Viral variants capable of utilizing both CXCR4 and CCR5, have been detected *in vivo* and are capable of infecting both T cells and monocyte derived cells [29]. Viral variants isolated during primary HIV infection have been found to be CCR5-tropic. In individuals exposed to a mixture of both X4-tropic and R5-tropic HIV variants, R5 utilizing HIV were first to be recovered in the peripheral blood [30]. X4 is slow to appear but its emergence has been associated with the clinical deterioration seen with rapid progressive to AIDS.

The initial attachment of HIV envelope with CD4 receptor and interaction with the chemokine co-receptors supports the process of viral entry into the cell as the conformational change in Env gp120 uncovers domains on gp41 that are needed for fusion of the viral envelope to the host cell membrane. Entry of the viral capsid containing the viral RNA genome is pH independent. Cyclophilins from the host cell associate with the p24 proteins to uncoat the genome [31]. The single stranded viral RNA genome is converted to complementary DNA by the viral reverse transcriptase enzyme (Fig. 17.2). The resulting cDNA is then copied to generate dsDNA, and is circularized into both covalently and noncovalently bound forms before being transported into the nucleus in a pre-integration complex with the viral matrix protein, Vpr, and integrase. The integrase contains the nuclear localization signal necessary for transport [32]. Integrase trims the ends of the viral DNA, cleaves the host genome, and catalyzes the integration of viral DNA into the host cell genome. During the early phase of transcription, transcripts of Tat, Rev and Nef are produced. This is followed by the production and export of unspliced large viral transcripts to the cytoplasm. Activated T cells support high levels of viral replication due to the availability of cellular enzymes. Viral structural proteins are synthesized during late phase transcription and newly synthesized viral genomic RNA is packaged into the capsid at the cell membrane. As the virus is assembled, the Gag and Gag-Pol

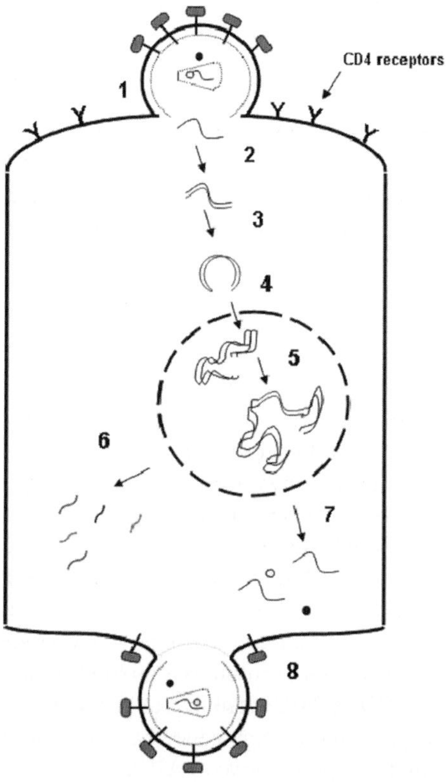

Fig. 17.2 Viral Replication Cycle. 1. Fusion of virus with host cell membrane following attachment to CD4 receptors and chemokine coreceptors 2. Uncoating of core, reverse transcription of viral RNA genome 3. Replication and circularization of cDNA genome 4. Transport into host cell nucleus 5. Integration into host cell genome 6. Early phase expression of HIV regulatory proteins (Tat,Rev,Nef) 7. Late phase production of structural proteins and replication of RNA genomes, followed by assembly of new virions. 8. Budding and cleavage of Gag and Gag-Pol polyproteins

polyproteins are processed by protease enzymes and the viral capsid buds through the membrane. During this process, the newly constructed virion takes up viral envelope proteins embedded at the site of budding. The viral budding involves the accumulation of lipid rafts in the membrane [20].

The cytopathic effects of HIV infection on CD4+ T cells result in CD4+ T cell loss. Multiple mechanisms contribute to cell death, including disruption of the integrity of the cell membrane due to budding of many viral particles, apoptosis induced by the viral gene products, and activation induced cell death. In addition, many uninfected CD4+ T cells are also killed due to the formation of multi-nucleated giant cells as a consequence of the fusion of HIV infected cells with uninfected cells. Eventually, the production of new CD4+ T cells is not sufficient to overcome the loss of CD4+ T cells in HIV infected individuals, a condition which leads to severe immune dysfunction.

During mucosal transmission of HIV, the virus gains entry through mucosal surfaces and infects susceptible target cells. Dendritic cells and infected CD4+ T cells contribute to further dissemination of the virus by transporting the virus to regional lymph nodes where it can spread to other cellular targets, resulting in a dramatic increase in the levels of infectious virus. Newly produced virions and HIV infected cells can disseminate the infection systemically to other lymphoid and non-lymphoid organs. During this primary acute stage of HIV infection, some patients develop flu-like symptoms within a few weeks to several months of initial exposure to the virus. At this point in the infection high HIV viral loads are detected in plasma. The viral population present in peripheral blood is quite homogeneous and is generally CCR5-tropic. The development of HIV-specific cellular and humoral immune responses at this stage leads to a reduction in the viral loads in peripheral blood as well as lymphoid tissues. HIV infected individuals can then enter a clinically asymptomatic stage that can last for a variable number of years, in some patients for more than 20 years. During this stage, new viruses are continually being produced at a low rate, but the magnitude of viral replication and damage to the immune system are limited by the anti-HIV host immune responses. Gradual loss of immune control during the course of infection can be attributed to the emergence of viral variants that evade HIV-specific host immune responses and the decreased capacity of the host to replenish the CD4+ T cell pool, leading to increased viremia, severe CD4+ T cell loss, and impaired immune functions and thus AIDS [33]. During this time the virus population becomes more heterogeneous, as mutations accumulate in the viral genomes. HIV infected patients with a decline in CD4+ T cell numbers below 350 cells in peripheral blood often have an increased susceptibility to opportunistic infections due to the impaired functions of both CD4+ and CD8+ T cells and blunted immune responses [34, 35]. AIDS patients are also more susceptible to neurological complications and several forms of cancer. Early intervention with highly active antiretroviral therapy [HAART] of HIV infected patients is beneficial in suppressing viral replication and enhancing immune functions, leading to a long clinically asymptomatic stage [36]. Since the lymphoid tissue in the body harbors most of the CD4+ T cells, it is a site for the dynamic interactions between the virus and the host immune system, resulting in many of the pathologic consequences of

HIV infection. Therefore, investigation of lymphoid tissues during HIV infection and during therapy will be important to arriving at a better understanding of the immunopathogenesis of the virus.

17.4 Pathogenic Effects of HIV Infection in the Gastrointestinal Mucosa

17.4.1 Structure of the Gastrointestinal Mucosa

Recent studies from a number of investigators have collectively shown that, regardless of the route of transmission, HIV causes massive disruption of the Gut Associated Lymphoid Tissue [GALT] very early in infection. Therefore, an understanding of the effects of HIV infection on GALT will be important in gaining insights into HIV pathogenesis. GALT harbors a significant portion of total lymphoid tissue in the body, i.e. >80% of the body's lymphocytes. Thus, GALT is an important player in providing the host immune defense against pathogens. In addition, the GI tract carries out the functions of processing and digesting food, absorbing nutrients and expelling waste. The GI tract is approximately 7.5 meters long [25 feet] and is divided into upper and lower sections. The upper section consists of mouth or buccal cavity, pharynx, esophagus, and stomach. The lower gastrointestinal tract includes the small intestine [duodenum, jejunum, and ileum] the large intestine or colon [ascending colon, transverse colon, descending colon and sigmoid flexure], the rectum, and the anus.

Histological examination shows that the small intestinal mucosa has simple columnar epithelium, submucosa, smooth muscle with inner circular and outer longitudinal layers, and serosa. The absorptive surface area of the small intestine is increased by plicae circulares, villi, and microvilli. Glandular epithelium is present along the whole length of the GI tract [goblet cells]. These cells secrete mucus that lubricates food particles and protects the cells from digestive enzymes. Villi are vaginations of the mucosa that increase the overall surface area of the intestine. The villi also contain a lacteal that is connected to the lymph system and aids in the removal of lipids and tissue fluid from the blood supply. Microvilli are present on the epithelium of a villus and serve to further increase the surface area over which absorption can take place (Fig. 17.3). The submucosa contains nerves, blood vessels and elastic fiber with collagen that stretch with increased capacity while maintaining the shape of the intestine.

Gastrointestinal mucosa and other mucous membranes provide a unique anatomical niche for an interface between a sterile internal environment and a contaminated external environment. Furthermore, they allow for the passage of selected materials through an otherwise impermeable barrier. Epithelial cells posses a polarity, having one side which faces the exterior environment and another which faces internally, or into the internal milieu. The epithelial cells require intimate contact with each other to maintain barrier functions while they require contact with the external

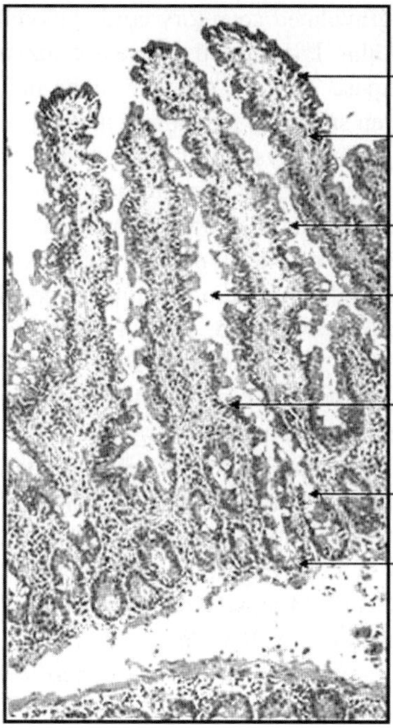

Fig. 17.3 H&E staining of jejunal villi. Goblet cells, Enterocytes and Paneth cells contribute to innate immunity while lymphocytes present in the Lamina Propria and intraepithelial areas provide acquired immunity

environment to perform functions such as nutrient absorption in the small intestine, and water reabsorption in the colon.

17.4.2 Immune Mechanisms in the GI Tract

Due to the constant exposure of the mucosa to the external environment, protection of the gut mucosa is orchestrated by an elaborate innate and acquired immune system. The renewal process involves stem cell proliferation in the crypts and maturation and differentiation of epithelial cells as they migrate along the villus axis. During this process, the epithelial cells mature into enterocytes, enteroendocrine cells, or goblet cells, most with a survival time of two to five days. Migration downwards into the crypts is accompanied by maturation into Paneth cells, with a survival time of 20 to 25 days. The bacterial load in the small intestine is 10^4 to 10^6 times lower than the colon. The innate mucosal defenses encompass both physical and chemical barriers like peristalsis, gastric acidity, mucous secretion and exfoliation of epithelium during regeneration.

Intestinal goblet cells secrete mucins that are central in the formation of a mucous barrier. Goblet cell numbers increase during inflammatory enteropathies like ulcerative colitis. They also secrete intestinal trefoil factor [mTFF3] that is involved in epithelial barrier formation. Absorptive enterocytes secrete IgA in addition to having antimicrobial activity via Histone H1 and Histone derived peptide fragments.

Paneth cells are granulated secretory epithelial cells that reside at the base of the crypts of Lieberkühn. Paneth cells secrete lysozymes, antimicrobial peptides, secretory phospholipase A2 [sPLA2] and alpha defensins. In addition, Paneth cell defensins have been shown to play a role in defense against viral particles [37].

The acquired immune system of the GALT develops during gestation such that at birth it is capable of responding to antigenic insult. The three parts and cells of the GALT are the lamina propria lymphocytes [LPL], intraepithelial lymphocytes [IEL] and Peyers patches lymphocytes. In normal adults, T cells are present in the lamina propria and epithelium. Lamina propria T cells are predominantly CD4+ αβTCR+. They are all α4β7+ or αEβ7+. They have an activated phenotype being L-selectinlo, CD69+, CD45RO+, CD25+, and Fas+. Lamina propria CD8+ T cells are also CD45RO+ and express high levels of αEβ7. Activated T cells are destined to die unless rescued by antigen or a common γ chain cytokine such as IL2, IL7 or IL15. Intestinal epithelial cells produce IL7 and lamina propria T cells express IL7 receptor. In the fetus, local differentiation of T cell progenitors in the intestine cause them to express markers normally associated with activation similar to the thymus. T cells expressing α4β7 extravasate into the gut following the binding to MAdCAM [Mucosa Addressin Cell Adhesion Molecule] which is expressed on endothelial cells. In normal adults IELs are predominantly CD3+CD8+ with a memory phenotype [CD45RO+]. These cells are mostly αEβ7+, binding to E cadherin expressed by gut epithelial cells. On activation, these IELs develop cytotoxic function [38, 39]. The phenotype of adult IELs is consistent with resting cytolytic T cells. Intestinal IELs do not express markers associated with cytolytic activity, such as granzyme B, perforin, FASL, and TNFα. After isolation, they upregulate these cytolytic effecter molecules. Intraepithelial TCRγδ+ T cells modulate epithelial growth and differentiation by producing keratinocyte growth factor [40, 41]. Mucosal γδ T cells can also influence the CD4+TCRαβ+ T cell dependent IgA B cell response.

The first organized Peyers patches appear at around 18 to 19 weeks of gestation. The numbers of Peyers patches increases from around 100 at birth to 250 in the mid teens and decreases to 100 at around 90 years of age. These lymphoid follicles bridge the mucosa and submucosa and distort the overlying dome epithelium, called the follicular dome epithelium. These areas contain specialized M cells that endocytose luminal contents and present the antigens to underlying lymphocytes and macrophages. The formation of M cells in Peyers patches is mediated in part by resident B lymphocytes.

Mucosal fluids contain high levels of IgA which contain a secretory component and a J chain that links two or more IgA monomers together [42, 43]. This is the result of a high number of IgA+ B lymphocytes localized in the lamina propria near the epithelial layer while the secretory component occurs at the basolateral side of the epithelial cells. Mucosa associated lymphoid tissue [MALT] are the major inductive sites for IgA responses in animals. They have three functions: to protect the mucosa against invasion and colonization, to block the uptake of undegraded dietary and microbial antigens, and to prevent the development of harmful immune responses [44, 45]. T lymphocytes and the variety of cytokines they secrete regulate

IgA synthesis in mucosal tissues and up to three to four grams of IgA are secreted daily into the GI tract [46]. The life span of the B-lymphocytes is five to nine days. With the help of local macrophages and dendritic cells, antigen specific B lymphocytes leave the Peyers patches and migrate through the systemic circulation, homing back to the mucosal effector site of the lamina propria. With T cell help, the B lymphocyte then matures into a plasma cell in the lamina propria.

Collectively, mucosal defenses mediated through immune cells and epithelial cells are critical in preventing the invasion and expansion of microbial pathogens and certain viruses, including HIV.

17.4.3 Pathogenic Effects of HIV Infection in the Gastrointestinal Mucosa

Gastrointestinal complications are commonly seen in patients with HIV-1 infections progressing to AIDS. During the early years of AIDS epidemic, HIV disease was frequently referred to as "Slim disease" [47]. Some GI complications appear in HIV infected patients prior to the development of advanced stages of immunodeficiency and in the absence of detectable enteric opportunistic infections, suggesting that the onset of enteropathogenic changes may occur early in infection and could be due to the direct effects of HIV in the GI mucosa [48, 49]. GALT is an early target of HIV that supports high levels of viral replication (Fig. 17.4). Despite advances in therapeutics, it continues to serve as an important viral reservoir and contributes to viral persistence [50, 51]. GALT harbors an abundance of activated effector memory CD4+ T cells that are CCR5+. These cells are highly susceptible to HIV

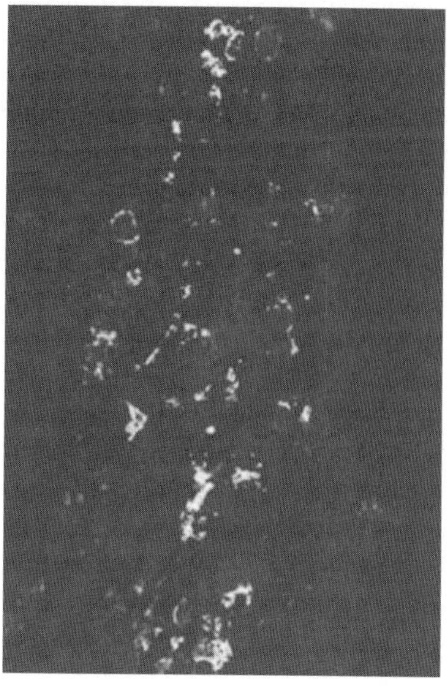

Fig. 17.4 Immunohistochemistry of cross-section of intestinal villi during HIV infection White staining indicates an infected macrophage or infected T cell in the jejunal mucosa

infection [52, 53]. Massive infection of these highly susceptible activated memory CD4+ T cells leads to severe CD4+ T-cell loss in GALT during primary HIV infection. The CD4+ T cell depletion persists throughout the course of infection in the absence of anti-retroviral therapy [54] (Fig. 17.5). Studies in Simian Immunodeficiency Virus [SIV] infected rhesus macaques have shown that the majority of extra-lymphoid CD4+ T effector memory T cells are productively infected within 10 days following exposure [52]. This is followed by a chronic state of immune activation, rapid regeneration and an immune deficiency in the absence of therapeutic interventions [55].

Recent studies have demonstrated that CD4+ T-cell restoration in GALT of HIV infected patients during highly active antiretroviral therapy [HAART] is incomplete

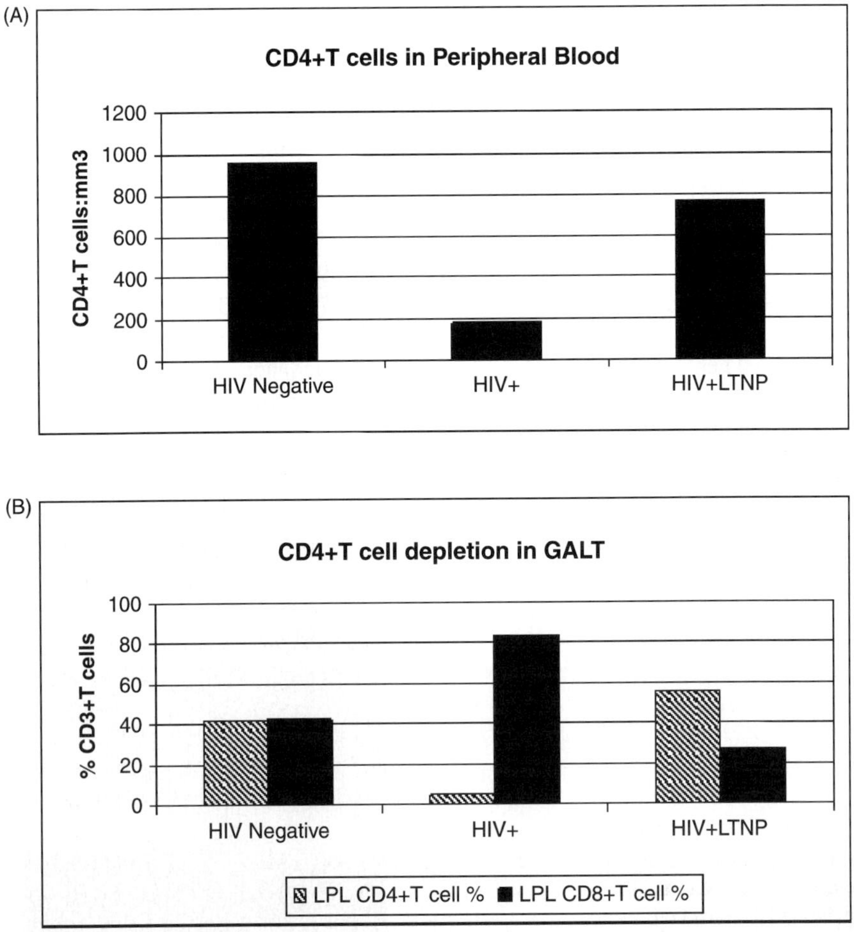

Fig. 17.5 T cell counts (**A**) CD4+ T cell numbers found in peripheral blood mononuclear cells in HIV negative subjects, HIV positive subjects, and HIV positive long term non-progressors. (**B**) Relative percentages of CD4+ and CD8+ T cells in lamina propria lymphocytes

and substantially delayed as compared to the peripheral blood compartment [50, 56]. This has been attributed to incomplete viral suppression in GALT during therapy. Increased levels of local inflammation and immune activation can continue to result in CD4+ T-cell killing and mucosal tissue damage [50, 57]. Therefore, GALT may play an important role not only in replenishment and maintenance of the viral reservoir but also may contribute to the continuing loss of CD4+ T cells in HIV infection. The dynamics of HIV infection in GALT may pose major challenges for complete mucosal immune restoration and eradication of viral reservoirs with current therapeutic strategies.

17.5 Clinical Complications in GI Mucosa During Progressive HIV Infection

HIV infected patients have experienced clinical complications that are attributed to the disruption of mucosal defenses in the GI tract. Clinical complications in the oral tissue of HIV infected patients have been well described and are reflective of the impaired immune functions. A number of opportunistic infections in the oral mucosa have been reported. One of the common infections in HIV infected patients is oral candidiasis, a condition which presents as white patches that can be atrophic or inflamed. Oral hairy leukoplakia may also be present and is caused by a co-infection with Epstein Barr virus (EBV) or reactivation of the latent EBV infection. Incidence of herpes simplex virus (HSV) infection is also common and presents as multiple aintal vesicles on the lip, buccal mucosa, and soft palate, often appearing ulcerative and crusty. Ulceration of the mucosa can be associated with several infectious and non-infectious etiologies. Apthous ulcers are painful ulcerations and can also be a common presentation. Dental manifestations and periodontal disease are rarely seen, but are important for general mucosal health. Linear gingival erythema with redness and swelling along gingival margins and a cratered appearance of the gingiva after healing are common symptoms. Necrotizing ulcerative periodontitis with loss of teeth may be seen in later stages of HIV infection.

In the esophagus, dysphagia and odynophagia are often encountered in HIV infected patients. Secondary infections like Candidiasis, cytomegalovirus (CMV), and HSV are frequently encountered. Neoplasms, gastric reflux disease and idiopathic motility diseases can complicate mucosal manifestations of HIV infection. Immune reconstitution following the initiation of HAART has been effective in resolving some of these complications. Additionally, treatment with prednisone or thalidomide may be warranted.

In the rectal mucosa, the occurrence of perirectal abscesses and fistulae, gonorrhea, chlamydia, lymphoma, and squamous cell dysplasia have been reported in HIV infected patients. Due to the generalized inflammatory response of the intestinal mucosa, redness of the anal rectal mucosa is common. Gastrointestinal bleeding in HIV infected patients is caused by CMV enteritis, Kaposi's sarcoma, and lymphoma. HIV-induced immune deficiency is also associated with the development of acalculous cholecystitis, cholangiopathy, papillary stenosis, persistence bile duct infection, pancreatitis, infections, tumors, CMV enteritis, intestinal lymphomas, and intusseption secondary to Kaposi's sarcoma.

Mucosal opportunistic infections are common among women and men. However, esophageal candidiasis is more common in women. The possible occurrence of female specific genitourinary manifestations of HIV is a subject of great interest and debate. Cervical dysplasia and neoplasm have been commonly found in women with advanced HIV infection. The center for disease control (CDC) has recently added invasive cervical carcinoma but not carcinoma *in situ* or cervical dysplasia to the list of AIDS-defining conditions. In summary, progressive HIV infection is associated with reactivation of latent infections, an increased susceptibility to various infections and pathologic changes in the mucosal tissues.

17.6 Histopathologic Changes in GALT During HIV Infection

Intestinal dysfunction is often detected in patients with HIV infection despite the success of HAART in improving the clinical status of these patients. HIV RNA, DNA, and Gag p24 protein are readily detected in intestinal biopsies of HIV infected patients in several studies [50, 56, 58, 59]. Investigations of HIV enteropathogenesis in humans are limited due to the difficulty in accessing gut mucosal tissue samples during the course of HIV infection. Simian immunodeficiency virus [SIV]-infected rhesus macaques provide an excellent animal model of AIDS not only for vaccine studies but also for the studies of HIV enteropathogenesis in a controlled experimental setting [60]. An acute enteropathy syndrome was observed in SIV-infected rhesus macaques during the primary acute stage of SIV infection as early as two weeks post-infection with signs of nutrient malabsorption, evidenced clinically by D-xylose malabsorption [61]. Animals had high levels of viral replication in both CD4+ T lymphocytes and macrophages in GALT [62, 63, 64]. These studies demonstrated that the onset of the disruption of intestinal structure and function occurred early during primary SIV infection, prior to the development of immunodeficiency, and were driven by the direct effects of viral replication and host response [64]. Histopathologic analysis of intestinal tissues detected dilated villus lacteals, excessive cellular debris, lymphoplasmocytic infiltrates, and cytoplasmic vacuoles in crypt epithelial cells [61]. Cellular injury and impaired functions of immature as well as differentiated epithelial cells in the GI tract may contribute significantly to the development of enteropathy [48, 65]. Recent studies of patients during early HIV infection reported that high levels of viral replication in GALT and marked CD4+ T-cell depletion correlated with decreased expression of genes regulating epithelial barrier maintenance and digestive and metabolic functions [48]. Occurrence of increased levels of inflammation, immune activation and apoptosis in GALT may contribute to gut mucosal damage. Similar changes were observed in the rectal mucosa [66]. These findings indicate that HIV-induced pathologic changes in GALT emerged at both the molecular and cellular levels very early during infection.

The magnitude of histopathologic and functional changes in GALT has been reported to increase during the course of HIV infection. These include inflammatory infiltrates with increased cellularity, increased expression of inflammatory cytokines, crypt hyperplasia, blunted villi, and decreased digestive enzyme activities. These changes have also been evaluated at the molecular level by DNA microarray analysis of intestinal biopsies of patients with chronic HIV infection. A marked decrease in the expression of genes involved in intestinal enterocyte digestive and absorptive functions as well as epithelial barrier function have been observed [67]. Genes associated with drug metabolism were also downregulated, potentially affecting the bioavailability of antiretroviral drugs in GALT, which may contribute to a lack of effective suppression of local viral replication [67].

17.7 Effects of HIV on CD4+ T Cell Populations in GALT

GALT supports robust viral replication in mucosal memory CD4+ T cells during primary HIV infection, leading to a severe depletion of the CD4+ T cells (Fig. 17.3) [54, 59]. Initially, HIV infection leads to vigorous proliferation and expansion of CD4+ T cells in GALT, as indicated by the expression of Ki67 [a cell proliferation marker] in 80% or more of the CD4+ T cells [54]. The killing of CD4+ T cells can be attributed to a massive infection of the majority of mucosal effector memory CCR5+CD4+ T cells as well as direct cytopathic effects of the viral infection, resulting in cellular necrosis [52]. In addition, some of the uninfected CD4+ T cells could be eliminated due to activation-induced cell death [bystander effect]. High levels of CCR5 and CXCR4 coreceptor expression on intestinal CD4+ T lymphocytes, compounded by their predominantly activated state, make these cells highly susceptible to the viral infection. This may explain the severity of CD4+ T-cell depletion seen in GALT. Insufficient production and reduced homing of CD4+ memory T cells may contribute to sustained CD4+ T-cell depletion during disease progression. The loss of CD4+ T cells was coincident with an increased prevalence of CD8+ T cells in GALT as seen by both flow cytometry and immunohistochemistry [67]. This dysregulation in T-cell homeostasis leads to increased levels of inflammatory cytokines in GALT and causes inflammation, chronic immune activation, tissue injury, and impaired functions [67]. Recruitment of CD4+ T cells from the periphery into GALT due to the ongoing viral replication and elevated chemokine production may provide new targets for the virus and cause a continuous loss of CD4+ T cells from the total pool in the body [68]. Presence of high levels of viral antigens in GALT may also contribute to the impaired functions of resident lymphoid and epithelial cells allowing for microbial translocation [69]. HIV Env, Tat, and Nef have been shown to directly cause cytopathic changes and impaired functions at both cellular and molecular levels in CD4+ T-cell lines and intestinal epithelial cell lines.

In contrast to the clinical course of disease progression in the majority of patients with HIV infection, a small subset of individuals, termed long-term HIV-1–infected nonprogressors [LTNP], remain clinically asymptomatic for more than 10 years in the absence of therapy. Their plasma viral loads are usually undetectable or very low, and CD4+ T-cell numbers in peripheral blood are in the normal range. Intestinal biopsies of LTNP have undetectable viral loads, normal CD4+ T-cell levels and high levels of virus-specific cellular responses in GALT [67]. Thus, suppression of viral replication and maintenance of CD4+ T-cell numbers and their function in GALT and peripheral blood correlate with better clinical outcome in LTNP and demonstrate the importance of the integrity of the mucosal immune system in the control of disease progression.

The CD4+ T-cell depletion in GALT persists through all stages of HIV infection in the absence of therapeutic interventions. Loss of CD4+ T cells in peripheral lymph nodes and blood is gradual, and disruption of lymph node architecture is seen only in advanced HIV infection. In contrast, the microenvironment of the GI tract is severely disrupted very early during primary HIV infection. The difference in the timing of the severe disruption of GALT and lymph node microenvironments may be partly attributed to the prevalence of activated memory CD4+ T cells in GALT [68, 69]. The effector memory CD4+ T cells constitute less than 20% of the total CD4+ T cells in lymph nodes and peripheral blood, whereas the majority of the CD4+ T cells are of naïve phenotype [53]. The lack of correlation between the peripheral blood T lymphocytes and intestinal mucosal T lymphocytes indicates that the higher numbers of circulating CD4+ T cells are not necessarily an accurate measure of intestinal mucosal immunity and pathologic changes occurring in GALT.

17.8 Viral Reservoir in GALT During HAART

Initiation of HAART in HIV infected patients has shown great promise in halting disease progression towards AIDS. The efficacy of HAART is measured by the suppression of viral loads and restoration of CD4+ T-cell numbers in peripheral blood. Long-term use of HAART is associated with the development of drug toxicities, metabolic abnormalities, and the emergence of drug-resistant variants. Therefore, current guidelines recommend delaying the initiation of HAART until patients reach CD4+ T cell counts below 350 cells/mm^3. While longitudinal follow-up of patients initiating HAART during primary HIV infection has demonstrated rapid suppression of viral loads, restoration of CD4+ T cells, and preservation of both HIV-specific CD4+ and CD8+ T-cell responses in peripheral blood, the kinetics of restoration of the gut mucosal immune system were substantially delayed and incomplete, independent of the time of initiation of HAART. This delay in mucosal immune restoration was attributed to incomplete suppression of viral replication in GALT. Gene expression profiling of intestinal biopsies from these patients showed an increased expression of genes regulating inflammation and immune activation but decreased expression of genes associated with mucosal repair and regeneration [50]. Thus it can be postulated that the regeneration of CD4+ T cells is dependant on the maintenance of a normal microenvironment in GALT.

Residual viral replication was detected in GALT of patients receiving HAART despite undetectable viral loads in the peripheral blood [50, 70, 71]. The lack of complete viral suppression may contribute to the continuous loss of CD4+ T cells from mucosal tissues during HAART and the delayed CD4+ T-cell restoration in GALT [50, 70]. Recent studies demonstrate a lack of appreciable decay of viral reservoirs of HIV-1 DNA-positive cells in GALT or peripheral blood monocytes in patients on HAART, despite undetectable plasma HIV RNA [70]. The number of potentially replication competent HIV-1 DNA positive cells in GALT was higher compared with peripheral blood. Thus, mucosal CD4+ T cells in GALT constitute an important viral reservoir, in addition to lymph nodes. Due to the constant exposure to antigenic and mitogenic stimuli in the GI mucosal compartment, it is an environment permissive to the activation of latently HIV-infected cells that may be trafficking to GALT. In addition, GALT may also support and maintain a low level of viral replication. These data also suggest that there is a continual replenishment of the CD4+ T-cell reservoir in the peripheral blood compartment during therapy. Persistence of latently infected CD4+ T-cell reservoirs is a major challenge for the eradication of HIV during therapy.

Studies utilizing the SIV model show that the initiation of antiviral therapy during primary SIV infection results in efficient suppression of viral replication and near complete restoration of CD4+ T cells in GALT of rhesus macaques [60]. The major mechanism of early CD4+ T-cell restoration in GALT involves trafficking of CD4+ T cells from the periphery to gut mucosa. Gene expression profiling of intestinal biopsies from these animals demonstrates decreased expression of inflammation and immune activation associated genes and increased expression of genes regulating mucosal repair and regeneration. Thus, it may be beneficial to start therapy earlier during the course of infection to maintain a mucosal microenvironment more conducive to the replenishment of CD4+ T cells.

17.9 Molecular Basis of Host Responses to HIV

High-throughput DNA microarray analyses of gene expression in lymphoid tissues of HIV infected patients has facilitated the identification of virus-induced molecular pathogenesis. Functional genomic analyses of host responses to HIV show that molecular processes regulating inflammation and immune activation are dominantly expressed in GALT and peripheral lymph nodes of therapy-naïve HIV-1 infected patients with a progressive clinical course [67, 72]. In GALT, dysregulation of gene expression associated with cell cycle regulation, lipid metabolism, and intestinal digestive and absorptive functions was observed. A sharp decrease in the expression of genes associated with intestinal epithelial barrier, nutrient digestive, and absorptive functions was observed in therapy-naïve chronically HIV-infected patients [67]. In addition, several of these processes were also impaired in GALT of LTNP. Multiple members of the cytochrome P450 family were also downregulated in both groups of patients, as well as in patients with acute infection [48, 67]. These findings suggest that the functions of drug metabolism and nutrient digestion were compromised in patients with HIV infection regardless of stage of infection. It is possible that the pathologic changes occurring during primary HIV infection cause long-lasting damage to the repair and regeneration processes of the intestinal epithelium as well as the gut mucosal immune system [48, 56, 58].

Investigation of the effects of HAART on GI mucosal gene expression in HIV-infected patients with varying degrees of mucosal immune restoration demonstrate the potential to identify molecular correlates of mucosal immune restoration and response to therapy. Comparative analyses of gene expression profiles from intestinal biopsies obtained prior to and at three months post-HAART were performed from two patients with HIV infection who had measurable restoration of mucosal CD4+ T cells and from two patients who displayed poor restoration of mucosal CD4+ T cells over the duration of therapy [50]. Patients with poor immune restoration had increased expression of inflammation, cell injury, apoptosis, and immune responses [50]. A substantial upregulation of genes associated with interferon response pathways were detected prior to and post-HAART, suggesting that chronic immune activation, inflammation and failure to control viral replication in mucosal tissues may be associated with an impaired ability to restore the GALT microenvironment. Genes associated with immune response and humoral and cellular defenses, as well as response to stress and wounding, were modulated during therapy. Patients with better immune restoration had a significant reduction in gene expression related to inflammation and immune response but increased expression of mucosal lymphocyte homing and repair and regeneration related genes. Molecular markers of immune activation and tissue injury dominated the expression profiles of patients failing to restore mucosal CD4+ T cells. These molecular changes in GALT preceded the changes in the mucosal CD4+ T-cell restoration at the cellular and tissue level. DNA microarray studies using the SIV model also support the observations made in patients with HIV infection during HAART. An understanding of the mechanisms of intestinal dysfunction in HIV infection will aid in the development of new therapeutic strategies that serve to repair and restore intestinal structure and function and alleviate nutritional complications in patients with HIV infection.

17.10 HIV Transmission and Genitourinary Mucosa

Globally, the majority of new HIV infections among women are acquired through heterosexual transmission. The structure of the female genital tract is described in Chapter 2. It is generally believed that Dendritic cells are involved in the uptake of HIV virions through the squamous epithelial layer of the vaginal mucosa within minutes, from where they migrate to the draining lymph nodes and then migrate to all tissues within days. High to detectable levels of HIV RNA have been found in semen samples from HIV infected men. Genital ulcers caused by several sexually transmitted diseases increase the risk of HIV infection in both women and men.

Exposure to HIV infection through rectal intercourse has a significantly higher likelihood to result in infection than vaginal intercourse, primarily because of increased trauma to the rectal mucosa. A significant proportion of heterosexual couples in previous studies have been found to engage in rectal intercourse at least some of the time. Although HIV can be recovered from cervical secretions, its infectivity appears to be less than that of semen. This may explain why HIV transmission from women to men is not as high.

Intercourse during menses has been suggested to increase the risk of HIV transmission to both men and women although there is as yet no virologic or epidemiological proof of such increased risk. The use of oral contraceptives has also been suggested as increasing the risk of infection, as has cervical ectopy. Anecdotal cases of female-to-female transmission have been reported. Although this mode of transmission is uncommon, data on lesbian sexual activity has not been studied on CDC AIDS case reports, preventing a full assessment of the role of such exposures in transmitting infections.

The gastrointestinal tract and the genitourinary tract share common embryonic origins, however they have distinct immunological features due to their physiological functions. The mucosal surfaces of both the male and female genital tracts are the entry sites of several sexually transmitted diseases. While the female genital tract has been extensively studied, the male genitourinary tract has limited information available with respect to immune responses at the mucosal surface. Human urine, seminal plasma and cervico-vaginal secretions contain IgG as the dominant isotype, in contrast to intestinal secretions that contain IgA. While the gut has special inductive lymphoid tissue like the Peyers patches, the presence of such tissue has not been demonstrated in the cervical mucosa. Additionally, while oral or rectal vaccination can generate both a local and a generalized immune response through the common mucosal system, vaccination at the vaginal site produces a local response only. Thus it is important that vaccination strategies target both the intestinal mucosal immune compartment and that of the genitourinary system.

17.11 HIV in Pulmonary Mucosa

Pulmonary manifestations of HIV infection are common in infants and children. The development of pulmonary complications is often the cause for morbidity and mortality in younger age groups. Acute bacterial pneumonia is often caused by pneumococcus, gram negatives and staphylococci. *Pneumocystis carinii* [PcP] is the most common agent causing death due to pulmonary infections in children with HIV. CMV co-infection with PcP is also common. Other chronic bronchopulmonary complications include lymphocytic interstitial pneumonitis and bronchiectasis. Tuberculosis [TB] is now increasingly noted as a common cause for acute pneumonitis in children, especially in areas endemic for TB infections. TB is recognized as one of the most important co-infections occurring with HIV in many parts of the world. Treatment of both HIV and TB in patients with advanced HIV infection has posed challenges of drug toxicity. Recent studies have demonstrated a decrease in pulmonary mucosa associated CD4+ T cell percentages, although to a lesser extent than in the GI mucosa. An in depth evaluation of the pathology of bronchopulmonary mucosa in HIV infection is under-investigated.

17.12 Antiviral Therapy

Initiation of HAART has resulted in a demonstrable decrease in HIV RNA loads and an increase in CD4+ T-cell counts in the peripheral blood of HIV infected patients [73]. HAART includes a combination of three or more Nucleoside and Non-nucleoside Reverse Transcriptase Inhibitors [NRTI and NNRTI], Fusion Inhibitors, and Protease Inhibitors [74]. Fusion Inhibitors are designed to prevent viral entry into the host-cell by interfering with the formation of the "hairpin" structure required for fusion to host-cell membranes. There are 19 licensed antiretroviral drugs in the latter three classes.

NRTIs and NNRTIs interfere with the activity of the reverse transcriptase [RT] by self-incorporation into newly synthesized DNA strands by HIV-1 reverse transcriptase, resulting in chain termination and an inhibition of genomic DNA synthesis. NRTI are nucleoside analogs that mimic cellular nucleosides. These include Zidovudine [AZT], Lamivudine [3TC], Stavudine [d4T] and Emtriva [FTC] [75, 76, 77, 78]. NNRTI inhibit RT by binding to a specific site in the enzyme [79]. The NNRTI include Nevirapine [Viramune], Efavirenz [Sustiva] and Delavirdine Mesylate [Rescriptor] [80, 81].

Protease inhibitors (PI), such as Atazanavir [Reyataz], Lexiva [Foramprenavir], Norvir [Ritonavir], Viracept [Nelfinavir] and Crixivan [Indinavir Sulfate] are known to inhibit HIV protease which leads to incomplete processing of viral protein precursors and gives rise to immature viral particles that are non-infectious [82]. The use of PI is limited due to the rapid emergence of resistant mutations and poor bioavailability.

Initiation of HAART has been shown to significantly prolong survival in HIV infected patients with central nervous system lymphoma compared to monotherapy. Side effects of HAART include metabolic complications such as diabetes, pancreatic lipodystrophy, lactic acidosis and drug resistance. [83, 84]. Several studies have proposed the early initiation of HAART as a way to preserve immune function, lower the viral set point and prevent the establishment of viral reservoirs. Concerns have been raised regarding the benefits of early treatment, which may inhibit the development of mature immunological responses and increase the risk of long-term medication toxicity [85, 86]. The long-term benefits of early versus delayed initiation of HAART remain largely under-investigated.

Structured Treatment Interruptions [STI] have been proposed as an alternative for patients experiencing drug-toxicity and an emergence of drug resistant mutations and lack of adherence [87]. STIs were explored to minimize drug related adverse effects and to boost the anti HIV immune response. However, recent studies do not support the use of STI, as patients consistently experienced virological and immunological failure shortly after the interruption of treatment and failed to develop HIV-specific immune responses. Thus, the best therapeutic strategy for the treatment of HIV infection needs to be continually investigated and improved.

The success of long-term HAART has greatly stimulated studies of viral reservoirs. The pool of latent HIV-1 positive T cells may be perpetuated by T cell homeostatic mechanisms for maintaining memory CD4+ T cells. Important HIV viral reservoirs include infected immune cells in GALT, tonsils and lymph nodes. It is possible that the bioavailability of drugs may be poor in certain tissues contributing to the persistence of viral reservoirs.

17.13 Role of Mucosal Immunity in Vaccine Development

Induction of mucosal immunity is an important factor in blocking transmission of HIV infection through sexual contact. Mucosal blockers can act as an early defense against systemic infection. If this were possible, it would be an appealing avenue to pursue. Forms of presentation of B and T cell epitopes capable of activating an entire set of inductive and effector immune elements of the mucosa are receiving increased attention.

Immunization of mucosal sites via trafficking throughout the mucosal immune system has been effective in other viral systems, especially those with respiratory targets. It remains to be shown that protection against HIV infection can be specifically conferred at the rectal or genital mucosa via trafficking through the course of the mucosal immune system as compared with local immunization. HIV vaccine development is discussed in detail in Chapter 18.

References

1. 2007, *AIDS Alert*, 22, suppl 1.
2. Evian C, Fox M, MacLeod W, Slotow SJ, & Rosen S. 2004, *S Afr Med J*, 94, 125.
3. [CDC]. 2006, *Centers for Disease Control and Prevention [CDC]HIV/AIDS Surveillance Report*, Volume 16.
4. Padma TV. 2005, *Nat Med*, 11, 695.
5. Qian HZ, Vermund SH, & Wang N. 2005, *Sex Transm Infect*, 81, 442.
6. Ruxrungtham K, Brown T, & Phanuphak P. 2004, *Lancet*, 364, 69.
7. Sharma VK & Khandpur S. 2004, *Natl Med J India*, 17, 310.
8. Tucker JD, Henderson GE, Wang TF, Huang YY, Parish W, Pan SM, Chen XS, & Cohen MS. 2005, *Aids*, 19, 539.
9. Kriebs JM. 2002, *J Perinat Neonatal Nurs*, 16, 1.
10. Regez RM, Kleipool AE, Speekenbrink RG, & Frissen PH. 2005, *Int J STD AIDS*, 16, 671.
11. Wig N. 2003, *Indian J Med Sci*, 57, 192.
12. Baggaley RF, Boily MC, White RG, & Alary M. 2006, *Aids*, 20, 805.
13. Zou S, Musavi F, Notari EPt, Fujii KE, & Dodd RY. 2005, *Transfusion*, 45, 1593.
14. Spira S, Wainberg MA, Loemba H, Turner D, & Brenner BG. 2003, *J Antimicrob Chemother*, 51, 229.

15. Hughes A & Corrah T. 1990, *Blood Rev*, 4, 158.
16. Kuznetsov YG, Victoria JG, Robinson WE, Jr., & McPherson A. 2003, *J Virol*, 77, 11896.
17. Russell RS, Liang C, & Wainberg MA. 2004, *Retrovirology*, 1, 23.
18. Ott DE, Coren LV, Kane BP, Busch LK, Johnson DG, Sowder RC, 2nd, Chertova EN, Arthur LO, & Henderson LE. 1996, *J Virol*, 70, 7734.
19. Zhu P, Liu J, Bess J, Jr., Chertova E, Lifson JD, Grise H, Ofek GA, Taylor KA, & Roux KH. 2006, *Nature*, 441, 847.
20. Nguyen DH & Hildreth JE. 2000, *J Virol*, 74, 3264.
21. Campbell GR, Pasquier E, Watkins J, Bourgarel-Rey V, Peyrot V, Esquieu D, Barbier P, de Mareuil J, Braguer D, Kaleebu P, et al. 2004, *J Biol Chem*, 279, 48197.
22. Khan MA, Aberham C, Kao S, Akari H, Gorelick R, Bour S, & Strebel K. 2001, *J Virol*, 75, 7252.
23. Goncalves J & Santa-Marta M. 2004, *Retrovirology*, 1, 28.
24. Garcia JV & Miller AD. 1991, *Nature*, 350, 508.
25. Bukrinsky M & Adzhubei A. 1999, *Rev Med Virol*, 9, 39.
26. Neil SJ, Eastman SW, Jouvenet N, & Bieniasz PD. 2006, *PLoS Pathog*, 2, e39.
27. Binette J, Dube M, Mercier J, Halawani D, Latterich M, & Cohen EA. 2007, *Retrovirology*, 4, 75.
28. Ullrich R, Schmidt W, Zippel T, Schneider T, Zeitz M, & Riecken EO. 1998, *Pathobiology*, 66, 145.
29. Suresh P & Wanchu A. 2006, *J Postgrad Med*, 52, 210.
30. Xiao H, Neuveut C, Tiffany HL, Benkirane M, Rich EA, Murphy PM, & Jeang KT. 2000, *Proc Natl Acad Sci USA*, 97, 11466.
31. Luban J, Bossolt KL, Franke EK, Kalpana GV, & Goff SP. 1993, *Cell*, 73, 1067.
32. Bouyac-Bertoia M, Dvorin JD, Fouchier RA, Jenkins Y, Meyer BE, Wu LI, Emerman M, & Malim MH. 2001, *Mol Cell*, 7, 1025.
33. Mellors JW, Rinaldo CR, Jr., Gupta P, White RM, Todd JA, & Kingsley LA. 1996, *Science*, 272, 1167.
34. Tough DF & Sprent J. 1995, *Immunol Res*, 14, 1.
35. Pantaleo G & Fauci AS. 1996, *Annu Rev Microbiol*, 50, 825.
36. Lederman MM. 2001, *AIDS*, 15 Suppl 2, S11.
37. Mackewicz CE, Yuan J, Tran P, Diaz L, Mack E, Selsted ME, & Levy JA. 2003, *AIDS*, 17, F23.
38. Blumberg RS, Yockey CE, Gross GG, Ebert EC, & Balk SP. 1993, *J Immunol*, 150, 5144.
39. Ebert EC. 1993, *Cell Immunol*, 147, 331.
40. Boismenu R & Havran WL. 1994, *Science*, 266, 1253.
41. Havran WL & Boismenu R. 1994, *Curr Opin Immunol*, 6, 442.
42. Kang YS, Calvanico NJ, & Tomasi TB, Jr. 1974, *J Immunol*, 112, 162.
43. Tomasi TB & Grey HM. 1972, *Prog Allergy*, 16, 81.
44. Czerkinsky C, Quiding M, Eriksson K, Nordstrom I, Lakew M, Weneras C, Kilander A, Bjorck S, Svennerholm AM, Butcher E, et al. 1995, *Adv Exp Med Biol*, 371B, 1409.
45. Eriksson K, Kilander A, Hagberg L, Norkrans G, Holmgren J, & Czerkinsky C. 1995, *Adv Exp Med Biol*, 371B, 1011.
46. Mestecky J. 1987, *J Clin Immunol*, 7, 265.
47. Serwadda D, Mugerwa RD, Sewankambo NK, Lwegaba A, Carswell JW, Kirya GB, Bayley AC, Downing RG, Tedder RS, Clayden SA, et al. 1985, *Lancet*, 2, 849.
48. Sankaran S, George MD, Reay E, Guadalupe M, Flamm J, Prindiville T, & Dandekar S. 2008, *J Virol*, 82, 538.
49. Verhoef K, Tijms M, & Berkhout B. 1997, *Nucleic Acids Res*, 25, 496.
50. Guadalupe M, Sankaran S, George MD, Reay E, Verhoeven D, Shacklett BL, Flamm J, Wegelin J, Prindiville T, & Dandekar S. 2006, *J Virol*, 80, 8236.
51. Dandekar S. 2007, *Curr HIV/AIDS Rep*, 4, 10.

52. Mattapallil JJ, Douek DC, Hill B, Nishimura Y, Martin M, & Roederer M. 2005, *Nature*, 434, 1093.
53. Picker LJ. 2006, *Curr Opin Immunol*, 18, 399.
54. Guadalupe M, Reay E, Sankaran S, Prindiville T, Flamm J, McNeil A, & Dandekar S. 2003, *J Virol*, 77, 11708.
55. Grossman Z, Meier-Schellersheim M, Paul WE, & Picker LJ. 2006, *Nat Med*, 12, 289.
56. Mehandru S, Poles MA, Tenner-Racz K, Horowitz A, Hurley A, Hogan C, Boden D, Racz P, & Markowitz M. 2004, *J Exp Med*, 200, 761.
57. Haase AT. 2005, *Nat Rev Immunol*, 5, 783.
58. Mehandru S, Poles MA, Tenner-Racz K, Manuelli V, Jean-Pierre P, Lopez P, Shet A, Low A, Mohri H, Boden D, et al. 2007, *J Virol*, 81, 599.
59. Mehandru S, Tenner-Racz K, Racz P, & Markowitz M. 2005, *J Allergy Clin Immunol*, 116, 419.
60. George MD, Reay E, Sankaran S, & Dandekar S. 2005, *J Virol*, 79, 2709.
61. Heise C, Vogel P, Miller CJ, Halsted CH, & Dandekar S. 1993, *Am J Pathol*, 142, 1759.
62. Lackner AA & Veazey RS. 2007, *Annu Rev Med*, 58, 461.
63. Veazey RS, Marx PA, & Lackner AA. 2001, *Trends Immunol*, 22, 626.
64. Hirsch VM & Lifson JD. 2000, *Adv Pharmacol*, 49, 437.
65. Douek D. 2007, *Top HIV Med*, 15, 114.
66. Kotler DP. 2005, *AIDS*, 19, 107.
67. Sankaran S, Guadalupe M, Reay E, George MD, Flamm J, Prindiville T, & Dandekar S. 2005, *Proc Natl Acad Sci USA*, 102, 9860.
68. Brenchley JM, Price DA, & Douek DC. 2006, *Nat Immunol*, 7, 235.
69. Brenchley JM, Price DA, Schacker TW, Asher TE, Silvestri G, Rao S, Kazzaz Z, Bornstein E, Lambotte O, Altmann D, et al. 2006, *Nat Med*, 12, 1365.
70. Poles MA, Boscardin WJ, Elliott J, Taing P, Fuerst MM, McGowan I, Brown S, & Anton PA. 2006, *J Acquir Immune Defic Syndr*, 43, 65.
71. Coffin JM. 1996, *AIDS*, 10 Suppl 3, S75.
72. Li Q, Schacker T, Carlis J, Beilman G, Nguyen P, & Haase AT. 2004, *J Infect Dis*, 189, 572.
73. Cohen OJ & Fauci AS. 2001, *Adv Intern Med*, 46, 207.
74. Sanchez R, Portilla J, Gimeno A, Boix V, Llopis C, Sanchez-Paya J, Merino E, de la Sen ML, Munoz C, Reus S, et al. 2006, *J Infect*, 22, Suppl 1.
75. Coates JA, Cammack N, Jenkinson HJ, Jowett AJ, Jowett MI, Pearson BA, Penn CR, Rouse PL, Viner KC, & Cameron JM. 1992, *Antimicrob Agents Chemother*, 36, 733.
76. De Clercq E. 1995, *J Med Chem*, 38, 2491.
77. Hazen R & Lanier ER. 2003, *J Acquir Immune Defic Syndr*, 32, 255.
78. Pollard RB, Peterson D, Hardy D, Pottage J, Murphy RL, Gathe J, Beall G, Rutkievicz V, Reynolds L, Cross AP, et al. 1999, *J Acquir Immune Defic Syndr*, 22, 39.
79. Ren J, Esnouf RM, Hopkins AL, Warren J, Balzarini J, Stuart DI, & Stammers DK. 1998, *Biochemistry*, 37, 14394.
80. Guay LA, Musoke P, Fleming T, Bagenda D, Allen M, Nakabiito C, Sherman J, Bakaki P, Ducar C, Deseyve M, et al. 1999, *Lancet*, 354, 795.
81. Marseille E, Kahn JG, & Saba J. 1998, *Aids*, 12, 939.
82. Venaud S, Yahi N, Fehrentz JL, Guettari N, Nisato D, Hirsch I, & Chermann JC. 1992, *Res Virol*, 143, 311.
83. Gulick RM, Ribaudo HJ, Shikuma CM, Lalama C, Schackman BR, Meyer WA, 3rd, Acosta EP, Schouten J, Squires KE, Pilcher CD, et al. 2006, *Jama*, 296, 769.
84. Struble K, Murray J, Cheng B, Gegeny T, Miller V, & Gulick R. 2005, *Aids*, 19, 747.
85. Blankson JN. 2005, *AIDS Read*, 15, 245.
86. Mauskopf J, Kitahata M, Kauf T, Richter A, & Tolson J. 2005, *J Acquir Immune Defic Syndr*, 39, 562.
87. Julg B & Goebel FD. 2006, *Infection*, 34, 186.

Chapter 18
Mucosal Vaccination Against HIV-1

Tom Evans

Abstract Sexual mucosal transmission of HIV-1 is the most common means of spread of HIV/AIDS throughout the world. Although it may be reasonable to assume that a vaccine that works to eliminate viral replication in the systemic lymphoid tissue may be partially protective, there is still a reasonable belief that a vaccine that engenders high level of immune defenses at mucosal surfaces will be needed in the quest for an effective HIV immunization regimen. Despite some positive advances, the fact remains that no candidate vaccine to date has elicited antibodies which can broadly neutralize a wide variety of primary HIV-1 virus isolates in humans. Moreover, the benefit of inducing either a quantitatively or qualitatively different response at the mucosal level has not been adequately addressed experimentally. Unfortunately, many of the same concerns are valid for the vaccines which have been developed to induce cytotoxic T cell responses to HIV, in an attempt to control HIV-1 replication in infected cells. Thus, all studies to date must be interpreted in the context of the failure of all human trials of HIV vaccines to date, and a lack of known correlates of immunity pending such a vaccine success.

This chapter thus focuses on non-human primate and human investigations of mucosal immunization for HIV-1. the pre-clinical evidence that mucosal routes of vaccination, or even mucosal induction of immune responses using systemic vaccination, provide complete or partial protection from HIV-1 infection or disease is evolving. Although most of the animal studies, especially in the macaque model, would imply that vaccination via the nose, rectum or vagina may be needed for optimum vaginal or rectal responses, that data is still weak and not substantiated by any human clinical trial. It is clear that we still do not know the optimum route, the desired antibody response, the qualitative or quantitative nature of the T cells and their homing markers that need to be induced, or the specific approach that will allow for development of an effective mucosal HIV vaccine. Indeed, there is still debate as to whether neutralizing antibodies or CTL or yet undefined immune parameters, either in mucosal or systemic compartments or both, may correlate with

T. Evans
Novartis Institute of Biomedical Research, Cambridge, Massachusetts, USA
e-mail: Tom.evans@novastis.com

M. Vajdy (ed.), *Immunity Against Mucosal Pathogens*,

protection against infection or disease. Nonetheless, most investigators in the HIV-1 vaccine field believe that mucosal responses will be needed and may be central to eventual success.

18.1 Introduction: There are Far more Questions than Answers

Sexual transmission of HIV-1 is the most common means of spread of HIV/AIDS throughout the world. Although a well-deserved emphasis has been focused on vaginal transmission, it is clear that other routes of infection, such as via penile surfaces, the rectum or the oral cavity, also deserve attention. The latter is especially important for protection during transmission by breast feeding in infants of infected mothers. In addition, the massive accumulation of virus that occurs in the gastrointestinal tract shortly after initial infection and dissemination, regardless of the initial route of transmission, has resulted in an appreciation of the need to protect this large mucosal lymphoid organ from depletion of CD4+ T cells [1]. It is quite clear that shortly following infection $CD4^+$ memory T cells are selectively and profoundly depleted in the gastrointestinal tract, that this depletion persist throughout the illness, and remains somewhat apparent even after successful antiretroviral therapy. Moreover, other studies have indicated that the respiratory tract [2] and vaginal tract [3] are also foci of early depletion of CD4+ T cells. In consequence, although it may be reasonable to assume that a vaccine that works to eliminate viral replication in the systemic lymphoid tissue may be partially protective, there is still a reasonable belief that a vaccine that engenders high level of immune defenses at mucosal surfaces will be needed in the quest for an effective HIV immunization regimen.

In order to achieve adequate control of HIV-1/AIDS, it is likely that a vaccine needs either to lower the viral load in the infected transmitting individual, to prevent infection in the recipient, or to decrease the rate of disease progression in the newly infected patient. The lowering of viral replication by use of therapeutic vaccination in chronically HIV-1-infected individuals is unproven to date, despite many misleading and "promising" Phase 1 or Phase 2 clinical trial results. New concepts in therapeutic vaccination emerging from cancer immunotherapy, as well as approaches to seemingly more tractable mucosal infectious diseases such as herpes viruses and human papillomavirus, may be applied in future therapeutic HIV-1 vaccine trials. However, the therapeutic vaccination approach will not be part of the further discussion in this chapter. This chapter thus focuses on some murine but primarily on non-human primate and human investigations of mucosal immunization for HIV-1.

For prevention of HIV-1 infection, most investigators agree that a high level of effective neutralizing antibody at the site of infection or against the gastrointestinal burst of replication will likely be needed. A number of studies have shown that passively administered IgG monoclonal antibodies can protect from mucosal simian/human immunodeficiency virus chimera (SHIV) challenges [4, 5]. There are also some studies of SHIV or SIV models of infection, in which the ability to elicit homologous virus neutralization has led to some degree of protection. However, the protection demonstrated has not always required that these vaccines be given by mucosal routes to induce mucosal IgG or IgA antibodies. Various mechanisms exist by which secretory IgA or transudated IgG, produced by mucosal or systemic immune responses, respectively, may protect from SIV or SHIV in such models [6].

Despite these positive advances, the fact remains that no candidate vaccine to date has elicited antibodies which can broadly neutralize a wide variety of primary HIV-1 virus isolates, and thus the benefit of inducing either a quantitative or qualitatively different response at the mucosal level has not been adequately addressed experimentally. Unfortunately, many of the same concerns are valid for the vaccines which have been developed to induce cytotoxic T cell responses to HIV, in an attempt to control HIV-1 replication in infected cells. That is, there is still lacking a robustly proven animal model to definitively answer the question of the need for a mucosal approach to provide protection through control of plasma viremia and replication in various lymphoid and non-lymphoid tissues. Thus, all of the studies cited below must be interpreted in the context of the failure of all HIV-1 vaccines attempted in humans to date, and a lack of known correlates of immunity, pending a vaccine success.

Despite this lack of a direct proof of the role of T cell responses as a correlate of immunity/protection against HIV-1 infection, a large number of investigations have shown that the T cell compartment, and the $CD8^+$ T cell compartment in particular, is critical in the control of SIV or HIV-1 viremia, and possibly in protection from disease. These include studies using T cell depletion, adoptive transfer of $CD8^+$ T cells in mice, observational studies in humans, and vaccine studies in non-human primates. Of potentially greater importance in developing an HIV-1 vaccine would be to apply the growing understanding of the role of mucosal T cell induction and trafficking. Although a general mucosal immune system has been postulated, and cross-talk may occur between some mucosal sites, there are also specific markers of T cell homing that are still under intense investigation (reviewed in [7]). However, the dynamic field of the specificity of those homing integrins and molecules and associated mucosal surface receptors is beyond the scope of this review.

The role of mucosal cytotoxic T cells (CTL), either MHC-I restricted or unrestricted, to provide protection from a variety of experimental pathogens that penetrate the mucosal barrier, is well documented. In a murine vaccinia-HIV-1-gp120 model it has been shown that rectal immunization with an HIV-1 peptide vaccine induced mucosal and systemic CTL and also provided protective immunity against rectal challenge, while systemic immunization using the same construct failed [8]. Moreover, selective induction of protective MHC-I restricted CTL in the intestinal lamina propia of rhesus macaques was shown following transient SIV infection of the colonic mucosa [9]. In rhesus macaques, these SIV-specific $CD8^+$ T cells with cytotoxic ability were also detected in the intraepithelial but not subepithelial compartment of the vaginal mucosa [10]. Indeed, in a recent study it was shown that the increased avidity of the seemingly protective colonic $CD8^+$ CTL was dictated by the route of immunization (i.e. rectal as opposed to systemic) and was also associated with colonic lamina propia $CD4^+$ T cell responses following rectal challenge with the pathogenic SHIV-ku-2 [11]. However, the lack of inbred rhesus colonies has resulted in an inability to do the appropriate transfer experiments to be assured of the protective specificity of the measured T cell response in such experiments.

In short, there are a number of fundamental unanswered questions as we enter 25 years of HIV-1 vaccine research. Do these vaccines need to induce immune responses at mucosal surfaces in order to have a beneficial effect? If so, can the use of systemically administered vaccines provide sufficient responses at the mucosal compartment to have some benefit? And if this is not true, then will mucosally *administered* vaccines be required to achieve the goal? If mucosal responses are needed and not induced by systemic vaccination, will a general mucosal vaccine suffice? That is, will there be a sufficient common mucosal immune system [12], or will there be compartmentalization of the mucosal immune response such that only some routes will protect? For example, does vaginal protection require administration by vaginal routes, or will iliac lymph node targeting, rectal, oral, or nasal immunization suffice? Although we have hints to all of these questions from animal models and a number of human studies, the definitive answer to any of the above questions is simply not known at this time.

18.2 Mucosal Routes of Immunization

The route of administration for induction of genital responses has been studied in a large number of systems, and not all of the data have been consistent. Nasal delivery has been prominent among the routes studied due to many positive characteristics. The doses of vaccines required have been relatively low, it lacks the low pH inactivation in the gut or vaginal tracts, and potent responses in multiple mucosal sites have been observed following nasal immunization [13, 14, 15]. In one murine study, intranasal immunization induced responses in a broader range of mucosal tissues than were observed following immunization by the vaginal, gastric, or rectal routes [16]. Likewise, intranasal vaccination of rhesus macaques with SIV p55gag combined with cholera toxin (CT) induced CD4+ and CD8+ T cell responses in both systemic and mucosal compartments, including the tonsil and mesenteric lymph nodes [17]. Specific antibody secreting B cells (ASC) were also found in the cervix, intestinal lamina propria, and nasal passages. In another study, intranasal immunization of humans with DNA-lipid complexes induced mucosal antibodies in genital and cervical lymph nodes [18].

In contrast to the intranasal studies cited above, the administration of CT to macaques either rectally or vaginally demonstrated that responses were limited to the compartment in which the immunization took place [19]. In humans, intranasal immunization with the nontoxic B subunit of CT (CTB) induced mucosal antibodies in cervicovaginal and rectal secretions as well as in peripheral blood [20], whereas the rectal or vaginal immunization with CTB again only induced local responses [21, 22].

Vaginal immunization to induce local vaginal antibody responses has yielded variable results. Following intravaginal immunization with poliovirus, women produce high levels of specific secretory IgA [23]. Likewise, immunization with Candida albicans also induced high specific IgA in vaginal secretions [24]. However, other studies have met with much more limited success [25, 26, 27]. An additional complication is added by the fact that the immune induction variation during vaginal vaccination requires assessment and timing to the different stages of the menstrual cycle [28]. Although the data from different laboratories has not been entirely consistent, the ability to induce generalized immunity outside the rectal or vaginal area by local administration of vaccines appears to be limited. How much protection may be afforded to small vs. large intestinal tissue when using rectal administration is still an open question, although there is recent intriguing supportive data for this concept from the Berzofsky research team at NIH [11].

Oral immunization has led to vaginal responses in some animal studies, but this route has been less potent in general than nasal approaches [29]. In one study, oral or intratracheal immunization with microparticles containing entrapped SIV induced protective immunity in systemically primed macaques against repeated intra-vaginal challenge with SIV [30]. In non-HIV studies in which priming versus no priming was compared, systemic priming has been necessary to achieve protection when using oral or IT delivered vaccines [31]. Oral immunization using SIV p55 gag and CT did not yield strong vaginal responses [32], despite eliciting a local response. Thus, oral vaccination has not been a preferred method for vaccine delivery in HIV models, except to deliver enteric modified bacterial vectors (such as *Salmonella*), which may prime for a systemic response after productive intestinal and lymph node infection.

Transcutaneous immunization has also been evaluated as potential immunization route to elicit mucosal CTL responses [7]. However, in these studies, the use of a potent mucosal adjuvant, such as either cholera toxin (CT) or *E. coli* labile toxin (LT) (with our without CpG), has been required, and appropriate control of the effects of these adjuvants in the transcutaneous immunizations by using multiple potential non-mucosal routes was not shown. Thus, these results at this point in time may be considered intriguing but require further confirmation.

18.3 Animal Studies and Mucosal Based Vaccine Approaches

The interest and potential need for mucosal vaccine development in HIV has led to a resurgence in interest and funding in the mucosal immunization field in general. The trial of Wright et al., carried out by the AIDS Vaccine Evaluation Group in the late 1990s and described below, was in many ways groundbreaking, and may have only been acceptable from a regulatory point of view because it took place in the HIV vaccine arena. Such trials in a regulatory environment with a lower benefit to risk ratio would likely require a great deal more pre-clinical toxicology than was initially required in HIV vaccine studies.

Many have questioned almost all of the initial early challenge studies in the SIV/SHIV model, as most were carried out with large infectious inocula that likely do not reflect the natural physiology in human transmission. Marthas et al. at UC Davis pioneered a method for infecting infant rhesus macaques using multiple feeding of low-dose SIV or SHIV placed in to milk, and were able to obtain infections in a majority of infants [33]. Likewise, investigators have used multiple low-dose inoculations via either the rectal route or the vaginal route to productively infect macaques [34, 35, 36]. With these models in hand, it is more likely that promising vaccine approaches can be evaluated in a manner more closely approximating the actual infectious events that occur in humans. In addition, it has become clear that the SHIV89.6P model that is highly pathogenic, but easy to protect against, may be of less value than studies using pathogenic SIV.

Some of the first studies that reinvigorated the mucosal field were those that showed that immunization protocols which protected monkeys from intravenous SIV challenge did not provide protection from mucosal challenge [37]. Shortly thereafter, protection from rectal challenge was achieved with a sub-infectious dose of attenuated SIV; however, the protection in those studies were shown likely to be due to a xenogeneic immune response to cellular antigens co-purified with the virus [38]. These UK-based investigators subsequently demonstrated that an SIV subunit vaccine, delivered to rhesus macaques by a non-mucosal route, but which targeted the regional nodes draining the rectum and vagina, was protective against rectal challenge [39]. This protection also then correlated with increased numbers of SIV-specific IgA-secreting cells in iliac lymph nodes. However, others have published that this regional targeting approach was not protective against vaginal transmission [40].

As these studies were being conducted and reported, other mucosal approaches continued to show promise. Live attenuated vaccines protected from the repeat oral challenge model of SIV in young macaques [41], and a SHIV attenuated vaccine protected from a virulent SIV vaginal challenge [42]. Multiple studies have now shown protection from rectal or vaginal challenge (with mostly SHIV viruses) using prime-boost combinations. These included those preformed by Ellenberger, Barouch, Amara, Rose [34, 43, 44, 45], and others. An ALVAC/protein prime-boost regimen was also shown to elicit CTL in the mucosal compartment and was partially protective against a SHIV mucosal challenge [46].

The natural tropism of adenoviruses for mucosal surfaces, and the fact that the widely used human adenovirus vaccine is mucosally delivered, sent this vector to the forefront of many disease vaccine programs, including those for tuberculosis, herpes simplex virus (HSV), as well as HIV [47]. Adenovirus vectors are also favored due to their ability to induce relatively robust and potent cellular and humoral responses following vaccination by multiple routes [48]. Early studies using an adenovirus SIV *env* vaccine administered by mucosal routes to rhesus macaques (intranasal or oral followed by intratracheal and intramuscular (IM) boosting) demonstrated an ability to elicit modest systemic and mucosal humoral responses, as well as potentially decreased viral burden following an intravaginal SIV_{mac251} challenge [49]. Later studies by these investigators also showed induction of cellular immunity and partial protection against a rectal challenge of the same SIV strain when using a multigenic (Env/Rev + Gag and/or Nef) adenovirus 5 vector prime (oral and intranasal boosted intratracheally) and IM protein boost [50]. However, the majority of studies performed with adenoviruses have used IM delivery, including the recent failed field trial of the Merck Ad5 vaccine encoding a polyprotein. Multiple studies are underway using adenoviruses of different serotypes or which have been genetically modified for mucosal delivery, mostly by the nasal route or prime-boost method, in HIV.

An AIDS vaccine based on an attenuated, recombinant vesicular stomatitis virus (VSV), administered by a combination of parenteral and mucosal routes, also proved effective in the rhesus SHIV-89.6PD mucosal challenge model [45]. In further studies using this construct Egan et al. showed that intranasal administration of this vaccine, which encoded HIV 89.6P envelope and SIV Gag (with alternate VSV G surface proteins used in the vector to avoid the neutralizing response), elicited markedly higher MHC class I T cell responses (measured by a combination of ELISPOT, tetramer staining, and CTL assays) than those achieved by the IM route [51]. The serum antibody titers to the HIV antigens were less affected by the route of administration, and the humoral IgG and IgA responses in the nasal wash, cervical and rectal secretions were detectable but disappointing, especially given the immunogenicity of the vaccine via the intranasal route. In the small number of experimentally infected monkeys, greater post-challenge titers appeared to arise in the intranasal group, although the level of protection did not vary between the intranasal and intramuscular administration. This study is, however, one of the few to show that an intranasal delivered vaccine can induce peripheral blood T cell responses superior to the same vaccine when given by the IM route.

Belyakov et al. also showed that mucosal CTLs specific for SIV virus can be induced by intrarectal immunization of macaques with a synthetic-peptide vaccine incorporating the LT(R192G) adjuvant [52], and that the response correlated with the level of T-helper response. After intrarectal challenge with pathogenic SHIV-Ku2, blood and intestinal viral titers, as well as local and peripheral $CD4^+$ T cells, were better preserved in intrarectally immunized than in subcutaneously immunized or control unvaccinated macaques. This same group has gone on to show differentiation between both the quantitative and qualitative response of mucosal (gut) CTL responses following rectal immunization. Much of this response and protective nature of rectal vaccination is hypothesized to function via the higher avidity and localization of the $CD8^+$ T cells induced [11, 53].

A further and intriguing mucosal approach has been developed by Andino et al. using poliovirus vectors encoding most of the genes in SIV. In early experiments using this novel system, rectal IgG and IgA were elicited in the majority of the cynomolgus macaques vaccinated via the intranasal route [54, 55]. Vaginal antibodies were detected in a minority of animals. Two immunizations were required, and two different Sabin serotypes were used as vectors. Of note, the investigators were then able to partially protect macaques from a pathogenic $SIV_{mac}251$ vaginal challenge. As with other vector systems in which either natural exposure or prior vaccination is likely to have elicited vector-neutralizing antibodies (e.g., adenovirus, parainfluenza virus, herpes viruses), the effect of preexisting immunity remains a significant hurdle for poliovirus-based vaccines.

There are a myriad of other reported vectors for use in mucosal protection. One of the more prominent among these is the alphavirus-based vector (Sinbis, Venezuelan Encephalitis Virus or Semliki Forest Virus). Although these have primarily been used parentally to date, they have been studied when administered by mucosal routes. Other promising approaches, espciecially for potential intranasal administration, include the use of herpes virus vectors (both amplicon based as well as attenuated constructs), rhinoviruses, adeno-associated viruses (AAV), and rabies viruses.

Although DNA, as a stand alone mucosal vaccination vector, has had a disappointing record [56], several investigators have used DNA for priming or boosting Initial work by Kalvinskis in the respiratory route led to responses that were intriguing, but this has not been followed by human trials [18]. The use of DNA transcutaneously may elicit mucosal T cell responses based on work in other systems [57], but this data is still far from definitive.

18.4 Mucosal HIV Vaccine Studies in Humans

In the 1990s the NIH-sponsored AIDS Vaccine Evaluation Group (which evolved into the HIV Vaccine Trials Network, or HVTN) carried out a number of innovative mucosal vaccine trials in humans. Two of the earliest studies involved the oral administration of an octomeric MN V3-loop peptide vaccine (United Biomedical) formulated as a polylactide microparticle prior to intramuscular boosting with the peptide adsorbed onto alum (AVEG 018) [58]. The peptide was given for three consecutive days at 1 mg on weeks 0, 4, and 24, followed by the intramuscular immunization at week 36. The vaccine was well-tolerated, but resulted in no significant humoral, cellular, or mucosal immune responses. Of note, in a prior trial using this same vaccine, immune responses were following three parental immunizations [59]. In a trial in which the intramuscular administration of the vaccine was followed by the oral boost (AVEG 023), the mucosal route again led to no detectable responses. The mucosal responses measured (but undetected) in these trials included assays for IgG and IgA antibodies in the feces, vaginal, parotid, salivary, seminal, and nasal wash secretions.

In another study a p17-24/Ty virus like particle (British Biotech) was given via the intramuscular route, and then boosted using the same particle given via either the oral or rectal routes (AVEG 019) [60]. Again, no meaningful responses were observed.

In the largest and most comprehensive study of HIV mucosal vaccination in humans performed to date (AVEG 027), a canarypox vCP-205 vaccine, which induced peripheral CTL responses when given IM in approximately 30% of healthy volunteers, was given by a variety of routes to 56 healthy subjects [61]. The non-replicating vectored vaccine encoded the p55Gag and p15 protease of the HIV-1 LAI strain, the gp120 of HIV-1 MN, and the gp41 of LAI, was administered IM, orally, intranasal, intravaginal, intrarectal, or in combination (IM + IR and IM + IN). These vaccinations were administered via the same route four times over 6 months, and then subjects in each arm were boosted with recombinant CHO-derived MNgp120 (VaxGen) at months 9 and 15. Of clinical design importance, a rabies glycoprotein canarypox vaccine construct (vCP65) was used as a control by each route (28 total subjects) to assess whether the immunogenicity of the HIV-1 encoded antigens would be potentially responsible in the event of suboptimal responses at mucosal sites.

The vaccines were well-tolerated despite the multiple routes and number of immunizations. Antibody responses to the canarypox vector were seen in all subjects and to the rabies glycoprotein in most of the vCP65 subjects after systemic administration. However, the HIV-1 specific responses generated by this vaccine administered mucosally or systemically were disappointing. After mucosal immunization, very few volunteers had responses to SF2 V3 loop, IIIB gp160, MN gp160, or LAI p24 proteins by any route, except for 2 of 7 who received the canarypox vaccine IM. The mucosal route also primed for gp120 binding antibody responses to the initial gp120 protein boost less effectively that did the systemic route (1/32 versus 10/14). No peripheral blood antibody-secreting cells (ASC) were detected following the canarypox immunizations, but IgG and IgA ASCs were detected following the protein boost (31/39 and 8/39 respectively). No mucosal antibody responses to HIV-1 were detected. The control rabies canarypox construct also elicited rabies antibody poorly when given mucosally. There were, however, mucosal antibodies detected (both IgG and IgA) to the parent canarypox vector (13/52 for IgA), including IgA in 8/11 of those who received intrarectal vaccinations.

Only sporadic CTL responses were measured in peripheral blood following any immunization (19% of volunteers), and this did not vary by route. However, there were detectable responses in 4 volunteers who received intrarectal delivery of the canarypox-vectored vaccine. The main conclusion that could be drawn was that the responses were suboptimal, but measurable (due to the frequent canarypox vector responses), and that the rectal route appeared to elicit the greatest local response, especially IgA to the canarypox and rabies antigens.

A number of studies have been now carried out using bacterial vectors encoding HIV-1 antigens. In a phase 1 study, a *Salmonella enterica* serovar *Typhimurium* was constructed to encode Gag via a type III secretion system [62]. Eighteen healthy adults received escalating doses orally beginning at 5×10^6 colony forming units (CFU) and reaching 1×10^6 CFU. Adverse events at low doses were in general mild, although at the highest doses diarrhea, fever, and abnormal transaminases were observed, and the highest dose arm had to be halted. No subject experienced bacteremia. Most of the subjects had immune responses to the *Salmonella* vector antigens, but responses to HIV-1 Gag were low and disappointing. No antibody was induced, and direct ex-vivo gamma interferon ELISPOT responses were minimal. Using an ex-vivo culture system, responses were observed in 4 of 10 volunteer receiving the three highest doses, but the actual number of cytokine secreting cells tended to be low, and persistence was not shown.

There has been an ongoing debate about whether the use of systemic vaccines can induce mucosal CTL responses, especially those in the cervico-vaginal or rectal tract (see discussion above). Some animal investigations have led to the thought that this is unlikely, including the extensive murine and nonhuman primate work by Belyakov and Berzofsky [63]. However, in one trial of a candidate canarypox vaccine, Musey selected subjects with known peripheral PBMC reactivity by chromium release and gamma interferon functional assays, and showed CTL activity in 5/7 in the blood compartment and in 4/7 in the rectal tract [64]. None of 5 placebo subjects had measurable responses in either compartment. In support of this concept, the protection from vaginal challenge by T-cell based vaccines given systemically in the SHIV 89.6P model also argues that mucosal vaccines, although possibly preferred, may not be necessary. However, because almost all of the related studies have focused on acute responses a few weeks after the final immunization, the notion remains that for long-term protection mucosal vaccination for induction of local memory may be beneficial or required.

18.5 Summary and Conclusion

At the time this chapter was compiled, the reported complete lack of efficacy of the Merck adenoviral-based IM vaccine (this vaccine had been partially effective in a SHIV mucosal challenge model) raises questions as to the validity or ability to interpret the majority of SHIV challenge studies. However, it is still unclear if the failure may be a SHIV strain-specific effect, an effect of low level prior immunity to the adenovirus construct, the inability to elicit functional mucosal responses in humans, the quantity or quality of the $CD4^+/CD8^+$ T cell response, lack of neutralizing antibody responses or a combination of these and other factors.

The pre-clinical evidence that mucosal routes of vaccination, or even mucosal induction of immune responses using systemic vaccination, provide complete or partial protection from HIV-1 infection or disease is still evolving. Although most of the animal studies, especially in the macaque model, would imply that vaccination via the nose, rectum or vagina may be needed for optimum vaginal or rectal responses, that data is still weak and not substantiated by any human clinical trial. The recent failure of the relatively immunogenic Ad5 multivalent vaccine does not leave answered the questions as to whether the failure was due to the level of response or the localization of response. It is clear that we still do not know the optimum route, the desired antibody response, the qualitative or quantitative nature of the T cells and their homing markers that need to be induced, or the specific approach that will allow for development of an effective mucosal HIV vaccine. Indeed, there is still debate as to whether neutralizing antibodies or CTL or yet undefined immune parameters, either in mucosal or systemic compartments or both, may correlated with protection against infection or disease. Nonetheless, 25 years into this epidemic, most investigators in the HIV-1 vaccine field believe that mucosal responses will be needed, and that directed mucosal vaccination maybe the best method available to achieve this goal.

References

1. Veazey, R.S., DeMaria, M.A., Chalifoux, L.V., Shvetz, D.E., Pauley, D.R., Knight, H.L., Rosenzweig, M., Johnson, R.P., Desrosiers, R.C., and Lackner, A.A. 1998, Science, 280, 427.
2. Vajdy, M., Veazey, R., Tham, I., deBakker, C., Westmoreland, S., Neutra, M., and Lackner, A. 2001, The Journal of Infectious Diseases, 184, 1007.
3. Veazey, R.S. 2007, Future, 1, 103.
4. Baba, T.W., Liska, V., Hofmann-Lehmann, R., Vlasak, J., Xu, W., Ayehunie, S., Cavacini, L.A., Posner, M.R., Katinger, H., and Stiegler, G. 2000, Nature Medicine, 6, 200.
5. Mascola, J.R., Stiegler, G., VanCott, T.C., Katinger, H., Carpenter, C.B., Hanson, C.E., Beary, H., Hayes, D., Frankel, S.S., and Birx, D.L. 2000, Nature Medicine, 6, 207.
6. Kozlowski, P.A., and Neutra, M.R. 2003, Current Molecular Medicine, 3, 217.
7. Belyakov, I.M., Ahlers, J.D., and Berzofsky, J.A. 2004, Expert Review of Vaccines, 3, S65.
8. Belyakov, I.M., Ahlers, J.D., Brandwein, B.Y., Earl, P., Kelsall, B.L., Moss, B., Strober, W., and Berzofsky, J.A. 1998, Journal of Clinical Investigation, 102, 2072.
9. Murphey-Corb, M., Wilson, L.A., Trichel, A.M., Roberts, D.E., Xu, K., Ohkawa, S., Woodson, B., Bohm, R., and Blanchard, J. 1999, The Journal of Immunology, 162, 540.
10. Lohman, B.L. 1995, The Journal of Immunology, 155, 5855.
11. Belyakov, I.M., Isakov, D., Zhu, Q., Dzutsev, A., and Berzofsky, J.A. 2007, The Journal of Immunology, 178, 7211.
12. Moldoveanu, Z., Russell, M.W., Wu, H.Y., Huang, W.Q., Compans, R.W., and Mestecky, J. 1995, Advances in Experimental Medicine and Biology, 371, 97.
13. Gallichan, W.S., and Rosenthal, K.L. 1995, Vaccine, 13, 1589.
14. Johansson, E.L., Rask, C., Fredriksson, M., Eriksson, K., Czerkinsky, C., and Holmgren, J. 1998, Infection and Immunity, 66, 514.
15. Russell, M.W., Martin, M.H., Wu, H.Y., Hollingshead, S.K., Moldoveanu, Z., and Mestecky, J. 2000, Vaccine, 19, S122.
16. Staats, H.F., Montgomery, S.P., and Palker, T.J. 1997, AIDS Research And Human Retroviruses, 13, 945.
17. Imaoka, K., Miller, C.J., Kubota, M., McChesney, M.B., Lohman, B., Yamamoto, M., Fujihashi, K., Someya, K., Honda, M., and McGhee, J.R. 1998, The Journal of Immunology, 161, 5952.
18. Klavinskis, L.S., Barnfield, C., Gao, L., and Parker, S. 1999, The Journal of Immunology, 162, 254.
19. Eriksson, K., Quiding-Järbrink, M., Osek, J., Möller, Å., Björk, S., Holmgren, J., and Czerkinsky, C. 1998, Infection and Immunity, 66, 5889.
20. Neutra, M.R., and Kozlowski, P.A. 2006, Nature Reviews. Immunology, 6, 148.
21. Kozlowski, P.A., Cu-Uvin, S., Neutra, M.R., and Flanigan, T.P. 1997, Infection and Immunity, 65, 1387.
22. Kozlowski, P.A., Cu-Uvin, S., Neutra, M.R., and Flanigan, T.P. 1999, The Journal of Infectious Diseases, 179, S493.
23. Ogra, P.L., and Ogra, S.S. 1973, The Journal of Immunology, 110, 1307.
24. Waldman, R.H., Cruz, J.M., and Rowe, D.S. 1972, The Journal of Immunology, 109, 662.
25. O'Hagan, D.T., Rafferty, D., McKeating, J.A., and Illum, L. 1992, Journal of General Virology, 73, 2141.
26. O'Hagan, D.T., Rafferty, D., Wharton, S., and Illum, L. 1993, Vaccine, 11, 660.
27. Thapar, M.A., Parr, E.L., Bozzola, J.J., and Parr, M.B. 1991, Vaccine, 9, 129.
28. Lu, F.X., Ma, Z., Moser, S., Evans, T.G., and Miller, C.J. 2003, Effects of Ovarian Steroids on Immunoglobulin-Secreting Cell Function in Healthy Women. pp. 944. American Society for Microbiology.
29. Challacombe, S.J., Rahman, D., and O'Hagan, D.T. 1997, Vaccine, 15, 169.

30. Marx, P.A., Compans, R.W., Gettie, A., Staas, J.K., Gilley, R.M., Mulligan, M.J., Yamshchikov, G.V., Chen, D., and Eldridge, J.H. 1993, Science, 260, 1323.
31. Tseng, J., Komisar, J.L., Trout, R.N., Hunt, R.E., Chen, J.Y., Johnson, A.J., Pitt, L., and Ruble, D.L. 1995, Infection and Immunity, 63, 2880.
32. Kubota, M. 1997, The Journal of Immunology, 158, 5321.
33. Van Rompay, K.K., Abel, K., Lawson, J.R., Singh, R.P., Schmidt, K.A., Evans, T., Earl, P., Harvey, D., Franchini, G., and Tartaglia, J. 2005, Journal of Acquired Immune Deficiency Syndromes, 38, 124.
34. Ellenberger, D., Otten, R.A., Li, B., Aidoo, M., Rodriguez, I.V., Sariol, C.A., Martinez, M., Monsour, M., Wyatt, L., and Hudgens, M.G. 2006, Virology, 352, 216.
35. McDermott, A.B., Mitchen, J., Piaskowski, S., De Souza, I., Yant, L.J., Stephany, J., Furlott, J., and Watkins, D.I. 2004, Journal of Virology, 78, 3140.
36. Otten, R.A., Adams, D.R., Kim, C.N., Jackson, E., Pullium, J.K., Lee, K., Grohskopf, L.A., Monsour, M., Butera, S., and Folks, T.M. 2005, The Journal of Infectious Diseases, 191, 164.
37. Marthas, M.L., Miller, C.J., Sutjipto, S., Higgins, J., Torten, J., Lohman, B.L., Unger, R.E., Ramos, R.A., Kiyono, H., and Mgghee, J.R. 1992, Journal of medical primatology, 21, 99.
38. Cranage, M.P., Baskerville, A., Ashworth, L.A.E., Dennis, M., Cook, N., Sharpe, S., Farrar, G., Rose, J., Kitchin, P.A., and Greenaway, P.J. 1992, Lancet (British edition), 339, 273.
39. Lehner, T., Wang, Y., Cranage, M., Bergmeier, L.A., Mitchell, E., Tao, L., Hall, G., Dennis, M., Cook, N., and Brookes, R. 1996, Nature Medicine, 2, 767.
40. Lu, X., Kiyono, H., Lu, D., Kawabata, S., Torten, J., Srinivasan, S., Dailey, P.J., McGhee, J.R., Lehner, T., and Miller, C.J. 1998, AIDS, 12, 1.
41. Otsyula, M.G., Miller, C.J., Tarantal, A.F., Marthas, M.L., Greene, T.P., Collins, J.R., Van Rompay, K.K.A., and McChesney, M.B. 1996, Virology, 222, 275.
42. Miller, C.J., McChesney, M.B., Lu, X., Dailey, P.J., Chutkowski, C., Lu, D., Brosio, P., Roberts, B., and Lu, Y. 1997, Journal of Virology, 71, 1911.
43. Amara, R.R., Villinger, F., Altman, J.D., Lydy, S.L., O'Neil, S.P., Staprans, S.I., Montefiori, D.C., Xu, Y., Herndon, J.G., and Wyatt, L.S. 2001, Science, 292, 69.
44. Barouch, D.H., Santra, S., Schmitz, J.E., Kuroda, M.J., Fu, T.M., Wagner, W., Bilska, M., Craiu, A., Zheng, X.X., and Krivulka, G.R. 2000, Science, 290, 486.
45. Rose, N.F., Marx, P.A., Luckay, A., Nixon, D.F., Moretto, W.J., Donahoe, S.M., Montefiori, D., Roberts, A., Buonocore, L., and Rose, J.K. 2001, Cell, 106, 539.
46. Pal, R., Venzon, D., Santra, S., Kalyanaraman, V.S., Montefiori, D.C., Hocker, L., Hudacik, L., Rose, N., Nacsa, J., and Edghill-Smith, Y. 2006, Journal of Virology, 80, 3732.
47. Gomez-Roman, V.R., and Robert-Guroff, M. 2003, AIDS AIDS Reviews, 5, 178.
48. Santosuosso, M., McCormick, S., and Xing, Z. 2005, Viral Immunology, 18, 283.
49. Buge, S.L., Richardson, E., Alipanah, S., Markham, P., Cheng, S., Kalyan, N., Miller, C.J., Lubeck, M., Udem, S., and Eldridge, J. 1997, Journal of Virology, 71, 8531.
50. Patterson, L.J., Malkevitch, N., Venzon, D., Pinczewski, J., Gomez-Roman, V.R., Wang, L., Kalyanaraman, V.S., Markham, P.D., Robey, F.A., and Robert-Guroff, M. 2004, Journal of Virology, 78, 2212.
51. Egan, M.A., Chong, S.Y., Rose, N.F., Megati, S., Lopez, K.J., Schadeck, E.B., Johnson, J.E., Masood, A., Piacente, P., and Druilhet, R.E. 2004, AIDS Research and Human Retroviruses, 20, 989.
52. Belyakov, I.M., Hel, Z., Kelsall, B., Kuznetsov, V.A., Ahlers, J.D., Nacsa, J., Watkins, D.I., Allen, T.M., Sette, A., and Altman, J. 2001, Nature Medicine, 7, 1320.
53. Belyakov, I.M., Kuznetsov, V.A., Kelsall, B., Klinman, D., Moniuszko, M., Lemon, M., Markham, P.D., Pal, R., Clements, J.D., and Lewis, M.G. 2006, Blood, 107, 3258.
54. Crotty, S., and Andino, R. 2004, Advanced Drug Delivery Reviews, 56, 835.
55. Crotty, S., Miller, C.J., Lohman, B.L., Neagu, M.R., Compton, L., Lu, D., Lu, F.X.S., Fritts, L., Lifson, J.D., and Andino, R. 2001, Journal of Virology, 75, 7435.

56. Lundholm, P., Leandersson, A.C., Christensson, B., Bratt, G., Sandström, E., and Wahren, B. 2001, Virus Research, 82, 141.
57. Belyakov, I.M., Hammond, S.A., Ahlers, J.D., Glenn, G.M., and Berzofsky, J.A. 2004, Journal of Clinical Investigation, 113, 998.
58. Lambert, J.S., Keefer, M., Mulligan, M.J., Schwartz, D., Mestecky, J., Weinhold, K., Smith, C., Hsieh, R., Moldoveanu, Z., and Fast, P. 2001, Vaccine, 19, 3033.
59. Gorse, G.J., Keefer, M.C., Belshe, R.B., Matthews, T.J., Forrest, B.D., Hsieh, R.H., Koff, W.C., Hanson, C.V., Dolin, R., and Weinhold, K.J. 1996, J Infect Dis, 173, 330.
60. Wahl, S.M., Smith, P.D., and Janoff, E.N. 1999, J. infect. Dis, 179, S397.
61. Wright, P.F., Mestecky, J., McElrath, M.J., Keefer, M.C., Gorse, G.J., Goepfert, P.A., Moldoveanu, Z., Schwartz, D., Spearman, P.W., and El-Habib, R. 2004, The Journal of Infectious Diseases, 189, 1221.
62. Kotton, C.N., Lankowski, A.J., Scott, N., Sisul, D., Chen, L.M., Raschke, K., Borders, G., Boaz, M., Spentzou, A., and Galán, J.E. 2006, Vaccine, 24, 6216.
63. Belyakov, I.M., and Berzofsky, J.A. 2004, Immunity, 20, 247.
64. Musey, L., Ding, Y., Elizaga, M., Ha, R., Celum, C., and McElrath, M.J. 2003, The Journal of Immunology, 171, 1094.

Part VI
Mucosal Vaccine Approaches

Chapter 19
Formulations and Delivery Systems for Mucosal Vaccines

Padma Malyala and Manmohan Singh

Abstract Mucosal vaccines offer great advantage and better patient compliance in many disease areas. Most vaccine delivery systems are generally particulate e.g. emulsions, microparticles, iscoms and liposomes, and mainly function to target associated antigens into antigen presenting cells (APC). Several key formulation and delivery approaches exist to deliver the antigens in combination with an adjuvant by the mucosal route. Some of these have also been evaluated in the clinic while many others are still in pre-clinical evaluation. The discovery of more potent adjuvants may allow further development of prophylactic vaccines against acute infections as well as therapeutic vaccines against chronic infectious diseases, some of which could be delivered mucosally. In addition, novel and creative delivery systems may also allow vaccines to be delivered mucosally in a more efficient manner.

Keywords Vaccine adjuvants · Immunostimulators · Mucosal delivery systems · Liposomes · Microparticles · Emulsions · LT mutants

19.1 Introduction

It is well known that majority of pathogens initially infect their hosts through mucosal surfaces. Therefore, the induction of mucosal immunity is likely to make an important contribution to protective immunity. In addition, mucosal administration, which avoids the use of needles, is becoming an increasingly attractive approach for the development of new generation vaccines. Although a number of vaccines are commercially available which control the spread of acute infections such as influenza [1], these vaccines induce serum immunity, but do not induce mucosal immunity at the site of infection in the nasal cavity. In addition, commercially available vaccines are ineffective for the induction of cytotoxic T lymphocyte (CTL) responses, which are responsible for killing virally infected cells [2]. As a consequence, work is currently underway to develop more effective mucosal vaccines

M. Singh
Novartis Vaccines, 4560 Horton Street, Emeryville, CA 94608, USA
e-mail: Manmohan.singh@novartis.com

M. Vajdy (ed.), *Immunity Against Mucosal Pathogens*,

that induce mucosal IgA responses through local administration, and also induce more potent systemic responses [3].

Most vaccines have traditionally been administered by intramuscular or subcutaneous injection. Mucosal administration of vaccines offers a number of important advantages; including ease of administration, reduced adverse effects induced through injection and the potential for frequent boosting. In addition, local immunization induces mucosal immunity at the sites where many pathogens initially establish infection of hosts while in general, systemic immunization has failed to induce mucosal IgA antibody responses. Oral immunization would be particularly advantageous in isolated communities, where access to health care professionals is difficult. Moreover, mucosal immunization would avoid the potential problem of infection due to the re-use of needles. Several orally administered vaccines are commercially available, which are based on live-attenuated organisms, including vaccines against polio virus, *Vibrio cholerae* and *Salmonella typhi*. In addition, a wide range of approaches is currently being evaluated for mucosal delivery of vaccines including many approaches involving non-living adjuvants and delivery systems.

19.2 Mucosal Adjuvants

The most potent mucosal adjuvants which are available for local immunization are heat labile enterotoxin from *Escherichia coli* (LT) and cholera toxin (CT) from *Vibrio cholerae*, and these molecules and their sub-units have shown some promise as intranasal adjuvants for e.g. influenza [4]. However, since the native toxins CT and LT are the causative agents, respectively for cholera and traveler's diarrhea, they are considered to be too toxic for use in humans. Therefore, several groups have focused on the development of detoxified mutants of LT and CT as mucosal adjuvants. Groups within Novartis Vaccines and Diagnostics Inc. (formerly Chiron Corporation) have focused on the development of LT mutants with reduced, or eliminated enzymatic activity, since it is the ADP-ribosylating enzymatic activity of LT and CT which causes abnormal intracellular accumulation of cAMP and excess fluid secretion from intestinal epithelial cells. Site-directed mutagenesis was used to replace single amino acids within the enzymatic A subunit of LT and mutants (LTK63 and LTR72) with reduced or eliminated enzymatic activity were developed [5]. LTK63 is completely devoid of ADP-ribosyltransferase activity and appears to be non-toxic both in vivo and in vitro, while LTR72 has residual enzymatic activity ($<1\%$ of the native LT) and has significantly reduced, but detectable, toxicity [5]. Both of these mutants have been previously shown to be potent mucosal adjuvants for antibody and CTL induction in a number of studies in mice [5]. In addition, LT mutants have been shown to be potent oral adjuvants for influenza vaccines and model antigens.

Nevertheless, due to the significant challenges associated with oral immunization, various alternative routes of immunization have been evaluated with LT mutants, including nasal, intravaginal and intra-rectal. Of these, intranasal immunization offers the most promise, both due to the potent responses induced by this route and due to the easy access and simple administration devices which already exist. On many occasions, the ability of LT mutants to induce potent antibody responses following intranasal immunization has been demonstrated [6]. In recent studies, LT mutant adjuvants have shown protection against challenge with *B. pertussis* [7], *S. pneumoniae* [8] and herpes simplex virus [9] following intranasal immunization and the induction of potent CTL responses. In addition, the potency of LT mutants was not affected by the presence of pre-existing immunity to the adjuvant [8]. The apparent safety of this approach in humans using wild type LT is strongly supportive of the approach using genetically detoxified LT mutants.

CpG, IC–31, Protollin, and Chitosan are other mucosal adjuvants used to target and enhance mucosal immune defenses [9, 10]. Chitosan has been used as both a mucosal adjuvant as well as a delivery system. CpG, a TLR agonist, has been used in conjunction with Tetanus Toxoid and Cholera Toxin antigens as a mucosal adjuvant and higher anti toxin titers have been reported by many research groups. Synthetic CpG-oligodinucleotides (ODN), containing immunostimulatory C poly G motifs is a novel class of adjuvant. These motifs are recognized by the innate immune system via the toll like receptor 9 (TLR9), and can induce broad adjuvant effects such as the direct activation of B cells, macrophages and dendritic cells as well as induction of IL-6 and IL-12 cytokine secretion [11]. CpG-ODNs contain a nuclease resistant phosphorothioate backbone, which can be co-administered with the vaccine antigen to induce specific immunity. It has been demonstrated that CpG-ODNs are safe and effective adjuvants with both parenteral and mucosal vaccine administration in various animal models and recent studies report the ability of CpG-ODNs to induce both systemic and humoral immunity upon mucosal application [11]. CpG motifs induce Th1 biased immune effects due to TLR9 signaling, which can be used to augment cell-mediated immunity [11]. In the development of an effective amebiasis vaccine, Ivory et al. developed a mucosal vaccine composed of the Entamoeba histolytica Galactose/N-acetyl-D-galactosamine inhibitable lectin (Gal-lectin) and CpG-ODN [11]. The Gal-lectin is a protein involved in parasite virulence and adherence and is known to activate immune cells, while CpG-ODN is a potent inducer of Th1-type immune responses. The group demonstrated that the intranasal administration of the vaccine resulted in strong Gal-lectin specific Th1 responses and humoral responses in gerbils [11].

The IC31 adjuvant, consisting of a vehicle based on the cationic peptide $KLKL_5$ KLK and the immunostimulatory oligodeoxynucleotide ODN1a, signaling through the TLR9 receptor, was found to promote highly efficient Th1 responses. The stimulatory effect of IC31 has been attributed to the ODN1a component leading to stimulation of antigen presenting cells through TLR9, triggering an intracellular cascade that eventually leads to secretion of proinflammatory cytokines and finally a Th1-response [12]. Upon ligation of TLR9, the production of IFN-α leading to Th1 response, and CTL activity was found to be induced in mice and signaling through TLR9 has thus been considered important for the generation of anti-bacterial and viral immune responses [12] The expression of TLR9 is selectively found in the endosomal compartment of plasmacytoid dendritic cells (pDCs) and B cells whereas TLR4 and the majority of the known TLRs are present on the surface of human myeloid DCs and monocytes [12]. The restricted distribution of TLR9 together with the intracellular location has questioned the applicability of targeting TLR9 through, e.g. soluble CpG-motifs [12]. However, recent clinical studies using various CpG motifs mixed with antigen have demonstrated the priming of a high number of CD8+ T cells specific for a melanoma peptide and an IFN-γ response after vaccination with an influenza vaccine [12]. These studies have emphasized the potential of TLR9 ligands, like IC31, as important immunomodulator for inclusion into human vaccines. The other characteristic feature of the IC31 adjuvant system is the cationic vehicle $KLKL_5KLK$ which is also shared with other successful tuberculosis (TB) vaccine formulations. Cationic DDA liposomes have, in this regard, previously been found to be efficient for priming anti-TB responses to both Ag85B-ESAT-6, a mycobacterial vaccine antigen, and HBHA, a Heparin Binding Haemagglutinin Antigen [12]. Both cationic liposomes and the cationic peptide, $KLKL_5KLK$, may provide efficient interaction with the negatively charged host cells and presumably also exert efficient delivery of antigen into the cell, a characteristic of cationic components, which have been exploited for the transfection of cells with plasmid DNA [12].

Agger et al. have reported augmentation of the immune response and protective efficacy of the mycobacterial vaccine antigen, Ag85B-ESAT-6 using IC-31 in mice [12]. The combination of Ag85B-ESAT-6 and IC31 exhibited significant levels of protection in the mouse aerosol challenge model of tuberculosis and a detailed analysis of the immune response generated revealed the induction of CD4+ T cells giving rise to high levels of IFN-γ secretion.

Protollin is a novel adjuvant comprising Proteosomes non-covalently complexed with LPS. Intranasal immunization of mice with Protollin combined with detergent-split influenza antigens (HA) or recombinant influenza hemagglutinin (rHA) enhanced serum IgG and mucosal IgA levels by up to 250-fold compared with immunization with the antigens alone [10] IFN-γ responses were also enhanced compared to the levels produced by splenocytes from mice immunized with antigen alone, while production of IL-5 was abrogated. Mice immunized with Protollin-rHA were completely protected against lethal challenge with influenza virus, demonstrating that Protollin is an effective mucosal adjuvant for prophylactic vaccines (Table 19.2) [13].

Chitosan, a naturally occurring biopolymer, has been used as a mucosal adjuvant as well as a bioadhesive and is discussed in detail in the section of mucosal delivery systems.

19.3 Mucosal Delivery Systems

19.3.1 Bioadhesive Formulations

Since the early 1980's, several groups have focused on bioadhesion as a concept to improve local and systemic drug delivery [14]. In general, bioadhesive delivery systems are designed to adhere to various tissue surfaces, mainly the mucosal epithelium. An alternative term, mucoadhesion is also used often to describe the interaction of a polymer delivery system and a mucosal site. Mucoadhesion appears to require a highly expanded and hydrated polymer network, which promotes an intimate molecular contact between the delivery system and the mucus layer [14]. The mechanisms of bio- or mucoadhesion can involve physical or chemical interactions, including electrostatic or hydrophobic bonding, van der Waal's forces or hydrogen bonding. Irrespective of the mechanisms involved, the main advantages of bioadhesive delivery systems include extended residence time at the site of action, local delivery to a selected site and enhanced interaction with the mucosal epithelium [14]. In a range of studies in recent years, several bioadhesive polymers have been described, including chitosans, methacrylic acids, starch, gelatin, hyaluronic acid and cellulose derivatives to enhance the absorption of co-administered protein drugs [15] Hyaluronic acid is a naturally occurring mucopolysaccharide consisting of residues of D-glucuronic acid and N-acetyl-D-glucosamine. Through the esterification of the carboxyl groups of hyaluronic acid with alcohols, biodegradable polymers have been developed, called HYAFF [16]. The HYAFF™ polymers can be used to make microspheres using a coacervation phase-separation process [17]. The HYAFF microspheres have strong bioadhesive properties and have been used for delivery of calcitonin and insulin following mucosal administration [15, 17]. However, HYAFF microspheres have not been used previously for mucosal delivery of vaccines.

The initial observations in mice offered significant encouragement that the bioadhesive microsphere delivery system may offer some benefit over administration of a formulation comprising of soluble haemagglutinin (HA) protein, with the potent adjuvant, LTK63. In addition, the results described here show that the potency of the HA + LT mutant combination can be enhanced by formulation into a bioadhesive microsphere delivery system (Figs. 19.1 and 19.2). Intranasal immunization with the bioadhesive microsphere formulation in pigs induced a significantly enhanced serum immune response in comparison to traditional intramuscular immunization.

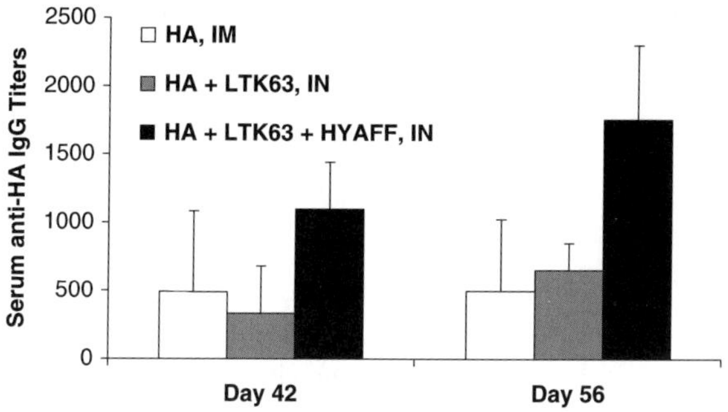

Fig. 19.1 Following two intranasal immunizations of mice four weeks apart, enhanced serum antibody responses were obtained with influenza vaccine (HA) and mucosal adjuvant LTK63 in combination with bioadhesive HYAFF microspheres (HA + LTK63 + HYAFF). For comparison, mice were also immunized with antigen alone (HA), antigen and microspheres (HA + HYAFF) or antigen plus adjuvant (HA + LTK63). Geometric mean titers (GMT) +/− standard error of the mean are presented for each group

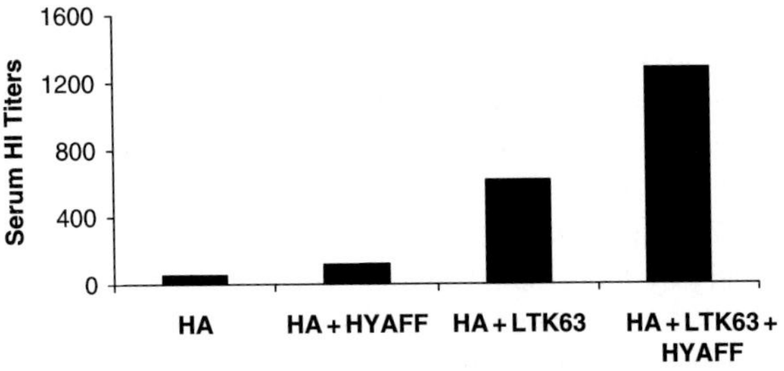

Fig. 19.2 Following two intranasal immunizations of mice four-week apart, enhanced serum hemagglutination inhibition (HI) antibody responses were obtained with influenza vaccine (HA) and mucosal adjuvant LTK63 in combination with bioadhesive HYAFF microspheres (HA + LTK63 + HYAFF). For comparison, mice were also immunized with antigen alone (HA), antigen and microspheres (HA + HYAFF) or antigen plus adjuvant (HA + LTK63). Mean HI Titers are presented for each group

In contrast, an alternative bioadhesive polymer, chitosan, which is a widely used pharmaceutical excipient [18], has shown encouraging results following intranasal immunization in a small animal model [18]. Nevertheless, chitosan and related molecules have previously been described as potent adjuvants or immunomodulatory compounds following parenteral/systemic immunizations [19]. Therefore, the adjuvant effect of chitosan following intranasal administration may not come solely from the bioadhesive properties of this polymer. It remains to be seen if simple 'bioadhesion' is enough to impart a potent adjuvant effect following mucosal delivery, but our data would seem to indicate that inclusion of an adjuvant active molecule may also be necessary to induce a potent response.

The bioadhesive microspheres may contribute to the immune response obtained due to one or more of the following reasons; (a) increased duration of retention in the nasal cavity, (b) greater interaction with the epithelium, (c) enhanced absorption, or (d) sustained release from the microspheres. Each of these effects may act upon the antigen, the adjuvant, or both. Although further studies are necessary to determine the mechanism of action of the formulation, it is notable that the microspheres alone were ineffective and the presence of a mucosal adjuvant was necessary for potent responses.

19.3.2 PLG Microparticles

In the recent past, we have focused on an alternative delivery system for vaccines consisting of microparticles prepared from the biodegradable, biocompatible polyesters, poly (lactide co-glycolide) (PLG). Microparticles represent an attractive approach to vaccine delivery since PLG has been used in humans for many years as resorbable suture material and as controlled-release drug delivery systems. [20, 21, 22] It has also been shown that microparticles ($\sim 1\ \mu m$) are taken up efficiently by antigen-presenting cells (APC) *in vitro*, and *in* vivo and are able to induce cytotoxic T lymphocyte (CTL) responses in rodents. [23, 24] We recently developed novel charged PLG microparticles prepared with an anionic surfactant as the particle stabilizer, capable of efficient adsorption of antigen onto their surface. [25, 26] This approach allows the preservation of antigen integrity and results in the presentation of multiple copies of the antigen to the APC, which is similar to the surface of a pathogen. We have also shown microparticles to be a potent vaccine delivery system for antigen from *Neisseria meningitides* serotype B (Men B) [27]. In previous studies, we have reported that immunization with a recombinant antigen from HIV adsorbed to PLG microparticles induces potent antibody and CTL responses in mice [28] and in non-human primates. [28]

Following intranasal administration of anesthetized mice with haemagglutinin (HA) from influenza virus entrapped in one of four microparticle resins (sodium polystyrene sulfonate, calcium polystyrene solfinate, polystyrene benzyltrimetylammonium chloride, or polystyrene divinylbenzene) sized to 20–45 μm enhanced anti-HA serum Haemagglutinin-inhibiting antibodies and nasal wash IgA antibodies were induced [29]. Importantly, this study showed that this immunization strategy reduced viral burden in the lungs following intranasal challenge of anesthetized mice with live virus. Interestingly, these resins induced enhanced serum IFN-γ levels following intranasal administrations, while the levels of IL-4, IL-2 and IL-6 remained unchanged, suggesting a Th1-type response (ibid).

To investigate a possible mechanism for the enhanced immune responses induced following intranasal immunizations with PLG-DNA, we localized and phenotypically identified the cells that expressed the HIV-1-gag protein in local and systemic lymphoid tissues. Following a single intranasal immunization with PLG-DNA expressing HIV-1 gag, we localized and identified the cells that expressed the encoded gene by immuno-fluorescent staining (unpublished data). In the immunostaining studies of cervical lymph i)nodes (CLN) and spleen, the majority of gag-expressing cells were $CD11b^+$, suggesting that this population is responsible for uptake and expression of DNA following intranasal immunization with PLG/DNA. Although CD11b is expressed by many cell populations, it is primarily considered a marker for tissue macrophages (Macs) and dendritic cells (DC), which are both professional antigen-presenting cells (APC) [29] [30]. However, compared to Macs, DC are more potent APC [31] Our data suggest that following IN immunization with DNA adsorbed onto PLG-microparticles, monocyte lineage cells, Macs and/or DC's, are involved in the uptake and expression of gag-DNA, since we detected both $CD11b^+$ and [31] naïve T cells *in vivo* is an important question, which needs further investigation. Our previous *in vitro* data showed that bone marrow-derived DC can take up PLG/DNA encoding HIV-1 gag and present it to a gag-specific T cell hybridoma [32].

19.3.2.1 Oral Immunization with PLG Microparticles

The most attractive route for mucosal immunization is oral, due to the ease and acceptability of administration through this route. However, due to the presence of acidity in the stomach, an extensive range of digestive enzymes in the intestine and a protective coating of mucus which limits access to the mucosal epithelium, oral immunization has proven extremely difficult with non-living antigens. However, novel delivery systems and adjuvants may be used to significantly enhance the responses following oral immunization.

In mice, oral immunization with PLG microparticles has been shown to induce enhanced mucosal and systemic immunity to entrapped compared to soluble antigens [33, 34, 35] In addition, mucosal immunization with microparticles induced protection against challenge with *Bordetella pertussis* [36, 37, 38, 39] *Chlamydia trachomatis* [40] and *Salmonella typhimurium* [41]. In primates, mucosal immunization with inactivated SIV in microparticles induced protective immunity against intravaginal challenge [42]. Also, in primates, mucosal immunization with microparticles induced protection against aerosol challenge with staphylococcal enterotoxin B [43]. Comparative studies have indicated that microparticles are one of the most potent delivery systems available for mucosal delivery of vaccines [44]. In recent studies, microparticles have also shown some promise for the mucosal delivery of DNA [45, 46]. The ability of microparticles to perform as effective adjuvants following mucosal administration is largely a consequence of their uptake into the specialized organized mucosa associated lymphoid tissue (oMALT) [47]. Although most of this work has described particle uptake following oral delivery, a recent paper described the uptake of microparticles into mice following intranasal delivery [48]. The potential of microparticles and other polymeric systems for mucosal delivery of vaccines was recently reviewed [13, 49] as was the use of a broader range of antigen delivery systems [50]. While microparticles have significant potential for mucosal delivery of vaccines, their potency may be improved by their use in combination with additional adjuvants. This is likely to be a pre-requisite for the development of effective oral vaccines, since the challenges should not be underestimated. Accumulated experimental evidence suggests that simple encapsulation of vaccines into microparticles is unlikely to result in the successful development of oral vaccines and improvements in the current technology are clearly needed [13].

19.3.3 Liposomal and Niosomal Formulations

Various liposomal formulations have been explored as mucosal delivery systems both with an antigen alone and with an incorporated adjuvant. Liposomal vaccine formulations are being explored in oral, and transdermal delivery systems. Liposomes enable formulation of both hydrophilic and hydrophobic molecules as they are comprised of hydrophilic and hydrophobic regions. The liposome lipid bilayers can adhere to the plasma membrane of the cognate mammalian cell and deliver their contents directly into the cytoplasm [51]. In transdermal vaccine delivery, vesicular systems such as liposomes (phospholipids-based artificial vesicles) and niosomes (non-ionic surfactant vesicles) are being tried to increase permeability. These penetration enhancers are biodegradable, non-toxic, amphiphilic in nature, and effective in the modulation of vaccine or drug release properties [52].

Liposomal delivery systems have been evaluated as an oral delivery system where gastrointestinal immunity is desired. For instance, due to the nature of the *Vibrio cholera* infection, oral vaccination, which induces intestinal immunity, is believed to be more effective than the immunity elicited by a parenteral vaccine.

In a study by Somroop et al. on developing a cholera vaccine, an oral vaccine made up of three antigens, i.e. heat-treated recombinant cholera toxin (CT), V. cholerae O1 lipopolysaccharide (LPS), and recombinant toxin-co-regulated pilus subunit A (rTcpA) was formulated using liposomes as an antigen delivery vehicle and unmethylated CpG-ODN as an adjuvant [51]. The kinetics of the specific antibody secreting cells (ASC) in the peripheral blood of vaccine primed-rats were studied in order to determine the mucosal memory response to the vaccine. Significant numbers of the ASC of all antigenic specificities were present in the lamina propria (LP) of the vaccinated rats on days 6 and 13 post-booster. The number of CT-ASC was highest on Day 4 after the boost, while the LPS-ASC and TcpA-ASC were at their peak on Day 3 post booster immunization. Table 19.1 below compares the mean of CT- ASC on day 3 between the placebo group and immunized group. [51].

Naito et al. used cationic liposomes as an oral delivery system for attenuated modified vaccinia virus Ankara, a vaccine vector (MVA) [53]. Their study explored whether cationic liposomes modified with tresyl monomthoxy poly ethylene glycol (TMPEG) are capable of shielding MVA expressing HIV glycoprotein (MVAIIIB/β-gal) from poxvirus-neutralizing antibodies in vitro and in vivo. They showed that, after modification using the optimized PEGylation method with cationic liposome complexes, the MVAIIIB/β-gal–TMPEG/liposome complexes retained infectivity in vitro. Furthermore, the same complexes provided an appreciable level of protection in vivo and elicited antigen-specific immune responses in mucosal and systemic tissues of BALB/c mice after repeated oral delivery, suggesting effective shielding against pre-existing poxvirus neutralizing antibodies for multiple deliveries (Table 19.1) [53]. Jain and Vyas used Mannosylated Niosomes as oral vaccine delivery system to protect the antigen, tetanus toxoid (TT), from gastrointestinal degradation [13, 54]. The coating of modified polysaccharide o-palmitoyl mannan also enhanced the affinity of the delivery system toward antigen-presenting cells

Table 19.1 Serum IgG antibody titers and Antigen-Specific Antibody Secreting Cells (ASC) using liposomes as mucosal delivery systems with different antigens and animal models using mucosal routes of administration [51, 53, 55]

Animal model	Route	Antigen	Functional readout	Formulation and dose	Titers	ASC
Mice	Oral	Ankara	HIV-Envelope-specific Antibody titers	MVA MVA/Liposome Complex	750 2200	– –
Rats	Oral	Cholera	Mean Antigen – specific Antibody Secreting Cells (ASC) on Day 4 for CT in peripheral blood	Placebo (CT+ *V. cholerae* O1 LPS+ Recombinant TcpA) combination formulated with liposomes using CpG-ODN.	– –	10 72
Rabbits	Nasal	TT	Mucosal IgA titers	TT Solution TT encapsulated in Liposome	31 64	– –

Table 19.2 Mucosal IgA antibody titers with Mannosylated Niosomes as Mucosal delivery system and Protollin as Mucosal adjuvant using TT and Influenza as antigens in rodent animal models [12, 54]

Animal model	Route	Antigen	Titer type	Formulation and dose	Titers
Rats	Oral	TT	Salivary Secretory IgA titers	TT/Alum	1.78
				TT/Plain Niosomes	56.23
				TT/Mannosylated Niosome	1000
Mice	Nasal	Influenza	Lung IgA titers	A/Beijing/IM	20
				A/Beijing/IN	24
				A/Beijing/Protollin/IN	3397

(APCs) of Peyer's patches. The formulation elicited better IgG levels and a significant mucosal immune response (SIgA) upon oral administration in comparison with intramuscularly- injected alum-adsorbed TT (Table 19.2) [54].

TT liposomes with CpG-ODN as an adjuvant were used as a nasal delivery system by Tafaghodi et al. to increase the systemic and mucosal immune responses in rabbits [55]. TT encapsulated in liposomes kept its intact structure, and its immunoreactivity was also completely preserved, as shown by the SDS–PAGE and ELISA methods. Although the highest serum IgG and antitoxin titers were observed in groups immunized with solution formulations ($P < 0.001$), the highest mucosal SIgA titers were achieved by liposomes encapsulated with TT (Table 19.1). CpG-ODN as an adjuvant was able to increase the serum IgG and antitoxin titers when co-administered with TT in solution ($P < 0.05$) or co-encapsulated with TT in liposomes ($P < 0.01$), but failed to increase the SIgA titers in nasal lavages. No hemolysis occurred upon incubation of liposomes with human RBCs. Also, no local irritation was observed in human volunteers after nasal administration of plain liposomes in this study. Intranasal administration of liposomes encapsulated with vaccines showed to be an effective way for inducing the mucosal immune responses [55]. Alcon et al. investigated the ability of biphasic lipid vesicles (as vaccine-targeting adjuvants) containing a bacterial antigen and CpG ODNs to induce systemic and mucosal immune responses in pigs. The results showed that while the protein, either alone or with CpG ODNs, did not induce mucosal immune responses, administration of antigen and CpG ODNs in biphasic lipid vesicles resulted in induction of both systemic and local antibody responses after immunization using a combined mucosal/systemic approach [56].

19.4 Summary

Future developments in adjuvants are likely to include the development of more site-specific delivery systems for both mucosal and systemic administration. In addition, the identification of specific receptors on APC's is likely to allow targeting of adjuvants for the optimal induction of potent, safe and specific immune responses.

The specific receptors on APC may be extra- or intracellular. If intracellular, then a means to promote uptake of the delivery system by the relevant cells may also be required for optimal efficacy. An interesting approach to targeting APC's has been described which involves co-expression of two linked proteins, with a targeting component and an adjuvant signal [57, 58, 59]. An alternative approach to vaccine targeting for CTL induction has also been described using a fusion protein with a bacterial toxin to deliver the antigen specifically to the Class I processing pathway [60, 61]. However, further developments of novel adjuvants will likely be driven by a better understanding of the mechanism of action of currently available adjuvants and this is an area of research that requires additional work.

Acknowledgments We would also like to thank all the members of the Vaccine Delivery Group at Novartis. Thanks are also due to Nelle Cronen for her help with the manuscript preparation.

References

1. Ghendon, Y. 1989, Adv. Exp. Med. Biol., 257, 37.
2. Riddiough, M.A., Sisk, J.E., and Bell, J.C. 1983, JAMA, 249, 3189.
3. Glezen, W.P. 1982, Epidemiol. Rev., 4, 25.
4. Tamura, S.I., and Kurata, T. 2000, Jpn. J. Infect Dis., 53, 98.
5. Barackman, J.D., Ott, G., and O'Hagan, D.T. 1999, Infect. Immun., 67, 4276.
6. Giuliani, M.M., Del Giudice, G., Giannelli, V., Dougan, G., Douce, G., Rappuoli, R., and Pizza, M. 1998, J. Exp. Med., 187, 1123.
7. Ryan, E.J., McNeela, E., Murphy, G.A., Stewart, H., O'Hagan, D., Pizza, M., Rappuoli, R., and Mills, K.H. 1999, Infect. Immun., 67, 6270.
8. Jakobsen, H., Bjarnarson, S., Del Giudice, G., Moreau, M., Siegrist, C.A., and Jonsdottir, I. 2002, Infect. Immun., 70, 1443.
9. Fraser, C.K., Diener, K.R., Brown, M.P., and Hayball, J.D. 2007, Expert Rev. Vaccines, 6, 559.
10. Jones, T., Cyr, S., Allard, F., Bellerose, N., Lowell, G.H., and Burt, D.S. 2004, Vaccine, 22, 3691.
11. Ivory, C.P., and Chadee, K. 2007, Infect. Immun., 75, 4917.
12. Agger, E.M., Rosenkrands, I., Olsen, A.W., Hatch, G., Williams, A., Kritsch, C., Lingnau, K., von Gabain, A., Andersen, C.S., Korsholm, K.S., and Andersen, P. 2006, Vaccine, 24, 5452.
13. Brayden, D.J. 2001, Eur. J. Pharm. Sci., 14, 183.
14. O'Hagan, D.T., Rafferty, D., Wharton, S., and Illum, L. 1993, Vaccine, 11, 660.
15. Illum, l., Farraj, N.F., Fisher, A.N., Gill, L., Miglietta, M., and Benedetti, L. 1994, J. Controlled Release, 29, 133.
16. Davide, C., Partrick, D., Marco, R., Paola, B., Giovanni, A., and David, F.W. 1998, Biomaterials, 19, 2101.
17. Richardson, J., Ramires, P., Miglietta, M., Rochira, L., Bacella, L., Callegaro, L., and Benedetti, L. 1995, Int. J. Pharm, 115, 9.
18. Jabbal-Gill, I., Fisher, A.N., Rappuoli, R., Davis, S.S., and Illum, L. 1998, Vaccine, 16, 2039.
19. Illum, L., Jabbal-Gill, I., Hinchcliffe, M., Fisher, A.N., and Davis, S.S. 2001, Adv. Drug Deliv. Rev., 51, 81.
20. O'Hagan, D., and Singh, M. 2003, Expert Review Vaccines, 2, 269.
21. Okada, H., and Toguchi, H. 1995, Crit. Rev. Ther. Drug Carrier Syst., 12, 1.
22. Putney, S.D., and Burke, P.A. 1998, Nat. Biotechnol., 16 [published erratum appears in Nat Biotechnol 1998 May;16(5):478], 153.
23. Maloy, K.J., Donachie, A.M., O'Hagan, D.T., and Mowat, A.M. 1994, Immunology, 81, 661.

24. Moore, A., McGuirk, P., Adams, S., Jones, W.C., McGee, J.P., O'Hagan, D.T., and Mills, K.H. 1995, Vaccine, 13, 1741.
25. Kazzaz, J., Neidleman, J., Singh, M., Ott, G., and O'Hagan, D.T. 2000, J. Control Release, 67, 347.
26. Otten, G.R., Schaefer, M., Greer, C., Calderon-Cacia, M., Coit, D., Kazzaz, J., Medina-Selby, A., Selby, M., Singh, M., Ugozzoli, M., zur Megede, J., Barnett, S., O'Hagan, D., Donnelly, J., and Ulmer, J.B. 2003, J. Virol., 77, 6087.
27. Newman, K.D., Elamanchili, P., Kwon, G.S., and Samuel, J. 2002, J. Biomed. Mater. Res., 60, 480.
28. Singh, M., Kazzaz, J., Chesko, J., Soenawan, E., Ugozzoli, M., Giuliani, M., Pizza, M., Rappuoli, R., and O'Hagan, D.T. 2004, J. Pharm. Sci., 93, 273.
29. Higaki, M., Takase, T., Igarashi, R., Suzuki, Y., Aizawa, C., Mizushima, Y., Heritage, P.L., Brook, M.A., Underdown, B.J., and McDermott, M.R. 1998, Vaccine, 16, 741.
30. Yan, M., Peng, J., Jabbar, I.A., Liu, X., Filgueira, L., Frazer, I.H., and Thomas, R. 2005, Immunol. Cell Biol., 83, 83.
31. Eyles, J.E., Carpenter, Z.C., Alpar, H.O., and Williamson, E.D. 2003, J. Drug Target., 11, 509.
32. O'Hagan, D., Singh, M., Ugozzoli, M., Wild, C., Barnett, S., Chen, M., Schaefer, M., Doe, B., Otten, G.R., and Ulmer, J.B. 2001, J. Virol., 75, 9037.
33. Challacombe, S.J., Rahman, D., Jeffery, H., Davis, S.S., and O'Hagan, D.T. 1992, Immunology, 76, 164.
34. Challacombe, S.J., Rahman, D., and O'Hagan, D.T. 1997, Vaccine, 15, 169.
35. Eldridge, J.H., Hammond, C.J., Meulbroek, J.A., Staas, J.K., Gilley, R.M., and Tice, T.R. 1990, J. Controlled Release, 11, 205.
36. Jones, D.H., McBride, B.W., Thornton, C., O'Hagan, D.T., Robinson, A., Farrar, G.H., Malone, R.W., Heritage, P.L., Brook, M.A., Underdown, B.J., and McDermott, M.R. 1996, Infect. Immun., 64, 489.
37. Shahin, R., Leef, M., Eldridge, J., Hudson, M., and Gilley, R. 1995, Infect. Immun., 63, 1195.
38. Conway, M.A., Madrigal-Estebas, L., McClean, S., Brayden, D.J., and Mills, K.H. 2001, Vaccine, 19, 1940.
39. Cahill, E.S., O'Hagan, D.T., Illum, L., Barnard, A., Mills, K.H., and Redhead, K. 1995, Vaccine, 13, 455.
40. Whittum-Hudson, J.A., An, L.L., Saltzman, W.M., Prendergast, R.A., and MacDonald, A.B. 1996, Nat. Med., 2, 1116.
41. Allaoui-Attarki, K., Pecquet, S., Fattal, E., Trolle, S., Chachaty, E., Couvreur, P., and Andremont, A. 1997, Infect. Immun., 65, 853.
42. Marx, P.A., Compans, R.W., Gettie, A., Staas, J.K., Gilley, R.M., Mulligan, M.J., Yamschikov, G.V., Chen, D., and Eldridge, J.H. 1993, Science, 260, 1323.
43. Tseng, J., Komisar, J.L., Trout, R.N., Hunt, R.E., Chen, J.Y., Johnson, A.J., Pitt, L., and Ruble, D.L. 1995, Infect. Immun., 63, 2880.
44. Ugozzoli, M., O'Hagan, D.T., and Ott, G.S. 1998, Immunology, 93, 563.
45. Jones, D.H., Corris, S., McDonald, S., Clegg, J.C., and Farrar, G.H. 1997, Vaccine, 15, 814.
46. Mathiowitz, E., Jacob, J.S., Jong, Y.S., Carino, G.P., Chickering, D.E., Chaturvedi, P., Santos, C.A., Vijayaraghavan, K., Montgomery, S., Bassett, M., and Morrell, C. 1997, Nature, 386, 410.
47. O'Hagan, D.T. 1996, J. Anat., 189 (Pt 3), 477.
48. Eyles, J.E., Spiers, I.D., Williamson, E.D., and Alpar, H.O. 2001, J. Pharm. Pharmacol., 53, 601.
49. O'Hagan, D. 1998, Adv. Drug Deliv. Rev., 34, 305.
50. Michalek, S.M., O'Hagan, D.T., Gould-Fogerite, S., Rimmelzwaan, G.F., and Osterhaus, A.D.M.E. 1999, Mucosal Immunology, P.L. Ogra, J. Mestecky, M.E. Lamm, W. Strober, J. Bienenstrock and J.R. Mcghee (Eds.), Academic Press, San Diego, 759.
51. Somroop, S., Tongtawe, P., Chaisri, U., Tapchaisri, P., Chongsa-nguan, M., Srimanote, P., and Chaicumpa, W. 2006, Asian Pac. J. Allergy Immunol., 24, 229.

52. Choi, M.J., and Maibach, H.I. 2005, Skin Pharmacol. Physiol., 18, 209.
53. Naito, T., Kaneko, Y., and Kozbor, D. 2007, J. Gen. Virol., 88, 61.
54. Jain, S., and Vyas, S.P. 2006, J. Liposome Res., 16, 331.
55. Tafaghodi, M., Jaafari, M.R., and Sajadi Tabassi, S.A. 2006, Eur. J. Pharm. Biopharm., 64, 138.
56. Alcon, V., Baca-Estrada, M., Vega-Lopez, M., Willson, P., Babiuk, L.A., Kumar, P., Hecker, R., and Foldvari, M. 2005, AAPS J., 7, E566.
57. Agren, L.C., Ekman, L., Lowenadler, B., and Lycke, N.Y. 1997, J. Immunol., 158, 3936.
58. Agren, L.C., Ekman, L., Lowenadler, B., Nedrud, J.G., and Lycke, N.Y. 1999, J. Immunol., 162, 2432.
59. Agren, L., Sverremark, E., Ekman, L., Schon, K., Lowenadler, B., Fernandez, C., and Lycke, N. 2000, J. Immunol., 164, 6276.
60. Goletz, T.J., Klimpel, K.R., Arora, N., Leppla, S.H., Keith, J.M., and Berzofsky, J.A. 1997, Proc. Natl. Acad. Sci. U.S.A., 94, 12059.
61. Cao, H., Agrawal, D., Kushner, N., Touzjian, N., Essex, M., and Lu, Y. 2002, J. Infect. Dis., 185, 244.

Index

D

N

O

P

R

S

T